Symposium Proceedings

PCI/FHWA
INTERNATIONAL SYMPOSIUM ON HIGH PERFORMANCE CONCRETE

October 20-22, 1997
New Orleans, Louisiana

Advanced Concrete Solutions for Bridges and Transportation Structures

Edited by
L.S. (Paul) Johal

PCI/FHWA
INTERNATIONAL SYMPOSIUM ON HIGH PERFORMANCE CONCRETE

Copyright © 1997
by Precast/Prestressed Concrete Institute

All rights reserved. No part of this publication may be reproduced in any form without the written permission of the Precast/Prestressed Concrete Institute.

ISBN-0-937040-55-X

Neither the Precast/Prestressed Concrete Institute nor the Federal Highway Administration is responsible for statements and/or opinions expressed by authors of papers.

PCI
PRECAST/PRESTRESSED CONCRETE INSTITUTE

175 W. Jackson Boulevard
Chicago, Illinois 60604
Phone: 312-786-0300
Fax: 312-786-0353
e-mail: info@pci.org
www.pci.org

PRINTED IN USA

STEERING COMMITTEE

James R. Hoblitzell
FHWA, Bridge Division

John M. Hooks
FHWA, Office of Technology Applications

Susan N. Lane
FHWA, Structures Division

Suneel N. Vanikar
FHWA, Office of Technology Applications

Wayne Aymond
Louisiana Department of Transportation and Development

Jerry L. Potter
Florida Department of Transportation

Thomas B. Battles
Precast/Prestressed Concrete Institute

John S. Dick
Precast/Prestressed Concrete Institute

Gary H. Munstermann
Precast/Prestressed Concrete Institute

COOPERATING ORGANIZATIONS

American Concrete Institute
American Segmental Bridge Institute
American Society of Civil Engineers
American Society for Concrete Construction
American Society for Testing and Materials
Asociacion Nacional de Industriales del Presfuerzo
    y la Prefabricacion, A.C.
Canada Centre for Mineral and Energy Technology
Canadian Portland Cement Association
Canadian Prestressed Concrete Institute
Composites Institute
Concrete Canada
Concrete Reinforcing Steel Institute
Expanded Shale, Clay and Slate Institute
Fédération Internationale de la Précontrainte
Instituto Mexicano del Cemento y del Concreto, A.C.
National Aggregates Association
National Ready Mixed Concrete Association
National Society of Professional Engineers
National Stone Association
Portland Cement Association
Post-Tensioning Institute
Wire Reinforcement Institute

TECHNICAL COMMITTEE

L. S. (Paul) Johal, Chair • Precast/Prestressed Concrete Institute

Shuaib H. Ahmad • North Carolina State University

John A. Bickley • John A. Bickley Associates, Canada

W. Vincent Campbell • Bayshore Concrete Products Corporation

Ramon L. Carrasquillo • University of Texas at Austin

Karen Cormier • T. Y. Lin International

Milo Cress • FHWA

Anat Y. Dabholkar • FDG, Inc.

Robert J. Desjardians • Cianbro Corporation

Charles W. Dolan • University of Wyoming

Catherine W. French • University of Minnesota

Diane Hughes • Sika Corporation

Gunnar M. Idorn • RAMBOLL-Consultant, Denmark

Paul Liles • Georgia State Department of Transportation

Colin Lobo • National Ready Mixed Concrete Association

M. Myint Lwin • Washington State Department of Transportation

V. M. Malhotra • CANMET/EMR, Canada

Juan Murillo • Parsons, Brinckerhoff Quade & Douglas

Charles K. Nmai • Master Builders, Inc.

Celik H. Ozyildirim • Virginia Transportation Research Council

Thomas J. Pasko • FHWA, Advanced Research

Basile G. Rabbat • Portland Cement Association

Mary Lou Ralls • Texas Department of Transportation

Kenneth B. Rear • W. R. Grace & Company

John P. Ries • Expanded Shale Clay and Slate Institute

Henry G. Russell • Henry G. Russell, Inc.

Scott A. Sabol • Delaware Transportation Institute

Surendra P. Shah • Northwestern University

Louis N. Triandafilou • FHWA

Paul Zia • North Carolina State University

TECHNICAL REVIEW PANEL

Shuaib H. Ahmad • North Carolina State University, Raleigh, NC

Atorod Azizinamini • University of Nebraska-Lincoln, Lincoln, NE

John A. Bickley • John A. Bickley Associates, Toronto, Ontario, Canada

Paul C. Breeze • Reid Crowther, Calgary, Alberta, Canada

Robert N. Bruce, Jr. • Tulane University, New Orlean, LA

Mark A. Bury • Master Builders, Inc., Cleveland, OH

Thomas D. Bush • University of Oklahoma, Norman, OK

W. Vincent Campbell • Bayshore Concrete Products Corporation, Cape Charles, VA

Ramon L. Carrasquillo • University of Texas at Austin, Austin, TX

Anthony P. Chrest • WALKER Parking Consultants, Kalamazoo, MI

George A. Christian • New York State Department of Transportation, Albany, NY

James L. Clarke • Master Builders, Inc., Mathews, NC

Karen Cormier • T.Y. Lin International, San Francisco, CA

Thomas Cousins • Virginia Technological University, Blacksburg, VA

Milo Cress • Federal Highway Administration, Lincoln, NE

Anat Y. Dabholkar • FDG, Inc., Arvada, CO

Thomas J. D'Arcy • The Consulting Engineers Group, Inc., San Antonio, TX

Robert J. Desjardians • Cianbro Corporation, Pittsfield, ME

Charles W. Dolan • University of Wyoming, Laramie, WY

Daniel Dorgan • Minnesota Department of Transportation, Roseville, MN

Catherine W. French • University of Minnesota, Minneapolis, MN

PREFACE

These Proceedings contain papers presented at the PCI/FHWA International Symposium on High Performance Concrete, held in New Orleans, Louisiana, October 20-22, 1997. The Symposium, sponsored by the Federal Highway Administration and organized by the Precast/Prestressed Concrete Institute, was conducted in conjunction with the 43rd PCI Annual Convention and Exhibition.

The International Symposium addresses research, design, construction, performance, and benefits of high performance concrete, with special emphasis on bridges and transportation structures. Although the concept of high performance concrete as a technology emerged only about 15 years ago, world-wide interest in this subject area has grown dramatically during the past decade. The potential economic advantages of high performance concrete with high strength and improved durability for bridges and transportation structures are very promising. Durability, and particularly chloride initiated reinforcement corrosion, is of major world-wide concern. Properly designed and executed high performance concrete has the potential to delay deterioration processes and thereby prolong the service life of a structure.

The purpose of the Symposium was to bring together experts in this specific field from around the world to discuss the most recent developments. The Proceedings reflect this collective knowledge by providing state-of-the-art information on research, concrete design, construction, materials, and quality control. In addition, research programs providing the background knowledge on code requirements and project histories are included.

PCI and FHWA would like to thank all the authors for their participation and for their cooperation in ensuring a high quality Symposium. We hope that the proceedings will form a valuable basis for further developments in this important field of civil and structural engineering, and in serving the public.

L. S. (Paul) Johal

Chicago, Illinois
August 1997

ACKNOWLEDGMENTS

The Symposium was sponsored by the Federal Highway Administration (FHWA) and organized by the Precast/Prestressed Concrete Institute (PCI). The Steering Committee comprising of Jim Hoblitzell, John Hooks, Sue Lane, Wayne Aymond, Jerry Potter, Tom Battles, John Dick, and Gary Munstermann provided the necessary guidance throughout the long planning process. The cooperating organizations were instrumental in publicizing and promoting the Symposium, which has been vital for its success. The PCI High Performance Concrete and Symposium Technical Committees played a significant role in reviewing abstracts and selecting papers for presentation and publication.

A deep sense of gratitude is expressed toward the speakers and their co-authors for their valuable time and effort to produce and present quality papers. The Technical Review Panel provided a most valuable service by reviewing the papers and providing comments in a timely manner.

Special thanks are due to a few PCI staff members. George Nasser willingly provided editorial help and reviews of selected material. Joe Hoyle helped in designing the cover. Chris Beckmann and Valorie Vieira, in their roles as Symposium coordinators, handled various details in an exemplary manner with commitment, devotion, and efficiency. They all deserve the gratitude of everyone who stands to benefit from the Symposium.

Finally, the overall support of PCI and FHWA Management is gratefully acknowledged.

L. S. (Paul) Johal

TABLE OF CONTENTS

*SESSION A (#4):   General History & Definition

Design Optimization of Precast Girder Bridges made with High-Performance Concrete ... 1
  Mostafa A. Hassanain • The University of Calgary, Calgary, Alberta, Canada
  Robert E. Loov • The University of Calgary, Calgary, Alberta, Canada

High Performance Concrete - Research and Forecast for Application in Bridge Structures in Poland ... 13
  Wojciech Radomski • Warsaw University of Technology, Warszawa, Poland

High-Strength Concrete in Bridges - History and Challenges ... 27
  Henry G. Russell • Henry G. Russell, Inc., Glenview, IL

High Strength Concrete in Spliced Prestressed Concrete Bridge Girders ... 39
  Zoubior Lounis • McGill University, Montreal, Quebec, Canada
  M. Saeed Mirza • McGill University, Montreal, Quebec, Canada

State-of-the-Art of HPC: An International Perspective ... 49
  Paul Zia • North Carolina State University, Raleigh, NC

Application of HPC in Infrastructure - An Overview in the Perspective of FIP and CEB ... 60
  Steinar Helland • Selmer ASA, Oslo, Norway

Use of HSC/HPC for Road Bridges in India ... 72
  S. A. Reddi • Gammon India Limited, Mumbai, India

*SESSION B (#5):   Materials & Mix Design

High-Performance Concretes Using Oklahoma Aggregates ... 84
  Thomas D. Bush, Jr., • University of Oklahoma, Norman, OK
  Bruce W. Russell • University of Oklahoma, Norman, OK
  Seamus F. Freyne • University of Oklahoma, Norman, OK

High Performance Concrete Mixture Proportion Optimization for Precast Concrete Using Statistical Methods ... 96
  James R. DeMaro • South Dakota School of Mines and Technology, Rapid City, SD
  M. R. Hansen • South Dakota School of Mines and Technology, Rapid City, SD
  Brian C. Anderson • South Dakota School of Mines and Technology, Rapid City, SD

Hydration Heat and Strength Development in High Performance Concrete ... 108
  Maria Kaszynska • Technical University of Szczecin, Szczecin, Poland

*Numbers in parenthesis refer to the sessions as shown in the Convention/Symposium Program

Early Age Shrinkage and Creep of High Performance Cement-Based Materials   118
   David A. Lange • Universaity of Illinois at Urbana-Champaign, Urbana, IL
   Leslie Struble • University of Illinois at Urbana-Champaign, Urbana, IL
   J. Francis Young • University of Illinois at Urbana-Champaign, Urbana, IL
   S. Altoubat • University of Illinois at Urbana-Champaign, Urbana, IL
   H.-C. Shin • University of Illinois at Urbana-Champaign, Urbana, IL
   H. Ai • University of Illinois at Urbana-Champaign, Urbana, IL

The Effect of Silica Fume Dosage on the Heat of Hydration of High Performance Mortars   124
   Roberto C. A. Pinto • Cornell University, Ithaca, NY
   Kenneth C. Hover • Cornell University, Ithaca, NY

High Performance Characteristics of Chemically Activated Fly Ash (CAFA)   135
   Thomas Silverstrim • By-Products Development Corporation, Collingdale, PA
   Hossein Rostami • By-Products Development Corporation, Collingdale, PA
   Yunping Xi • Drexel University, Philadelphia, PA
   Joseph Martin • Drexel University, Philadelphia, PA

Shrinkage Cracking in High Performance Concrete   148
   Surendra P. Shah • Northwestern University, Evanston, IL
   W. Jason Weiss • Northwestern University, Evanston, IL
   Wei Yang • Northwestern University, Evanston, IL

High-Performance Prefabricated Silica Fume Concrete for Infrastructure   159
   Per Fidjestol • Elkem ASA Materials, Kristiansand, Norway
   Tony Kojundic • Elkem Materials, Inc., Pittsburgh, PA

The Chloride Penetration Resistance of Concrete Containing High-Reactivity Metakaolin   172
   R. Doug Hooton • University of Toronto, Toronto, Ontario, Canada
   K. Gruber • Engelhard Corporation, Iselin, NJ
   A. M. Boddy • University of Toronto, Toronto, Ontario, Canada

The Effectiveness of Calcium Nitrite in Producing High Performance Concrete   184
   Eric P. Steinberg • Ohio University, Athens, OH
   Kenneth B. Edwards • Ohio University, Athens, OH

*SESSION C (#10):   Laboratory Research & Future Direction

High Early Strength Concrete for Prestressed Concrete Applications   194
   Shuaib H. Ahmad • North Carolina State University, Raleigh, NC
   Paul Zia • North Carolina State University, Raleigh, NC

Reinforced Membrane Elements with Concrete Strength up to 100 MPa   206
   Thomas T. C. Hsu • University of Houston, Houston, TX
   Li-Xin Zhang • McDermott Engineering Houston, Houston, TX

*Numbers in parenthesis refer to the sessions as shown in the Convention/Symposium Program

Mechanical Properties and Structural Response of High-Performance Concrete Beams   218
    G. Rosati • Politecnico of Milan, Milan, Italy
    S. Cattaneo • Politecnico of Milan, Milan, Italy
    M. Marazzini • Politecnico of Milan, Milan, Italy
    A. Meda • Politecnico of Milan, Milan, Italy

Concrete Mixture Optimization Using Statistical Mixture Design Methods   230
    Marcia J. Simon • Federal Highway Administration, McLean, VA
    Eric S. Lagergren • National Institute of Standards and Technology, Gaithersburg, MD
    Kenneth A. Snyder • National Institute of Standards and Technology, Gaithersburg, MD

Influence of Silica Fume on Chloride Resistance of Concrete   245
    R. Doug Hooton • University of Toronto, Toronto, Ontario, Canada
    P. Pun • University of Toronto, Toronto, Ontario, Canada
    Tony. Kojundic • Elkem Materials, Inc., Pittsburgh, PA
    Per Fidjestol • Elkem ASA Materials, Kristiansand, Norway

Development of High Performance Grouts for Corrosion Protection of Post-Tensioning Tendons   257
    Andrea J. Schokker • The University of Texas at Austin, Austin, TX
    B. D. Koester • The University of Texas at Austin, Austin, TX
    Michael E. Kreger • The University of Texas at Austin, Austin, TX
    John E. Breen • The University of Texas at Austin, Austin, TX

Permeability of High Performance Concrete: Rapid Chloride Ion Test Versus Chloride Ponding Test   268
    John J. Myers • The University of Texas at Austin, Austin, TX
    Wissam E. Touma • The University of Texas at Austin, Austin, TX
    Ramon L. Carrasquillo • The University of Texat at Austin, Austin, TX

*SESSION D (#11): Quality Concepts & Construction Techniques

Curing of High Performance Concrete - An Overview   283
    Changqing (Max) Wang • The University of Calgary, Calgary, Alberta, Canada
    Walter H. Dilger • The University of Calgary, Calgary, Alberta, Canada
    Wib S. Langley • Jacques Whitford & Associates, Dartmouth, Nova Scotia, Canada

High Performance Concrete in Precast Concrete Tunnel Linings; Meeting Chloride Diffusion and Permeability Requirements   294
    A. John R. Hart • CCL Joint Venture, Guelph, Ontario, Canada
    John Ryell • Trow Consulting Engineers, Ltd., Brampton, Ontario, Canada
    Michael D. A. Thomas • University of Toronto, Toronto, Ontario, Canada

Design and Construction Demands on Concrete that Necessitated a Very Low Water-Cement Ratio and High Slump   308
    Robert W. LaFraugh • Wiss, Janney, Elstner Associates, Inc., Seattle, WA

*Numbers in parenthesis refer to the sessions as shown in the Convention/Symposium Program

Use of Quality Systems to Consistently Produce Durable High Strength Concrete Railroad Ties 320
    Dave Millard • CXT, Inc., Spokane, WA

Developing High Performance Concrete Specifications for Highway Bridge Construction – Experience of the Ontario Ministry of Transportation 328
    Hannah C. Schell • Ontario Ministry of Transportation, Downsview, Ontario, Canada
    Beata Berszakiewicz • Ontario Ministry of Transportation, Downsview, Ontario, Canada
    Alan K. C. Ip • Ontario Ministry of Transportation, Downsview, Ontario, Canada

High Performance Silica Fume Grout for Post-Tensioning Ducts 343
    Michael Sprinkel • Virginia Transportation Research Council, Charlottesville, VA

High-Performance Concrete Bridges: The Canadian Experience 355
    Denis Mitchell • McGill University, Montreal, Quebec, Canada
    Pierre-Claude Aitcin • University of Sherbrooke, Sherbrooke, Quebec, Canada
    John A. Bickley • Concrete Canada, Toronto, Ontario, Canada

Quality Control & Quality Assurance Program for Precast Plant Produced High Performance Concrete U-Beams 368
    John J. Myers, The University of Texas at Austin, Austin, TX
    Ramon L. Carrasquillo, The University of Texas at Austin, Austin, TX

*SESSION E (#16):   Structural Design & Concepts

Seismic Behavior of High Strength Concrete Filled Tube (CFT) Columns 383
    A. El-Remaily • University of Nebraska-Lincoln, Lincoln, NE
    Atorod Azizinamini • University of Nebraska-Lincoln, Lincoln, NE
    M. Zaki • University of Nebraska-Lincoln, Lincoln, NE
    F. Filippou • University of California-Berkeley, Berkeley, CA

Tests of Two High Performance Concrete Prestressed Bridge Girders 394
    Catherine French • University of Minnesota, Minneapolis, MN
    Theresa Ahlborn • Michigan Technological University, Houghton, MI

Prestressed I-Girder Design Using High Performance Concrete and the New AASHTO LRFD Specifications 406
    M. Myint Lwin • Washington State Department of Transportation, Olympia, WA
    Bijan Khaleghi • Washington State Department of Transportation, Olympia, WA
    Jen-Chi Hsieh • Washington State Department of Transportation, Olympia, WA

Time-Dependent Effects in High Performance Concrete Bridge Members 419
    Xiaoming Huo • University of Nebraska, Omaha, NE
    Maher K. Tadros • University of Nebraska, Omaha, NE

*Numbers in parenthesis refer to the sessions as shown in the Convention/Symposium Program

Shear Limit of HPC Bridge I-Girders 431
    Zhongguo (John) Ma • University of Nebraska, Omaha, NE
    Maher K. Tadros • University of Nebraska, Omaha, NE
    Mantu Baishya • University of Nebraska, Omaha, NE

Shear Behavior of Pretensioned I-Shaped Girders made with High Strength Concrete 443
    Bruce W. Russell • The University of Oklahoma, Norman, OK
    James H. Allen, III • Coreslab Structures, Inc., Oklahoma City, OK

The Influence of Tensile Strength of High Strength Concrete on the Behavior of Deflected
Reinforced Concrete Beams in Serviceability Conditions 455
    Salvatore Russo • Venice Institute University, Venice, Italy

*SESSION F (#17): Precast Concrete Fabrication & Transportation

Implementation of High Strength Concrete Research for Prestressed Girders –
A DOT's Perspective 467
    Daniel Dorgan • Minnesota Department of Transportation, Roseville, MN

Application of High Performance Concrete in Two Bridges in New Hampshire 475
    Michelle L. Juliano • New Hampshire Department of Transportation, Concord, NH
    Christopher M. Waszczuk • New Hampshire Department of Transportation, Concord, NH

Fabrication of Prestressed Concrete Beams for Two High Performance Concrete Bridge
Projects in Texas 488
    Burson Patton • Texas Concrete Company, Victoria, TX

Design and Analysis of Pretensioned/Post-Tensioned Long-Span High Performance
Concrete I-Beams 504
    Lisa Carter Powell • P. E. Structural Consultants, Austin, TX

Covington High Performance Concrete Bridge 515
    Chuck Prussack • Central Pre-Mix Prestress Company, Spokane, WA

Applications of a New High Performance Polyolefin Fiber Reinforced Concrete in
Transportation Structures 521
    V. Ramakrishnan • South Dakota School of Mines & Technology, Rapid City, SD

Design and Construction of High Performance Concrete Bridges on 407 Express Toll Route .. 533
    Hari K. Jagasia • Ontario Transportation Capital Corporation, Thornhill, Ontario, Canada

*Numbers in parenthesis refer to the sessions as shown in the Convention/Symposium Program

*SESSION G (#20): Structural Performance & Code Requirements

Design of Tension Lap Splices in High Strength Concrete   543
   Atorod Azizinamini • University of Nebraska-Lincoln, Lincoln, NE

Investigation of Allowable Compressive Stresses for High Strength, Prestressed Concrete   554
   Bruce W. Russell • The University of Oklahoma, Norman, OK
   Joo Pin Pang • The University of Oklahoma, Norman, OK

Instrumentation and Measurements - Behavior of Long-Span Prestressed High Performance Concrete Bridges   566
   Ned H. Burns • The University of Texas at Austin, Austin, TX
   Shawn P. Gross • The University of Texas at Austin, Austin, TX
   Kenneth A. Byle • The University of Texas at Austin, Austin, TX

Treatment of High-Strength Concrete in U.S. Codes   578
   S. K. Ghosh • Portland Cement Association, Skokie, IL

Use of High-Performance Concrete in a Bridge Structure in Virginia   590
   Jose Gomez • Virginia Transportation Research Council, Charlottesville, VA
   Thomas E. Cousins • Virginia Technological University, Blacksburg, VA
   Celik Ozyildirim • Virginia Transportation Research Council, Charlottesville, VA

Longitudinal Seismic Response of Precast Spliced Girder Bridges   599
   Jay Holombo • University of California-San Diego, La Jolla, CA
   M. J. Nigel Priestley • University of California-San Diego, La Jolla, CA
   Frieder Seible • University of California-San Diego, La Jolla, CA

Evaluation of Long-Term Behavior of High Performance Prestressed Concrete Girders   612
   John F. Stanton • University of Washington, Seattle, WA
   Marc O. Eberhard • Unviersity of Washington, Seattle, WA
   Paul Barr • University of Washington, Seattle, WA
   Elizabeth A. Fekete • University of Washington, Seattle, WA

Deformation Behavior of Long-Span Prestressed High Performance Concrete Bridge Girders   623
   Shawn P. Gross • The University of Texas at Austin, Austin, TX
   Kenneth A. Byle • The University of Texas at Austin, Austin, TX
   Ned H. Burns • The University of Texas at Austin, Austin, TX

*SESSION H (#21): Project Profiles & Case Histories

Research and Utilitzation of High Performance Concrete in North Carolina   635
   Azam Azimi • North Carolina Department of Transportation, Raleigh, NC
   Paul Zia • North Carolina State University, Raleigh, NC

*Numbers in parenthesis refer to the sessions as shown in the Convention/Symposium Program

Application of High Performance Concrete in Giles Road Bridge, Nebraska   646
    Xiaoming Huo • University of Nebraska, Omaha, NE
    Maher K. Tadros • University of Nebraska, Omaha, NE

Use of High Performance Concrete in Highway Bridges in Washington State   657
    M. Myint Lwin • Washington State Department of Transportation, Olympia, WA

Application of 100 MPa High-Strength, High-Fluidity Concrete for a Prestressed Concrete Bridge with Span-Depth Ratio of 40   669
    Kenro Mitsui • Takenaka Corporation, Chiba, Japan
    T. Yonezawa • Takenaka Corporation, Chiba, Japan
    M. Tezuka • Technical Research Institute Oriental Construction Co., Ltd., Tochigi, Japan
    M. Kinoshita • Takemoto Oil & Fat Co., Ltd., Aichi, Japan

Virginia's Bridge Structures with High Performance Concrete   681
    Celik Ozyildirim • Virginia Transportation Research Council, Charlottesville, VA
    Jose Gomez • Virginia Transportation Research Council, Charlottesville, VA

Texas High Performance Concrete Bridges - Implementation Status   691
    Mary Lou Ralls • Texas Department of Transportation, Austin, TX
    Ramon L. Carrasquillo • University of Texas at Austin, Austin, TX

Colorado Showcase on HPC Box-Girder Bridge: Development and Transfer Length Tests   705
    P. B. Shing • University of Colorado, Boulder, CO
    D. Cooke • University of Colorado, Boulder, CO
    D. M. Frangopol • University of Colorado, Boulder, CO
    M. A. Leonard • Colorado Department of Transportation, Denver, CO
    M. L. McMullen • Colorado Department of Transportation, Denver, CO
    W. Hutter • Colorado Department of Transportation, Denver, CO

HPC Bridge Showcase Project in Alabama   717
    J. Michael Stallings • Auburn University, Auburn, AL
    David Pittman • Auburn University, Auburn, AL

*Numbers in parenthesis refer to the sessions as shown in the Convention/Symposium Program

Proceedings of the PCI/FHWA
International Symposium on High Performance Concrete
New Orleans, Louisiana, October 20-22, 1997

DESIGN OPTIMIZATION OF PRECAST GIRDER BRIDGES MADE WITH HIGH-PERFORMANCE CONCRETE

Mostafa A. Hassanain
Ph.D. Candidate, Department of Civil Engineering
The University of Calgary
Calgary, Alberta, Canada T2N 1N4

Robert E. Loov
Professor, Department of Civil Engineering
The University of Calgary
Calgary, Alberta, Canada T2N 1N4

ABSTRACT

The potential economic advantages from the utilization of high-performance concrete (HPC) with high strength and improved durability for precast, prestressed concrete highway bridge girders are quite promising. In spite of this, relatively little has been done regarding the implementation of HPC for bridges when compared to buildings. The objective of this paper is to describe a research program currently being carried out at the University of Calgary to study the impact of utilizing HPC for the production of precast, pretensioned girders on the design of slab-on-girder, continuous-span bridges. The cost effectiveness of the use of HPC, as compared to normal-strength concrete, for bridge girders is illustrated.

INTRODUCTION

In a recent survey conducted by the Precast/Prestressed Concrete Institute (PCI) High-Strength Concrete Committee, the average 28-day concrete strength used in the precast concrete industry was found to be 41 MPa (6000 psi).[1] The maximum strength that producers felt they can reliably supply was 59 MPa (8500 psi). These strengths are much less than those in the 100 MPa (14,500 psi) range that have been achieved for many buildings. Increased material costs and additional quality control requirements seem to deter producers from utilizing concretes of much higher strength. More research is needed to provide a comfort level in the industry that higher strength concretes can be economically and confidently produced.

The term high-performance concrete (HPC) has emerged in the past few years to describe concrete that exhibits higher compressive strength and improved durability. One advantage of utilizing HPC in precast bridge girders may be the reduction in the overall costs leading consequently to less expensive bridges. Although the basic concrete cost per unit volume is increased, primarily due to increased mix constituent and quality control costs, this may be partially or fully offset by the reduced quantities of concrete required as a result of the proportionate increase in girder capacity. The amount of non-material costs associated with an HPC girder (e.g., the labour, transportation and erection costs) is the same as that for a normal-strength concrete girder; the cost savings come from the reduction in the number of girders. Another advantage of using HPC may be increased span lengths resulting in fewer piers. From a practical point of view, longer spans and/or fewer girders allow engineers an economical alternative to bridges made from competing materials. An important additional factor which increases the attractiveness of HPC, especially in corrosive environments, is the added durability due to its lower permeability. Increased durability results in longer-lasting bridges. It is apparent, therefore, that HPC should enable the design and construction of more efficient, more cost effective and more durable prestressed concrete highway bridges.

Several researchers have investigated the impact of HPC on the design of precast bridge girders.[2,3] They focused on simply-supported bridges. No similar studies, however, have been carried out on continuous-span bridges which are increasingly called for in bridge design philosophy. Moreover, extensive cost effectiveness analysis of the use of HPC, as compared to normal-strength concrete has not been adequately addressed in such a way as to provide the incentive for precast concrete producers to utilize concretes of higher strength. A research program has started at the University of Calgary to accomplish this end. A brief description of the research methodology is given herein along with some cost effectiveness analyses. This paper is an extension of the work previously described in reference 4.

RESEARCH PROGRAM METHODOLOGY

Assumptions
The following assumptions were made:
1. Design conforms to the Ontario Highway Bridge Design Code (OHBDC).[5]
2. The draft Canadian Highway Bridge Design Code (CHBDC) live load specifications shown in Figure 1[6] are used.
3. Tendon profile and locations of critical sections are as shown in Figure 2.
4. Unshored construction is used.
5. 75 mm (3 in)-thick future wearing surface will be placed on top of the concrete deck.
6. Each barrier has a cross-sectional area of 0.3 m^2 (3.23 ft^2). This load is distributed equally among the girders.
7. Effective prestressing force after losses is 0.8 of the prestressing force at transfer.
8. Concrete strength at transfer is 0.75 of the 28-day strength.
9. Seven-wire, low-relaxation, 13 mm (0.5 in) strands having a tensile strength of 1860 MPa (270 ksi) are used.

Structural System and Method of Analysis
Bridges built of cast-in place, reinforced concrete deck slabs on precast, prestressed

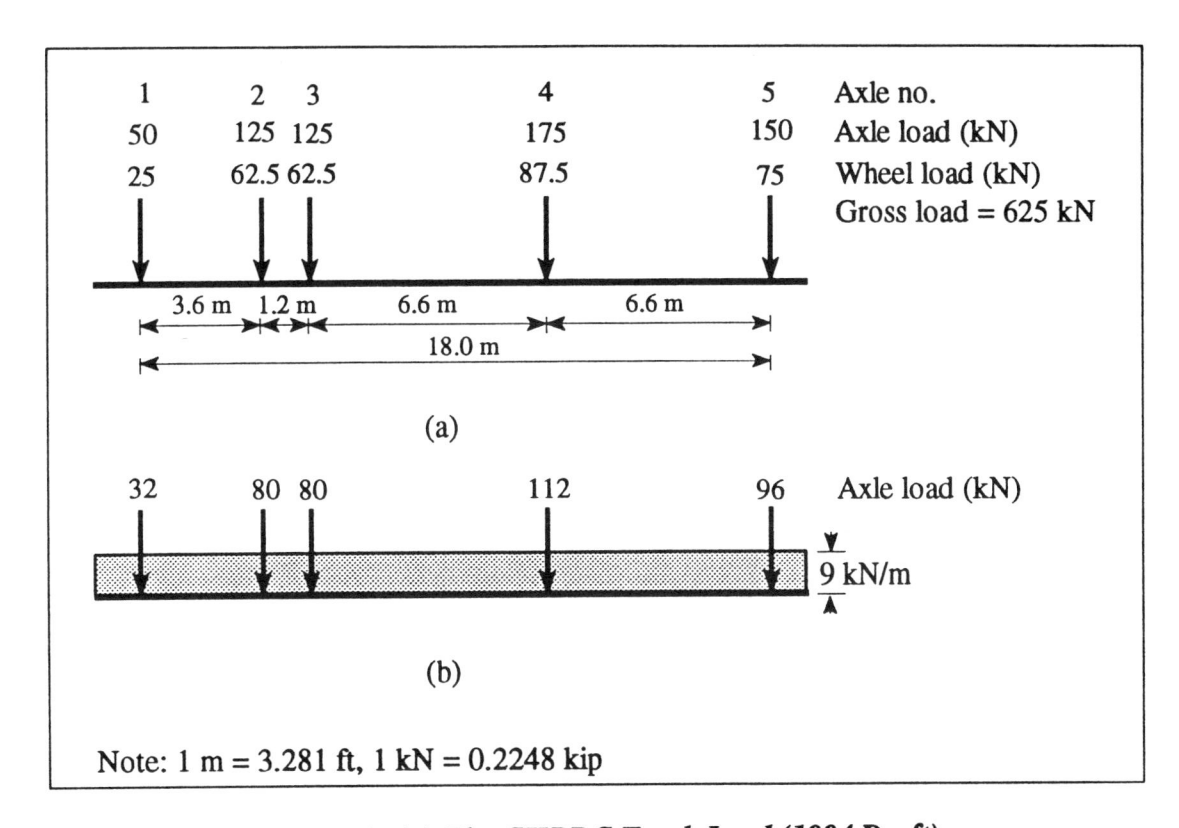

Figure 1. (a) The CHBDC Truck Load (1994 Draft)
(b) The CHBDC Lane Load (1994 Draft)

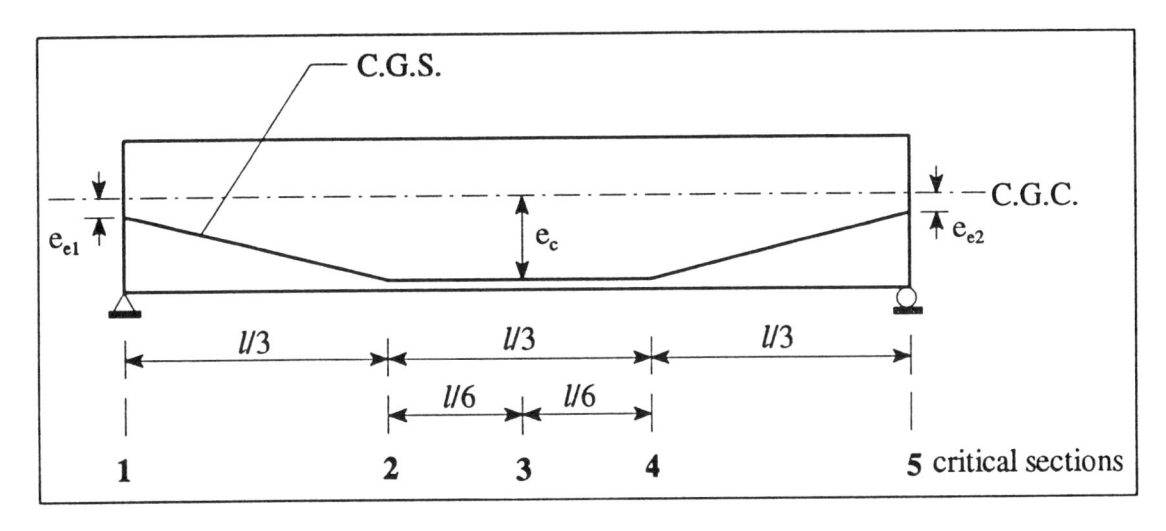

Figure 2. Tendon Profile and Locations of Critical Sections

concrete I-girders represent about half of all prestressed concrete bridges that have been constructed during the last 40 years in Canada[7] and the United States.[8] Therefore, this structural system has been selected for this investigation.

In order to account properly for the construction sequence of slab-on-girder bridges, the moments at different cross sections are computed as follows: Initially, the precast girders are placed on the substructure and the analysis under the girder and slab self-weights is carried out as for simple beams. At this stage, the stresses are determined using the girder section only. Subsequently, continuity is achieved by casting concrete between the ends of the girders and adding reinforcing bars over the piers. The analysis for the live load and pavement weight is then carried out as for continuous beams. Composite action of the precast girder and cast-in-place slab must be considered in evaluating the stresses produced by these loads.

The OHBDC permits a simplified method of analysis in determining the live load longitudinal moments in continuous-span girder bridges. Recently, Jaeger et al.[9] have reported that this method is valid for the positive moment regions, but is inaccurate for negative moments. Because of the uncertainties that could be associated with the OHBDC method, the semicontinuum method[10], which is approved by the OHBDC as a refined method of analysis, has been used for live load analysis. In this method, a slab-on-girder bridge is represented by discrete longitudinal members and a continuous transverse medium. This idealization is a closer representation of this type of bridge.

Continuity of slab-on-girder bridges requires that positive moments developed over piers due to creep of the girders and the effects of live load on remote spans be considered. Positive moments due to creep under sustained loads; i.e., prestressing and dead loads, are partially counteracted by the negative moments due to differential shrinkage between the cast-in-place deck slab and the precast girders. The Trost-Bažant age-adjusted elastic modulus method has been used to compute the prestress and dead load restraint moments. As recommended by the OHBDC, the restraint moments induced by differential shrinkage have been reduced by 60 percent to take into account the presence of creep.

Formulation of the Optimal Design Problem

General — Design of prestressed concrete girder bridges on a trial and error basis is commonly used. However, it is a tedious and time consuming process. Mathematical optimization techniques provide systematic procedures by which optimal designs can be obtained with substantially less time and effort. The formulation of an optimal design problem requires identification of design variables for the structural system, an objective function that needs to be minimized, and design constraints that are imposed on the system. Once the design problem has been formulated, it is transcribed into the following standard nonlinear constrained optimization model: Find the set of n design variables contained in the vector b that will minimize the objective function represented by Equation (1) subject to the constraints represented by Equations (2) through (4).

$$f(b) = f(b_1, b_2, ..., b_n) \tag{1}$$

$$g_i(b) \leq 0, \quad i = 1, ..., m \tag{2}$$

$$h_i(b) = 0, \quad i = 1, ..., k \tag{3}$$

$$b_i^l \leq b_i \leq b_i^u, \quad i = 1, ..., n \tag{4}$$

where m = number of inequality constraints; k = number of equality constraints; b_i^l and b_i^u are lower and upper bounds on the ith design variable, respectively.

Many numerical methods have been developed to solve the general nonlinear programming problem described above. The methods start from an initial design provided by the user which is iteratively improved until the optimum is reached. Many of these methods have been incorporated into general-purpose design optimization software packages. One such package is IDESIGN (Interactive *DESIGN* Optimization of Engineering Systems).[11] IDESIGN is widely used in the United States and elsewhere. It has been used in this study to solve the nonlinear programming problem. The optimization process consists of cycling between two distinct phases defined as analysis and optimal design in an iterative fashion until the optimum is reached.

Design Variables — For standard precast I-girders, cross-sectional dimensions are known and become preassigned parameters instead of design variables. CPCI girders[12] are used for this investigation. However, there is no loss of generality because it has been shown previously[13] that the use of standard CPCI girders or standard AASHTO-PCI girders result in practically identical optimum solutions. Composite action of the precast girder and cast-in-place slab is assumed, and thus the slab thickness is taken as a design variable. Other design variables are the required amounts of prestressed and non-prestressed flexural reinforcements in the girders and slab, and the tendon profile defined by the eccentricities at midspan and piers as shown in Figure 2. In this investigation, the girder concrete strength is also taken as a design variable. In cases where the concrete strength has a given fixed value, b_i^l and b_i^u in Equation (4) are both set to be equal to that value.

Objective Function — The relevant objective function in the design of bridges with a fixed number of spans is the minimum superstructure cost. This assumes that the cost of piers and abutments is relatively unaffected by changes in the number of girders. This objective function is taken as the material (concrete and steel) costs plus overhead and waste, in addition to the labour, transportation and erection costs. Slab formwork cost may vary slightly with the change in the number of girders. For example, as the number of girders decreases, the actual slab formwork area increases; however, the labour cost of placing this formwork decreases in a way that may partially or fully compensate for the increase in material cost. Therefore, it was decided to exclude the slab formwork cost from the objective function. Furthermore, costs of some other items that are relevant to the superstructure such as wearing surfaces, barriers, diaphragms and drains were not included in the objective function because these costs do not influence the optimum number of girders. In the present investigation, the objective function is defined as the superstructure cost / deck area:

$$COST = [\, n_g C_g + C_c V_c + C_s (\, m_s + n_p (m_{sn}+m_{sp}) + m_p \,) \,] / WL \qquad (5)$$

where n_g = number of girders; n_p = number of positive moment connections (at piers); C_g = cost of girder (including cost of materials, production, transportation and erection); C_c = cost of concrete in slab per unit volume; C_s = cost of non-prestressed steel per unit of mass; V_c = volume of concrete in slab; m_s, m_{sn}, m_{sp}, m_p = mass of non-prestressed steel in slab, negative moment steel in slab at each pier, positive moment connection steel at each pier, and positive moment steel in girders, respectively; W = width of the bridge; and L = total length of the bridge.

Design Constraints — The objective function is minimized under all relevant flexural constraints on serviceability and safety as well as practical constraints according to the OHBDC. Other constraints (e.g., shear) could easily be added. However, in general, they are expected to have marginal effect on the design and are left to be checked at the final design stage. It should be mentioned here that the OHBDC does not have explicit criteria for limiting deflections in prestressed concrete bridges. In all of the cases investigated for this paper, the active constraints (governing design criteria) were observed. For brevity, only these active constraints are presented here.

At the time of prestress transfer, concrete stresses at the top face of the girder due to its own weight should be within the allowable limit at all critical sections (Tensile stresses are assumed positive and compressive stresses are assumed negative):

$$-\frac{P_i}{A} + \frac{P_i e}{S_t} - \frac{M_{dg}}{S_t} \leq 0.2 \sqrt{f'_{ci}} \qquad (6)$$

where P_i = initial prestressing force; A = girder cross-sectional area; e = tendon eccentricity; S_t = girder cross section modulus with respect to the top surface; M_{dg} = moment due to self-weight of the girder; and f'_{ci} = girder concrete strength at the time of transfer. At the bottom face of the girder:

$$-\frac{P_i}{A} - \frac{P_i e}{S_b} + \frac{M_{dg}}{S_b} \geq -0.6 f'_{ci} \qquad (7)$$

where S_b = girder cross section modulus with respect to the bottom surface.

At service conditions after continuity is achieved, concrete stresses at the bottom face of the girder should satisfy the allowable limit at all critical sections:

$$-\frac{P_e}{A} - \frac{P_e e}{S_b} + \frac{M_{dg} + M_{ds}}{S_b} + \frac{M_{da} + M_l}{S_{bc}} \leq 0.2 \sqrt{f'_c} \qquad (8)$$

where P_e = effective prestressing force after losses; M_{ds} = moment due to slab weight; M_{da}

= moment caused by the additional dead load applied after the slab hardens, M_l = live-load moment; S_{bc} = section modulus of composite transformed section with respect to the bottom face of the girder; and f'_c = 28-day concrete strength of girder. This approximate equation assumes that most losses have occurred before the girder becomes composite.

A limit on the tendon eccentricity has been imposed to ensure adequate concrete cover:

$$e \leq e_{max} \tag{9}$$

where e_{max} = maximum practical eccentricity that can be accommodated within the girder.

Ultimate limit state provisions require the factored negative moment at any pier, M_f, not to exceed the flexural strength of the section, M_r:

$$M_f \leq M_r \tag{10}$$

The OHBDC requires a minimum reinforcement area, A_s^+, of 1.50 times the nominal depth of the girder to be provided in the bottom flanges over the piers:

$$A_s^+ \geq 1.50 \, d \tag{11}$$

where d = nominal depth of the girder, and the units of the constant 1.50 and d are in mm.

The OHBDC also requires the deck slab to have a minimum thickness of 225 mm (8.9 in):

$$t \geq 225 \, mm \tag{12}$$

where t = thickness of the deck slab. And according to the empirical design method for slabs allowed by the OHBDC, the ratio of centre-to-centre girder spacing, S, to the thickness of the slab shall not exceed 15:

$$\frac{S}{t} \leq 15 \tag{13}$$

ECONOMIC ANALYSIS

Case Study
In order to study the economic advantages from the utilization of HPC for precast, pretensioned girders in continuous slab-on-girder bridges, a series of optimal designs was generated for a two-equal-span continuous bridge. The cross section of the bridge is shown in Figure 3. It has three lanes with an overall width of 12 m (39.4 ft). Each span is 28 m (92 ft) in length. This span length is slightly more than what is required to cross a typical three-lane urban roadway in order to allow for possible future road widening. The girders are CPCI type 1400[12] and are spaced at a distance S. Strength of the deck slab

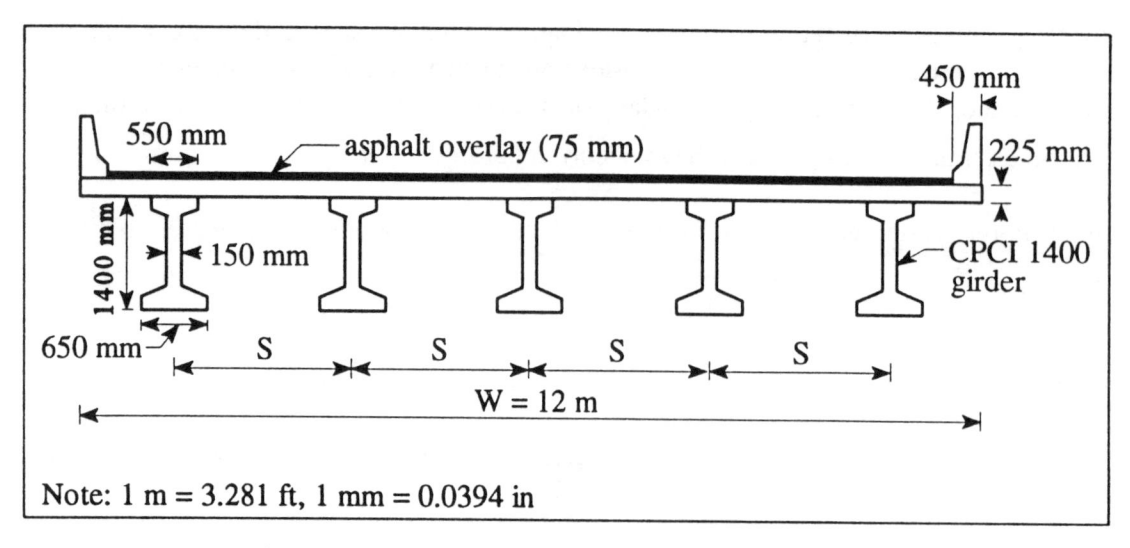

Figure 3. Cross Section of the Bridge Investigated

concrete was fixed at 35 MPa (5075 psi). Transverse spacings of the girders were varied as 2.5, 3.0 or 4.0 m (8.2, 9.8 or 13.1 ft). These spacings were chosen because they optimized the design of the exterior and interior girders when 5, 4 or 3 girders are used, respectively. Exterior girders are generally required to have a capacity greater than or equal to the interior girders. Design is optimized by increasing the girder spacing until the service load stresses in the exterior girders are equal to those in the interior ones. It was noted that the optimal deck slab thickness for transverse spacings of 2.5 and 3.0 m always corresponds to the minimum thickness permitted by the OHBDC (Equation (12)). At larger spacings, the slab thickness increases and the constraint represented by Equation (12) ceases to be active, while that represented by Equation (13) becomes active. It should be mentioned that the first two spacings are less than the 3.7 m (12.1 ft) maximum limit imposed by the OHBDC when designing slabs according to the empirical design method, while the third spacing exceeds this limit.

Cost Estimation

Unit cost estimates were obtained from local precast concrete producers. They are expressed here in Canadian dollars. Because the girder concrete strength was taken as a design variable that can change at every iteration in the optimization process, it was necessary to have a continuous function for the concrete mix cost. In order to accomplish this end, several groups of current concrete cost data from local producers and relative cost data from the literature were compiled and normalized with respect to the cost of a 40 MPa (5800 psi) concrete mix, which has been assumed to be \$95/m^3 (\$73/yd^3) including overhead and waste. Using regression analysis, the curve and expression shown in Figure 4 were obtained. They relate the girder concrete strength and the concrete mix cost ratio. Labour and curing cost an additional \$34/m^3 (\$26/yd^3). Prestressing strands cost \$1.32/m (\$0.40/ft), while epoxy-coated reinforcing bars cost \$1.68/kg (\$0.76/lb) for material, labour, overhead and waste. Transportation and erection costs depend on the distance between the precasting plant and the construction site, and the condition of the

$$Y = 0.936 + (X / 100)^3$$
Normalized with respect to the cost of a 40 MPa mix

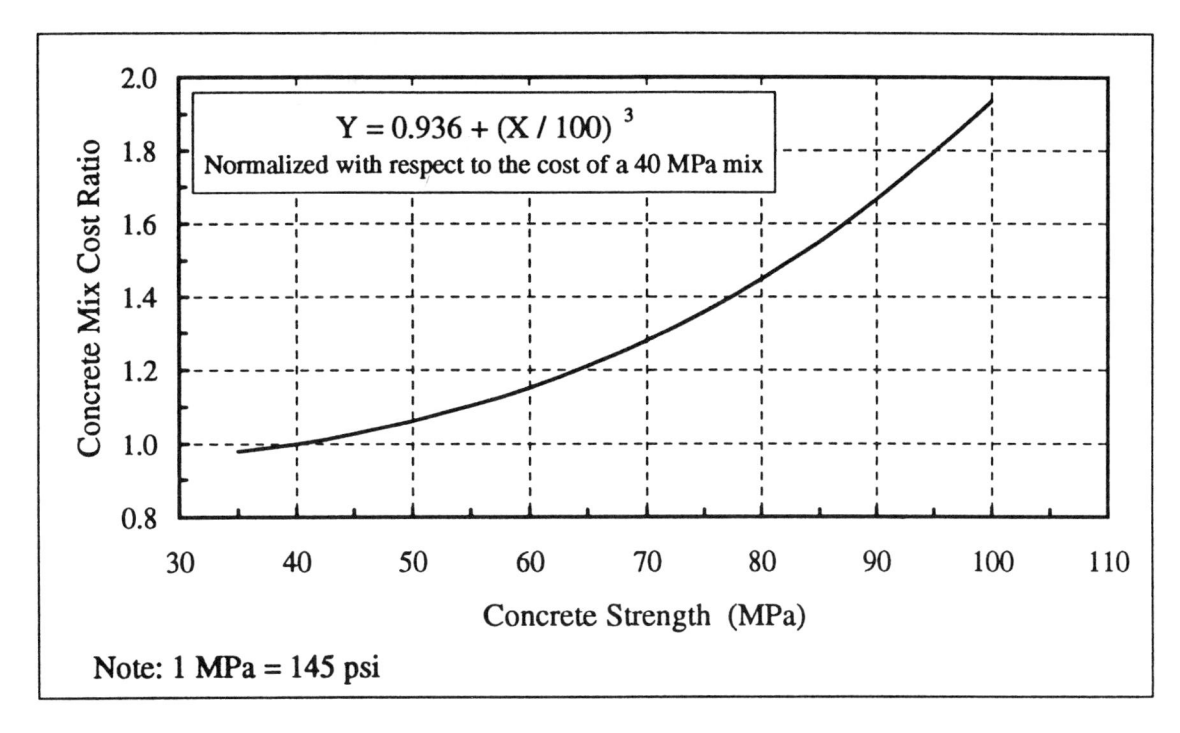

Figure 4. Concrete Mix Cost Ratio vs. Concrete Strength

site among other factors. In this investigation, two levels of transportation and erection costs were considered. Low transportation and erection costs were taken as $6000 per girder, while high transportation and erection costs were assumed to be $12,000 per girder.

Cost Comparisons

Table 1 contains the results obtained from a series of optimal designs performed for the above case study. Review of the data in the table reveals that as a result of the increase in

Table 1 - Results of the generated series of optimal designs

Number of Girders	Optimum Concrete Strength (MPa)	Minimum Superstructure Cost ($/m²)	
		Low Transportation and Erection Costs Utilized	High Transportation and Erection Costs Utilized
3	72.1	130	183
4	51.6	140	212
5	43.4	163	252

Note: 1 MPa = 145 psi, $1/m² = $0.093/ft²

girder capacity, the increase in the concrete cost associated with the use of HPC can be fully offset by the reduced number of girders. For example, at low values for transportation and erection costs, using concrete with a normal strength of 43.4 MPa (6293 psi) would require 5 girders at a total superstructure cost of $163/m² ($15/ft²). Only 3 girders would be required if the concrete strength specified was 72.1 MPa (10,455 psi) resulting in a total cost of $130/m² ($12/ft²). Thus, a saving of about $33/m² ($3/ft²) is achieved. For this two span bridge, the superstructure cost savings would have exceeded $22,000. Furthermore, the achieved savings increase dramatically for girders with higher transportation and erection costs; for this case, the total savings would have exceeded $46,000. Figure 5 illustrates graphically the information contained in Table 1. From the

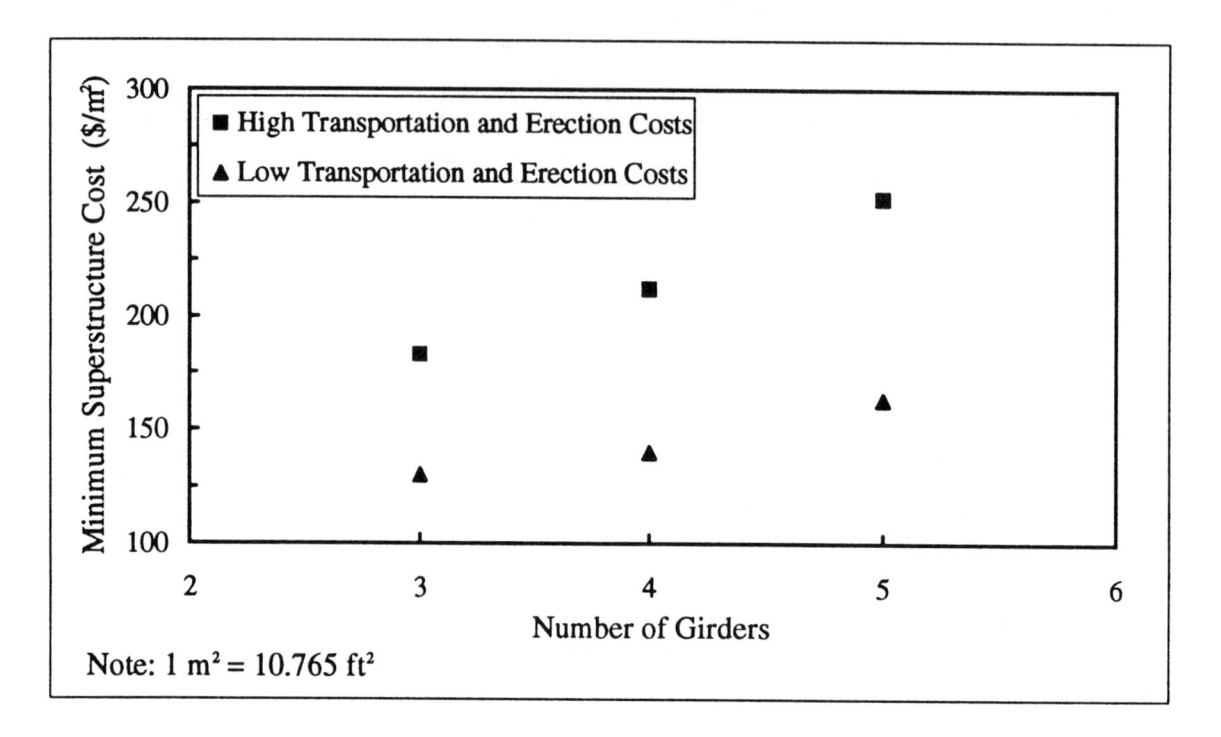

Figure 5. Minimum Superstructure Cost vs. Number of Girders

figure, it is clear that as the number of girders decreases as a result of specifying concretes of higher strength, the total superstructure cost also decreases. It can also be noted that the reduction in the superstructure cost associated with high transportation and erection costs for the above case is much higher than that associated with lower transportation and erection costs. It should be emphasized that the above costs exclude the costs of such items as slab formwork, wearing surfaces, barriers, diaphragms and drains.

It is worth mentioning that the trend of cost reduction in Figure 5 has prompted the trial of a design involving only 2 girders spaced at 6.0 m (19.7 ft). It was found that such a design would require the girders to be made with concrete of a strength of 134 MPa (19,430 psi). This concrete strength is beyond the current practical range. Therefore, the case of 2 girders was left for further investigation.

CONCLUSION

The potential economic advantages from the utilization of HPC with high strength and increased durability for the production of standard bridge I-girders are quite promising. The results presented in this paper illustrate the cost effectiveness associated with the utilization of HPC for girders. As a result of the increase in girder capacity, the increase in the concrete cost associated with the use of HPC can be fully offset by the reduced number of girders. It is apparent that HPC should enable the design and construction of more economical prestressed concrete highway bridges.

ACKNOWLEDGEMENT

The authors gratefully acknowledge the financial support for this study provided by Concrete Canada (The Network of Centres of Excellence on High-Performance Concrete), and Alberta Economic Development and Tourism.

REFERENCES

1. Dolan, C. W. and LaFraugh, R. W., "High Strength Concrete in the Precast Concrete Industry," *PCI Journal*, V. 38, No. 3, May-June 1993, pp. 16-19.

2. Rabbat, B. G. and Russell, H. G., "Optimized Sections for Precast Prestressed Bridge Girders," *PCI Journal*, V. 27, No. 4, July-August 1982, pp. 88-104.

3. Russell, B. W., "Impact of High Strength Concrete on the Design and Construction of Pretensioned Girder Bridges," *PCI Journal*, V. 39, No. 4, July-August 1994, pp. 76-89.

4. Hassanain, M. A. and Loov, R. E., "The Effect of Increasing the Concrete Strength of Bridge Girders on Span Capability and Spacing," *Proceedings of the 1997 Annual Conference of the Canadian Society for Civil Engineering*, Sherbrooke, Quebec, Canada, V. 7, pp. 359-368.

5. *Ontario Highway Bridge Design Code* (3rd edition), Ministry of Transportation, Ontario, 1992.

6. Dorton, R. A., "Development of Canadian Bridge Codes," *Developments in Short and Medium Span Bridge Engineering '94*, The Fourth International Conference on Short and Medium Span Bridges, Halifax, Nova Scotia, Canada, pp. 1-12.

7. Kulka, F. and Lin, T.Y., "Comparative studies of medium-span box girder bridges with other precast systems," *Canadian Journal of Civil Engineering*, V. 11, September 1984, pp. 396-403.

8. Dunker, K. F. and Rabbat, B. G., "Performance of Prestressed Concrete Highway Bridges in the United States - The First 40 Years," *PCI Journal*, V. 37, No. 3, May-June 1992, pp. 48-64.

9. Jaeger, L. G., Bakht, B. and Aly, A., "Revisiting Simplified Analysis of Multi-Span Girder Bridges," *Proceedings of the 1996 Annual Conference of the Canadian Society for Civil Engineering*, V. IIa, Edmonton, Alberta, Canada, pp. 431-442.

10. Jaeger, L. G. and Bakht, B., "Bridge analysis by the semicontinuum method," *Canadian Journal of Civil Engineering*, V. 12, No. 3, September 1985, pp. 573-582.

11. Arora, J. S., *IDESIGN User's Manual, Version 3.5.2*, Technical Report No. ODL-89.7, Optimal Design Laboratory, College of Engineering, The University of Iowa, Iowa, June 1989.

12. *Design Manual* (3rd edition), Canadian Prestressed Concrete Institute, Ottawa, 1996.

13. Lounis, Z. and Cohn, M. Z., "Optimization of Precast Prestressed Concrete Bridge Girder Systems," *PCI Journal*, V. 38, No. 4, July-August 1993, pp. 60-78.

HIGH PERFORMANCE CONCRETE - RESEARCH AND FORECAST FOR APPLICATION IN BRIDGE STRUCTURES IN POLAND

Wojciech Radomski
Warsaw University of Technology
Institute of Roads and Bridges
Al. Armii Ludowej 16, 00-637 Warszawa, Poland

ABSTRACT

The needs concerning improvement in durability of existing and new bridges in Poland are briefly presented. Application of high performance concrete (HPC) to improve the bridge durability is analysed in the light of technical and economical aspects according to the Polish conditions, including the programme of construction of a new network of motorways. Current state and material possibilities to produce HPC in Poland are presented. Special attention is paid to the research on HPC carried out currently in Poland. Design problems concerning the structural applications of HPC are listed and discussed, focussing on bridge structures. Concluding remarks include a summary of research on HPC in Poland and its potential future applications for bridge structures.

INTRODUCTION

It is widely known that high performance concrete (HPC) is used in various structural applications not only because of its high compressive strength but also because of other qualities. According to Y. Malier[1], a review of structures built throughout the world up to 1990 reveals that the use of HPC would be economically justifiable in only 15 to 25 percent of them if high compressive strength were the only criterion, which seems to be true even at present. For instance, in the case of bridge structures, HPC is mainly used because of its high short-term strength and improvement in durability compared with normal strength concretes. The use of HPC leads also to reduction of the total concrete volume, up to more than 30 percent in some cases[2], and allows to reduce the number of precast girders or to increase their span in composite bridges[3], compared with application of normal concrete. Although the unit price of HPC is generally higher than that of normal strength concrete, the use of HPC is economically profitable in many applications.

However, it should be emphasized that HPC is also known to be more brittle than normal concretes. Therefore, it requires to be reinforced in a special manner in structural applications using traditional reinforcing bars or, very rarely, using fibre reinforcement, for example in Japan, where carbon fibre has been used in bridge columns for retrofitting against seismic forces[4].

HPC has been extensively studied during the last decade in several countries, mainly in Canada, France, Germany, Japan, Norway, and U.S.A. Many experiments prior to the structural applications have been performed on this concrete. However, the use of HPC throughout the world is still evidently less than that of normal strength concretes and therefore HPC may be considered as relatively new material. Morover, a lack of standard or code requirements concerning design of structures using HPC contributes to the limited use of the material.

The Polish experience concerning the applications of HPC, including bridge structures, is rather poor so far, mostly because of economical reasons existing up to 1990. However, an evident growth of interest concerning HPC and its recent use in various types of structures can be noticed. It has been reflected among others in the special issues of the Polish engineering journals and increasing number of papers prepared for national and international conferences. In bridge engineering, this growth of interest results mostly from the needs to improve the durability of structures and the new technical and economical problems required to be solved in connection with the development of road network in Poland, including the construction of new motorway network. Moreover, the increased use of modern bridge construction methods in Poland during the last years, for example, cast-in-place balanced cantilever or incremental launching, are also very favorable factors to the increased applications of HPC. Its high short-term strength allows to advance a progress in the construction works.

DURABILITY PROBLEM

About 29,000 bridge structures with a total lentgh of about 540 km (340 miles) are located on the public road network in Poland. About 22,000 of them with a total length of obout 380 km (240 miles), i.e. about 75% of the total number and about 70% of the total length, are concrete bridges. The total number of railway bridges in Poland is about 8,800. About 5,500 (i.e. about 60%) are concrete and masonry bridges.

The bridge stock in Poland is relatively old. About 50% of highway bridges and about 80% of railway bridges have been more than 50 years in service. According to the last report published by the General Directorate of Public Roads[5], nearly 20% of the total number of highway bridges are structurally deficient or functionally obsolete and therefore require to be repaired, strengthened or modernized. The situation described above may be considered similar to many other countries in the world, including highly developed ones. In some of them, the situation is even more critical.

The above information indicates that durability of existing bridges, especially constructed with the use of normal strength concrete, has shown to be highly insufficient.

A progress in the research on various non-conventional materials and their applications in concrete bridges to improve the durablity has been observed during the past several years. It relates mostly to testing and the practical use of several types of fibrous concrete, synthetic materials, sealing of existing exposed concrete surfaces using various chemical products and above all in the area of HPC.

HPC IN POLISH BRIDGE ENGINEERING

Current State and Potential Needs

Polish experience concerning the applications of HPC in bridge structures is very poor so far. The ordinary B30, i.e. concrete with the standard compressive strength 30 MPa (4.35 ksi), to B35 concretes are generally used for reinforced cast-in-place concrete structures, while B35 to B40 concretes are used for precast prestressed (mostly pretensioned) elements. However, improvement in concrete quality has been noticed in the past few years. For instance, the superstructure of Border Bridge in Cieszyn, 760 m (2,492 ft) long, constructed in 1991 using incremental launching, located between Poland and Czech Republic, used B45 concrete. The superstructure of the motorway bridge over Vistula River near Toruń, 955 m (3,131 ft) long, with the three central spans of 130 m (426 ft), constructed using cast-in-place cantilever balance, located in the central part of Poland, is being built with the strength not less than B50.

After a relatively long period of rather limited development of bridge engineering in Poland, when about 90% of concrete bridge structures have been constructed with the use of precast elements, an evident trend to apply the other erection methods has ben observed recently. This trend results from the new needs related to the development of the highway network in Poland and the necessity to construct many new bridges, including bridges over the big Polish rivers. According to 'Drogownictwo' report[6], it is necessary to construct by the year 2005, fourteen relatively large bridges with a total span length of 6,600 m (21,640 ft), excluding the approach structures. A major part of these bridges will be constructed of concrete or as composite with steel girders (beam or truss) and the concrete deck.

The most important task for bridge engineering in Poland in the near future is construction of a number of bridges in connection with the Project of Motorway Construction. This Project also includes three new transeuropean motorways A1 (Nord-South, Helsinki-Łódź-Budapest), A2 (West-East, Berlin-Warsaw-Moscow) and A4 (West-East, Berlin-Cracow-Lvov) with a total length of nearly 2,000 km (1,240 miles), which should be constructed by the year 2010. It also requires more than 500 bridge structures with a total length of more than 35 km (2.2 miles) and about 1,600 culverts. Many other bridge structures are or will be constructed in the next few years in connection with the general development of road network in Poland.

The basic criteria for designing the motorway bridge structures has been defined as[7]: durability, minimum manitenance cost, structural solutions enabling easy inspections, replacement of bearings, expansion joints, etc. Most tasks to be accomplished in the next few years demand development of modern construction methods and assurance of high durability. One of the ways to improve durability is the use of HPC. It is demonstrated among others by the projects submitted in December 1996 for competition concerning a new bridge over Vistula River in Płock (central Poland). From fourteen projects, six concrete bridges of various structural systems have been proposed. In five of them, B60 concrete as a minimum has been proposed. One of the award winning projects, concrete cable-stayed bridge with concrete B60, is shown in Figure 1.

The trend to use HPC is also endorsed by the bridge structure in Chabówka (south of Poland), 268 m (879 ft) long, constructed by incremental launching, completed in October 1996. The superstructure of this bridge is made of concrete B60.

Material Availability

This problem has been considered in detail by Z. Jamroży[8]. Generally, there are no major obstacles in obtaining HPC in Poland because of the following factors:

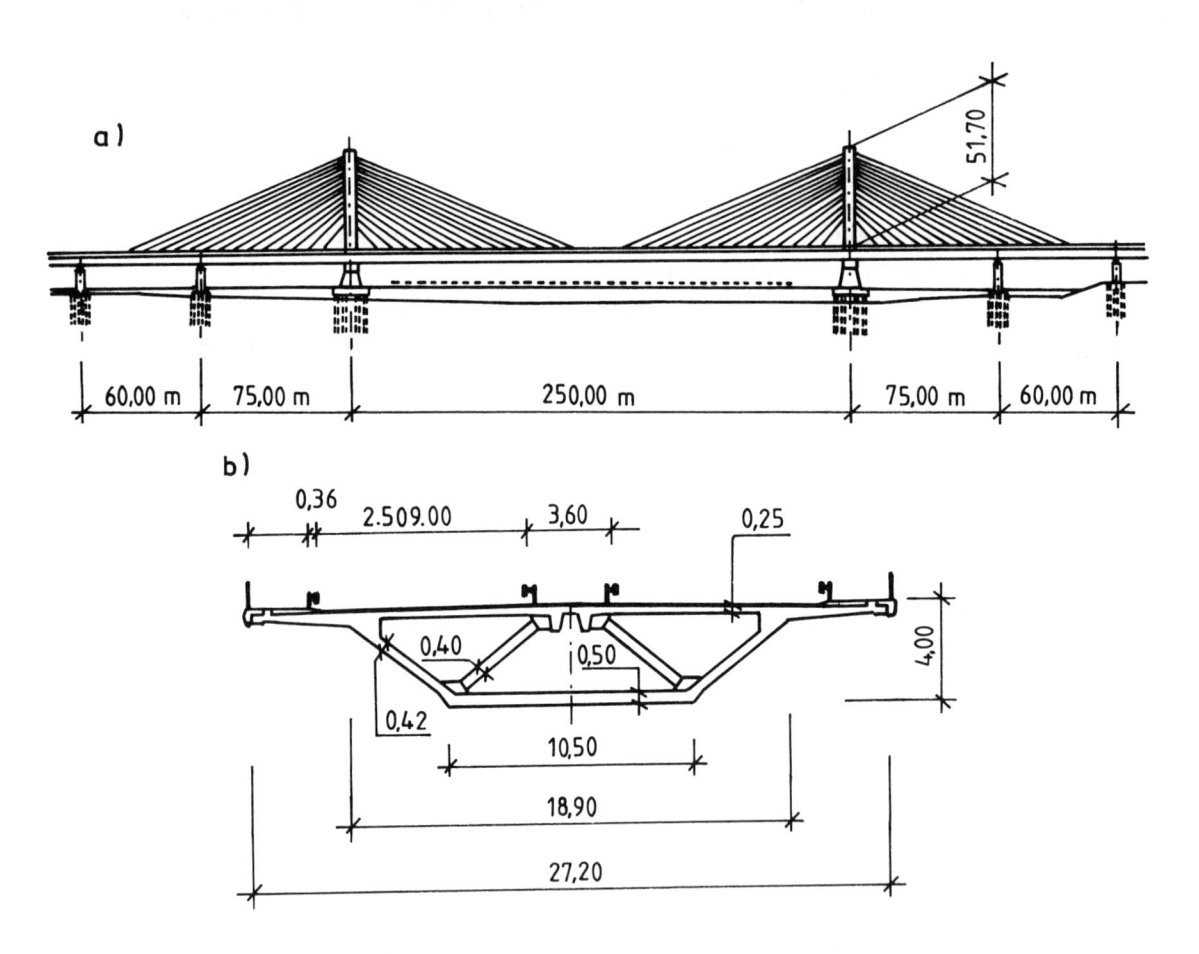

Figure 1. Cable-Stayed Concrete Bridge Over Vistula River in Płock. Competition Project. Elevation (a) and Cross Section (b). (Note: 1 m = 3.3 ft)

(a) Several types of high quality cement are made in Poland.
(b) Deposits of good quality granite, basalt, carbonate and rounded coarse aggregate quality are available in Poland.
(c) Several types of Polish superplasticizers of a ralative high quality and many superplasticizers from abroad are available.
(d) Microsilica is also available in Poland but its quality is not very high.

Taking into account some special requirements for bridge construction process, direct influence of climate and environmental factors, heavy traffic, durability, etc., the use of HPC seems justified using microsilica and superplasticizers as well as other high quality admixtures and additives (e.g. retarder if necessary) from abroad and cement and aggregate from Poland.

Economical Problems

The economical comparison between concretes B40 and B60 performed in the construction of a viaduct in Warsaw (208 m (682 ft) long, completed in 1992) has shown that the total unit cost of concrete B60, including transport and quality control costs, is about 16% higher[9].

A particular analysis concerning the costs of normal strength and HPC was performed by J. Brzezicki and J. Kasperkiewicz[10]. They also compared the relevant costs in some other countries. The results of this analysis are summarized in Table 1. All the costs are referred to the cost of cement. The average cost of 1 ton of cement is different depending on the country (Poland about $45, Japan - $90, Great Britain - $100, Sweden - $85).

Table 1 - Aproximate Relative Cost of Constituent Materials For Concrete

Material	Poland	Japan	Great Britain	Sweden
Cement	1.0	1.0	1.0	1.0
Natural aggregate	0.08 - 0.15	0.13	0.09 - 0.11	0.095
Crushed aggregate	0.09 - 0.14	0.13	0.10	0.095
Microsilica	1.1	10.1	2.0	1.6
Superplasticizers	12 -25	16 -25	9 - 14	4 - 8
Air Entraining Adm.	30 - 50	22.7	6.8	23.8
Concrete	0.7 - 1.5	1.1 - 1.3	0.6 -1.0	1.2

It can be seen that the unit cost of concrete ranges from 0.6 to 1.5 of the unit cost of cement. However, the decisive factor seems to be the cost of microsilica and admixtures.

Silica fume in Poland is relatively inexpensive but its quality is lower than that from some other countries.

The cost of concrete itself in bridge structures in Poland is approximately 15 % -20 % of the total construction cost. Therefore, the use of HPC instead of conventional concrete increases the total construction cost by a maximum of 10%. The gain resulting from much better durability justifies this additional cost. It should be noted, however, that the total volume of concrete is generally less when HPC is used.

Design and Research Problems

HPC has considerably distinct characteristics compared with normal concrete. Particularly in the design stage, the increased brittleness of the material requiring a special reinforcement (especially the transversal one) should be kept in mind. Unfortunately, the new concrete classes (beginning from 60 MPa concrete) are beyond the scope of Polish and many other national or international standards or codes and design methodologies. However, a number of tests performed on HPC specimens or structural members recently, provide additional data for design[11].

Structural application of HPC demands that extensive tests be performed prior to its use. Several trial mixes to make choice of the optimum mix are recommended. The following characteristics of HPC should be tested in particular: quality of constituent materials, compressive strength, shear strength, bond strength between concrete and reinforcing bars, hydration and thermal effects and their influence on the mechanical properties of the material, brittleness of the material and the influence of the reinforcement on the behaviour of the structural members under various types of loading, permeability, freeze-thaw resistance and chemical resistance. There is very limited information on the fatigue strength of HPC. For practical purposes, the comparative tests related to Polish and other materials are recommended.

POLISH RESEARCH ON HIGH PERFORMANCE CONCRETE

Research Fields

Research on HPC is more advanced than its structural applications. Investigations on HPC that have been carried out recently in Poland can be classified into the following groups.

(a) Technological research focused on studying the influence of admixtures and additives, aggregate and other factors on the properties of HPC, using Polish and imported products for HPC. Studies concerning optimization of HPC mix using various constituent materials and additives.
(b) Material research focused on some physical phenomena, mostly thermal conditions, in HPC and their influence on its properties, especially the strength development.
(c) Research on bond strength between steel reinforcement and HPC matrix.
(d) Strength tests of HPC structural elements, mostly beams, subjected to different type of loading.

Research classified above into group (a) concerns mainly the applicability of various Polish and other materials and products for HPC and therefore is only briefly presented below. Research classified into groups (b), (c) and (d) is of more general importance and therefore is presented below in greater detail.

Technological Research

Extensive studies with the compressive strength ranging from 60 MPa (8.7 ksi) to 100 MPa (14.5 ksi) have been performed recently by Rajski[12,13]. An analytical model for concrete mix design has been developed and verified experimentally. Concrete mix design of 36 MPa (5.2 ksi) strength after one day has also been recently studied[14] using commercially available constituent materials and admixtures and conventional technique. The factors that were expected to affect the early compressive strength and workability are the water-cemet ratio, total aggregate-cement ratio and curing temperature. Results indicated that superplasticizer causes significant air entraining and retarding effects. The amount of air is directly related to cement content in the concrete mix. The results also showed that it is difficult to obtain such high early strengths without heating the concrete up to 40-50°C (104-122°F).

The effect of superplasticizer and silica fume on workability of HPC has been also studied by J. Szwabowski and J. Gołaszewski[15]. The results showed that workability of HPC depends on the dosage, time of introducing and origin of superplasticizer and the amount of silica fume. An increase of the dosage of superplasticizer, retardation of its introducing into the concrete mix and presence of silica fume improve the workability.

Hydration and Thermal Effects

The thermal stresses caused by self warming of concrete resulting from exothermic process of cement hydration are significant problems during design and construction stages of various structures, especially massive ones. Many investigations have been performed for normal strength concretes, whereas it is not sufficiently recignized as yet in HPC.

Increased amount of cement and presence of silica fume in HPC causes thermal effects that are of prime interest. The relevant research has been performed recently by M. Kaszyńska[16], using a special equipment set to provide the adiabatic conditions of hardening. The mix proportions for tested specimens are listed in Table 2. All the constituent materials, except superplasticizers, are from Poland.

To check the influence of different initial temperatures on hydration thermal effects and the compressive strength of concrete, the following initial temperatures have been applied: 10°C ([50°F], mix designation M10 and MP10, respectively), 20°C ([68°F], M20 and MP2O), and 35°C ([95°F], M35 and MP35).

Table 2 - Mix Proportions For Tested Concrete

Mix Designation	Cement PC45 Kg/m³	Natural Aggregate kg/m³	Water dm³/m³	Fly Ash kg/m³	Microsilica kg/m³	Super-plasticizer dm³/m³
M	450	1786	158	-	30	9
MP	315	1730	158	135	30	9

Note: 1 kg/m³ = 0.062 lb/ft³, 1 dm³/m³ = 1 l/m³ = 0.0063 gal/ft³

These temperatures correspond to the summer and autumn conditions of concreting in Poland. The reference specimens cured in the ambient temperature in the laboratory, i.e. 18±2°C (64±4°F), have shown the compressive strength more than 70 MPa (10.2 ksi) after 28 days. Test results and their analysis are published elsewhere[19]. Some selected results concerning the development of the early strength depending on the initial temperatures and presence of fly ash in the mixes are presented in Figures 2 and 3.

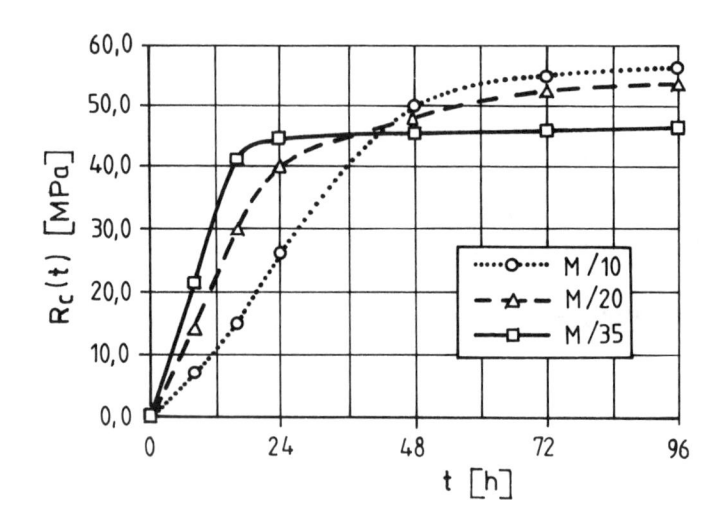

Figure 2. Development of Compressive Strength of Concrete M Cured in Adiabatic Conditions. (Note: 1 MPa = 0.145 ksi)

These test results show effect of initial temperture and fly ash on the development of compressive strength. The investigations have also shown that heat evolution depends strongly on the initial temperature. These effects are radically different from those observed in the normal strength concretes and therefore relations concerning these concretes should be verified in case of HPC.

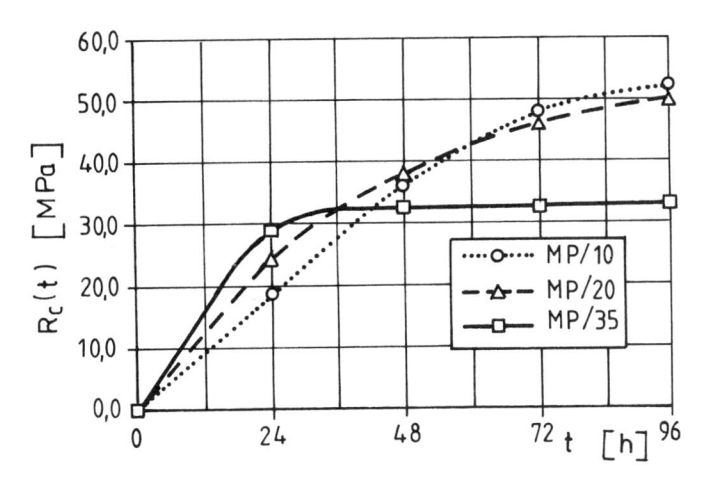

Figure 3. Development of Compressive Strength of Concrete MP Cured in Adiabatic Conditions. (Note: 1 MPa = 0.145 ksi)

Bond Strength

One of the most important factors concerning structural applications of HPC is bond between the concrete and steel reinforcement. This problem has been recently tested by J. Kozicki and D. Ulańska[17]. The comparative studies using normal-strength concrete B30 and HPC B90 and more have been performed. The test results have been also compared with those obtained by other authors and some models for bond strength have been verified. Investigations have been performed on ribbed bars subjected to axial pulling out at anchorage equal to five bar diameters. The influence of concrete compressive (f_c) and tensile ($f_{ct,sp}$) strengths, bar diameter (d) and its form of ribbing (f_R) as well as concrete cover thickness (c) on the bond strength (τ_{du}) have been studied in detail. Some selected results, indicating the differences between model of bond concerning normal strength concrete and HPC are shown in Figures 4 and 5, where the results obtained by other researchers are also presented for comparison.

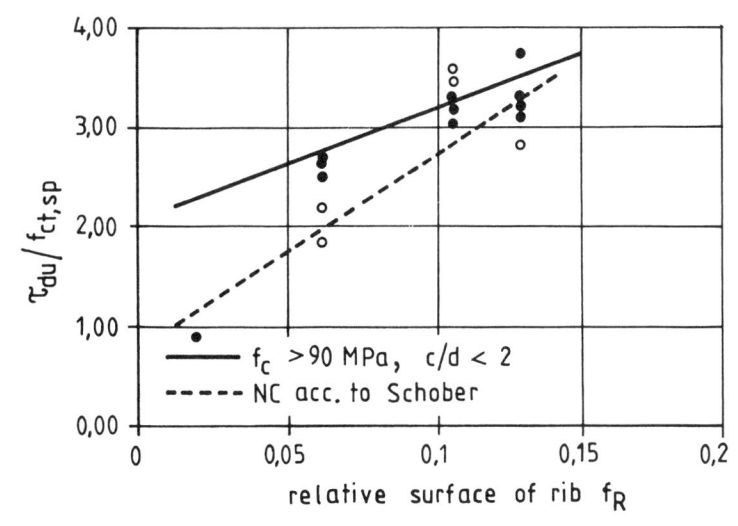

Figure 4. Relation Between Bond Strength and Relative Surface of Rib of the Bar.

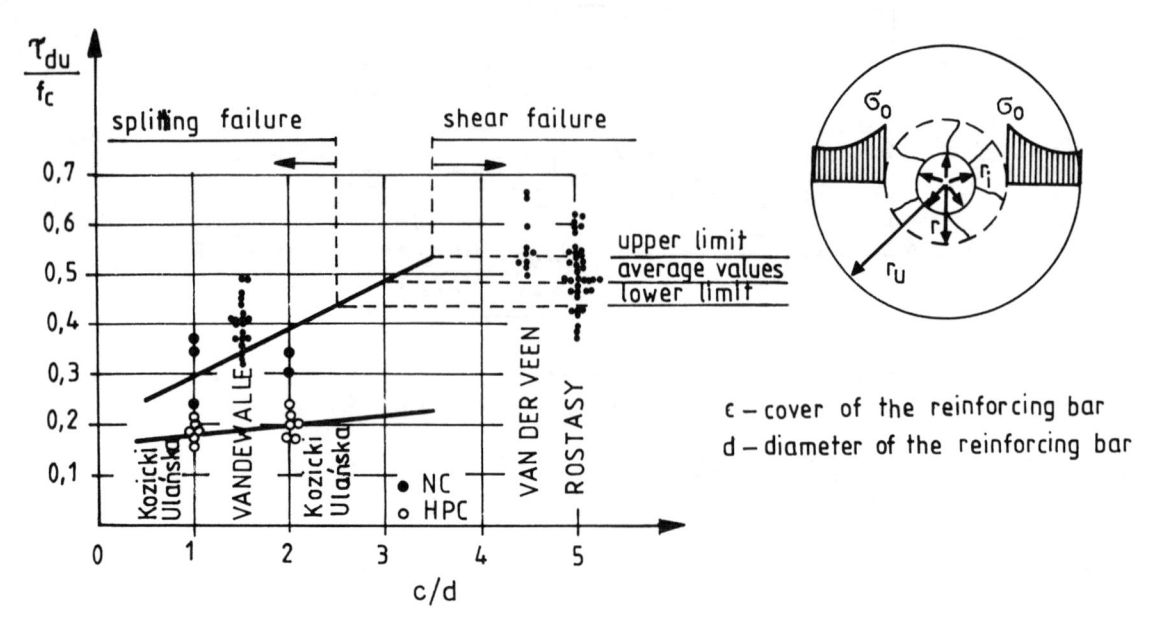

Figure 5. Type of Failure Depending on Parameter c/d and Strength Characteristics of Concretes as well as Model for Analytical Approach

Testing of Structural Elements

A. Kosińska, and A.B. Nowakowski[18] recently tested the reinforced concrete beams made of normal strength concrete B25 and HPC B70 using all materials and additives produced in Poland. Every beam, reinforced as shown in Figure 6, was subjected to pure torsion up to failure.

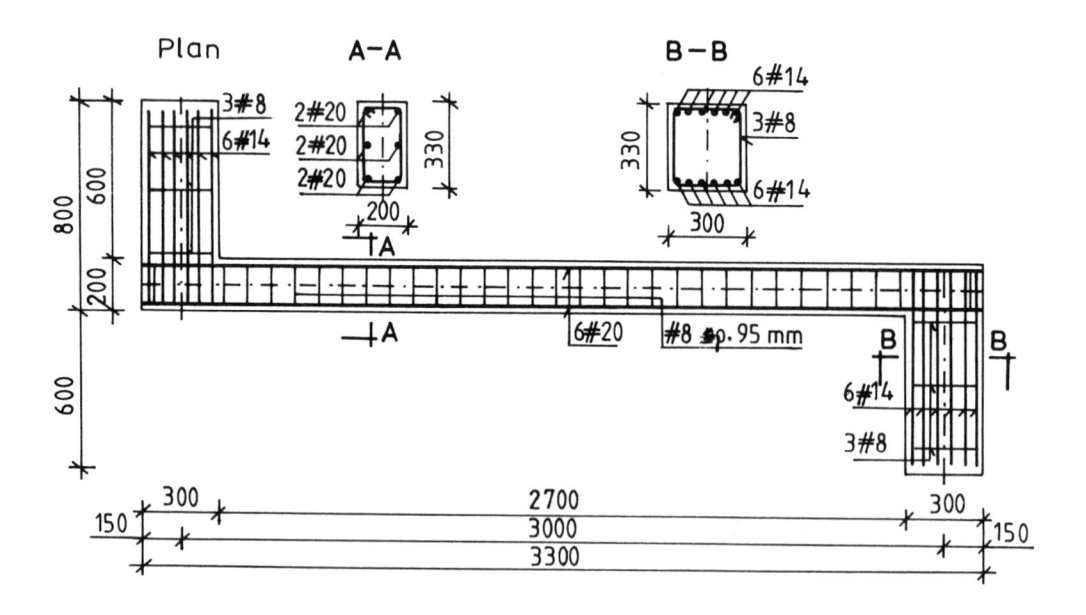

Figure 6. Beam Tested Under Pure Torsion. (Note: 10 mm = 0.394 in)

As expected, torsional stiffness of the beams does not depend on the quality of concrete as shown in Figure 7.

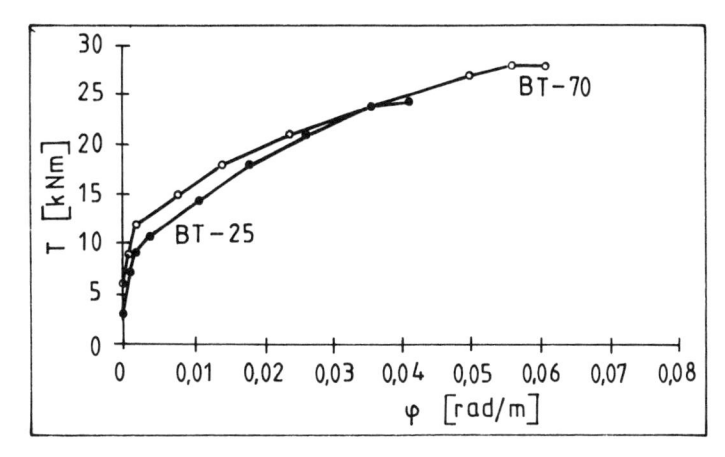

Figure 7. Rotation versus Torsional Moment for NC and HPC Beams.
(Note: 1 kNm = 737.6 lbf/ft)

Tensile stresses in longitudinal reinforcing bars and the stirrups located 0.55 m (0.168 ft) from the support versus torsional moment are shown in Figures 8 and 9, respectively. It can be seen that in both cases the tensile stresses in reinforcement are less in the case of HPC than in the case of normal concrete when the value of torsional moment is the same. It indicates the better effectiveness of reinforcement in HPC due to probably a better bond.

Calculations performed according to Eurocode 2 revealed that average safety factor is equal to 1.20 in case of NC and 1.40 in case of HPC beams. Test results indicate that HPC beams subjected to pure torsion up to failure are more crack resistant, and therefore more durable, and have higher carrying capacity than NC beams. The rotational stiffness of beams is almost the same regardless of the concrete strength and quality.

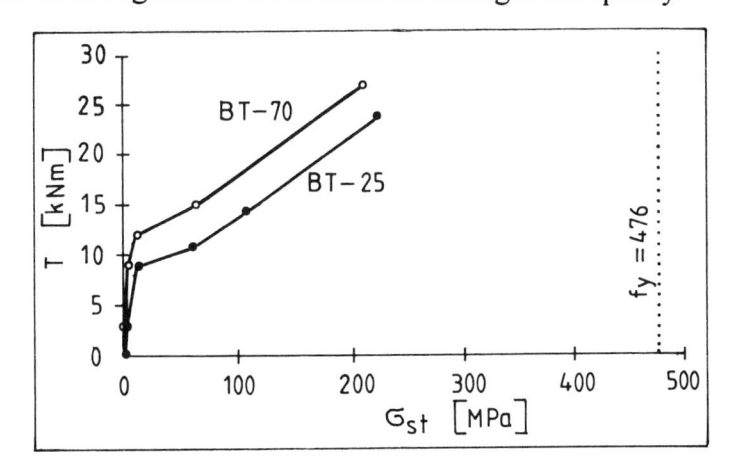

Figure 8. Tensile Stresses in Longitudinal Reinforcement versus Torsional Moment for NC and HPC Beams.
(Note: 1 MPa = 0.145 ksi, 1 kNm = 737.6 lbf/ft)

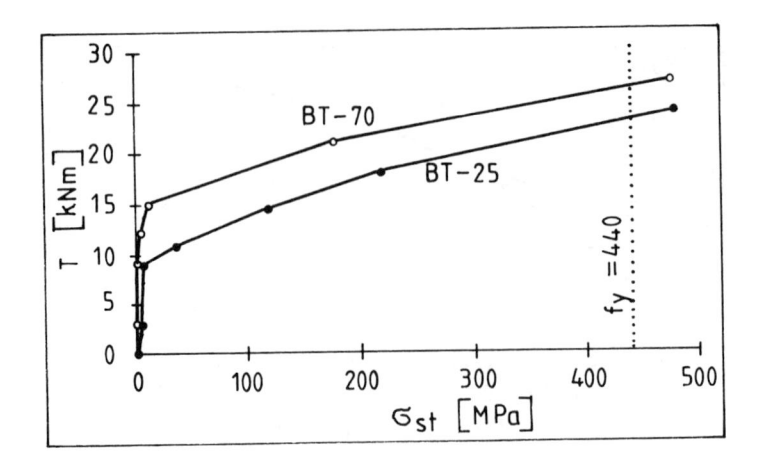

Figure 9. Tensile Stresses in Stirrups versus Torsional Moment.
(Note: 1 MPa = 0.145 ksi, 1 kNm = 737.6 lbf/ft)

CONCLUDING REMARKS

The following conclusions are based on the research review described above:

1. The development of road network in Poland, especially construction of motorways, demands construction of a number of modern bridge structures with high durability. HPC is perceived to be a very viable solution to improve the durability.
2. The modern construction methods, for example, cast-in-place cantilever balance or incremental launching also justifies the use of HPC because of its high short-term strength.
3. There are no severe material or technological obstacles to producing and developing HPC in Poland.
4. Research on HPC prior to its structural application are at present intensively performed in Poland. Fundamental research on HPC is also being performed.
5. Under new economic conditions, transfer of modern technology and research, including HPC, from highly developed countries to Poland, has made substantial progress in recent years.

REFERENCES

1. Malier, Y., "The French Approach to Using HPC", Concrete International, July 1991, pp. 38-32.

2. Malier, Y., Brazzilier, D., and Roi, S., "The Bridge of Joigny", Concrete International, May 1991, pp. 40-42.

3. Russell, B. W., "Impact of High Strength Concrete on the Design and Construction of Pretensioned Girder Bridges", PCI Journal, July-August 1994, pp. 76-89.

4. Mwamila, B. L. M., Mashima, M., and Nakai, H., "Fiber Reinforced Concrete in Japan", Published by the Faculty of Engineering, Osaka University, Japan, V. 30, 1989, pp. 199-214.

5. "General Directorate of Public Roads - Report on the Bridge State on the Public Roads", Drogownictwo, No.1, January 1995, pp. 18-25 (in Polish).

6. "Report on the State of the Large Road Bridges", Drogownictwo, No. 6, June1993, pp. 121-124 (in Polish).

7. Ryżyński, A., "Technical Conference - Engineering Structures in the Construction Programme of the Motorways in Poland", Drogownictwo, No. 8, August 1993, pp. 188-190 (in Polish).

8. Jamroży, Z., "On the Necessity and Possibilities of Introducing of High Performance Concrete in Poland", Inżynieria i Budownictwo, No. 9, September 1993, pp. 361-362 (in Polish).

9. Kowalski, R., and Lewandowska, S., "Has the High Quality Concrete To Be More Expensive?", Przegląd Budowlany, No. 8-9, August-September 1992, pp. 345-347.

10. Brzezicki, J., and Kasperkiewicz, J., "On Costs of High Performance Concretes", Inżynieria i Budownictwo, No. 9, September 1993, pp. 386-387 (in Polish).

11. Radomski, W., "Some Analytical and Technical Problems Connected with Application of High-Strength Concrete", Inżynieria i Budownictwo, No. 9, September 1993, pp. 370-373 (in Polish).

12. Rajski, O., "Proposal of the Proportioning Method of High Quality Concretes" Inżynieria i Budownictwo, No. 2, February 1996, pp 101-102 (in Polish).

13. Rajski, O., "Some Technologicsl Factors of HPCs", Inżynieria i Budownictwo, No. 6, June 1996, pp. 346-350 (in Polish).

14. Wawrzeńczyk, J., Szymczyk, M., and Wąż, S., "Concrete Mix Design of 36 MPa after 1 Day", Proc. XLI Sci. Conf. of KILiW PAN and KN PZITB, Krynica 1996, Vol. 6, pp.103-110 (in Polish).

15. Szwabowski, J., and Gołaszewski, J., "The Effect of Superplasticizer and Silica Fume on Workability of High Performance Concrete", Cement-Wapno-Beton, V. I/LXIII, No. 6, June 1996, pp. 212-214 (in Polish).

16. Kaszyńska, M., "The Effect of Temperature on the Heat of Hardening and Strength of High Performance Concretes", Cement-Wapno-Beton, V. I/LXIII, No. 6, June 1996, pp. 221-223 (in Polish).

17. Kozicki, J., and Ulańska, D., "The Bond between Steel Bars and High Strength Concrete Matrix", Proc. XLI Sci. Conf. of KILiW PAN and KN PZITB, Krynica 1996, Vol. 4, pp. 101-108 (in Polish).

18. Kosińska, A., and Nowakowski, A.B., "The Experimental Investigations on Reinforced High-Strength Concrete Beams Subjected to Pure Torsion", ibid., pp. 93-100 (in Polish).

HIGH-STRENGTH CONCRETE IN BRIDGES—HISTORY AND CHALLENGES

Henry G. Russell
Henry G. Russell, Inc.
Glenview, Illinois, USA

ABSTRACT

The first part of the paper contains a brief history of high-strength concrete in bridges and illustrates that high performance concrete in the disguise of high-strength concrete has been used for many years. The second part of the paper presents some challenges that need to be addressed if the full potential of high-strength concrete is to be achieved. These challenges include ways to effectively use higher strength concretes; greater consideration of prestress losses, vertical deflections, lateral stability and heat of hydration; development of design specifications applicable to concrete strengths greater than 69 MPa (10,000 psi); improved quality control; use of later ages for release strengths and specified strengths and limitations of prestressing bed capacities.

INTRODUCTION

High-strength concrete in the United States is generally considered to be concrete with a compressive strength of 41 MPa (6000 psi) or greater. In precast, prestressed concrete bridges, girders with specified compressive strengths of 34 MPa and 41 MPa (5000 and 6000 psi) at 28 days have been used for many years. However, because mix proportions are generally dictated by release strengths, concrete strengths at 28 days are frequently in excess of the specified 28-day value. Actual strengths of 55 MPa (8000 psi) are often achieved, yet not considered in design. Consequently, the industry has been using high-strength concretes for many years without formal recognition. The performance of these high-strength concrete girders has been excellent. In recent years, the industry has begun to take advantage of the availability of the higher strength concretes and several bridges. have been designed and built with high-strength concrete. The first part of this paper summarizes specific applications of high-strength concrete in bridges. The second part of the paper describes some challenges that need to be addressed in design and production, if the full utilization of high-strength concrete is to be achieved in the future.

HISTORY

Early Developments

Walnut Lane Bridge, built in Philadelphia in 1949, was the first major prestressed concrete bridge in North America. It was a three span structure with a main span of 48.8 m (160 ft). Walnut Lane Bridge consisted of 13 cast-in-place post-tensioned girders with a web thickness of 180 mm (7 in.) and a depth of 2.00 m (6 ft 7 in.) According to Zollman,[1] the specification called for a 50-mm (2-in.) slump concrete with a compressive strength of 37 MPa (5400 psi) at 28 days. Because it was difficult to place the 50-mm (2-in.) slump concrete, the slump was gradually increased to 65, 75 and 90 mm (2-1/2, 3 and 3-1/2 in.) This made it more difficult to achieve the strength which required 28 days instead of 15 days with the 50-mm (2-in.) slump concrete. Nevertheless, 37 MPa (5400 psi) concrete in 1949 must be considered high-strength concrete—thus making Walnut Lane Bridge the first high-strength concrete bridge in North America.

In 1951, Art Anderson[2] achieved concrete compressive strengths of 69 MPa (10,000 psi) with "no-slump" concrete in precast members. Concrete strengths much beyond 21 MPa (3000 psi) were a rarity in North America at that time. Anderson also reported that 28-day strengths exceeding 52 MPa (7500 psi) were routinely achieved without steam curing. In the late 1950's, Dr. Anderson's company—Concrete Technology Corporation—produced precast beams with a concrete compressive strength of 62 MPa (9000 psi) for a 21-story building.[3] At about the same time, they also developed a bulb-tee cross section which combined pretensioning and post-tensioning—a concept that could still become increasingly important with high-strength concrete.

Washington State Bridges
In the early 1980's, several bridges[4-6] were built in the State of Washington using concrete with specified strengths at 28 days in excess of 41 MPa (6000 psi). Tower Road Bridge, shown in Fig. 1, has a span length of 49 m (161 ft) and was reported to be the first major bridge application in the United States where 62 MPa (9000 psi) concrete at 28 days was specified for the girders. The cross section comprised six modified bulb-tee girders with a depth of 1.5 m (5 ft) spaced at 1.6 m (5 ft 5-1/2 in.). The girders were post-tensioned to obtain continuity—an application of Anderson's concept of a post-tensioned high-strength concrete bulb tee. The specified compressive strength of the 115-mm (4-1/2-in.) thick deck was 41 MPa (6000 psi). Use of the 62 MPa (9000 psi) concrete allowed the use of relatively shallow girders, thereby minimizing roadway approach costs.

Braker Lane Bridge
Braker Lane Bridge[7] over Interstate I-35 at Austin, Texas, was built in 1986 using Texas Type C girders with a depth of 1015 mm (40 in.). The girders have a span length of 25.9 m (85 ft) and are spaced at 2.56 m (8.4 ft). Required compressive strengths were 51 MPa (7400 psi) within 17 hours and 66 MPa (9600 psi) at 28 days. This allowed a design with increased girder spacing and, consequently, fewer girders compared to a design with normal strength concrete. For example, at a 25.9-m (85-ft) span length, girders with a concrete compressive strength of 41 MPa (6000 psi) would require a girder spacing of 1.52 m (5 ft). The release strength at 17 hours controlled the concrete mix design. The actual concrete compressive strengths at 28 days averaged 92.4 MPa (13,400 psi) with a coefficient of variation of 7.4%. Although it did not receive a great deal of publicity, Braker Lane Bridge represented a significant step forward in the actual usage of high-strength concrete in bridge girders outside Washington State.

Figure 1 - Tower Road Bridge

East Huntington Cable-Stayed Bridge
The East Huntington Bridge,[8] built in 1984, is a segmental prestressed concrete cable-stayed bridge over the Ohio River between the states of West Virginia and Ohio. The main span is 274.3 m (900 ft). In a cable-stayed bridge, the superstructure is a long compression member designed to resist the horizontal components of the forces in the cable stays. Also, in long-span bridges, the superstructure weight is the predominant load for which the structure must be designed. Consequently, the selection of high-strength concrete results in a lighter structure to resist the longitudinal compressive forces. In the case of the East Huntington Bridge, high performance concrete was selected because of its compressive strength as well as its improved durability and higher tensile strength. Specified concrete compressive strength for the superstructure elements was 55 MPa (8000 psi) at 28 days. Actual strengths averaged 68 MPa (9900 psi) at 28 days. In competitive bidding against a steel box-girder alternate design, the concrete option was bid at 10 million US dollars less than the steel alternate.

Annacis Cable-Stayed Bridge
The Annacis Bridge,[9, 10] also named the Alex Frazer Bridge, was built in 1986 and crosses the Frazer River near Vancouver, British Columbia, Canada. The cable-stayed portion of the structure consists of five continuous spans with a center span of 465 m (1526 ft). The composite superstructure consists of twin 25.3-m (83-in.) deep steel I-beams of constant depth, transverse floor beams that taper in depth, and a composite precast concrete deck with a cast-in-place overlay. The precast deck panels have a specified compressive strength of 55 MPa (8000 psi) at 56 days. Composite action between the precast concrete elements and the steel framework is achieved through shear studs welded to the floor beam top flanges. Deck panels are integrated with the steel superstructure by cast-in-place strips of concrete with a specified strength of 55 MPa (8000 psi) at 56 days. A cost comparison between a composite concrete deck and an orthotropic steel deck indicated that the total cost with a concrete deck was substantially less.

Normandy Bridge
Completed in 1994, the Normandy Bridge is a cable-stayed bridge across the River Seine in Normandy, France. It has a main span length of 856 m (2808 ft) which makes it the world's longest cable-stayed bridge.[11] The center 624 m (2047 ft) of the main span superstructure consists of a steel box girder. The remaining portion of the main span and the back spans are concrete box sections. The pylons consist of composite steel and concrete boxes. By increasing the concrete strength from 40 MPa (5800 psi) to 60 MPa (8700 psi), the pylon wall thickness was reduced from 600 mm to 400 mm (24 to 16 in.) and the thickness of the bottom flange of the concrete box girder in the side spans was reduced from 200 mm to 180 mm (8 to 7 in.).[12]

Lacey V. Murrow Bridge
The Lacey V. Murrow replacement floating bridge [13] is 2012 m (6600 ft) long and was opened to traffic in 1993. The bridge consists of 20 prestressed concrete pontoons connected to form a continuous structure. The structural design required a compressive strength of 45 MPa (6500 psi) at 28 days. In addition, the concrete was required to have a maximum chloride permeability of 1000 coulombs and a shrinkage not exceeding 400 millionths at 28 days. The latter two requirements controlled the mix design and actual strengths exceeded 69 MPa (10,000 psi). The concrete was high performance for strength, permeability and shrinkage.

Northumberland Strait Crossing
The Northumberland Strait Crossing,[14, 15] shown in Fig. 2, is a 12.9-km (8-mile) long bridge that links Prince Edward Island with New Brunswick in Canada. Completed in 1997, the bridge is designed for a service life of 100 years. The main portion of this bridge consists of 43 spans each 250 m (820 ft) long. The spans consists of two continuous precast, prestressed concrete variable depth cantilevered girders with a length of 95 m (312 ft) each and a drop in span of 60 m (197 ft). All of the substructure and superstructure components for the main spans were precast and floated out and erected using a large floating crane. The main girder was precast segmentally using a balanced cantilever approach. Total weight for the 190-m (625-ft) long girder is approximately 8200 tonnes (8820 tons). Because of the aggressive environment in the Northumberland Strait, the majority of concrete used for the superstructure has low permeability and a compressive strength of 55 MPa (8000 psi). For some piers, the ice shields utilize concrete with a compressive strength of 80 MPa (11,600 psi) to resist abrasion damage. The Northumberland Strait Crossing, also known as The Confederation Bridge, represents the largest usage of high-strength concrete in a prestressed concrete bridge in North America.

FHWA Showcase Projects
The Federal Highway Administration, in cooperation with state highway departments, has a current program to construct demonstration showcase bridges[16] with high performance concrete. The objective is to advance the use of high performance concrete in order to achieve economy of construction and enhance long-term performance. Projects are

Figure 2 - The Confederation Bridge

underway in Alabama, Colorado, Georgia, Indiana, Nebraska, New Hampshire, North Carolina, Ohio, Texas, Virginia, Washington, and Wisconsin. Actual bridges are already constructed in Nebraska, New Hampshire, Texas, Virginia and Washington. Durability and economical long-term maintenance have become goals along with economical construction costs. The demonstration bridges are being constructed in a variety of climates and with different cross sections for the prestressed concrete girders.

The advantages of using high performance concrete in prestressed concrete girders occur because the higher concrete strengths allow the use of larger prestressing forces. This means that a given girder size can be designed for a longer span length. Alternatively, for a fixed span length, fewer girders are required for the same width of bridge, or shallower girders may be used. In all cases, the use of higher strength concrete can result in lower initial costs. There are, however, strength limitations beyond which the use of higher strength concretes is not beneficial. The strength limitation varies from 69 to 97 MPa (10,000 to 14,000 psi) depending on the cross-sectional shape. Beyond these strength levels, sufficient prestressing force cannot be incorporated into the cross section to take advantage of the higher concrete strength. The use of 15-mm (0.6-in.) diameter strand is advantageous in this regard. At the present time, concrete strengths up to 69 MPa (10,000 psi) should be considered in the design of all long-span prestressed concrete girders with a minimum design strength of 55 MPa (8,000 psi).

CHALLENGES IN DESIGN

Strength Limitations
Several research studies[17-20] on the use of high-strength concrete in prestressed concrete

girders have concluded that the use of high-strength concrete results in a larger span capability for the same depth of section. Alternatively, for the same span length, fewer girders are needed for the same width bridge or shallower girders may be used.

The use of high-strength concrete in long-span simply supported precast, prestressed concrete beams was investigated in a series of parametric studies by Zia, et. al.[17] Their results indicated that longer span lengths can be achieved with higher strength concretes. However, when the compressive strength was increased beyond a certain strength level, there was little or no benefit to be gained. The strength level at which the use of higher strength concrete was not beneficial varied from 55 to 83 MPa (8,000 and 12,000 psi) depending on the cross-sectional configuration and the prestressing force. Zia also found that shallower sections with higher strength concretes could be used in place of larger sections with lower strength concretes.

In a study at the University of Texas, Castrodale[18] determined the maximum span lengths that could be used with different girder types, girder spacings and girder concrete compressive strengths. He showed that maximum span lengths can be increased through the use of higher strength concretes. However, for all cross sections analyzed, the rate of increase in span length decreased as concrete compressive strength increased. Reduced benefits were achieved at the higher strength levels.

The results of Zia and Castrodale were further confirmed by Russell[19] in a parametric study comparing girder spacings and span lengths for four different cross sectional shapes. The results clearly showed that the use of high-strength concrete can allow longer bridge spans for concrete girders made with standard shapes or can decrease the number of girders required for a given design case. Russell concluded that, through the use of high-strength concrete, longer spans can be achieved using shallower sections in place of deeper cross sections. The author recommended that the use of high-strength concrete should be employed to maximize span lengths, increase girder spacings or reduce the depth of structures.

The advantages of going beyond about 69 MPa (10,000 psi) concrete strength in standard solid sections are limited because sufficient prestressing force cannot be incorporated into the cross section to take advantage of the higher strength concretes.[20] More prestressing force can be incorporated by using closer strand spacing, larger diameter strand, higher grade strand, post-tensioning, less assumed prestress losses and girders with a large bottom flange. For the industry to take full advantage of the availability of higher strength concretes, one or more of the means listed above must be utilized. Otherwise, the industry will never go much beyond the use of 69 MPa (10,000 psi) compressive strength concrete.

Prestress Losses
When compared to the properties of conventional strength concretes, high-strength concrete has a higher modulus of elasticity, a higher tensile strength and reduced creep. The higher modulus of elasticity results in less elastic shortening in prestressed concrete

girders at time of release for the same stress level. This reduction in shortening may be offset by the use of higher prestress levels with high-strength concrete. The higher tensile strength does not directly affect prestress losses but allows high-strength concrete girders to be designed for a higher permissible tensile stress. This, in turn, results in higher service load design moments.

Creep per unit stress of high-strength concrete is lower than creep for conventional strength concretes.[21] Thus, the direct substitution of a high-strength concrete in place of a lower strength concrete will result in less prestress losses. However, the utilization of a higher level of prestress will offset the reduction in creep per unit stress. The magnitude of the net result will depend on the reduction in creep per unit stress and the increase in stress level. Shrinkage of high-strength concrete used in prestressed concrete girders is equal to or less than the shrinkage of conventional strength concretes. This may result more from the effect of temperature during curing than from changes in the concrete constituent materials.

Since current methods for calculating prestress losses are based on conventional strength concretes, their applicability with high-strength concrete needs to be assessed. Available information indicates that prestress losses will be lower with high-strength concrete. Consequently, a reduction in prestress losses means that for the same amount of initial prestress, a greater force is available for design at service load. Since the amount of force available at service load controls the design of long-span girders, reduced prestress losses will be beneficial in the more effective utilization of high-strength concrete.

Vertical Deflections
The initial camber at release of the prestressing strands and subsequent changes in camber are determined by the modulus of elasticity, creep and shrinkage of the concrete. Since high-strength concrete has different properties from conventional strength concrete, an assessment of this on prediction of initial camber and long-term deflections is needed. Figure 3 shows a comparison of calculated midspan deflection versus time for 83 MPa (12,000 psi) compressive strength BT-72 girder with three different span lengths. The calculations were made using a step-by-step time-dependent analysis, assuming that the deck was cast at 83 days and became effective at 90 days. Deflections ranged from a camber of 75 mm (3 in.) to a sag of 100 mm (4 in.) depending on the span. It should be noted that the longest span girder does not have the highest initial camber to offset the subsequent deflection during and after the deck is cast. It is unlikely that a sag of 100 mm (4 in.) would be acceptable. Consequently, there may be deflection requirements that limit the span lengths for which high-strength concrete can be used. It is also necessary to assess, through field measurements, the accuracy of the predicted deflections and the variability of the deflections. It is anticipated that the deflections will be more variable with longer span lengths.

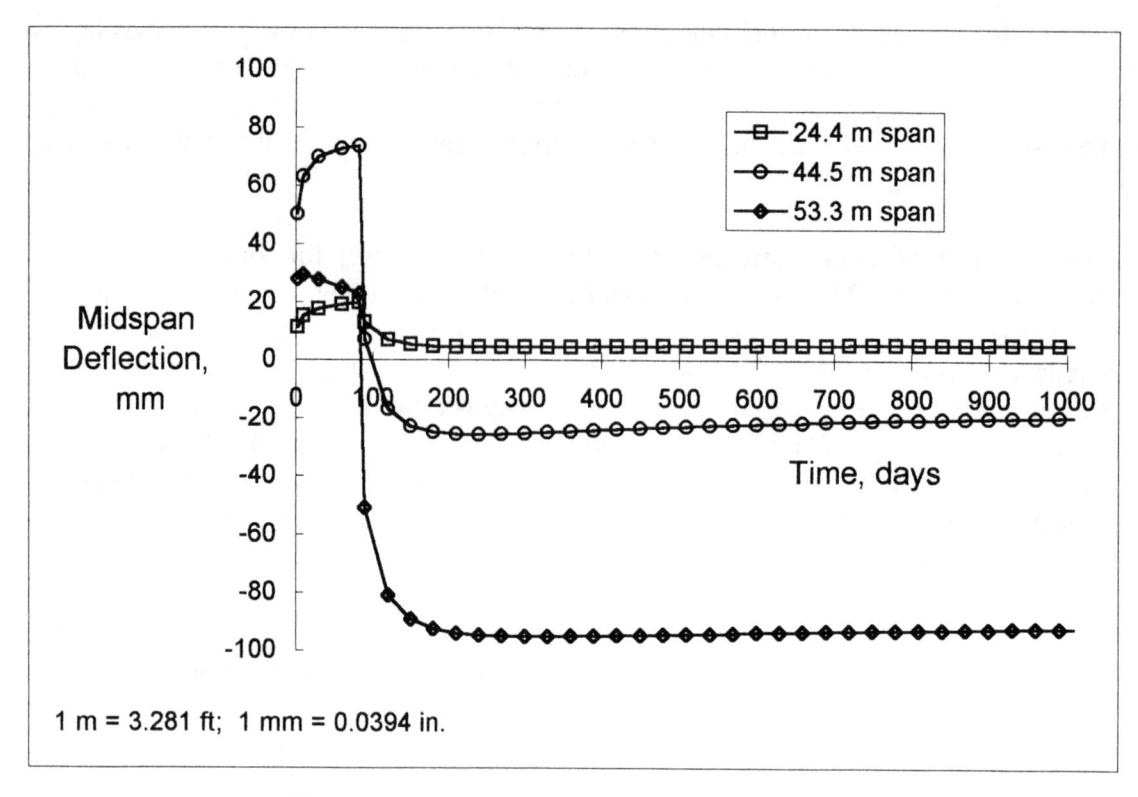

Figure 3 - Midspan Deflections Versus Time

Design Specifications

The AASHTO *Standard Specifications for Highway Bridges*[22] states that:
> "The design of precast prestressed members ordinarily shall be based on $f_c' =$ 5,000 psi. An increase to 6,000 psi is permissible where, in the Engineer's judgement, it is reasonable to expect that this strength will be obtained consistently. Still higher concrete strengths may be considered on an individual area basis. In such cases, the Engineer shall satisfy himself completely that the controls over materials and fabrication procedures will provide the required strengths."

This statement may have been applicable when it was first introduced into the *Standard Specifications*. Today, most engineers have satisfied themselves that strengths of at least 41 MPa (6000 psi) can be obtained consistently. A 1995 survey of precast concrete producers, made in connection with development of the *PCI Bridge Design Manual*, indicated that specified strengths for bridge beams ranged from an average low of 35 MPa (5000 psi) to an average high of 52 MPa (7500 psi). In some locations, specified strengths of 59 MPa (8500 psi) were reported.

The *LRFD Bridge Design Specifications*[23] states that "Concrete strengths above 10 ksi shall be used only when physical tests are made to establish the relationships between the concrete strength and other properties." This statement does not prohibit the use of concrete with a compressive strength above 69 MPa (10,000 psi). However, it does very little to encourage the future use of strengths above 69 MPa (10,000 psi) as the amount of testing to cover all aspects of design can be overwhelming. It does, however, represent a significant step forward compared to the 41 MPa (6000 psi) implied limit in the *Standard Specifications*. Nevertheless, the applicability of *LRFD Specifications* to concretes with compressive strengths in excess of 69 MPa (10,000 psi) needs to be assessed.

CHALLENGES IN PRODUCTION

Concrete Production

The production of high-strength prestressed concrete girders in a prestressed concrete plant requires special quality control procedures that possibly may not be used for lower strength concretes. Since the production of high-strength concrete requires the optimal use of all constituent materials, it is extremely important that the quality of these materials not vary during the production cycle of the girders. All materials used in actual girder construction must be the same as those used in the preproduction trial mixes. Substitution of alternate materials is not acceptable with high-strength concrete. It is also important that sufficient trial batches be performed in the plant prior to girder production so that the fabricator has confidence and experience in achieving the required strengths. The precaster must, therefore, be prepared to implement quality control procedures beyond those used in the production of conventional strength concrete members.

Heat of Hydration

Control of concrete temperature during the curing cycle is extremely critical. Since high-strength concretes contain more cementitious material than used in lower strength concretes, the heat generated during hydration is greater. It is, therefore, important that the temperatures of the girders be controlled during the curing period. When radiant or steam curing methods are utilized, temperature control of the heat and the steam must be based on the temperature of the concrete and not on the temperature of the enclosure surrounding the girders. It is possible for the temperature of the concrete to be 11 to 17° C (20 to 30° F) hotter than that of the surrounding air or steam. This may require modification of the normal plant production procedures. If not properly controlled, it is possible for the water in the concrete to boil.

It is also important that the concrete cylinders used to determine concrete strength at release of the prestressing strands undergo the same temperature regime as the concrete in the girder. Placing the cylinders inside the same enclosure does not necessarily provide this situation. It is, therefore, necessary to go to a match-curing procedure to determine strength of the concrete at release. The effect of heat curing in the prestressing bed is to increase the early strength gain at the expense of relatively lower strengths at later ages.

Strength Versus Age

The traditional approach in precast concrete production has been to achieve a 24-hour turnaround in the prestressing bed. This has meant that the required concrete strength at release of the prestressing strands has been achieved at a concrete age between 12 and 18 hours. Consequently, concrete mix proportions are dictated by the strength at release of the strands. This has, generally, ensured that the concrete strength at 28 days is more than adequate. The effect of increasing the early strength means that the strength development at later ages will be less. With higher strength concrete and the necessity to achieve higher strength at later ages, it may be more appropriate to consider a slower strength gain and extend the release age to later ages such as 48 hours. This would help achieve a greater strength gain at later ages and achieve closer agreement between design and actual strengths at both strand release and 28 days. It has been stated that doubling the bed time will significantly increase the girder costs. The challenge to the fabricator is to find ways to allow a longer bed time without significantly increasing the total cost of the bridge.

Another alternative for consideration is to specify the strength at an age other than 28 days. For many years, the building industry has used an age of 56 or 90 days for cast-in-place high-strength concrete. This allows for the later age strength gain to be used in selection of concrete mix proportions. Since the effect of heat curing is to slow down and decrease the later age strength development, the use of later age strength would allow the fabricator to take advantage of the strength gain after 28 days. The 56-day or 90-day strength could be specified in the contract documents with shipping approved whenever the strengths are attained.

Bed Capacity
The efficient use of high-strength concrete requires that the girders be prestressed to the maximum allowable amount. This will result in the total prestressing force being greater than for lower strength concretes. It is possible to incorporate 99 - 15-mm (0.6-in.) diameter strands in the lower flange and webs of a Texas U-beam. When stressed to 75% of 1.86 GPa (270 ksi), a total force of 19.3 MN (4.35 million pounds) must be resisted at the ends of the bed. When draped or harped strands are used in I-shaped girders, the uplift forces can be greater than the hold down locations can resist.

Lateral Stability
The optimization of girder cross sections has resulted in cross sections that are more slender than used previously, The use of high-strength concrete in these girders can lead to longer span lengths and more initial camber at release. The combination of more slender girders, longer span lengths and more camber increases the likelihood of lateral instability during handling, shipping and erection of high-strength concrete girders.

SUMMARY AND CONCLUSIONS

Although high-strength concrete has been used in several bridges, it is only recently that it has become recognized as a means to achieve longer span lengths, fewer girders or shallower sections. On some structures, high-strength concrete was selected to provide a more durable structure in addition to providing a higher compressive strength for structural design. These structures used high performance concrete by today's definition.

At the present time, barriers exist to the use of concretes with compressive strengths above 69 MPa (10,000 psi). The *LRFD Specifications* [23] requires special physical tests if concrete strengths above 69 MPa (10,000 psi) are to be used. At the same time, the advantages of going beyond 69 MPa (10,000 psi) concrete strength in standard sections are limited because sufficient prestressing force cannot be incorporated into the cross section to take advantage of the higher strength concretes even with the availability of 15-mm (0.6-in.) diameter strands.

As higher strength concretes are used, greater consideration will need to be given to beam shape, prestress losses, vertical deflections, lateral stability, heat of hydration, quality control, later ages for release and specified strengths and prestressing bed capacity.

REFERENCES

1. Zollman, C. C., "Reflections on the Beginnings of Prestressed Concrete in America - Part 1: Magnel's Impact on the Advent of Prestressed Concrete," *PCI Journal*, Vol. 23, No. 3, May/June 1978, pp. 21-48.
2. Anderson, A. R., "An Adventure in Prestressed Concrete," *PCI Journal*, Vol. 24, No. 5, September/October 1979, pp. 90-113.
3. Anderson, A. R., "An Adventure in Prestressed Concrete," *PCI Journal*, Vol. 24, No. 6, November/December 1979, pp. 79-93.
4. "Mountain View Road Bridge," *PCI Journal*, Vol. 29, No. 2, March/April 1984, pp. 148-151.
5. "Tower Road Bridge," *PCI Journal*, Vol. 29, No. 2, March/April 1984, pp. 144-147.
6. "Fancher Road Bridge," *PCI Journal*, Vol. 29, No. 2, March/April 1984, pp. 166-171.
7. Durning, T. A. and Rear, K. B., "Braker Lane Bridge - High Strength Concrete in Prestressed Bridge Girders," *PCI Journal*, Vol. 38, No. 3, May/June 1993, pp. 46-51.
8. "Hybrid Girder in Cable-Stay Debut," *Engineering News Record*, November 15, 1984, pp. 32-36.
9. Taylor, P. R. and Torrejon, J. E., "Annacis Bridge - Design and Construction of the Cable-Stayed Span," *Quarterly Journal of the Federation International de la Precontrainte*, Vol. 4, 1987, pp. 18-23.
10. "Stayed Girder Reaches a Record with Simplicity," *Engineering News Record*, May 22, 1986, pp. 26-28.
11. "Up and Away to a World Record," *Engineering News Record*, September 19, 1994, pp. 76-82.
12. Lane, S. N. and Podolny, W. Jr., "The Federal Outlook for High Strength Concrete Bridges," *PCI Journal*, Vol. 38, No. 3, May/June 1993, pp. 20-33.
13. Lwin, M. M., Bruesch, A. W. and Evans, C. F., "High-Performance Concrete for a Floating Bridge," *Conference Proceedings, Fourth International Bridge Engineering Conference, San Francisco, California, August 28-30, 1995*, Vol. 1, Transportation Research Board, Washington, D.C., 1995, pp. 155-162.

14. Lester, B. and Tadros, G., "Northumberland Strait Crossing: Design Development of Precast Prestressed Bridge Structure," *PCI Journal*, Vol. 40, No. 5, September/October 1994, pp. 32-44.
15. *FHWA Study Tour of Northumberland Strait Crossing Project (NSCP)*, Report Facilitator Russell, H. G., FHWA, U. S. Department of Transportation, Report No. FHWA-PL-96-022, 1996, 70 pp.
16. Smith, D. C., "The Promise of High-Performance Concrete," *Public Roads*, Autumn 1996, pp. 31-40.
17. Zia, P., Schemmel, J. J. and Tallman, T. E., "Structural Applications of High-Strength Concrete," Report No FHWA/NC/89-0006. Center for Transportation Engineering Studies, North Carolina State University, Raleigh, June 1989, 330 pp.
18. Castrodale, R. W., Kreger, M. E. and Burns, N. E., "A Study of Pretensioned High-Strength Concrete Girders in Composite Highway Bridges—Design Considerations," Report 381-4, University of Texas for Transportation Research, 1988.
19. Russell, B. W., "Impact of High Strength Concrete on the Design and Construction of Pretensioned Girder Bridges," *Journal of the Precast/Prestressed Concrete Institute*, Vol. 39, No. 4, July/August 1994, pp. 76-89.
20. Russell, H. G., Volz, J. S. and Bruce, R. N., "Optimized Section for High-Strength Concrete Bridge Girders," FHWA, U. S. Department of Transportation, Report No. FHWA-RD-95-180, 1995, 165 pp.
21. Zia, P., Leming, M. L. and Ahmad, S. H., "High Performance Concretes, A State-of-the-Art Report," Report No. SHRP-C/FR-91-103, Strategic Highway Research Program, National Research Council, Washington, D.C., 1991.
22. *Standard Specifications for Highway Bridges*, sixteenth edition, American Association of State Highway and Transportation Officials, Washington, D.C., 1996, 677 pp.
23. *AASHTO LRFD Bridge Design Specifications*, first edition, American Association of State Highway and Transportation Officials, Washington, D.C., 1994.

HIGH STRENGTH CONCRETE IN SPLICED PRESTRESSED CONCRETE BRIDGE GIRDERS

Z. Lounis, NSERC Post-Doctoral Fellow and M. S. Mirza, Professor
McGill University
Department of Civil Engineering and Applied Mechanics
Montreal, QC, Canada H3A 2K6

ABSTRACT

This paper illustrates the benefits of using high strength concrete in spliced post-tensioned precast I-girder bridges. The combination of high-strength concrete and post-tensioning increases the span length and girder spacing capabilities of standard precast I-girders, improves their durability and yields economical design of bridge superstructures. The combination of high strength concrete and splicing by post-tensioning enables the precast I-girder system to achieve longer spans and become far more competitive with steel plate girders and cast-in-place concrete box girder systems. The maximum girder spacing capability of the modified AASHTO-PCI bulb-tee 72 section is determined for girder concrete strengths varying between 40 MPa (5.8 ksi) to 70 MPa (10 ksi). This study shows that in general, the required post-tensioning force is controlled by the compressive stress constraint, while the required pretensioning force in the field and pier segments is controlled by the tensile stress constraints. An example illustrates the practical benefits of high strength concrete in reducing the superstructure cost.

INTRODUCTION

The use of high strength concrete in bridge girders is relatively less common than in heavily loaded columns of multistory buildings. High strength concrete provides higher compressive strength, modulus of elasticity and tensile strength than normal strength concrete. Compressive strengths of up to 80 MPa (12 ksi) can be achieved through the use of low water-cement ratios with high range water reducers and pozzolanic materials (e.g. fly ash, ground granulated blast-furnace slag or silica-fume). High strength concrete provides improved freeze-thaw durability, lower permeability which reduces the rate of penetration of oxygen and thus the rate of corrosion of the steel reinforcement.

In bridge girder design, high strength concrete enables increasing the span or girder spacing capability of standard precast I-sections which results in longer spans, or fewer girders and reduced dead load. Moreover, high early strength concrete is often required to expedite the daily production cycle of pretensioned girders in bridge construction under tight schedules. The required initial concrete strength increases with the required prestressing force. The initial one-day concrete strength f_{ci}' may be in the order of 70 percent of the 28-day strength f_c'. The impact of high strength concrete on simply supported pretensioned bridge girders has been investigated by several authors[1,2,3,4,5]. For this bridge system, the span length or girder spacing capability is often limited by the allowable tensile stress under full service loads. The impact of increasing the concrete strength is limited because the corresponding allowable tensile stress increases at a much lower rate than the allowable compressive stress. In addition, the benefits of high strength concrete are limited by the girder cross-section capability to contain the required pretensioning strands. This latter limitation can be overcome by using 15 mm (0.6 in.) strands instead of the conventional 13 mm (0.5 in.) strands, or adopting smaller strand spacings. In order to take full advantage of the increasing concrete strength with time, a 56-day concrete strength has been recommended by some authors[4] as the reference design strength assuming that the application of full live load is deferred for at least 56 days.

The trend in prestressed concrete bridge construction is changing from simple to continuous span girders. Continuity allows greater spans, elimination of costly and difficult to maintain deck joints and reduction of intermediate piers for safety. The use of continuous post-tensioned bulb-tee girders allows taking full advantage of the benefits of high strength concrete and thus enables increasing the span length or girder spacing capability of the existing bulb-tee sections. Prestressed concrete girders have better long-term performance than other structural systems and they outperform the concrete decks[4]. Hence, the introduction of continuity by post-tensioning eliminates cracking and enhances the deck durability besides minimizing the life-cycle costs. Moreover, the splicing of precast segments leads to long span girders and enables to overcome the problems of transporting and erecting long and heavy girders. The splices can be located either at the supports, or away from the supports (field splices).

The main objective of this paper is to investigate the impact of high-strength concrete on the girder spacing or span length capability of spliced post-tensioned bulb-tee girders. For this bridge system, the girder spacing, or span length capability is often controlled by the allowable compressive stress under full working loads[6]. Hence, the impact of high strength concrete is more significant for the spliced post-tensioned system than for the conventional pretensioned system. This investigation also includes the determination of the required girder concrete strength for span lengths varying between 40m and 55 m and the required post-tensioning and pretensioning forces.

SPLICED PRESTRESSED CONCRETE BRIDGE I-GIRDERS

The splicing of precast I-girders using longitudinal post-tensioning is becoming a competitive alternative solution for medium span bridges throughout North America. The combination of this splicing technique with high strength concrete increases the span length capability of precast I-girders and overcomes the problems associated with the transportation and erection of long and heavy precast girder segments. Spans of up to 75 m (246 ft.) can be achieved by this technique which makes precast I-girder systems far more competitive compared with steel plate girders and cast-in-place concrete box girder alternatives.

The service life performance of this system is much better than that of the conventional continuous pretensioned girder system because cracking can be eliminated by providing the appropriate post-tensioning force. This investigation is based on a two-span continuous system consisting of post-tensioned precast concrete spliced bulb-tee girder segments with the field splices being located near the contraflexure points due to the dead loads shown in Fig.1. The girders act compositely with a reinforced concrete slab deck of 200 mm (7.9 in.) thickness. The girders are erected as two-simply supported field segments and a double cantilever segment. Each field segment has a length equal to 80% of the span length and is simply supported at the abutment and on temporary supports. The pier segment is erected as a double cantilever with a length equal to 40% of the span length. A 600 mm (24 in.) wide cast-in-place joint with a concrete strength of 30 MPa is assumed to permit access for coupling of conduits and thorough vibration of the concrete. The investigation is based on the modified standard AASHTO-PCI bulb-tee 72 shown in Fig. 2 which is obtained by widening the flanges and webs by 23 mm (0.9 in.) in order to accommodate the post-tensioning ducts. This bulb-tee section has been found to be an efficient section for use in spliced post-tensioned bridge girders[6]. The girder spacing capability is determined for span lengths of 45, 50, and 55 m (148, 164, 180 ft.) and girder concrete strengths of 40, 50, 60 and 70 MPa (5.8, 7.3, 8.7, and 10 ksi). However, the maximum length of the field segments is dictated by the different maximum transportable lengths and weight requirements of the various states and provinces.

For this system, post-tensioning can be implemented in single or multiple stages; however, more than two-stage post-tensioning is not recommended. The principal inconvenience of this system is the accommodation of the post-tensioning tendon anchorages. This problem has been successfully overcome by incorporating the anchorages in the end diaphragms, thus avoiding the anchorage zone reinforcement congestion in the girder web and also reducing the weight of the girders by eliminating the end blocks.

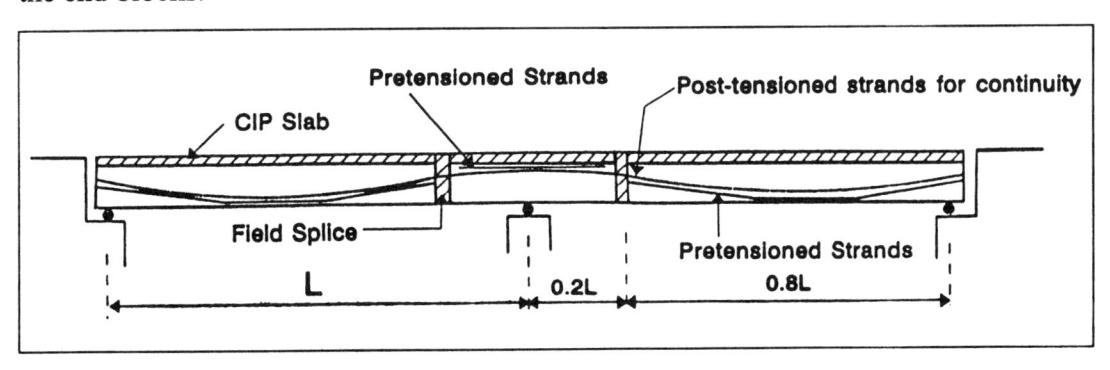

Figure 1. Two-Span Spliced Prestressed Concrete Bridge Girders

The precast segments and the completed continuous bridge girders are checked at the following stages: (1) Release of pretensioning force in the field and pier segments; (2) Erection of precast girders on abutments, piers and temporary supports; (3) Placement of the cast-in-place reinforced concrete deck and joints; (4) Application of post-tensioning and removal of the temporary supports; and (5) Application of the superimposed dead and live loads.

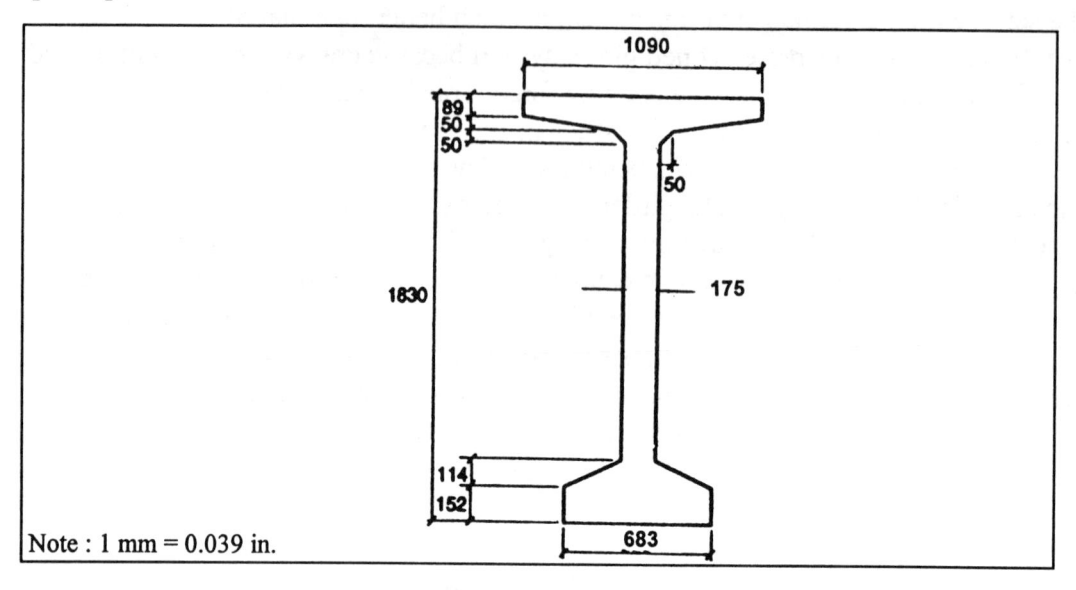

Figure 2. Modified AASHTO-PCI Bulb-Tee 72

IMPACT OF HIGH STRENGTH CONCRETE

Design Requirements and Assumptions

The bridge design complies with all requirements of the AASHTO Code and is based on the new AASHTO live load, load distribution factor, impact factor, and effective flange width formula. A superimposed dead load (wearing surface, parapet) of 2.5 kN/m^2 (52 psf) is assumed. Deflected pretensioned strands are assumed for the field segments, while straight strands are used for the pier segments with maximum feasible tendon eccentricities using a minimum concrete cover of 70 mm (2.75 in.) for the extreme row of strands. The pretensioning steel consists of 13 mm (0.5 in.) diameter, 1860 MPa (270 ksi), low relaxation strands with a spacing of 50 mm (2 in.). The parabolic post-tensioning tendons consist of 15 mm (0.6 in.) diameter, 1860 MPa (270 ksi) low relaxation strands. The allowable stresses at transfer are $0.25\sqrt{f_{ci}'}$ ($3\sqrt{f_{ci}'}$ psi) and less than 1.38 MPa (200 psi) for tension and $0.6 f_{ci}'$ for compression. The allowable tensile stress under dead load and 80% of the live load is $0.5\sqrt{f_c'}$ ($6\sqrt{f_{ci}'}$ psi), while the allowable compressive stress under full dead and live loads is $0.45 f_c'$. The girder concrete strength at transfer f_{ci}' is taken as 70% of the 28-day concrete strength f_c'. The ductility requirement is satisfied by ensuring that the relative neutral axis depth at the ultimate limit state (c/d_e) is less than 0.42. This ductility constraint may be critical at the pier section, particularly for long spans and high strength concrete. The cast-in-place reinforced concrete slab thickness was assumed to be constant and equal to 200 mm (7.9 in.). The slab concrete strength was taken as 30 MPa (4.4 ksi) at 28 days.

Maximum Girder Spacing/Span Length

A computer program was written to determine the maximum girder spacing that the modified AASHTO-PCI bulb-tee 72 can sustain for span lengths of 45, 50, and 55 m (148, 164, 180 ft.) and concrete strengths of 40, 50, 60 and 70 MPa (5.8, 7.3, 8.7, and 10 ksi) and the results are illustrated in Fig.3 and Table 1. The figure shows a quasi-linear increase of the girder spacing capability with the concrete strength for the three spans investigated. This is due mainly to the fact that the compressive stress under full working loads is the governing design requirement up to a certain limiting concrete strength. Beyond this limit, the compressive stress constraint becomes inactive and the girder spacing capability increases at a slower rate due to the fact that the tensile stress at service becomes the governing constraint.

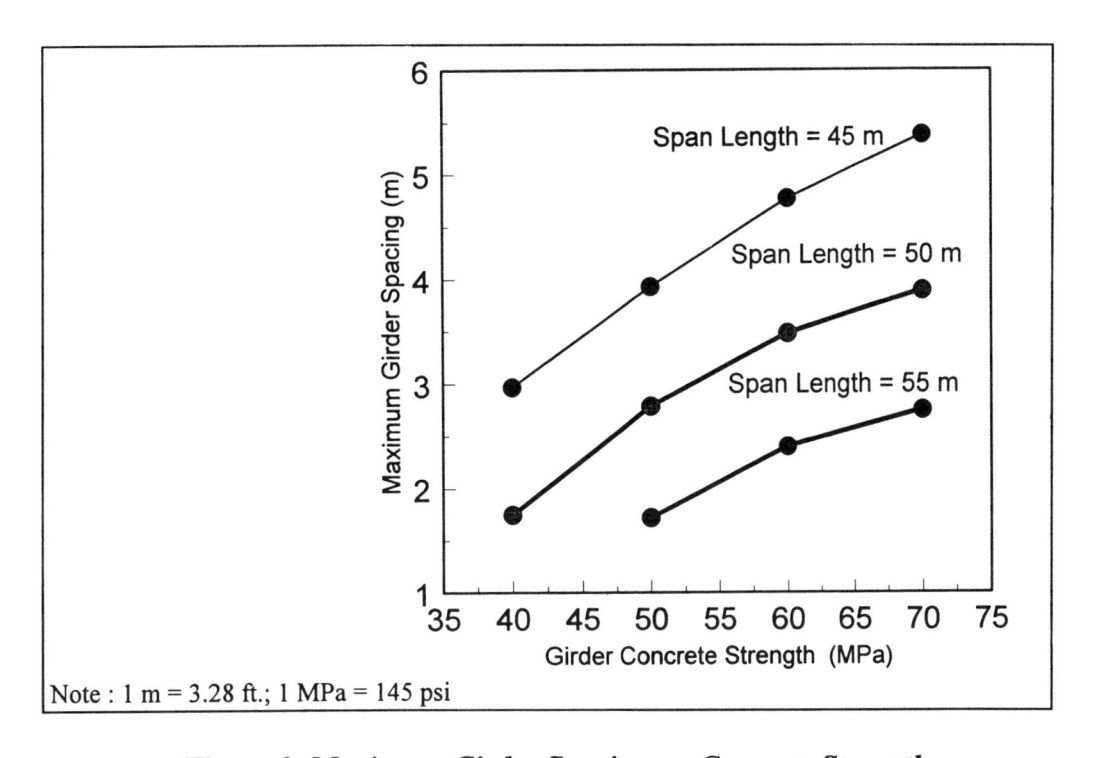

Figure 3. Maximum Girder Spacing vs. Concrete Strength

For the same concrete strength, the girder spacing capability decreases by about 30% for a 5 m (16 ft.) increase in the span length. This investigation shows that for the span length of 55 m (180 ft.), the use of normal strength concrete with $f_c' = 40$ MPa (5.8 ksi) is not feasible and high strength concrete is required. It should be pointed out that for girder spacings of more than 3.6 m (11.8 ft.), it is necessary to increase the slab thickness to achieve an adequate ratio of the girder spacing to slab thickness.

In general, the required effective post-tensioning is controlled by the compressive and/or tensile stress constraints under full working loads. The required pretensioning force in the field segment is controlled by the tensile stress constraint at the stage of placement of the slab deck. For concrete strengths above 60 MPa (8.7 ksi), the pretensioning force in the field segment is also controlled by the capacity of the modified bulb-tee 72 section to contain the required number of 13 mm (0.5 in.) strands; however, this constraint can be overcome by using 15 mm (0.6 in.) strands. The required pretensioning force in the pier segment is controlled by the tensile stress constraint at transfer at the end of the pier segment. The values of the effective post-tensioning force and effective pretensioning force in the field segment corresponding to the maximum girder spacings in Fig. 3 are summarized in Table 1. This table shows that the effective post-tensioning and pretensioning forces increase with increasing concrete strengths.

Table 1- Maximum Girder Spacing and Effective Post-Tensioning and Pretensioning Forces

Span Length (m)	Concrete Strength (MPa)	Maximum Girder Spacing (m)	Post-tensioning Force (kN)	Pretensioning Force in Field Segment (kN)	Pretensioning Force in Pier Segment (kN)
45	40	2.97	4096	2855	2568
	50	3.93	5174	3381	2568
	60	4.77	5865	4114	2568
	70	5.38	5863	4962	2568
50	40	1.75	3286	2853	2349
	50	2.79	4345	3521	2624
	60	3.48	4767	4332	2877
	70	3.89	5103	5198	2929
55	40	-	-	-	-
	50	1.72	3378	3378	2624
	60	2.40	4156	4328	2202
	70	2.75	4601	5338	843

Note: 1 m = 3.28 ft.; 1 MPa = 145 psi; 1 kN = 0.225 kips.

Minimum Required Girder Concrete Strength

The above program was used to determine the minimum concrete strength required to achieve a feasible bridge design. The variations of the required concrete strength with the span length are shown in Fig. 4 for the case of a girder spacing of 2.5 m. The required girder concrete strengths are 30, 36.5, 47 and 62 MPa (4.4, 5.3, 6.8 and 9 ksi) for span lengths of 40, 45, 50, and 55 m (130, 148, 164, 180 ft.), respectively. Using a linear regression analysis, the following approximate relationship between the required concrete strength and span length can be established for this system (where f_c' in MPa and span length L in m):

$$f_c' = 2.13\,L - 57 \tag{1}$$

or in terms of the U.S. customary units (where f_c' in ksi and span length L in ft.):

$$f_c' = 0.094\,L - 8.27 \tag{2}$$

This figure shows that for this spliced bridge girder system and using the modified AASHTO-PCI bulb-tee 72, high strength concrete is required for spans above 45 m (148 ft.). Moreover, the above standard section cannot be used for spans beyond 55 m (180 ft.) even with very high strength concrete such as 70 MPa (10 ksi); thus, the maximum feasible span for this section is 55 m (180 ft.). However, high strength concrete may be required even at spans below 45 m (148 ft.) if shallower sections are required due to a depth limitation.

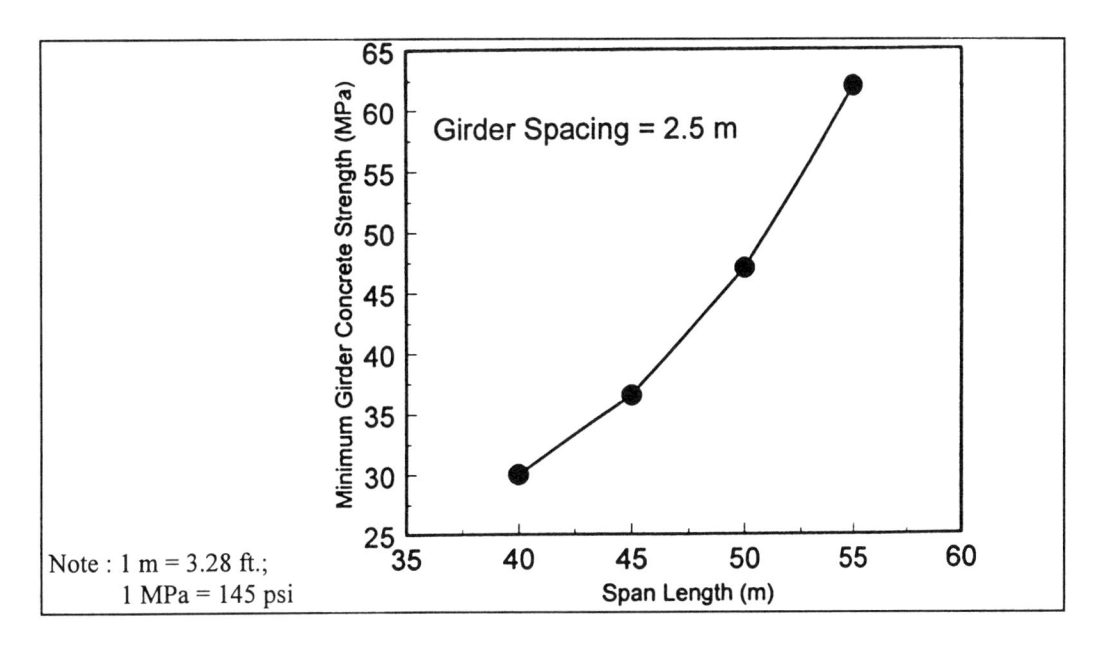

Figure 4. Minimum Required Girder Concrete Strength

Required Post-Tensioning and Pretensioning Forces

The required post-tensioning and pretensioning forces are controlled by the compressive and tensile stresses at service and transfer as described above. In Figure 5, the required post-tensioning forces are determined for span lengths of 45 m (148 ft.) and 50 m (164 ft.), girder spacings of 2 m (6.6 ft.) and 2.5 m (8.2 ft.), and for girder concrete strengths of 40, 50, 60, and 70 MPa (5.8, 7.3, 8.7, and 10 ksi). For a span of 45 m (148 ft.), the required post-tensioning force decreases considerably when the girder concrete strength is increased from 40 MPa (5.8 ksi) to 50 MPa (7.3 ksi), and then it becomes nearly constant for $f_c' \geq 50$ MPa (7.3 ksi). This is due to the fact that for $f_c' \leq 50$ MPa (7.3 ksi), the service load compressive stress constraint is the governing design criterion, and for $f_c' \geq 50$ MPa (7.3 ksi), the service tensile stress constraint becomes the controlling design criterion.

Figure 5 also shows that for the 50 m (164 ft.) span length and girder spacing of 2 m (6.6 ft.), the required post-tensioning force is insensitive to the concrete strength, which indicates that the compressive stress constraint is not a governing requirement.

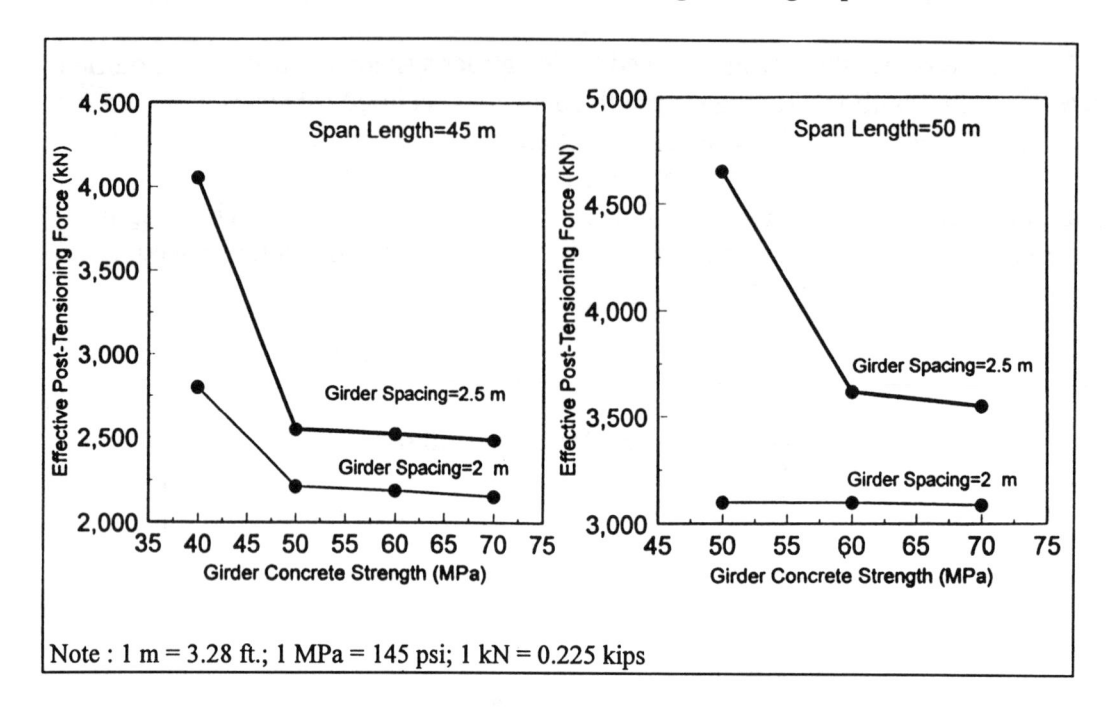

Figure 5. Required Effective Post-Tensioning Force

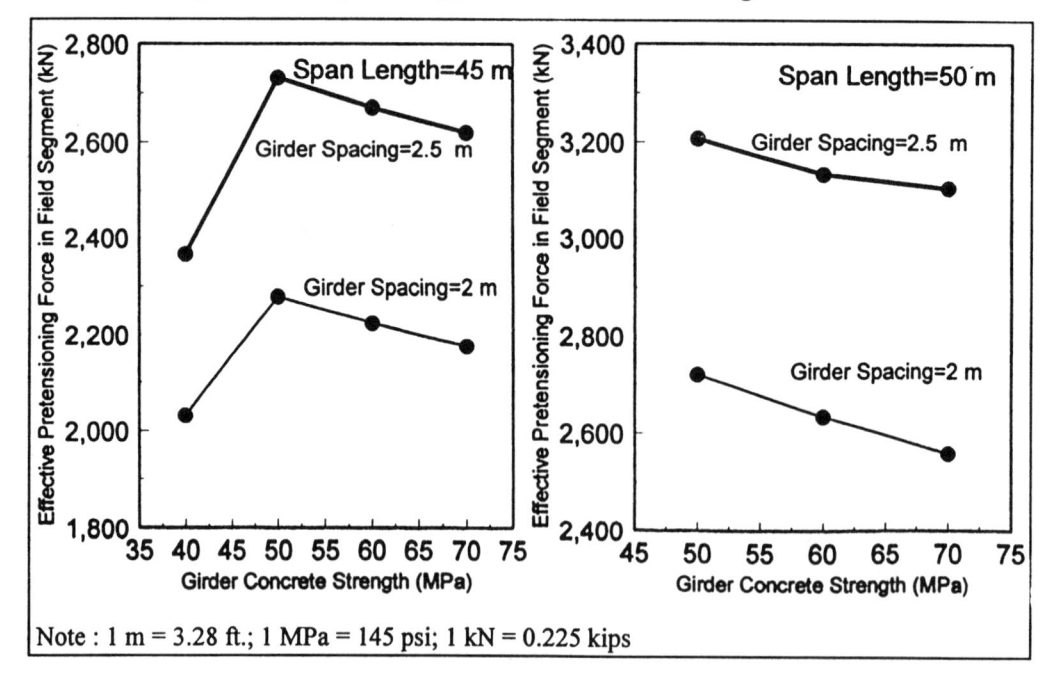

Figure 6. Required Effective Pretensioning Force in the Field Segment

In Figure 6, the required effective pre-tensioning forces in the field segment are determined for span lengths of 45 m (148 ft.) and 50 m (164 ft.), girder spacings of 2 m (6.6 ft.) and 2.5 m (8.2 ft.), and for girder concrete strengths of 40, 50, 60, and 70 MPa (5.8, 7.3, 8.7, and 10 ksi). Figure 6 shows that the required pretensioning force increases with the concrete strength up to 50 MPa (7.3 ksi), and then decreases quasi-linearly beyond that concrete strength. For $f_c' \leq 50$ MPa (7.3 ksi), the required pretensioning force in the field segment increases to compensate for the considerable decrease in the post-tensioning force (Fig. 5). However, the total prestressing force (post-tensioning and pretensioning forces) decreases with increasing concrete strength for all spans.

Design Example

The example is a three-lane highway bridge with a total width of 12 m (40 ft.), and consisting of two equal span continuous post-tensioned girders with a span length L=50 m (164 ft.) as shown in Fig. 1 using the modified AASHTO-PCI bulb-tee 72 in Fig. 2. The girders are first erected as two simply supported field segments of 40 m (131 ft.) (i.e. 0.8L) and a double cantilever segment of 20 m (65.5 ft.) (i.e. 0.4L). Using normal strength concrete with $f_c' = 40$ MPa (5.8 ksi), Table 1 (or Fig.3) shows that the maximum feasible girder spacing is 1.75 m (5.74 ft.). Hence, the bridge requires 7 girders at a spacing of 1.7 m (5.6 ft.) and slab overhangs of 900 mm (3 ft.). Using Table 1, the required effective post-tensioning force is 3286 kN (739 kips), while the required pretensioning forces in the field and pier segments are 2853 kN (641 kips) and 2349 kN (528 kips), respectively.

Alternatively, if high strength concrete is used, for example $f_c' = 50$ MPa (7.3 ksi), Table 1 (or Fig.3) shows that the maximum feasible girder spacing is 2.79 m (9.1 ft.). Hence, the bridge requires 5 girders at a spacing of 2.50 m (8.2 ft.) and slab overhangs of 1000 mm (3.3 ft.). Using Fig.5, the required effective post-tensioning force is 4651 kN (1046 kips), while the required pretensioning forces in the field and pier segments are 3207 kN (721 kips) and 500 kN (112 kips), respectively.

This example shows that high strength concrete enables the elimination of two lines of girders, which will result in considerable savings in the cost of the superstructure.

CONCLUSIONS

The use of high strength concrete and the splicing of precast girder segments by longitudinal post-tensioning increases the span length and girder spacing capability of standard precast I-girders, improves their durability and yields economical designs of bridge superstructures. The combination of high strength concrete and splicing by post-tensioning enables the precast I-girder system to achieve longer spans and become far more competitive with steel plate girders and cast-in-place concrete box girder systems. The positive impact of high strength concrete in this bridge system is more significant than for conventional pretensioned girders because the compressive stress is in general the governing design criterion. Moreover, the introduction of post-tensioning enables standard precast I-girders to overcome the geometrical constraint associated with the capacity of the section to accommodate the required number of 13 mm (0.5 in.) pretensioning strands. This study shows that in general, the required post-tensioning force is controlled by the compressive stress constraint. The required pretensioning force in the field segment is controlled by the tensile stress at the placement of the deck, while the required pretensioning force in the pier segment is controlled by the transfer tensile stress at the segment ends.

Acknowledgment

The financial support of the Natural Sciences and Engineering Research Council (NSERC) of Canada is gratefully acknowledged.

References

1. Russell, H. G. (Editor), *High-Strength Concrete*, Special Publication SP-87, American Concrete Institute, Detroit, MI, 1985, 278 pp.
2. Durning, T. A., and Rear, K. B., "Braker Lane Bridge-High Strength Concrete in Prestressed Bridge Girders," PCI JOURNAL, V. 38, No. 3, 1993, pp. 46-51.
3. Russell, B. W., "Impact of High Strength Concrete on the Design and Construction of Pretensioned Girder Bridges," PCI JOURNAL, V.38, No.3, 1993, pp. 76-89.
4. Dolan, C. W., Ballinger, C. A., and LaFraugh, R. W., " High Strength Prestressed Concrete Bridge Girder Performance," PCI JOURNAL, V.38, No.3, 1993, pp. 88-97.
5. Russell, H. G., Volz, J. S., and Bruce, R. N., "Applications and Limitations of High-Strength Concrete in Prestressed Bridge Girders," *Fourth International Bridge Engineering Conference*, 1995, 169-180.
6. Lounis, Z., Mirza, M. S., and Cohn, M. Z., " Segmental and Conventional Precast Prestressed Concrete I-Bridge Girders," *ASCE Journal of Bridge Engineering*, V.2, No.3, 1997, pp. 1-10.
7. AASHTO *LRFD Bridge Design Specifications*, First Edition, American Association of State Highway and Transportation Officials, Washington, D.C., 1994.

STATE-OF-THE-ART OF HPC:
AN INTERNATIONAL PERSPECTIVE

Paul Zia
Distinguished University Professor, Emeritus
North Carolina State University
Raleigh, N.C., U.S.A.

ABSTRACT

This paper presents an overview of the rapid development of HPC at the international level, focused on three areas of interest: (1) definitions, (2) applications in bridges and pavements, and (3) codes and standards.

INTRODUCTION

Although the concept of high performance concrete (HPC) as a technology emerged out of the laboratory no more than 15 years ago, worldwide interest in this technology has grown dramatically during the past decade. While publications on the subject were rare in the beginning of the 1980's, it is difficult today to cope with the massive stream of information in the technical literature. Table 1 illustrates the trend of the growing interest by the number of publications listed in the two annotated bibliographies on HPC for highway applications compiled by Leming et al.[1] and Zia et al.[2] It should be noted that if the publications on building applications are also included in the tabulation, the number would be even more staggering. Today, concrete of up to 83 MPa (12,000 psi) specified strength can be readily obtained in most parts of the world if the concrete is required in large quantities and basic materials are locally available. Under these conditions, it can be justified economically to import any special materials, technology, and facilities for production and control, if necessary.

Table 1 – Number of Publications on HPC for Highway Applications

Year	No. of Publications
1974 – 1977	97
1978 – 1981	118
1982 – 1985	238
1986 – 1989	471
1990 – 1993	554

To give a glimpse of the rapid development of HPC at the international level, this paper will focus on three areas of interest: (1) definitions, (2) applications, and (3) codes and standards. The paper is based in large part on a recently updated state-of-the-art report prepared for FHWA[3]. The HPC developments in the U. S. and Canada are not included since they are covered by a number of other papers in this symposium. An earlier version of this paper was presented at the FHWA HPC Showcase Conference in Omaha, Nebraska in the fall of 1996.

DEFINITIONS

A broad definition of HPC that seems to have been accepted by the engineering community is that the concrete must achieve certain performance requirements or characteristics for a given application that otherwise cannot be obtained from normal concrete routinely produced as a commodity product. A similar but slightly more specific definition of HPC has been given by the ACI TAC Subcommittee on HPC. The committee defines HPC as "*concrete which meets special performance and uniformity requirements that cannot always be achieved routinely by using only conventional materials and normal mixing, placing, and curing practices. The requirements may involve enhancements of characteristics such as placement and compaction without segregation, long-term mechanical properties, early-age strength, toughness, volume stability, or service life in severe environments.*" These qualitative definitions are perhaps useful for identifying HPC but are not usable for design purposes.

Based on the results of the SHRP projects[4], the FHWA developed a classification of HPC according to different levels of performance characteristics[5]. Such a classification would enable design engineers to select appropriate performance criteria of HPC for different highway applications in different environmental conditions.

Since compressive strength can be easily quantified and higher concrete strength is often (but not always) associated with other desirable performance characteristics of concrete, it becomes convenient to define HPC in terms of compressive strength. Table 2 shows the limits of concrete strengths for HPC specified by the various countries. One can see that almost every country except the U.S.A. specifies an upper limit of the compressive strength. Furthermore, a minimum strength level of about 50 MPa (7,000 psi) is specified by most countries and the lowest strength level of 41 MPa (6,000 psi) is accepted in the U.S.A.

APPLICATIONS

Perhaps the most comprehensive listing of HPC projects published to date is in the FIP/CEB Report No. 222, November 1994[6]. The listing contains 116 entries for more than 120 projects reported from Europe, North America, Asia, and Australia. It covers high-rise buildings, precast and prestressed components, slabs and pavements, bridges, floating structures, offshore gravity-based structures, repairs, special structures and structures where impermeability of concrete or special durability is required. Due to space limitations, our focus will be placed on bridges and pavements.

Table 2 – Definition of HPC by Strength

Code	Strength, MPa (ksi)
ACI	≥ 41 (≥ 6)
CEB	50-100 (7-15)
Norway	44-94 (6-14)
Finland	60-100 (8-15)*
Japan	50-80 (7-12)
Germany	65-115 (9-17)*
The Netherlands	65-105 (9-15)*
Sweden	60-80 (8-12)*
France	50-80 (7-12)

*Based on cube strength

It is quite interesting to note that a survey[7] conducted by FIP/CEB in 1990 indicated that in the U.S.A., HPC was used far more in buildings than in bridges. Of the total of 27 building projects reported, 22 were located in the U.S.A. As early as 1965, HPC was used in the U.S.A. and by 1989 concrete with strength as high as 110 MPa (16,000 psi) had been used.

On the other hand, the same report indicated that HPC was used abroad more in bridges than in buildings. Of the total of 26 bridges reported, only 8 were in the U.S.A. All the bridges were of long-span structures, mostly prestressed box girders and cable-stayed bridges. The concrete strength used for the bridges was generally lower than that used for the buildings. It is noted, however, that lightweight concrete with 69 MPa (10,000 psi) strength was used in a German bridge in 1978. Let us review some of the developments in the various countries.

Japan

In Japan, several railway bridges using prestressed concrete trusses with HPC were built in the 1970's. The objectives were to reduce dead load for transporting, to minimize deflection under live load, and to eliminate noise and vibration problems caused by the running of the trains. Specified concrete strength ranged from 60 to 80 MPa (8,700 to 11,600 psi), but the average strength achieved in the field was over 90 MPa (13,000 psi). Water/cement ratio was at 0.23 and 0.30

Also in Japan, a 40 m (131 ft) long single-span post-tensioned box beam was built in 1993 as a pedestrian bridge between two research laboratory buildings. Due to height limitation, the span-depth ratio of the beam was kept at 40. There was a problem of vibration due to such a large span-depth ratio. To overcome the problem, flat steel bars with dampers were mounted on the underside of the beam. A very flowable "self-compacting" concrete with a water-cementitious materials (w/cm) ratio of 0.20 was used. Specified concrete strength was 102 MPa (14,800 psi) and the actual strength achieved was 122 MPa (17,700 psi). The slump was 25 ± 2 cm (9.8 ± 0.8 in.) and the slump flow was 60 ± 5 cm (23.6 ± 2 in.).

France

Three HPC cable-stayed bridges were built in France in the past 10 years. HPC was used for structural efficiency and durability. The Pertuiset bridge with a span of about 110 m (360 ft) over the Loire River used flowable high strength concrete (HSC) with slump over 20 cm (7.9 in.) and w/cm ratio of 0.33. About 7.5% silica fume was used. The concrete strength achieved in 16 hours was 33 MPa (4,800 psi) and it reached 80 MPa (11,600 psi) at 28 days. For the same reason of structural efficiency and durability, HPC was also used for both Elorn and Nomandie bridges constructed more recently.

The Joigny bridge was constructed as an experimental bridge in France 8 years ago to demonstrate the ability to produce high strength concrete without using silica fume in a commercial ready-mixed concrete plant. Using high strength concrete resulted in 30% savings of the amount of concrete used which, in turn, reduced dead load by 24% for the substructure. This bridge has been instrumented for monitoring the long-term performance.

Norway

In Norway, a major concern for concrete bridges has been the resistance to chloride penetration. To improve the chloride resistance of concrete, it has been a general requirement followed by designers since 1989 that w/cm ratio must not exceed 0.40 and silica fume must be used. In addition, it is unique that lightweight HSC has been used far more widely in Norway than in any other country.

High strength concrete of 85 to 90 MPa (12,300 to 13,000 psi) has been used for highway pavements to provide the necessary abrasion resistance since in Norway almost every car is equipped with steel studded tires.

Germany

In Germany, the Deutzer bridge crossing the Rhine river close to Cologne was built in 1978. The bridge was a free cantilever construction with three spans of 132 m, 185 m, 121 m (433 ft, 607 ft, 397 ft). LWHSC was used for the central 61 m (200 ft) of the center span and NWHSC was used for the rest of the bridge. The specified strength for both concretes was 55 MPa (8,000 psi), but the mean strength obtained in the field was 69 MPa (10,000 psi) for the normal weight concrete and 73 MPa (10,600 psi) for the lightweight concrete. The amount of cement used was 400 kg/m^3 (673 lbs/cy) for both concretes. No mineral admixture was used for the NWHSC but an additional 50 kg/m^3 (84 lbs/cy) fly ash was used for the LWHSC. The w/c ratio was 0.43 for NWHSC and the w/cm ratio was 0.38 for LWHSC. The reported workability varied from 40 to 46 cm (15.7 to 18.1 in.) for the NWHSC and from 38 to 43 cm (15 to 16.9 in.) for the LWHSC.

Denmark

The Great Belt Crossing in Denmark is a major tunnel and bridge connection now under construction. This massive project of US$7 billion includes two single track railway tunnels, each of 8,000 m (26,250 ft) in length between the islands of Sprogoe and Zealand, and a parallel road bridge (East Bridge) of 6,800 m (22,310 ft). The central part of this bridge is a suspension bridge with a main span of 1,624 m (5,328 ft) and pylons of 254 m (833 ft) in height. The islands of Sprogoe and Funen will be connected by the West Bridge, which is a combined road and railway bridge with a total length of 6,600 m (21,650 ft). The structure consists mainly of 110 m (361 ft) long precast concrete girders. The specified design service life is 100 years, so the concrete durability is a major design consideration. HPC was used for pylons, precast girders, and precast tunnel segments. The concrete strength achieved ranged from 50 to 65 MPa (7,250 to 9,430 psi) and the chloride diffusion coefficient was less than 2 x 10^{-12} m^2/s.

The HPC mixture proportions used for the Great Belt Crossing project are shown in Table 3. To ensure very low chloride permeability, limits for w/cm ratio and the amounts of water, fly ash, and silica fume were all specified.

Table 3 – HPC Mixture Proportions for the Great Belt Crossing

	Type A*	Type B*
W/CM	<0.35	<0.40
FA/CM	>10%	>10%
SM/CM	5~8%	5~8%
FA + SM	<25%	<25%
Tot. Water	<135 L/m^3	<140 L/m^3

*Type A used for precast tunnel segments
*Type B used for pylons and precast girders

China

HPC has also been used widely in China for major highway and railway bridges in the past 10 years. They include prestressed continuous rigid frame bridges, arch bridges, cable-stayed and suspension bridges. In 1986, HPC of 80 MPa (11,600 psi) was used for the prestressed concrete girders of a railway bridge on the Hengyang-Quangzhou Line. Since 1988, 10 major bridges have been completed and 3 are still under construction.

Span ranges varied from 45 to 452 m (148 to 1,483 ft) and concrete strength varied from 50 to 75 MPa (7,250 to 10,900 psi).

CODES

Modulus of Elasticity

Let us now look at the various national codes. One area with considerable variation is the modulus of elasticity for HPC as shown in Table 4. Both the CEB/FIP Model Code 90 and the Norwegian Code express the modulus of elasticity as being proportional to the 0.3 power of the compressive strength. When concrete strength exceeds 100 MPa, a special test is required to determine the modulus. The Norwegian Code can be used for both normal- and lightweight concretes. It is interesting to note that the Finnish Code specifies a constant value for the modulus of elasticity.

Table 4 – Modulus of Elasticity of Different Codes

Code	Modulus of Elasticity (MPa)
CEB/FIP	$E_c = 11000 (f_c')^{0.3}$ for $50\ MPa \leq f_c' \leq 100\ MPa$
Norway	$E_c = 9500 (f_c')^{0.3} (\rho/2400)^{1.5}$ ρ = Concrete Density in kg/m^3 for f_c' up to 85 MPa (Cube) FOR $85\ MPa < f_c' \leq 105\ MPa$, special test is required.
Finland	$E_c = 38,700\ MPa$ for $60\ MPa < f_c' \leq 100\ MPa$ (Cube)
Netherlands	$E_c = 35,900 + 40 f_c'$ for f_c' up to 105 MPa (Cube)
Germany	E_c values are tabulated for f_c' = 65 to 95 MPa (Cube) FOR f_c' = 105 MPa, 115 MPa (Cube), special test is required.
Sweden	E_c values are tabulated for f_c' up to 80 MPa (Cube)

The Netherlands Code treats the modulus of elasticity as a linear function of the compressive strength of concrete, while the German and Swedish Codes tabulate the values of modulus of elasticity for different classes of HPC. Note that all these three codes use cube rather than cylinder for testing compressive strength. Furthermore, an upper limit on the compressive strength is specified by each of these six codes.

Stress-Strain Curve

With respect to the stress-strain curve used for design, there are variations among these codes. Both the CEB/FIP MC 90 and the German Code use a curve of similar shape, but specify different values for ε_o, ε_u, and E_c as shown in Fig. 1. The stress-strain curves specified by the Norwegian and Swedish codes are also similar in shape as shown in Fig. 2. However, the values for ε_o, ε_u, and E_c are all different. The Finnish and Netherlands codes specify yet another two different stress-strain curves as shown in Fig. 3. The different values for ε_o and ε_u can be found in References 6 and 8.

Even though these curves may be quite different in shape, their effect on the design of flexural members is probably insignificant since the flexural strength is normally governed by the tension reinforcement. However, when these curves are applied to the design of compression members, the differences in their shapes may produce appreciable differences in the load carrying capacity.

Tensile Strength

The relationships between the tensile strength and the compressive strength of HPC as specified by the various national codes are shown in Table 5. A functional relationship is specified by the Norwegian and the Netherlands codes. Note again that the equation given by the Norwegain code can be used also for lightweight concrete.

CEB/FIP

$$\varepsilon_o = [2.0 - 0.005(f_c' - 50)]/1000$$

$$\varepsilon_u = [2.5 + 2(1 - f_c'/100)]/1000$$

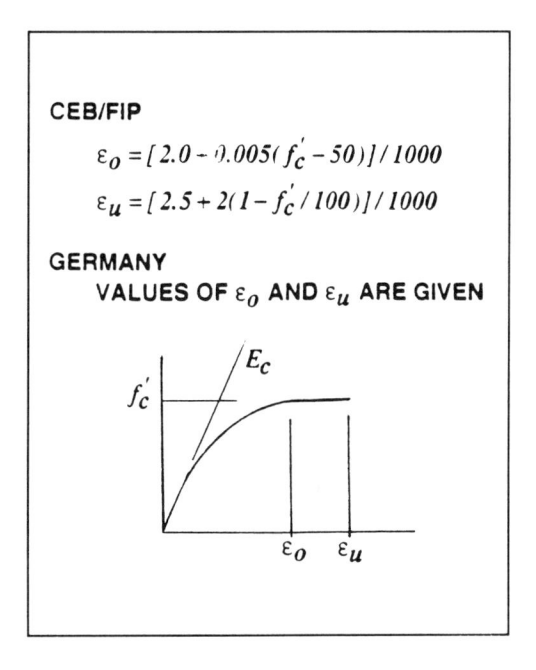

GERMANY
   VALUES OF ε_o AND ε_u ARE GIVEN

Figure 1 – Stress-Strain Curves of CEB/FIP Model Code and German Code

NORWAY

$$\varepsilon_o = [0.004 f_c' + 1.9]/1000$$

$$\varepsilon_u = (2.5\varepsilon_o E_c / f_c' - 1.5) f_c' / E_c$$

SWEDEN

$$\varepsilon_o = 0.002$$

$$\varepsilon_u = [0.3 + 0.7(\rho/2400)]0.0035$$

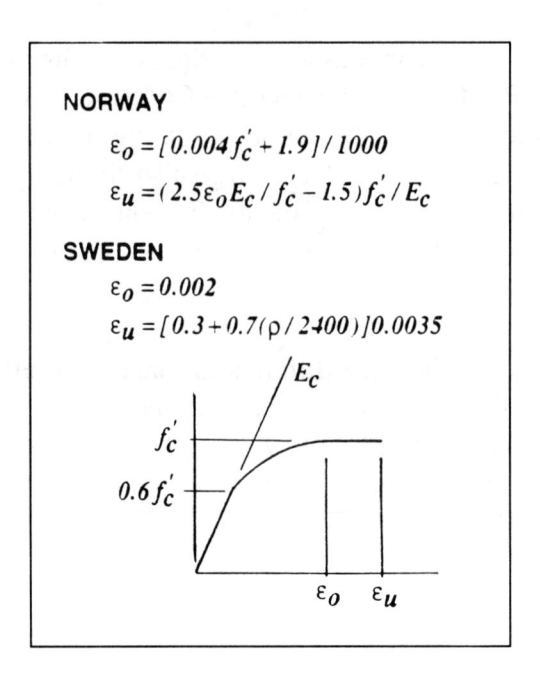

Figure 2 – Stress-Strain Curves of Norwegian and Swedish Codes

FINLAND
VALUES OF ε_o AND ε_u ARE GIVEN

THE NETHERLANDS
VALUES OF ε_o AND ε_u ARE GIVEN

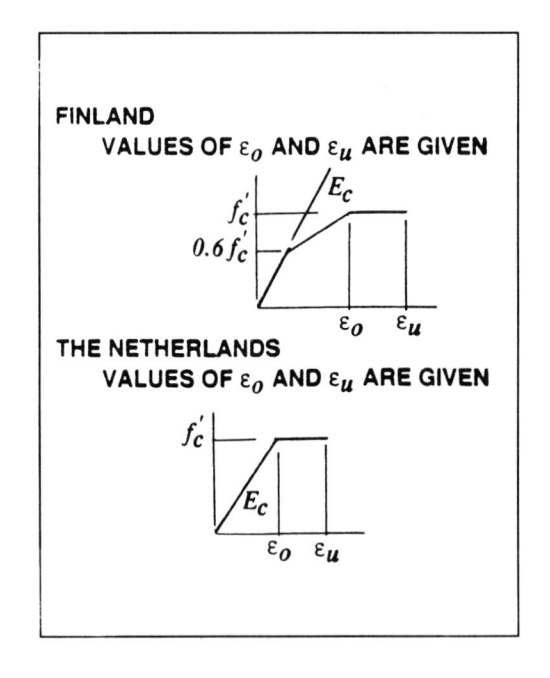

Figure 3 – Stress-Strain Curves of Finnish and Netherlands Codes

Table 5 – Relationships between the Tensile and Compressive Strengths of HPC

Code	Tensile Strength vs Compressive Strength
Norway	$f_t' = k(f_c')^{0.6}(0.3 + 0.7\rho/2400)$ f_t' is constant for $f_c' > 85$ MPa
The Netherlands	$f_t' = 0.7(3 + 0.02 f_c')$
CEB/FIP	$f_c' = 50 \quad 60 \quad 70 \quad 80 \quad 90 \quad 100$ $f_t' = 3.3 \quad 3.7 \quad 4.0 \quad 4.4 \quad 4.7 \quad 5.0$
Finland	$f_c' = 70 \quad 80 \quad 90 \quad 100$ $f_t' = 3.3 \quad 3.5 \quad 3.7 \quad 3.9$
Germany	$f_c' = 65 \quad 75 \quad 85 \quad 95 \quad 105 \quad 115$ $f_t' = 4.3 \quad 4.6 \quad 4.8 \quad 5.0 \quad 5.2 \quad 5.3$
Sweden	$f_c' = 60 \quad 70 \quad 80$ $f_t' = 2.5 \quad 2.6 \quad 2.65$

The relationships between the tensile strength and the compressive strength of HPC are given in tabulated values by the CEB/FIP model code and the national codes of Finland, Germany, and Sweden. Comparing these tabulated values, it appears that the CEB/FIP model code and the German code allow much higher values than the other codes.

Design for Shear

With respect to design for shear, code provisions for concrete contribution to shear capacity are quite different in the Norwegian, German, Swedish, and Netherlands codes. Each code provision does account for the effects of concrete strength and the amount of longitudinal steel provided.

CONCLUSIONS

Based on such a brief overview, several conclusions may be drawn:

(1) The worldwide interest in HPC and its applications has been growing in a very rapid pace in the past several years. HPC has become widely accepted practically on all continents. Much of the applications of HPC remains in the areas of long-span bridges and high-rise buildings.

(2) In terms of design, increasing emphasis is being placed on concrete durability rather than on its strength. In many applications, high strength concrete is used only because of its high durability quality rather than the need for its strength.

(3) The use of high strength concrete for highway pavements in Norway is unique. No other countries have used HSC for pavements.

(4) The development of "self-compacting" concrete in Japan is a very significant step in the quality control of concrete placement. The ability to pump such "flowable" concrete for a long distance without segregation minimizes the manual handling of concrete in trasporting, placement, and consolidation. Such a mechanized operation, along with automated batching and mixing of concrete, will enable the achievement of automation of the concreting process.

(5) There has been an enormous amount of research performed on durability of concrete but without much correlation, largely because the property is "material specific" and dependent on test methods. There is an urgent need for new and improved test methods that would provide more consistent correlation between the laboratory and field results so that the data on durability can be better quantified. More research is needed to develop a rational design methodology for durability.

(6) Much research continues to be focused on the mechanical properties of high- and very-high-strength concretes and their structural applications. The results of the research are being incorporated into various national codes of practice. However, more information is needed on the behavior of the concrete at its early age and its relationship to the long-term performance.

REFERENCES

1. Leming, M. L., Ahmad, S. H., Zia, P., Schemmel, J. J., Elliott, R. P., and Naaman, A. E., "High Performance Concretes: An Annotated Bibliography 1974-1989," SHRP-C/WP-90-001, Strategic Highway Research Program, National Research Council, Washington, D.C., 1990, v, 403 pp.

2. Zia, P., Ahmad, S. H., and Leming, M. L., "High Performance Concretes: An Annotated Bibliography (1989-1994)," Publication No. FHWA-RD-96-112, Federal Highway Administration, McLean, VA., June 1996, iv, 337 pp.

3. Zia, P., Ahmad, S. H., and Leming, M. L., "High Performance Concretes: A State-of-the-Art Report (1989-1994)," Final Report prepared under Purchase Order No. DTFH61-94-P-00959, Federal Highway Administration, McLean, VA., September 1996, 250 pp.

4. Zia, P., Leming, M. L., Ahmad, S. H., Schemmel, J. J., Elliot, R. P., and Naaman, A. E., "Mechanical Behavior of High Performance Concretes, Volume 1: Summary Report," SHRP-C-361, Strategic Highway Research Program, National Research Council, Washington, D.C., October 1993, xi, 98 pp.

5. Goodspeed, C. H., Vanikar, S., and Cook, R. A., "High-Performance Concrete Defined for Highway Structures," *Concrete International*, February 1996, Vol. 18, No. 2, pp. 62-67.

6. Joint CEB-FIP Working Group on High Strength/High Performance Concrete, *Application of High Performance Concrete*, CEB Bulletin No. 222, Lausanne, Switzerland, November 1994, 66 pp.

7. Joint FIP-CEB Working Group on High Strength Concrete, *High Strength Concrete: State of the Art Report*, CEB Bulletin No. 197 (FIP SR 90/1), Federation Internationale de la Prescontrainte, London, England, August 1990, 61 pp.

8. Holand, I. and Helland, S., "CEB-FIP Working Group on High-Strength/High-Performance Concrete," *Fourth International Symposium on the Utilization of High Strength/High Performance Concrete*, Paris, France, May 29-31, 1996, Vol. 3, pp. 1251-1256.

APPLICATION OF HPC IN INFRASTRUCTURE
-
AN OVERVIEW IN THE PERSPECTIVE OF FIP AND CEB

Steinar Helland
Head, Concrete Technology Dept.
Selmer ASA
Oslo, Norway

ABSTRACT

There are many proposals for definitions of this type of concrete. However, some of the proposals for HPC are very wide and might be summarized as "good concrete", or concrete" according to expectations".

The FIP/CEB[1] definitions are somewhat more restrictive and reflect in essence a concrete with a "low-porosity" paste. They define high strength concrete (HSC) as: "*concrete with a cylinder strength above 60 MPa and up to 130 MPa (≈ 9 - 19 ksi), the practical upper limit for concretes with ordinary aggregates. It also includes lightweight aggregate concrete with a cement paste of similar properties*". FIP/CEB similarly regards high performance concrete (HPC) as material with a water-binder ratio (w/b) less than 0.40

This paper presents work that has taken place within the two organizations during recent years flavored with some of my own experience with this material.

FIP / CEB

FIP, Fédération Internationale de la Précontrainte and CEB, Comité Euro-International du Béton, are two international bodies presently in the process of merger. The merged organization under the name FIB, Fédération Internationale du Beton, is expected to be operative in 1998. FIP/CEB presently acts as the umbrella organization for some 50 national concrete associations worldwide. PCI is the US member group. The objective of both FIP and CEB is to promote the use of concrete and to push the frontiers of technology. Their main work takes place in technical commissions which produce various reports.

Their main document is the so-called "Model Code", which was last revised in 1991. This Code is intended for codewriters worldwide and forms the main reference for national and regional concrete codes in a great part of the world. It has a dominating influence in Europe.

Another main activity is to initiate and organize international congresses and symposiums. The last FIP/PCI congress was held in Washington DC in 1994, the last symposium in Johannesburg half a year ago. HSC/HPC has been a main field of activity for these organizations during the last 1½ decades.

FIP/CEB PUBLICATIONS ON HSC/HPC

In 1990 Ippatti of Finland presented a bibliography on the international published reports on HSC/HPC, Fig 1[2]. While this material was hardly produced under conditions other than in the laboratory before 1980, the number of reports on this exiting family of concretes has exploded during the last years.

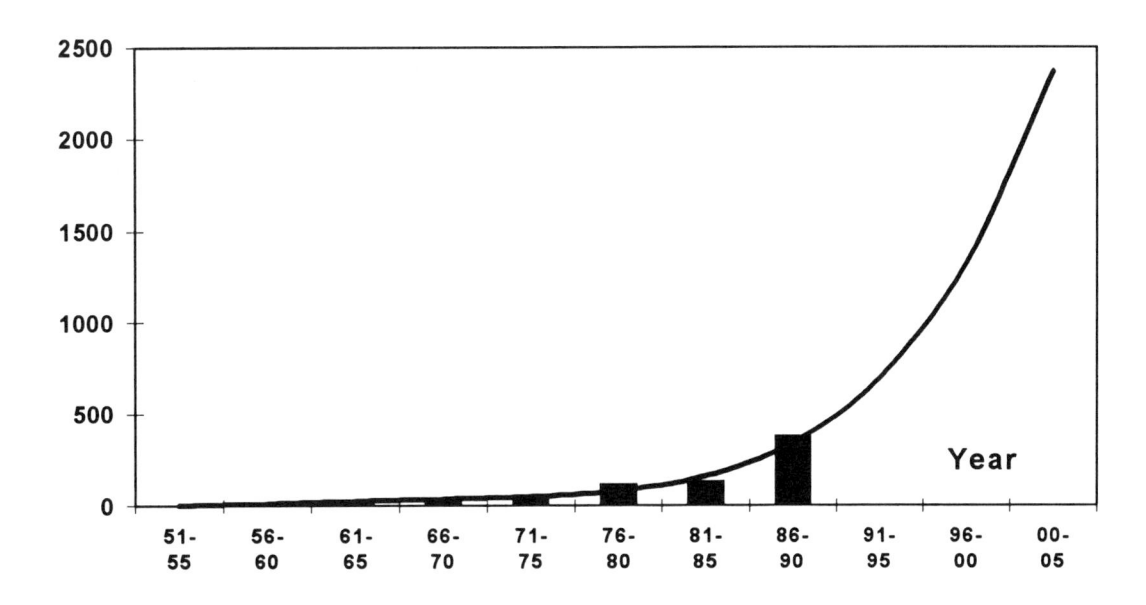

Figure 1. Number of internationally published reports on HSC according to Ippatti[2]

To digest all this experience and to give authoritative guidance to the industry, FIP and CEB formed joint working groups to prepare international consensus reports. These groups included representatives from Japan, Australia, Mid-East, Europe and North America.

The first of these documents was the 1988 "state-of-the-art" report "Condensed silica fume in concrete"[3]. This document has during the last years formed the scientific basis for European standardisation of this very efficient additive to HSC/HPC. In 1990 FIP and CEB worked out a "state-of-the-art" report on HSC[1]. It does cover both design, material and execution aspects. This document then formed the basis for the 1995 "HPC - Recommended Extension to the Model Code 90"[4]. It gives detailed advice and rules for the structural use of concrete grades a characteristic cylinder strength of up to 100 MPa.

Presently CEN, Comité Européen de Normalisation (the standardization body in charge of producing the new unified European standards), has decided to extend the coming version of the design standard CEN-EN 1992 to cover concrete grades of up to 100 MPa ($\approx$ 14.5 ksi). The FIP/CEB extension to the Model Code is the main reference for this work.

In 1994, FIP/CEB published the "Application of HPC"[5].

Table 1 - International standards covering HSC[5]

Country	Document	Max. char. Compressive strength MPa 1 ksi = 6.9 MPa	Test specimen	Notes and special requirements for LWA etc.
International	CEB-FIP MC-90 Addendum to MC from 1995	80 100	Cylinder 150/300 mm	LWA to be considered
Norway	NS 3473 / 1992	105 94	Cube 100 mm Cyl 150/300	LWA considered to LC 85 f_{ck} 105 (ρ/ρ_2)$^{1.5}$
Finland	Rak MK B4 1983/84 suppl. 1989	100	Cube 150 mm	LWA not considered
USA	ACI 318-89	No maximum strength specified; Limits for some design parametres	Cylinder 6/12 inch 152/304 mm	LWA considered; Strength is determined from tests
Canada	CSA A23.1, 23.2, 23.3 1994	No maximum strength specified Limits (80 MPa) for some design parametres	Cylinder 150/300 mm 100/200 mm	LWA considered; Strength is determined from tests
Japan	Specification for HSC	80	Cylinder 100/200	-
Germany	Supplement to DIN 1045, DIN 488 and DIN 1055	115	Cubes 200 mm	-
Sweden	BBK 79	80	Cubes 150 mm	-
The Netherlands	Supplement to NEN 6720, NEN 5950	105	Cubes 150 mm	-

This paper gives a detailed description of the use of this material in 16 examples worldwide and key-figures of another 100 projects. In addition it gives a summary of all national and international codes on HSC/HPC and also an overview of research taking place on the subject worldwide at that time.

Closely related to this subject, FIP published in 1996 the "state-of-the-art" report on "Durability of concrete structures in the North Sea"[6]. This summarizes the infield performance of some 24 concrete structures constructed during the last 20 years with a material composition close to the technical frontier concerning HPC/HSC.

Presently, an FIP/CEB commission works on similar documents on Lightweight Aggregate Concrete (LWAC). In these documents, which are expected to be issued in 1998/1999, both High Strength and High Performance LWAC will be incorporated.

It is my hope that the cited FIP/CEB consensus-documents also will help other regions and countries presently upgrading their local standards.

HSC/HPC SYMPOSIUMS

FIP has acted as the main sponsor for four international symposiums on HSC/HPC. The first was arranged in Stavanger, Norway in 1987, the second in Berkeley, California in 1990, the third in Lillehammer, Norway in 1993 and the fourth in Paris, France in 1996. The fifth is scheduled for Norway in 1999.

In total, the proceedings from these symposiums represent some 450 technical reports[7-10]. At the closing session of the Paris event last year, I had the privilege of summarizing the worldwide trends in utilizing this material as it was reported during the symposium. Some of the main observations were:

Field of application

Traditionally, application in bridges and offshore structures seems to have represented the spearhead to penetrate the market for HSC/HPC. However, while the concrete industry has been able to compete successfully with steel for offshore gas and oil installations until recently by developing and applying state-of-the-art technology concerning strength, durability and density, this trend has changed. New exploration technology based on submerged and/or floating concepts have favoured steel, and unfortunately some of the previous locomotive-effect of the offshore industry to boost the HPC/HSC technology has dried out.

Today, the use of this material in bridges dominates the applications all over the world and generates the locomotive effect for further development of this technology. 75% of the projects described at the Paris symposium thus represented such structures.

While the use of HSC/HPC just a few years ago was an exclusive technology known only in a few countries, this symposium clearly demonstrated that it is becoming more accepted in the market on all continents, with a possible exception of Africa.

Durability -Strength

For almost all projects reported in Paris, improved durability performance was cited as a prime reason for applying this dense material. In particular the improved resistance to the ingress of chlorides is a highly appreciated property.

The partly submerged bridge "Shore Approach" at the western coast of Norway was constructed in 1982 with w/b ratio of < 0.38 combined with 8% of silica fume. It thus represents one of the oldest structures exposed to a harsh marine environment where the new principles of HPC were applied. Helland[10] reported on the full-scale long-term performance of the structure. Based on extensive investigations on the object, the optimistic expectations for such a concrete have been confirmed.

The project, together with a great number of other field-exposed marine structures in Norway, has also been part of the background to develop a more realistic version of Fick's 2nd Law of diffusion for ingress of chlorides. This model has been reported[11] and gives new arguments to the conclusion that only modern HPC will withstand a chloride rich environment. The investigation further concluded that the diffusion coefficient in the Fick's law must be replaced by a time dependent number as the material resistance increases considerable with time.

However, there is a growing concern about using dense concrete for only durability reasons without taking advantage of the enhanced strength properties resulting from the low porosity.

The trend is that the durability aspects open up for the technology. The designer is gradually forced to include in his design the improved strength properties that come as a "free" bonus to be able to compete in the market with cost-effective concepts.

Designing with HSC

The reports are similar from all parts of the world. Designers used to work with normal concrete qualities are confronted with both formal and practical challenges when they consider HSC. By just copying traditional concepts and geometrical shapes, the engineer ends up in a corner. Russell et al[10] clearly illustrated this in their study on how to replace normal strength with HS concrete in highway bulb-tee girders with standard AASHTO-PCI geometry. They concluded "..., *the maximum concrete compressive strength is in the range of 62 to 69 MPa (≈ 9 - 10 ksi). Above this strength level, sufficient prestressing force cannot be introduced into the cross section to take advantage of any higher compressive strengths*".

Actually, in my opinion the concrete designers are in the same phase as the shipbuilders were in the post WW-II period when they shifted from rivets to welding. As a result of new available technology, the design concept changed dramatically. The concrete industry must face a similar innovative change if they want to benefit from HSC in structural design.

To cope with this challenge, several of the Paris-reporters saw a way out of the dilemma by use of steel tubes filled with HSC as structural members. In this way they have both obtained the desirable confinement and avoided too congested reinforcement.

Two spectacular projects were reported with this concept. One is from a Canadian experimental pedestrian bridge to be built in Sherbrooke. In this structure, 200 - 300 MPa ($\approx$ 29-43 ksi) mortar is poured into 150 mm (6 in.) diameter stainless steel tubes of 4.5 m ($\approx$ 15 ft.) length. These members are then assembled at the site to form a space-truss for the structure spanning some 70 m, Aitcin et al[10].

Zihua[10] referred to a world record bridge crossing the Yangtze River in China. This 420 m long arch structure has been under construction since 1994 and the main element in the arch is also a space-truss with 402 mm ($\approx$ 1.5 in.) steel tubes filled with 60 MPa ($\approx$ 9 ksi) concrete. An impressive activity for large scale testing and instrumentation of the long-term behavior, is an integrated part of the project.

Concrete technology

The great majority of projects reported, were within the FIP definition, i.e. with strengths between 60 and 130 MPa ($\approx$ 9-19 ksi). To achieve the desired properties, most every project applied silica fume. On some projects even 3-powder binders like cement, fly ash or granulated slag and silica, were used to tailor the mix. Fig 2 gives the correlation between w/b and strength. As might be noticed, even thresholds like w/b lower than 0.30 seem to have been passed in full-scale production in many parts of the world. These reports are indeed remarkable as such mixes are not possible to batch, transport or cast in a controlled way by traditional means.

Low w/b mixes tend to contain high amounts of cement with a high curing temperature as a result. The attitude to high peak temperatures did indeed vary from country to country.

While some regions put extreme requirements to avoid detrimental effects (f.x. < 12°C ($\approx$ 54 F) between two sections), the majority of codes today have a clause of maximum 65 to 70°C ($\approx$ 150-160 F) to avoid long term strength and permeability loss due to, among other effects, delayed formation of ettringite.

A US bridge project was reported where the peak temperature had reached 93°C without much concern about the consequences for properties other than short term strength[10].

I feel that the industry should be very conscious if it bases construction on such elevated temperatures if it does not have sound documentation for the long-term performance for the actual binder combination.

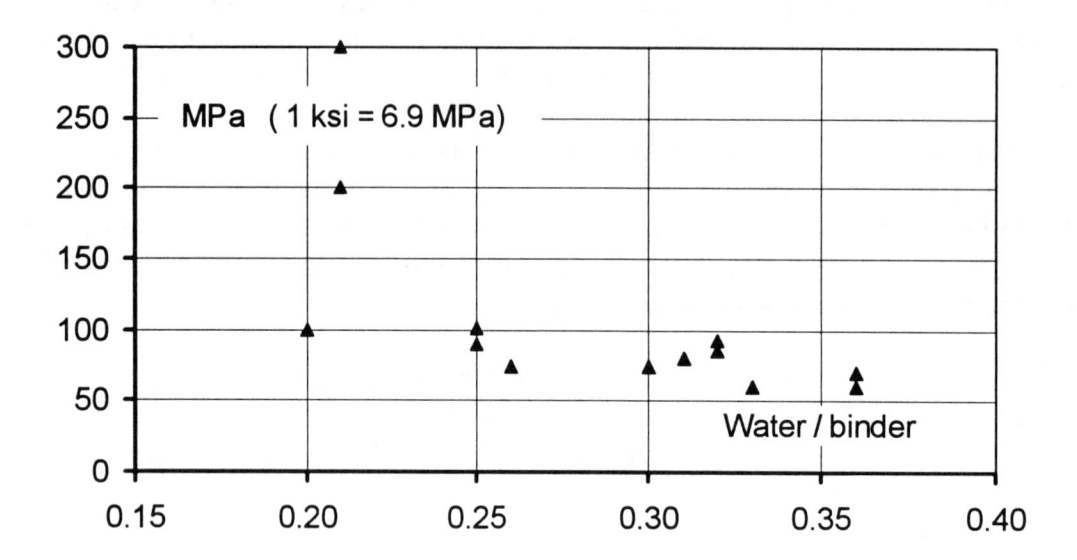

Figure 2. Correlation between strength and w/b ratio for a number of projects reported at the Paris Symposium, Helland[10]

SOME PERSONAL EXPERIENCE FROM NORWAY

Having been involved in full-scale use of HPC/HSC since the late 1970's, I have gained some tough experience and developed some strong points of view on aspects I consider crucial for the successful application of this new generation of concrete.

Competence

Norway was the first nation in the world to have HSC with characteristic cube strengths up to 105 MPa ($\approx$ 15 ksi) incorporated in its code of design, NS 3473, in 1989[12]. In the same year, our Public Roads Administration revised its general specification for highway construction and put an upper limit on the w/b-ratio of 0.40 combined with a demand for silica fume[13]. This was to build up defence lines against the alarming observations of chloride ingress in older bridges. Our options at that time were either to abandon concrete as a material to be exposed to marine environment, or to specify the promising new HPC with low w/b and addition of silica fume. We thus had to introduce HPC on the market before a major part of the building industry had time to familiarise itself with this new technology.

During the last few years a maximum w/b ratio of 0.38 combined with > 5% silica fume has also appeared as a typical specification for the offshore gas and oil installations in the North Sea[14].

According to the FIP/CEB definitions, all concrete installations built in Norway in the 1990s for the oil and gas-fields in the North Sea and most highway structures are thus built with HSC/HPC. This amounts to about 20-25 % of our total domestic concrete production.

I was myself involved in this process and still consider these documents to be satisfactory, but we missed one very important aspect. We did not make a sufficiently strict definition of the required level of competence for those allowed to work with this material.

For steel structures it is well accepted that personnel involved in welding of different qualities of steel must posses certificates proving their proficiency. The concrete industry lacks tradition of certification for workers, engineers or designers. Thus, when Norwegian clients, design offices and contractors observed that the new generation of "hi-tech" concrete qualities were standardized, they considered them to be free to use by anybody.

The result has been that Norway has acted as a full-scale laboratory and training field for the industry for a number of years. Initially in this period, we experienced an unacceptable number of problems in handling these delicate qualities for the inexperienced participants in the market. In particular cracking in the plastic phase was a phenomena all new participants in this field had to face and learn to control the hard and expensive way.

What other industry would ever open its doors to new technology without anchoring it to a competence requirement?

Today in our complex world, most professions have a system of certification based on personal practical and theoretical competence to protect society against rascals.

In Norway we have actually had a requirement since 1986 in our standard for material and execution, NS 3420, for leading workers and supervisors to have a 3 year practical and theoretical education with a public diploma as craftsmen. This is well implemented for ironworkers and carpenters, but due to tradition, it is more or less ignored for concrete-workers. For engineers the standards require an undefined "special skill in concrete technology".

General competence is a balanced mixture of practical skill and theoretical understanding. Practical skill can be obtained during daily work by conscientious workers. The observation of honeycombing and other visual macrodefects will be enough to improve their techniques. The feedback loop is short and I am not too concerned about this part of our workers' competence that takes care of the "outside" qualities of structures.

What is more difficult to obtain are the theoretical aspects of the profession and in particular the understanding of deterioration mechanisms and advanced concrete technology representing the "inner" qualities. The only way of learning this is through traditional lecturing on the school-bench.

Our readymix-industry has faced this situation and put in specific requirements for documented competence with emphasis on HSC/HPC. Their plants are by law subject to certification by a public body. Eight years ago this industry organised a modular system of courses with examination and diplomas. At the same time the certification agency required documentation of this theoretical competence for all operators as well as supervisors. Merely to be able to obtain the necessary certification of the plant and stay in business, the Norwegian readymix industry had to go through a massive process of schooling. This process probably represents the major push in recent years to lift the competence in concrete technology in our country. Generally speaking, this part of the industry now represents a stronghold that is to ensure quality. This is an example where a bureaucratic public body is in charge of following up a requirement.

For the contractors and designers, their competence, or lack of such, in the general understanding of modern "hi-tech" concrete technology like HSC/HPC, results in functional or non-functional structures for the client. Even if such shortcomings do not result in structural collapse, the normal consequence will be increased maintenance costs in the future. Maintenance problems are however not defined as a field of priority for the public building control. The various clients are therefore considered as those with prime responsibility for checking that these clauses in NS 3473/3420 are fulfilled.

Again, due to tradition, clients very seldom put emphasis on elements other than the lowest bid price. Any process of pre-qualification based on level of competence or checking of the actual proficiency for the design or construction crew during the job would be an exception. Surprisingly, this is also the case for major public clients.

Since there is very little outside pressure on this part of the industry to educate their staff in this field, the process of raising the level of competence is too slow as it actually represents quite a heavy investment for employers.

In Norway, the Association of General Contractors (LBA) has organised a course in the theoretical aspects of the profession in a similar way as the ready-mix-industry with special emphasis on HSC/HPC and the "inner" qualities of the concrete. The duration is 40 hours on the school-bench and some 20 hours of homework. For those passing the final examination, the reward is a diploma documenting formal competence. The Standardization Board (NBR) further accepts this module as a valid entry concerning the requirements presently formulated in NS 3420. For craftsmen in adjacent professions such as carpenters and ironworkers, this addendum gives them the formal qualifications as concreters as well. Similarly for those with the education of an engineer, this satisfies the requirement in the Standard for "special skill in concrete technology". In 1991, the board of LBA determined that this training programme should be a matter of priority to counteract society's criticism concerning inadequate quality standards in the industry.

The response by employees to this offer has been very positive. Workers are by nature curious and they are proud of their profession.

courses on grown-up blue-collar has so far not been the case and companies to get their

than 450 have passed the ndidates, so we expect soon to bitions.

uropean concrete standards for onsider my main contribution that the level of proficiency "hi-tech" materials like he European system for possible to link these

materials. In Norway, structures and for structural we are energetically building wing scepticism to concrete,

per structures. To achieve construction with some his challenge is to apply anced durability performance

ures built with modern HPC s field documentation will other parts of the world in high quality concrete.

However, too little attention has been focused on introducing such qualities as standardized materials to the industry.

My personal opinion is that the real threshold and challenge for the industry in the near future is to educate and man the projects to be able to deal with such concrete.

The frontrunners represented at this symposium probably represent teams that are doing their homework, but as soon as HSC/HPC becomes an integrated part of the national standards, and if the standards do not include specific requirements to the personnel's competence, everybody will be allowed to work in this field on the lowest-bid basis.

The projects presented here in New Orleans represent "hi-technology". This is in sharp contrast to most traditional concreting which certainly might be labeled as "low-tech".

If the industry allows "hi-tech" projects to be executed by "low-tech" teams, society soon will accuse us of being irresponsible rascals.

REFERENCES

1. FIP/CEB Bulletin d'Information no 197 "High Strength Concrete". State-of-the-art-report 1990, available through CEB secreteriat EPF Lausanne, Case Postale 88, CH-1015 Switzerland, Fax 6935060

2. Ari Ipatti, "A Bibliography on High Strength Concrete 1930-1990", Imatran Voima Oy, Concrete and Soils Laboratory, SF-01600 Vantaa, Finland

3. FIP state-of-the-art-report 1988 "Condensed Silica Fume in Concrete", available through FIP secreteriat - Institution of Structural Engineers, 11 Upper Belgrave Street - London SW1X 8BH, UK, Fax 0171 2354294

4. FIP/CEB Bulletin d'Information no 228 "High Performance Concrete - Recommended Extentions to Model Code 90", 1995, available through CEB secreteriat EPF Lausanne, Case Postale 88, CH-1015 Switzerland, Fax 6935060

5. FIP/CEB Bulletin d'Information no 222 "Application of High Performance Concrete" 1994, available through CEB secreteriat. EPF Lausanne, Case Postale 88, CH-1015 Switzerland, Fax 6935060

6. FIP state-of-the-art-report 1996 "Durability of Concrete Structures in the North Sea", available through FIP secreteriat - Institution of Structural Engineers, 11 Upper Belgrave Street - London SW1X 8BH, UK, Fax 0171 2354294

7. Proceedings from "The First International Symposium on the Utilization of High Strength Concrete", Stavanger, Norway 1987. Available from Norwegian Concrete Association, P.O. Box 2312 Solli, Oslo, Norway

8. Proceedings from "The Second International Symposium on the Utilization of High Strength Concrete", Berkeley, USA 1990. ACI (SP-121)

9. Proceedings from "The Third International Symposium on the Utilization of High Strength Concrete", Lillehammer, Norway 1993. Available from Norwegian Concrete Association, P.O. Box 2312 Solli, Oslo, Norway

10. Proceedings from "The Fourth International Symposium on the Utilization of High Strength /High Performance Concrete", Paris, France 1996. Available from Presse de l'ENPC 49, rue de l'Université, 75007 Paris, France

11. Helland S., Maage M., Carlsen J., Vennseland Ø., Paulsen E. "Service Life Prediction of Existing Concrete Structures Exposed to Marine Environment", ACI Materials Journal Vol. 93 No 6 Nov.-Dec. 1996 pp 602-608

12. NS 3473E "Concrete Structures - Design rules" (English version), revision 1992, Norges Standardiseringsforbund, P.O.Box 7020 Homansbyen 0306, Oslo - Norway, Fax +47 22464457

13. "Prosesskode-2" 1989 (In Norwegian), Vegdirektoratet - Håndboksekretariatet, P.O.Box 8109 Dep. Oslo 1, Norway

"Concrete Structural Materials", (In english) Offshore general technical specifications - Norsk Hydro document NHT-S51-24, 1992, Norsk Hydro a.s 0240 Oslo, Norway

USE OF HSC/HPC FOR ROAD BRIDGES IN INDIA

S.A. REDDI
Dy. Managing Director, Gammon India Limited, Mumbai, India

ABSTRACT - The paper presents an exhaustive review of the use of High Strength Concrete (HSC)/High Performance Concrete (HPC) for road bridges in India. Major historical landmarks right from the 1940s till date in the bridge engineering field in India are discussed. The paper also presents a broad overview about various specifications/properties of different ingredients of concrete as had been used in the construction of Indian road bridges. Recent development in the use of HPC for some of the outstanding bridges in the world are highlighted. The paper concludes with remarks on the future scenario.

DEFINITION OF HIGH PERFORMANCE CONCRETE (HPC) - Traditionally, HSC/HPC has been defined by its high compressive strength. Concrete with a strength of 35 MPa (5 ksi) was considered HSC in the Fifties while 40 MPa (6 ksi) or above was regarded high strength in the Sixties. Today high strength probably means a strength of 50 MPa (7.5 ksi) and above.

HPC may be defined as concrete with :

(i)   a maximum water cement ratio of 0.35
(ii) a minimum durability factor of 80 percent (ASTM), and
(iii) a minimum strength criteria of either
      21 MPa (3 ksi) at 4 hours (VES),
      34 MPa (5 ksi) at 24 hours (HES), or
      69 MPa (10 ksi) at 28 days (VHS).

Very Early Strength (VES) concrete is used for repairs to components of bridges and for precast members to facilitate early removal of formwork. High Early Strength (HES) concrete enables prestressing of tendons at 24 to 48 hours after concreting. Very High Strength (VHS) concrete have applications in all types of bridges.

Other High Performance requirements include :

- Ease of placement, consolidation
- Long term mechanical properties
- Volume stability
- Longer life in severe environments.

HISTORICAL LANDMARKS IN INDIA - Outstanding concrete bridges built during the Forties include :
* Coronation Bridge, Siliguri-82m.(270ft.) main span (Fig.1).
* Napier Bridge in Madras. This 55 years old marine structure is in excellent condition.
* Three Nos. PSC Railway Bridges near Siliguri.

Fig.1 : Coronation Bridge

The Railway Bridges built in 1948 consist of precast, PSC `I' beams of 18m (60 ft.) spans. The tendons were painted with bitumen and wrapped with waterproof paper, in lieu of ducts ! Zero slump concrete of grade 40 MPa (6 ksi) was used. Prototype girder was tested near Mumbai. A loaded open wagon was derailed and made to skid on the beams, with no distress to the beams.

The first precast PSC road bridge in India (Palar bridge) has 23spans of 28.35m (90ft.). Each span consists of 4 `U' beams (Fig.2). After the concrete attained a cube strength of 31mpa (4.5 ksi), tendons werethreaded through the preformed holes and prestressed. R.C. deck slab was cast on top of the `U'.

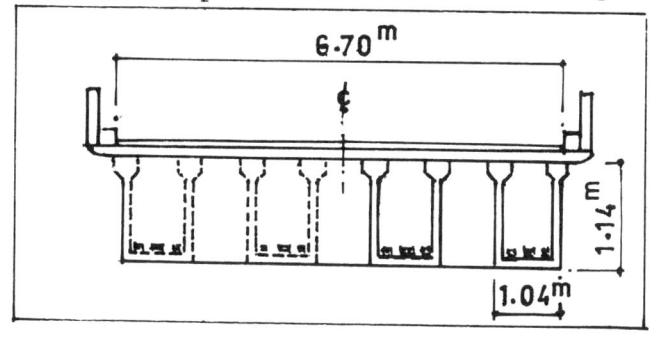

Fig.2 C.S. of the deck of Palar bridge

In the Sixties, 3062m (10043 ft.) long Sone Bridge in Bihar was constructed (Fig.3) with 93 spans of 33m (108 ft.) and concrete grade 42 MPa (6.2 ksi). All the 465 beams were precast in a yard 6 Km (4 miles) away, transported on a trailer and erected by a pair of mechanised rail mounted gantries. Statistical quality control techniques were introduced for the first time in the country. Zero slump concrete was used.

Fig.3 : Sone Bridge

Barak bridge at Silchar has a central span of 122m (400ft). The uniformity of concrete strength was a critical factor for evaluation of precamber during construction.

The Bassein Creek Bridge, Mumbai (Fig.4) with continuous deck of 361.6m (1186 ft.), used concrete of strength 46MPa (6.8ksi) to restrict the depth of girders.

Fig.4 : Bassein Creek Bridge

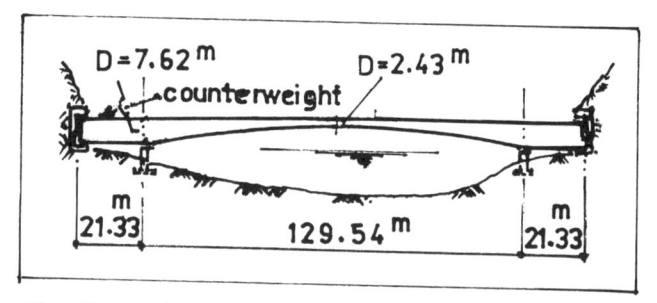

The Lubha Bridge (Fig.5) also in Assam has main cantiliver span of 130m (430ft.), with concrete counterweights at both ends. This involved tunnelling into rock.

Fig. 5 Lubha bridge in Assam

PRECAST SEGMENTAL BRIDGES - One of the longest river bridges was built across the Ganges at Patna in the Seventies(Fig.6). This 5575m (3.5Miles) long bridge consists

Fig.6 : Ganga Bridge Patna

of 46 spans of about 121m (400ft) each. The cellular R.C. piers were of 35MPa (5ksi) concrete and the superstructure of 45MPa (6.5 ksi) concrete. The mix was designed for High Early Strength of 20MPa (3ksi) at 6 hours and 35MPa (5ksi) at 48 hours, for early removal of formwork and stressing of tendons. The actual 28 days strength exceeded 70MPa (10ksi). Precast epoxy glued segmental construction was adopted using about 60 tonnes of epoxy (Fig.7).

Fig.7 : Ganga Bridge Patna - Deck Erection

The Ganga Bridge Buxar, 1122m (3680ft.), Narmada Bridge at Zadeshwar, 1350m (4428ft.) and Krishna Bridge have used precast segmental system. The 540m (1771ft.) long Krishna bridge consists of 18 spans of 30m (98ft.) each with expansion joints at 180m (590ft.) centres. The 3 cell box deck rests on Teflon bearings over solid ellptical pier (Fig.8). The precast units were

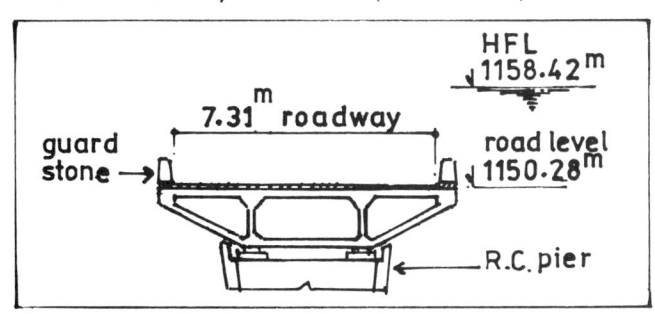

Fig.8 C.S. of deck of Krishna bridge at Deodurg

matchcast vertically, rotated, launched, jointed with epoxy mortar and prestressed. The Morhar bridge in Bihar with Precast steam cured PSC beams were used for High Early Strength of 35MPa (5ksi) in 12 hours. The 420m (1378ft.) long bridge was completed in six months.

PRESENT SCENARIO - The awareness of durability requirements and re-appraisal of Standards and Codes after the collapse of some spans of Mandovi Bridge resulted in the IRC Special publication SP-33 in 1989.The minimum grade of concrete for concrete components of the bridge for severe exposure condition is now 40MPa (6ksi).

The 2.5Km.(8200ft.)long (precast psc) Pamban Bridge(1984-88) (Fig.9) focused on durability.The minimum grade of concrete for PSC is 45MPa (6.5 ksi). Beams were steam cured for strength of 35MPa (5ksi) at 12 hours. Initially prestressed at 24 hrs, the beam was shifted to the stacking yard. Final stressing, grouting was completed at seven days and the beam launched.

Fig.9 :Pamban Bridge

Special specifications were evolved to facilitate effective grouting, using low w/c ratio of 0.40. The grout temperature was maintained at about 25°C by using chilled water. Reciprocating positive displacement type grout pumps were used for grouting. The grout was mixed using colcrete mixers to obtain a colloidal mix.

Akkar Bridge in Sikkim (Fig.10) was the first All-Concrete cablestayed bridge in India. The deck slab is integral with transverse beams spaced at 3m centres. The slab-cum-beam system is monolithic with longitudinal girders having a depth of 800mm. Precast RC panels formed permanent formwork.

The piers and deck slab of the Second Hooghly Bridge with cablestay span of 457m (1500ft.) necessitated high grade

Fig.10 :Akkar Bridge

concrete, 50MPa (7.4ksi). The tolerance specified for the deck slab was +5mm, in a deck width of 32m (105 ft.), involving design of special High Performance Concrete mix.

Four major bridges were constructed in Nepal utilising HPC. The RC piers with heights upto 20m (66 ft.) were concreted in one pour with no construction joints. The 30m (98ft.)span girders were precast using 45 MPa (6.6ksi) grade concrete and launched in position.

HIGH SLUMP CONCRETE - The concrete mixes are now designed with high workability. For the reconstruction of Mandovi Bridge in Goa, the piles were of 40MPa (6ksi) grade underwater concrete with slump of 200mm (8inch). For Raoli Flyover in Mumbai, the construction sequence for the box girders necessitated placing 45MPa (6.6 ksi)concrete in layers without cold joints and retardation of upto 6hours was realised with the use of retarding plasticisers.

For the precast elevated box girders for MRTS in Chennai, concrete of 45MPa (6.6ksi) grade was specified with 3days strength of 35MPa (5ksi) and 7days strength of 40MPa (6ksi).

The third Godavari Rly.Bridge is under construction. The bowstring arch bridge consists of 28 PSC spans of 92.55m (304ft.). The RC arches are of 45MPa (6.6ksi), braced laterally by precast RC struts. To reduce imbalance of concreting on either side, stringent tolerances for concrete density with permissible variations of 5% only were specified. Strengths of upto 70MPa (10ksi) were realised, with standard deviation of 3.6MPa (0.5ksi).

UNDERWATER HIGH STRENGTH CONCRETE - During the construction of the Ganga Bridge at Patna one of the 12m dia. (39ft.) caissons sunk to a depth of 55m (180ft.) had developed cracks. After detailed investigations, it was decided to construct a fresh caisson of smaller diameter inside the crack caisson. The heavily reinforced steining of 40mpa (6 ksi) concrete was concreted underwater, to a depth of 55m (180ft.). This perhaps is the deepest underwater structural concrete for a bridge anywhere in the World. The high performance relates to its ability to be placed underwater maintaining the required structural integrity with a concrete strength of 40MPa (6ksi).

For the Second Hooghly Bridge (Fig.11),the main foundations consisted of 23m (75ft.) dia. multi-cellular RC caissons.

Fig.11 : Second Hooghly Bridge

Fig.12 :Panvel Nadhi Viaduct

During the service life of the bridge, the caissons are kept empty, to minimise the dead load. Concrete of liquid retaining structure grade was used for the caisson steining, with water bars at construction joints. The bottom plugging utalised special grades of colcrete/concrete. The caisson was pumped dry after bottom plugging.

KONKAN RAILWAY BRIDGES - The 760Km. (475 miles) long new railway line completed in 1997 includes 43 major and 1670 minor bridges. Almost all are of precast, PSC construction, concrete grade upto 45MPa (6.6ksi). Both pretensioned and post-tensioned systems are used. For Panvel Nadhi Viaduct, 423m (1410ft.) long, the 70m (230ft.) tall piers are of slipformed concrete. The continuous deck in PSC was incrementally launched (Fig.12.)

HIGH STRENGTH/HIGH PERFORMANCE CONCRETE MATERIALS

Coarse Aggregates - In India, both gravel and crushed aggregates are available. Wherever available, natural gravel is preferred for preparation of High Performance Concrete. I.S.Code 383 on aggregates permits the use of both gravel and crushed stone. The reduced water demand by natural gravel results in reduced cement content, a reduction in the w/c ratio for the same workability, resulting in higher strength. The author has been using natural gravel successfully for prestressed concrete bridges.

For a bridge in Norway recently completed, the following mix was used to achieve the concrete grade of 75Mpa (11ksi), thanks to the use of gravel :

Cement	: 475 Kg.	CSF	: 40 Kg.
Admixtures	: 6.5 Kg.	Water	: 180 Lt.
Sand :0-8 mm	: 1080 Kg.	Gravel	: 720 Kg.
Slump	: 240-260mm	A/C ratio	: 3.6

Fine Aggregates - Sand or crushed aggregates conforming to IS:383 are used. In the coastal areas, natural sand is dredged from the creek. Such material, even after washing, is invariably contaminated by chlorides. In such situations, crushed aggregates should be used. Concrete for dams in India is being produced with crushed aggregates.

Cement, Admixtures - The 53 grade cement is adequate for concrete strengths upto 70MPa (10ksi). For Indian conditions retarders, plasticizers and superplasticizers are relevant. These admixtures are now being selectively used to avoid cold joints and to increase workability of concrete.

Superplasticisers are used to produce flowing concrete with high slump (150 to 200mm) for heavily reinforced bridges and where adequate vibration cannot be realised.

Condensed Silica Fume (CSF) - In view of its extreme fineness and high silica content, silica fume is a highly effective pozzolanic material and is used in concrete to improve its various properties including compressive strength. It also reduces permeability, protecting the reinforcement from corrosion. The silica fume, after collection, is condensed in order to facilitate handling.

The use of CSF increases water demand due to increased surface area of fine particles. This is mitigated by the use of superplasticizers. The use of CSF and superplasticizers are obligatory for HPC, especially beyond strengths of 80MPa (12ksi). CSF is presently imported.

Role of Supervision - This is crucial for realising High Performance Concrete. Some construction companies have qualified, trained personnel for realising concrete upto 100MPa (15ksi). However, there is a tremendous amount of misinformation and misunderstanding regarding the concrete mix design in India. This needs to be corrected.

In this context, the example of concrete mix design for the Ganga Bridge, Patna may be cited. A testing laboratory was established at the project site and an excellent degree of control (Standard Deviation:3.5MPa) was exercised, besides most economical design, resulting in saving in cement and costs compared with conventional practice elsewhere in the country. The following mixes were successfully adopted.

Characteristic strength	35 MPa	45 MPa
W/C ratio	0.42	0.36
A.C. ratio	5.6	4.0
Cement Consumption per cu.m.	340 Kg.	450 Kg.

The bridge was built, using Ordinary 10/7 mixers. The mix was designed by weight and converted into volume during actual execution. Such economic mix was possible, as the mix design was based on the author's innovations.

Role of Indian Codes and Specifications - A review of concrete in more than 1000 bridges built in the last 50 years reveals that the characteristics strength of concrete adopted for bridge construction remains static in the range of 40-50MPa (6-7.5ksi), vide Table 1. A number of factors have contributed to the state of affairs.

TABLE 1 : CONCRETE GRADES FOR TYPICAL BRIDGES SURVEYED

Name of the Bridge	Mpa (ksi)	Year
1. Railway Bridges	40 (6)	1948
2. Sone Bridge at Dehri-on-Sone	42 (6.2)	1963
3. Bassein Creek Bridge near Mumbai	48 (7)	1969
4. Khalidiyah Bridge, Iraq	45 (6.6)	1984
5. Pamban Bridge, Tamil Nadu	45 (6.6)	1985
6. Raoli Flyover Bridge, Mumbai	45 (6.6)	1992
7. Third Godavari Bridge	45 (6.6)	1996

Cement of quality was available in the Forties and Fifties. With the introduction of State controls, the cement quality in the Sixties and Seventies suffered. After the subsequent decontrol, cement now available satisfies ASTM requirements. Admixtures of good quality are now being produced in the country.Unfortunately, the Indian Codes are rather restrictive and at best only grudgingly admit their use.

The introduction of IS:102626, on mix design has retarded development of higher grades of concrete. Though meant to provide guidelines to the uninitiated, it is taken as the Bible and designing the mix as per the document is widespread. The publication's stumbling block comes in the form of the higher current margins based on specified standard deviations; this acts as license for poor quality concrete!

No developed country in the World has a National Standard or Guidelines for concrete mix design. This is normally left to professional bodies such as ACI, BRE etc. For the Ganga Bridge at Patna, the contractors were able to optimise the concrete mix with low cement consumption, primarily because of the absence of IS Guidelines in the early Seventies.

The Indian Roads Congress (IRC) Codes are conservative. The maximum grade of concrete is only 55MPa (8ksi) with the target mean strength specified as 69MPa (10ksi). The acceptance criteria given in the Indian Codes are not capable of being implemented in practice and also inhibits use of higher grades of concrete. The IRC Codes limit the maximum permissible stress in reinforced concrete is 8.5MPa (1.25ksi) in compression and 11.5 MPa(1.7ksi)in flexture. For PSC the maximum permitted compressive stress is limited to 20MPa (3ksi), limiting the characteristic strength used in the designs to 55MPa (8ksi). If the constraints are removed, strength of 80MPa (12ksi) may be used.

PSC bridges have been constructed with 28 days strength upto 50MPa (7ksi) with 80% strength required at the time of prestressing. Typically, precast girders and cantilever segments can be after 48 to 72 hrs. of concreting. and the mix is designed for strength of 35MPa (5ksi) at 48 hours. However, the Codes do not permit their use due to conservative provisions regarding creep and shrinkage.

HPC - INTERNATIONAL SCENARIO - HPC has been advantageously used for a number of bridges all over the World. Reasons for using HPC in bridges include economy, slender members, reduced weight and extended service life. Some bridges using HPC is given in Table 2:

TABLE 2 : SOME BRIDGES USING HPC

Name	Year	Strength	Other Data
Laval, Canada	1992	70MPa (10ksi)	0.30,495 Kg/cu.m
Mirabel, Canada	1993	80 (12)	
Perluiset, France	1988	80 (12)	
Joigny, France	1983	78 (11.5)	Without CSF
Elorn, France	1994	97 (14)	
Great Belt, Denmark	1997	70 (10)	One million cu.m
CNT, Japan	1993	122 (18)	W/C = 0.2

HPC is also used for bridge substructures and pylons of cablestay bridges. For the 400m (1312ft.) span Elorn Bridge in France, the RC pylons are 117m (384ft.) high and are of 80MPa (12ksi). This is a high strength application for a RC bridge member for the first time in the World.

Economics of HPC - The use of HSC/HPC improves concrete properties and reduces costs. The unit cost of HSC/HPC is obviously higher, but is offset by the reduced quantity. The concrete. The cost of prestressing strands remains unchanged. The real saving comes from the reduction in non-material costs associated with the girders, including reduction in cost of labour, transportation and erection costs and overheads due to reduced number of girders.

Codes and Regulations for HPC - Most National Standards are applicable to concrete strengths upto about 60MPa (9ksi). However, some Standards include HPC (Table No. 3)

IMPACT ON CONSTRUCTION METHODS - Construction with HPC requires a greater degree of control. The use of automated batching plants with good quality pan mixers or turbo mixers are required. The w/c ratio control is extremely crucial for HPC and as such, the metering device is required to very sensitive with variations not exceeding one percent.

At the same initial workability, slump loss may be higher in concrete with superplasticisers. As they are invariably used

TABLE NO.3 : SOME NATIONAL STANDARDS INCLUDING HPC

Country	Specifications	Maximum Strength		
CEB/FIP	MC-90	80 MPa	(12 ksi)	Cylinder
Norway	NS-373 : 1992	105 "	(15.5ksi)	Cube
Finland	MK B4 : 1984	100 "	(15 ksi)	Cube
Japan	HSC Specs.	80 "	(12 ksi)	Cylinder
Germany	DIN 1045 (S)	115 "	(17 ksi)	Cube
Sweden	BBK 79	80 "	(12 ksi)	
Netherlands	NEN 6720 (S)	105 "	(15.5ksi)	Cube

for HPC, the time elapsed between mixing and placement of concrete should be kept to the minimum.

High frequency vibrators assist in proper compaction of concrete. As the w/c ratio is extremely low, preventive measures are necessary to protect the fresh concrete from the wind and sun, with curing for a minimum of seven days. All the measuring and testing equipments shall be of appropriate quality and shall be calibrated initially and at periodical intervals during the operation of the equipment. The cube moulds shall be of the dimensions and tolerances specified in the Codes.

FUTURE PROSPECTS - The technology and materials for HPC exists in India. However, there is at present no incentive to go in for HPC, because of the restrictions of Indian Codes and Specifications. There is an urgent need to remove the restrictive clauses and incorporate the provisions relating to HPC in the existing Codes. The manufacture of CSF of the requisite purity is required to be initiated by the Ferro Alloy Industries. Till then CSF is imported. Training programmes should be initiated,both by the Owners as well as Constructors for mix design for HPC.

With the above refinements it is possible to gradually increase the strength of concrete to 80 Mpa (12 ksi) in the first instance and to 100MPa (15ksi)before the end of the millennium. Already concrete of 60MPa (9ksi) is used for a Nuclear Reactor in South and residential towers in Mumbai.

ACKNOWLEDGEMENTS - The author is grateful to Gammon India Limited for affording the experience on HPC Technology for over 35 years. Most of the examples (since 1961) cited in the paper are borne out of the author's personal experience.

HIGH-PERFORMANCE CONCRETES USING OKLAHOMA AGGREGATES

Thomas D. Bush, Jr., P.E.	Bruce W. Russell, P.E.	Seamus F. Freyne
Asst. Professor of CEES	Asst. Professor of CEES	Graduate Research Asst.
University of Oklahoma	University of Oklahoma	University of Oklahoma
Norman, Oklahoma, U.S.A	Norman, Oklahoma, U.S.A	Norman, Oklahoma, U.S.A.

ABSTRACT

The State of Oklahoma and its DOT have recognized the need to develop high-strength and high-performance concretes using materials that are available locally. A three year study is currently underway to identify suitable materials. This paper presents results indicating coarse aggregate type and grading can substantially affect compressive strength and elastic modulus; some significant differences in compressive strength were also observed from use of different cement sources.

INTRODUCTION

A comprehensive research study has been undertaken at the University of Oklahoma to identify local materials suitable for production of high-strength concrete, and will culminate in the production and testing of HS/HPC (high-strength/high-performance concrete) bridge girders with compressive strengths up to 100 MPa (14,000 psi). The research program is examining cements, coarse aggregate types and sources, fine aggregate gradings, and addition of chemical and mineral admixtures to identify materials and proportions most suitable for producing HS/HPC in Oklahoma.

This paper focuses on: 1) suitability of locally available cements, and 2) suitability of various coarse aggregates. In the cement study, concretes with three different water-cement ratios (w/c) were tested, varying only the cement source for a given w/c. The coarse aggregate study utilized a single mixture proportion, and varied only the source and/or grading of coarse aggregate. Eight cements and four coarse aggregates were included in the studies. This paper will concentrate on the extent to which compressive strength and modulus of elasticity were affected by cement source and coarse aggregate.

EXPERIMENTAL PROGRAM

Cement Study
The effects of cement type and source on the properties of HPC were studied at two compressive strength levels. Eight different cements were tested, including Types I, II, and III, from a total of six sources (cement plants). The lower strength mixture contained approximately 7 sacks of cement per cu. yd. (387 kg/m^3; 650 lb/cyd) and the higher strength mixture contained about 8.5 sacks of cement per cu. yd. (464 kg/m^3; 780 lb/cyd). At each strength level, all other parameters were held constant, i.e., proportions, type of

coarse and fine aggregate, and type and dosage of chemical admixtures. Chemical admixtures included a conventional water reducing and retarding admixture coupled with a high range water reducing admixture (HRWR). Mineral admixtures were not included in these mixtures. All specimens were moist cured at 23 ± 1.7°C (73.4 ± 3°F).

Proportions (saturated surface dry condition) for the 7 sack and 8.5 sack mixtures are shown in Table 1. The w/c's for the mixtures were 0.40 and 0.34, respectively. For the lower strength mixture, the coarse aggregate met an AASHTO #67 gradation, with a nominal maximum particle size of 19 mm (3/4 in.). The higher strength mixture contained coarse aggregate with a 9.5 mm (3/8 in.) maximum particle size. Both of the limestone aggregates came from the same quarry in central Oklahoma. The same fine aggregate, a river sand, was used for all batches at both strength levels. The conventional water reducer/retarder was a hydroxylated organic compound, and possessed slight air entraining capabilities. The HRWR was a modified naphthalene sulfonate. Chemical admixture dosage rates used for the two mixture classes are contained in Table 1.

Table 1 Mixture Proportions (per m^3) for Cement Study

	7 Sack Mixtures	8.5 Sack Mixtures
Water (kg)	154	157
Cement (kg)	387	464
Coarse Aggregate (kg)	1,053[a]	1,009[b]
Fine Aggregate (kg)[c]	795	753
Conventional WR (mL)	773	889
HRWR (mL)	3,017	4,177
w/c	0.40	0.34
Sand Fraction of Total Aggregate by Wt.	0.43	0.43
HRWR Dosage (mL/kg cement)	7.81	9.02
Notes: a. #67 crushed limestone, max. particle size = 19 mm b. "3/8 in. chips" crushed limestone, max. particle size = 9.5 mm c. "Dover Sand," Fineness Modulus=2.50		

1 kg/m^3 = 1.686 lb/yd^3 ; 1 fl.oz. = 29.6 ml ; 1 fl.oz/yd^3 = 38.7 ml/m^3, 1 in.= 25.4 mm

Cement samples from production facilities in Oklahoma, southern Kansas, western Arkansas, and northern Texas were represented in the eight cement test group. Compound compositions of the cements, as provided by the suppliers, are shown in Table 2. Cements are designated by Type (I, II, or III) followed by an identification number (.1 through .8).

The values of Blaine fineness presented in the Table 2 were independently obtained using an air permeability apparatus in accordance with ASTM C 204.[1] These values agreed well with the values of fineness provided by the cement suppliers. For the most part, chemical compositions for the cements were similar to one another, with the exceptions of lower C_3A contents for the Type II cements, and higher fineness and lower C_2S for the Type III cement.

Table 2 Compound Composition (%) and Fineness of Cements

	Designation							
	I.1	I.3	I.4	I.6	II.2	II.5	II.7	III.8
C_3S	59.6	54.4	58	58	58.8	54	56	57.1
C_2S	-	18.4	-	18	-	21	22	13.8
C_3A	10.6	11.4	13.4	9	5.1	8	4	10.5
C_4AF	-	-	-	8	-	10	12	-
Blaine Fineness (cm^2/gm)	3480	3390	3390	3690	3470	3610	3600	5490

- data not available

Coarse Aggregate Study

Four sources of coarse aggregate were tested in HPC mixtures to determine their suitability for producing high-strength HPC. The aggregates included crushed *limestone* from central Oklahoma, crushed *rhyolite* from central Oklahoma, crushed *granite* from southwestern Oklahoma, and *river gravel* (predominantly crushed, but with some uncrushed larger particles) from southeastern Oklahoma. Each coarse aggregate was used in mixtures in two conditions: 1) a "standard" grading, and 2) the commercial (as-received) grading. The standard grading was selected to fall within the specifications for AASHTO #7 material. Use of the standard grading isolated differences in strength due to the coarse aggregate materials themselves, such as mineralogy, surface texture, and particle shape. Mixtures containing coarse aggregate with the commercial grading were compared to the companion mixtures with standard grading to examine potential strength differences related to the grading.

Measured properties of the aggregates and their gradings are shown in Tables 3 and 4, respectively. The commercial limestone was finer than the other three coarse aggregates (Table 4), and had a maximum particle size of approximately 9.5 mm (3/8 in.). The rhyolite, granite, and gravel had maximum size particles of 15.9 mm (5/8 in.). To achieve the standard grading, each coarse aggregate was separated into different sizes and recombined in the required amounts. The commercial 9.5 mm (3/8 in.) limestone material was augmented with larger material from the same quarry to achieve the standard grading.

Table 3 - Properties of Coarse Aggregates

Property	Limestone LI	Rhyolite RH	Granite GN	River Gravel GV
Bulk Specific Gravity (SSD)	2.67	2.71	2.62	2.59
Percent Absorption (SSD)	1.2	1.4	0.5	1.3
DRUW, Standard (kg/m^3)	1590	1510	1520	1605
DRUW, Commercial (kg/m^3)	1620	1520	1500	1630

DRUW = Dry Rodded Unit Weight; 1 lb/ft^3 = 16.02 kg/m^3

Table 4 - Coarse Aggregate Gradings (Percent Passing)

Sieve Size	Commercial Grading				Standard Grading
	Limestone LI	Rhyolite RH	Granite GN	Gravel GV	All
19.05 mm (3/4 in.)	100	100	100	100	100
12.5 mm (1/2 in.)	100	91.5	92.6	91.2	91
9.5 mm (3/8 in.)	94.2	62.4	48.5	67.5	59
4.75 mm (#4)	16.4	8.7	0.9	11.2	2
2.36 mm (#8)	4.7	2.7	0.4	1.9	0

Table 5 - Mixture Proportions (per m^3) for Coarse Aggregate Study

Water (kg)	177
Cement (kg)	474
Flyash (kg)	166
w/(c+p)	0.28
Conventional Water Reducer (mL)	1255
High Range Water Reducer (mL)	2925
Fine Aggregate	496-617
Coarse Aggregate	957-1040
Coarse Aggregate Content, Dry Rodded Volume per Concrete Volume	0.64

1 kg/m^3 = 1.686 lb/yd^3 ; 1 fl.oz. = 29.6 mL ; 1 fl.oz/yd^3 = 38.7 mL/m^3

All mixtures (both commercial and standard gradings) in the coarse aggregate study used the proportions shown in Table 5. The mixtures had a constant *volume of DRUW aggregate per unit concrete volume* equal to 0.64, and were proportioned by absolute volume. Since each aggregate had a slightly different bulk specific gravity, minor differences in absolute coarse and fine aggregate weights were required to achieve the constant ratio of 0.64. The mixtures all contained flyash and had w/(c+p) (ratio of water to cement plus pozzolans) of 0.28. A total of 16 mixtures were batched, two for each coarse aggregate material (limestone, rhyolite, granite, gravel) in each grading condition (commercial, standard). Compressive strengths at 1, 3, 7, and 28 days were determined, and modulus of elasticity was measured at 7 and 28 days. All specimens were wet cured at 23 ± 1.7°C (73.4 ± 3°F).

RESULTS AND DISCUSSION

Results of the Cement Study

Measured compressive strengths at 1, 3, 7, and 28 d are contained in Figs. 1 and 2. Each reported strength represents the average strength of three, 100 x 200 mm (4 x 8 in.) cylinders. Cylinders were made in conformance with ASTM C 192[2] and tested following ASTM C 39[3]. Overall the average 28 day strength of the 7 sack concrete was 60.2 Mpa (8,740 psi) with an average standard deviation of 5.13 MPa (744 psi). For the 8.5 sack concrete mixtures, the average 28 day strength was 71.7 MPa (10,410 psi) with average standard deviation of 4.73 MPa (686 psi).

In Fig. 1 (7 sack mixtures), the 28 day strengths for all eight mixtures are very near the average value of 60.2 MPa (8,740 psi) except for mixture I.1-7, which possessed a 28 day strength of 49.3 MPa (7,150 psi), about 18% less than the average of the other 7 sack mixtures. Concrete mixture I.4-7 achieved the highest 28 day strength of 64.3 MPa (9,330 psi), which is only 7% greater than the average of all 7 sack concretes tested.

Figure 2 depicts the concrete strengths of the 8.5 sack mixtures for all ages. From these charts, it is apparent that the highest 28 day strength was achieved by mixture I.3-8.5. Its 28 day strength was 81.4 MPa (11,810 psi) which is about 13% greater than the average 28 day strength for these eight mixtures. Of the remaining seven mixtures, only one 28 day strength fell below 68.9 MPa (10,000 psi); an average 28 day strength of 64.2 MPa (9,320 psi) was measured for mixture II.2-8.5.

Figures 1 and 2 also illustrate the differences in strength gain with age. In general, the concrete made with Type III cement gained strength more rapidly (due to the cement's greater fineness) than concretes made with the Type I and II cements. Comparing the 7 sack concretes, the Type III concrete made 34.5 MPa (5,010 psi) at 1 day, whereas the Type I concretes achieved an average of 23.2 MPa (3,370 psi) and the Type II concrete achieved an average of 21.6 MPa (3,140 psi) at one day. Comparing the 8.5 sack mixtures, the concrete made with Type III cement achieved a 1 day strength of 37.4 MPa (5,430 psi), whereas the concretes made with Type I cement averaged a 1 day strength of 29.6 MPa (4,300 psi).

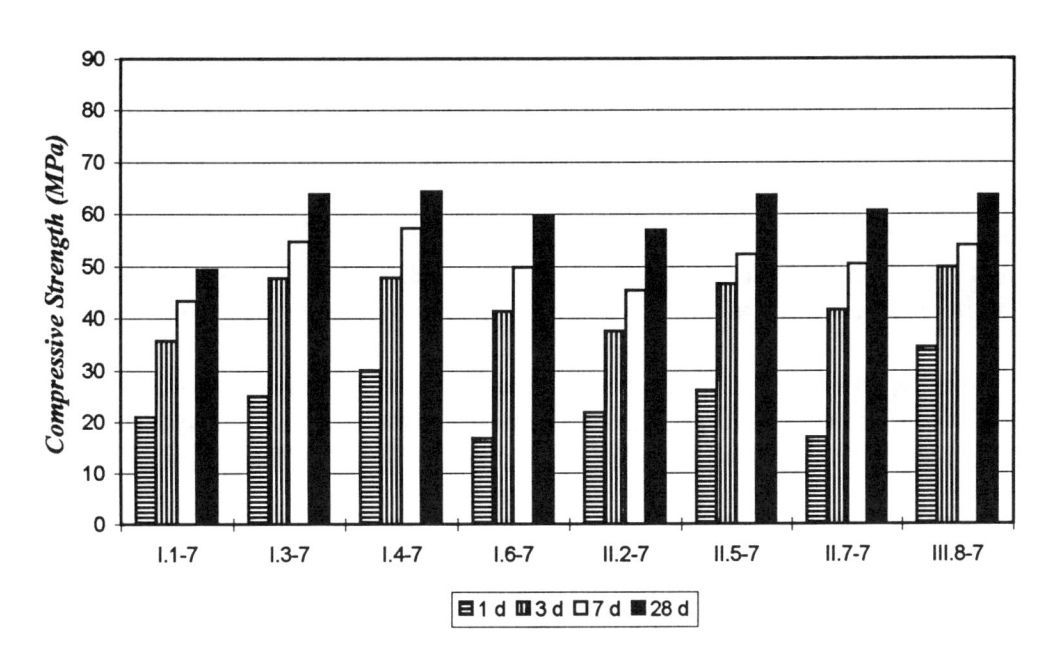

1 ksi = 6.89 MPa
Figure 1 Strength Gain for 7 Sack Mixtures

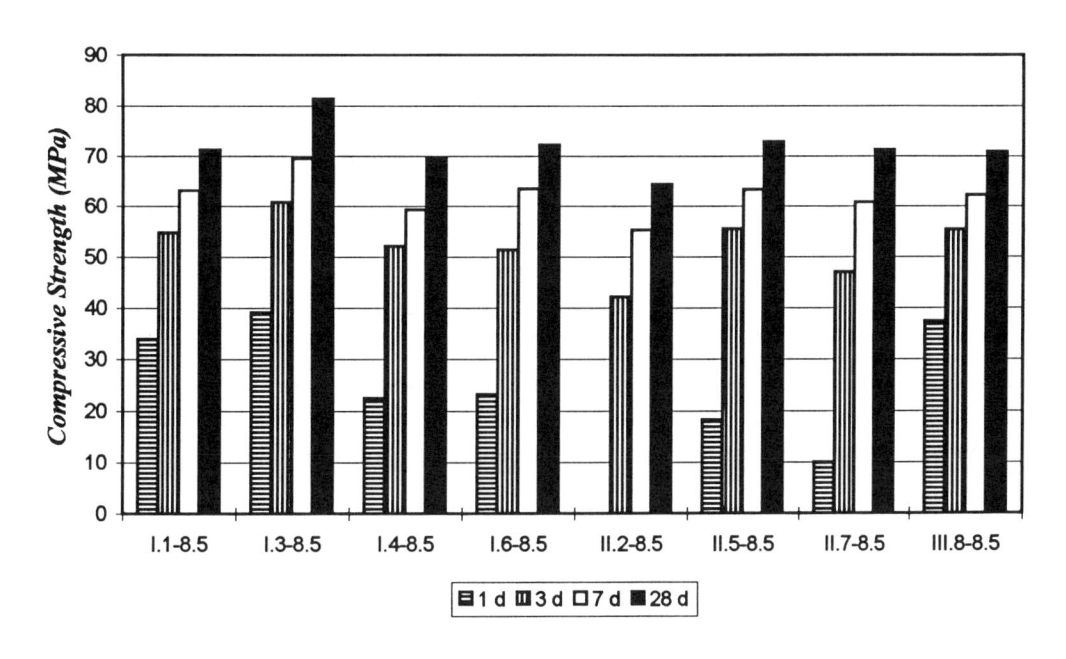

1 ksi = 6.89 MPa
Figure 2 Strength Gain for 8.5 Sack Mixtures

Although the rapid strength gain achieved by using Type III cement is apparent from the data, the Type II cements do not appear to substantially delay strength gain when compared to Type I cements. This is discovered by examining the strength gain with age for the different mixtures of concrete, and evidenced by the overlapping concrete strengths from the Type I and Type II cements, especially at early ages. This indicates that very few real differences exist between these Type I cements and Type II cements. The exception to this observation is for 1 day compressive strengths of the 8.5 sack concretes made with Type II cements (Fig. 2). Mixture II.7-8.5 achieved lower 1 day strengths than the other cements, and mixture II.2-8.5 did not have sufficient strength to be tested at 1 day. From Table 2 that lists cement chemical compositions and finenesses, the fineness of the Type II cements is essentially the same as the Type I cements. The fineness of the Type II cements average 3560 cm^2/g whereas the Type I cements have an average fineness of 3490 cm^2/g. The Type II cements are distinguished only by reduced amounts of C_3A which should result in reduced heat of hydration.

Overall, the variations in 28 day concrete strengths between the different cements appear to be slight, and the concrete strengths tend to group tightly around the average for each class of concrete. To compare the effects of cement on concrete strengths, 90% confidence intervals were computed for the 28 day strengths of each mixture. A confidence interval is a statistical measure that rates the probability that the true average strength for the population (for a given mixture) will fall within a specified range. For the 7 sack mixtures, these comparisons indicated that some measurable differences in concrete strength can be attributable to the cement used to make the concrete. For example, the concretes from mixture I.1-7 and mixture II.2-7 fell measurably below the average value for strengths in this mix class. Little statistical difference was found to exist between the 28 day strengths for mixtures I.3-7, I.4-7, II.5-7, II.7-7 and III.8-7.

Ninety percent confidence intervals for concretes from the 8.5 sack mix class are shown in Fig. 3. Only one mixture (II.2-8.5), possessed 28 day strengths significantly less than the average of the other mixtures. Conversely, the cement used in mixture I.3-8.5 could hold high potential for the manufacture of high-strength concrete. Taken as a whole, comparisons indicate that one cement, designated I.3, could be more suitable than the other cements for the production of high-strength concrete. Two cements, II.2 and possibly I.1 may be less suitable for producing high-strength concrete.

In reviewing the data, questions could be raised as to whether the observed differences in concrete strength result from variations in cement source, or rather the differences were produced by naturally occurring variations in the data. To address this concern, three separate casts were made using the same cement source and the same mix proportions. The concrete strengths from these three casts were then compared at 1, 7, and 28 days. From these three casts, the average 28 day strength varied no more than 3.5% and the 90% confidence limits were overlapping, indicating that nearly identical concrete strengths were produced when the same cement was used. As an extension, the consistency demonstrated by these three casts indicates the variations in strength observed between concretes made with different cements were likely caused by the variations in cement, as all other constituent materials and procedures were held constant. Additional data is being obtained to further verify these observations.

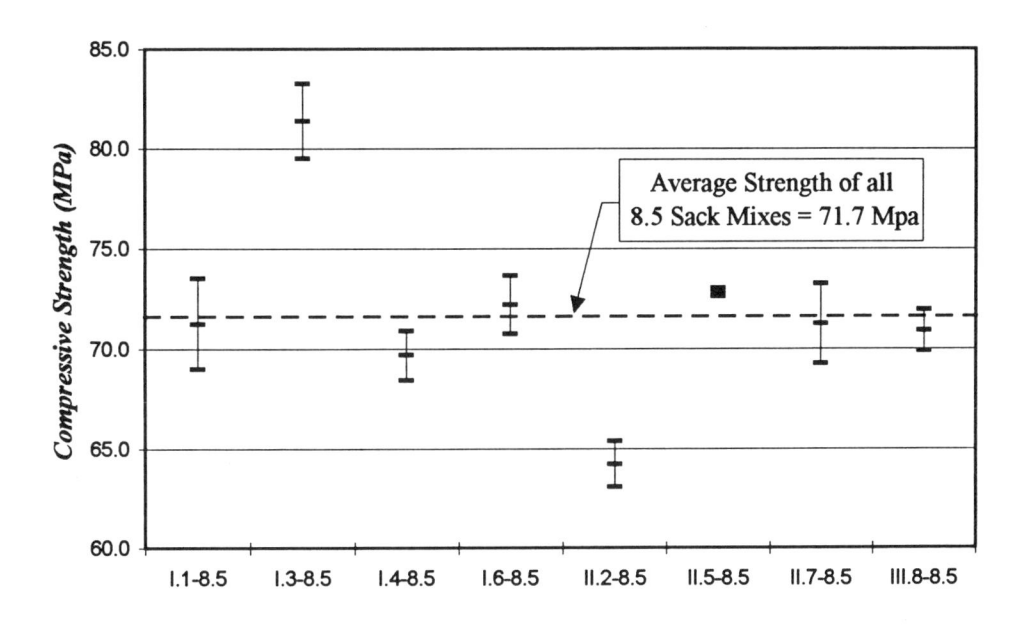

Figure 3 Ninety Percent Confidence Intervals for 8.5 Sack Mixtures - Cement Study

1 ksi = 6.89 MPa

The modulus of elasticity at 28 days is reported for each mixture in Table 6. Elastic modulus was determined using a compressometer in accordance with ASTM C 469[4]. The average from two cylinders was used for each modulus data point reported.

Table 6 Elastic Moduli - Cement Study

Mixture Class	Elastic Modulus (GPa)								
	I.1	I.3	I.4	I.6	II.2	II.5	II.7	III.8	Average
7 Sack	41.4	42.8	43.1	41.0	37.2	42.1	41.0	41.7	41.3
8.5 Sack	41.0	42.4	43.1	39.3	39.0	42.1	42.4	37.9	40.9

1 ksi = 6.89 MPa

For the 7 sack mix class, the modulus of elasticity averaged 41.3 GPa (5,990 ksi) for all casts with a standard deviation of 1.69 GPa (245 ksi), or 4.1%. For the higher strength concrete, the modulus of elasticity averaged 40.9 GPa (5,930) ksi with standard deviation of 1.80 GPa (261 ksi), or 4.4%. Interestingly, the average modulus of elasticity was slightly lower for higher strength (8.5 sack) concretes, 40.9 GPa vs. 41.3 GPa. This observation is contrary to the accepted equations for elastic modulus that relate E_c as a function of concrete strength. No strong trend was observed that relates the elastic modulus to concrete strength for the cement study. Instead, both classes of concrete, the 7 sack mixture and the 8.5 sack mixture, possess roughly equivalent elastic moduli. This is possibly a reflection that the same coarse aggregate (crushed limestone) was used in roughly the same amounts for both mix classes, even though the 7 sack mix class used a larger aggregate than the 8.5 sack mixtures, 19 mm vs. 9.5 mm.

Results of the Coarse Aggregate Study

Measured compressive strengths at 1, 7, and 28 days are shown in Table 7. Average results from six 100 x 200 mm (4 x 8 in.) cylinders (three per batch) were used for each data point reported. Designations in the table indicate the aggregate type (LI=limestone, RH=rhyolite, GN=granite, GV=gravel) and grading (s=standard, c=commercial). Compressive strengths at 28 days ranged from 64.8 to 85.2 MPa (9,410 to 12,360 psi).

Ninety-five percent confidence intervals were computed for 28 day strengths of all mixtures. When comparing average strengths for mixtures, if the 95% confidence intervals did not overlap, the resulting strengths were considered to be significantly different, i.e., not due to chance.

Table 7 Compressive Strengths, Coarse Aggregate Study

	\multicolumn{5}{c}{Standard Grading}	\multicolumn{5}{c}{Commercial Grading}
	LIs	RHs	GNs	GVs	Ave	LIc	RHc	GNc	GVc	Ave
1 d Strength (MPa)	23.3	25.3	30.2	24.6	25.8	26.9	28.8	23.6	22.1	25.4
7 d Strength (MPa)	59.7	64.2	69.0	56.6	62.4	69.9	62.6	61.9	51.7	61.6
28 d Strength (MPa)	73.7	78.8	83.7	70.0	76.5	85.2	75.9	76.1	64.8	75.5

1 ksi = 6.89 MPa

Strengths of mixtures containing aggregates with the standard grading ranged from 70.0 MPa (10,160 psi) to 83.7 MPa (12,150 psi) at 28 days. The average compressive strength was 76.5 MPa (11,090 psi) and the average standard deviation was 1.45 MPa (210 psi). Confidence intervals for the 28 day strengths of these mixtures did not overlap indicating significant quantifiable differences can be attributed to the coarse aggregate. The highest strength was observed for the mixtures containing granite (9% above the average), followed by rhyolite, limestone, and gravel. The limestone aggregate produced strengths slightly below the average, while the gravel resulted in strengths about 9% lower than the average. The lower strength of the mixtures with river gravel is consistent with the reduced bond associated with the smooth particle surfaces. The granite, rhyolite, and limestone have rougher textures, possibly leading to higher strengths. Granite and rhyolite, being dense, hard aggregates, produced the highest strength mixtures for the standard grading.

Mixtures with the commercial (as-received) grading yielded 28 day compressive strengths ranging from 64.8 MPa to 85.2 MPa (9,410 psi to 12,360 psi). The average 28 day strength was 75.5 MPa (10,960 psi) with an average standard deviation of 2.00 MPa (290 psi). As with the standard grading, strengths of the river gravel mixtures were found to be lower than for mixtures containing the other aggregates. Granite and rhyolite mixtures had the same (statistically similar) intermediate strength of 76 MPa. However, unlike for the standard grading, the mixtures with limestone produced the highest strength. The dramatically increased strength of the limestone mixtures (nearly 16% as compared to the standard grading) is likely due to the smaller aggregate size in its commercial grading. Smaller coarse aggregate particles tend to produce less microcracking in the transition zone, leading to improved strength. The commercial limestone was a nominal 9.5 mm (3/8 in.) maximum size material, while the other three aggregates were nominal 15.9 mm (5/8 in.) in size. If similar strength differences are exhibited by granite and rhyolite, then strengths approaching 100 MPa (14,000 psi) may be possible with these mixture proportions.

With regard to compressive strength, it is evident that the limestone, rhyolite, and granite aggregates are all suitable for production of HPC with strengths in excess of 69 MPa (10,000 psi). The granite and rhyolite appear to hold better potential for producing mixtures of even higher strengths, especially if gradings utilize smaller average particle sizes (such as 9.5 mm, similar to the commercial limestone). The limestone produced mixtures with laboratory strengths in excess of 85 MPa (12,300 psi) when used in the smaller (9.5 mm) maximum particle size, but achieved lower (10,700 psi) strengths when used at the standard grading.

Elastic moduli at 7 and 28 days are shown for the mixtures in Table 8. Modulus of elasticity (ASTM C 469[4]) was obtained by averaging the results from four 100 x 200 mm (4 x 8 in.) cylinders. Measured values ranged from a low of 40.0 GPa to a high of 44.4 GPa (5,800 to 6,450 ksi). At 28 days, granite produced the highest modulus (44.4 GPa) for the standard grading; however, large differences in modulus were not observed between mixtures containing limestone, rhyolite, or granite at a given age and for a given grading. Mixtures containing river gravel had moduli substantially lower than for the other aggregates (ranging from 34.8 to 40.4 GPa), for both commercial and standard gradings, at 7 and 28 days.

Unlike in the cement study, moduli generally increased from age 7 days to age 28 days, slightly more so for the mixtures with the standard aggregate grading. Increases were on the order of 4-7% for mixtures with commercial gradings and 9-10% for mixtures with the standard grading. Mixtures with rhyolite aggregate exhibited essentially the same modulus (42.0 GPa (6,100 ksi)) for both gradings and ages. The general trend of increase in modulus with age can be seen in Fig. 4. The solid lines indicate modulus predicted from the equation, $E_c = 3,320\sqrt{f_c} + 6900$ *MPa* ($40,000\sqrt{f_c} + 1.0 \times 10^6$ *psi*), recommended for high-strength concrete[5,6] and the AASHTO[7] equation, $E_c = 4,730\sqrt{f_c}$ *MPa* ($57,000\sqrt{f_c}$ *psi*). The ACI Committee 363 equation for modulus of high strength concrete was conservative for the mixtures tested.

Table 8 Modulus of Elasticity, Coarse Aggregate Study

	Standard Grading					Commercial Grading				
	LIs	RHs	GNs	GVs	Ave	LIc	RHc	GNc	GVs	Ave
7 d E_c (GPa)	40.0	41.7	40.3	36.5	39.6	40.7	42.4	41.0	34.8	39.7
28 d E_c (GPa)	43.4	42.0	44.4	40.0	42.5	42.4	42.0	42.7	37.2	41.1

1 ksi = 6.89 MPa

$$E_c = 4730\sqrt{f'_c}\ MPa$$

$$E_c = 3{,}320\sqrt{f'_c} + 6900\ MPa$$

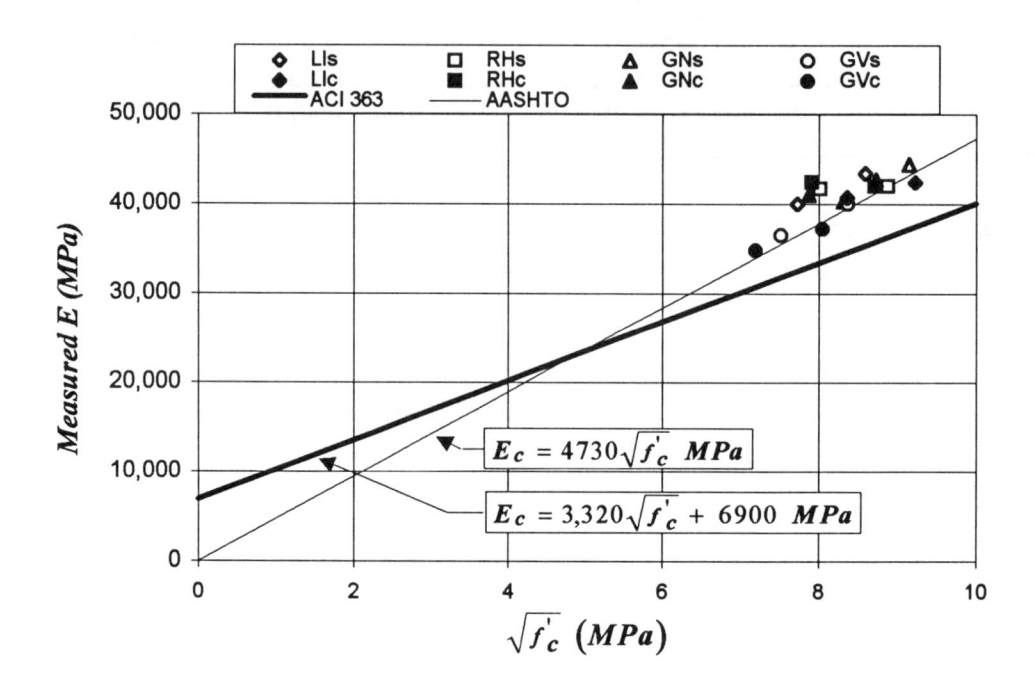

1 ksi = 6.89 MPa

Figure 4 Modulus of Elasticity - Coarse Aggregate Study

CONCLUSIONS

1. These data indicate that while strengths from mixtures using cements from different sources were generally closely grouped, cement source and type can affect concrete compressive strengths. Therefore, it is important to test individual cements to determine their suitability for applications in high-strength concrete.

2. In this study, concretes made with Cement I.3 possessed higher strengths than the concretes made with other cements. Conversely, concretes made with Cements I.1 and II.2 possessed lower strengths when compared to the concretes made with other cements. These results demonstrate that for the cements studied, the cement source can influence the compressive strengths of plain concrete, and that

engineers, constructors and transportation agencies should be aware of the possible differences in performance of HS/HPC.

3. For the cement study, modulus of elasticity was largely unaffected by increases in concrete strength. Instead, E_c remained unchanged as did most of the constituent materials for the two concrete strengths, indicating that the E_c in this strength range may be affected more by the constituent materials than by concrete strength.

4. For the coarse aggregate study, when using a standard grading, granite aggregate produced higher strengths than rhyolite and limestone; mixtures with river gravel achieved strengths substantially lower than for the other aggregates. The river gravel does not appear well suited to producing concrete with strengths in excess of 70 MPa.

5. Limestone aggregate with the commercial grading achieved the highest compressive strength of all mixtures tested in the coarse aggregate study. This result is believed to be closely tied to reduced microcracking in the transition zone due to the smaller particle sizes of the commercial limestone aggregate. It is expected that even higher strengths would be achieved using granite or rhyolite in gradings with smaller particle sizes.

6. Granite aggregate produced mixes with slightly higher modulus of elasticity than the limestone and rhyolite aggregates. River gravel mixtures produced moduli substantially lower than the average for all aggregates. A modest increase in modulus was observed from 7 to 28 days for all mixtures containing aggregates except rhyolite, which remained unchanged.

7. Overall, the expression *$E_c = 3,320\sqrt{f'_c} + 6900$ MPa $(40,000\sqrt{f'_c} + 1,000,000$ psi)* underestimated the elastic modulus by 10 to 20 percent for both the cement study and the coarse aggregate study.

REFERENCES

1. Fineness of Hydraulic Cement by Air Permeability Apparatus (C 204), V. 4.01, *ASTM*, Philadelphia, 1995, pp. 161-169.

2. Making and Curing Concrete Test Specimens in the Laboratory (C 192), V. 4.02, *ASTM*, Philadelphia, 1995, pp. 116-122.

3. Compressive Strength of Cylindrical Concrete Specimens (C 39), V. 4.02, *ASTM*, Philadelphia, 1995, pp. 17-21.

4. Static Modulus of Elasticity (C 469), V. 4.02, *ASTM*, Philadelphia, 1995, pp. 241-244.

5. Carrasquillo, R.L., A.H. Nilson, and F.O. Slate. Properties of High Strength Concrete Subject to Short-Term Loads, *ACI Journal*, V. 78, No. 3, May-June 1981, pp. 171-178.

6. ACI Committee 363, State-of-the-Art Report on High-Strength Concrete (ACI 363R-92), American Concrete Institute, Detroit, 1992.

7. Standard Specification for Highway Bridges, 15th edition, *American Association of State Highway and Transportation Officials*, Washington, D.C., 1992.

HIGH PERFORMANCE CONCRETE MIXTURE PROPORTION OPTIMIZATION FOR PRECAST CONCRETE USING STATISTICAL METHODS

James R. DeMaro
Graduate Student
South Dakota School of Mines and Technology
Rapid City, SD USA

Dr. M. R. Hansen
Associate Professor
South Dakota School of Mines and Technology
Rapid City, SD USA

Brian C. Anderson
Graduate Student
South Dakota School of Mines and Technology
Rapid City, SD USA

ABSTRACT

South Dakota Concrete Products (SDCPC), makes precast concrete pipe, box culverts, and prestressed/precast beams for bridges. SDCPC uses 27.6 Mpa (4000 psi), 34.5 Mpa (5000 psi), and 41.1 Mpa (6000 psi) concrete in their production and batches the concrete using a central mix operation.

The South Dakota Department of Transportation (SDDOT), requires that compressive strengths must be at least 10% over design strength at seven days. The compressive strengths that the company was achieving were normally within these limits, but were variable and unreliable.

A five variable statistical experimental design was organized to determine the effect of the main components on performance, strength and economy of the mixture.

Research Objective & Scope

The objective of this project was to optimize three mixture proportions. There were three design strengths to consider, 27.6 Mpa (4000 psi), 34.5 Mpa (5000 psi), and 41.1 Mpa (6000 psi). The mixture proportions must have consistent slump, strength, and air content from 4.5% to 6%. For precast and prestressed products the SDDOT allows no more then 15% (1:1 replacement) fly ash. For this project the new mixes would use up to 15% addition of fly ash, for economy. The lab work at South Dakota School of Mines and Technology (SDSM&T) would be conducted first where all appropriate ASTM standard concrete tests would be conducted. Compressive strength tests would be conducted at one, three, seven, and twenty-eight days. Cylinders would also be put on a thermocouple data collection machine to record the early temperature of the concrete versus time.

At the end of the lab trial mixes the data was analyzed, and an appropriate mixture proportion for the design strength was chosen. The chosen mixture proportions were then repeated in the lab to see if the predicted results could be achieved. After the mixture proportions were stabilized in the lab, field testing was conducted. For the field test, the mixture proportions chosen were made at the SDCPC plant, with the batching operation normally used. The cost of the original mixes were as follows: 27.6 Mpa (4000 psi) ($39.00/cy), 34.5 Mpa (5000 psi) ($44.00/cy), and 41.1 Mpa (6000 psi) ($50.00/cy).

Batching and Mixing Sequence
The concrete batches were mixed according to the standard practice used at the plant, which was to add all of the sand and rock and mix for 1 minute, add all of the air entraining admixture and water and mix for 1 minute, add the cement and mix for 1 minute. At this time the high range water reducer (HRWR) was added and a total mix time of fifteen minutes was achieved. All ASTM field test were conducted and cylinders were cast for strength tests.

Materials Used
All materials used for this research were provided by local suppliers for use in the SDSM&T lab. The ASTM Type I/II Cement was manufactured by Dacotah Cement and supplied in 42.64 Kg (94 lb.) bags. The 1.9 cm (3/4 inch) crushed limestone coarse aggregate was supplied by Pete Lien and Sons, from their quarry located in Rapid City, SD. The river sand was supplied by Birdsall Sand and Gravel of Rapid City, SD, which came from their plant in Creston, SD. The air entraining admixture consisted of an aqueous solution of neutralized resin acids and rosin acids. The HRWR consisted of Naphthalenesulfonate Formaldehyde Copolymer, in aqueous solution. The fly ash came from Coal Creek Station power plant located in Beaulah, ND. The fly ash was a class F, and was tested by the Energy and Environmental Research Center of the University of North Dakota. These materials were all stored in covered containers in the SDSM&T concrete lab. A sieve analysis of the fine and coarse aggregate was conducted on the material used at the school, and compared to a test conducted on the materials used by SDCPC.

Experiment Design
It was decided that the best approach to the amount of work needed to conduct this research, was to use statistical methods.[1] The process involved thirty-three different trial mixture proportions, made eleven at a time. For the thirty-three trials, five independent variables were chosen. The first problem was to chose the five variables and their rate of change for the experiment. Since SDCPC was having trouble with compressive strength, slump, bug holes in the finished product, and wanted to incorporate fly ash into their mixture proportions, these were the major considerations for the project. The five variables chosen were cement content, fly ash (percentage replacement), water to cementitious materials ratio, HRWR dosage, and aggregate blend (percent coarse to sand). The cement content for the original SDCPC standard mixture proportions had a 385 kg/ m^3 (650 pcy) average. Using this as the mid point, the range for the cement content was 350 to 950 pcy in 150 pcy increments. The original average for their w/c ratio was 0.33. Therefore the range for the w/c ratio was 0.27 to 0.39 with 0.03 increments. A range from 266 to 621 ml/cwt (9 to 21 oz/cwt) was chosen for the HRWR dose, with increments of 88.7 ml (3 oz/cwt). The last variable was chosen to try to evaluate the problem SDCPC was having with bug holes in their finished product. The range for the

aggregate blend was from 50 to 70 percent coarse aggregate with 5 percent increments. These variables and their ranges are summarized in Table 1.[2]

Table 1 - Variables Chosen For This Experiment

CEMENT (pcy)	FLY-ASH (%)	W/C	HRWR (oz/cwt)	AGG BLEND (% coarse)	CODED UNITS
350	0	0.27	9	50	-2
500	5	0.30	12	55	-1
650	10	0.33	15	60	0
800	15	0.36	18	65	1
950	20	0.39	21	70	2

(1 oz = 29.57 ml, 1 pcy = .593 kg/m^3)

After the five variables and their rates of change were chosen, a method had to be devised to determine which variables to change for each mix. For this a 5 factor central composite design broken into 3 groups of 11 runs each was used. (1) A coded unit is assigned to each value of each variable. To go from coded units to real units, simply use the following expression. A=C*d+M where:

```
A = actual value
C = coded value (-2 To +2)
d = actual amount of  1 coded unit
M = middle value of a variable
```

This method was ideally suited for this experiment, and was used to design the thirty-three different mixes.

Mixture proportions for the first eleven mixes (33-1 through 33-11)

The materials were stored in the SDSM&T concrete laboratory in covered containers. A 1000 gram sample of the coarse and fine aggregates was dried for 24 hours to find the actual moisture content of the aggregates on mix day. The water required for these eleven mixture proportions was then adjusted for moisture content of aggregates. After generating the eleven mixture proportions (33-1 through 33-11), they were then produced in a random order in the lab. The mix proportions for the thirty three mixes are listed in Tables 2, 3, and 4.

Table 2 - Mixture Proportions For Mixes 33-1 Through 33-11

MIX #	CEMENT pcy	FLY ASH pcy	FINE pcy	COARSE pcy	WATER pcy	HRWR oz/cwt	AEA oz/cwt
1	425	75	1457	1781	148	12	2
2	760	40	950	1820	284	12	2
3	425	75	1133	2105	147	18	2
4	760	40	1260	1540	283	18	2
5	760	40	1282	1566	236	12	2
6	425	75	1123	2085	178	12	2
7	760	40	997	1851	235	18	2
8	425	75	1444	1764	177	18	2
9	585	65	1209	1814	211	15	2
10	585	65	1209	1814	211	15	2
11	585	65	1209	1814	211	15	2

(1 oz = 29.57 ml, 1 pcy = .593 kg/m^3)

Table 3 - Mixture Proportions For Mixes 33-12 Through 33-22

MIX #	CEMENT pcy	FLY ASH pcy	FINE pcy	COARSE pcy	WATER pcy	HRWR oz/cwt	AEA oz/cwt
12	680	120	996.8	1851	237	12	2
13	475	25	1444	1764	178	12	2
14	680	120	1282	1566	236	18	2
15	475	25	1123	2085	177	18	2
16	475	25	1133	2105	148	12	2
17	680	120	1260	1540	285	12	2
18	475	25	1457	1781	147	18	2
19	680	120	980	1820	284	18	2
20	585	65	1209	1814	211	15	2
21	585	65	1209	1814	211	15	2
22	585	65	1209	1814	211	15	2

(1 oz = 29.57 ml, 1 pcy = .593 kg/m^3)

Table 4 - Mixture Proportions For Mixes 33-23 Through 33-33

MIX #	CEMENT pcy	FLY ASH pcy	FINE pcy	COARSE pcy	WATER pcy	HRWR oz/cwt	AEA oz/cwt
23	315	35	1369	2053	114	15	2
24	855	95	1050	1574	309	15	2
25	650	0	1209	1814	211	15	2
26	520	130	1209	1814	212	15	2
27	585	65	1225	1837	173	15	2
28	585	65	1194	1790	251	15	2
29	585	65	1209	1814	213	9	2
30	585	65	1209	1814	211	21	2
31	585	65	1511	1511	212	15	2
32	585	65	907	2166	212	15	2
33	585	65	1209	1814	212	15	2

(1 oz = 29.57 ml, 1 pcy = .593 kg/m^3)

Test Results

The thirty-three mixes were all tested for slump, percent of entrained air, initial concrete temperature, and unit weight according to applicable ASTM's. There were also twelve cylinders made for each of the thirty-three mixes. These cylinders were tested for compressive strength at 1, 3, 7, and 28 days. Also one cylinder from each batch was installed with a thermo-couple on a Digi-Strip data acquisition machine. The strip chart recorder collected temperature readings for the concrete at fifteen minute intervals for the first twenty-four hours. A summary of these tests and material costs (using the raw material costs listed in Table 1) per cubic yard are listed in Tables 5 through 7.

Table 5 Test Data For Mixes 33-1 Through 33-11

MIX	SLUMP (in)	AIR %	1 DAY (psi)	3 DAY (psi)	7 DAY (psi)	28 DAY (psi)	COST ($/cy)
1	0.0	3.9	1890	4380	5130	6660	32.85
2	2.0	3.7	2670	5270	6460	7780	44.61
3	0.0	2.6	3480	4790	6600	8350	33.72
4	8.5	10.5	1750	2700	6940	5090	46.16
5	1.5	4.2	3400	4830	5850	7660	44.86
6	0.0	3.7	2230	3720	5170	6400	32.66
7	2.0	5.5	3480	4970	6110	7520	46.30
8	0.0	3.2	2720	4670	6210	7560	33.66
9	3.0	6.9	2090	3180	4420	5610	39.18
10	2.0	4.7	2590	4060	5170	6520	39.18
11	2.5	7.0	2190	3580	4570	5490	39.18

(1 in = 2.54 cm, 100 psi = .69 Mpa)

Table 6 Test Data For Mixes 33-12 Through 33-22

MIX	SLUMP (in)	AIR %	1 DAY (psi)	3 DAY (psi)	7 DAY (psi)	28 DAY (psi)	COST ($/cy)
12	4.0	5.7	2980	4810	5670	6960	43.01
13	3.0	7.5	1490	2550	3500	4480	33.85
14	7.3	9.7	1510	2190	2780	4380	44.57
15	2.5	6.7	1830	3000	3840	5210	34.72
16	0.0	3.0	2690	4970	5930	6760	33.90
17	8.5	5.2	1870	3300	4040	5270	42.87
18	0.8	5.9	2190	3880	4500	6090	34.91
19	8.0	7.8	1710	2980	3180	4550	44.32
20	5.5	9.6	1350	2860	3360	4520	39.18
21	6.5	7.2	1270	2310	2980	4060	39.18
22	3.5	7.1	2030	3200	4100	5290	39.18

(1 in = 2.54 cm, 100 psi = .69 Mpa)

Table 7 Test Data For Mixes 33-23 Through 33-33

MIX	SLUMP (in)	AIR %	1 DAY (psi)	3 DAY (psi)	7 DAY (psi)	28 DAY (psi)	COST ($/cy)
23	0.0	4.8	950	2450	3460	4750	28.38
24	6.0	8.2	3260	4910	5390	7060	49.99
25	3.5	4.8	2190	4220	4930	6660	40.64
26	5.5	7.0	1190	2840	3700	5170	37.73
27	0.0	2.4	4610	6190	7100	9550	39.34
28	9.0	7.0	1010	2230	2980	3940	39.03
29	3.0	5.2	1970	3540	4570	6070	37.97
30	7.5	9.0	1830	3620	4000	5650	40.40
31	6.3	8.4	1350	3200	4380	5610	39.24
32	2.0	4.4	2090	4240	5430	6500	39.32
33	3.3	6.6	1550	3540	4420	5610	39.18

(1 in = 2.54 cm, 100 psi = .69 Mpa)

Statistical analysis trial number 1

After all of the results from the 33 mixes were gathered a statistical analysis of this data was done. Since the modeling was quite extensive a computer program was used. The program that was chosen was MINITAB for Windows release 10 Xtra. Each of the variables for the experiment had to be modeled for each of the important responses which were slump, air, and 7 day strength. To do this a response surface regression fit was used. For this process the five variables were entered as the responses and the slump, air, and 7 day strength were used as the factors. The computer then generated the coefficients for an equation for each variable in terms of slump, air, and 7 day strength.

Next a linear regression of cost was done, again using MINITAB. A spread sheet was used to generate a mix design to give the required strength at minimum cost from the linear regression. The inputs for the spread sheet include 7 day strength, air %, and required slump. The equation is as follows: (2)

$$\text{cost} = \text{constant} + C1*7day + C2*air\% + C3*slump + C4*7day^2 + C5*air\%^2 + C6*slump^2 + C7*7day*air\% + C8*7day*slump + C9*air\%*slump.$$

It was decided to use a 10.16 cm (4 inch) slump and 6% air as standard inputs and use the strengths 30.36 Mpa (4400 psi), 37.95 Mpa (5500 psi), and 45.54 Mpa (6600 psi) as outputs since the requirements are 10% over design strength within 7 days. The three mixture proportions generated by the spread sheet are listed in Table 8.

Table 8 Mix Proportions For Mixes 4400, 5500, and 6600 For First Trial

MIX ID	CEMENT lb/cy	FLY ASH lb (1:1 replacement)	W/C	HRWR oz/cwt	AGG % coarse	AEA oz/cwt	ROCK lb/cy	SAND lb/cy	COST $/cy
4400	674	10.0	0.33	13.3	61.6	2.0	1840	1150	$41.89
5500	798	9.2	0.32	14.8	62.1	2.0	1760	1070	$46.98
6600	975	6.0	0.31	15.6	64.1	2.0	1680	940	$53.74

(1 oz = 29.57 ml, 1 pcy = .593 kg/m^3, 1 lb = .454 kg)

These mixes were all made in one day with the help of the same people as the original 33 mixes. The mixes were made in the lab with an ambient temperature approximately equal to the ambient temperature for the 33 mixes. The normal ASTM tests were performed on all mixes and the concrete temperature vs. time for the first 24 hours was also taken. Table 9 presents the ASTM test values for these mixes.

Table 9 ASTM Test Values For Mixes 4400, 5500, and 6600 For First Trial

MIX ID	AIR %	SLUMP (in)	AMBIENT TEMP (F)	CONC. TEMP (F)	UNIT WT. (pcf)
4400	6.2	1.50	62	60	148
5500	6.4	1.75	62	60	146
6600	7.0	4.25	62	61	142

(1 in = 2.54 cm, 1 pcf = 16.01 kg/m^3)

Compressive strength tests were conducted at 1, 3, and 7 days. The seven day compressive strength breaks were conducted with unsatisfactory results. The 30.36 Mpa (4400 psi) mix was considerably over its design strength, while the 37.95 Mpa (5500 psi) mix was approximately 11% over design strength. The 45.54 Mpa (6600 psi) mix was approximately 20% under its design strength. The values are listed in Table 10.

Table 10 1, 3, and 7 Day Compressive Strengths For Mixes 4400, 5500, and 6600

MIX ID	1 DAY (psi)	3 DAY (psi)	7 DAY (psi)
4400	1990	4550	5770
5500	2210	4670	5650
6600	2470	4650	5670

(100 psi = .69 Mpa)

The data must be analyzed with the variables as the factors and the slump, air, and seven day strength as the responses. The three mixes made from this analysis all had the same strengths at 1, 3, and 7 day breaks. Also the w/c ratio for the three mixes was almost identical. The data was analyzed using the reverse approach and three new mixture proportions selected.

Statistical analysis number 2
For the second trial, a response surface regression was again used but with the 7 day strength entered as the response, and cement content, fly ash %, HRWR dose, w/c, agg %, and air % as the factors.

A spread sheet was designed to minimize the cost while meeting a design strength, by changing the values for cement, fly ash, w/c, HRWR. The value for the agg % was held constant at 60% and the air content was also held constant at 6%. The equation for this analysis is as follows: (2)

$$\begin{aligned}
7\text{day} = &\text{constant} + C1^*\text{cement} + C2^*\text{flyash} + C3^*\text{wc} + C4^*\text{hrwr} + C5^*\text{agg} + \\
&C6^*\text{air} + C7^*\text{cement}^2 + C8^*\text{flyash}^2 + C9^*\text{wc}^2 + C10^*\text{hrwr}^2 + \\
&C11^*\text{agg}^2 + C12^*\text{air}^2 + C11^*\text{cement}^*\text{flyash} + C12^*\text{cement}^*\text{wc} + \\
&C13^*\text{cement}^*\text{hrwr} + C14^*\text{cement}^*\text{agg} + C15^*\text{cement}^*\text{air} + \\
&C16^*\text{flyash}^*\text{wc} + C17^*\text{flyash}^*\text{hrwr} + C18^*\text{flyash}^*\text{agg} + \\
&C19^*\text{flyash}^*\text{air} + C20^*\text{wc}^*\text{hrwr} + C21^*\text{wc}^*\text{agg} + C22^*\text{wc}^*\text{air} + \\
&C23^*\text{hrwr}^*\text{agg} + C24^*\text{hrwr}^*\text{air} + C25^*\text{agg}^*\text{air}
\end{aligned}$$

The variables that would be changed in this spread sheet have allowed ranges of cement 178-569 kg/m^3 (300-1000 pcy), fly ash (5-15%), w/c (.25-.35), and HRWR 295-473 ml/cwt (10-16 oz/cwt). The strengths were 30.36 Mpa (4400psi), 37.95 Mpa (5500psi), and 45.54 Mpa (6600 psi) at seven days as explained earlier. The three mixture proportions generated by the spread sheet are listed in Table 11.

Table 11 Mix Proportions For Mixes 4401, 5501, and 6601 For Second Trial

MIX ID	CEMENT lb/cy	FLY ASH lb (1:1 replacement)	W/C	HRWR oz/cwt	AGG % coarse	AEA oz/cwt	ROCK lb/cy	SAND lb/cy	COST $/cy
4401	448	5.0	0.32	16.0	60	2.0	1977	1318	$33.17
5501	584	5.0	0.30	14.6	60	2.0	1877	1251	$38.25
6601	643	5.0	0.28	13.3	60	2.0	1838	1225	$40.31

(1 oz = 29.57 ml, 1 pcy = .593 kg/m^3, 1 lb = .454 kg)

These mixes were all made in one day with the help of the same people as the original 33 mixes. The mixes were made in the lab with an ambient temperature approximately 50 degrees F. The normal ASTM tests were performed on all mixes and the concrete temperature vs. time for the first 24 hours was also taken. Table 12 presents the ASTM test values for these mixes.

Table 12 ASTM Test Values For 4401, 5501, and 6601 Mixes For Second Trial

MIX ID	AIR %	SLUMP (in)	AMBIENT TEMP (F)	CONC. TEMP (F)	UNIT WT. (pcf)
4401	4.6	0.00	50	46	138
5501	4.0	0.00	50	47	148
6601	5.0	0.00	50	43	144

(1 in = 2.54 cm, 1 pcf = 16.01 kg/m^3)

Table 13 1, 3, and 7 Day Compressive Strengths For Mixes 4401, 5501, and 6601

MIX ID	1 DAY (psi)	3 DAY (psi)	7 DAY (psi)
4401	1390	2350	4570
5501	1590	5370	6420
6601	2190	5810	6760

(100 psi = .69 Mpa)

The compressive strength test results from these mixes were very good with all mixes meeting the ten percent over design strength at seven days. All the mixes were approximately twelve percent over design strength except for the 37.95 Mpa (5500 psi) mix. This mix was approximately 20 percent over design. The only problem was that all of the mixes had zero slump. These mixes will be made again, with the HRWR dose raised in an attempt to raise the slump.

Mixes redone with HRWR dose adjusted
For the 30.36 Mpa (4400 psi) mix the cement content was changed to 279 kg/m^3 (470 pcy) and the water was left alone in an attempt to raise the strength slightly, for a better factor of safety. This changed the w/c ratio from .32 to .31 for this mix. The only other change that was made was in the HRWR dose. The dose for the 30.36 Mpa (4400 psi) mix was changed from 473-591 ml/cwt (16 to 20 oz/cwt). The dose for the 37.95 Mpa (5500 psi) mix was changed from 432-591 ml/cwt (14.62 to 20 oz/cwt). The dose for the 45.54 Mpa (6600 psi) mix was changed from 393-561 ml/cwt (13.3 to 19 oz/cwt). The mix proportions are listed in Table 14.

Table 14 Mix Proportions For 4402, 5502, and 6602 Mixes With Redosing

MIX ID	CEMENT lb/cy	FLY ASH lb (1:1 replacement)	W/C	HRWR oz/cwt	AGG % coarse	AEA oz/cwt	ROCK lb/cy	SAND lb/cy	COST $/cy
4402	470	5.0	0.31	20.0	60	2.0	1977	1318	$34.73
5502	584	5.0	0.30	20.0	60	2.0	1877	1251	$39.24
6602	643	5.0	0.28	19.0	60	2.0	1838	1225	$41.45

(1 oz = 29.57 ml, 1 pcy = .593 kg/m^3, 1 lb = .454 kg)

These mixes were all made in one day with the help of the same people as the original 33 mixes. The mixes were made in the lab with an ambient temperature approximately 77 degrees F. The normal ASTM tests were performed on all mixes and the concrete temperature vs. time for the first 24 hours was also taken. Table 15 presents the ASTM test values for these mixes.

Table 15 ASTM Test Values For 4402, 5502, and 6602 Mixes With Redosing

MIX ID	AIR %	SLUMP (in)	AMBIENT TEMP (F)	CONC. TEMP (F)	UNIT WT. (pcf)
4402	4.0	0.00	77	73	150
5502	3.1	3.00	77	72	154
6602	2.8	1.50	77	73	154

(1 in = 2.54 cm, 1 pcf = 16.01 kg/m^3)

As before compressive strength tests were conducted at 1, 3, and 7 days, and all results were satisfactory as shown in Table 16.

Table 16 1, 3, and 7 Day Compressive Strengths For Mixes 4402, 5502, and 6602

MIX ID	1 DAY (psi)	3 DAY (psi)	7 DAY (psi)
4402	2150	2570	6680
5502	2110	3040	6800
6602	3100	3640	7060

(100 psi = .69 Mpa)

Field Studies

The plant verification was performed at the South Dakota Concrete Products Company plant, located in Rapid City, South Dakota. As recommended in the lab results, the mix proportions for mixes 4401, 5501, and 6601 were used for the first trial. All materials were the same as used in the lab, and were mixed in the central mix operation. The charging sequence was done according to SDCPC normal procedures. The mix proportions for the first trial mixes (4403, 5503, and 6603) are listed in Table 17.

Table 17 Mix Proportions For First Trial Mixes 4403, 5503, and 6603

MIX ID	CEMENT lb/cy	FLY ASH lb (1:1 replacement)	W/C	HRWR oz/cwt	AGG % coarse	AEA oz/cwt	ROCK lb/cy	SAND lb/cy	COST $/cy
4403	427	22.5	0.34	20.0	60	2.0	1977	1318	$33.97
5503	556	29.3	0.31	20.0	60	2.0	1877	1251	$40.13
6603	611	32.2	0.29	20.0	60	2.0	1838	1225	$42.86

(1 oz = 29.57 ml, 1 pcy = .593 kg/m^3, 1 lb = .454 kg)

For this trial the mixes were all made by SDCPC personnel, and delivered to the testing site. The ASTM tests that were conducted were slump, air percentage, concrete temperature, and unit weight. Also twelve 10.16cm X 20.32cm (4" X 8") cylinders, using plastic molds were made for all mixes so that compressive strength data at 1, 3, and 7 days could be collected. These cylinders were stored at the plant according to SDCPC normal practices. The results from the ASTM tests on the plastic concrete are listed in Table 18.

Table 18 ASTM Test Results For Mixes 4403, 5503, and 6603

MIX ID	AIR %	SLUMP (in)	AMBIENT TEMP (F)	CONC. TEMP (F)	UNIT WT. (pcf)
4403	9.8	7.50	70	52	145
5503	7.5	8.50	70	58	146
6603	9.0	10.00	70	58	144

(1 in = 2.54 cm, 1 pcf = 16.01 kg/m^3)

The cylinders for all of the mixes were picked up 24 hours after they were cast and transported to the South Dakota School of Mines and Technology concrete laboratory for testing. The results for the 1, 3, and 7 day compressive strength tests are listed in Table 19.

Table 19 Compressive Strength Test Results For Mixes 4403, 5503, and 6603

MIX ID	1 DAY (psi)	3 DAY (psi)	7 DAY (psi)
4403	2980	3130	3280
5503	4380	4570	4870
6603	4280	4310	4340

(100 psi = .69 Mpa)

The design requirement for the compressive strength, of 10 % over design in 7 days was not met for this trial. This was probably due to the fact that during mixing, mix 4402 was overdosed with water by 2.37 kg/m^3 (4 pcy), mix 5502 was overdosed by 2.37 kg/m^3 (4 pcy), and mix 6602 was overdosed by 1.18 kg/m^3 (2 pcy). The slump for all three mixes was not within SDDOT standards of no more than 19 cm (7.5 in). Also, the air content for all three mixes was high

Field Study Trial 2
For trial 2 a decisions were made to lower the W/C for all mixes to help the strength gain, and to also lower the HRWR, and air doses to help with the high slumps, and air contents. The W/C was lowered by .02 for all mixes, and the HRWR dose was lowered by 88 .72 ml/cwt (3 oz/cwt) for all mixes. Also the AEA dose for all three mixes was lowered by 29.57 ml/cwt (1 oz/cwt). The mix proportions for mixes (4404, 5504, and 6604) are listed in Table 20.

Table 20 Mix Proportions For Trial 2 Mixes 4404, 5504, and 6604

MIX ID	CEMENT lb/cy	FLY ASH lb (1:1 replacement)	W/C	HRWR oz/cwt	AGG % coarse	AEA oz/cwt	ROCK lb/cy	SAND lb/cy	COST $/cy
4404	427	22.5	0.32	17.0	60	1.0	1977	1318	$33.57
5504	556	29.3	0.29	17.0	60	1.0	1877	1251	$39.60
6604	611	32.2	0.27	17.0	60	1.0	1838	1225	$42.29

(1 oz = 29.57 ml, 1 pcy = .593 kg/m^3, 1 lb = .454 kg)

For trial 2 the mixes were once again all made by SDCPC personnel, and delivered to the testing site. The ASTM tests that were conducted were slump, air percentage, concrete temperature, and unit weight. Also twelve 10.16 cm X 20.32 cm (4" X 8") cylinders were made for each mix so that compressive strength data at 1, 3, and 7 days could be collected. These cylinders were stored at the plant according to SDCPC normal practices. The results from the ASTM tests are listed in Table 21.

Table 21 ASTM Test Results For Mixes 4404, 5504, and 6604

MIX ID	AIR %	SLUMP (in)	AMBIENT TEMP (F)	CONC. TEMP (F)	UNIT WT. (pcf)
4404	7.5	3.00	55	60	143
5504	6.4	6.00	55	60	144
6604	7.0	7.00	55	60	144

(1 in = 2.54 cm, 1 pcf = 16.01 kg/m^3)

The cylinders for all of the mixes were again picked up 24 hours after they were cast and transported to the South Dakota School of Mines and Technology concrete laboratory for testing. All results for the 1, 3, and 7 day compressive strength tests were satisfactory and are listed in Table 22.

Table 22 Compressive Strength Test Results For Mixes 4404, 5504, and 6604

MIX ID	1 DAY (psi)	3 DAY (psi)	7 DAY (psi)
4404	4080	4380	4770
5504	4460	4890	5460
6604	5000	5270	6510

(100 psi = .69 Mpa)

Conclusions And Recommendations

1. The field data gathered for this research indicated satisfactory results for mixes 4404, 5504, and 6604. The compressive strength data for mix 4404 was above the 10 % over design strength required. The compressive strength data for mixes 5504, and 6604 was not exactly 10 % over design strength, but was close enough to be within workable limits. The slumps for mixes 4404, 5504, and 6604 were a little high, but were still within workable limits. The percentage of entrained air for all three mixes was also within workable limits. The proportions can be changed slightly as production continues.

2. At this time it is recommended that mixes 4404, 5504, and 6604 should be the mixes utilized by SDCPC on a daily basis. The fly ash should be stored in a bin similar to the storage bin used for the cement. The fly ash should not be stored in bags for use since it will attract moisture in this manner.

3. Recommended Batching Sequence
   All fine and coarse aggregate should be introduced and mixed for 2 minutes.
   Add all cement and fly ash and mix for 1 minute.
   Add all water with AEA and mix for 1 minute.
   Immediately add HRWR and mix for the remaining fifteen minutes.

4. The cost of SDCPC standard mixes at the beginning of the project were $39.00/cy 27.6 Mpa (4000 psi), $44.00/cy 34.5 Mpa (5000 psi), and $50.00/cy 41.1 Mpa (6000 psi). The cost for the recommended mixes are $29.99/cy (Mix 4404),

$33.50/cy (Mix 5504), and $34.91/cy (Mix 6604). These costs are based on the raw material prices listed in Table 1. The resulting savings are $10.01/cy, $10.50/cy, and $15.09/cy respectively.

5. Insure that the water is measured accurately at all times. The HRWR dose for mixes 4404, 5504, and 6604 should be adjusted as needed to achieve the slump required. This should not affect the compressive strength of the mixes. The initial concrete temperature is also critical and must be carefully controlled year round. If these steps are followed SDCPC should achieve the consistent concrete required for their operation.

6. Additional research is recommended to improve slump & air contents and reduce admixture dose.

7. Further research is also needed to simplify the statistical model and to reduce the number of variables with insignificant coefficients.

Acknowledgments
The author would like to thank the following individuals and institutions who have contributed to the completion of this research:
1. John Luciano, Master Builders, for personal help, and
2. South Dakota Concrete Products Company, for sponsoring the SDCPC Fellowship.

References
1. Luciano, J., and Bobrowski, G., "Using Statistical Methods to Optimize High Strength Concrete Performance," *Transportation Research Record 1284*, pp. 60-69.
2. DeMaro, J., "Mix Design Optimization for Precast Concrete Using Statistical Methods," Master of Science Thesis, SDSM&T, 1996.

HYDRATION HEAT AND STRENGTH DEVELOPMENT IN HIGH PERFORMANCE CONCRETE

Maria Kaszyńska Ph.D.(Eng.)
Institute of Civil Engineering
Technical University of Szczecin, Poland.

ABSTRACT

While designing and building of massive engineering structures, the important problem encountered is thermal stresses evolving from self-heating of large concrete masses in the result of exotermic process of cement hydration. For determining the values of these stresses both in the precise and aproximated manner what is needed a set of data to characterize the properties of concrete hardening in massive structure. Data referring to cement hydration heat allow to determine the temperature distribution in concrete mass. Data regarding the mechanical properties of hardening concrete allow to judge safely the danger of structure damaging. This paper presents analyses and results of simultaneous investigation of hydration heat and compressive strength of early age concrete hardening in adiabatic conditions such as it prevails inside concrete mass.

INTRODUCTION

Resultant state of stresses which occurs in massive structures is produced by combined reaction of external (operational) loads and internal loads induced by the heat of cement hydration. In structures with low massiveness, the internal stresses are influenced predominantly by the dampness fields while the effect of thermal fields may be neglected. In structures with medium massiveness, both the effect of dampness and the effect of thermal fields should be taken into account. In structures with high massiveness, the thermal stresses predominate over other kinds of internal stresses. These stresses may, at given stage of construction of a concrete structure, exceed the tensile strength of concrete used, leading thereby to cracks and fractures in structure even during its construction. Sustained period of elevated temperature action in the course of setting and hardening of concrete in a block has a decisive impact on the development of mechanical properties of concrete. Although this problem has already been the subject of many publications[1] which relate to normal concrete, it is still not sufficiently known for the high performance concrete (HPC). The cement hydration process in the presence of superplasticizer and silica fume leads to a formation in HPC which differs qualitatively from that of normal concrete.

It is believed that in case of HPC, where increased amounts of high grade cement and active silica are applied, the hydration heat effects on thermal stresses may be already substantial in case of structures with even a medium massiveness. While investigating the self-heating in concrete mixtures, Hegger [2] found out that columns with a cross-section of 1200 mm x 1200 mm (47.24 in. x 47.24 in.) made in B85 concrete revealed a much greater thermal shock than those made in B45 concrete. The maximum temperature in B85 concrete reached 69.7°C, and in B45 concre te it was 61.1°C, at air temperature was 7°C to 12°C. Mirambell and others [3] registering the self-heating temperature in the pedestrian bridge constructed of high-strength concrete (W/C = 0.3) determined the maximum temperature of 76.3°C in the cross section, after 20-hours hardening, whereas the air temperature was 7°C to 12°C. In turn Cook and others [4] analysing the distribution of temperatures and stresses in concrete blocks with dimensions 1000 mm x 1000 mm x 1000 mm (39.37 in. x 39.37 in. x 39.37 in.) found out that the maximum temperatures in blocks made in high-performance concrete (90 MPa, 120 MPa) (13.04 ksi, 18.84 ksi) were not higher than those in an normal concrete (35 MPa) (5.07 ksi) block.

Another problem having been highlighted by researchers is a decrease in the strength of concrete used in a structure, caused by its elevated self-heating. Hegger [2] reckons that the decrease reaches 10 to 15 per cent for B85 concrete compared with the concrete which hardens at 28 days at the normal temperature $T_0=20°C$. He further suggests that a diminishing coefficient be introduced while determining the relevant characteristic strength.

A few of the conclusions drawn by different researchers and presented above are a good evidence on how contradictory the opinions regarding the hydratation heat and its influence on the thermal stresses in high-performance concrete are. In this paper, the results of a research on the hydration heat and compressive strength of high-performance concrete cured in adiabatic conditions, i.e. those which prevail inside a massive structure, are presented. The purpose of this research was to determine a relationship between the amount and kinetics of heat generation versus compressive strength of concrete containing silica fume, fly ash and superplasticizer, when cured in varying thermal conditions.

EXPERIMENTAL PROCEDURE

Assumptions

A temperature distribution in concrete, produced by the action of internal or external source of heat, is described by the Fourier - Kirchhoff equation :

$$\frac{\partial T}{\partial \tau} = a_T \nabla^2 T + \frac{1}{c_b \gamma} \frac{\partial Q}{\partial \tau} \qquad (1)$$

where: T - temperature in given point of concrete [°C]
τ – time of the process [h]
a_T - coefficient of temperature equalization (diffusion) in concrete [m²/h]
c_b - specific heat of concrete [kJ/kg °C]
γ - density of concrete [kg/m³]
Q - heat of cement hydration in concrete [kJ/kg]

The kinetics of hydration heat emission is described by so called "source function", $W(\tau)$. This function determines the density of power of heat of hydration of cement in concrete and is related to the heat of hydration by means of the following formula :

$$Q(\tau) = \int_0^\tau W(\tau)d\tau \qquad (2)$$

The course of the function $W(\tau)$ depends on time, temperature of curing at given point of concrete and on the volume and kind of cement used. In the literature we can find many attempts to describe mathematically the source functions and the amount of heat emitted within a concrete block [5]. Appropriate determination of the source function value has a direct impact on the accuracy of the heat transfer equation solution and this, on the other hand, conditions on correct solution to the problem of thermal stresses in concrete. All mathematical formulas which attempt to present analytically the heat of hydration and the source function are based on coefficients determined experimentally and are dependent upon many parameters. There is no possibility to determine a one, universal, function which would describe the course of thermal effects in concrete and which would be independent of the type of binder, the composition of concrete or the temperature of hydration. This is why properly conducted experiments are so vital when determining the value of the source function for given kind of concrete

The heat of hydration is most frequently investigated on the samples of cement grout or mortar cured under isothermal conditions. Such investigations allow a researcher to determine basic relationships which describe the hydration process to evaluate the effect of mineralogical composition of concrete or the effect of admixtures used, etc. An increase in the temperature of concrete self-heating, which has been found during investigations carried out on constructions of such objects indicates that there are quasi-adiabatic conditions of curing within a concrete block.In the investigations carried out by the author the hydration heat and the mechanical properties of concrete were analyzed and eveluated under adiabatic conditions which yield upper estimates of the amount of heat emitted within concrete [5]

Research apparatus

For the purposes of investigations, a special test setup was designed and constructed. The schematic diagram of this apparatus is shown in Fig. 1. The basic system used for investigating the heat of hydration of cement in concrete under adiabatic conditions is composed of a calorimetric container, an ultrathermostat, a purpose-built regulator and a recorder. Two temperature sensors were used for measuring temperatures in the interior of concrete sample (T_b) and in its surroundings (T_o). A temperature difference ($T_b - T_o$) measured between the concrete sample and its surroundings makes an error signal to be transmitted to the regulator which is used for continuous controling the heating device. The floating control so performed has allowed us to maintain the temperature of the sample surroundings within the limits : $T_o = T_b \pm 0{,}01\ °C$, that is, to provide adiabatic conditions of curing for the sample.

To achieve possibly highest accuracy in measuring the temperature rise in the concrete cured, cylindrical specimens with diameter Φ = 240 mm (9.45 in.) and hight h = 300 mm (11.81 in.) were taken for investigating the heat of hydration. The specimens for testing the compressive strength, sized at 100 mm x 100 mm x 100 mm (3.94 in. x 3.94 in. x 3.94 in.) were cured in an additional calorimetric cointainer at a temperature controlled by the temperature of the specimen used for testing the heat of hydration. The temperature control in the additional calorimetric container was effected analogically to that in the basic system. The system was designed so as to provide the course of temperature, with an accuracy of ±0,5°C, for the specimens destined for mechanical tests.

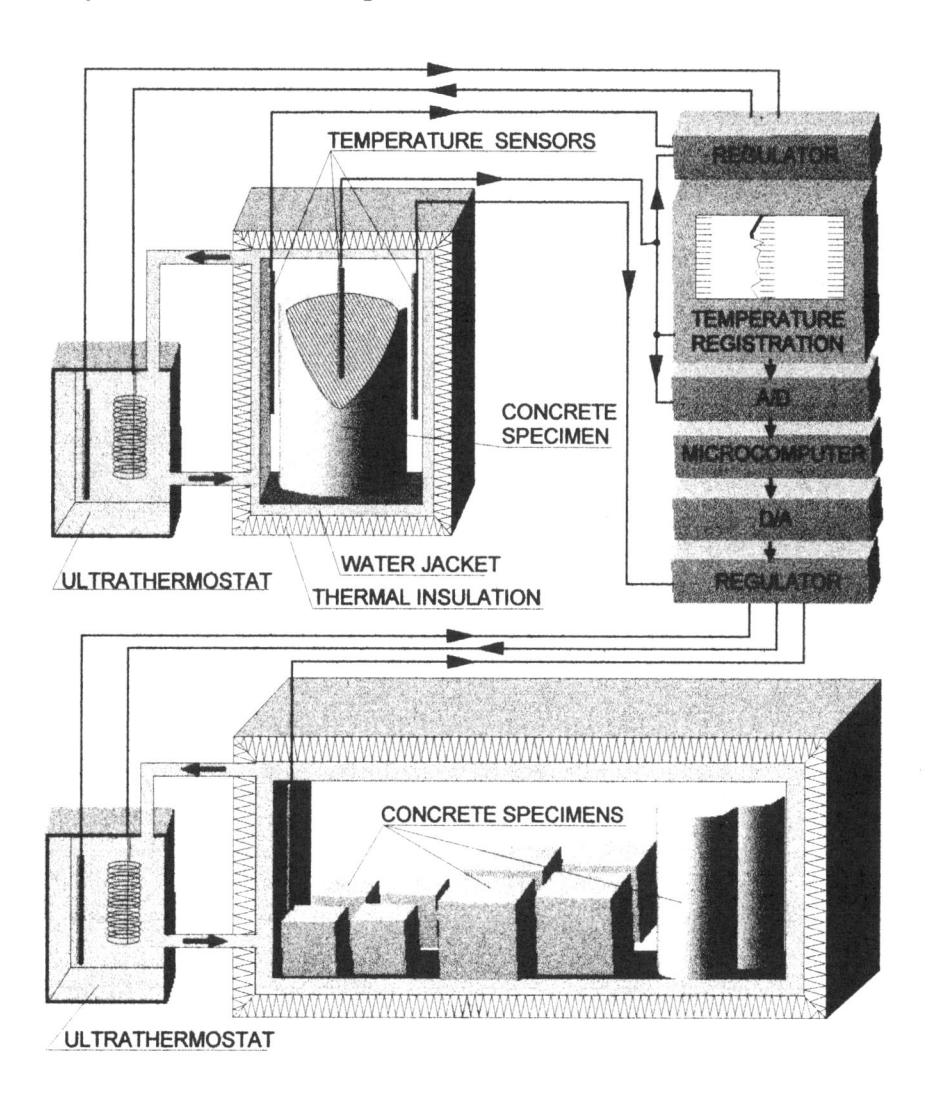

Fig. 1. Schematic diagram of the apparatus

Research method
The research on the cement hydration heat of concrete was carried out in a calorimetric apparatus which ensured cylindrical specimens adiabatic curing conditions. The temperature in a given sample was registered continously over 7 days of concrete hardening. On the basic of the temperature vs. time curves recorded during the tests, the

amounts of heat of hydration of cement in concrete $Q(\tau)$ were calculated from the formula (3) and the values of the source function $W(\tau)$ were calculated from the relationship (4):

$$Q(\tau) = \Delta T(\tau)\frac{c_b\gamma}{C} \qquad (3)$$

$$W(\tau) = \frac{dQ}{d\tau} \approx \frac{Q(\tau_{i+1}) - Q(\tau_i)}{\tau_{i+1} - \tau_i} \qquad (4)$$

The concrete compressive strength was examined after 8, 16 hours and 1, 2, 3, 7 days of curing in adiabatic conditions, as well as after 1, 3, 7, 28 days of curing in laboratory conditions. To determine the influence of diverse initial temperatures on both the cement hydration thermal effects and concrete strength, the following initial temperatures of the concrete mixtures were assumed: $T_0=20^0C$, and additionally $T_0=10^0C$ and $T_0=35^0C$ as the characteristic temperatures prevailing in autumn/winter and summer concrete pouring periods respectively.

Materials

For the purpose of investigation there has been taken concrete characterized by constans value of coefficient W/C = 0.35. The composition of concrete mixtures are specified in Table 1.

Table 1 - Concrete mixture composition

Concrete type	Cement [kg/m³]	Aggregate [kg/m³]	Water [dcm/m³]	Fly Ash [kg/m³]	Silica Fume [kg/m³]	Superplasticizer [dm/m³]
M	450	1786	158	-	30	9
MFA	315	1730	158	135	30	9

The M symbol is assigned to mixtures based on CP45 Portland cement with addition of silica fume, and FM6 superplasticizer fabricated by Addiment. As regards the MFA mixtures, 30 per cent of cement was replaced with fly ash. The following initial temperatures of the concrete mixtures were assumed: $T_0=20^0C$ (mixtures M/20 and MFA/20), and $T_0=10^0C$ (mixtures M/10 and MFA/10), and $T_0=35^0C$ (mixtures M/35 and MFA/35).

RESULTS OF INVESTIGATION

The characteristic values derived from the research on the concrete self-heating (ΔTmax), generated amount of hydration heat (Q), the maximum value of the source function (*W*max), as well as the time (tw_{max}), are summarised in Table 2.

Table 2 - Characteristic values of concrete self-heating research results

Concrete mixture	T_o [°C]	T_{max} [°C]	ΔT_{max} [°C]	Q_{max} [kJ/kg]	W_{max} [W/kg]	t_{Wmax} [h]
M/10	8	61	53	288	10.3	15
M/20	20	71	51	277	12	10
M/35	37	80.9	43.9	239	13.9	5
MFA/10	12	60.5	48.5	264	7.6	18
MFA/20	19.1	62.8	43.7	238	8.5	12
MFA/35	36.4	75.8	39.4	214	9.1	5

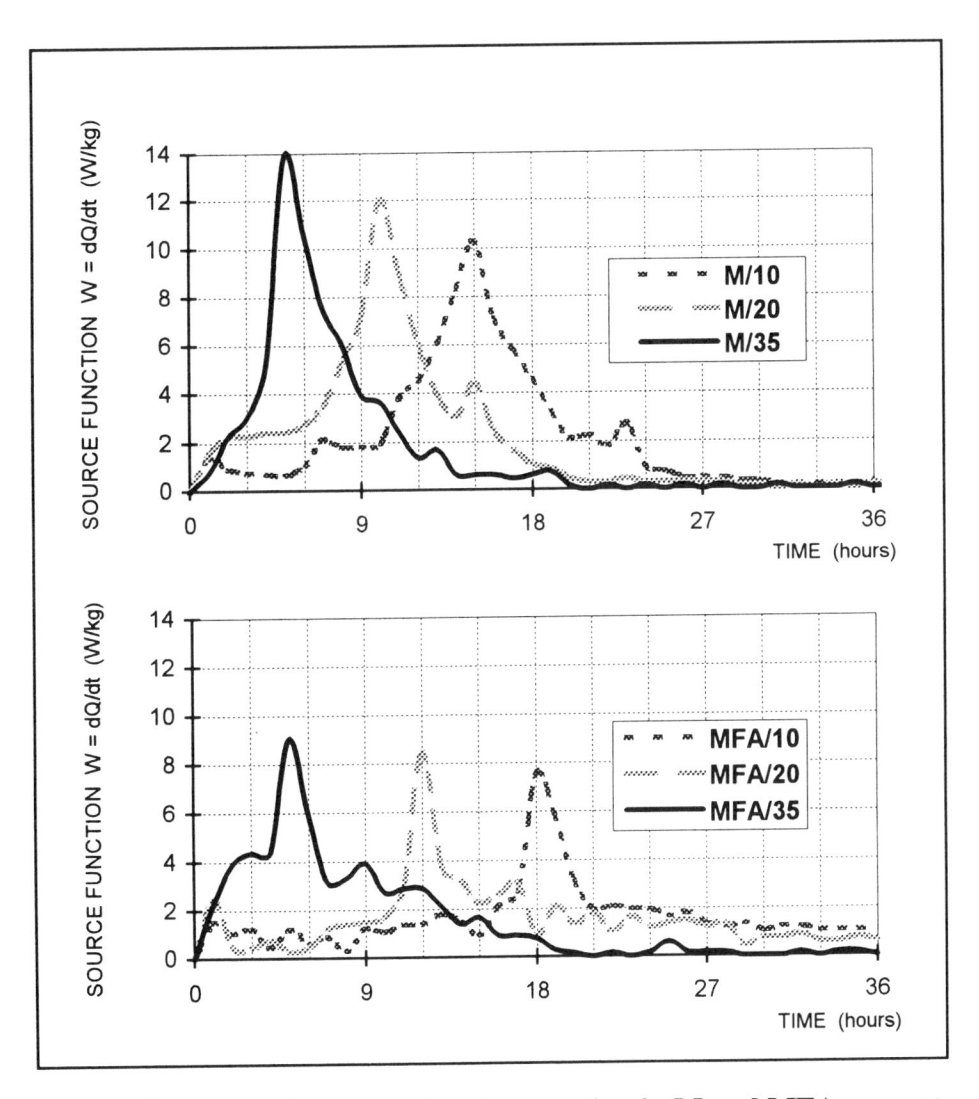

Figure 2. Kinetics of hydration heat generation in M and MFA concrete for varied initial temperature

The self-heating of concrete reached very significant values (44°C to 53°C) depending on the initial temperature of the concrete mixture, (maximum temperature of 61°C to 81°C) regardless of the fact that the amounts of generated specific heat (i.e. per 1 kg) were not greater than those valid for normal concrete. Similar values of the self-heating were registered for the high-performance concrete during construction of structures [2,3,4].
Figure 2 shows the graphs on the source function $W(t) = dQ/dt$ which reflect the kinetics of heat generation during the cement hardening in M concrete which does not contain fly ash and in MFA concrete with fly ash.

The research on the ordinary concrete [5,6] hydration heat conducted previously revealed that the concrete mixture initial temperature does not have any effect on the whole amount of heat emitted in concrete, although it does affect the hydration process kinetics.

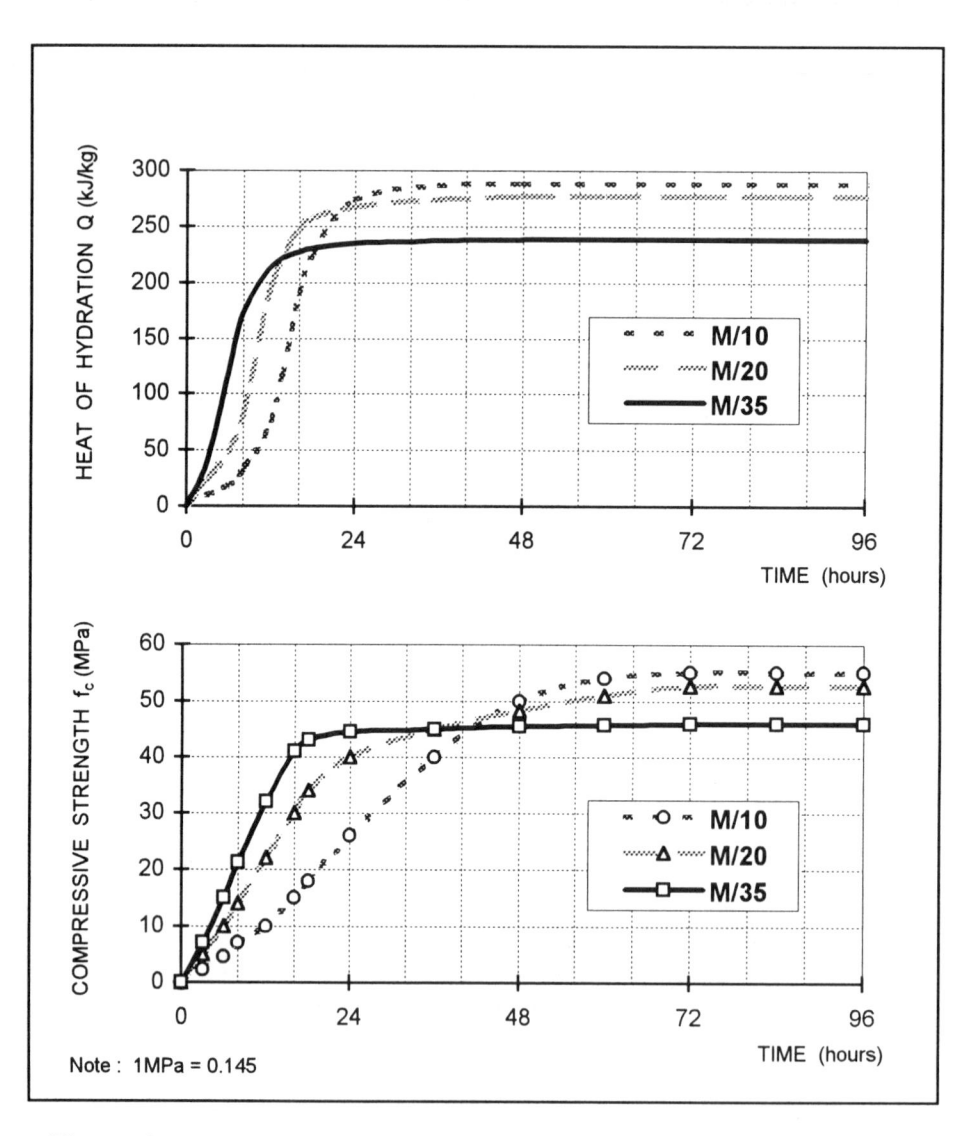

Figure 3. Hydration heat and compressive strength of M concrete cured in adiabatic conditions.

In the case of high-performance concrete it was found out that both the kinetics and amount of emitted hydration heat depend equally upon the concrete's initial temperature. The source function reaches its maximum values in mixtures (M/35 and MFA/35) prepared at 35°C. The greatest amount of heat is emitted during cement hydration in the mixtures (M/10 and MFA/10) prepared at 10°C.

Figures 3 & 4 present the test results on the amount of generated hydration heat and on compressive strength concerning the M and MFA concrete grades at an early stage of their curing in the adiabatic conditions.

With respect to concrete grades with and without fly ash, appoximately the same relationships were obtained, although the ash admixture defers all the effects, and the process of heat emission and strength development is more slowly at first.

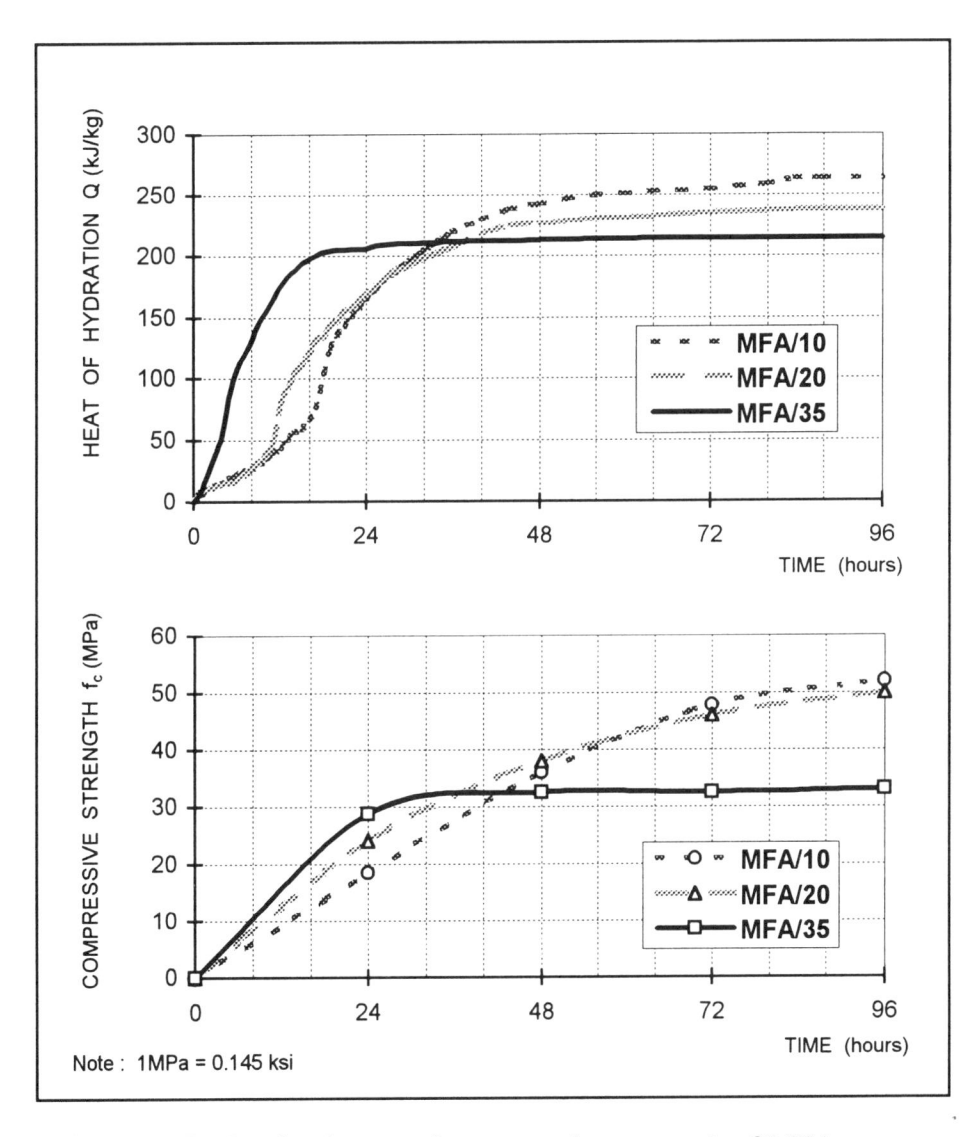

Note: 1MPa = 0.145 ksi

Figure 4. Hydration heat and compressive strength of MFA concrete cured in adiabatic conditions.

A high temperature of the self-heating influences positively the ash reactivity in mixtures MFA/10 and MFA/20, and in the course of the pozzolana reaction, a higher strength increase rate of concrete with fly ash is observed. In case of high-performance concrete, a relatively low W/C ratio causes a certain amount of cement involved not to hydrate at all. When the initial temperature of the concrete mixture is elevated, the hydration process developes higher which, in turn, restricts the heat generation and strength development. At a lower initial temperature, the hydration process developes more slowly and according to the existent research, a higher cement contents hydrates which results in a higher self-heating rate and in a longer strength development period.

The test results allowed the establishment of correlations between the amount of cement hydration heat emitted and concrete strength in the initial stage of hardening in adiabatic conditions. Fig.5 shows diagrams of function $f_c = f(Q)$ for concrete without fly ash for varied initial temperatures.

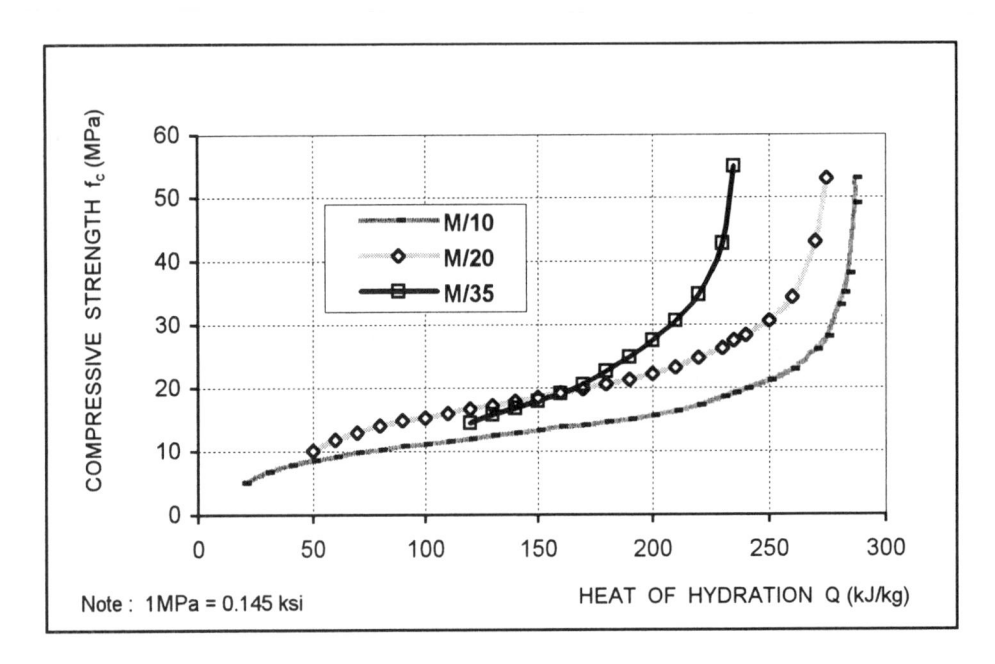

Figure 5. Diagrams of function $f_c = f(Q)$ for concrete M

CONCLUSIONS

The self-heating of high-performance concrete reaches substantial values due to the presence of cement therein, but the specific heat emission rate per kg is similar to that for ordinary concrete grades. A low W/C ratio results in an incomplete hydration due to lack of water required for that process. The research has proved that there is a certain relationship between the hydration progress, expressed by the amount of emitted hydration heat, and the development of mechanical properties of concrete which hardens in varying thermal conditions. In the case of concrete grades without fly ash contents, the heat emission progresses faster than the strength development while curing concrete under the adiabatic conditions. An addition of fly ash into MFA mixtures worsens the kinetics and reduces the amount of heat and original concrete strength. However, after seven days of curing, the strength of concrete with ash admixture is approximately the same as that of ash-free concrete. The tests have shown a very strong influence of the initial temperature of concrete mixture upon the course of heat generation processes and strength development in high-performance concrete grades, the course itself being different compared with the one pertinent to normal concrete grades. Both the amount of emitted heat and concrete strength, at initial stages of curing, increase as the initial temperature of concrete mixtures rises and they reach their maxima at $T_0=35^0C$. In the course of hydration process, both the amount of emitted heat and compressive strength are highest for concrete grades prepared at $T_0=10^0C$.

A high temperature of the self-heating at the begining of hardening worsens the hydration process and reduces the subsequent concrete strength. The research results have revealed that for high-performance concrete, no relationships pertinent to normal concrete grades may be implemented without verifying them first. While applying high-performance concrete for massive structures, some research is needed on a given mixture which is intended to harden under conditions simulating the curing process of concrete in the real structure.

REFERENCE

1. Springenschmid, R., " Thermal Cracking in Concrete at Early Ages", Procedings of the International RILEM Symposium, E & FN SPON London, 1995.
2. Hegger, J., "High strength concrete for a 186m high office building in Frankfurt, Germany", High-Strength Concrete 1993, Lillehammer, Norway, pp.504-511.
3. Mirambell, E., Calmon, J.L., Aguado, A., "Heat of hydration in high-strength concrete:case study", High-Strength Concrete 1993, Lillehammer, Norway. pp.554-561.
4. Cook, W. D., Miao,B., Aitcin,.P.C., Mitchell,D.,"Thermal stresses, large high strength concrete columns" ACI Materials Journal, V. 89, No.1, 1992.
5. Kaszyńska, M.,"Early-Age Fly Ash Concrete Properties in Adiabatic Cured Conditions"., CONCRETE 2000, Dundee, Scotland, E & FN SPON 1993, pp.672-678.
6. Kaszyńska, M., Matyszewski, T., "Influence of Temperature on Properties of Fly Ash Concrete in Massive Structures".Procedings: Tenth International Ash Use Symposium. EPRI Washington, 1993. pp.49-1.

EARLY AGE SHRINKAGE AND CREEP OF
HIGH PERFORMANCE CEMENT-BASED MATERIALS

D. Lange, L. Struble, F. Young, S. Altoubat, H.-C. Shin, H. Ai
University of Illinois at Urbana–Champaign
Urbana, Illinois, U.S.A.

ABSTRACT

Work at the FAA Center of Excellence (COE) for Airport Pavement Research at UIUC includes two projects that focus on high performance concrete. The experimental approach utilizes paste, mortar, and concrete specimens to investigate the bulk behavior of concrete. One of the projects deals with high performance concrete (HPC) for thin overlays and the other studies the role of fiber reinforced concrete (FRC) for airport pavement applications. Both projects include experimental investigation of early age drying shrinkage and creep because serviceability and durability can be compromised by poorly designed materials that lead to early failure in pavements. If early problems are avoided, HPC and FRC can be excellent materials that offer superior mechanical properties and long-term durability. This paper discusses findings from both on-going studies.

INTRODUCTION

High performance concrete usually uses a very low water-to-cement ratio (w/c) and considerable replacement of portland cement by silica fume or other supplementary cementing material to provide higher strength and lower permeability. The strength of ordinary concrete (w/c ~ 0.45) is about 50 MPa (7000 psi), the strength of HPC (w/c 0.30) is about 70 MPa (10,000 psi), and the strength of high performance mortar (DSP, w/c 0.20) may be as high as 200 MPa (30,000 psi). Water permeability of ordinary concrete is around 10^{-12} m/s, whereas permeability of HPC has been estimated to be from 10^{-13} to 10^{-16} m/s. With the low permeability of HPC comes a considerable improvement in long-term durability. This is probably the most important benefit to be gained by use of HPC in airport pavements. Other benefits include improved abrasion resistance and better bonding to substrate concrete. However, the low permeability may lead to self-desiccation, resulting in shrinkage cracking.

FRC using steel fibers is studied under a separate project within the FAA COE activity. FRC is a proven material with greater resistance to cracking and improved tensile, flexural, and fatigue strengths. The improved properties of FRC may allow use of thinner and larger pavement slabs with fewer joints. The largest volume application of FRC has been in airport pavements, both as overlays to repair existing pavement and as full depth pavement construction. The benefit of FRC in airport pavements is in superior crack resistance and, therefore, superior durability.

Overlays can exhibit curling, either due to drying shrinkage or differential thermal expansion between the overlay and the underlying pavement. Creep mechanisms serve to significantly mitigate the development of critical stresses at early ages. For example, our study has shown that creep reduces by half the stresses that would be estimated from free shrinkage strain. For this reason, it is important to characterize creep along with drying shrinkage and thermal expansion. Constitutive models of HPC and FRC material behavior are being developed for use in other FAA COE efforts that utilize finite element analysis for pavement analysis.

This research attempts to understand shrinkage and creep at a fundamental level. Shrinkage and creep result from structural changes in C-S-H microstructure that occur under stresses created by drying and applied loads. Experiments on paste, mortars, and concretes have been conducted. Free shrinkage of paste, mortars, and concrete specimens has been measured under various curing and drying environments. A test of larger concrete specimens has been developed to measure shrinkage and creep simultaneously starting one day after casting. Our study intends to link changes in paste and mortar to bulk behavior in concrete.

This paper describes two FAA COE projects (HPC and FRC) that are closely related. Together, they characterize material properties and address the fundamental modeling of overlay behavior. The end objective of these projects is to establish new knowledge of HPC and FRC materials, and assimilate the material properties in models of pavement behavior.

EXPERIMENTAL PROGRAM

The two projects together provide several lines of inquiry that are being pursued for HPC and FRC materials. The inquiry sequences together as follows:

1. Drying shrinkage of paste.
2. Drying shrinkage of mortars.
3. Composite analysis to link shrinkage of paste and mortars.
4. Thermal expansion of pastes.
5. Elastic modulus of pastes.
6. Bond strength measurements.
7. Drying shrinkage and creep of unrestrained and restrained concrete.
8. Computational modeling of pavement response.

Materials
Pastes, mortars, and concretes are investigated in this work. The paste specimens are configured as small beams measuring 2.5 x 10 x 100 mm. The mortars are configured as 25 x 25 x 250 mm beams. The concrete specimens are "dog-bone" specimens where the cross-section within the 0.6 m gage length is 75 x 75 mm. The ends of the dog-bone specimen flare out and fit within grips that provide restraint during the test.

The pastes and mortar series vary in w/c from 0.2 to 0.4 and include 0%, 6%, 18%, and 24% silica fume mix designs. The concrete investigation focuses on more conventional range of w/c from 0.46 to 0.56 without use of silica fume. The pastes and mortar series include two different aggregate volume fractions of 20% or 60%.

Drying shrinkage
The paste and mortar specimens used during the drying experiments are wet cured for either two days or eight weeks. The drying shrinkage measurements are being carried out to long ages under two different humidity conditions, 11% RH and 50% RH. Figure 1 shows typical results showing the effect of silica fume content on drying shrinkage of paste. Drying shrinkage occurs even after 1400 hours of drying, although the rate of

shrinkage is low. The very fine pore structure of these low porosity pastes makes diffusion of moisture from the paste very slow, and it may not be practical to reach true equilibrium. Although equilibrium is not achieved, the maximum shrinkage recorded for a paste (w/c 0.20) with no silica fume was 0.4% and with 12-24% was 0.2%. The maximum drying shrinkage of mortar (w/c 0.25; 60% volume of sand) with no silica fume was 0.15% and with 12-24% was 0.13%.

Composite analysis linking paste to mortar

The drying shrinkage of pastes and mortars are being linked using a composites modeling approach developed by Hansen[1]. A simple composites model, however, does not account for microcracking that may occur due to residual stress in the material due to restraint from the aggregate. Furthermore, the simple model does not allow for variation in elastic modulus in the paste due to moisture gradients and change of moisture with time. We anticipate extension of the existing model to better account for moisture changes and residual stresses in mortar.

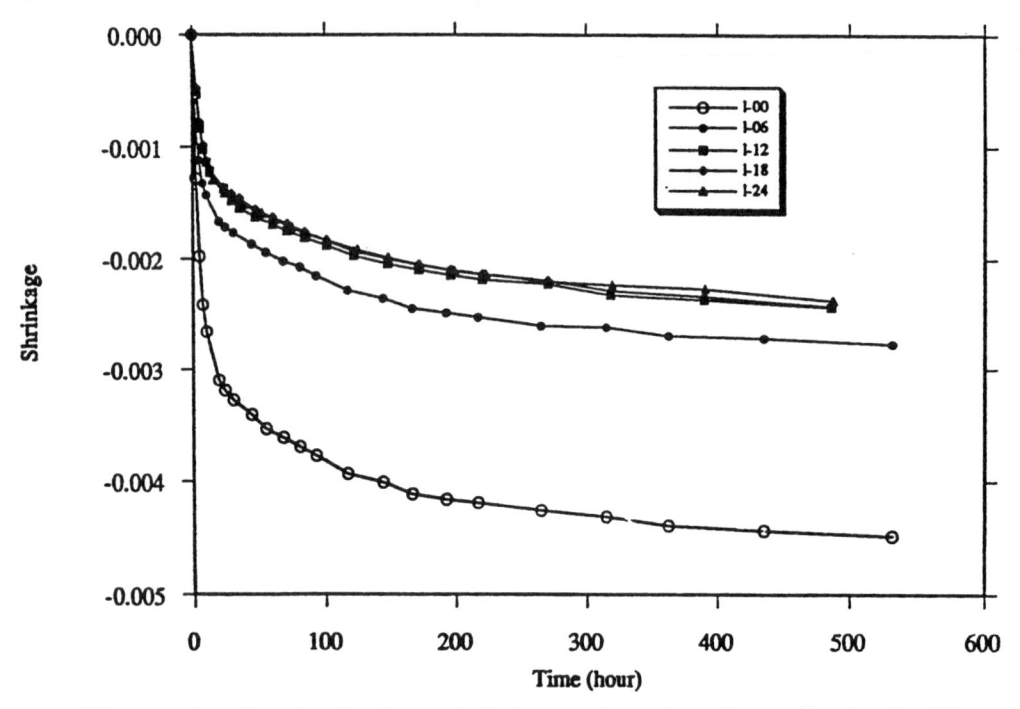

Figure 1. Paste drying shrinkage showing the effect of silica fume (w/c 0.20, silica fume content from 0-24%; shrinkage is $\Delta l/l$)

Thermal expansion

Thermal expansion of paste samples immersed in a water bath was measured in an apparatus that elevated the bath temperature from 25°C to 50°C. The values of thermal expansion coefficient using saturated paste samples (w/c = 0.20 or 0.30) were $5-10 \times 10^{-6}$ per °C, in the range we expected. The results suggest that there is no significant effect of silica fume on thermal expansion. The experiments also revealed a previously unknown behavior. We expected a linear expansion as a function of temperature in tests of small paste samples. Indeed, we observe linear expansion during the first hour of heating from 25°C to 50°C, but then a contraction of the sample occurs as it is held at 50°C during the next several hours. Several physical explanations, including polymerization of calcium silicate hydrate phases, are being considered.

Elastic modulus of paste

Elastic modulus of pastes are measured on small beam samples in flexure. The elastic modulus of a low w/c paste (w/c 0.20-0.30) is about 9 GPa (1.3×10^6 psi), and increases to 15 GPa (2.1×10^6 psi) if dried at 50% RH. Silica fume content has little effect on the elastic modulus of paste. The modulus is quite sensitive to moisture condition, and it is important to ensure saturated condition of the samples during the test. The dependence of elastic modulus on moisture content has a profound consequence for modeling large samples and real pavement sections. If the paste modulus changes with moisture condition, then our model must account for moisture gradient through the thickness of the sample and moisture content as a function of time.

Bond strength

Bond strength measurements are being tested using two methods. We are using the Iowa Shear Test and ASTM C 1245 tensile test methods to characterize bond. These tests are being conducted on candidate HPC overlay mixes and other mix designs derived from mixes used at Champaign's Willard Airport during a recent pavement overlay project. The results of the bond strength study will feed directly into computational models of overlay performance. As an example of the results obtained, the shear bond strength of an HPC overlay on an ordinary concrete base after 28 days of moist curing was 3.5 MPa (500 psi) and the tensile bond strength was 1.5 MPa (220 psi).

Shrinkage and creep of concrete

Shrinkage and creep of restrained concrete is being studied under the FRC project, and the findings will be equally applicable to HPC overlays. These experiments are focused on early-age behavior, starting with testing 18 hours after casting until about 7 days after casting. In this project we have seen that creep acts as an important stress relaxation mechanism. Engineers have been known to estimate internal stress due to drying shrinkage by measuring free shrinkage of concrete and applying Hooke's Law to compute stress. This approach ignores creep, and is, in fact, a gross simplification of the problem. We have found that creep reduces by half the stress estimated from free shrinkage. Figure 2

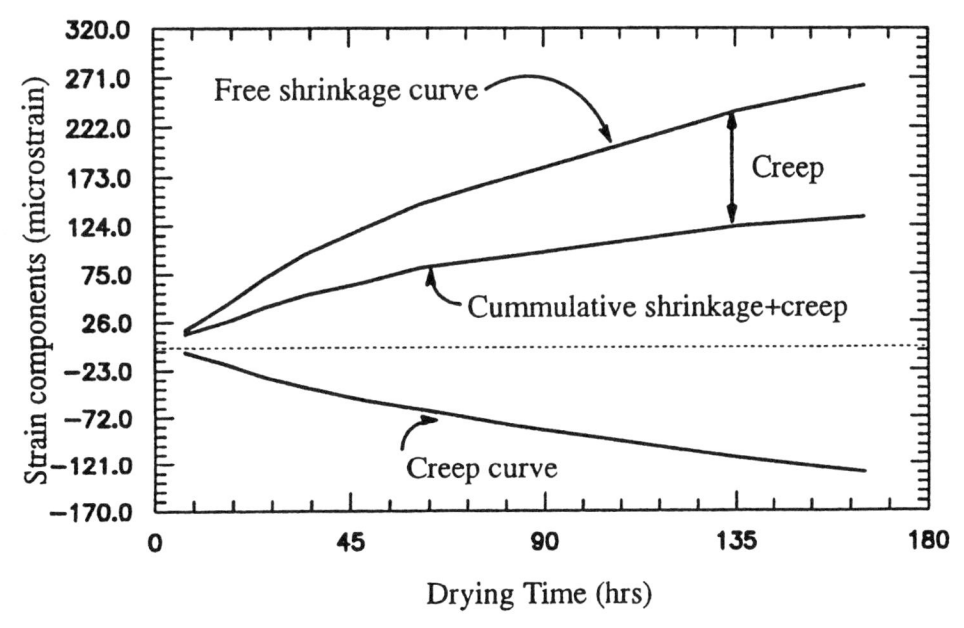

Figure 2. Creep strain resolved from free shrinkage specimen and restrained (cumulative shrinkage + creep) specimen

illustrates the relationship between free shrinkage and creep in early-age FRC (w/c 0.48-0.56). Figure 3 shows how the stresses due to drying of a restrained specimen increase as w/c increases, up to 50% of 7-day tensile strength. Clearly, creep is a factor that must be included in the computational model of overlay performance.

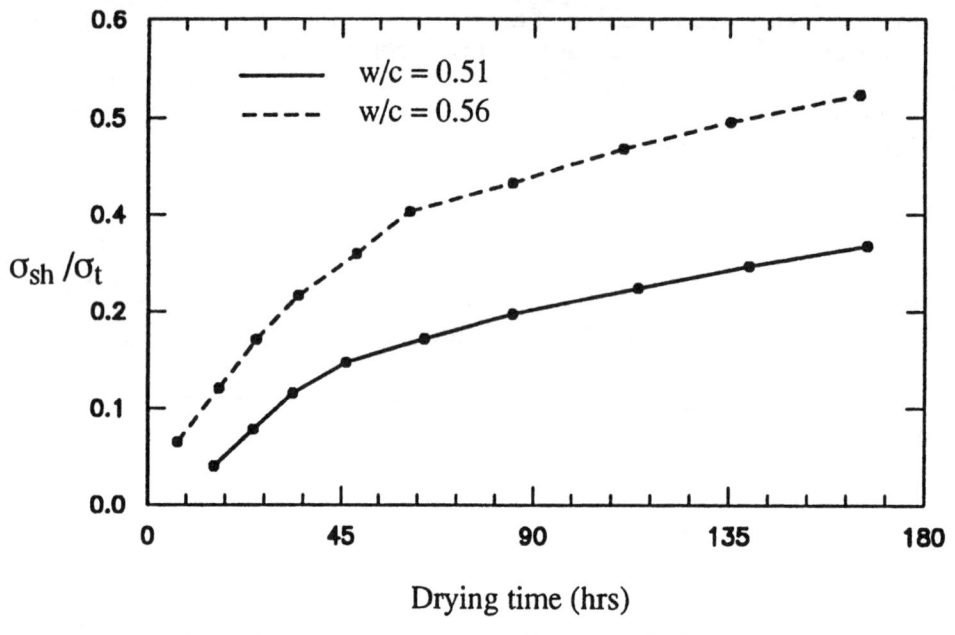

Figure 3. Shrinkage stress normalized to 7-day tensile strength

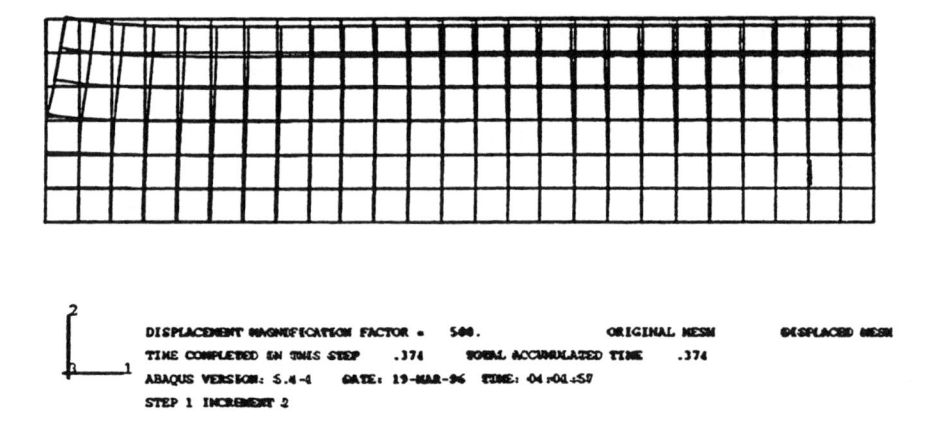

Figure 4. Finite element model of pavement overlay debonding and curling due to thermal and moisture gradients.

Modeling pavement behavior

A computational model will incorporate findings regarding drying shrinkage, elastic modulus, thermal expansion, bond strength, and creep. All of these material properties are critical to a comprehensive understanding of the overlay performance. The model will be a 2-dimensional finite element model that assess potential for curling under various environmental conditions. A typical result is shown in Figure 4. Temperature and moisture gradients are the sources of stress that cause the debonding and curling of the overlay. Thickness and joint spacing recommendations will be extracted from the modeling effort. A more sophisticated, 3-dimensional pavement model is being developed by other researchers within the FAA COE. Ultimately, our materials projects will provide constitutive modeling parameters for the 3-D computational model to provide a more robust solution concerning overlay performance.

SUMMARY

This paper has provided a brief overview of a comprehensive experimental program underway to assess performance of HPC and FRC in terms of fundamental material behavior. The scope of the materials under test, experimental approach, and analytical strategy have been discussed. The goal of the experimental work is to define material properties of high performance paste, mortar, concrete and bond between HPC and concrete substrates. These material parameters will be input to models of structural behavior of pavement overlays. The FAA COE goal for the HPC and FRC projects is to produce modeling capability to predict performance and provide criteria for design of overlay thickness and joint spacing.

ACKNOWLEDGMENT

This paper was prepared from a study conducted in the Center of Excellence for Airport Pavement Research which is funded in part by the Federal Aviation Administration under Research Grant Number 95-C-001. The opinions expressed in this paper are those of the authors and do not necessarily reflect the official views and policies of FAA.

REFERENCES

1. Hansen, W., "Constitutive Model for Predicting Ultimate Drying Shrinkage of Concrete," *J. Am. Ceram. Soc.*, **70** [5] 329-332 (1987).

THE EFFECT OF SILICA FUME DOSAGE ON THE HEAT OF HYDRATION OF HIGH PERFORMANCE MORTARS

Roberto C. A. Pinto
Graduate Research Fellow
Cornell University
Ithaca, NY, U.S.A.

Kenneth C. Hover
Professor
Cornell University
Ithaca, NY, U.S.A.

ABSTRACT

The heat of hydration of high performance mortars with different doses of silica fume was assessed. Silica fume was introduced as an additional cementitious material, and as a partial cement replacement. The temperature rise, total heat liberated, and the rate of heat liberation were monitored by a quasi-adiabatic calorimeter. Compressive strength was also measured at 90 days. Silica fume was found to increase compressive strength and to greatly influence the heat development characteristics of the mixtures. The maximum temperature rise was increased, and the heat liberated curves were greatly modified as a result of the inclusion of silica fume.

INTRODUCTION

The heat evolved during cement hydration is an important factor in the development of both strength and thermal stresses in concrete structures. When heat is generated via chemical activity more rapidly than it is dissipated to the environment, high internal concrete temperatures, harmful thermal gradients, or both, can develop. The risk of cracking is increased when there is a 20°C (68°F) temperature difference between the warmest and coolest portions of the member[1,2]. This is more likely to happen when the outer face of the section is exposed to lower temperatures.

Various techniques can be applied to reduce the maximum temperature and/or thermal gradients within a concrete section, such as protective measures to reduce heat loss on the surface[3], encourage heat loss in the interior, or modification in the mixture characteristics to reduce early heat generation. These methods reduce the temperature difference by either protecting the outer face from rapid cooling or by decreasing the temperature rise in the interior of the concrete[3]. A combination of these methods is common.

The study of heat development becomes particularly important in high performance mixtures where increased cement contents are not uncommon. Pozzolanic materials with lower, or slower heat generation characteristics, such as ground granulated slag or fly ash, are often used in such mixtures. In addition to modifying hardened concrete characteristics such as strength and permeability, these mineral admixtures frequently reduce the temperature rise at early ages[4].

Silica fume is a pozzolanic material widely used in the manufacture of high-performance concrete mixtures. It increases paste strength due to its pozzolanic activity, and the bond between paste and aggregate[5] However, unlike other pozzolanic materials, silica fume influences the early hydration of cement by accelerating the hydration of the tricalcium silicate[6-8], primarily responsible for early heat generation, and by reacting at the early stages of cement hydration[6]. These factors combine to cause rapid early strength gain, an important characteristic to faster demolding and/or prestressing of precast elements, and a different heat development behavior as compared to mixtures without silica fume.

The purpose of this study was to investigate the effect of different dosages of silica fume on the heat characteristics of a high-performance mortar. Silica fume was used in two modes: as a partial cement replacement (reducing the cement content), and as an additional cementitious material (maintaining the cement content, but reducing the water-cementitious materials ratio).

The maximum temperature rise, heat evolved, and rate of heat generated were monitored for 72 hours in mortar samples. Compressive strength was obtained from samples stored in laboratory conditions.

EXPERIMENTAL METHOD

The study consisted of two set of experiments. In the first, cement content was held constant, while silica fume was added to a control mortar mixture as 5% and 10% of the cement mass. The water-cement ratio was 0.33 for all mixtures, whereas the water-cementitious material ratio varied from 0.33 to 0.27. In the second set of experiments, silica fume replaced part of the cement content at two levels: 10%, and 20% of cement mass. In this case, the water-cementitious materials ratio remained 0.33 for all mixtures, with variations of water-cement ratio as more cement was replaced by silica fume.

Sample nomenclature indicates the mode of introduction and the amount of silica fume in the mixture. For example mixture A-5% refers to a 5% addition of silica fume, while mixture R-20% stands for a mixture with 20% replacement of cement by silica fume.

ASTM C 33 concrete sand with a fineness modulus of 2.78, and an ASTM C 150 Type I cement were used in all mixtures. A naphatalene sulfonate-type superplasticizer (Daracem-100© by W. R. Grace) with a solids content of 40% was introduced to the silica fume mixtures in different amounts to achieve a flow (ASTM C 109) of within 20% of the control mixture. Condensed silica fume, manufactured by Elkem Materials was introduced as a dry powder. Table 1 shows the mixture proportions, and the mortar flow values obtained.

A total of five mixtures were prepared in a 0.1 m³ (3.5 ft³) drum mixer in the laboratory. For each batch, five 102 mm x 204 mm (4 in. x 8 in.), and one 152 mm x 305 mm (6 in. x 12 in.) cylinders were filled with mortar (according to ASTM C 192) for compressive strength tests and heat development measurement, respectively. The 204 mm (8 in.) specimens were sealed and stored at ambient temperature for the first day (around 27°C - 81°F), then removed from the molds and immersed in lime saturated water (25°C - 77°F) until being tested for long-term compressive strength at 90 days after casting.

Table 1 - Mix proportions (per m³)

		control	A-5%	A-10%	R-10%	R-20%
cement	kg	749	731	749	665	579
water	kg	247	242	245	244.	241
fine aggregate	kg	1251	1222	1142	1233	1211
silica fume	kg	-	36	74	72	144
superplasticizer	mL	-	4.7	9.8	3.0	9.9
water/cement		0.33	0.33	0.33	0.37	0.42
water/cementitious materials		0.33	0.31	0.30	0.33	0.33
silica/cement (by mass)	%	-	5	10	10.8	24.9
flow (ASTM C 109)		50	46	58	58	64

Note: 1 kg/m³ = 1.69 lb/yd³

The heat development from the 152 x 305 mm (6 x 12 in.) cylinders was monitored for 72 hours, by a QD0612 Qdrum calorimeter, manufactured by Digital Site Systems, Inc. The Qdrum calorimeter is a large insulated drum equipped with thermal sensors. A data recording system attached to the calorimeter monitors sample temperature, and the calorimeter heat loss through the drum walls. The unit was programmed to record these values at 15 minute intervals. The total heat generated in this quasi-adiabatic condition was obtained by the summation of the actual heat retained by the sample itself plus the heat lost through the walls.

Under true adiabatic conditions no heat would be lost from the specimen. In this case, some heat (about 30% of the total liberated) was lost through the walls of the calorimeter. This heat loss was recorded, however, and was included in all subsequently reported values of total heat generated. Hence the term "quasi-adiabatic" used here. Specimens placed in the calorimeter experienced a thermal environment similar to that experienced by concrete at the center of a concrete element of a total thickness around 1.2 meters[9] (4 ft).

RESULTS

Figure 1 shows the time-temperature profiles recorded for the control mixtures and mixtures with addition of silica fume. Figure 2 shows the time-temperature profiles for the control mixture and mixtures with silica fume as a partial cement replacement. Table 2 summarizes the maximum temperature rise, and the temperature rise per mass of mortar, shown both in absolute terms and normalized to the control mixture. The time when the maximum temperature was observed is also shown.

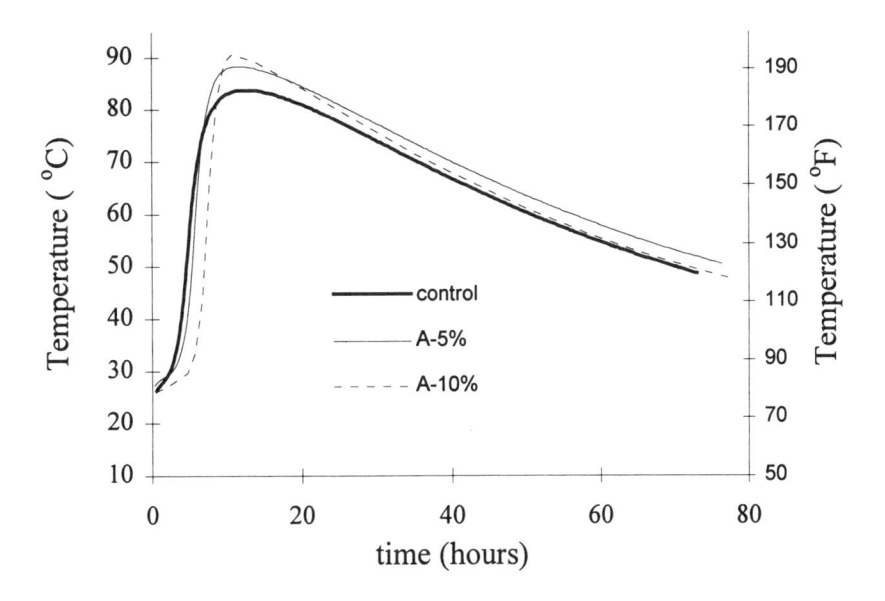

Figure 1 - Time-temperature profiles for mixtures with addition of silica fume

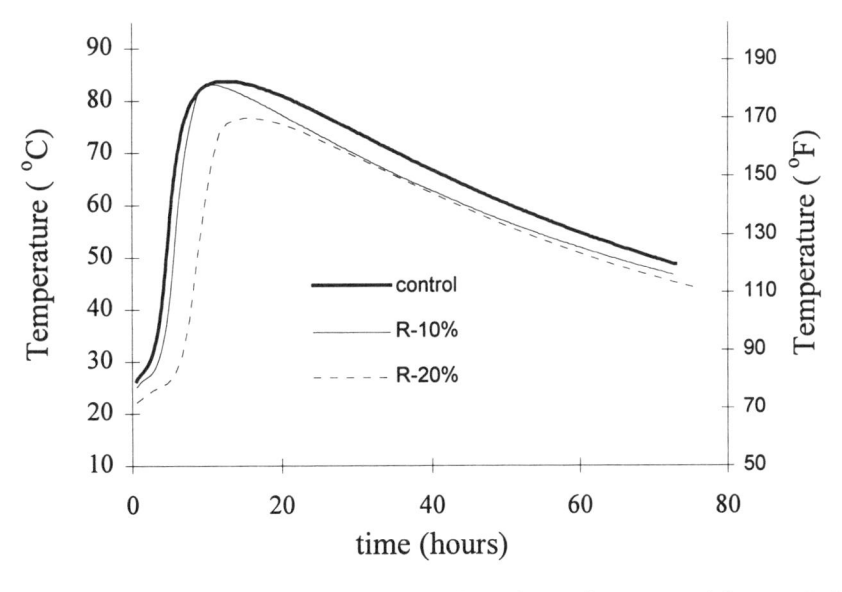

Figure 2 - Time-temperature profiles for mixtures with partial cement replacement by silica fume

Table 2 - Temperature rise of mixtures

Mixture	Observed maximum sample temperature rise (°C)	Observed sample temperature rise per mass of mortar (°C/kg)	Relative increase in sample temperature rise (base control mixture)	Time at maximum temperature (hours)
control	57.4	4.58	-	13
A-5%	61.0	4.87	+6.3%	12
A-10%	64.7	5.16	+12.7%	11
R-10%	57.9	4.78	+4.4%	11
R-20%	54.6	4.90	+7.0%	15

Note: 1°C/kg = 0.82°F/lb

Figure 3 shows the development of the quasi-adiabatic heat generation per mass of mortar over the first three days for the control mixture, and mixtures with addition of silica fume. Similarly, Figure 4 displays the results obtained for the mixtures with silica fume as a partial cement replacement

The rate of heat liberated per mass of mortar is shown in Figures 5 and 6, for the mixtures studied.

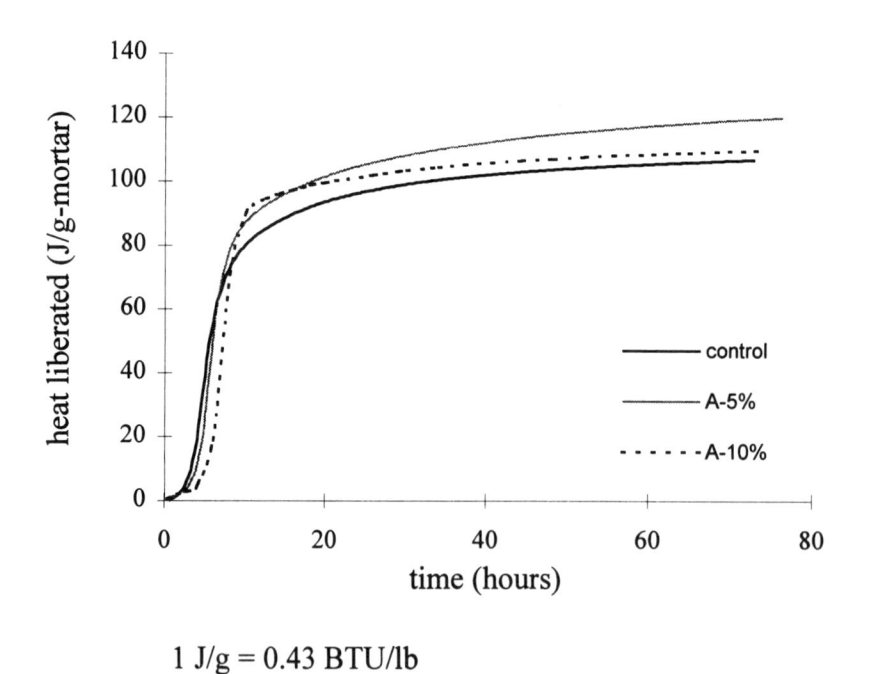

1 J/g = 0.43 BTU/lb

Figure 3 - Heat development for mixtures with addition of silica fume

1J/g = 0.43 BTU/lb

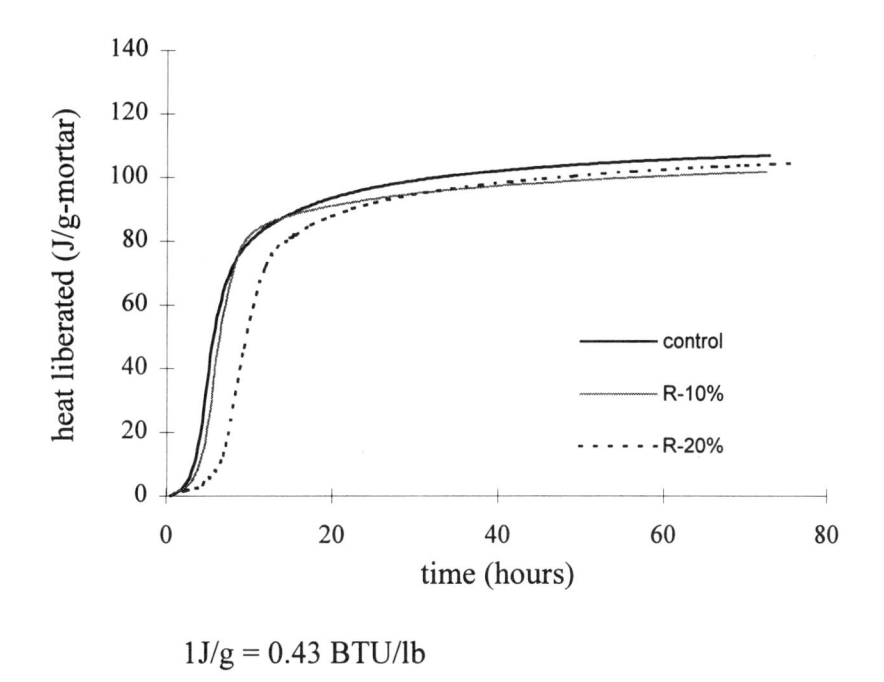

Figure 4 - Heat development for mixtures with partial cement replacement by silica fume

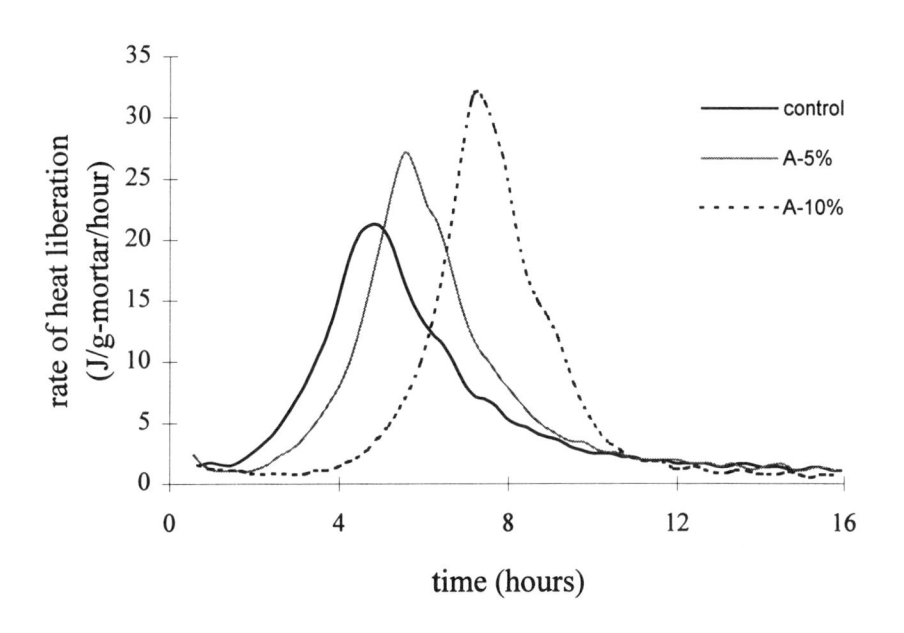

1J/g/hour = 0.43 BTU/lb/hour

Figure 5 - Rate of heat liberation for mixtures with addition of silica fume

Figure 6 - Rate of heat liberation for mixtures with partial cement replacement by silica fume

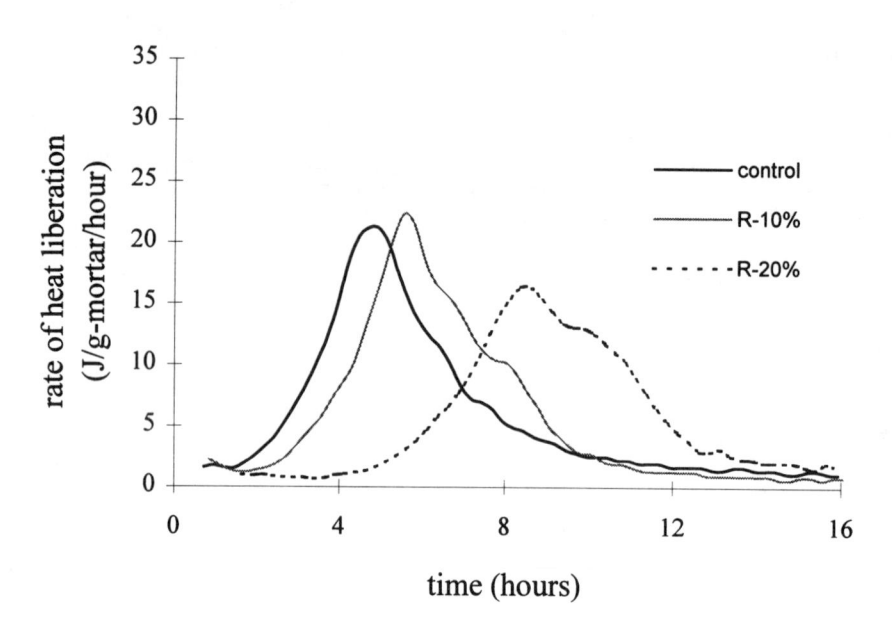

1J/g/hour = 0.43 BTU/lb/hour

The compressive strength results at 90 days for the 102 x 204 mm cylinder specimens cured at laboratory conditions are presented in Table 3.

Table 3 - Compressive strength results

Mixture	Lab specimens 90 days (MPa)	Relative increase in strength (base control mixture)
control	60.3	-
A-5%	75.0	24.4%
A-10%	77.8	29.0%
R-10%	63.9	6.0%
R-20%	70.8	17.4%

Note: 1 MPa = 0.145 ksi

DISCUSSION

The set of mixtures in which silica fume was used as an additional material (A-5%, and A-10%) experienced the highest temperature rise per mass of mortar. The temperature rise for the A-10% mixture was almost 0.6°C/kg-mortar (0.49°F/lb-mortar) greater than

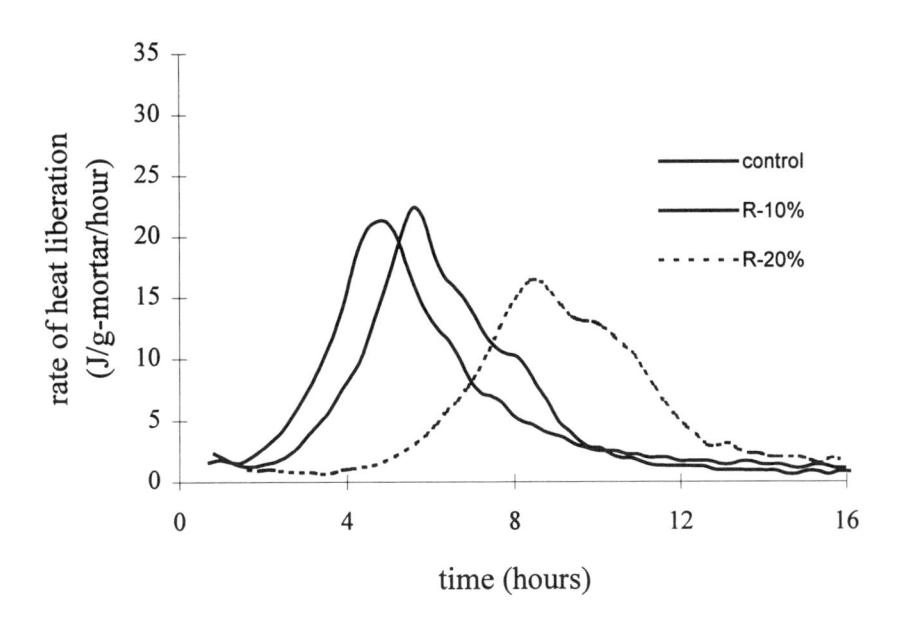

Figure 6 - Rate of heat liberation for mixtures with partial cement replacement by silica fume

1J/g/hour = 0.43 BTU/lb/hour

The compressive strength results at 90 days for the 102 x 204 mm cylinder specimens cured at laboratory conditions are presented in Table 3.

Table 3 - Compressive strength results

Mixture	Lab specimens 90 days (MPa)	Relative increase in strength (base control mixture)
control	60.3	-
A-5%	75.0	24.4%
A-10%	77.8	29.0%
R-10%	63.9	6.0%
R-20%	70.8	17.4%

Note: 1 MPa = 0.145 ksi

DISCUSSION

The set of mixtures in which silica fume was used as an additional material (A-5%, and A-10%) experienced the highest temperature rise per mass of mortar. The temperature rise for the A-10% mixture was almost 0.6°C/kg-mortar (0.49°F/lb-mortar) greater than for the control mixture (an increase of 13%). Since the calorimeter walls simulate the insulation provided by about 0.6 meters (2 ft.) of concrete cover, the results shown here indicate that for mass concrete, a mixture like the silica fume mixtures used here would likely cause more problems with thermal gradients than a mixture without silica fume.

These observed higher maximum temperatures at increasing levels of silica fume addition were a consequence of the increased chemical activity during this period, being in agreement with observations of an acceleration effect of silica fume on early cement hydration[6-8].

In this regard, the curves shown in Figure 5 are particularly interesting. The peak on the rate of heat evolved curves was higher for the mixtures with silica fume addition, occurring before the maximum sample temperatures were observed (around 11-12 hours in this case). These higher peaks indicate the increase of chemical activity for the mixtures with silica fume addition, with more heat being liberated during the first 12 hours.

Figure 5 also shows that the initial period of small chemical activity, often called the induction or dormant period[10], was extended for the mixtures with higher silica fume content. This observed extension of the dormant period cannot be attributed solely to the presence of silica fume in the mixture, since superplasticizer in different amounts was introduced for the mixtures with silica fume. In fact, it was observed for the particular brands tested that both admixtures affect differently the extent of the dormant period, as indirectly measured by the time of initial set[11]. While superplasticizer retarded the initial set, silica fume accelerated it.

Despite these differences in dormant period, it can be noticed from Figure 5 that when silica fume was added to the mixture, with enough superplasticizer to maintain workability (as measured by mortar flow), heat liberation is delayed, but once it starts, it occurs proportionally more rapidly, releasing more heat per time, as discerned by the higher peaks.

When silica fume was used as a partial cement replacement, as in mixtures R-10%, and R-20%, the expected acceleration in early cement hydration due to the silica fume presence was counteracted by the increase in the water-cement ratio, as more cement was replaced by silica fume. The final result was a combination of both acceleration and dilution effects. At higher levels of replacement, the maximum temperature rise per mass of material increased. The dormant period was extended at higher levels of replacement, as seen in Figure 6.

The extension of the dormant period observed in the mixture R-20% might have been affected by the greater flow value of this mixture (>20% of the control mixture). More retardation may have occurred due to the higher superplasticizer content.

Clearly, it would have been useful to test compressive strength specimens that had experienced the same temperature history as those in the calorimeter, unfortunately the only results available are for the 90 days, 25°C cylinders as shown in Table 3. Due to the small size of these specimens, it is unlikely that they experienced the high temperatures developed in the calorimeter.

Therefore, the results in Table 3 serve to characterize the effect of the silica fume on compressive strength in relatively thin sections in which heat build-up is unlikely. In this study the introduction of silica fume at the levels tested here increased the laboratory cured compressive strength of the mortars, no matter its form of utilization, as an additional cementitious material, or as a replacing material. The effect of high temperature on compressive strength would be of greater influence on the in-situ strength development in larger members.

The choice of whether to use silica fume in high-performance mixtures, as either an additional cementitious material, or as a partial cement replacement, would depend on the acceptability of the effects of the silica fume on both temperature rise and other required properties such as compressive, tensile or flexural strength permeability, or frost resistance. From the results presented here, the introduction of 10% silica fume, for instance, in a high-performance mortar mixture increased 13% the observed maximum temperature rise. The same amount of silica fume as a replacing material caused an increase of 4% on the maximum temperature rise of the mixture. No matter how used, the silica fume led an increase in mortar temperature.

While the results presented here were obtained for mortar mixtures, similar behavior is expected to occur in concrete mixtures. The introduction of coarse aggregate in a fixed amount for all mixtures studied (i.e. same mortar fractions) would surely change the potential compressive strength and maximum temperature rise values, but it should not significantly modify mixture behavior relative to the control mixture.

In massive structures made of high-performance silica fume concrete, care should be taken to monitor interior concrete temperature rise, and the risk of early cracking. Although the introduction of silica fume may allow the utilization of smaller sections, due to an increase in compressive strength, there is a need to closely monitor the development of potentially harmful thermal gradients.

CONCLUSION

- The introduction of silica fume to a high-performance mixture not only influences compressive strength, but also modifies its heat development characteristics.
- Silica fume presence accelerates heat liberation, resulting in an increase in the temperature rise.

- The choice of whether to use silica fume in high-performance mixtures, as either an additional cementitious material, or as a partial cement replacement, would depend on the acceptability of the effects of the silica fume on temperature rise and compressive strength, or other required concrete property.

ACKNOWLEDGMENT

Funding for the present study was provided by CNPq - Conselho Nacional de Pesquisa e Desenvolvimento - Brazil, and Cornell University. The authors would also like to thank W. R. Grace & Co. and B. R. DeWitt Inc. for supplying materials.

REFERENCES

1. FitzGibbon, M. E., "Large Pours for Reinforced Concrete Structures," *Concrete (London)*, Vol. 10, No.3, 1976, p. 41.

2. FitzGibbon, M. E., "Large Pours - 2 Heat Generation and Control," *Concrete (London)*, Vol. 10, No.12, 1976, pp. 33-35.

3. Neville, A. M., *Properties of Concrete*, 4th Ed., John Wiley & Sons, New York, 1196, 844 pp.

4. ACI Committee 207, *Mass Concrete*, ACI 207.1R-87, American Concrete Institute, 44 pp.

5. Malhotra, V. M., Ramachandran, V. S., Feldman, R. F., and Aïtcin, P., *Condensed Silica Fume in Concrete*, CRC Press, 1987, 221 pp.

6. Cheng-yi, H., and Feldman, R. F., "Hydration Reactions in Portland Cement-Silica Fume Blends," *Cement and Concrete Research*, Vol. 15, 1985, pp. 585-592.

7. Kurdowsky, W., and Nocun-Wczelik, W., "The Tricalcium Silicate Hydration in the Presence of Active Silica," *Cement and Concrete Research*, Vol. 13, 1983, pp. 341-348.

8. Wu, Z., and Young, J. F., "The Hydration of Tricalcium Silicate in the Presence of Colloidal Silica," *Journal of Material Science*, Vol. 19, 1984, pp. 3477-3486.

9. Johnson, G. L., "Concrete Testing on the Kroch Library Project: Thermometry, Maturity and Calorimetry," Master of Science Thesis, Cornell University, 1993.

10. Bye, G. C., *Portland Cement - Composition, Production and Properties*, Pergamon Press, 1983, 149 pp.

11. Pinto, R. C. A., and Hover, K. C., "Effect of Silica Fume and Superplasticizer Addition on Setting Behavior of High-Strength Mixtures," paper accepted for publication in Transportation Research Board Record.

HIGH PERFORMANCE CHARACTERISTICS OF CHEMICALLY ACTIVATED FLY ASH (CAFA)

Thomas Silverstrim and Dr. Hossein Rostami,
By-Products Development Corporation
Collingdale, Pennsylvania, USA

Dr. Yunping Xi and Dr. Joseph Martin,
Drexel University, Department of Civil and Architectural Engineering
Philadelphia, Pennsylvania, USA

ABSTRACT

Chemically Activated Fly Ash (CAFA) concrete is a new development in fly ash cementitious material technology. CAFA concrete is produced using conventional concrete mixing and forming techniques. CAFA requires dry curing at elevated temperatures of 50°C-93°C (130°F-200°F) making it feasible for production of precast concrete products. CAFA concrete has high performance properties including: rapid strength gain (up to 90% of 28 day compressive strength in 24 hours), high ultimate strengths (over 124 MPa (18,000 psi)), and excellent acid resistance. CAFA concrete is resistant to chemical attack, such as, sulfuric (H_2SO_4), nitric (HNO_3), hydrochloric (HCl), and organic acids. Recent work includes testing for alkali silica reactivity, chloride permeability, microstructure, and concrete pipe manufacture. CAFA compositions were studied using SEM Micrographs and EDS (Energy Dispersive Spectroscopy). The microstructure following long-term exposure to strong acid was also investigated. CAFA has been used to create High Performance CAFA Concrete Pipe. The results of this work are included. The cost of materials of CAFA are competitive with the cost of materials of portland cement concrete of over 40 MPa (6,000 psi). Further work is needed for utilization of CAFA in high performance structural applications, such as, creep testing, prestressing tendon creep, performance with reinforcing steel, alkali-silica reactivity with reactive aggregates, and application specific testing. Near term applications of CAFA concrete are, block, pipe, burial vaults, median barriers, sound barriers, overlaying materials, and chemical resistant products. It is envisioned that with further development CAFA will be suitable for high strength construction products, such as, bridge beams, prestressed members, concrete tanks, highway appurtenances, and other concrete products.

INTRODUCTION

In 1996, the United States consumed approximately 760 million metric tons (850 million short tons) of coal for electric generation and industrial use. This resulted in the generation of about 80 million metric tons (90 million short tons) of coal combustion by-products comprising fly ash, bottom ash, and boiler slag. Of this 80 million metric tons of material, about 45 million metric tons (50 million short tons) are fly ash. In 1996, 27% of fly ash was reused and the remaining 73% was landfilled or surface impounded. Fly ash has been researched for the past six decades including significant work on various methods of alkali activation. Chemically Activated Fly Ash (CAFA) uses a blend of chemicals to produce unexpected results. This paper summarizes information from prior publications regarding the high performance characteristics of CAFA. CAFA cement shows great promise in increasing the quality, durability, and reducing the cost of high performance concrete products.

According to ASTM C 618, there are two main classifications of fly ash: Class F and Class C. This classification is often cited by transportation departments for material specification. Class F fly ash has a minimum combined silicon oxide, aluminum oxide, and iron oxide content of 70%. The particles are classified as an aluminosilicate glass which exhibit pozzolanic reactivity in the presence of alkali, but do not themselves exhibit cementitious properties when mixed with water. Class C fly ash has a combined silicon oxide, aluminum oxide, and iron oxide content greater than 50%. The material is a calcium aluminosilicate and exhibits cementitious properties when exposed to water along with pozzolanic reactivity.[1] Over 50% of Class C fly ash is recycled while less than 10% of Class F fly ash is recycled.

FLY ASH AND ALKALI ACTIVATION IN CONCRETE

Fly ash is a commonly used cementitious component of concrete due to its pozzolanic activity and high specific surface area. Over 6.6 million metric tons (7.4 million short tons) of fly ash were utilized in concrete in 1994.[2] The addition of fly ash to cement mixtures is well established and a great body of research on the subject is available. Possible benefits from the use of fly ash in concrete include better economics, increased ultimate strength, better chemical resistance, and a number of other property improvements.[1]

At the same time, the conversion of fly ash into cementitious material through the use of chemical mechanisms other than portland cement has generated considerable interest. Majiling and Roy present the hydrothermal transformation of Class F fly ash in the presence of lime to produce a new reactive fly ash cement.[3] The use of alkali to catalyze the hardening properties of pozzolanic materials is well researched. Roy and Silsbee's state of the art report on ceramic-based composites includes a variety of technologies including the alkali activation of latent hydraulic materials, such as, blast furnace slag, to produce ceramic-based composites.[4] Shi reports on alkali activated lime fly ash pastes with high levels of fly ash in an alkali activated systems.[7] Considerable work has been conducted on alkali activated slags with and without fly ash.[7-11] Slags typically contain about 50% calcium and alkali activation catalyzes hydration reactions. Other work has been conducted by Davidovits in which alkali activation of dehydrated phyllosilicates (i.e. metakaolin) with calcium glasses forms hardened products. The hardening mechanism is described as a polycondensation reaction. This chemistry was patented and applied by Lone Star to create Pyrament© cement. The geopolymer binder creates concrete with high strength and good chemical resistance. In addition, use of the binder in place of portland cement results in a reduction of CO_2 emissions by 80% compared to Portland cement.[12, 13, 14]

On a related front, there are a group of commercially available cement materials based on silicate. These cements are "polymerized" through a flouride activation process and have high chemical resistance. The technology dates back to the 1940's and the best information on the mechanism is available from Iler.[15] Results from prior publications is included below detailing the high performance concrete properties and the excellent chemical resistance.

HIGH PERFORMANCE CAFA MATERIALS

According to ACI Committee on High Performance Concrete (HPC) A2E01, HPC must improve upon the following parameters: a) ease of placement and consolidation without affecting strength, b) long term mechanical property, c) early high strength, d) volume stability, and e) longer life in severe environments. A state of the art SHRP report was prepared on High Performance Concrete (HPC). HPC was defined based on the following criteria [16]:

    (1) Maximum water-cementitious ratio of 0.35;
    (2) Minimum Durability Factor of 80% based on ASTM C 666, Procedure A;
    (3) Minimum Strength Criteria of either:
        (a) 21 MPa (3,000 psi) within 4 hours (Very Early Strength),
        (b) 34 MPa (5,000 psi) within 24 hours (High Early Strength), or
        (c) 69 MPa (10,000 psi) within 28 days (Very High Strength).

Work on HPC indicates that concrete with compressive strength ranging from 69 MPa (10,000 psi) to 138 MPa (20,000 psi) has improved the concrete durability.[16-19]

Typical CAFA concrete and mortar formulations used in the experimental work are described in Table 1 and Table 2.

The preparation of CAFA samples involves first mixing the dry ingredients. The activating solution is added to the dry ingredients and mixed until homogenous. This fresh mixture is placed in molds or shapes as desired. Vibration and compaction can be employed as necessary based on the workability of the mixture. The workability of the mixture can be altered by changing the amount of fly ash in the binder and by changing the ratio of binder to aggregate. This product is cured at temperatures between 40°C (104°F) and 120°C (248°F). The curing time is dependent upon several factors. In general, the strength of CAFA increases as the curing time increases. Proper mix designs with adequate curing create CAFA concrete with high performance properties.[20-23]

Table 1- Typical Mix Design for CAFA concrete

Component	kg/m^3	lb./yd^3
CAFA Activating Solution	136	300
Class F Fly Ash	409	900
ASTM C 33 Fine Aggregate	500	1100
Coarse Aggregate- #66	682	1500

Table 2- Typical Mix Design for CAFA mortar

Component	kg/m^3	lb./yd^3
CAFA Activating Solution	273	600
Class F Fly Ash	591	1300
ASTM C 33 Fine Aggregate	773	1700

EFFECT OF CAFA BINDER COMPOSITION ON STRENGTH

CAFA binder is defined as the mixture of the fly ash with the activating chemical. This mixture plays the same fundamental role in CAFA materials as Portland cement in traditional concrete. Based on traditional concrete definitions, CAFA mixtures have very low water to cementitious solids ratios. The percentage of total water within CAFA binder is a better guide to strength prediction of CAFA. Figure 1 indicates CAFA strength versus the total water content of the binder. These tests were conducted with mortar mixtures, using Class F fly ash, C 33 sand, and cured in the same conditions. Specimen sizes were 5.08 cm x 10.16 cm (2" x 4") specimens cured for 18 hours. The viscosity of the mortar mixture was held constant for several mixtures by the addition of aggregates. The total water content was to calculate available total water content. The relationship between water cement ratio in which the surface saturated dry calculations for aggregates were used showed a poor correlation. In further experiments, water from saturated aggregates correlated with the strength of CAFA using dry aggregates and added water.

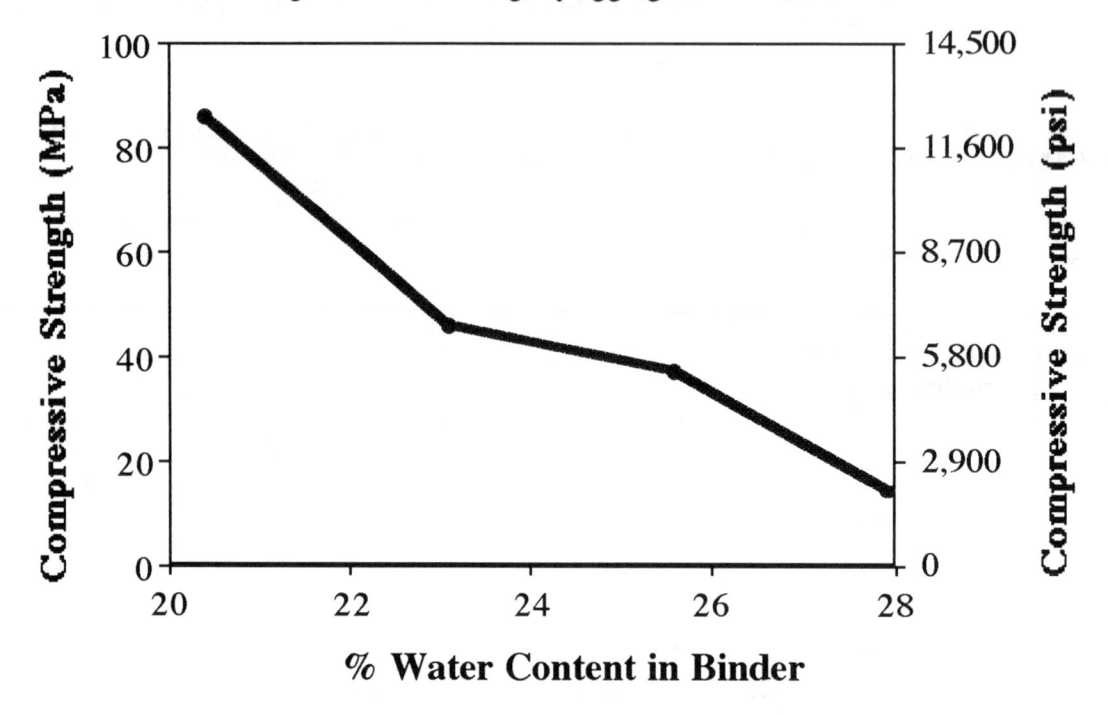

Figure 1- Compressive Strength of CAFA Versus Water Content of Binder.

The other primary factor of the strength of CAFA is the % weight of activating solution in the composition. Figure 2 indicates the strength of CAFA vs. the % of activating solution in the total mix.

The tests were conducted using CAFA mortar compositions with similar concentration of activating solution, and the same fly ash, curing conditions, and aggregates. Based on Figures 1 and 2, it can be seen that higher concentration alkali activating solution in greater quantities results in higher strength.

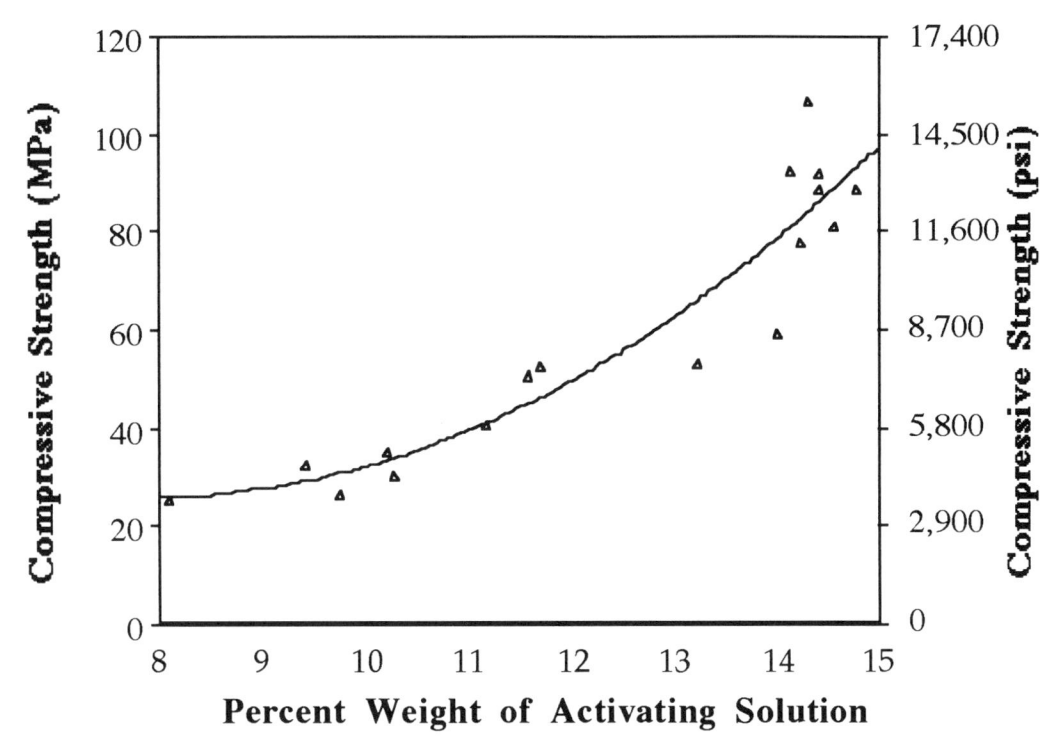

Figure 2- Percent Weight of Activating Solution Versus Compressive Strength

COMPRESSIVE STRENGTH DEVELOPMENT

CAFA binder is used to create CAFA concrete with high performance properties. As reported previously, the compressive strength development of CAFA increases as a function of curing time and temperature. Increasing temperature results in more rapid strength development. The ultimate strength of CAFA is a function of curing time and temperature. The time to reach maximum compressive strength is reduced as the temperature increases as shown in Figure 3.

Another significant difference is the strength development of CAFA over time. CAFA strength after 1 year is within 5% of the strength after 7 days. The hardening mechanism of CAFA involves rapid development of strength which levels within a short period of time. Testing of CAFA for drying shrinkage are underway according to ASTM 596 using CAFA mortar. Preliminary results indicate a strain of less than 50 μ in./in. over 56 days starting one day after curing. This indicates excellent dimensional stability of the early age cement as compared to Portland cement mortars.

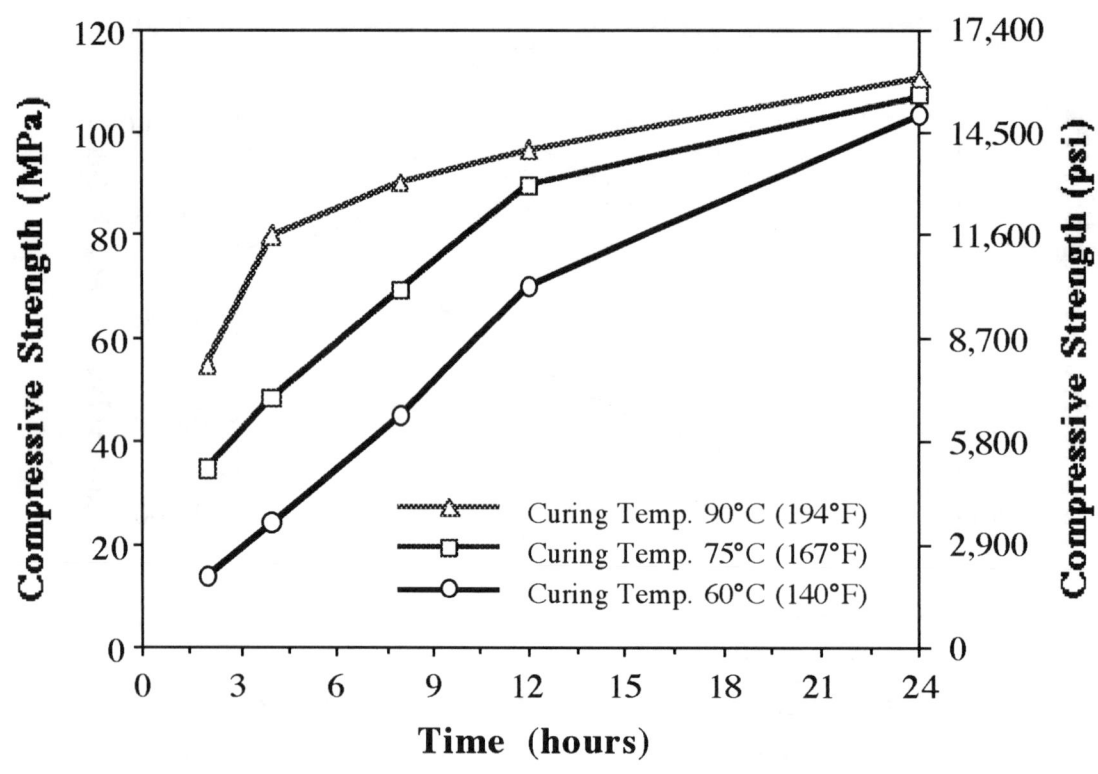

Figure 3- Compressive Strength Development Various Time at Varying Temp. [22]

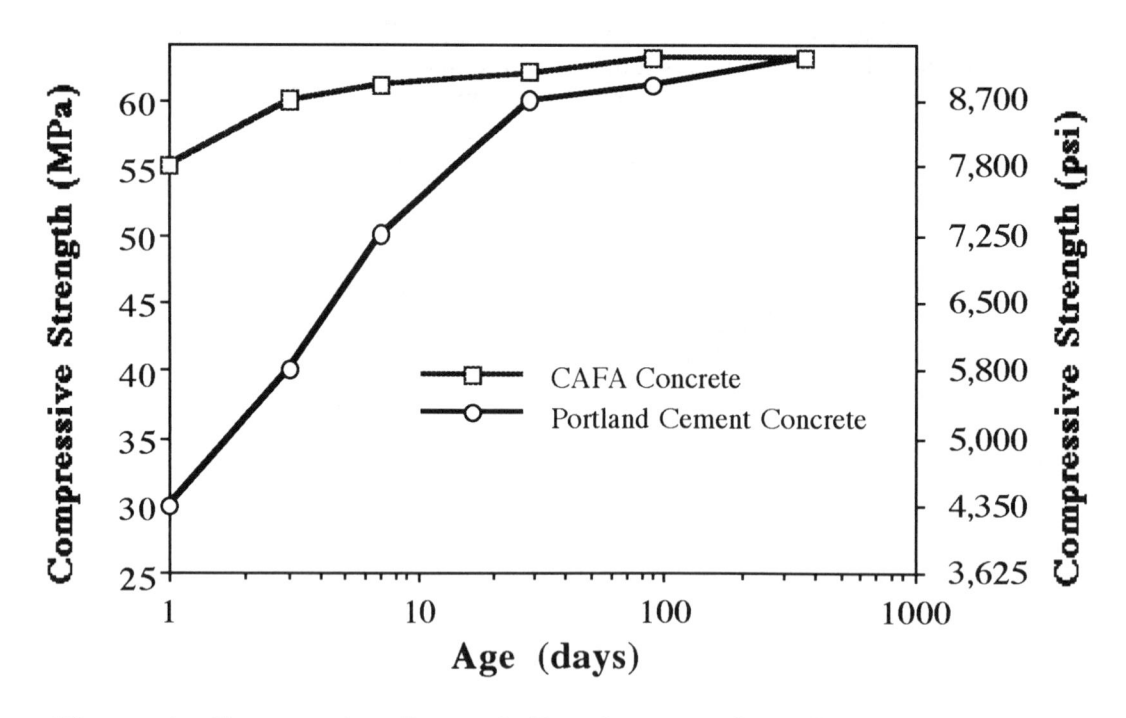

Figure 4- Compressive Strength Development Over Long Time Period

RESISTANCE TO FREEZING AND THAWING

The weathering action of freezing and thawing on CAFA concrete has been evaluated.[22] The results from testing of CAFA with air entrainment and without tested based on ASTM C 666 are summarized. The samples were evaluated based on weight loss. The results indicate the CAFA matrix has resistance to freeze and thaw. In addition, CAFA mortar has more resistance than CAFA concrete. It was found that commonly used air entraining admixtures increased the weight loss of CAFA concrete. The results based on the weight loss of the specimens exposed to freezing and thawing is shown in Figure 5.

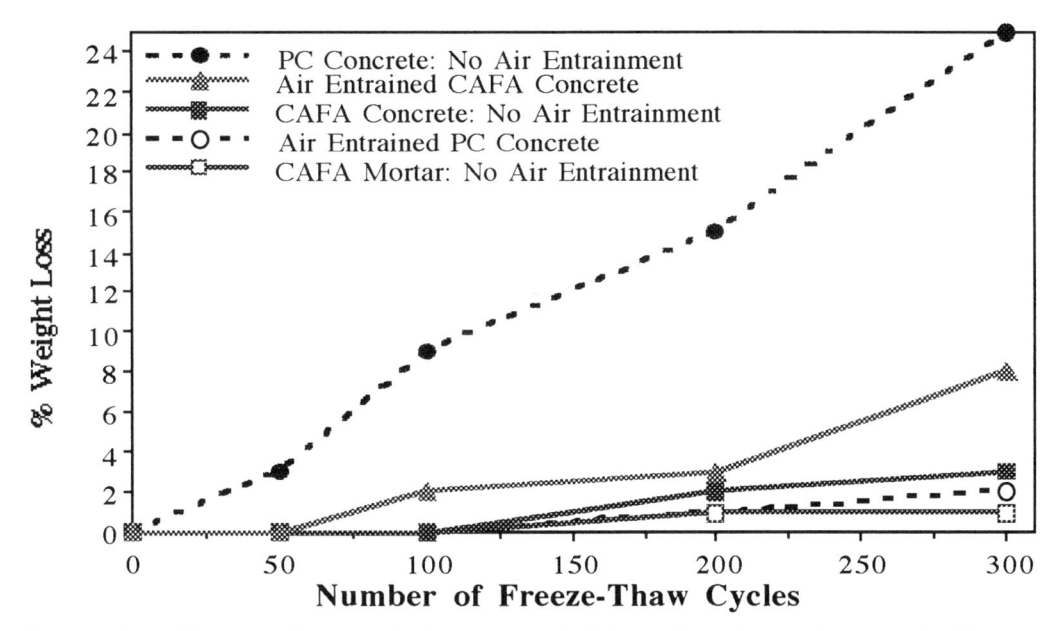

Figure 5- Weight Loss of Concrete Subjected to Freezing and Thawing

RESISTANCE TO ACID ATTACK

The ability of CAFA to withstand aggressive chemical environments has been reported.[22] The results from continued testing are presented. CAFA mortar cylinders 5.08 cm x 10.16 cm (2" x 4") and 7.62 cm x 15.24 cn (3" x 6"), and prismatic 5.08 cm x 3.81 cm x 20.32 cm (2"x 1 1/2"x 8") samples have been immersed in 20% solutions of Hydrochloric Acid, Sulfuric Acid, Nitric Acid, and 100% Acetic Acid for over 1 year. Weight loss, change in flexural strength, and change in compressive strength were determined based on ASTM C 267. The weight loss over 1 year is shown in Figure 6. For comparison, Portland cement concrete has poor chemical resistance. The best compositions use silica fume to reduce permeability and reduce the rate of deterioration. Based on a failure criterion of 25% weight loss, the best portland cement compositions will fail after 35 days in 5% sulfuric acid and 220 days in 1% sulfuric acid. For nitric and hydrochloric acid, Portland cement concrete fails within 7 days.[24] Portland cement has no resistance to the strong acid solutions used in these experiments. The chemical resistant properties of CAFA are more similar to silicate cements than portland cement. Based on the results, CAFA has potential for chemical resistant applications including sewage pipes, chemical resistant overlays, containment structures, septic tanks, marine applications, etc.

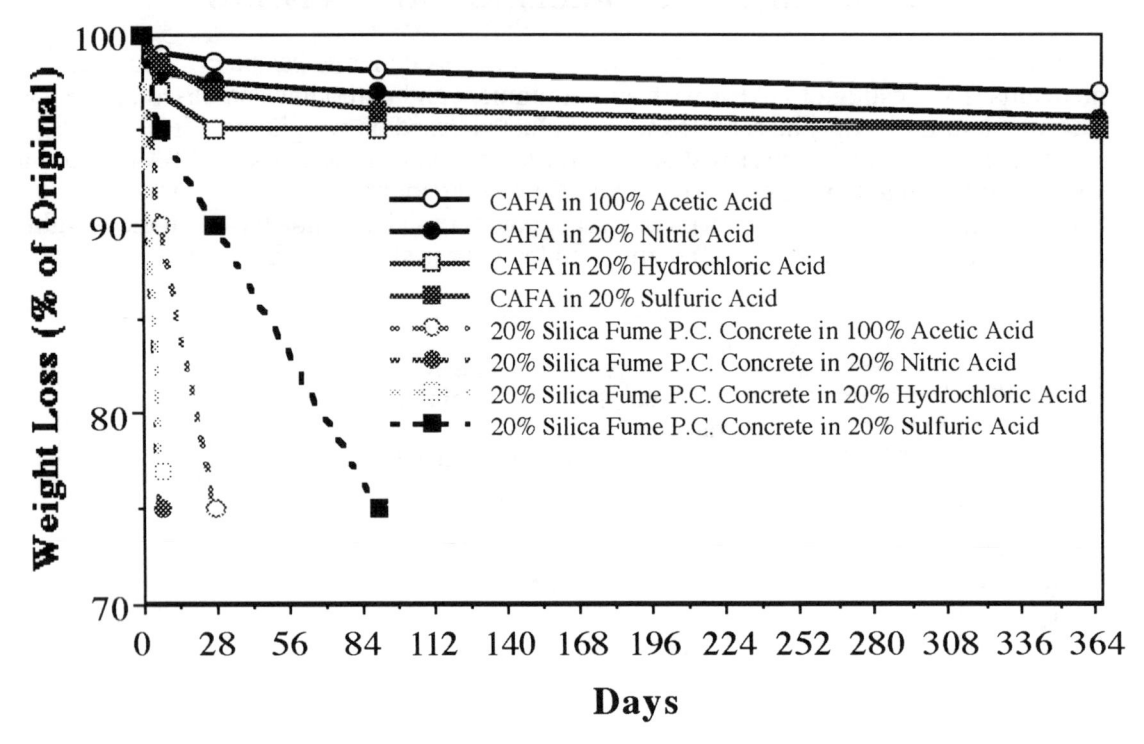

Fig. 6- Weight Loss of CAFA and Silica Fume Concrete in Acid Immersion

ALKALI SILICA REACTIVITY

CAFA concrete is an alkali activated system with concentrations of sodium and/or potassium which far exceed permissible levels in portland cement. Naturally, there were concerns raised as to the potential for deleterious alkali-silica reactivity. Dr. Yunping Xi at Drexel University conducted experiments on alkali-silica reactivity based on ASTM C 227. 2.54 cm x 2.54 cm x 25.4 cm (1"x 1"x 10") prismatic samples of CAFA mortar were prepared with crushed pyrex glass replaced for C 33 sand. Samples were immersed in 1 N NaOH at 38°C (100°F) and tested for length change with a comparator Samples exhibited no indications of alkali-silica reactivity as shown in Figure 7.

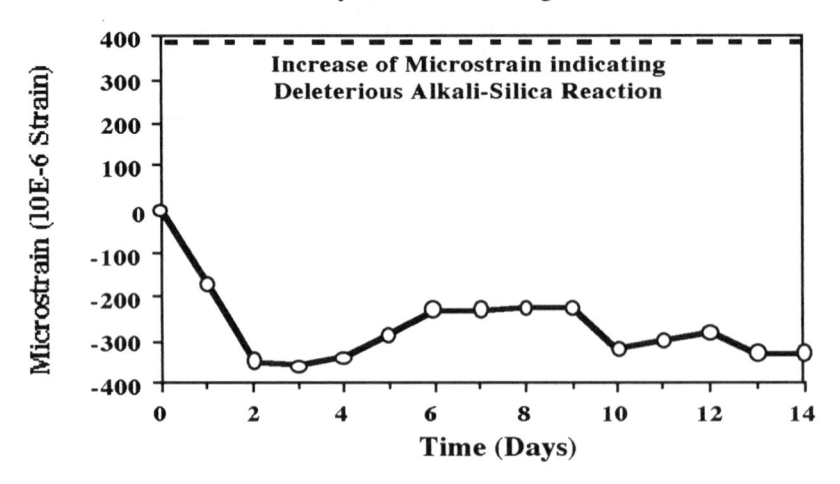

Figure 7- Alkali Silica Reactivity Testing

SEM AND EDS INVESTIGATION OF CAFA

CAFA was investigated with Scanning Electron Microscope (SEM) and Energy Dispersive Spectroscopy (EDS) with the help of the R. J. Lee Group in Monroeville, PA. Investigations were conducted on hardened CAFA and CAFA immersed in 70% Nitric acid for 90 days. SEM analysis was performed on petrographic thin sections. Polished thin sections were prepared from each of the CAFA samples. Each CAFA sample was cut and then polished using increasing finer diamond grit to achieve a final polish of 1000 grit. The polished surfaces were then impregnated with an ultra-violet (UV) epoxy dye, repolished, attached to a glass slide, and back ground to approximately 20 micrometer thickness, again using increasing finer diamond grit to a final 3 micron diamond grit polish. The results can be seen in Figure 7 and Figure 8. It can be seen from the EDS that the sodium from the alkali activation process is solubilized in the low pH environment. The structure of the CAFA was intact as seen by the levels of aluminosilicate in the EDS and the appearance in the SEM.

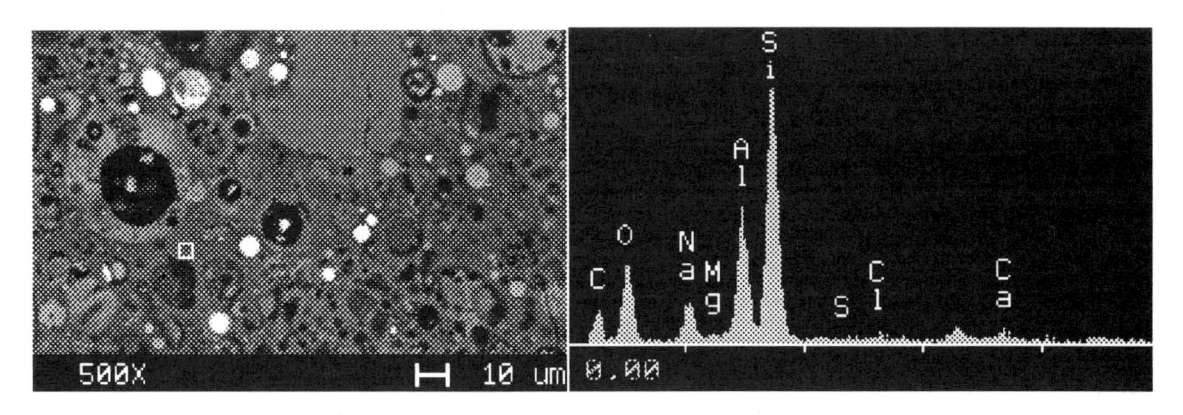

Figure 8: SEM and EDS of CAFA Cured at 60°C (140°F)

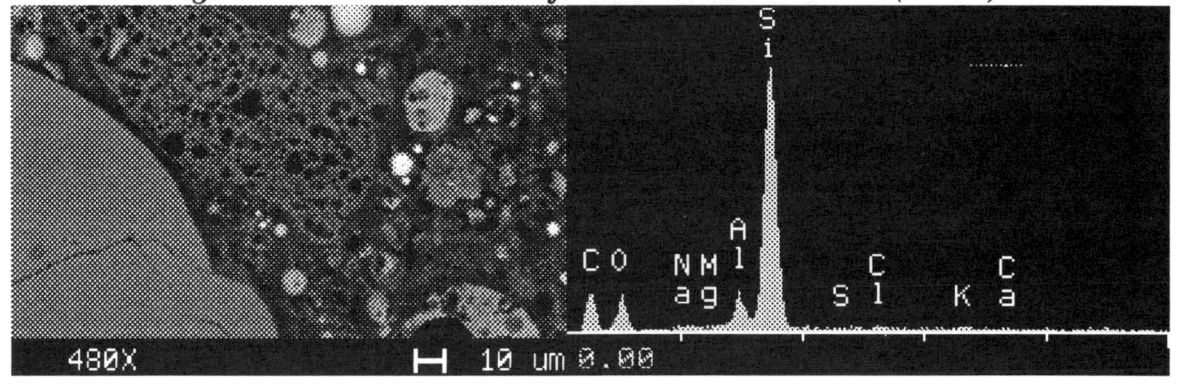

Figure 9: SEM and EDS of CAFA Cured at 60°C (140°F) and Immersed in HNO_3 for 90 Days

Experiments in the laboratory are helpful for evaluating properties. However, the most important aspect of concrete research is to be sure that the material can be used to create a product. The manufacturability of CAFA concrete for the production of concrete pipe was investigated.

HIGH PERFORMANCE CONCRETE PIPE PRODUCTION

Recently, CAFA technology has been applied to the production of dry cast concrete pipe. The results indicate high performance mechanical and chemical properties. By-Products Development Corp. (BPDC) working with Drexel University's Department of Civil and Architectural has teamed with Eastern Shore Concrete Products (ESCP) in New Holland, PA for the application of CAFA technology to the production of CAFA. High Performance concrete pipe. The pipe were produced with ESCP's International Concrete Pipe Machinery Packerhead system. This equipment is standard for pipe production in the industry and can be seen in Figure 10. The portland cement in ESCP concrete mix designs was replaced with CAFA. CAFA consists of an alkali activating solution and fly ash. CAFA pipe were produced flawlessly by the expert crew at ESCP. Fine tuning of the process was fast and painless. The crew quickly figured out the best production procedure.

The pipe were produced with International Concrete Pipe Machinery BiDi- Bi directional packerhead system. 2.4 m. long x 30 cm (8 ft. long x 12") diameter pipe were created. In order to satisfy the dry heat curing requirement, a polyethylene shrink wrap was wrapped around the pipe prior to curing. The pipe were then placed in steam at 65°C (150°F) for 7 hours and 54°C (130°F) for 13 hours. The pipe were removed and placed outside at ambient temperatures of approximately 10°C (50°F).

After 14 days, the pipe were tested for external load crushing strength by the three edge bearing method according to ASTM C 497. Core samples were taken from the pipe and tested according to ASTM C 39. The absorption test was conducted based on ASTM C 497, Test Method A. The permeability test was conducted as described in ASTM C 497. The result of this testing can be seen in Table 3.

The chemical analysis of the fly ash used in the demonstration can be seen in Table 4. The fly ash was acquired from JTM Industries, Allentown, PA.

Figure 10- Removal of the Jacket from a 2.4 m. (8') long by 30 cm (12") dia. CAFA Pipe

Table 3- CAFA Pipe Test Results (ASTM C 497)

Load to Produce 0.254 mm (0.01") Crack	4000 kg (8,750 pounds)
Ultimate Crushing Strength	5100 kg (11,250 pounds)
Core Sample Crushing Strength	59 MPa (8,500 psi)
Absorption Testing	7.0%
Permeability Testing	Negative

Table 4- Chemical Analysis of Fly Ash in CAFA Pipe

Constituent	Amount of Total
SiO_2	49.10%
Al_2O_3	28.41%
Fe_2O_3	14.89%
CaO	1.64%
MgO	0.83%
SO_3	0.44%
LOI	2.82%

FUTURE WORK

The investigation of CAFA for use as a structural material is only at the beginning stage. The properties which have been observed to date are encouraging. Based on current information, CAFA testing indicates the following:
- Excellent dimensional stability, i.e., very low drying shrinkage and sulfate expansion
- High pull-out strength of steel reinforcing bar
- Enhanced protection of reinforcing steel
- Good flexural toughness with fiber reinforcement
- Resistance to de-icing salts and marine exposure (ongoing)
- Good high temperature stability (suitable for refractory applications)
- Low alkali-silica reactivity

Other research which needs to be conducted includes:

- creep properties
- alkali-silica reactivity with reactive aggregate
- performance under prestressing forces
- exposure to biogenic sulfide

CONCLUSION

CAFA has excellent potential for specialty cement applications. The matrix can be used in chemical resistant, and high strength construction applications. The pipe investigation shows that a high performance pipe product can be created from the matrix. The micrographs show a dense alkali aluminosilicate microstructure resistant to extremely aggressive chemical attack can be created. This stone-making process has a composition similar to naturally occurring alkali feldspars. The best opportunities for CAFA are in specialty cement applications, such as, chemical resistant products, wastewater products, corrosion rehabilitation, refractory, and high strength concrete applications. Overall, CAFA is an excellent material for manufacturing concrete products especially because of its dense microstructure, rapid strength development, and durability.

ACKNOWLEDGMENTS

The authors are very grateful to the professionals at Eastern Shore Concrete Pipe Co. who were invaluable in applying their knowledge of dry casting to CAFA production. Special thanks to Bill Hecker, Bob Perrone, and others for their invaluable assistance. The authors would also like to thank the R.J.Lee Group for efforts on the SEM and EDS and the parties who have helped support the research including the Civil and Architectural Engineering Dept. of Drexel University, Drexel University, PQ Corporation, FMC Corporation, Environmental Protection Agency, Department of Energy, and the Ben Franklin Technology Center of Southeastern Pennsylvania.

REFERENCES

1. Helmuth, Richard, "Fly Ash in Cement and Concrete", Portland Cement Association, Skokie, IL, 1987, p. 17.

2. Tyson, S., "Ash at Work", American Coal Ash Association, Alexandria, VA, March,1996, 40 pp.

3. Majiling, J., and Roy, D. M., "The Potential of Fly Ash for Cement Manufacture," *Am.Ceramics Soc. Bullet.*, 72 (10), 1993, pp. 167-175.

4. Roy, D.M. and M.R. Silsbee,"Novel Cements", *The Construction Specifier*, Volume 47, Number 12, 1994, p. 66.

5. Gravitt, B., R. Heitzmann, J. Sawyer, "Hydraulic Cementand Compositions employing same", U. S. Patent # 4,997,484, Issued March 5, 1991.

6. Roy, A., P. Schilling, H. Eaton, "Alkali Activated Class C Fly Ash Cement", U. S. Patent # 5,435,843, Issued July 25, 1995.

7. Shi, Caijin, "Early Microstructure Development of Activated Lime-Fly Ash Pastes", *Cement and Concrete Research*, Vol. 26, No. 9, 1996, pp. 1351-1359.

8. Freidin, K, Erell, E., "Bricks Made of Coal Fly Ash and Slag, Cured in the open Air", *Cement and Concrete Composites*, 1995.

9. Bijen, J., and H. Waltje, "Alkali Activated Slag-Fly Ash Cements", *Fly Ash, Silica Fume, Slag, and Natural Pozzolans in Concrete*, 1989 Trondheim Conference, Vol. 2, 1989, pp. 835-842.

10. Shi, Caijin,"Strength, Pore Structure and Permeability of Alkali-Activated Slag Mortars", *Cement and Concrete Research*, Vol. 26, No. 12, 1996, pp. 1789-1799.

11. Kutti, T., "Alkali Activated Slag Mortar", *Chalmers Tekniska Hogskola*, Gotenborg, Sweden, Vol. 12, No. 35, 1990, pp. 553-562.

12. Davidovits, Joseph, " Recent Progresses in concretes for nuclear and uranium waste containment", *Concrete International*, V. 16, No. 12, Dec. 1994, pp. 53-58.

13. Davidovits, Joseph, "Geopolymer cements to minimize carbond dioxide greenhouse warming", Ceram. Trans., Vol. 37, No. Cement-Based Materials: Present, Future, and Environemntal Aspects, 1993, pp. 165-182.

14. Davidovits, Joseph, M. Davidovits, "Geopolymer: Room temperature ceramic matrix for composites", Ceram. Eng. Sci. Proc., Vol. 9, No. 7-8, 1988,

pp. 835-41.

15. Iler, Ralph K, <u>The chemistry of silica : solubility, polymerization, colloid and surface properties, and biochemistry</u>, Wiley, New York, NY, 1979, pp.856.

16. Goodspeed, C., Vanikar, S., and Cook, P., "High - Performance Concrete (HPC) Defined for Highway Structures", Federal Highway Administration, Washington D.C., Version 1, 1995, 16 pp.

17. Zia, P., Leming, M.L., Ahmad, S.H., "High Performance Concretes: A State-of-the-Art Report," SHRP-C/FR-91-103, Federal Highway Administration, Washington D.C., 1991.

18. Mokhtarzadeh, A., Kriesel, R., French, C., and Suyder, M., Mechanical Properties and Durability of High-Strength Concrete for Prestresses Bridge Girders", Transportation Research Record, Minneapolis, MN, 1995, 4 pp.

19. Streeter, D., "Developing High Performance Concrete Mix for New York State Bridge Decks", Transportation Research Record, Albany, NY, 1996, 5 pp.

20. Silverstrim, T., H. Rostami, J. Larralde, A. Samadi, "Chemically Activated Fly Ash Cementitious Material", U.S. Patent #5,601,643, Issued February 11, 1997.

21. Rostami, H. and Silverstrim, T. "Chemically Activated Fly Ash (CAFA); A New Type of Fly Ash Based Cement'" Proceedings of the Thirteenth Annual International Pittsburgh Coal Conference, Pittsburgh, PA, 1996, pp. 1074-1079.

22. Rostami, H. and Silverstrim, T. "Chemically Activated Fly Ash Concrete", Proceedings of the Twelfth International Symposium on Use and Management of Coal Combustion By Product (CCBs), Orlando, FL, Vol. 2, No. 53, 1997, pp. 1-10.

23. Silverstrim, T., Rostami, H., Clark, B. and Martin, J., "Microstructure and Properties of Chemically Activated Fly Ash Concrete", Presented to Nineteenth International Cement Microscopy Association, Cincinnati, OH, March, 1997, 18 pp.

24. Durning, T.A. and M.C. Hicks, "Using Microsilica to Increase Concrete's Resistance to Aggressive Chemicals", Concrete International, March, 1991, pp. 42-48.

SHRINKAGE CRACKING IN HIGH PERFORMANCE CONCRETE

Surendra P. Shah
Walter P. Murphy Professor Of Civil Engineering
Director Of The NSF Center For Advanced Cement Based Materials
Northwestern University
Evanston, Illinois U.S.A

W. Jason Weiss
Research Assistant
Northwestern University
Evanston, Illinois U.S.A

Wei Yang
Research Associate
Northwestern University
Evanston, Illinois U.S.A

ABSTRACT

Early age cracking often occurs in concrete when volumetric changes caused by water loss (drying shrinkage) are prevented. A long term investigation of early age shrinkage cracking is currently being conducted at the National Science Foundation Center for Advanced Cement Based Materials. The goals of this study include: determining the parameters which influence shrinkage cracking, investigating material improvements which prevent or limit shrinkage cracking, and developing computer models capable of predicting the age of first cracking. This paper summarizes the results of several recent investigations focused on understanding and characterizing early age cracking in high performance concrete. Experimental trends are presented for several material compositions to describe the role of various percentages of silica fume and a newly developed shrinkage reducing admixture on mechanical properties. A model is presented which predicts the age of first cracking through the use of fracture mechanics concepts in conjunction with coupling the effects of shrinkage stress and creep relaxation. Finally, a comparison is made between the theoretical predictions and experimental observations.

INTRODUCTION

In addition to higher strength, the potential for increased durability has fueled the use of higher performance concretes. Pozzolanic materials and high-range water reducing agents have been used in the production of these high performance materials. Despite increased strength, these concretes demonstrate an increased sensitivity to early age cracking which may limit more widespread use.[1] In addition to being unsightly, cracks weaken concrete structures and provide a means for water and salt ingress, thereby accelerating premature deterioration. Shrinkage cracking is a major concern in large flat structures including highway pavements, bridge decks, industrial floors, and parking garages due to the large surface area which is exposed to drying.

One common method used in comparing the shrinkage of different concrete mixes is to measure the change in length of a concrete prism which is free to shrink as water is lost to the environment. While free shrinkage is useful in comparing mixes of different compositions, it can not by itself predict if concrete will crack in service. In service, restraint arising from rigid inclusions, reinforcement, specimen thickness, or subgrade friction often limits the volumetric changes. This restraint results in the development of tensile stresses which, if high enough, can lead to cracking. The prediction of shrinkage cracking due to restraint is a complex phenomenon dependent on the interaction of several factors including: free shrinkage, creep relaxation, material stiffness, fracture resistance, environmental conditions, time dependence, and degree of restraint.

This paper will summarize findings from several recent investigations concerned with assessing the influence of silica fume[1,2] and a newly developed shrinkage reducing admixture[3,4] on early age cracking. Experimental results will be presented to illustrate the effect of mix proportions on compressive strength, free shrinkage, creep, and brittleness. In addition, the potential for restrained shrinkage cracking is assessed using the ring test.[5,6] Finally, a computer automated fracture mechanics model is described which couples creep and shrinkage stresses to predict the age of first cracking.

INFLUENCE OF SILICA FUME

Recently, the addition of silica fume has been used to produce concrete of increased strength and durability. Silica fume increases the strength of concrete due to pozzolanic reactions and increased particle packing density[7]. In addition to higher strength, the addition of silica fume to concrete has been proposed as one method to substantially reduce the permeability of concrete due to reduced pore size[8]. Despite the decrease water to binder ratios, it has been proposed that silica fume concretes may show increased free shrinkage due to pore refinement[7].

Mix Proportions
The mix proportions used for investigating the influence of silica fume on restrained cracking are provided in Table 1. A typical micro-concrete was used throughout this study in which the aggregates consisted of a 9 mm coarse aggregate and a 3 mm natural river sand. To eliminate the influence of aggregate volume, the paste volume was maintained constant (33%) in each mix. All

specimens were cured at 20°C and 50% R.H. after demolding. Restrained shrinkage (ring) specimens were demolded after 6 hours, while specimens used to measure mechanical properties and free shrinkage were demolded at 24 hours. As anticipated, the addition of silica fume and the reduction of the water to binder ratio resulted in a substantial increase in compressive strength (Table 1).

Table 1 - Mix Proportions For Silica Fume Comparison [1]

	NSC	HSC-SF-10%	HSC-SF-15%
Water/Binder	0.40	0.32	0.29
Densified Silica Fume/Cement	0.00	0.10	0.15
Fine Agg./Binder	2.00	1.82	1.74
Coarse Agg./Binder	2.00	1.82	1.74
Plasticizer/Binder	0.000	0.016	0.020
Paste Volume/Total Volume	0.33	0.33	0.33
Aggregate Volume/Total Volume	0.67	0.67	0.67
28 Day Compressive Strength (MPa)	53.4	74.6	86.5

Free Shrinkage and Water Loss

The influence of silica fume on free shrinkage was measured using 100 mm x 100 mm x 400 mm prisms in which the deformation of a 250 mm gage length was measured. In addition, weight changes due to water loss were measured for the silica fume specimens. Figure 2 illustrates that while minor differences in free shrinkage were measured (Fig. 2a), a significant increase in weight loss was noticed with a higher proportion of silica fume (Fig. 2b). These results appear to demonstrate that free shrinkage is not directly proportional to water loss.

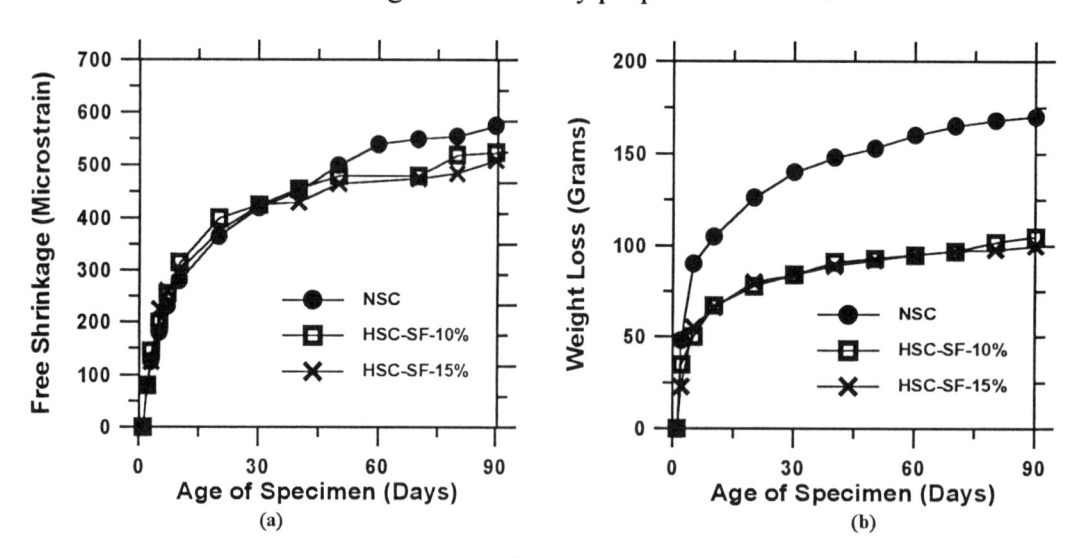

Figure 2. The Effect of Silica Fume[1] on (a) Shrinkage and (b) Weight Loss

Creep

The development of tensile stress in young concrete can be significantly reduced at early ages by creep relaxation[9]. To determine the role of silica fume on early age creep, specimens were loaded on the third day to 40% of the compressive strength. Figure 3 provides representative results of these creep tests in which a significant reduction in specific creep (creep per unit stress) can be observed with the addition of silica fume.

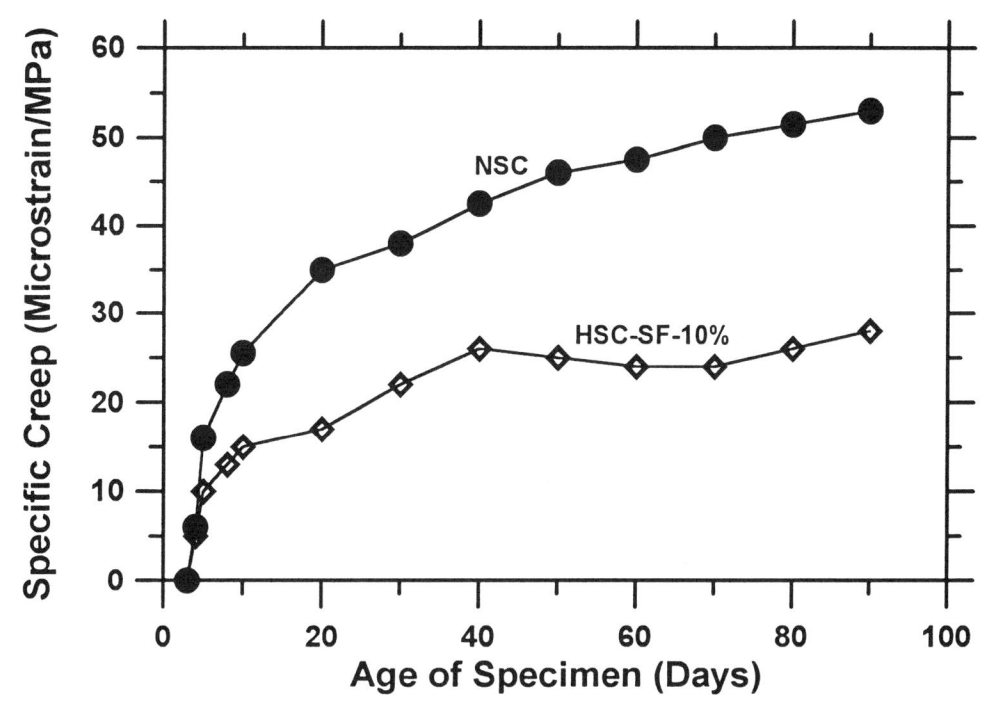

Figure 3. The Effect of Silica Fume on Specific Creep[1]

Fracture Parameters/Brittleness

As the strength of concrete increases, the failure becomes more brittle. Fracture mechanics provides one method which can be used to quantify the brittleness of a specimen. Concrete is classified as a quasi-brittle material because it exhibits a significant amount of prepeak stable crack growth (Δa). A lower prepeak crack extension is characteristic more brittle material. For example, a classically brittle material will exhibit no crack growth prior to the peak load at which time the crack will propagate in an unstable manner. Using the Two-Parameter Fracture Model[10], John and Shah[11] have shown the increase of compressive strength to be accompanied by a decrease in the effective crack ratio (Figure 4a*). These results were confirmed in this test as shown in Figure 4b** which shows a consistently

---

*Effective Crack Ratio $= \dfrac{a_c - a_o}{D - a_o}$ (%)

**Effective Crack Extension $= a_c - a_o$

lower effective crack length in the higher strength specimen. This increase in brittleness may contribute to earlier ages of cracking in higher strength concrete.

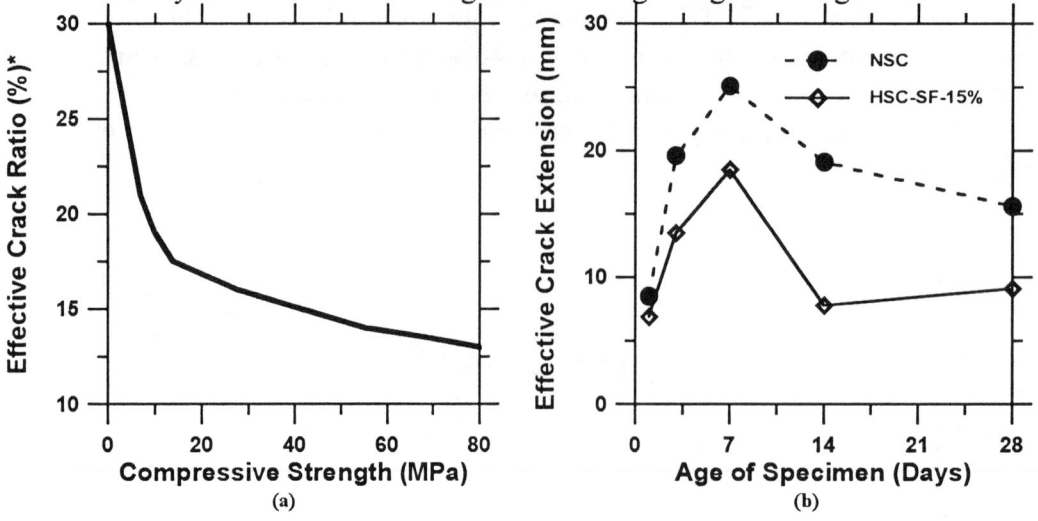

Figure 4. The Effect of Increasing Compressive Strength on Brittleness [11,2]

Restrained Shrinkage Cracking

The restrained ring test has been proposed as one method to directly evaluate the susceptibility of a concrete to shrinkage cracking when it is restrained[5,6]. In this test concrete is cast around a rigid steel ring. As the concrete dries it attempts to shrink, but the inner ring restrains this movement resulting in the development of tensile stresses. If these tensile stresses exceed the materials resistance capability, cracking results. Despite substantial increases in compressive strength (Table 1), high strength concrete is more susceptible to early age shrinkage cracking as shown in Table 2. This increase in early age cracking can be attributed to the combination of several factors in higher strength concrete including: higher initial rates of shrinkage, higher overall shrinkage, lower creep relaxation, higher material stiffness, and increased brittleness.

Table 2 - Results of Restrained Ring Testing [1]

	NSC	HSC-SF-10%	HSC-SF-15%
28 Day Compressive Strength (MPa)	55.3	74.6	87.2
Age of First Cracking (days)	10	8	6
Crack Width at 50 Days (mm)	0.45	0.60	0.90

INFLUENCE OF SHRINKAGE REDUCING ADMIXTURES

One method which has recently been investigated to reduce shrinkage includes the use of a non-expansive organic shrinkage reducing admixture (SRA)[3,12]. It is now thought that shrinkage occurs in part due to a meniscus which is formed as water is removed from the rigid C-S-H skeleton. This meniscus creates forces which compress the C-S-H skeleton resulting in a volumetric change. Balogh[13] has proposed the use of SRA to reduce shrinkage by altering the surface tension of water, which reduces the force generated by the meniscus. Reducing free shrinkage results in a reduction in the tensile stress which is developed in the restrained specimen.

Mix Proportions

The mix proportions for the micro-concrete used in studying the effectiveness of the SRA are provided in Table 3. Specimens were demolded at 24 hours and stored in an environmental chamber at 40% R.H. and 30°C. The addition of the shrinkage reducing admixture exhibited no significant difference in stiffness or brittleness as shown in Figure 5, however, a slight retardation in strength development and a 16% reduction in 28 day strength was observed (Table 3).

Table 3 - Mix Proportions For Shrinkage Reducing Admixture Comparison[4]

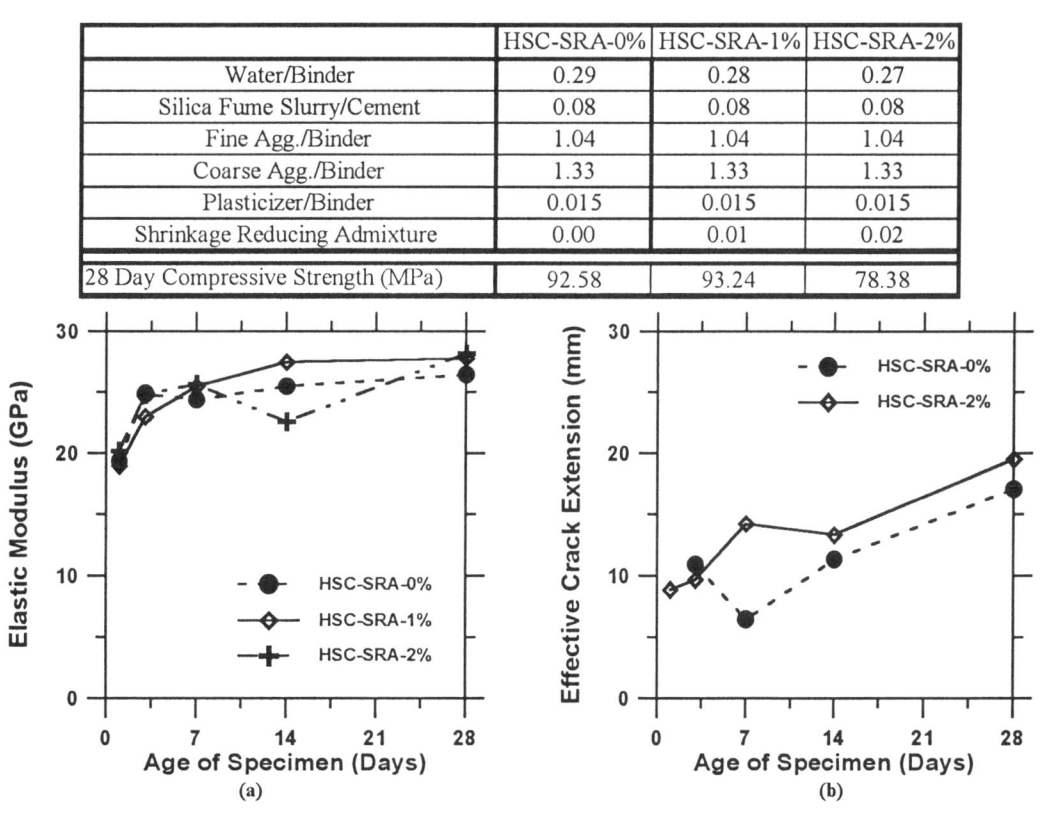

	HSC-SRA-0%	HSC-SRA-1%	HSC-SRA-2%
Water/Binder	0.29	0.28	0.27
Silica Fume Slurry/Cement	0.08	0.08	0.08
Fine Agg./Binder	1.04	1.04	1.04
Coarse Agg./Binder	1.33	1.33	1.33
Plasticizer/Binder	0.015	0.015	0.015
Shrinkage Reducing Admixture	0.00	0.01	0.02
28 Day Compressive Strength (MPa)	92.58	93.24	78.38

Figure 5: The Effect of Shrinkage Reducing Admixture[4] on (a) Elastic Modulus and (b) Effective Crack Extension

Free Shrinkage
Figure 6 illustrates that a significant reduction in free shrinkage can be observed with the addition of the shrinkage reducing admixture. It is also interesting to note that the addition of 2% SRA results in a lower initial rate of shrinkage which may play a significant role in reducing the potential of early age cracking.

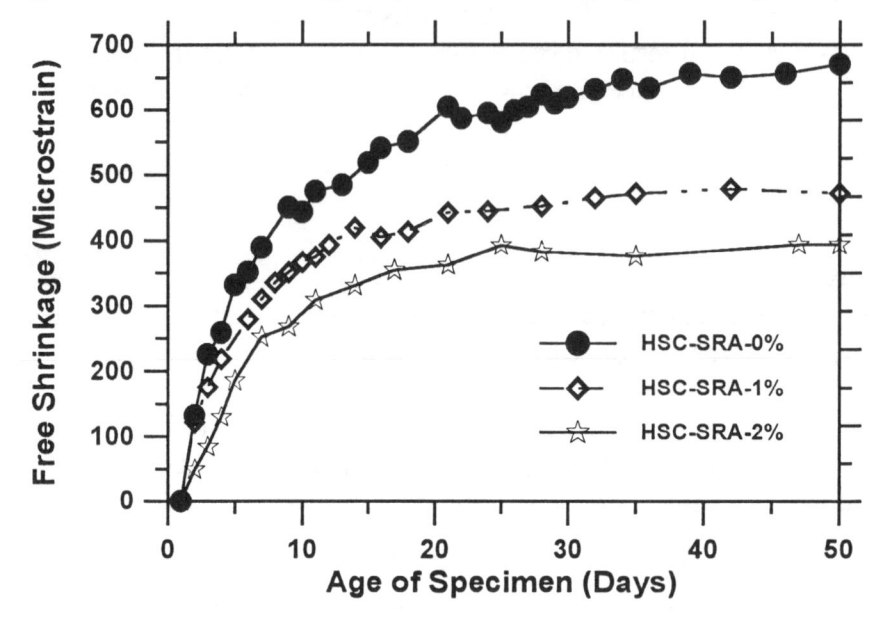

Figure 6. Comparison of Free Shrinkage With The Addition of A Shrinkage Reducing Admixture (SRA)[4]

Restrained Shrinkage Cracking
Again, the ring test was used to determine the effect of restraint on concrete cracking. Using the results from three specimens, it was observed that the age of first cracking can be extended from 3.2 days (HSC-SRA-0%) to 11.7 days (HSC-SRA-2%) with the addition of 2% SRA by weight of cement. A 1% addition of SRA showed little improvement in the age of first cracking (4.6) which may be attributed to the fact that little difference in the rate of free shrinkage exists at very early ages (1 to 4 days). By delaying the age of cracking the average crack opening at 50 days was reduced from 1.21 mm (HSC-SRA-0%) to 0.56 mm (HSC-SRA-2%).

THEORETICAL MODELING

Predicting early age shrinkage cracking is complicated due to time dependent material behavior and coupled creep and shrinkage effects. Recently Shah et al.[2] have outlined the use of a fracture mechanics model which incorporates creep relaxation, shrinkage stress development, and the time dependent development of material properties. This paper provides a brief conceptual overview of this model while more details can be found in literature. [2,14]

Stress Analysis

A stress analysis subroutine has been developed to compute the stress generated in a specimen caused by coupling creep and shrinkage. The differential strain ($d\varepsilon(t,\xi)$) can be expressed as a function of time (ξ) and age of loading (t) in the equation:

$$d\varepsilon(t,\xi) = d\varepsilon_\sigma(\xi) + d\varepsilon_c(t,\xi) + d\varepsilon_s(\xi) \tag{1}$$

where $d\varepsilon_\sigma(\xi)$ is the differential elastic strain at loading, $d\varepsilon_c(t,\xi)$ is the creep strain at time $t > \xi$, and $d\varepsilon_s(\xi)$ is the differential shrinkage strain. Expressing these differential strains as a function of time, age of loading, elastic modulus, and creep, the following equation can be obtained for total strain, $\varepsilon(t)$:

$$\varepsilon(t) = \int_0^t \left[\left(\frac{1}{E_\sigma(\xi)} + \frac{1}{E_C} \phi(t,\xi) \right) \frac{d\sigma(\xi)}{d\xi} + \alpha(\xi) \right] d\xi \tag{2}$$

in which $E_\sigma(\xi)$ is the tangent modulus, E_c is the 28 day modulus, $\phi(t,\xi)$ is a creep coefficient, and $\alpha(\xi)$ is the shrinkage rate at time ξ. Elastic modulus is determined from experiments while the CEB-FIP creep model[15] is used to define the remaining material properties. By setting $\varepsilon(t)$ equal to zero (definition of complete restraint) Equation 2 can be solved by discretizing the time integral and using numerical integration. Stress development as a function of time can be calculated using Equation 2 as shown in Figure 7 (HSC-SRA-0%). Figure 7 clearly shows that early age creep can significantly reduce the developed strain because creep and shrinkage are often similar in magnitude and specimens are exposed to loading at a very early ages.

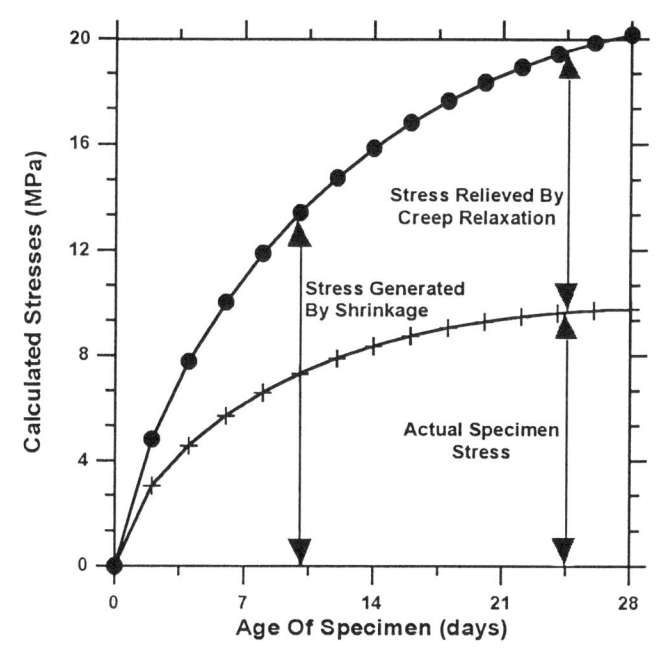

Figure 7. Predicted Stress Development Due to Creep Relaxation[4]

Failure Criteria

After calculating the specimen stress, a failure criterion can be applied to predict the age of first cracking. The failure criterion chosen for this test is based on fracture mechanics because it can be used to account for brittleness, specimen geometry, specimen size, conventional reinforcement, fiber reinforcement, and prepeak crack growth. The fracture mechanics approach chosen for this model is based on R-Curve concepts. The R-Curve (solid line-Fig. 8) represents the materials resistance as a function of crack growth (Δa). The geometry dependent R-Curve defined by Ouyang and Shah[16] was used. The G-Curve (dashed line-Fig. 8) is the rate of strain energy release with respect to crack growth. Increasing the specimens stress results in an increase in energy release which causes the G-Curve to move from position (b) to position (c). The intersection of the G and R curve gives the amount of stable crack growth (Δa) which has occurred. As this crack continues to grow, the initial slope of the G curve continues to increase until the G-curve reaches position (d) where the energy balance conditions are met. The energy balance condition exists when equations 3 are satisfied. At this point (d) the crack propagates in an unstable fashion resulting in material failure.

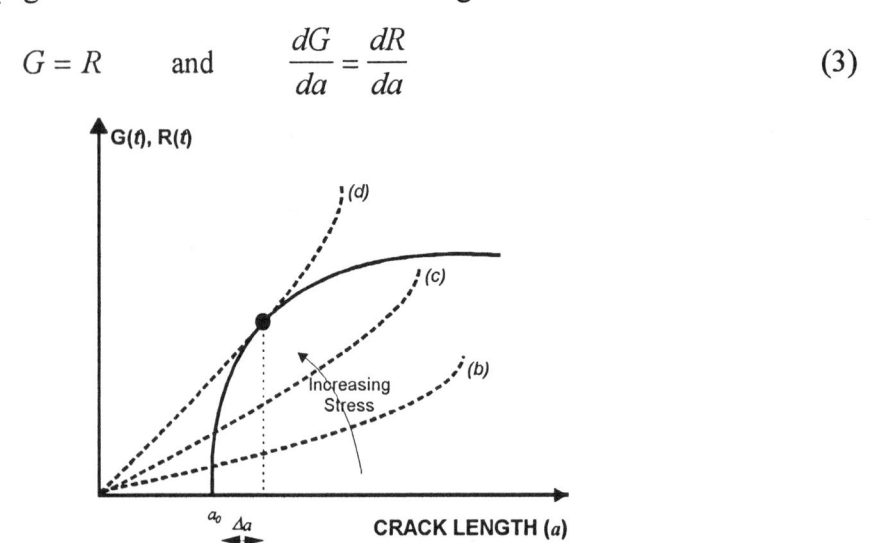

$$G = R \quad \text{and} \quad \frac{dG}{da} = \frac{dR}{da} \qquad (3)$$

Figure 8. Fracture Mechanics Failure Criteria: G and R Curves

Comparison of Theoretical and Experimental Results

Once this failure criterion is met the age of unstable crack propagation (age of first cracking) is output. Results obtained from this model are presented in Figure 9. From this figure it can be seen that a favorable correlation exists between experimental observations and theoretical predictions. The computer model is consistent in predicting the observed trend of higher strength materials exhibiting earlier ages of first cracking. In addition, the model predicts a significant increase in the age at first cracking with the use of the SRA as was evidenced in the experiments.

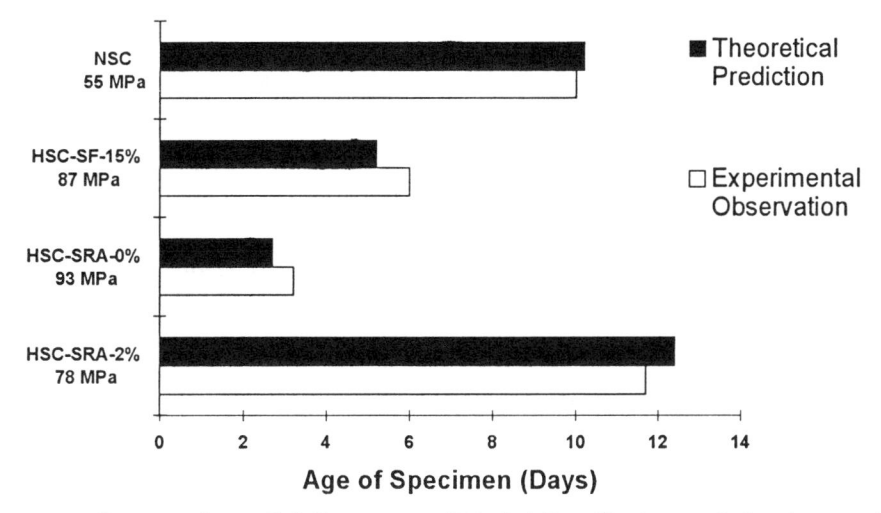

Figure 9. Comparison Of Computer Model Predictions of the Age at First Cracking vs. Experimental Observations For Restrained Ring Specimens [2,4]

CONCLUSIONS

While a long term investigation is still under way, several conclusions have been obtained from these studies including:

1. Despite increases in early age and overall strength, high performance concrete is more susceptible to early age cracking than normal strength concrete.

2. Specific creep is significantly reduced with increasing silica fume content.

3. The use of 2% shrinkage reducing admixture (SRA) by weight of cement results in a 42% reduction in free shrinkage at 50 days.

4. Delaying shrinkage cracking results in a significant decrease in crack opening.

5. Favorable comparison is observed between experimental observations and the values predicted using the fracture mechanics based model.

6. Due to the significance of creep relaxation in determining specimen stress, experimental data is necessary to confirm theoretical creep predictions.

ACKNOWLEDGMENTS

The support of Arco Chemical Company, W.R. Grace and Company, and the National Science Foundation Center for Advanced Cement Based Materials is gratefully acknowledged.

CONVERSION FACTORS

0.227 Pounds Force (lb) = 1 Newton (N)
0.006895 Pound Per Square Inch (psi) = 1 Megapascal (MPa)
0.03937 Inch (in) = 1 Millimeter (mm)
0.0022 Pounds Weight (lb) = 1 Gram

REFERENCES

1. Weigreink, K., Marinkunte, S., and Shah, S. P. " Shrinkage Cracking of High Strength Concrete," ACI Materials Journal, Sept-Oct, Vol. 93, No. 5, 1996, pp. 409-415

2. Shah, S. P., Ouyang, C., Marikunte, S., Yang, W., and Becq-Giraudon, E. (in press) "A Fracture Mechanics Model for Shrinkage Cracking of Restrained Concrete Ring", ACI Materials Journal, to appear.

3. Shah, S. P., Karaguler, M. E., and Sarigaphuti, M., " Effects of Shrinkage Reducing Admixture on Restrained Shrinkage Cracking of Concrete", ACI Materials Journal, Vol. 89, No. 3, May-June 1992, pp. 88-90.

4. Weiss, W. J., "Shrinkage Cracking in Restrained Concrete Slabs: Test methods, material Compositions, Shrinkage Reducing Admixtures, and Theoretical Modeling", MS Thesis, Northwestern University, Evanston, IL, USA

5. Grysbowski, M., and Shah, S. P., "Shrinkage Cracking of Fiber Reinforced Concrete", ACI Materials Journal, March-April, Vol. 87, No. 2, 1990, pp. 138-148.

6. Carlson, R., W., and Reading, T., J., "Model of Studying Shrinkage Cracking In Concrete Building Walls", ACI Structures Journal, July/Aug, Vol. 85, No. 4, 1988, pp. 395-404.

7. de Larrad, F., Acker, P., and Le Roy, R. "Chapter 3- Shrinkage Creep and thermal Properties" in High Performance Concrete and Applications, ed. Shah, S.P., and Ahmad, S.H., Edward Arnold, Great Britain, © 1994, pp. 65-114

8. Hooton, R. D., "Influence of Silica Fume Replacement on Physical Properties and Resistance to Sulfate Attack, Freezing and Thawing, and Alkali-Silica Reactivity", ACI Materials Journal, Vol. 90, No. 2, March-April 1993, pp. 143-151.

9. Kovler, K., "Testing System for Determining the Mechanical Behavior of Early Age Concrete Under Restrained and Free Uniaxial Shrinkage", Materials and Structures, Vol. 27, No. 10, pp. 324-330.

10. Jenq, Y. S., and Shah, S. P., "A Two Parameter Fracture Model for Concrete", Journal of Engineering Mechanics, Vol. 111, No. 4, 1985, pp. 1227-1241.

11. John, R., and Shah, S.P., "Fracture Mechanics Analysis of High Strength Concrete", Journal of Materials In Civil Engineering, Vol. 1, No. 4, 1989, pp. 185-198.

12. Shoya, M., and Sugita M., "Application of Special Admixture to Reduce Shrinkage Cracking of Air Dried Concrete," Hachinohe Institute of Technology, Hachinohe, Japan, 1-11.

13. Balogh, A., "New Admixture Combats Concrete Shrinkage", Concrete Construction, July 1996, pp. 546-551.

14. Yang, W., Wang, K., and Shah, S.P., "Predictions of Shrinkage Cracking Under Coupled Shrinkage and Creep Loads", Materials for the New Millenium, Proceedings of the Fourth Materials Engineering Conference, Washington, D.C., November 1996, Vol. 1, Ed. K.P. Chong, pp564-573.

15. Müller, H. S. "New Prediction Models for Creep and Shrinkage of Concrete", ACI SP 135-1, 1994, pp. 1-19

16. Ouyang, C., and Shah, S. P., "Geometry Dependent R-Curve for Quasi-Brittle Materials" Journal of the American Ceramics Society, Vol. 74, No. 11, 1991, pp. 2831-2836

High-Performance Prefabricated Silica Fume Concrete for Infrastructure

Ass. Prof. Per Fidjestøl, MSc, FACI,
Technical Manager
Elkem ASA Materials, Kristiansand, Norway

Tony Kojundic
Business Manager
Elkem Materials Inc., Pittsburgh, PA

ABSTRACT

The last ten years have been intensive in terms of infrastructure development in the Scandinavian countries. This has coincided with developments in concrete technology and improved production technology, much of it learned in the offshore concrete industry, giving exciting structures with high durability. One of the new developments in materials technology is the use of high-performance concrete (HPC) with silica fume and appropriate chemical admixtures.

There are benefits to be gained by silica fume addition in prestressed infrastructure construction. Aside from the general benefits of high strength and durability, deleterious effects of accelerated curing such as reduced chloride resistance, lower final strength, and risk of delayed ettringite formation can be reduced or eliminated.

INTRODUCTION

Silica fume was first collected in Kristiansand, Norway, in 1947. Investigations into the properties of the material and its uses began promptly. In 1976 a Norwegian standard permitted the use of silica fume in blended cement. Two years later the direct addition of silica fume into concrete was permitted by Norwegian standards. The first major placements of ready-mixed silica fume concrete in the United States were done by Norcem in 1978. The first publicly-bid American project using silica-fume concrete was done by the Corps of Engineers in late 1983 (Kinzua Dam, PA). The use of silica fume as a concrete addition is getting increasingly commonplace. Annually an estimated five million m^3 of silica fume concrete are placed on a global basis, (more than 1/3 of this in Scandinavia).

Silica fume has been used to provide concrete with very high compressive strength, with very high levels of durability, or both. In the United States it is now often used to produce concretes with reduced permeability for applications such as parking structures and bridge decks.

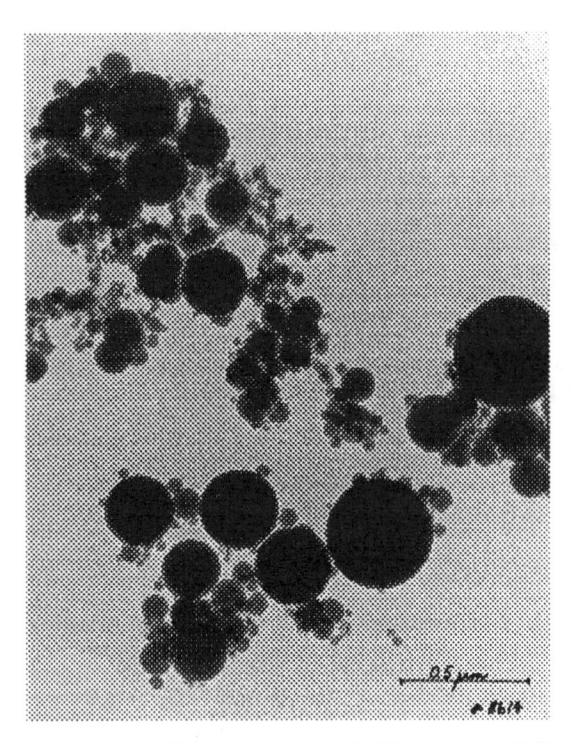

Figure 1. TEM picture of silica fume (Elkem Materials)

Concrete production technologies have been evolving. While many bridges and tunnels are built using cast-in-place techniques, increasingly prefabrication technology is utilized. Three of the largest infrastructure projects in the nineties rely heavily on prefabrication; namely the Great Belt link in Denmark, the Oresund link between Denmark and Sweden, and Northumberland bridge in Canada.

In addition, the techniques for prefabricating such various components as bridge beams, tunnel lining elements and culverts are developing. Even such seemingly commonplace products as paving blocks and curbs can be improved by high performance silica fume concrete (e.g. 125 MPa (17500 psi) paving blocks for wear resistance produced by OC Østraadt of Sandnes, Norway).

EFFECT OF SILICA FUME ON CONCRETE PROPERTIES

Silica fume Fundamentals

General Description - Silica fume (Also known as microsilica or condensed silica fume[1]) in its basic form is a gray (near white to near black) powder. The primary silica fume particle is spherical and has a mean diameter of about 0.15 microns. Figure 1 shows a transmission electron microscope (TEM) photograph of silica fume. The small particle size of the material gives a surface area of 15000-25000 m^2/kg. The primary particles will often be present in groups of several particles, fused together at the intersections. Silica fume is which satisfies the established standards for the material[2] contain more than 85 % silicon dioxide.

Other amorphous silica products are occasionally confused with silica fume. These products come in two main forms:

- Products that are manufactured, like precipitated or fumed silica. While these products offer the potential of performing well in concrete, they are typically very expensive and complicated to use.
- Product found in nature, i.e. pozzolanic materials such as diatomaceous earth, calcined clay (metakaolin), trass, rice husk ash etc.

These other pozzolanic products are different from silica fume, both in chemical and physical composition. The experience gained from research into the performance of silica fume can therefore not be credited towards the usability of these materials.

Product Forms - Silica fume is available commercially in several forms.

- <u>As-produced silica fume</u> -- Silica fume as collected is an extremely fine powder.
- <u>Slurried silica fume</u> -- Silica fume in a water-based slurry.
- <u>Densified silica fume</u> -- Dry, densified silica-fume products are also available. These products have high enough bulk density enough to be transported economically.
- <u>Pelletized silica fume</u> -- As-produced silica fume may also be pelletized by mixing the silica fume with a small amount of water and often a little cement. Typically used for off-grade or excess material. Pelletizing is not a reversible process -- the pellets are too hard to break down easily during concrete production.

General effect on fresh concrete

The small size of the silica fume particle means that the material has a large specific surface area. This surface area has great effects on the properties of fresh concrete mixes with silica fume. At a typical dosage of 8% silica fume by cement weight, between 50,000 and 100,000 microspheres are added for each cement particle, and this addition causes a increase in internal surface by a factor of 5 to 6.

The large increase in surface area gives a corresponding increase in internal surface forces with a consequent increase in the cohesiveness of the concrete. As the concrete is more cohesive, it is less susceptible to segregation than regular concrete, even in a flowing concrete, which is also useful for high fluidity grouts and pumped concrete mixes. A consequence of the cohesiveness is that a silica fume concrete will produce virtually no bleed water. This means that voids due to internal bleed pockets (under aggregate and reinforcement) are avoided, but the lack of a water layer on the surface means that the concrete is susceptible to plastic shrinkage cracking due to early drying of the surface.

General Effect On Hardened Concrete

As the concrete hardens, the chemical action of the silica fume sets in. The silica fume reacts with calcium hydroxide (from the cement hydration) to produce calcium silicate hydrates (CSH). Thus the amount of binder is increased, and the number of sites where binder is formed is widely distributed (as are the silica fume particles), which both increases the strength and reduces the permeability by densifying the matrix of the concrete. A silica fume concrete can be very homogenous (due to reduction in segregation and the many nucleation points for hydration) and very dense and have greatly improved strength and impermeability[3].

It has been found that the relatively porous interface, rich in portlandite (calcium hydroxide), that surrounds aggregate particles in normal concrete is virtually absent in silica fume concrete[4]. This improved interfacial zone means that the bond between aggregate and matrix is improved in HPC with silica fume.

USING HIGH PERFORMANCE SILICA FUME CONCRETE

For Fresh Concrete Properties

Silica fume will ease the production of concrete by allowing more fluid mixtures and more intense vibration without segregation[5]. The lack of bleed water also means that the weak surface layer sometimes caused by bleeding can be avoided.

Silica fume has been used to reduce the heat of hydration in large pours

High Strength Concrete Construction

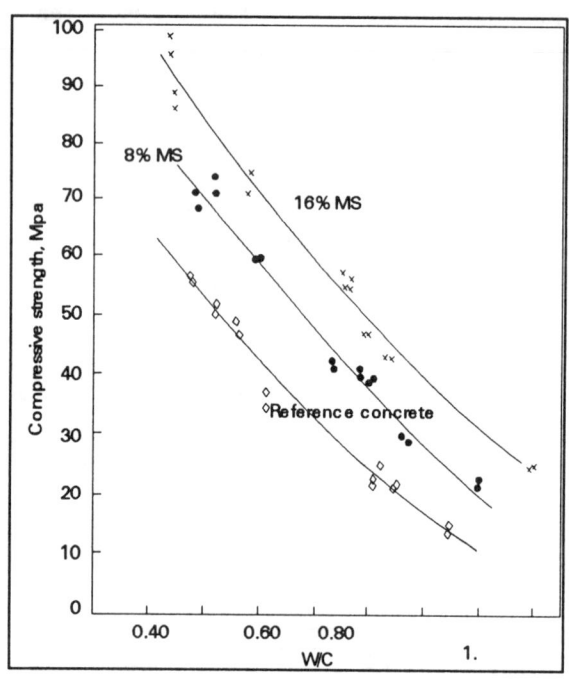

Figure 2. Effect of silica fume on compressive strength.

Many reports are available which show that silica fume added to a concrete mix will increase the strength of that mix (fig. 2)[6]. The actual strength increase will depend upon numerous factors, among others: the type of mix, type of cement, amount of silica fume, use of water reducing admixtures, aggregate properties and curing.

With proper mixture proportioning concretes of very high strengths can be produced using normal readymix facilities. In the USA 100 to 130 MPa concrete has been used in tall buildings, and silica fume concrete was recently used in the construction of the worlds tallest concrete building, Petronas Towers in Kuala Lumpur, Malaysia.

Recent infrastructure construction in Scandinavia and France have used more than 60 MPa for design strength, allowing longer spans and more slender structures together with increased durability. These structures include the impressive cable stayed bridge Pont de Normandie in France[7], the world record cantilever Varodd bridge in Kristiansand[8], Norway (260 m span), and the new candidate for this title, the 302 m span high strength lighweight concrete bridge currently under construction in

Durable Infrastructure

Abrasion Resistance -- High strength silica fume concrete shows greatly improved resistance to abrasion and erosion. Though this application is mostly overlooked in literature, abrasion resistance has become an important area of use for silica fume concrete. The large repair project on the Kinzua dam (PA), after 10 years[9], shows excellent performance of the silica fume concrete. The Norwegian practice of using steel studded tires on cars during winter means extreme wear of the road surface. In recent years, high strength silica fume concrete has become the paving material of choice for high wear-resistance. One example is a 95-MPa concrete on a 10 km stretch of highway north of Oslo, Norway. It has been found that the use of such paving reduces the wear by a factor of 5-10 compared to high-quality black-top[10].

Permeability -- An important effect of silica fume addition to concrete is reduction in the permeability of concrete. Permeability is important for two reasons: the parameter describes how rapid deleterious substances may enter the concrete and it describes how easily material can be leached from the concrete. Figure 3 from Sandvik[11] shows the effect of replacing cement by silica fume on the permeability of concrete at a constant water/cementitous materials ratio.

Life Cycle Considerations - The lifetime of a concrete structure exposed to an aggressive environment can be illustrated as shown in Figure 4: The concrete cover protects against attack for a certain time (the initiation period), then the attack progresses during the propagation period until the attack is severe enough that the serviceability of the structure is threatened. This conceptual view is especially useful for looking at corrosion of steel reinforcement.

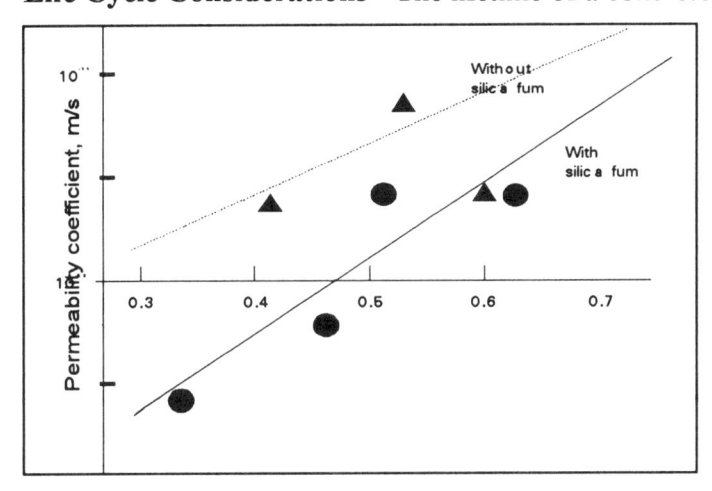

Figure 3. Permeability plotted against a modified water/cementitious materials ratio (w/(c+3SF)), which is commonly (in Norway) considered to give similar 28 day strength

Traditionally only the initiation period has been included in the lifetime, recently, however, the availability of high performance silica fume concretes with high electrical resistivity has opened the door for including a part of the propagation period in the lifetime, a very important step, especially where cracked structures are considered[13].

Corrosion Resistance

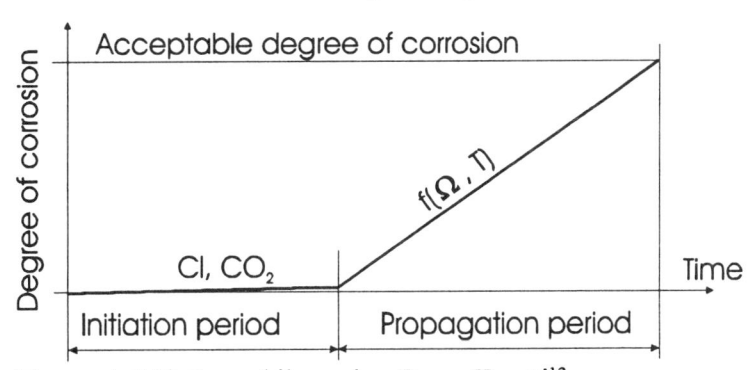

Figure 4. Lifetime philosophy. From Tuutti[12]

In high-performance concrete, corrosion initiation will only be due to chlorides. Such quality concretes should not be susceptible to carbonation.

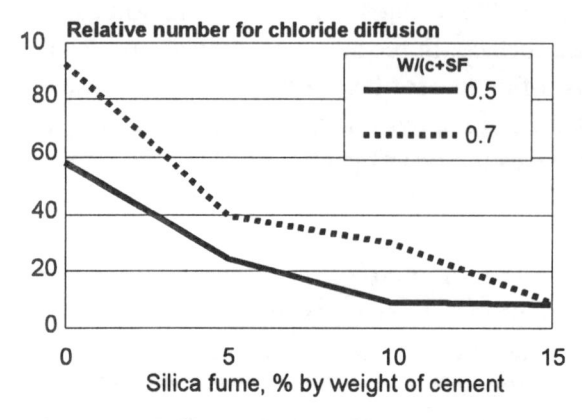

Figure 5. Effect of silica fume on chloride transport into paste.

Chloride normally enters from the surface of the concrete, and silica fume has been well documented to reduce chloride ingress. Figure 5[14] shows the result of tests on chloride ingress into cement pastes of varying silica fume contents.

Hooton & al[15] compared typical mixes, both "ordinary" and high performance with and without silica fume. An additional variable was the introduction of a curing temperature cycle similar to the procedure often used in the precast industry in North

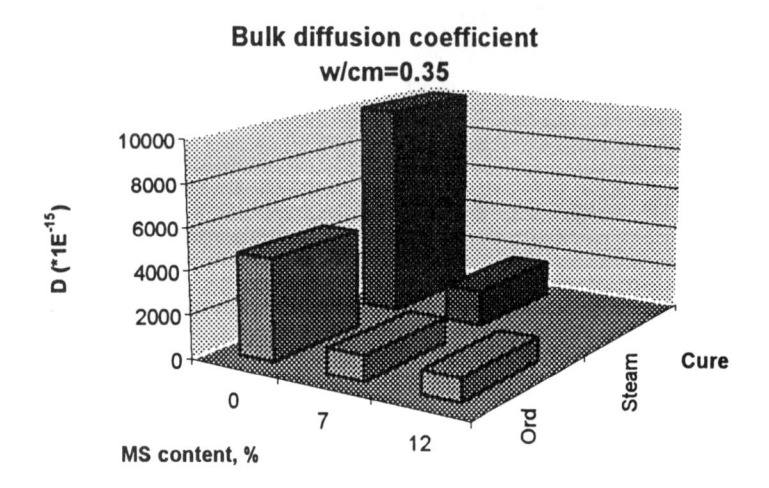

Figure 6. Chloride diffusion vs. silica fume and curing regime. w/cm=0.35

America. Figures 6 and 7 show some of the results, as determined by 90-day ponding. Not only does the use of silica fume cause a significant reduction in chloride transport in every case, but it also alleviates the detrimental effect of heat curing on chloride transport that virtually all other investigators report.

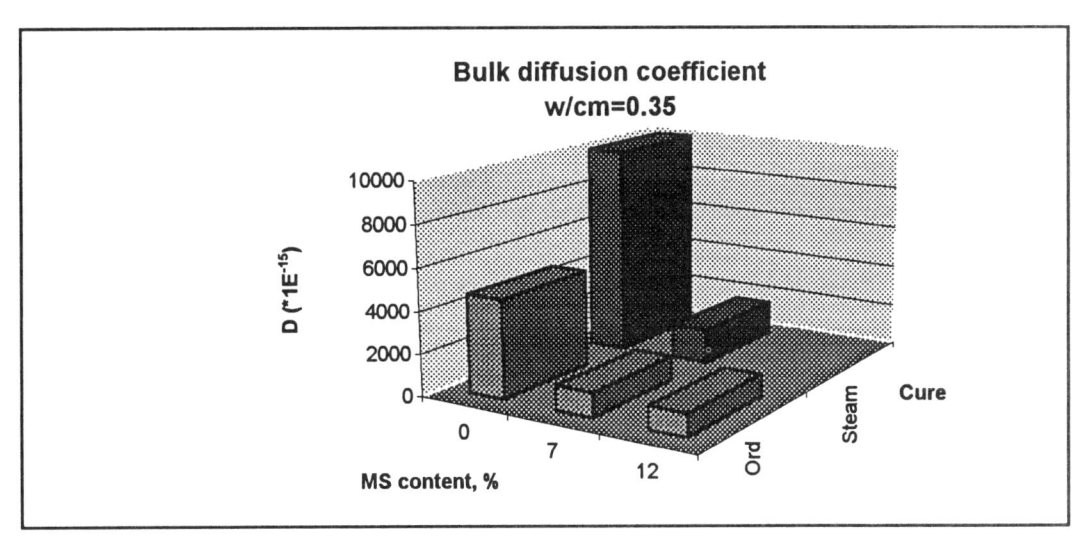

Figure 7. Silica fume and curing temperature, w/cm=0.45

Corrosion rate

Silica fume works in several ways to reduce the risk of corrosion. The improved impermeability properties of silica fume concrete means greatly reduced rate of chloride penetration in marine structures and structures exposed to deicing salts. Silica fume concrete also has very high electrical resistivity, thereby greatly diminishing the rate of corrosion, should it be initiated[16,17,18,19,20,21].

Silica fume provides a large increase in the electrical resistance of concrete. The electrical resistance of concrete is essential to determine the rate of any initiated corrosion. High resistivity in the concrete severely limits the corrosion current that can flow, and thus the corrosion rate will be small. It is reportedthat a limiting resistance on the order of 3-600 Ohm-meter is sufficient to prevent corrosion, and such resistivities are frequently exceeded by silica fume concrete, even in very wet conditions.

Frost Resistance

In studies[22] to determine the effect of silica fume addition on frost resistance, it was found that by increasing the dosage of air entrainer and adding a plasticizer it was easy to achieve the desired levels of air in the mix. In hardened concrete the spacing factor and the stability of the bubbles improved compared to the reference concrete. Very long term tests [23] show salt scaling resistance of air-entrained silica fume concrete to be similar to that of ordinary concrete.

Reactive Aggregates

Deterioration to reaction between alkalis and silica in aggregates (ASR) in concrete requires:

a) High alkali content in the mix (normally supplied by the cement)
b) Reactive aggregates, containing reactive silica.

c) Available water.

Well dispersed silica fume reacts with the available alkalis in the fresh concrete, forming alkali silicates. This binds the alkalis which would otherwise attack reactive siliceous aggregates. Since silica fume contributes to a reduced permeability, the amount of available water is reduced. These two factors combine to reduce the susceptibility of silica fume concrete to ASR.

This aspect of silica fume concrete was one of the first to be commercially utilized on a large scale. Since 1979 all concrete in Iceland has been made with about 7% silica fume, a remedy which, together with other improvements in construction procedures, ended the previously huge ASR problems [24],[25].

Sulfate Resistance

A major study was started in Oslo during the first years of testing of silica fume in concrete. This involved submerging specimens in acidic sulfate rich groundwater leached from the alun shale in Oslo. 12 and 20 year results are available from this trial[26]. These results show that the silica fume concrete (15% MS, w/c=0.6) performed as well as the mixes with sulfate resistant cement (at w/c=0.45). This is confirmed in the unpublished data after 30 years and by laboratory tests[27],[28].

The good performance of the silica fume mixes in a sulfate environment can be attributed to several factors[29], the most important are:

- the refined pore structure and thus the reduced passage of harmful ions[30], and

- the lower calcium hydroxide content which leads to reduced formation of gypsum (and consequently of ettringite).

Delayed Ettringite Formation

Results[31] from recent research confirm that initial curing of concrete at temperatures higher than 60 degrees C can lead to a volume expansion in specimens stored in water. This expansion is caused by delayed ettringite formation. The volume increase coincide with an extensive compressive strength reduction. Use of silica fume (5%) suppresses the expansion, and the use of low alkali cement reduces the expansion. Unfortunately, higher silica fume dosages were not tested.

PRACTICAL APPLICATIONS OF PREFABRICATED SILICA FUME CONCRETE IN INFRASTRUCTURE

Producing Silica fume Concrete

The production of silica fume concrete does not differ much from the production of conventional concrete, except for the need for adequate dispensing equipment and appropriate revisions to mixture design. The use of silica fume will often give greater flexibility in the choice of design strength. Often, durability specifications will require the use of silica fume, and the high strength then obtained can be used in actual design.

General

The first systematic use of silica fume concrete in prefabricated infrastructure construction dates back to about 1982. For several years, until a merger stopped this part of production, virtually all concrete from a major Norwegian producer contained about 5% silica fume. The components produced included: DT-elements, deep I-beams, hollow-core elements, culvert elements, railway sleepers etc. The structures built include bridges, wharves and parking structures. A survey is being established that will investigate the performance of these products.

Below, some projects that illustrate the use of silica fume concrete are presented.

Great Belt Connection

The Great Belt Link (Storebælt-forbindelsen) links the two large Danish islands of Zealand and Funen. This is a very large project, the concrete volume is estimated to about 1.060.000 m^2. Construction commenced in July 1989, and the complete project will be finished at the end of 1998. The connection consists of three main elements:

EASTERN TUNNEL: Two parallel tunnels, 8 km (5 miles) in length. The tunnels are lined with a total of about 60 000 prefabricated elements, while the annulus is filled with cement grout containing silica fume.

EASTERN BRIDGE: The multilane highway runs from the Western Bridge across Sprogø onto a suspension bridge. The total length of this bridge is about 7 km (4.4 miles). The towers for the suspension bridge and the supports for the approaches were all prefabricated in segments on shore and assembled at sea.

WESTERN BRIDGE: The Western Bridge has a total length of about 6.5 km (4 miles) and connects Funen to Sprogø. There are two parallel bridges, one for rail and one for road traffic. Both supports and bridge beam are made up from prefabricated sections, the main load-bearing system is made up from about 60 spans of 110.4 m (356 ft) each.

Table 1. Some requirements to the mixture for Storebelt

w/cm	max.	0.35 (A), 0.45 (B)
Cement	min.	300 kg/m^3
Fly ash	min.	15 % of total cementitious
Silica fume		4 - 8 % of total cementitious
FA + SF	max.	25 % of total cementitious

The project is required to maintain its serviceability with a minimum amount of maintenance and repair throughout the 21st Century. This means the strictest requirements to materials and structural design. Two basic concrete mixes have been developed, termed A and B. Type A is to be used in the most exposed situations, such as the tunnel lining:

Since this project was started, several other large scale prefabrication schemes have been started. Notable among these are Northumberland Bridge in Canada and Øresund connection between Sweden and Denmark.

Bergsøysundet Bridge

Table 2. Key Figures, Bergsøysundet

Total length	914 m
Floating length	830 m
No. of pontoons	7
Span between pontoons	105 m
Pontoon dimensions	20 x 34 x 5.8 m
Concrete volume pontoons	4800 m³ (LWC)

1m=, 1m³,=

Table 3. Mix design for the pontoons

Component	kg/m³
Cement P30-4A (ASTM Type 1)	420
Silica fume slurry	40 (20 kg dry)
Sand 0-5	700
Lightweight aggregate (Liapor)	550

1kg/ m³=

Bergsøysundet Bridge in Northwest Norway is the first floating bridge to be built with anchoring only at the end-points. The bridge uses 7 pontoons in lightweight concrete, the superstructure is steel. The bridge opened for traffic in July 1992. The bridge elements were prefabricated in Stavanger, and towed to the site for final assembly. A similar structure, also with lightweight concrete, was used for the Nordhordland bridge, connecting Bergen to the northern islands.

The pontoons were built by Norwegian Contractors.

Other structures/products

A number of prefabricated bridges using deep I beams have been built with silica fume concrete over the last 15 years. Typically these are bridges with 20-30 m spans. Concrete design strengths are 55-65 MPa, and the typical mixture design will have about 5% silica fume at a w/cm of 0.35-0.42.

Similar mixtures have been used for virtually all railroad sleeper production for the last 10-12 years, and for DT elements and rectangular beam elements for parking structures and wharves.

CONCLUSIONS

Silica fume is a mineral additive that gives great benefits to the production of high performance concrete. The benefits include high (early and final) strength, excellent chloride and corrosion resistance and resistance to external and internal (DEF) attacks. High performance silica fume concrete have been used in precast concrete structures ranging from the mundane to the very advanced.

While the benefits of silica fume concrete are relatively easy to obtain, the potential for problems related to the lack of bleed water in high performance concrete should be noted. In short, proper curing procedures are required.

REFERENCES

[1] Final Report. Siliceous By-products In Concrete. 73-sbc Rilem Committee.

[2] ASTM C1240

[3] DIAMOND S. The microstructures of cement paste in concrete. 8th International Congress on Chemistry of Cement, Rio de Janeiro, Brazil, Vol 1, pp 122-147, 1986.

[4] Bentur,A. Goldman, Cohen, MD. The contribution of the transition zone to the strength of high quality silica fume concretes. Proc Mat.research Society Symposium Boston 1987. Vol. 11,1988. **Lnr 1766**

[5] Referanse om vibreringsevne og lite segregering. Wallevik???

[6] Sellevold, E and Radjy,F.

[7] Referanse til Pont de Normandie fra Paris møte

[8] Varoddbroa fra Norsk betongdag

[9] Luther, M. D. and Halczak,W. Long-term performance of silica fume concretes in the USA exposed to abrasion-erosion or cavitation -- With 10-year results for Kinzua Dam and Los Angeles River, Fifth CANMET/ACI International Conference on Fly Ash, Slag, Silica Fume and Other Natural Pozzolans, SP-153, American Concrete Institute, Detroit, 1995, pp. 863-884.

[10] Helland,S. High strength concrete used in highway pavements. Proc. Second international symposium on utilisation of high strength concrete. W.Hester. ed. Berkeley 1990.

[11] Sandvik,M Silikabetong, Herdevarme og Egenskapsutvikling. FCB/SINTEF rapport STF65 A85041 1983

[12] Tuutti, K. Doktoravhandling

[13] Fidjestöl, P and Tuutti, K. The importance of chloride diffusion. Proc. RILEM meeting on chloride initiated corrosion in concrete. Olivier and Nilsson ed. To be published 1997.

[14] Gautefall,O.; Effect of Condensed Silica Fume on the Diffusion of Chlorides through Hardened Cement Paste. Paper# SP 91-48, Second CANMET/ACI International Conference on the Use of Fly Ash, Silica Fume, Slag and Natural Pozzolans in Concrete, Madrid 1986

[15] Hooton,RD, Pun, P, Kojundic,T and Fidjestol,P. Influence of Silica Fume on Chloride Resistance of Concrete

[16] Wolsiefer, JT Sr., Silica fume concrete: A solution to steel reinforcement corrosion in concrete, Durability of Concrete -- Second International Conference, SP-126, American Concrete Institute, Detroit, 1991, pp. 527-558

[17] Petterson, K. Chloride Threshold Value and the Corrosion Rate in Reinforced Concrete. Proc. Nordic Seminar on Corrosion of reinforcement: Field and laboratory Studies for Modelling and Service Life. Report TVBM-3064. University of Lund, 1995. K. Tuutti, ed.

[18] Fidjestøl, P and Frearson, J. High-Performance Concrete Using Blended and Triple Blended Binders. Proc. ACI international Conference on High Performance Concrete. Singapore 1994. ACI SP-149. VM Malhotra ed.

[19] Berke, NS, Dallaire, MP and Hicks, MC Plastic, mechanical, corrosion and chemical resistance properties of silica fume (microsilica) concretes, Proceedings, Fourth CANMET/ACI International Conference Fly Ash, Silica Fume, Slag and Natural Pozzolans in Concrete, Istanbul, 1992, SP-132, American Concrete Institute, Detroit, 1992.

[20] Gautefall, O and Vennesland, Ø., Elektrisk motstand og pH-nivå. Modifisert portlandcement, delrapport 5, (Electrical ristivity and pH level. Modified Portland Cement project. Part report 5) Sintef Report # STF 65 A85042, Trondheim 1985. (In Norwegian)

[21] Zhang, Min-hong and Gjørv, O E., Effect of silica fume on pore structure and chloride diffusivity of low porosity cement pastes. Cement and Concrete Research, Vol. 2,1 No. 6, 1991.

[22] Okkenhaug K. and Gjorv O.E. Influence of condensed silica fume on the air -void system in concrete. FCB/SINTEF, Norwegian Institute of Technology, Trondheim, 1982. Report STF65 A82044.

23. Fidjestøl, P. Salt-scaling resistance of silica fume concrete. Proc. Int. Symp on utilization of by-products in concrete. Milwaukee 1992. Naik ed.

[24] Asgeirsson H. and Gudmundsson G. Pozzolanic activity of silica dust. Cement and Concrete Research, 1979, 9, 249-252.

[25] Asgeirsson, H. Silica fume in cement and silane for counteracting of alkali-silica reactions in Iceland. Cement and concrete research. Vol. 16, no. 3, 1986

[26] Fiskaa O.M. Betong i Alkunskifier. Norwegian Geotechnical Institute, Oslo, 1973, Publication 101. (In Norwegian).

[27] Mather K. Factors affecting the sulphate resistance of mortars. Proc. 7th International Conference on Chemistry of Cements, 1980, 4, 580-585.

[28] Fidjestøl, P. Concrete for low sulfate concentrations. Proc. Concrete for the 90's. I.Hinczak, ed. Leura, NSW 1990.

[29] Fidjestøl, P. The benefit of microsilica-based additives in concretes exposed to aggressive environments. Proc. 4th international conference on the durability of buildings, materials and construction. Singapore- 4DBMC 1987

[30] Popovic K. et al. Improvement of mortar and concrete durability by the use of condensed silica fume. Durability of Building Materials, 1984, 2, 171-186.

[31] Rønne, M & al. Chemical stability of LWAC exposed to high hydration generated temperatures. Proceedings of the International Symposium on Structural Lightweight Aggregate Concrete, Sandefjord, Norway 1995

THE CHLORIDE PENETRATION RESISTANCE OF CONCRETE CONTAINING HIGH-REACTIVITY METAKAOLIN

R.D. Hooton (Professor, University of Toronto, Toronto, Ontario, Canada)
K. Gruber (Engelhard Corporation, Iselin, N.J., USA)
A.M. Boddy (Undergraduate, University of Toronto, Toronto, Ontario, Canada)

ABSTRACT

In this study, the chloride penetration resistance of concrete containing high-reactivity metakaolin was investigated. Metakaolin is a processed and heat treated kaolinite clay which has pozzolanic properties and high internal surface area. Six concretes were cast at water to cementitious materials ratios of 0.3 and 0.4 with 0, 8 and 12% by mass replacement of portland cement by high-reactivity metakaolin. The concretes were tested for 1) Strength, 2) Bulk Diffusion, 3) AASHTO T259 Ponding, 4) Water Sorptivity, 5) Rapid Chloride Permeability, and 6) Resistivity.

It was found that strength increased at all ages with increasing contents of metakaolin and decreasing water-to-cementitious materials ratio. The results from all of the other experimental work, except the sorptivity test, showed that higher metakaolin content and lower w/cm ratio decreased diffusion, permeability, and conductivity and increased resistivity. Overall, the conclusion from this research is that high reactivity metakaolin will substantially increase a concrete's resistance to chloride ingress.

INTRODUCTION

The term durability refers to a concrete's ability "to withstand the processes of deterioration to which it can be expected to be exposed"[1]. For example, inflowing water containing soluble chloride ions can depassivate reinforcing steel and lead to corrosion, a deterioration mechanism which can severely shorten the service life of reinforced concrete projects such as parking, bridge, and marine structures.

Research is being directed to the production of high quality impermeable concretes which inhibit salts from reaching and corroding the reinforcing steel. These often include lower water-to-cementitious materials ratio (w/cm) and use of one or more mineral admixtures and supplementary cementing materials.

One of the newest supplementary cementing materials to be developed in North America is high-reactivity metakaolin. High purity kaolin clay is treated by controlled thermal activation to drive off the water "bound in the interstices of the kaolin", so "the structure collapses, resulting in an amorphous aluminosilicate", effectively converting the material to the metakaolin phase[2]. The end result is an almost 100 percent reactive pozzolan, which will chemically combine with calcium hydroxide to form calcium silicate and calcium aluminate hydrates[3].

Because of its white color, high-reactivity metakaolin does not darken concrete as silica fume typically does, making it suitable for color matching and other architectural applications[4]. From other research it has been concluded that mixes containing high-reactivity metakaolin yield comparable performance to silica fume mixes in terms of strength, permeability, chemical resistance, and drying shrinkage resistance[3,4,5].

The objective of this study was to investigate the chloride penetration resistance provided by concrete containing high-reactivity metakaolin at various replacement rates and water-to-cementitious material ratios. This was accomplished by measuring the chloride ingress using a variety of test procedures into the concrete mixtures.

EXPERIMENTAL

Mix Design
Six air-entrained concrete mixtures were cast using 0, 8, and 12% by mass replacement of cement by High Reactivity Metakaolin (HRM) at both w/cm of 0.30 and 0.40. The high-reactivity metakaolin used was manufactured by MetaMax. The portland cement used was Type I/II low-alkali manufactured by Lafarge Canada. The fine aggregate had a density of 2700 kg/m^3, an absorption of 1.4%, and a fineness modulus of 2.56. The coarse aggregate was a 10 mm crushed limestone with a density of 2670 kg/m^3 and an absorption of 1.67%. The volumetric mix designs are summarized in Table 1 along with fresh concrete properties. All mixtures contained a Type A water-reducer, a naphthalene sulfonate-based superplasticizer and an air entraining agent. Fine aggregate contents were adjusted to maintain yield.

Casting
The mixer used was an Eirich, 150 liter flat pan mixer. The dry materials were placed in the mixer first and blended for two minutes before the water was added to ensure a homogeneous dispersion of all the materials throughout the mix. In each case, the water reducer was combined with the water and the air entraining agent was spread across the sand before these components were put into the mixer. The superplasticizer was added after some initial mixing to minimize absorption by the cement.

For each mixture, three slabs, 75 mm deep by 250 mm by 350 mm (3 x 10 x 14 inch), were cast and then compacted on a vibrating table. Eight 100 x 200 mm (4 x 8 inch) cylinders were also cast in plastic molds for compressive strength.

Curing
All of the concrete was initially covered with wet burlap and plastic for 20 hours. After demolding, the slabs were cured for 5 days in a lime water tank, and

Table 1 - Summary of Mix Designs

Mixture	AB1	AB2	AB3	AB4	AB5	AB6
w/cm	0.4	0.4	0.4	0.3	0.3	0.3
Portland Cement	100%	92%	88%	100%	92%	88%
HRM	0	8%	12%	0	8%	12%
Ingredients:						
Cement (kg/m^3)	380	350	334	460	423	405
HRM (kg/m^3)	-	30	46	-	37	55
Coarse Agg. (kg/m^3)	1100	1100	1100	1100	1100	1100
Fine Agg. (kg/m^3)	655	647	641	619	608	601
Water (kg/m^3)	152	152	152	138	138	138
25 XL Water Reducer (mL/100 kg)	325	325	325	325	325	325
SPN Superplasticizer (mL/100 kg)	400	600	800	700	900	1100
MicroAir Air Entrainer (mL/100 kg)	40	40	40	40	40	40
Fresh Properties:						
Slump (mm)	135	125	200	170	90	145
Air Content (%)	9.5	7	6	7.5	5	5
Plastic Density (kg/m^3)	2275	2353	2424	2346	2459	2473

Notes: 12.0 kg/m^3 = 1.685 pcy; 25.4 mm = 1.0 in; 1.0 ml/100 kg = 0.015 fl.oz/100 lb.

subsequently stored in an enclosure at room temperature, approximately 23°C, and 50% relative humidity. The compressive strength cylinders were immediately placed into limewater at 23°C after being removed from their molds, and were left there until their respective test times.

Sample Coring

All tests used 100 mm diameter samples, six of which were cored with a diamond drill from the formed face of each slab at 25 days of age. The cores for the water sorptivity, resistivity, bulk diffusion, and rapid chloride permeability tests were then cut with a diamond saw to a thickness of 50 mm, with the top 25 mm (closest to the finished face) discarded. The AASHTO T259 ponding test cores were left at a thickness of 75 mm. After sealing their sides, core samples were then placed in a 50% relative humidity enclosure at ambient temperature until the specific sample preparation for each test was initiated at approximately 28 days of age.

TEST PROCEDURES

Bulk Diffusion Test

The apparent diffusion coefficient for each mix was determined using the Bulk Diffusion Test[6]. Each 50 mm thick, 100 mm diameter sample was epoxy coated on all sides except the test face. The samples were then vacuum saturated as per ASTM C1202. They were sealed in plastic containers with the test face exposed to 1.0 mol/L NaCl solution at 23°C. Because the samples are initially saturated and then submerged in the NaCl solution, this test measures the chloride transport solely as a result of pure diffusion. The test does not, however, correct for chloride binding effects.

Two replicates from each mix were tested for each time period of 28 and 90 days. Upon removal from solution, the specimens were profile ground on a milling machine fitted with a 50 mm diameter diamond drill bit (Drill bit rotation = 320 rpm; Cross bed speed = 50 mm/min). The profile grinding was completed by first discarding the surface 0.5 mm of the sample, and then collecting the powdered samples from ten to twelve 0.5 mm layers at various depths to a maximum depth between 20 and 25 mm. When grinding was completed, the powdered samples were placed in a 105°C oven for 24 hours. They were then put through a nitric acid digestion process. The digestion process involved adding 35 mL of distilled water and 7 mL of nitric acid (1 part distilled water to 1 part nitric acid) to the powdered sample, stirring for 10 seconds, resting for a 4 minute reaction period, and then boiling the solution on a hot plate for one minute. Upon cooling, the samples were filtered to remove excess solids. The remaining liquid solutions were titrated using silver nitrate using an autotitrator to determine the total chloride content.

Using these results, a plot of chloride content versus depth was created to determine the coefficient of apparent diffusion, D_a. This plot, which is often referred to as a chloride concentration profile, was fit using Fick's Second Law of diffusion according to Crank's solution. These equations are listed as (1) and (2) respectively:

$$\frac{dc}{dt} = D_a \frac{d^2c}{dx^2} \quad (1)$$

$$C(x,t) = C_o \left[1 - \mathrm{erf}\left(\frac{x}{2\sqrt{D_a t}} \right) \right] \quad (2)$$

where D_a is the apparent diffusion constant (m²/s), C_o is the surface chloride concentration (% of concrete mass), $C(x,t)$ is the chloride concentration (% of concrete mass) at a certain depth, x (mm) and time, t (s), and erf() is the standard error function.

After obtaining the titration results for chloride content with depth, the chloride concentrations determined at all depths were reduced by the base or background chloride concentration. The base chloride concentration represents the chloride in the original mix ingredients. The data was then fit using Crank's solution to determine the unknown variables, D_a and C_o. This was accomplished using Jandell-Table Curve, a software program which can find the best curve fit using a non-linear, least-squares method. The best fit was determined by adjusting both of the unknowns to achieve the highest coefficient of determination, r^2, for the solution. A typical chloride profile is shown in Figure 1. It should be noted that typically the first profile value at approximately 1 mm depth is not included in the calculation of C_o and D_a. It is affected by the higher volume of paste at the surface.

Sample AB3R   w/cm=0.4   88%OPC   12%Metakaolin

1.0mol/L NaCl @ 23°C for 28 days

$D_a = 0.5863 \times 10^{-11}$ m²/s $C_o = 0.58\%$ $r^2=0.9974$

Figure 1 - Typical 28 day Bulk Diffusion Test Chloride Profile

It must be noted that Fick's second law does not perfectly represent the diffusion process that occurs, since it does not include the effects of chloride binding, or how D_a will change with time. Still, it has been found that the equation and the resulting D_a

and C_o values reflect the measured chloride profiles very well, and hence, allow for a solid comparison to be made of the diffusion properties of various concrete mixes[6].

Modified AASHTO T259 90 Day Pond Test
Two samples from each concrete mix were also tested using a modified version of the AASHTO T259 90 day pond test. After coring, the samples were stored in a 50% relative humidity, 23° enclosure. At 28 days of age, a bituthene rubber membrane was wrapped around the circumference of the 100 mm diameter, 75 mm high cores with a 20 mm dam of this membrane left above the formed face of the samples and sealed with silicone caulking. Then a 15 mm depth of 3% NaCl solution was ponded on the tops (formed faces) of the samples. This upper "pond cell" was then covered with plastic wrap to minimize evaporation of the solution. The samples were subsequently placed back on the wire rack in the 50% relative humidity enclosure for 90 days.

After the 90 days of ponding, the samples were profile ground in a similar manner to the bulk diffusion test samples; however, powdered samples were taken to a maximum depth of approximately 19 mm for the AASHTO T259 specimens.

It should be noted that the samples were ground in 0.5 mm layers, rather than the 12 mm standard thickness specified by AASHTO since using 12 mm layers gives an unrepresentative and crude depiction of the actual penetration profile[6].

The "diffusion" coefficient for this test was termed the combined coefficient (D_c) since the test measures the combined effects of initial sorption and wicking (toward the 50% relative humidity lower face), as well as pure diffusion.

Water Sorptivity Test
The property defined as the rate of penetration of water into unsaturated concrete is the sorptivity[7]. This parameter (S) is defined as the slope of the line relating the one-dimensional absorption of water to the square root of time as shown in Figure 2. The relationship can be expressed in equation form as:

$$i = S \times t^{1/2} + k \qquad (3)$$

where i is the volume of absorbed water per unit area of suction surface (mm^3/mm^2 = mm), t is the elapsed absorption time (min), S is the rate of absorption or sorptivity ($mm/min^{1/2}$) and k is a constant.

After sealing the sides of the disk, a conditioning regime consisting of 3 days in a 50°C oven, followed by 4 more days at 50°C but sealed in a polyethylene food storage container. This procedure was found to condition samples to about a 1% moisture content, similar to many field exposures.

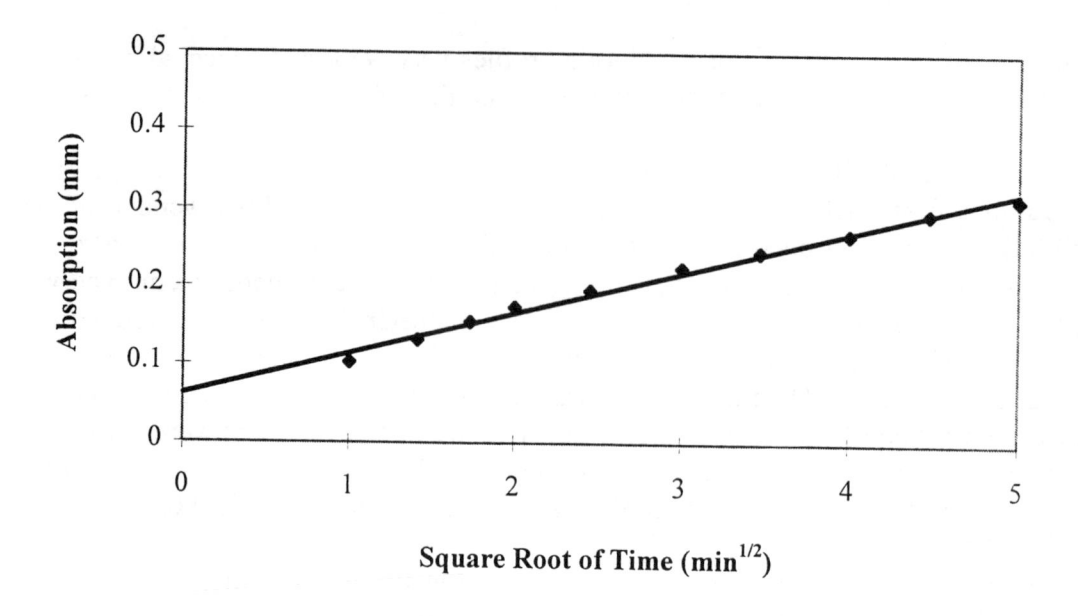

Figure 2 - Sorptivity Relationship - Absorption vs. Square Root of Time

At 28 days of age, each specimen's initial mass was recorded then the test face was exposed to water. After each of 1, 2, 3, 4, 6, 9, 12, 16, 20, and 25 minutes, the core's mass was measured.

Resistivity and Rapid Chloride Permeability Tests

The Rapid Chloride Permeability Test (RCPT) test procedure was applied at 28 days of age to 2 replicates from each mix and followed the standard, as per ASTM C1202 (similar to AASHTO T277).

After vacuum saturation, the resistivity was measured first. The resistivity test applied to the concrete samples were based on the DC potential approach developed by Monfore[8]. The DC potential was switched between 4 and 6 volts every 30 seconds for 15 minutes and voltage readings were taken each time across the specimen and across the resistor. This alternating pattern was employed to eliminate the polarization effects that are caused by the passage of a direct current.

The (RCPT) was applied to the specimens immediately after completion of the resistivity test. In addition to measuring the total charge passed after 6 h, the charge passed after 30 minutes was recorded and multiplied by 12 to obtain a 6 h equivalent value[6]. Since one of the criticisms of this test is the potential heating of solutions when high currents develop, the extrapolation to 6 hours using the 30 minute value practically eliminated the concern.

RESULTS AND DISCUSSION

The average compressive strengths obtained for each mixture are presented in Table 2.

It is evident from the results presented in Table 2 that an increased content of high-reactivity metakaolin greatly improves strengths at all ages. The increase in strength caused by 8% replacement was larger than that achieved by reduction of w/cm from 0.40 to 0.30.

Bulk Diffusion (D_a) values are shown in Table 3, and Figures 3 and 4, for both 28 and 90 day ponding periods. All of the 90 day values are lower than the 28 day values as would be expected since diffusion coefficients decrease with age.

Table 2 - Average Compressive Strengths

Mixture	Average Compressive Strength (MPa)		
	7 day	28 day	56 day
1 - 0.4 w/cm, 0% HRM	23.7	31.8	36.9
2 - 0.4 w/cm, 8% HRM	52.4	63.3	61.2
3 - 0.4 w/cm, 12% HRM	60.3	66.9	70.4
4 - 0.3 w/cm, 0% HRM	48.1	59.2	62.6
5 - 0.3 w/cm, 8% HRM	67.1	77.5	79.8
6 - 0.3 w/cm, 12% HRM	78.5	83.7	85.9

Table 3 - Bulk Diffusion and T259 Test Results - Apparent and Combined Diffusion Coefficients and Surface Concentrations

Ponding Time Span	D_a (m^2/s x 10^{-12}), C_o (% of concrete mass)				D_c (m^2/s x 10^{-12}), C_o (% of concrete mass)	
	28 days		90 days		T259 - 90 days	
Mixture	D_a	C_o	D_a	C_o	D_a	C_o
1 - 0.4 w/cm, 0% HRM	16.71	0.32	11.31	0.33	6.10	0.27
2 - 0.4 w/cm, 8% HRM	8.25	0.69	5.73	0.58	4.08	0.39
3 - 0.4 w/cm, 12% HRM	5.86	0.58	3.52	0.53	2.97	0.32
4 - 0.3 w/cm, 0% HRM	9.64	0.42	5.67	0.44	4.99	0.40
5 - 0.3 w/cm, 8% HRM	3.76	0.52	2.88	0.49	2.06	0.46
6 - 0.3 w/cm, 12% HRM	2.95	0.50	2.76	0.48	1.58	0.41

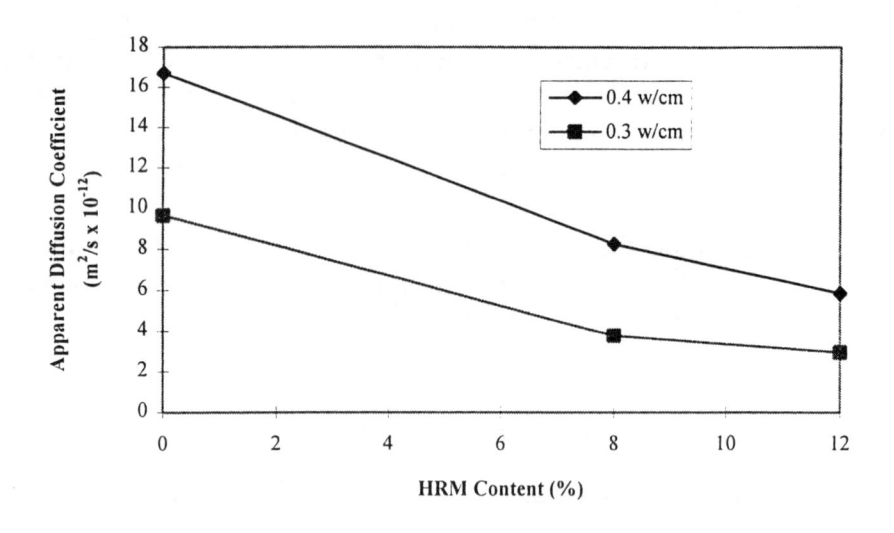

Figure 3 - 28 Day Bulk Diffusion Results - Apparent Diffusion Coefficient vs. HRM Content

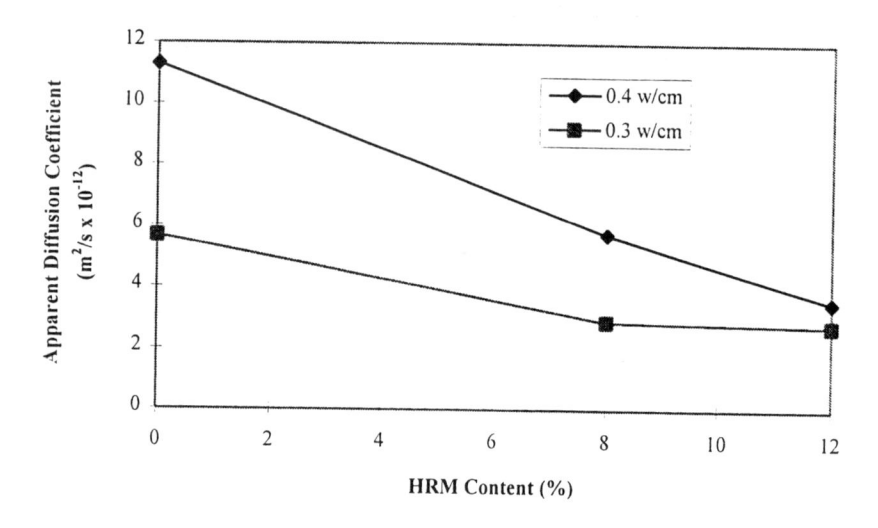

Figure 4 - 90 Day Bulk Diffusion Results - Apparent Diffusion Coefficients vs. HRM Content

Replacing 8% cement with high reactivity metakaolin (HRM) in a 0.40 w/cm concrete improved diffusion characteristics as much or more than a reduction of w/cm to 0.30. The best performance was exhibited by the 0.30 w/cm, 12% HRM concrete.

The modified AASHTO T259 chloride penetration results, expressed as a combined diffusion result (D_c), are also shown in Table 3 and Figure 5. The observations are similar and the rankings the same as those found for the bulk diffusion tests.

Resistivity values and RCPT values are also given in Table 4. Increasing levels of high reactivity metakaolin resulted in large reductions in RCPT and large increases in resistivity. As expected with both tests, the portland cement concretes were not

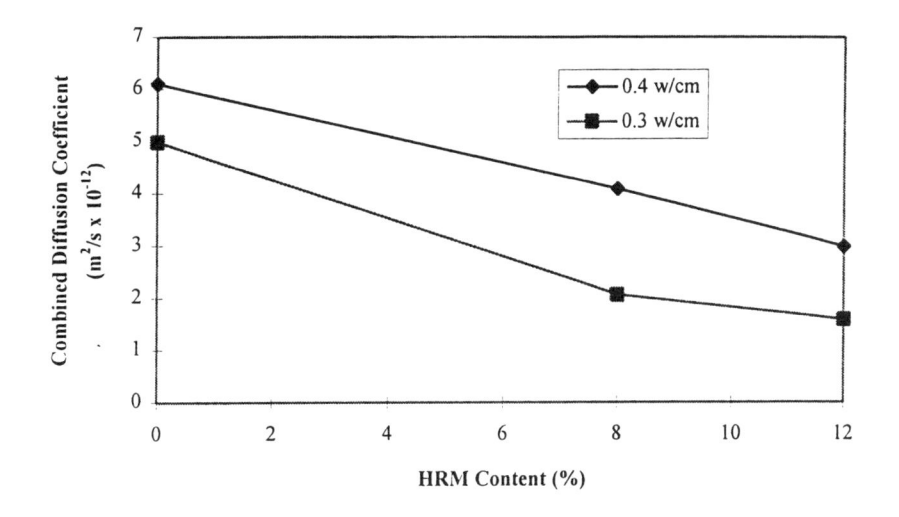

Figure 5 - Modified AASHTO T259 - Combined Diffusion Coefficient vs. HRM Content

Table 4 - Average Resistivity, RCPT and Sorptivity Results

Mixture	Resistivity (ohm-cm)	RCPT Coulombs 6 h values*	RCPT Coulombs 0.5 h x 12 values	Sorptivity (mm/min$^{1/2}$)
1 - 0.4 w/cm, 0% HRM	9960	2770	2210	0.075
2 - 0.4 w/cm, 8% HRM	34910	560	510	0.050
3 - 0.4 w/cm, 12% HRM	72260	310	290	0.052
4 - 0.3 w/cm, 0% HRM	11370	2350	1990	0.046
5 - 0.3 w/cm, 8% HRM	51960	400	373	0.048
6 - 0.3 w/cm, 12% HRM	99360	230	221	0.048

* standard ASTM C1202 test results

greatly improved by a reduction in w/cm. This is thought to be due to porous interfacial transition zones (ITZ) around aggregate particles. The porosity of the ITZ is thought to be reduced by the pozzolanic reaction of the HRM, accounting for the significant improvement in resistivity. The 30 minutes x 12 RCPT values were found to be lower than the standard 6 h test values for the portland cement concretes which exhibited the highest coulomb values. This shows the effect of correcting for heating of the solutions during the test.

Sorptivity values are also given in Table 4. At w/cm = 0.40, HRM reduced sorptivity values but the 0.30 w/cm values were essentially the same. This is likely a limitation of the initial moisture conditioning.

In Table 5, all of the test results for the concretes have been ranked from best (ranking = 1) to worst (ranking = 6). In general, all of the tests, are consistent in ranking the 0.40 w/cm portland cement concrete the worst and the 0.30 w/cm, 12% HRM concrete the best. The only differences are that the RCPT and resistivity tests ranked the 0.40 w/cm, 12% HRM concrete better than the 0.30 w/cm, 8% HRM concrete better than the 0.30 w/cm, 8% HRM concrete. The general agreement provides confidence that the rapid indicator tests can be used to rank resistance to chloride ingress.

Table 5 - Ranking by Different Test Methods

Mixture	RCPT	Resistivity	$D_{a\,28\,day}$	$D_{a\,90\,day}$	$D_{c\,90\,day}$ (T259)
1 - 0.4 w/cm, 0% HRM	6	6	6	6	6
2 - 0.4 w/cm, 8% HRM	4	4	4	4=5	4
3 - 0.4 w/cm, 12% HRM	2	2	3	3	3
4 - 0.3 w/cm, 0% HRM	5	5	5	4=5	5
5 - 0.3 w/cm, 8% HRM	3	3	2	2	2
6 - 0.3 w/cm, 12% HRM	1	1	1	1	1

CONCLUSIONS

1. For concretes with the same water-to-cementitious materials ratio, compressive strengths increased dramatically at all ages with increased replacement percentage by mass of high-reactivity metakaolin.

2. The 28 day and 90 day chloride ponding tests showed that both increasing the high-reactivity metakaolin content and decreasing the water-to-cementitious materials ratio resulted in a significant decrease of the apparent diffusion coefficient. The effect of replacing 8% cement with HRM at w/cm = 0.40 was larger than that of reducing the w/cm from 0.40 to 0.30.

3. The lowest apparent diffusion coefficient was obtained with the 0.3 water-to-cementitious materials ratio with 12% replacement by mass of high-reactivity metakaolin after both 28 and 90 days of ponding.

4. The modified AASHTO T259 combined diffusion coefficients showed a similar trend to the apparent diffusion coefficients

5. While at w/cm = 0.40, sorptivity values were reduced by incorporating HRM. At w/cm = 0.30 no trend in sorptivity was found since, for the initial moisture conditioning used here, results may have reached a lesser limit or terminal value.

6. High-reactivity metakaolin exhibited a dramatic increase in reactivity.

7. The Rapid Chloride Permeability Test results indicated that permeability decreases with a decrease in water-to-cementitious materials ratio and decreases significantly with an increase in replacement of cement with high reactivity metakaolin.

Overall, the ranking of all the concretes by each of the test methods (except sorptivity) was the same. The 0.40 w/cm portland cement concrete performed the worst, while the 0.30 w/cm, 12% HRM concrete performed the best.

ACKNOWLEDGEMENTS

Partial funding and the High Reactivity Metakaolin for the research was provided by Engelhard Corporation, Iselin, N.J. Additional funding was provided by the Ontario Centre for Materials Research.

REFERENCES

1. Neville, A.M., "Properties of Concrete", Fourth Edition. John Wiley & Sons Inc., New York, 1995.

2. Kuennen, T., "Metakaolin Might", Concrete Products, May 1996, pp. 106-110.

3. Caldarone, M.A., Gruber, K.A., and Burg, R.G., "High-Reactivity Metakaolin: A New Generation, Mineral Admixtures", Concrete International, November 1994, pp. 37-40.

4. Balogh, A., "High-Reactivity Metakaolin", reprinted from Concrete Construction, July, 1995, 3 pp.

5. Marsh, D., "An Alternative to Silica Fume?", Concrete Products, 1994.

6. McGrath, P.F., "Development of Test Methods for Predicting Chloride Ingress into High Performance Concrete", Doctor of Philosophy Thesis, Department of Civil Engineering, University of Toronto, 1996.

7. Kelham, S., "A Water Absorption Test for Concrete", Magazine of Concrete Research, V. 40, No. 143, 1988, pp. 106-110.

8. Monfore, G.E., "The Electrical Resistivity of Concrete", Journal of the PCA Research and Development Laboratories, May 1968, pp. 35-47.

The Effectiveness of Calcium Nitrite in Producing High Performance Concrete

Eric P. Steinberg, Ph.D., P.E.
Associate Professor
&
Kenneth B. Edwards, Ph.D., P.E.
Assistant Professor
Ohio University
Department of Civil Engineering
Stocker Center
Athens, OH 45701

ABSTRACT

High performance concrete (HPC) provides higher durability and if necessary, higher strength. HPC leads to higher initial costs which are often offset by longer life and reduced repair costs. To improve the durability of prestressed concrete bridge girders, the Ohio Department of Transportation allows the use of the corrosion inhibitor calcium nitrite as an alternative to epoxy coated reinforcement. The objective of the work discussed in this paper was to determine the effectiveness of calcium nitrite. This was done by removing concrete samples from a total of 48 prestressed concrete box beam bridges in two different geographical locations. The concrete samples were then chemically tested for nitrite and chloride content. The testing for the nitrite was to determine if the corrosion inhibitor was adequatly dispersed and if any of the nitrite was depleted over time. Chloride contents were determined because of its effect of inducing corrosion. Test results showed the nitrite still exists in sufficient quantities to mitigate corrosion of the reinforcement. The results also showed possible unfavorable dispersion of the nitrite. Chloride concentrations were sufficiently high to induce corrosion.

INTRODUCTION

High performance concrete (HPC) is designed to provide greater durability and higher strength, if necessary. The increased durability of conventional concrete can be enhanced by decreasing the permeability of the concrete, using chemical admixtures, and improving quality control. Improved durability of concrete is of enormous importance because it can reduce replacement and repair costs by extending the life of the structure. Though initial costs of HPC exceed conventional concrete, these costs can be offset by extended life and reduced quantities when the HPC also incorporates higher strength. Safety is also increased by improving the durability of concrete.

Calcium nitrite has been specified by the Ohio Department of Transportation's (ODOT) Bureau of Bridges as an alternative to epoxy coated reinforcement since 1986. The purpose of the research work described herein was to evaluate the effectiveness of this chemical admixture in providing a HPC by studying the field performance of this concrete.

BACKGROUND

Concrete typically has a high alkalinity with a pH of about 12.5 to 13.5[1]. This condition, present within concrete, causes a passive ferrous oxide film to form on the surface of the iron and prevents corrosion. This passive oxide film is destroyed if the pH falls below a value of approximately 11. The oxide layer then becomes porous and active. The passive ferrous oxide film is also destroyed in the presence of chloride ions that are typically introduced to the concrete system by de-icing salts or seawater. The amount of chloride required to initiate this corrosion depends on the pH of the solution in contact with the steel, and comparatively small quantities of about 0.6 to 1.2 kg/m^3 of concrete (1.0 to 2.0 lb/yd^3) are required[2].

The calcium nitrite additive provides protection for the reinforcement by the nitrite ion oxidizing the ferrous metal ion to form an insoluble ferric oxide coating on the surface of the reinforcement. This insoluble coating prevents corrosion from continuing at this location in the reinforcement. If the chloride ions are able to react with the ferrous oxide layer, the ferrous chloride complex (rust) moves away from the steel surface and exposes more ferrous oxide at this location. The ferrous oxide is then susceptible to continued reaction with the chloride ions to form more rust, or it reacts with the nitrite ions to form the insoluble ferric oxide coating[3]. A typical commercial solution of a calcium nitrite corrosion inhibitor contains 30% (by weight) calcium nitrite.

Literature related to the calcium nitrite corrosion inhibitor has shown the additive to be beneficial in reducing the rate of corrosion in reinforcement. One report concluded that calcium nitrite provides significant protection of reinforcement even in adverse chloride environments when compared to concrete without calcium nitrite[4]. A more recent laboratory research project showed that the calcium nitrite did not significantly delay the onset of corrosion, but it substantially reduced the severity of the corrosion[5]. These results agree with the process in which calcium nitrite protects reinforcement. However, these studies on calcium nitrite, which evaluated corrosion protection by either electrical and/or visual methods, were performed in the laboratory.

The North Carolina Department of Transportation (NCDOT) has implemented a procedure developed by the W.R. Grace Company to determine the amount of nitrite in hardened concrete[6]. However, this method used to determine the amount of the nitrite ion in hardened concrete has not been standardized. Accuracy of this method ranges from 80 to 100% nitrite recovery.

SAMPLING

Bridge Selection
The northeast section of Ohio receives an appreciable amount of lake effect snow leading to a higher use of road salt. Therefore, it was chosen as one of the geographical sampling

regions. In order to compare the results from two different geographical areas, the southwest section of Ohio was also chosen because this area would have weather conditions most opposing to the northeast. Each area contained 24 bridges from which samples were removed.

To account for the effectiveness of the calcium nitrite corrosion inhibitor over various years of serviceability, a range of 12 years in which the bridges were constructed was considered. Four bridges from each geographical region were selected during various years from 1985 to 1993. Documentation that verified the inclusion of the calcium nitrite in the bridges was also used to select the bridges.

Location of Samples
Four samples were removed from each bridge to account for any uneven distribution of the calcium nitrite. The location of each sample within a bridge was selected depending on the condition of the beams, the accessibility, and the areas most susceptible to corrosion. Signs of corrosion included staining due to either water or rusting, spalling, or cracking. In most cases, the beams were in good condition with no actual sign of corrosion. However, this was not always the case. For instance, in Fig. 1, a large section of concrete spalled from the fourth interior beam of bridge ADA-52-13.63. If any signs of possible corrosion were noticed on a bridge, then that area was selected as the location for one of the samples. These sections were the most critical and were determined first.

Figure 1. Spalled Section of ADA-52-13.54

If there were no noticeable signs of corrosion on a bridge, the location of the samples was then determined by accessibility. An example of an obstacle is depicted in Fig. 2 where the bridge (ERI-269-08.03) was too low to get underneath, and access to the east side of the bridge was blocked. Therefore, all the samples had to be removed from the edge of the first beam from the west.

If the bridge did not show any signs of corrosion and was fully accessible, the sample locations were usually as follows. Sample one was located in the edge of an exterior beam. Sample two was then located in the bottom of that same exterior beam. Sample three was taken from the bottom of an interior beam located approximately halfway across the width of the bridge. Finally, sample four was located in the edge of

Figure 2. Inaccessibility of ERI-269-08.03

the exterior beam on the opposite side of the bridge from sample one. The samples were typically taken in an end span as far from the abutment as the water, roadway, and embankment would permit.

Each sample was taken near the stirrup location so that the effect of the chloride on the reinforcement could be determined. Prestressing strands were avoided due to their high stress and critical function of the bridge.

Method of Sample Removal
The samples were removed using an accessory rotary hammer with a 16 mm (5/8") drill bit. The drill was used to remove approximately 6 mm (1/4") of the surface to avoid contaminating the sample with anything that may be on the surface of the beam and to remove any mortar rich concrete. Each hole was drilled approximately 100 mm (4") deep. The concrete powder from the drilling process was collected with a vacuum motor and a small bag and filter. To reduce the amount of sample loss, a container was used to catch the powder from the hole. The bag from each unit was then removed and placed into a sealed container labeled with all pertinent information. The final step in the sampling retrieval was filling the drilled holes with a two-part epoxy to ensure that the integrity of the beam was not sacrificed. Fig. 3 shows the set-up for the procedure used to sample the bridges. A complete description can be found in Gamble[7].

CHEMICAL ANALYSES
Nitrite
The procedure used for determining the nitrite content was NCDOT's chemical procedure #C-20.0. This procedure was based on the W. R. Grace Procedure #809, which was developed in 1980 at the Washington Research Center. The method contained three parts to calculating the amount of calcium nitrite present in a hardened concrete sample: preparation of standard calibration curves for the nitrite ion, nitrite extraction and determination for a sample, and calculation of calcium nitrite in liters per cubic meter of concrete. A detailed explanation of the procedure can be found in Gamble[7].

Figure 3. Sampling Procedure

Chloride
The procedure used for determining the chloride content was AASHTO method: T 260-93. In this research, the total chloride ion content determination using the potentiometric titration procedure was followed.

RESULTS

Figs. 4 and 5 provide the results of the calcium nitrite concentrations for the bridges in the northern and southern regions, respectively. The solid lines shown on the figures connect the average concentration values for each sample year. The average concentrations did not show any consistent trend. Both figures show scatter in the data. This scatter could be caused by the samples containing different proportions of aggregate and/or variation in the distribution of the calcium nitrite.

Fig. 4 provides the calcium nitrite test results for the northern region bridges. The data points that show no calcium nitrite in 1984 and 1986 are the results of samples removed from control bridges that did not contain the corrosion inhibitor. One data point for 1993 that shows no calcium nitrite was from a sample of a bridge that was supposed to have the corrosion inhibitor. In fact, samples removed from other beams of the same bridge revealed that the calcium nitrite existed at these locations. The sample lacking the inhibitor, which was removed from the under side of a beam at the centerline of the bridge near the northern abutment, shows that unfavorable variation in the distribution of the corrosion inhibitor may be possible.

Fig. 5 shows the nitrite test results for the southern region bridges. The data point showing no calcium nitrite in 1984 represents the results of samples from another control bridge that did not contain the corrosion inhibitor. The data point showing no calcium nitrite in 1986 is for a sample taken from a bridge that contained the corrosion inhibitor. This was verified by other samples from the same beam and other beams of the bridge. This again showed the possible unfavorable distribution of the corrosion inhibitor.

4.95 L / m³ = 1 gal / yd³

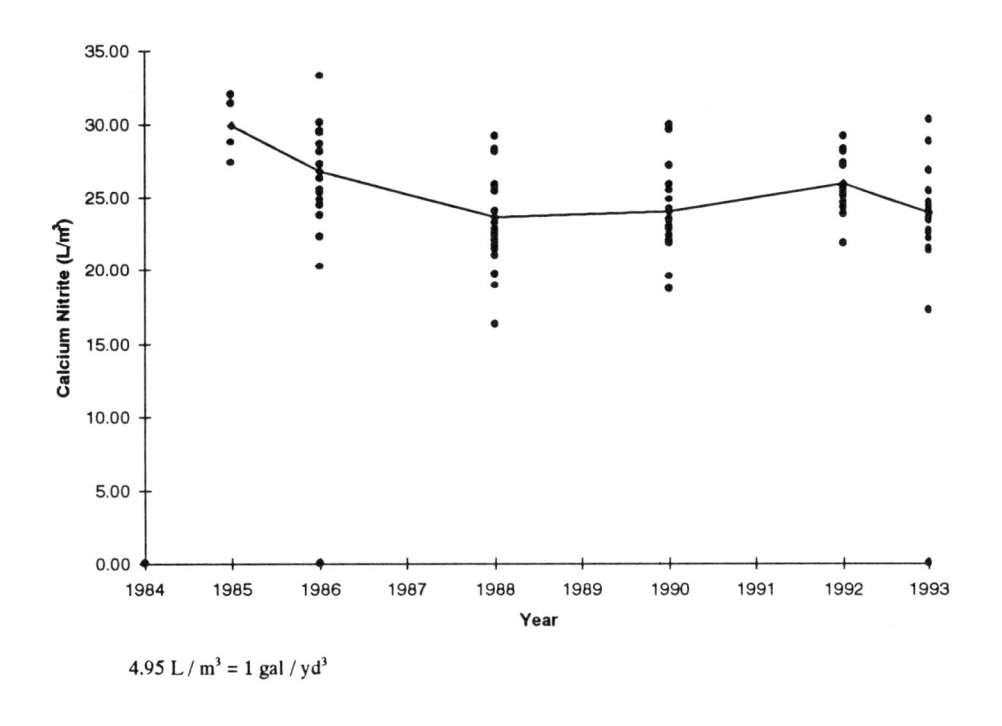

Figure 4. Calcium Nitrite versus Year of Construction (Northern Region)

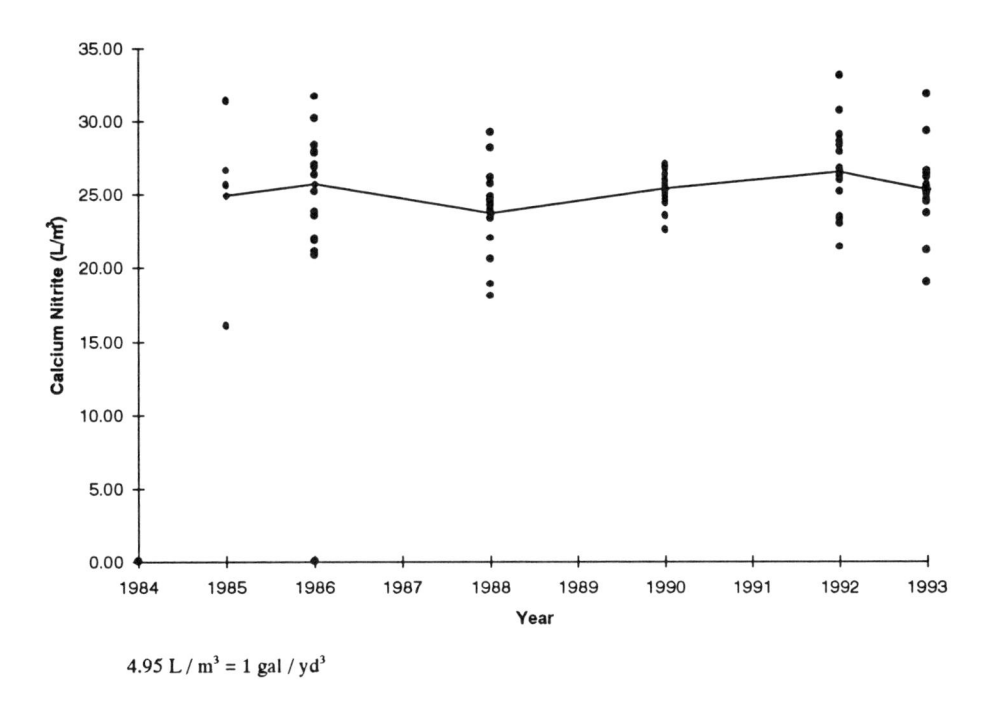

4.95 L / m³ = 1 gal / yd³

Figure 5. Calcium Nitrite versus Year of Construction (Southern Region)

The results from the chloride testing are shown in Figs. 6 and 7. The figures show scatter in the data that was expected. Some of the scatter in the data is undoubtedly caused by variation in the permeability of the concrete at the location where the sample was removed, the amount of chloride in the form of road salt that was applied to the bridge, the drainage of runoff from the bridge, and the weather conditions at each specific bridge site. The solid lines in both figures connect the average chloride concentrations for each year. Linear regression was then performed on the average values to establish the broken line and the equations.

Fig. 6 displays the chloride results for the entire northern region. The general trend is an increase in chloride content for the older bridges. This is because the older bridges have been exposed to road salt for longer periods. The extreme chloride contents shown in 1984 were taken from a spalled and a cracked region on the underside of a beam toward the center of the width of the bridge. The spalled region appeared to be caused from freeze-thaw action. The crack, however, appeared to be from corrosion of the prestressing strand because it followed the location of the of the strand. Water staining was also observed at the crack location. This bridge did not contain the calcium nitrite.

The highest chloride value shown in 1988 was also taken from a cracked region on the underside of a beam toward the center of the bridge's width. The highest values in 1993 were from samples taken from the sides of two edge beams near the expansion joint at the bridge's north abutment. These regions were severely damaged and corrosion was definitely occurring. The detail of this region contained high early strength cast-in-place (CIP) concrete over dowel bars that anchored the prestressed beams to the abutment. A galvanized steel drip edge was also placed in this region at approximately 14" below the top of the prestressed beam. The location of the severe damage was above the drip edge. The CIP concrete was also in worse condition than the concrete of the prestressed beam. Both of these bridges contained the calcium nitrite.

Fig. 7 shows the results of the chloride testing for the bridges in the southern region. In general, the data shows lower values of chloride content as compared to the northern region. This was expected due to the difference in road salt usage. However, the year 1990 somewhat contradicts this conclusion. Examination of the data for the top six chloride contents for 1990 revealed that the samples were from two bridges that had heavy traffic volume. The average daily traffic (ADT) for these two bridges were 19,320 and 20,830. The average of the ADT for all the southern region bridges was less than 7,000. The higher ADT for the previous two bridges may cause an increase in the amount of road salt applied to them resulting in higher chloride contents. The extremely high chloride content shown in 1986 bridges may also be due to higher than average ADT (12,853 for the bridge which contained this sample). Though some of the bridges had staining and one actually had a spalled section with the prestressing strand exposed, the chloride contents of these specific bridges did not show higher chloride values than the other bridges sampled. The bridge with the exposed strand showed signs of the corrosion, but the spalled concrete was likely due to differential movement between the beams and not corrosion of the strand.

1 kg / m³ = 0.59 lb / yd³

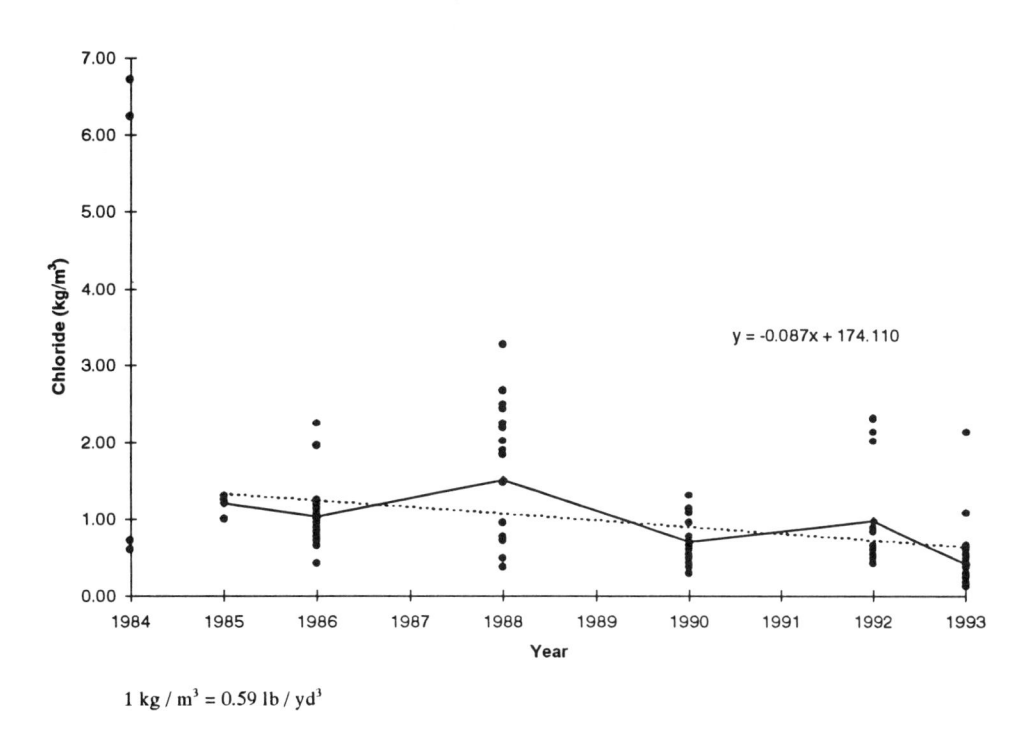

Figure 6: Chloride Content versus Year of Construction (Northern Region)

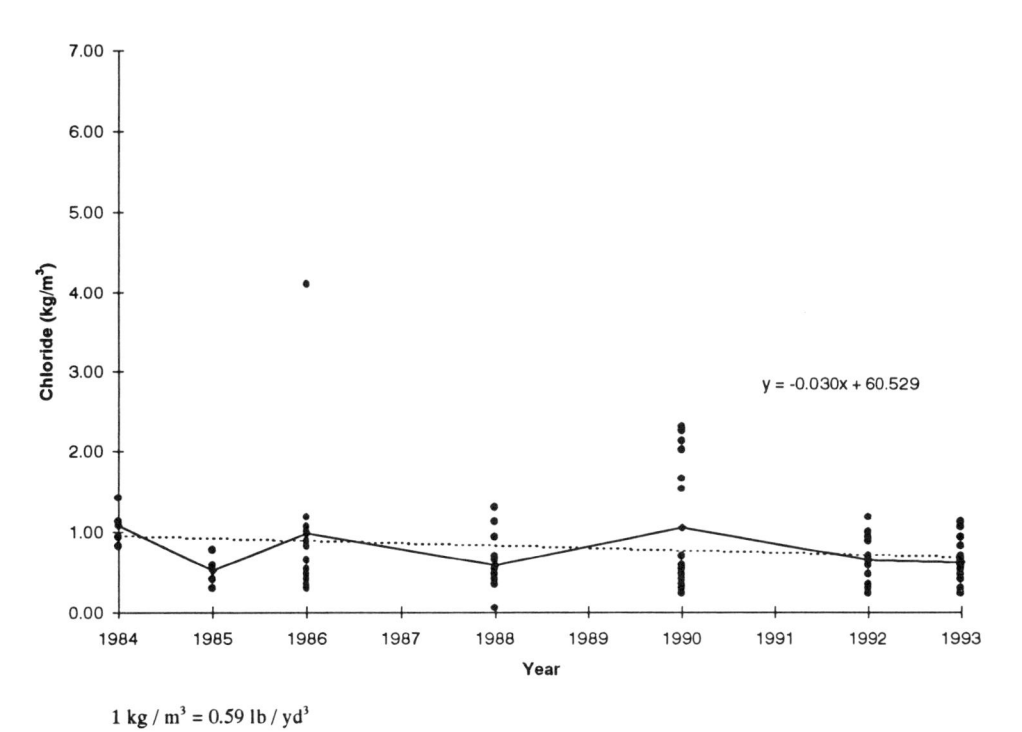

1 kg / m³ = 0.59 lb / yd³

Figure 7: Chloride Content versus Year of Construction (Southern Region)

The amount of chloride ions detected in each sample were typically above 0.6 kilogram per cubic meter, suggesting that without the presence of the calcium nitrite, there would likely be the initiation of corrosion. With the inclusion of calcium nitrite, 7.6 kilograms of chloride per cubic meter initiates corrosion, according to the manufacturer's literature[3]. The values of chloride ion content are significantly lower than this threshold and therefore suggest that corrosion has not begun in the bridges.

A rough estimate to the time of corrosion can be calculated using the results of this study if it is assumed that the rate of chloride intrusion continues at a linear rate and that depletion of the corrosion inhibitor does not occur for some other unknown reason. From Fig. 6, the chloride content is approximately 1.4 kg/m³ for a bridge approximately 10 years old. Thus, chloride is added to the northern bridges at approximately 1.4 kg/m³ per 10 years. Therefore, assuming corrosion does not begin until a value of 7.6 kg/m³ is reached, the time to corrosion becomes an additional 44 years for the northern region as shown in Eqn. 1.

$$Add.\ Yrs.\ Corrosion = \frac{7.6 - 1.4}{\frac{1.4}{10\ years}} = 44\ years \qquad (1)$$

For an assumed chloride content of 1.0 kg/m³ in 10 years for the southern region, the estimation becomes 66 additional years. These results show that calcium nitrite is very effective in producing a HPC by making the concrete much more durable.

CONCLUSIONS

Overall the test results for the nitrite typically show values in excess of 20 L/m³ (4 gal/yd³). Though the corrosion inhibitor may distribute unfavorably, it appears to be remaining in sufficient quantities to be effective against corrosion.

The chloride contents of the samples were found to be high enough in some locations to induce corrosion. Chloride levels were generally higher in the northern region in comparison to the southern region, as expected. Locations of damage, such as staining, cracking, and spalling, typically had high chloride contents.

It was also concluded that the corrosion inhibitor is very effective in producing a high performance concrete by extending the time to corrosion of the reinforcing steel.

ACKNOWLEDGMENTS

The authors would like to thank the Ohio Department of Transportation (ODOT) and the Federal Highway Administration (FHWA) for providing financial support for this project. The authors would also like to thank Joanne Gamble, former graduate research assistant, for her assistance in this work and the following ODOT individuals for their technical support: Vik Dalal, Bridge Research Engineer; Lloyd Welker, Bridge Engineer;

Dave Moore, Bridge Rating Engineer; and the District Bridge Engineers.

REFERENCES

1. Derucher, K., Korfiatas, G., and Ezeldin, A., *Materials for Civil & Highway Engineers*, Prentice Hall, Englewood Cliffs, NJ, 1994.

2. Mindess, S. and Young, J., *Concrete*, Prentice Hall, Englewood Cliffs, NJ, 1981.

3. Berke, N. and Rosenberg, A., Technical Review of Calcium Nitrite Corrosion Inhibitor in Concrete. In *Transportation Research Record 1211*, TRB, National Research Council, Washington D.C., 1989, pp. 18-27.

4. Virmani, Y., Clear, K., and Pasko, T., *Time-to-Corrosion of Reinforcing Steel in Concrete Slabs. Vol. 5: Calcium Nitrite Admixture or Epoxy-Coated Reinforcing Bars as Corrosion Protection System*. Report No. FHWA-RD-83-012, FHWA, U.S. Department of Transportation, Washington, D.C., 1983.

5. Pfeifer, D., Landgren, J., and Zoob, A., *Protective Systems for New Prestressed and Substructure Concrete*. Report No. FHWA-RD-86-193, FHWA, U.S. Department of Transportation, Washington, D.C., 1987.

6. Jeknavorian, A., Chin, D., and Saidha, L., Determination of a Nitrite-Based Corrosion Inhibitor in Plastic and Hardened Concrete. *Determination of the Chemical and Mineral Admixture Content of Hardened Concrete*, ASTM STP 1253, S. Kosmatka and A. Jeknavorian, Eds., American Society for Testing and Materials, Philadelphia, 1995.

7. Gamble, J., *Field Evaluation of Calcium Nitrite and Chloride in Ohio Prestressed Concrete Box Beam Bridge Girders*, Thesis presented in partial fulfillment of Master of Science in Civil Engineering, Ohio University, Athens, OH, August, 1996.

8. *Standard Test for Specific Gravity of Soils*, ASTM D854-92, American Society for Testing and Materials, V. 4, Philadelphia, PA, 1994.

HIGH EARLY STRENGTH CONCRETE FOR PRESTRESSED CONCRETE APPLICATIONS

Shuaib H. Ahmad and Paul Zia
Civil Engineering Department
North Carolina State University
Raleigh, North Carolina 27695-7908 U.S.A.

ABSTRACT

One category of High Performance Concretes (HPC) is high early strength concrete. Concretes which have superior early-age properties are needed to reduce the time required for producing precast/prestressed structural members. The objectives of this investigation were to: (i) develop mixture proportions for producing high early-strength (HES) concretes with a target strength of 5000 psi (35 Mpa) within 18 hours after the addition of water, (ii) demonstrate the feasibility of producing HES concretes in the prestressing plant and determine the mechanical properties of such plant produced concretes and (iii) study the flexural behavior of prestressed flexural members using HES concretes and check the validity of current ACI equations for transfer length.

INTRODUCTION

Use of concretes with higher compressive strengths enable greater prestressing in precast members; however, this requires concrete strengths of at least 5000 psi (34 MPa) at release of the prestressing force. Currently under a normal 18 hour production cycle, concrete strengths of 3500 to 4000 psi (24.5 to 28 MPa) are achieved in most prestressing plants. To achieve 5000 psi (34 MPa) strengths, the detensioning operation of the prestressed tendons is delayed by a couple of days.

Through the utilization of high-early strength (HES) concretes, the prestressing industry would be able to reduce the time required for producing precast/prestressed structural members. High-early strength (HES) concrete is one category of high performance concretes which may be defined in a broad sense as concretes with improved material characteristics, placeability and service life. In this study, conventional materials, such as Type I cement and common aggregates, are utilized to facilitate the adaptation of the mixture proportions by a typical prestressing yard.

The objective of this investigation is to develop mixture proportions for producing high-early strength (HES) concretes with target strengths of 5000 psi (34 MPa) within 18 hours after the addition of water. It also seeks to demonstrate the feasibility of producing high-early strength (HES) concretes in the prestressing plant and to determine the mechanical properties of such plant-produced concretes. Finally, this investigation studies the flexural behavior of prestressed beams using HES concretes and checks the validity of the current equation for transfer length in the ACI Code.

EXPERIMENTAL INVESTIGATION

The testing program included tests for the fresh or plastic concrete and for hardened concrete. Laboratory and field concretes were each tested to determine the conformance of their observed strengths to the target strengths of 5000 psi (34 MPa) within 18 hours and to determine other mechanical properties. A total of 176 specimens was tested.

Constituent Materials

A Type I cement was used for all phases of the program The cement was of low alkali content and met the requirements of ASTM C150. All of the concrete mixes used crushed granite as coarse aggregate (ASTM 33). Natural yellow sand was used as fine aggregate (ASTM C33). The corrosion inhibitor (CI) was a calcium nitrite solution with a retarder. The high range water reducers (HRWR) were naphthalene-based. The air-entraining agents (AEA) were of neutral vinyl resin.

Mixture Proportions

To gain a strength of 5000 psi (34 MPa) within 18 hours after the addition of water, steam curing was employed. For producing the laboratory mixture (LAB-1), no accelerators were used.

To obtain the field mix, the LAB-1 mix was slightly modified by utilizing the type of HRWR and AEA which were available at the prestressing yard. Two different mixture proportions were developed during the site phase of this research. These were identical except that a corrosion inhibitor was added to the second mix to ascertain the effect of this additive on the material properties of hardened concrete. In the laboratory, only one method of curing was used. Three different procedures of curing were used for the field specimens.

Testing

The compressive strength of 4 x 8 in. (100 x 200 mm) cylinders were determined according to ASTM C39, except that an unbonded capping system was used. The unbonded capping system consisted of steel restraining rings with neoprene pad inserts. The elastic modulus was determined by two methods. In the field, 6 x 12 in. (150 x 300 mm) cylinders were tested, and 4 x 8 in. (100 x 200 mm) cylinders were tested in the laboratory. The modulus of rupture (MOR) beam specimens were tested by loading at third-points over a 12 in. (30.5 cm) clear span (ASTM C78). The testing for the splitting tensile strength was done with 4 x 8 in. (100 x 200 mm) cylinders (ASTM C496). The cylinders were loaded to failure at a rate of 7500 pounds (33.4 kN) per minute. The shrinkage and creep tests were done with 4 x 8 in. (100 x 200 mm) cylinders. All other aspects of the testing conformed to ASTM C512. To study the flexural behavior of prestressed beams with HES concrete, two prestressed beam specimens were cast in the prestressing plant and tested in the laboratory at NCSU. To study the transfer length of strand in HES concretes, one additional beam specimen was cast in the prestressing plant.

TEST RESULTS AND DISCUSSION

Table 1 - Laboratory results for compressive strengths with LAB-1 mixture proportions.

Test Age	Strength[a] psi (MPa)		
	Batch 1	Batch 2	Batch 3
17 hours	5240 (36.13)	5160 (35.58)	5190 (35.79)
3 days	5740 (39.58)	5720 (39.44)	5880 (40.54)
7 days	6330 (43.65)	6520 (44.96)	6500 (44.82)
28 days	7750 (53.44)	7910 (54.54)	7870 (54.26)

[a] Average of three replicate 4 x 8 in. (100 x 200 mm) specimens.

Table 2 Field concrete mix proportions for NI/NS/WC test specimens.

Aggregate	Target[b]	Actual	Moisture
Cement (Type I), pcy (kg/m^3)	870 (517.2)	860 (511.3)	-
#67 Granite, pcy (kg/m^3)	1750 (1040.4)	1760 (1046.4)	2.3 %
Natural Sand, pcy (kg/m^3)	1039 (617.7)	1040 (618.3)	3.8 %
HRWR (Daracem), oz/cy (l/m^3)	87 (3.37)	82 (3.18)	-
AEA (MBVR), oz/cy (l/m^3)	5 (0.19)	5 (0.19)	-
DCI-S, oz/cy (l/m^3)	0 (0)	0 (0)	-
Water, pcy (kg/m^3)	217 (129.0)	209 (124.3)	-
W/C	0.34	0.33	-
Slump, in (cm)	7.0 (17.8)	4.5 (11.4)	-
Air, %	4.5	3.5	-
Temperature, °F (°C)	-	66 (19)	-

[b] All weights adjusted for actual moisture content.

Material Properties

During the laboratory phase of this investigation, compression testing was performed on specimens from three batches, all with the mix proportions of LAB-1 (**Table 1**). Three replicate 4 x 8 in. (100 x 200 mm) cylinders were used for testing at ages of 17 hours, 3 days, 7 days and 28 days. The results shown in **Table 1** are the averages of three replicate specimens.

The mixture proportions for high-early strength (HES) concrete developed in the laboratory were used in the prestressing yard to obtain field concrete strengths of 5000 psi (34 MPa) within 17 hours. Two mix proportions were used for the field concrete, differing only in the addition of a corrosion inhibitor (DCI-S) to the second mix. The field testing included compressive strength, flexural strength, splitting tensile strength and elastic modulus testing.

Table 3 - Average material properties for the two field concrete mixture proportions with three curing procedures produced at the prestressing yard.

		NI[d]/NS[e]/WC[f]	CI/NS/WC	NI/ST[g]/WC	CI/ST/WC	NI/ST/AC[c,h]
Compression[a]	psi (MPa)					
17 hours		2080 (14.3)	1660 (11.4)	5020 (34.6)	4860 (33.5)	4400 (30.3)
3 days		5030 (34.7)	3880 (26.8)	5610 (38.7)	5910 (40.7)	5910 (40.7)
7 days		6560 (45.2)	5560 (38.3)	5980 (41.2)	6420 (44.3)	6580 (45.5)
28 days		8250 (56.9)	6910 (47.6)	7040 (48.5)	7550 (52.1)	7160 (49.4)
Elastic Modulus[a]	ksi (GPa)					
17 hours		2879 (19.9)	3062 (21.1)	3318 (22.9)	3333 (23.0)	3259 (22.5)
3 days		3802 (26.2)	3786 (26.1)	3714 (25.6)	3833 (26.4)	3802 (26.2)
7 days		4352 (30.0)	4305 (29.7)	3960 (27.3)	4147 (28.6)	4057 (28.0)
28 days		4557 (31.4)	4354 (30.0)	4314 (29.7)	4547 (31.4)	4479 (30.9)
Flexure[b]	psi (MPa)					
17 hours		360 (2.48)	255 (1.76)	550 (3.79)	550 (3.79)	465 (3.21)
28 days		565 (3.90)	645 (4.45)	710 (4.90)	655 (4.52)	550 (3.79)
Split Tensile[a]	psi (MPa)					
17 hours		230 (1.59)	175 (1.21)	270 (1.86)	340 (2.34)	285 (1.97)
7 days		415 (2.86)	385 (2.65)	315 (2.17)	350 (2.41)	345 (2.38)
28 days		545 (3.76)	415 (2.86)	480 (3.31)	435 (3.00)	365 (2.52)

a   Average of three replicates.
b   Average of two replicates.
c   Specimens removed from steam at age of 14 hours due to detensioning of casting bed.
d   No corrosion inhibitor.
e   No steam curing
f   Water cured
g   Steam cured 11-12 hrs
h   Air cured

In total, five sets of concrete mixes were produced in the prestressing plant. The variables were the curing conditions and the use of corrosion inhibitor. In **Table 2** an example of the target and the actual mix proportions used for each of the five mixtures is shown. A summary of the results for compressive strength, elastic modulus, flexural strength and split cylinder strength is shown in **Table 3**. These results are the average of the number of replicate specimens tested.

Compressive Strength - The use of the corrosion inhibitor has the effect of retarding the compressive strength gain of young concretes (t ≤ 3 days). After the age of 3 days, its retarding effect is relatively smaller for steam-cured concretes as compared to concretes without steam curing. **Figure 1** shows a comparison of the observed strengths for the field concretes with the predictions as per the equation recommended by ACI 209[1]. The figure shows that the ACI prediction underestimates the strength of steam cured HES concretes at early ages (t ≤ 3 days). For HES concrete without steam curing, and with and without DCI-S corrosion inhibitor, the strength predictions as per ACI 209 underestimated the measured values at each age tested.

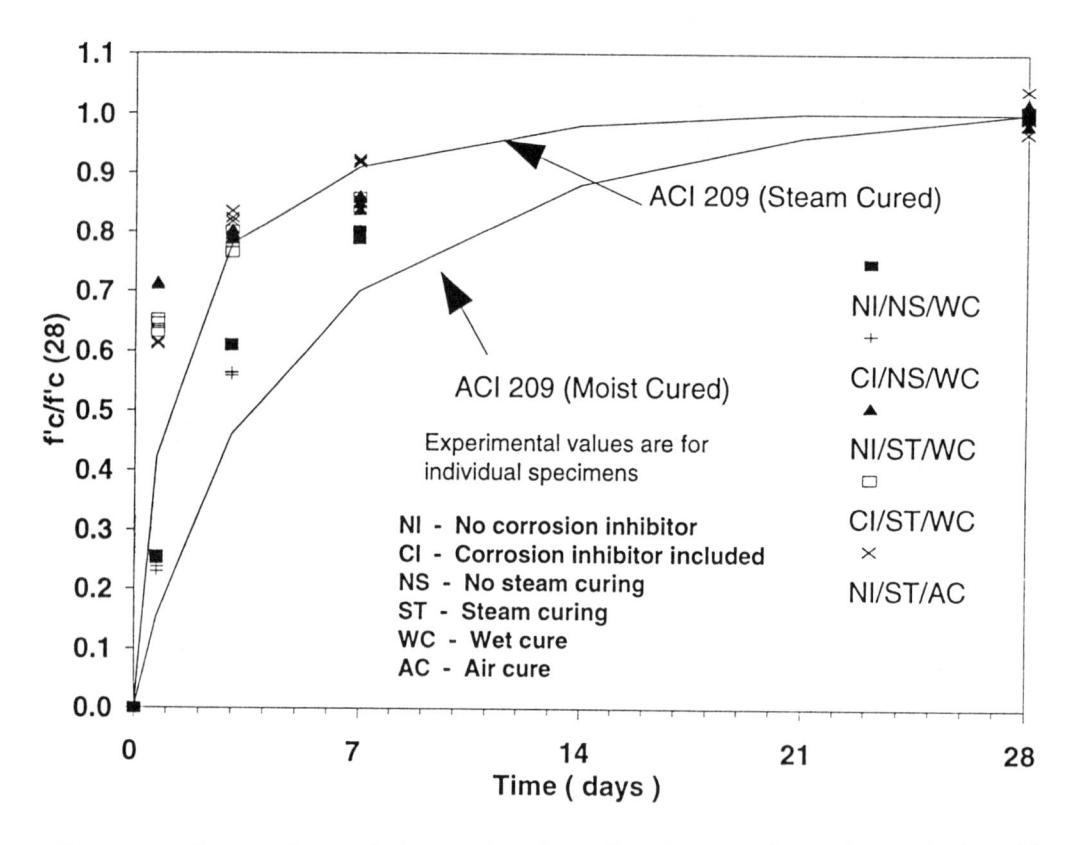

Figure 1 Comparison of observed and predicted strength vs. time relationship

Elastic Modulus - The elastic modulus of concrete depends on factors including the strength, stiffness, fraction of the coarse aggregate, unit weight of concrete and matrix of the concrete. Generally the modulus of elasticity is expressed as proportional to the square root of the compressive strength of the concrete.

The experimental results were compared with some equations used for predicting the modulus of elasticity. These equations are:

$$E_c = 33W^{1.5}\sqrt{f'_c} \qquad (1)$$

$$E_c = 1{,}000{,}000 + 40{,}000\sqrt{f'_c} \qquad (2)$$

$$E_c = W^{2.5}(f'_c)^{0.325} \qquad (3)$$

The above equations are as per the recommendations of ACI Code 318-89[2] (Eqn 1), ACI 363[3] (Eqn 2), and Ahmad and Shah[4] (Eqn 3). The value used for the unit weight of concrete (W) is 145 pcf (22.83 kN/m³) as was determined in the laboratory.

At all test ages, the ACI 318 predictions overestimated the results except for concretes with DCI-S and without steam curing. The predictions were within ± 8% of the observed values for 80% of the specimens tested. The test results show that the equation proposed by Ahmad and Shah[4] predicts the value within 10% at all ages for concretes without steam curing and at ages greater than 3 days for concretes with steam curing. At the age of 17 hours, all equations overestimated the elastic modulus of steam cured concretes by a minimum of 12%. It appears that for steam cured concretes, the predictions as per the equation of ACI 363 are closest to the observed strengths.

The results show that the use of corrosion inhibitor has no significant influence on the elastic modulus (E_c), since E_c primarily is a function of the aggregate type and its percent content. Furthermore, it can be seen that steam curing has much less effect on the increase in the elastic modulus with time as compared to the increase in compressive strength with time.

Stress-Strain Relationship - Stress-strain relationships were determined for four replicate specimens at ages of 7 and 28 days respectively. The results show that at the age of 7 days, the strains at the maximum compressive stress values (peak strains) are equal or slightly greater than the values at an age of 28 days. For HES concrete without corrosion inhibitor, the peak strains at 7 and 28 days age are 2400 and 2500 microstrains for concretes without steam curing and 2200 and 1500 microstrains for concretes with steam curing. The peak strains for HES concretes with corrosion inhibitor at 7 and 28 days age are 1600 and 1550 microstrains without steam curing and 1700 and 1800 microstrains for concretes with steam curing. The results also indicate that in the initial portion of the stress-strain curve, the stress-strain relationship of younger concretes is very comparable to that of mature concrete. However, near the maximum strength stage, the stress-strain relationship of younger concretes is relatively more non-linear as compared to the response of the mature concretes.

Results of the effects of steam curing on the stress-strain relationship of HES concrete at 28 days show that steam cured HES concretes without corrosion inhibitor exhibit a 28 day compressive strength of about 85% and peak strains of about 60% of the non-steam cured specimens. Also the indication is that the stress-strain relationship for these concretes are similar up to 5000 psi (34.45 MPa); for concretes with and without DCI-S, the stress-strain relationship up to a strain of 500 microstrains (or a stress of 2000 psi or 13.79 MPa) is similar for non-steam cured and steam cured specimens.

Flexural Strength - Two replicate 4 x 4 x 17.5 in. (100 x 100 x 445 mm) beams were tested at 17 hours and 28 days. The results were compared with empirical equations of ACI 318-89[2] (Eqn 4), ACI 363[3] (Eqn 5) and Ahmad and Shah[4] (Eqn 6) which are:

$$f_r = 7.5\sqrt{f'_c} \qquad (4)$$

$$f_r = 11.7\sqrt{f'_c} \qquad (5)$$

$$f_r = 2.3 (f'_c)^{2/3} \qquad (6)$$

The ACI-318 equation provided a better prediction of the observed results than the other two equations, however, the difference at all test ages were as high as 21%. The differences between predictions and experimental results ranged from -5% to +7% for 70 % of the specimens tested. Differences between measured and predicted values were almost 90% for some specimens with the ACI 363 equation and up to 66% for the equation proposed by Ahmad and Shah[4].

Splitting Tensile Strength - Two replicate 4 x 8 in. (100 x 200 mm) cylinders were tested at 17 hours, 7 days and 28 days for each of the five field concretes. Some of the notable empirical equations for estimating the splitting tensile strength include the equations of ACI 318-89[2] (Eqn 7), ACI 363[3] (Eqn 8), Oluokun[5] (Eqn 9) and Ahmad and Shah[4] (Eqn 10), which are:

$$f_{ct} = 6\sqrt{f'_c} \qquad (7)$$

$$f_{ct} = 7.4\sqrt{f'_c} \qquad (8)$$

$$f_{ct} = 1.38 (f'_c)^{0.69} \qquad (9)$$

$$f_{ct} = 4.34 (f'_c)^{0.55} \qquad (10)$$

The empirical equations of ACI 318 and ACI 363 significantly overpredicted the splitting tensile strength of high early strengths at all ages. The equations of Ahmad and Shah, and Oluokun are quite similar, and in this investigation the predictions as per Oluokun's equation are compared with the observed results since Oluokun's equation was developed for early-age concretes.

In this investigation the equation proposed by Oluokun[5] was used with a correction factor of 0.94 to convert the 4 x 8 in. (100 x 200 mm) cylinder strengths to 6 x 12 in. (150 x 300 mm) cylinder strengths[6]. The predictions for the non-steam cured specimens at very early ages were close to observed values, but the predictions overestimated the observed values for the steam cured specimens at early ages. The equation also overpredicted the results for cylinders at 28 days age.

Shrinkage and Creep Behavior - The shrinkage and creep behavior were determined for the air cured high-early strength concrete without DCI-S corrosion inhibitor.

The shrinkage value was determined using two companion 4 x 8 in. (100 x 200 mm) cylinders. The average of the measurements from both sides of each cylinder were used to calculate the shrinkage strain for the specimen. The average shrinkage strain was determined to be 1060×10^{-6} in/in for steam cured concrete. The observed shrinkage was higher than the mean value of 730×10^{-6} in/in as per recommendation of ACI Committee 209[1]. It should be noted that the specimens for shrinkage testing were stored in an area with a relative humidity below 40%. Since the specimens were tested under conditions with lower humidity than those generally existing in service and reported by ACI 209, this could have attributed to larger measured shrinkage strains for these test.

The creep testing was also conducted using two replicate 4 x 8 in. (100 x 200 mm) cylinders. The cylinders were loaded to 1035 psi (7.13 MPa) and the measurements recorded. The average of the creep values for cylinders showed an initial strain of 300×10^{-6} in/in and a final strain of 950×10^{-6} in/in at an age of 104 days. The ultimate creep coefficient was determined to be 3.17 which is within the range of 1.30 to 4.15 reported by the ACI Committee 209[1]. The higher than average creep strains observed could be attributed to the high cement content (870 pcy or 517.2 kg/m^3) used for HES concretes.

Beam Behavior

Prestressed beams were fabricated to study the behavior of HES concrete in prestressed concrete flexural members. Two beams were cast for flexural testing up to failure, and one beam was cast to determine the transfer length of the prestressing strands. The 42-day compressive strengths of the specimens were f_c =7500 psi.

Load-Deflection Behavior - Two 12 ft (3.66 m) prestressed beams were tested under flexure. The cross section of each beam was 8 x 12 in (20.3 x 30.5 cm) with an effective depth of 8 in (20.3 cm). Three 1/2 in. (12.7 mm) diameter strands were each prestressed with an initial force of 30,980 pounds (138.1 kN) giving an initial prestressing stress (f_{pi}) of 202,480 psi (1396 MPa). The beams were simply supported at 6 in (15.2 cm) from each end with a free span of 11 feet (3.35 m).

The beams were loaded until the first flexural crack was observed. The beams were then unloaded until the crack closed and then loaded to reopen the crack. The load at which the crack reopened was noted to determine the effective prestress. The effective prestressing stress (f_{se}) was computed to be 175,300 psi (1209 MPa). The strain at the level of the strands and 2 inches (5 cm) from the top fiber were monitored by embedment gages and recorded. The results show a loss of prestressing force of 14% at the test age of 42 days.

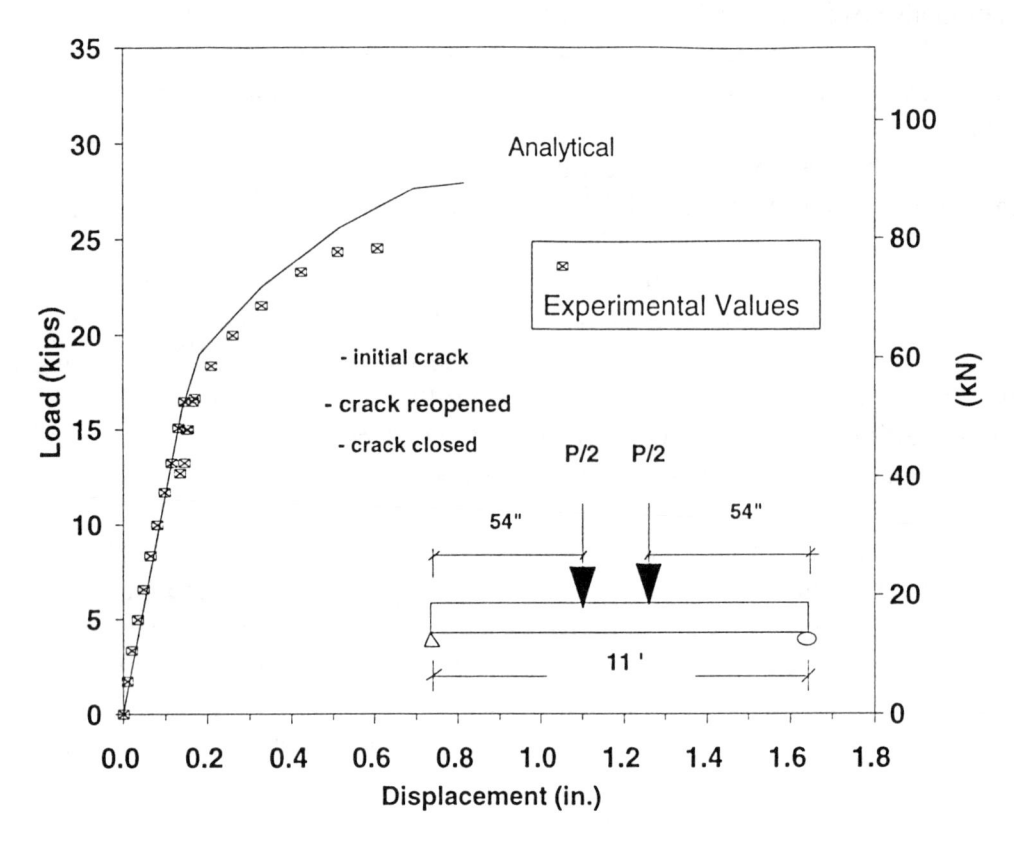

Figure 2 Load vs. midspan deflection for HES S-1 prestressed beam.

Figures 2 and **3** show comparisons of the observed and predicted load vs. midspan deflection response of beams HES S-1 and HES S-2 respectively. For the prediction of the load-deflection, the model proposed by Ahmad and Shah[4] was utilized. The major cause of difference between measured and predicted ultimate load and the behavior after the initial cracking stage could be attributed to the occurrence of premature shear failure in the test specimens.

An initial flexural crack occurred and developed in the constant moment region of the beam. At the failure load, the cracking had progressed approximately 7 inches (17.8 cm) from the extreme tensile fiber toward the extreme compression face.

At failure of the beams, the strains from the embedment gages at the level of the prestressing strands are on the order of 1350 microstrains, which translates into a stress of 37.8 ksi (260.6 MPa) using a modulus of elasticity of prestressing steel of 28,000 ksi (193.06 GPa). The prestressing steel reached a stress in excess of 213 ksi (1469 MPa) at a load of 24 kips (107 kN) which is the sum of the effective prestress of 175,300 psi (1209 MPa) and an additional stress of 37.8 ksi (260.6 MPa) due to loading. A higher steel stress could not be achieved due to the shear cracking which caused premature failure.

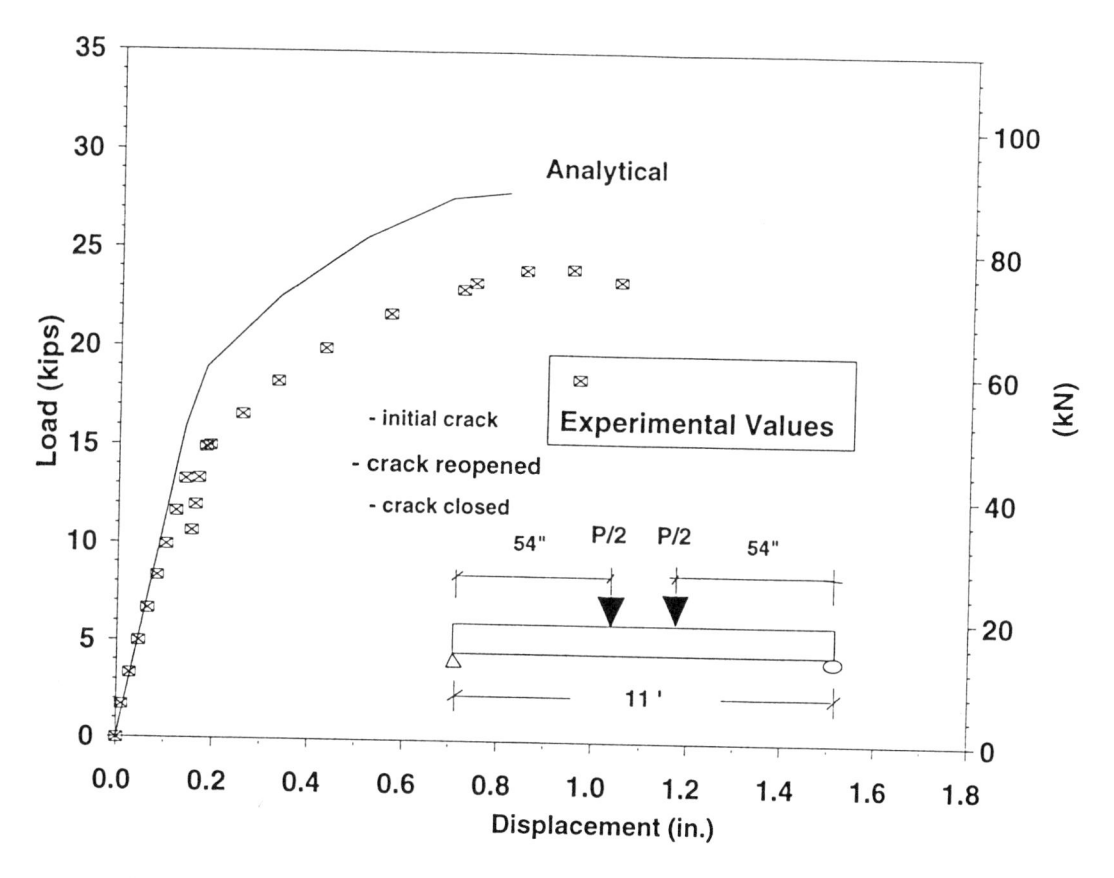

Figure 3 Load vs. midspan deflection for HES S-2 prestressed beam.

At failure, the maximum compressive concrete strains did not reach 0.003. This can be attributed to the premature shear failure of the beams. An analysis of the beam shows the nominal shear capacity (V_n) to be 11,290 lbs (50.34 kN). At an applied load of 24,000 lbs (107.02 kN) and a self-weight of 1076 lbs (4.80 kN) of the beam, the imposed shear force is computed to be 12,538 lbs (55.91 kN).

Transfer Length - One prestressed beam was cast in order to determine the transfer length of the prestressing strands. The beam had a length of 84 in. (213 cm) and a cross section of 12 x 5-1/2 inches (30.5 x 14.0 cm) with a 3-1/2 in. (8.89 cm) effective depth. Four 1/2 in. (12.7 mm) Grade 270 Low Relaxation strands were cast in the beam. The stress in the concrete at the strand level was computed by using the calculated loss of prestressing force. The initial stress in the tendon was measured to be 202 ksi (1396 MPa) resulting in an effective stress of 185 ksi (1279 MPa). The resulting concrete stress was 2167 psi (14.94 MPa).

The results show that the stress in concrete had stabilized at approximately 30 inches (76.2 cm) from the free end, hence the stress in the prestressing steel should be nearly constant at a lesser distance. This discrepancy is due to the edge effect present in all specimens.

The ACI Code equation[2] for predicting transfer length is:

$$L_t = 50 \, d_b \qquad (11)$$

where d_b = Nominal diameter of the prestressing strand.

Based on a linear regression analysis of research data published before 1977, Zia and Mostafa[7] proposed this equation for the transfer length of prestressing strands:

$$L_t = [\, 1.5 \, (f_{si}/f_{ci}) \, d_b \,] - 4.6 \qquad (12)$$

where f_{si} = initial stress in strand before losses, ksi
f_{ci} = compressive strength of concrete at transfer, ksi
d_b = nominal diameter of prestressing strand, inches

Later, Cousins, Zia and Johnston[8] proposed equations from additional published data for coated and uncoated prestressing strands to predict transfer and development length. The transfer length equation for uncoated strands was:

$$L_t = 0.5 \frac{U'_t \sqrt{f_{ci}}}{B} + \frac{f_{se} A_s}{\pi d U'_t \sqrt{f_{ci}}} \qquad (13)$$

where U'_t = 6.7 for uncoated strand
B = 300 psi/inch (81.5 kPa/mm)
f_{ci} = compressive strength of concrete at transfer, psi
f_{se} = effective prestress, psi
A_s = area of prestressing strand, in^2
d = nominal strand diameter, inches.

The equation proposed by Cousins et al. was not for high-early strength concretes.

The observed transfer length of about 30 in (76.2 cm) is a little greater than the prediction of 25 inches (63.5 cm) by the ACI Code equation and falls between the predictions of 26.6 inches (67.7 cm) by equation 4.12 and 39.5 inches (100.2 cm) by equation 4.13. Therefore, it appears that a reasonable estimate of the transfer length of prestressing strands in high-early strength concretes can be made by use of the ACI Code 318-89 equation, and other equations available in literature.

SUMMARY AND CONCLUSIONS

From the results of this investigation, the following conclusions can be drawn for the high-early strength concretes tested.

1. Concrete with a compressive strength of 5000 psi (34 M Pa) in 18 hours can be achieved with high quality aggregates and Type I cement using curing techniques currently in use in the precast/prestressed concrete industry.

2. The predicted values as per ACI 363 for elastic modulus are within ±8% of the observed values for 80% of the specimens tested.
3. The ACI 318 equation for predicting modulus of rupture range from -5% and +7% for 70% of the observed results of the high-early strength concretes tested.
4. The empirical equations of ACI 318 and ACI 363 significantly overpredicted the splitting tensile strength of high-early strength concretes at all ages regardless of the curing method used.
5. The average shrinkage strains of the high-early strength concretes tested are larger than those reported in literature for conventional concretes. This can be attributed to the abnormally low humidity level (below 40%) during the tests.
6. The ultimate creep coefficient for high-early strength concretes are larger than for conventional concretes, and can be attributed to the higher cement content.
7. The ACI Code procedure for shear strength gives an acceptable prediction of the shear capacity of flexural members using high-early strength concrete.
8. The transfer length for prestressing strands in members utilizing high-early strength concretes can be adequately predicted (within ±17%) by the ACI Code equation.

REFERENCES

1. ACI Committee 209, "Prediction of Creep, Shrinkage and Temperature Effects in Concrete Structures, SP-27 American Concrete Institute, Detroit, Michigan, 1971, pp.51-93.
2. ACI Committee 318, "Building Requirements for Reinforced Concrete (ACI 318-89)," American Concrete Institute, Detroit, Michigan, 1989.
3. ACI Committee 363, "State-of-the-Art Report on High Strength Concrete," ACI Journal, Proceedings Vol. 81, No. 4, July-August 1984, pp.364-411.
4. Ahmad, S.H., and Shah S. P., Structural Properties of High Strength Concrete and its Implications for Precast Prestressed Concrete, PCI Journal, Vol. 30, No.6, November-December 1985, pp. 92-117.
5. Oluokon, F. A., Prediction of Concrete Tensile Strength from its Compressive Strength: Evaluation of Existing Relations for Normal Weight Concrete, ACI Materials Journal, Vol. 88, No. 3, May-June 1991, pp. 302-309.
6. Maccaferri, R., Mechanical Properties of High Performance Concrete, Master of Science Thesis, North Carolina State University, 1993.
7. Zia, P. and Mostafa, T., Development Length of Prestressing Strands, PCI Journal, Vol. 22, No. 5, September-October 1977, pp. 55-65.
8. Cousins, T., Zia, P., and Johnston, D., Transfer and Development Length of Epoxy Coated and Uncoated Prestressing Strand, PCI Journal, Vol. 35, No. 4, July-August 1990, pp. 92-102.

Proceedings of the PCI/FHWA
International Symposium on High Performance Concrete
New Orleans, Louisiana, October 20-22, 1997

REINFORCED MEMBRANE ELEMENTS WITH CONCRETE STRENGTH UP TO 100 MPA

Thomas T. C. Hsu
Professor
Dept. of Civil & Environmental Engineering
University of Houston, Houston, Texas, USA

Li-Xin Zhang
Senior Structural Engineer
McDermott Engineering Houston
Houston, Texas, USA

ABSTRACT

Fourteen full-size reinforced panels (or membrane elements) with concrete strength of 100 MPa (14,500 psi) were recently tested using the universal panel tester. Five of these panels were tested under tension-compression to study the effect of concrete strength on the constitutive laws of concrete in compression. Comparing panels with concrete of 100 MPa (14,500 psi) in this research with those of 42 MPa (6,000 psi) and 65 MPa (9,500 psi) tested previously, the softening coefficient was found to be inversely proportional to the square root of concrete strength ($\sqrt{f'_c}$). As a result, we were able to generalize the existing softening coefficient for normal strength concrete to include concrete strengths of up to 100 MPa (14,500 psi). Applying this generalized softening coefficient, we were also able to show that the existing softened-truss models can correctly predict the behavior of the remaining nine test panels subjected to pure shear.

SOFTENED TRUSS MODELS

Brief Description

Wall-type and shell-type reinforced concrete structures, such as offshore platforms, nuclear containment vessels, shear walls, shell roofs and I-girders, can be visualized as assemblies of membrane elements (panels) subjected to in-plane stresses. The key to rational analyses of the behavior of these whole structures is by thoroughly understanding the behavior of reinforced concrete membrane elements isolated from such structures.

A **softened-truss model (STM)**[1,2] has been developed at the University of Houston to predict the response of reinforced concrete membrane elements subjected to in-plane shear and normal stresses. The model is based on the three fundamental principles of mechanics of materials, namely, the stress equilibrium, the strain compatibility and the constitutive laws of materials. Thus, the STM can be generally used to predict the load-deformation relationship throughout a loading history.

The phenomenon of "softening of concrete" was discovered in 1972 when Robinson and Demorieux[3] observed that a reinforced concrete panel subjected to compression in one direction was softened by tension in the perpendicular direction. This biaxially softening phenomenon was later quantified by Vecchio and Collins[4], who proposed a softened stress-strain relationship of concrete in compression. This softened stress-strain relationship was then used in combination with Mohr's stress and strain circles to predict the behavior of elements subjected to shear[5].

Vecchio and Collins' model, known as the "compression field theory," was derived from the rotating angle, defined in the principal directions of cracked concrete. This rotating-angle model was later generalized at the University of Houston into the **rotating-angle softened-truss model (RA-STM)**[6] which was applicable to both shear and torsion. This type of rotating-angle models had the advantage of simplicity, but was incapable of predicting the "concrete contribution" (V_c) to shear resistance.

In order to derive V_c, a **fixed-angle softened-truss model (FA-STM)** was developed in 1996 at the University of Houston[7,8]. In this model the equilibrium and compatibility equations, as well as the materials laws, were developed in the principal directions of the applied stresses. Because this derivation involved the shear stress and shear strain of concrete, the fixed-angle model was more complex, but also more powerful.

Rotating-Angle vs. Fixed-Angle

A reinforced concrete membrane element subjected to in-plane stresses is shown in Fig. 1(a). The applied stresses, $\tau_{\ell t}$, σ_ℓ and σ_t, are oriented in the ℓ-t coordinate system of the longitudinal and transverse steel. The applied principal stresses, σ_2 and σ_1, for the element are defined in the principal 2-1 coordinate system as shown in Fig. 1(d). The angle between the 2-1 coordinate and the ℓ-t coordinate of steel bars is defined as the **fixed-angle** α_2, because this angle does not change when the three applied stresses, $\tau_{\ell t}$, σ_ℓ, and σ_t, increase proportionally.

Before cracking, the principal stresses in the concrete, Fig. 1 (b), coincide with the applied principal stresses σ_1 and σ_2. When the principal tensile stress σ_1 reaches the tensile strength of concrete, cracks will form and the concrete will be separated by the cracks into a series of concrete struts in the 2-direction as shown in Fig. 1(f). If the element is reinforced with different amounts of steel in the ℓ- and t-directions [i.e., $\rho_\ell f_\ell \neq \rho_t f_t$ in Fig. 1(c)], the directions of the post-cracking principal stresses in **concrete** will "rotate away" from the principal directions of the proportionally increasing applied stresses. These new directions of the post-cracking principal stresses in concrete are defined by the d-r coordinate system shown in Fig. 1(e). The angle between the post-cracking principal d-r coordinate and the ℓ-t coordinate of steel bars is defined as the **rotating-angle** α.

In the **RA-STM** the direction of cracks is assumed to coincide with the principal d-direction in the cracked concrete, Fig. 1(g). The derivations of all the equations are based on the d-r coordinate. In the **FA-STM**, however, the direction of the cracks is assumed to coincide with the principal 2-direction of the applied stresses, Fig. 1(f). All the equilibrium and compatibility equations are derived based on the 2-1 coordinate.

Fixed-Angle Softened-Truss Model

The equilibrium and compatibility equations are derived utilizing the principle of transformation for the stresses and strains in the concrete[1,2,7]:

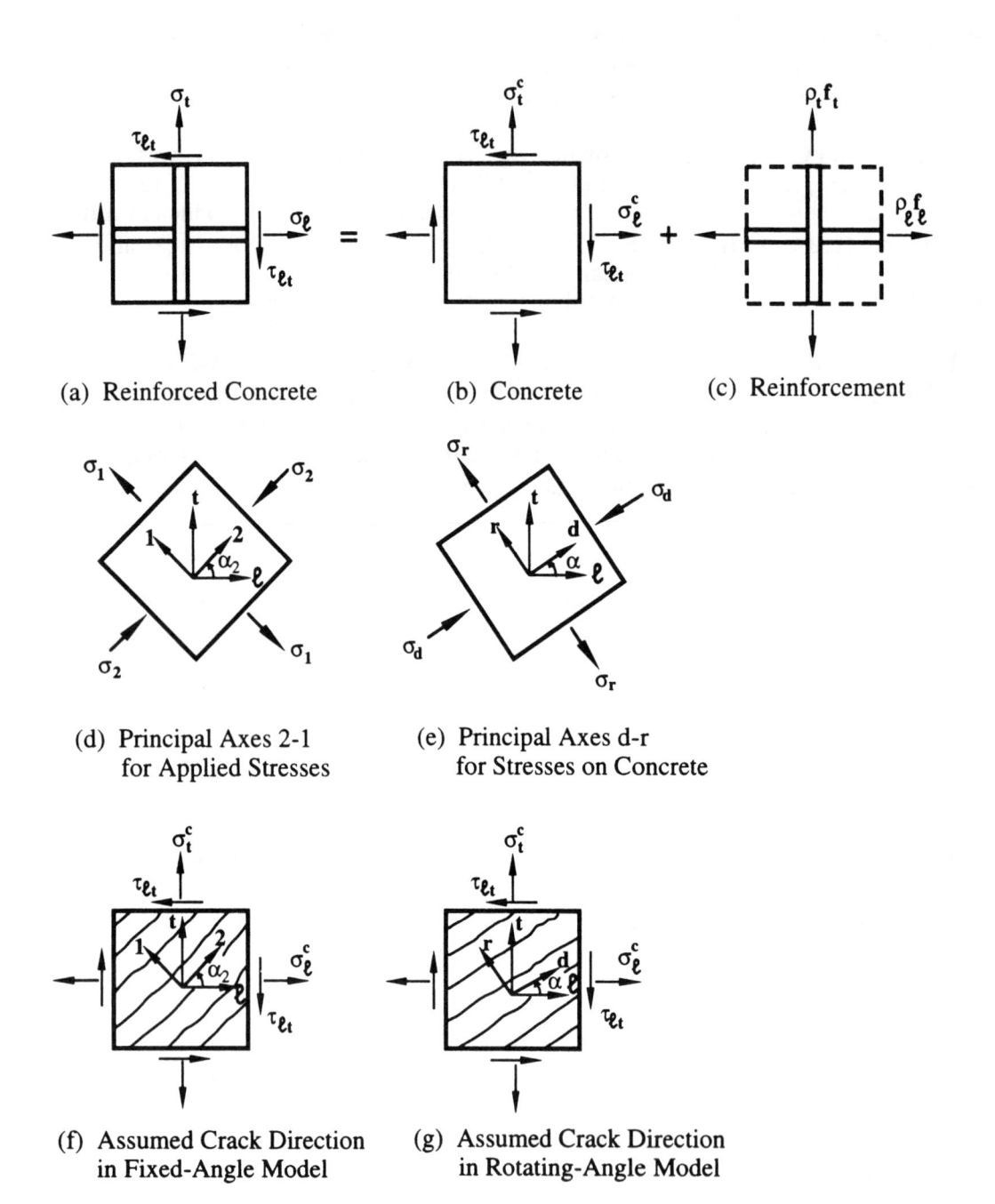

(a) Reinforced Concrete (b) Concrete (c) Reinforcement

(d) Principal Axes 2-1 for Applied Stresses

(e) Principal Axes d-r for Stresses on Concrete

(f) Assumed Crack Direction in Fixed-Angle Model

(g) Assumed Crack Direction in Rotating-Angle Model

Fig. 1 Reinforced Concrete Membrane Elements Subjected to In-plane Stresses

Equilibrium Equations

$$\sigma_\ell = \sigma_2^c \cos^2\alpha_2 + \sigma_1^c \sin^2\alpha_2 + \tau_{21}^c\, 2\sin\alpha_2\cos\alpha_2 + \rho_\ell f_\ell \quad (1)$$

$$\sigma_t = \sigma_2^c \sin^2\alpha_2 + \sigma_1^c \cos^2\alpha_2 - \tau_{21}^c\, 2\sin\alpha_2\cos\alpha_2 + \rho_t f_t \quad (2)$$

$$\tau_{\ell t} = (-\sigma_2^c + \sigma_1^c)\sin\alpha_2\cos\alpha_2 + \tau_{21}^c(\cos^2\alpha_2 - \sin^2\alpha_2) \quad (3)$$

Compatibility Equations

$$\varepsilon_\ell = \varepsilon_2 \cos^2\alpha_2 + \varepsilon_1 \sin^2\alpha_2 + \gamma_{21} \sin\alpha_2 \cos\alpha_2 \qquad (4)$$

$$\varepsilon_t = \varepsilon_2 \sin^2\alpha_2 + \varepsilon_1 \cos^2\alpha_2 - \gamma_{21} \sin\alpha_2 \cos\alpha_2 \qquad (5)$$

$$\gamma_{\ell t} = 2(-\varepsilon_2 + \varepsilon_1)\sin\alpha_2 \cos\alpha_2 + \gamma_{21}(\cos^2\alpha_2 - \sin^2\alpha_2) \qquad (6)$$

where the symbols are explained in the List of Notations.

The fixed-angle softened-truss model degenerates into the rotating-angle softened-truss model when the 2-1 coordinate becomes the d-r coordinate, the fixed-angle α_2 becomes the rotating-angle α, and the shear stress and shear strain, τ_{21}^c and γ_{21}, become zero in Eq. (1) to (6).

Constitutive Laws

The four constitutive laws required in the fixed-angle softened-truss model are based on the smeared crack concepts. An average (or smeared) stress means an average value of the stresses from a crack to midway between two cracks; an average (or smeared) strain is measured from the displacement over a length that traverses several cracks, thus including the crack widths. The four average stress-strain curves proposed by Belarbi and Hsu[9,10], Pang and Hsu[7,11] and Hsu and Zhang[12] are given in Fig. 2 (a) to (d) as follows:

Fig. 2 (a): average compression stress-strain curve of concrete.
Fig. 2 (b): average tensile stress-strain curve of concrete.
Fig. 2 (c): average shear stress-strain curve of concrete.
Fig. 2 (d): average stress-strain curve of steel bars embedded in concrete.

Application of Softened-Truss Models to High-Strength Concrete

An accurate prediction of the panel behavior by the softened-truss models depends on a good set of constitutive laws for concrete and steel in membrane elements (or panels). The constitutive laws of panels given in Fig. 2 are applicable only to normal strength concrete (42 MPa or 6,000 psi). These constitutive laws were modified for medium-high strength concrete (65 MPa or 9,500 psi) by Zhang[13]. This paper reports the tests of five panels to determine the constitutive laws of 100 MPa (14,500 psi) concrete. Nine additional panels with 100 MPa (14,500 psi) concrete were also tested to check the applicability of the softened truss models in predicting the behavior of panels subjected to pure shear. As a result, the softened-truss models are now applicable to panels made of concrete with compressive strength up to 100 MPa (14,500 psi).

TESTS OF 100 MPA CONCRETE PANELS

Fourteen full-size reinforced panels of 100 MPa (14,500 psi) concrete were recently tested[14]. The fourteen panels are divided into three series: VE-, VA-, and VB-series.

The dimensions and steel arrangements for panels in VE-series are shown in Fig. 3(a). In this series the longitudinal reinforcement (ℓ-direction) was placed perpendicular to the direction of the applied compressive loads (2-direction), giving a α_2 angle of 90°. For convenience, panels with α_2 angle equal to 90° will be called "90° panels." Panels in VA-

and VB-series were subjected to pure shear in the ℓ-t coordinates of the steel bars. In these two series, the longitudinal and transverse reinforcements in the ℓ-t coordinates are oriented at an angle of 45° to the principal 2-1 coordinates of the applied stresses ($\alpha_2 = 45°$) as shown in Fig. 3(b). Panels with α_2 angle equal to 45° will be called "45° panels."

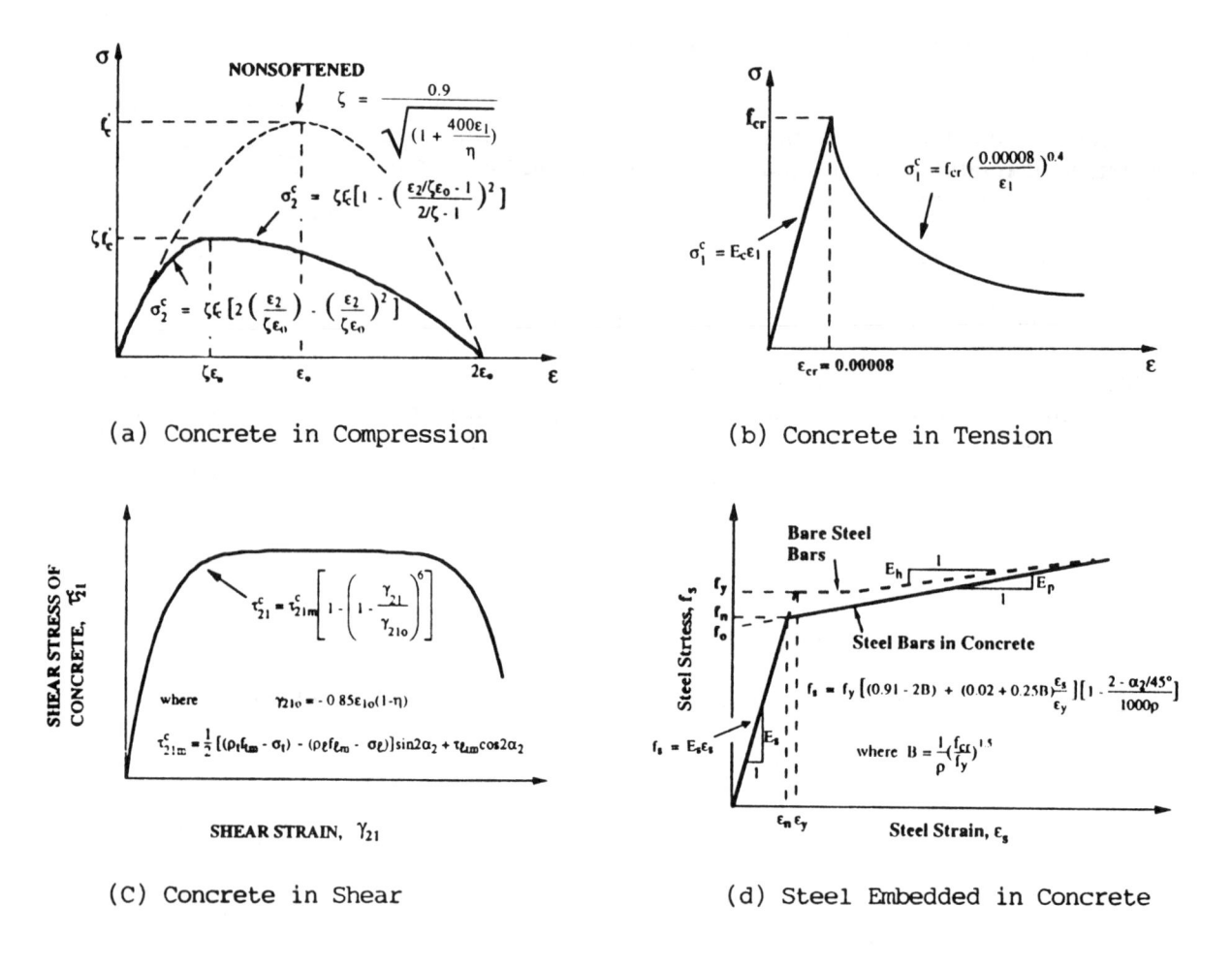

Fig. 2 Constitutive Laws of Concrete and Steel Bars

(a) Concrete in Compression

(b) Concrete in Tension

(c) Concrete in Shear

(d) Steel Embedded in Concrete

90° Panels (Series VE)

The tests on the five 90° panels in the VE-series are designed to study the fundamental constitutive laws of concrete and steel bars. The compressive loading and the tensile loading were applied sequentially with the tensile loading first. The principal tensile strain ε_1 in the horizontal direction was first increased up to a point where a desired value was achieved. The principal compressive strain ε_2 was then gradually increased in the vertical direction until failure, while the servo-control system maintained the desired principal tensile strain ε_1 constant. The main variable in this series was the principal tensile strains ε_1 of 0.0065, 0.010, 0.020, and 0.030, for panels VE0, VE1 (and VE2), VE3, and VE4, respectively.

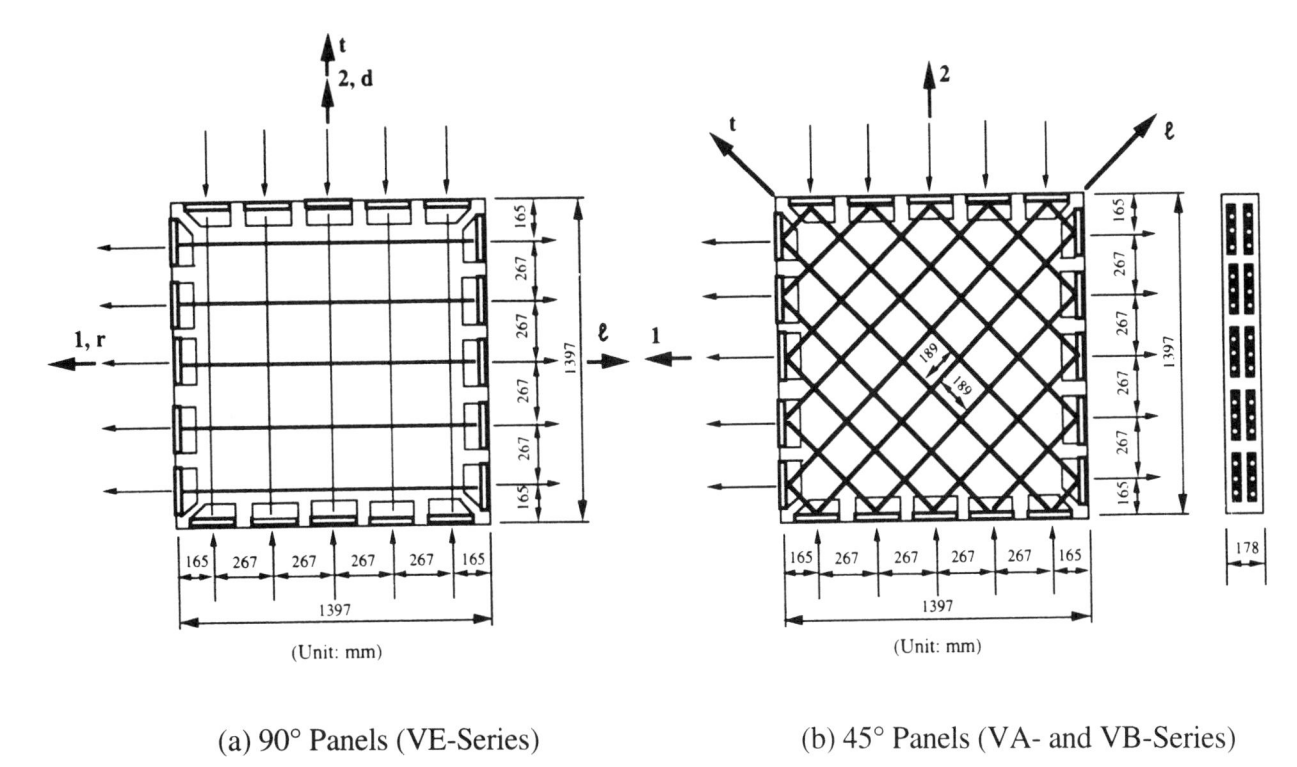

(a) 90° Panels (VE-Series)   (b) 45° Panels (VA- and VB-Series)

Fig. 3. Dimensions and Coordinates of Test Panels

45° Panels (Series VA and VB)

The five 45° panels in VA-Series -- VA0, VA1, VA2, VA3 and VA4 -- were reinforced equally in the ℓ- and t-directions with 0.60%, 1.20%, 2.39%, 3.59% and 5.24% of steel, respectively. They were subjected to pure shear loading in the directions of steel bars by applying equal magnitudes of principal compressive and tensile stresses. The configuration and loading of the four 45° panels in VB-series was the same as that in the VA-series, except that the steel ratios in the longitudinal direction (ℓ-axis) were greater than those in the transverse direction (t-axis). The main variable of panels in VB-series was the ratio of the transverse steel force to the longitudinal steel force, η ($= \rho_t f_{ty}/\rho_\ell f_{\ell y}$), which was designed to be 0.546, 0.332, and 0.189 corresponding to test panels VB1, VB2, and VB3, respectively. The η ratio of panel VB4 was the same as that of panel VB2, but the steel percentage of VB4 was only one-half the steel percentage of VB2.

SOFTENED COEFFICIENT FOR CONCRETE

The softened coefficient ζ in Fig. 2 (a) for normal strength concrete of 42 MPa was found to be a function of the tensile strain ε_1 and the parameter η[7,8]:

$$\zeta = \frac{0.9}{\sqrt{(1 + \frac{400\varepsilon_1}{\eta})}} \qquad (7)$$

Eq. (7) can be generalized to include high strength concrete, if the constant 0.9 is replaced by a factor $R(f_c')$ which is a function of concrete cylinder strength f_c' :

$$\zeta = \frac{R(f_c')}{\sqrt{(1 + \frac{400\varepsilon_1}{\eta})}} \qquad (8)$$

The $R(f_c')$ values in Eq. (8) are calculated from the measured values of ζ and ε_1 obtained from three sources: (1) VE-series of 90° panels with high strength concrete of 100 MPa (14,500 psi)[14], (2) HE-series of 90° panels with medium-high strength concrete of 65 MPa (9,500 psi)[13], and (3) E-series of 90° panels with normal strength concrete of 42 MPa (6,000 psi)[10]. Relating $R(f_c')$ to the concrete cylinder strengths f_c' in Fig. 4 shows that $R(f_c')$ is inversely proportional to $\sqrt{f_c'}$:

$$R(f_c') = \frac{5.8}{\sqrt{f_c'}} \qquad (9)$$

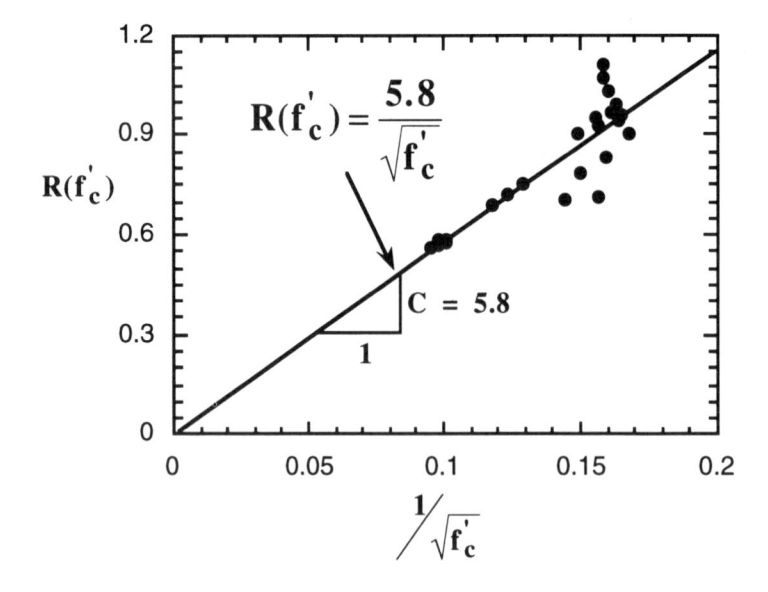

Fig. 4. Relationship Between $R(f_c')$ and $1/\sqrt{f_c'}$

Substituting Eq. (9) into Eq. (8), the equation of stress-softened coefficient ζ for concrete strength up to 100 MPa becomes

$$\zeta = \frac{5.8}{\sqrt{f_c'(\text{MPa})}} \frac{1}{\sqrt{(1 + \frac{400\varepsilon_1}{\eta})}} \quad (10)$$

ANALYSIS OF 45° TEST PANELS

Since 45° test panels are subjected to pure shear, their behavior are best described by the applied shear stress vs. shear strain curves ($\tau_{\ell t}$-$\gamma_{\ell t}$ curves) shown in Figs. 5 and 6. The theoretical curves are calculated by the equilibrium and compatibility equations, (1) to (6), and the constitutive laws in Fig. 2, except that the generalized softening coefficient, Eq. (10), is used in Fig. 2 (a).

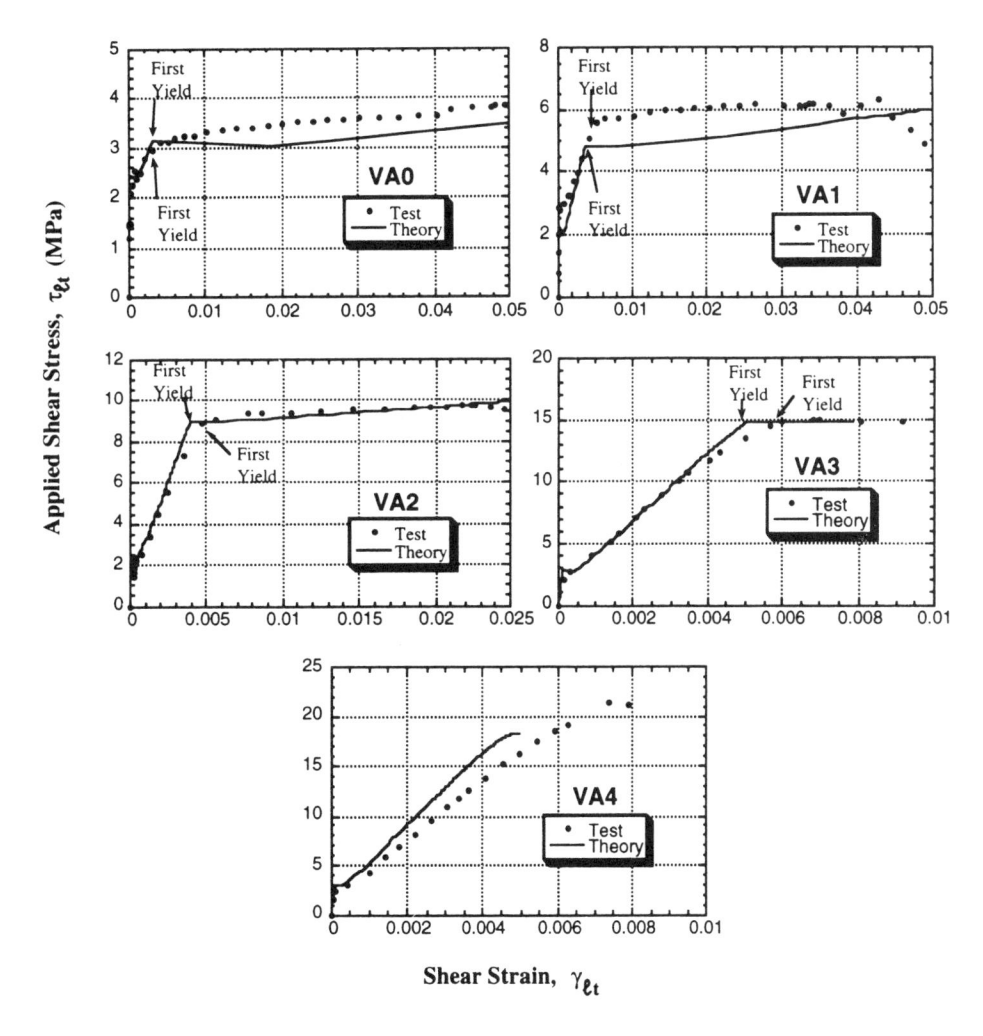

Fig. 5 Shear Stress vs. Shear Strain Curves of Panels in VA-Series Predicted by FA-STM

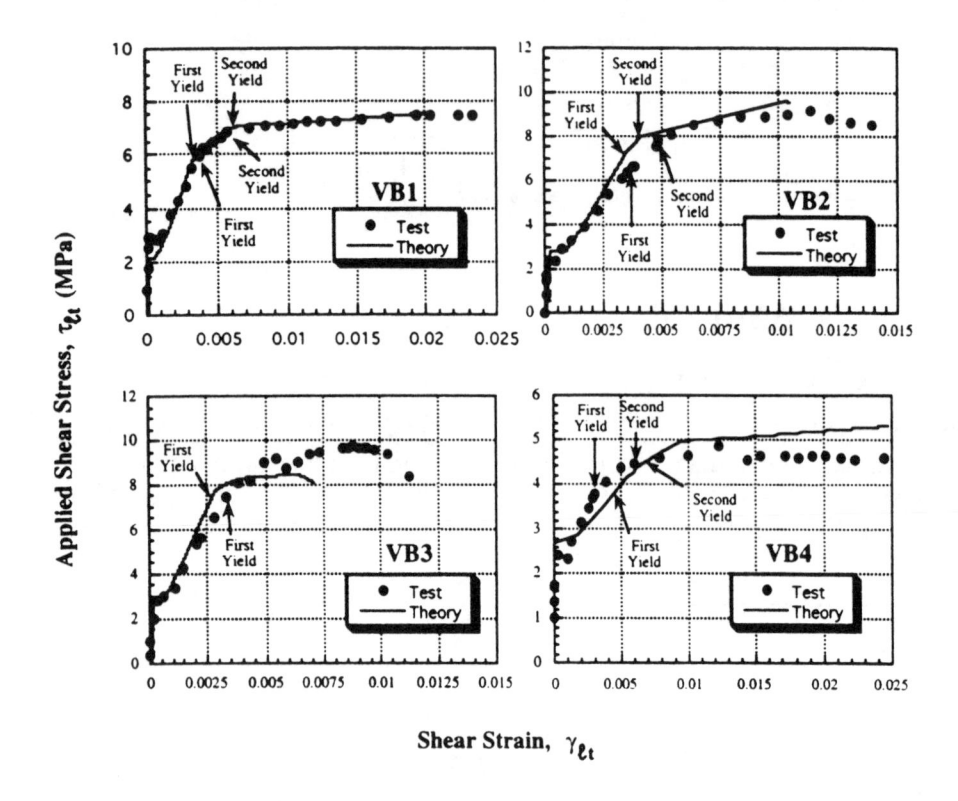

Fig. 6 Shear Stress vs. Shear Strain Curves of Panels in VB-Series Predicted by FA-STM

Panels in VA-Series --- The steel ratios of panels in Fig. 5 are the same in the ℓ- and t-directions. For these panels, all the governing equilibrium and compatibility equations, Eq. (1) to (6), as well as the constitutive laws of materials, Fig. 2, in the FA-STM become identical to those in the RA-STM. Consequently, the prediction of VA-series by these two models are the same.

The theoretical $\tau_{\ell t}$-$\gamma_{\ell t}$ curves predicted by the FA-STM and RA-STM, together with the experimental curves, are shown in Fig. 5 for the five panels in VA-series. There are good agreements between the theoretical predictions and the experimental results. The four panels VA0 to VA3 are under-reinforced, while the panel VA4 is over-reinforced. Thus, Fig. 5 shows that the softened truss models are applicable to both these two modes of failure.

Panels in VB-Series --- In this series of panels, Fig. 6, the steel ratios in the ℓ-direction are larger than those in the t-direction. Therefore, the predictions by FA-STM and RA-STM are different. For simplicity, only the prediction of the more powerful FA-STM is presented here.

The theoretical and experimental $\tau_{\ell t}$-$\gamma_{\ell t}$ curves for the four panels in Fig. 6 demonstrate excellent agreement between theoretical predictions and experimental results.

This agreement is particularly significant because the four panels have widely different η ratios of transverse-to-longitudinal steel force. Even panel VB3, which is partially under-reinforced in the t-direction (low η value of 0.189), provides a reasonable theoretical $\tau_{\ell t}$-$\gamma_{\ell t}$ curve that terminates well beyond the first yield of steel.

CONCLUSIONS

The behavior of reinforced concrete membrane elements with normal strength concrete can be predicted by the rotating-angle softened-truss model (RA-STM) and the fixed-angle softened-truss model (FA-STM). This paper shows that with a minor modification these models are applicable to elements made of concrete up to 100 MPa (14,500 psi).

The only required modification of the constitutive laws is the softening coefficient ζ in the stress-strain curve of concrete in compression. As derived from our panel tests, the softening coefficient, Eqs. (10), is inversely proportional of the square root of concrete strength ($\sqrt{f'_c}$), in addition to being a function of the principal tensile strain (ε_1) and the parameter η.

ACKNOWLEDGMENT

This research was supported by National Science Foundation Grants No. MSS-9114543 and No. CMS-9213707. These generous supports are gratefully acknowledged.

REFERENCES

1. Hsu, T. T. C., <u>Unified Theory of Reinforced Concrete,</u> CRC Press, Inc., 1993, 336 pp.

2. Hsu, T. T. C., "Toward a Unified Nomenclature for Reinforced Concrete Theory," <u>Journal of Structural Engineering,</u> ASCE, Vol. 122, No. 3, March 1996, pp. 275-283.

3. Robinson, J. R. and Demorieux, J. M., "Essais de Traction-Compression sur Modeles d'Ame de Poutre en Beton Arme," <u>IRABA Report,</u> Institut de Recherches Appliquees du Beton de L'Ame, Part 1, June 1968, 44 pp; "Resistance Ultimate du Beton de L'ame de Poutres en Double Te en Beton Arme," Part 2, May 1972, 53 pp.

4. Vecchio, F., and Collins, M. P., "Stress-Strain Characteristic of Reinforced Concrete in Pure Shear," <u>IABSE Colloquium, Advanced Mechanics of Reinforced Concrete,</u> Delft, Final Report, International Association of Bridge and Structural Engineering, Zurich, Switzerland, 1981, pp. 221-225.

5. Vecchio, F. J. and Collins, M. P., "The Response of Reinforced Concrete to In-Plane Shear and Normal Stresses," <u>Publication No. 82-03</u> (ISBN 0-7727-7029-8), Department of Civil Engineering, University of Toronto, Toronto, Canada, 1982, 332 pp.

6. Hsu, T. T. C., "Softened Truss Model Theory for Shear and Torsion," <u>Structural Journal of the American Concrete Institute,</u> Vol. 85, Nov. 6, Nov.-Dec. 1988, pp. 624-635.

7. Pang, X. B. and Hsu, T. T. C., "Fixed-Angle Softened-Truss Model for Reinforced Concrete," Structural Journal of the American Concrete Institute, Vol. 93, No. 2, Mar.-Apr. 1996, pp. 197-207.

8. Hsu, T. T. C. and Zhang, L. X., "Nonlinear Analysis of Membrane Elements by Fixed-Angle Softened-Truss Model," Structural Journal of the American Concrete Institute, Vol. 94, No. 5, Sept.-Oct. 1997.

9. Belarbi, A. and T. T. C. Hsu, "Constitutive Laws of Concrete in Tension and Reinforcing Bars Stiffened by Concrete," Structural Journal of the American Concrete Institute, Vol. 91, No. 4, July-Aug. 1994, pp. 465-474.

10. Belarbi, A. and T. T. C. Hsu, "Constitutive Laws of Softened Concrete in Biaxial Tension-Compression," Structural Journal of the American Concrete Institute, Vol. 92, No. 5, Sept.-Oct. 1995, pp. 562-573.

11. Pang, X. B. and Hsu, T. T. C., "Behavior of Reinforced Concrete Membrane Elements in Shear," Structural Journal of the American Concrete Institute, Vol. 92, No. 6, Nov.-Dec. 1995, pp. 665-679.

12. Hsu, T. T. C. and Zhang, L. X., "Tension Stiffening in Reinforced Concrete Membrane Elements," Structural Journal of the American Concrete Institute, Vol. 93, No. 1, Jan.-Feb. 1996, pp. 108-115.

13. Zhang, L. X., "Constitutive Laws of Reinforced Elements with Medium-High Strength Concrete," Master's Thesis, Department of Civil and Environmental Engineering, University of Houston, Houston, TX, 1992, 245 pp.

14. Zhang, L. X., "Constitutive Laws of Reinforced Elements with High Strength Concrete," Ph.D. Dissertation, Department of Civil and Environmental Engineering, University of Houston, Houston, TX, 1995, 303 pp.

LIST OF NOTATIONS

E_c = elastic modulus of concrete
E_s = elastic modulus of bare steel bars
f'_c = cylinder compressive strength of concrete
f_{cr} = tensile strength of concrete
f_ℓ, f_t = average steel stresses in longitudinal (ℓ-) and transverse (t-) directions, respectively
f_s = stress in mild steel; f_s becomes f_ℓ or f_t, when applied to the longitudinal and transverse steel, respectively
f_y = yield strength in bare steel bars
α = rotating-angle; angle between the principal compressive stress of concrete (d-axis) and the longitudinal steel bars (ℓ-axis)
α_2 = fixed-angle or steel bar angle; angle between the applied principal compressive stress (2-axis) and the longitudinal steel bars (ℓ-axis)
ε_{cr} = tensile cracking strain of concrete, taken as 0.00008

$\varepsilon_\ell, \varepsilon_t$	=	average strains in ℓ- and t-directions, respectively, of the steel bars
$\varepsilon_d, \varepsilon_r$	=	average concrete tensile strains in the principal d- and r-direction, respectively
ε_s	=	strain in mild steel; ε_s becomes ε_ℓ or ε_t, when applied to the longitudinal and transverse steel, respectively
$\varepsilon_2, \varepsilon_1$	=	average normal strains in the 2- and 1-directions, respectively
γ_{dr}	=	average shear strain in d-r coordinate
$\gamma_{\ell t}$	=	average shear strain in ℓ-t coordinate
γ_{21}	=	average shear strain in 2-1 coordinate
γ_{21m}	=	average shear strain corresponding to τ^c_{21m}
η	=	ratio defined as $(\rho_t f_{ty} - \sigma_t)/(\rho_\ell f_{\ell y} - \sigma_\ell)$. $\eta = \rho_t f_{ty}/\rho_\ell f_{\ell y}$ under pure shear
ρ	=	steel ratio, ρ becomes ρ_ℓ or ρ_t, when applied to the longitudinal and transverse steel, respectively
ρ_ℓ, ρ_t	=	steel ratios in the longitudinal (ℓ-) and transverse (t-) directions, respectively
σ_d, σ_r	=	average concrete compressive stresses in the principal d- and r-directions, respectively
σ_ℓ, σ_t	=	applied normal stresses in the ℓ- and t-directions, respectively, of steel bars
σ_2, σ_1	=	applied principal stresses in the 2- and 1-directions, respectively
σ^c_2, σ^c_1	=	average normal stresses of concrete in the 2- and 1-directions, respectively
$\tau_{\ell t}$	=	applied shear stress in ℓ-t coordinate
τ^c_{21}	=	average shear stress of concrete in the 2-1 coordinate
τ^c_{21m}	=	the maximum shear stress of concrete in the 2-1 coordinate
ζ	=	softened coefficient of concrete in compressive stress-strain curve

MECHANICAL PROPERTIES AND STRUCTURAL RESPONSE OF HIGH-PERFORMANCE CONCRETE BEAMS

G. Rosati, Ph.D.
Assistant Professor
Politecnico of Milan, Italy

S. Cattaneo
M.S. Student
Politecnico of Milan, Italy

M. Marazzini
Architect
Milan, Italy

A. Meda
M.S. Student
Politecnico of Milan, Italy

ABSTRACT

The load-displacement response of a concrete-like specimen shows a size dependence. An identification procedure is presented that allows to evaluate a constitutive behavior without structural effects. The results are applied to the determination of the mechanical properties of high-performance fiber-reinforced cement-based materials. With this objective, three-point bending tests and tensile tests were performed on specimens made of different materials, by means of through a closed-loop testing machine.

INTRODUCTION

In many engineering fields, the study of composite materials is nowadays an important topic[1,2]. Among these materials, high performance cementitious composites are specifically suited to Civil Engineering, owing to their durability, abrasion resistance, early strength and volume stability[3]. Compared to normal-strength concretes, high-performance concretes are more expensive and their use increases the initial cost of a structure, but - bearing in mind the service life - there are many economical advantages, such as a decrease in dimensions, weight, maintenance, not to speak of the lower environmental impact[4].
It is well recognized that high-strength concrete is less ductile than normal-strength concrete[5], and the introduction in the matrix[6] of an adequate amount of dispersed fibers is instrumental in improving the toughness of the material and - in the end - the ductility of a structure.
Some structural elements are so important to the performance of the whole structure that

their initial integrity must be safeguarded at any cost, without compromising efficiency. Pylons carrying major bridges, and underwater tunnel liners may be nearly unrepairable and irreplaceable if they suffer major deterioration.

Water and salt are among the most aggressive substances which endanger the durability of concrete structures. In order to obtain long-term durability, a solution is to produce concrete having very low permeability. The permeability of concrete is related to its pore structure, and diffusion and transport processes must be fully understood if the improvement of durability is an issue.

In this paper, the mechanical behaviour of a plain and fiber-reinforced high-performance concretes (f_c=130-150 MPa (18.8-21.7 ksi)) is examined. Direct tensile tests and three-point bending tests are discussed and an identification procedure that allows to determine stress-strain and stress-displacement curves unaffected by any undesired structural effect is presented.

EXPERIMENTAL ANALYSIS

To obtain the correct mix-design of the DSP (Densified with small particles) micro-concrete more then 140 bending tests (three point bending) were performed on prismatic elements 40x40x160 mm (1.57x1.57x6.30 in.) controlled by CMOD (crack-mouth opening displacement). The broken pieces were subjected to compression tests.

The three-point bending tests were also used for the selection of the superplasticizer. The most efficient was the polyacrylate-based dispersing agent, for which it was possible to get materials with a lower porosity. This is demonstrated in Fig.1a, where the nominal tensile strenghts, as a function of two superplasticizers (acrilic and melaminic) are compared. A significantly poorer structural response was obtained with the sulphonated naphthalene formaldehyde condensate superplasticizer.

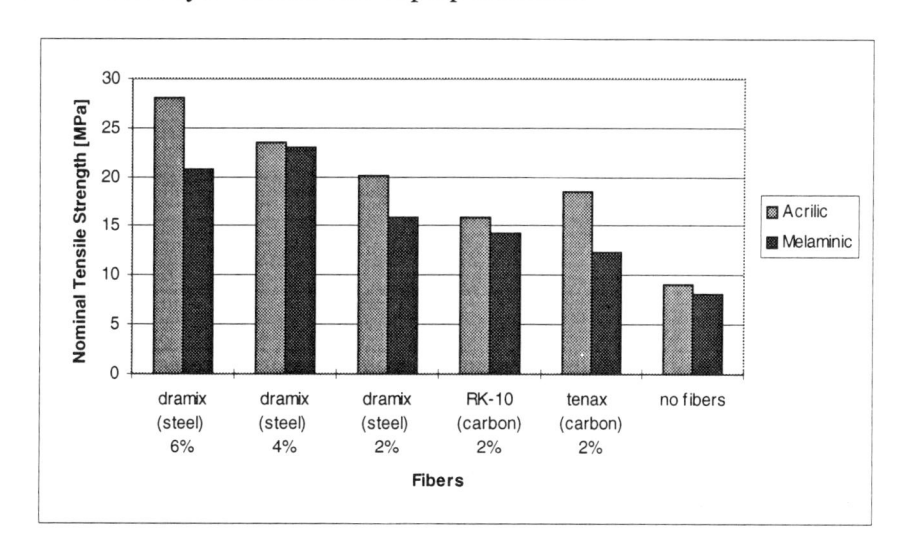

Fig. 1a. Influence of superplasticizer on nominal tensile strengths

After determining the optimal mix-design (Tab. 1), and the ideal fiber (steel) and its contents, curing tests were made, with the results shown in Fig. 1b, to display the independence from this factor.

A new series of tests was carried out in order to determine the Young's modulus, Fig. 2, and better know the mechanical behavior of this material.

To carry out the tensile and bending tests, a suitably modified 100 kN (22481 lb.) electro-mechanical Instron test machine was used. The main characteristics of this machine are a minimum speed of 2 μm per hour (7.9 10^{-5} in. per hour), three control channels, one of them external for choosing the feedback signal for a stable control of the test, closed loop control with integral and derivative gain (for possibility of removing the effect of finite axial stiffness of the machine to avoid failure due to instability).

Table 1. Ideal mix-design

MIX-DESIGN	NO FIBERS	FIBERS (2/4% in volume)
cement 52.5	696 kg/m^3	667 kg/m^3
silica fume (SF/C)	174 kg/m^3 (25%)	167 kg/m^3 (25%)
quartz 0.06-3.2 mm	1300 kg/m^3	1246 kg/m^3
superplasticiser (SP/C)	14.4 kg/m^3 (2.1%)	13.8 kg/m^3 (2.1%)
water (W/C+SF)	196 kg/m^3 (22.5%)	187 kg/m^3 (22.4%)
steel fibers (v$_f$ %)	-	162/324 kg/m^3 (2.1/4.2%)

Note: 1m = 3.2808 ft., 1 Kg = 2.2046 lb., 1 mm = 0.039 in.

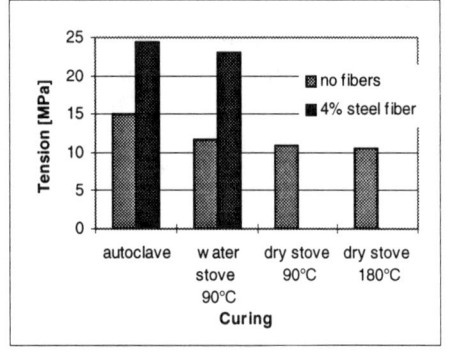

Note: 1 MPa = 0.145 ksi

Fig. 1b. Influence of curing on tensile and compressive strengths

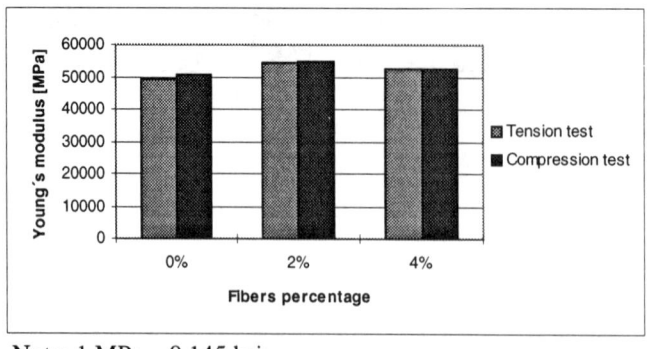

Note: 1 MPa = 0.145 ksi

Fig. 2. Influence of fiber content on tensile and compressive modulus

In tensile tests shown in Fig. 3a, the displacements were measured using four LVDT's, having a sensitivity of 0.2 µm (7.9 10^{-6} in.), set across the notch on a 50 mm (1.97 in.) base, arranged at 90° around the specimen axis. The relative shift of the machine heads was measured at four points. On the specimen portion outside the notched zone, resistance measuring instruments (strain gauges) were set in two diametrically opposite points. The most important modification made to the press consisted in the addition of four dynamometric bars, adjustable by screw/nut-thread/thrust-bearing, that connected the lower plate of the actuator with the fixed cross piece. The high-sensitivity loading cell was connected to the cross piece, so that it could directly measure the load acting on the specimen. The bars made it possible to intervene at each loading step on any undesired rotation, assuring uniformity of displacement within a 1 µm (3.9 10^{-5} in.) tolerance. In this way, it was possible to impose a step-by-step uniform crack opening through the critical cross section of the specimen.

In bending tests, as shown in Fig. 3b, the feedback signal was the crack opening displacement, measured by a clip-gauge placed at the bottom edge astride the midspan section.

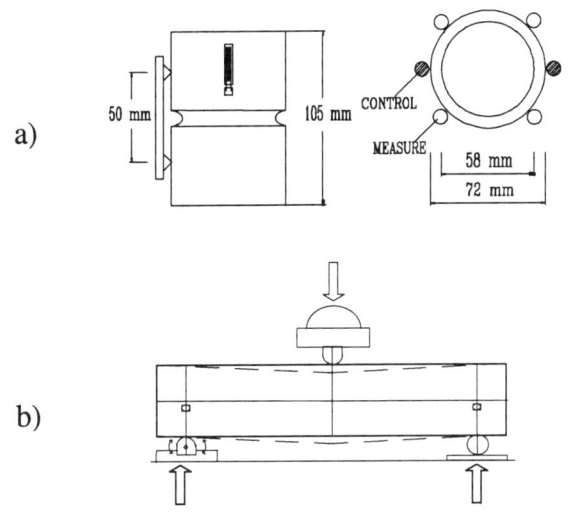

Note: 1 mm = 0.039 in.

Fig. 3. Specimens in tension (a) and in bending (b)

About it, bending, tension and compression tests were realized:
- for the bending tests prismatic specimens with dimensions 100x100x400 mm (3.94x3.94x15.74 in.) were utilized;
- for tension tests, cylinders with a diameter of 72 mm (2.83 in.) and height of 105 mm (4.13 in.), were used;
- for the compression one the two half pieces of the specimen obtained from the flexure test were used again.

The bending and tension test results (with fiber contents 0-4% by volume) are showed in Fig. 4.

MODELING

A mathematical model is formulated, based on both a non-linear elastic analysis for the solid state and a cohesive analysis for the cracked state. The elastic analysis is formulated in orthogonal hyperbolic coordinates to model the notch geometry and accounts for a plastic region in front of the notch-tip to predict the pre-peak non-linear behavior. The cohesive analysis first investigates the final localizing process (aggregate debonding) and then, on the basis of this investigation, considers the early localization (coalescence), otherwise difficult to identify.

The proposed model provides a closed-form solution for the following aspects: notch-sensitivity and strength scaling laws, effective non-linear stress - strain relation, extension of the microcracked region and effective stress - crack opening relation.

The simultaneous application of correct testing procedures and this model allows us to predict more objective values for the material properties.

Experimental evidence shows that the peak load, in a specimen in a tensile or bending test, is reached when a localization of deformation is already developed.

This means that a process zone and, subsequently, a crack are formed in a critical zone of the specimen.

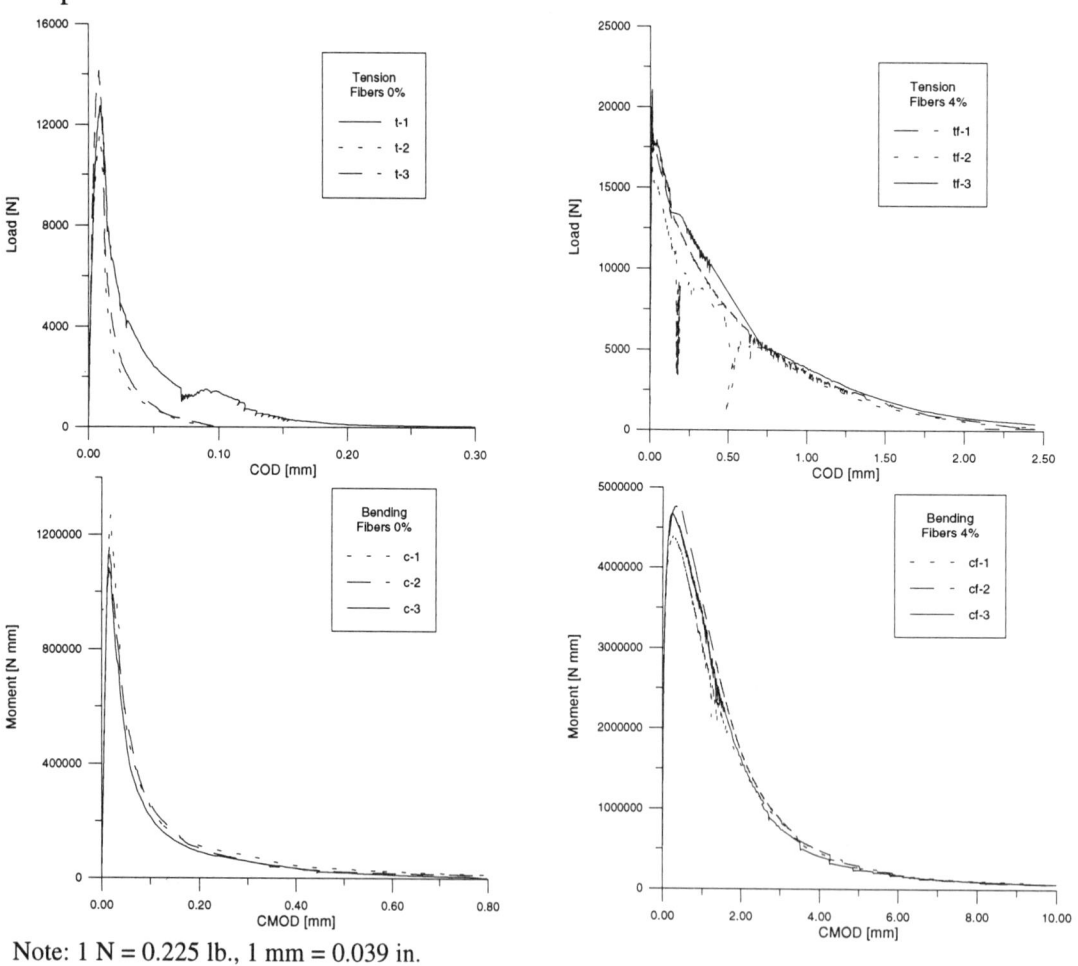

Note: 1 N = 0.225 lb., 1 mm = 0.039 in.

Fig. 4. Experimental curves in tension (top) and in bending (bottom)

Concrete under tension

A strength criterion suggests that the onset of a crack in direct tension is due to achievement of the true tensile strength of the material and a notch effect is produced. In this condition:

$$f_{ct} = \sigma_{Nt} \alpha_{K\rho small} \tag{1}$$

where f_{ct} is the true tensile strength of the material, σ_{Nt} is the stress in the specimen evaluated with the classical beam theory at the load level of the onset of cracking (90% of the nominal strength) and $\alpha_{K\rho small}$ is the stress concentration factor, function of the load and the singularity geometry, defined, in according with Neuber's theory[7], by:

$$\alpha_{K\rho small} = 1 + \sqrt{(a/\rho)/(1+(a/4t))} / (1+\sqrt{\rho'/\rho}) \tag{2}$$

where a is the net area radius, t is the notch depth, ρ is the notch radius and ρ' is half of the maximum aggregate size (the stresses acting on its side-surfaces can be uniformly distributed).
To describe the behavior of cracked concrete under tension, in the purely cohesive phase, a hyperbolic stress σ_w - crack opening w relation is adopted:

$$\sigma_w = f_{ct} (1 - w/w_c) / (1 + k w / \phi_a) \tag{3}$$

where ϕ_a is the maximum aggregate size, w_c is the stress-free crack width (experimentally evaluated), and k is a coefficient related to the specific fracture energy G_F.
The value of k can be determined using a best-fitting procedure and consequently G_F is the definite integral of Eq.3, evaluated from zero to w_c.

Concrete under bending

The same tension strength f_{ct} can be obtained by the relation:

$$f_{ct} = \sigma_{Nt} \alpha_{SM} \tag{4}$$

applied to bending specimen without notch, where σ_{Nb} is the 50% of the nominal strength and α_{SM} is given by

$$\alpha_{SM} = 1 + \sqrt{0.663(a/\rho')/(1+0.166 a/(t+t_0))} \tag{5}$$

with a is the ligament depth, ρ' is the same parameter of Eq.2, t is the crack depth and t_0 is the unknown value of the material dominant defect.
In this case t is set to zero because the equation is applied to the onset of cracking.

The crack depth t, during its propagation, can be related to the crack opening w_m by the expression:

$$t = w_m(1+k_t)/(w_u + k_t w_m) \qquad (6)$$

where w_u is the bending stress-free crack width and k_t is a coefficient that depends on material properties, Fig. 5.

The coefficient k_t can be determinate using a optical method (such as moiré technique) to measure crack opening displacements versus crack penetration.

According to the previous considerations, the response of a single-notched concrete beam in bending is governed by both an elastic and a cohesive contribution, produced by the stress-transfer through the cracked surface.

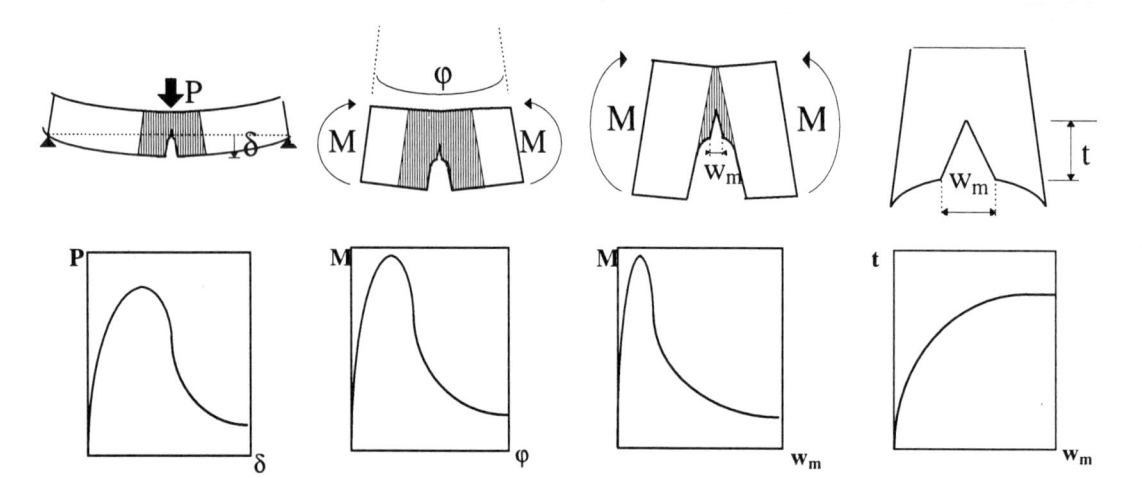

Fig. 5. Physical phenomena (top) and experimental curves (bottom) in a bending test.

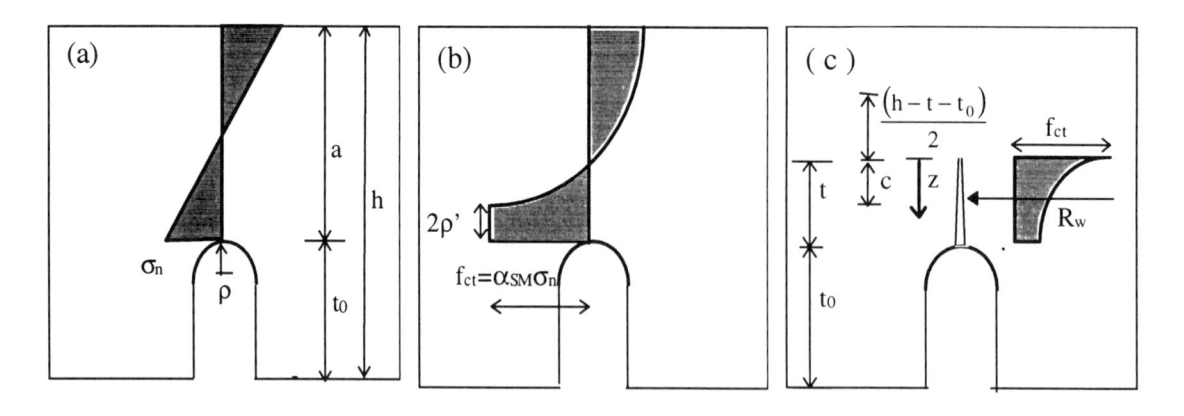

Fig. 6. Nominal (a) and effective (b) stresses for t=0 and M=M$_{cr}$; cohesive stresses (c) for t>0.

During crack propagation (t>0), the mid-span notched element of the beam, regarded as a perfectly elastic material, can be considered as subjected both to bending moment, due to external loading, and to a force R_w, which is the resultant of the cohesive stresses keeping the crack closed. Thanks to this scheme, the notch is no longer deep t_0, but $t+t_0$, where t is the depth of the crack at the analyzed loading step, Fig. 6. In this way it is possible to consider the crack as a pointed notch, with a concentration factor α_{SM} according to Neuber's theory.

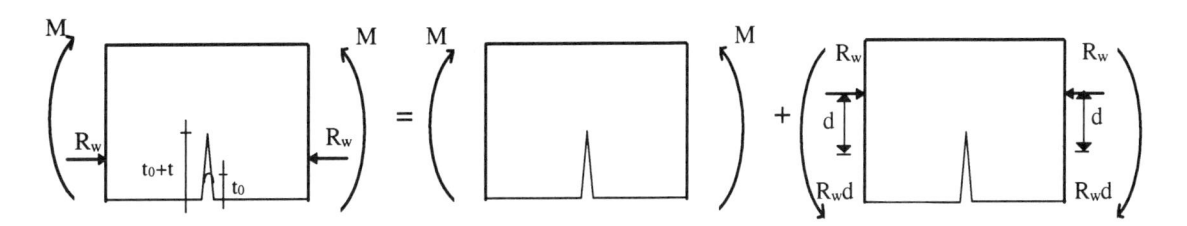

Fig. 7. Principle of superposition applied to the midspan notched element in bending.

Applying the principle of superposition, Fig. 7, the maximum stress provoked at the crack tip by the applied bending moment M, acting on the center of the ligament, is equal to the sum of three effects: the cohesive compression force R_w, applied on the center of the ligament, its transfer bending moment $R_w d$ and the elastic contribution of the ligament. As a result, the following stress equivalence at the crack tip can be obtained:

$$6M/(ba^2)\alpha_{SM} = R_w/(ba)\alpha_{SC} + 6R_w d/(ba^2)\alpha_{SM} + f_{ct} \qquad (7)$$

where b is the cross section width, $d=a/2+c$ is the distance of the center of the cohesive stresses from the middle of the ligament, c is the distance of the center of the cohesive stresses from the crack tip and:

$$\alpha_{SC} = 1 + \sqrt{0.367(a/\rho')/(1+0.815a/(t+t_0))} \qquad (8)$$

is the concentration factor at the apex of a pointed-notch for the compressive stresses[8]. Assuming that the crack opening displacement w is linearly distributed along the crack and referring to the local coordinate z, with origin at the crack tip, its distribution through the crack length is:

$$w = w_m \cdot z/t \qquad (9)$$

where w_m is the maximum crack width or crack opening displacement (c.o.d). at the distance t (crack depth) from the tip.
The cohesive stresses transmitted through the crack faces are governed by Eq.3, so that the resultant of the cohesive stresses R_W can be determined as:

$$R_w = b\int_0^t \sigma_w dz \qquad (10)$$

and the bending moment $R_W d$ (with respect to the center of the ligament) can be determined as:

$$R_w d = b\int_0^t \sigma_w(z+a/2)\,dz \quad (11)$$

The applied bending moment is obtained as the sum of an elastic and a cohesive contribution, according to the stress equivalence at the crack-tip, Eq.7:

$$M = M_e + M_c = ba^2 f_{ct}/(6\alpha_{SM}) + (R_w a \alpha_{ST}/(6\alpha_{SM}) + R_w d) \quad (12)$$

Referring to the midspan beam element with length L_1, which could be considered as equal to the cross section depth h, the midspan rotation $\Delta\phi$ is, Fig. 8:

$$\Delta\varphi = ((\varepsilon_A^+ + \varepsilon_B^+)\cdot L_1/2 + (\varepsilon_A^- + \varepsilon_B^-)\cdot L_1/2 + w_m)/(h-t_0) \quad (13)$$

where the strain at the top edge and at the notch-tip level in the cross section A, at a distance $L_1/2$ from the notched section B so as to be considered notch insensitive, are:

$$\varepsilon_A^+ = (6M\cdot(1-L_1/L))/(bh^2\cdot E_c) \quad (14)$$
$$\varepsilon_A^- = (12M\cdot(1-L_1/L)\cdot(h/2-t_0))/(bh^3\cdot E_c) \quad (15)$$

where L is the beam span; the strain at the notch-tip and at the top edge in the notched section B are:

$$\varepsilon_B^- = \sigma_w(w_m)/E_c \quad (16)$$
$$\varepsilon_B^+ = \sigma_{TOT}/E_c \quad (17)$$

where σ_{TOT} is the effective top-edge longitudinal stress due to both bending and compression.

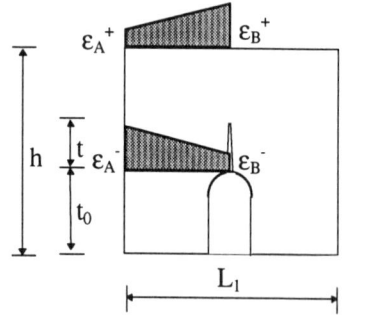

Fig. 8. Longitudinal strains at the top edge level (+) and at the notch-tip level (-).

Finally, the midspan deflection δ, neglecting, for the sake of simplicity, both punching and shear effects, is:

$$\delta = \Delta\varphi L/4 + (M(1-L_1/L)(L-L_1)^2)/(12E_c J) \quad (18)$$

DISCUSSION

The proposed model can predict the structural response quite accurately. In order to identify the basic theoretical parameters, the flexural behavior can be expressed by means of suitable functional relation:

$$F(M, w_m, t, \delta) \qquad (19)$$

In order to identify the "best" values of the governing parameters, the same approach presented in [Rosati Schumm '91][9] has been adopted here. The functional relationship (19) has been split into a set of two-variable relationship:

$$\begin{aligned}(M\text{-}w_m)_{th} &= (M\text{-}w_m)_{exp} \\ (t\text{-}w_m)_{th} &= (t\text{-}w_m)_{exp} \\ (4M/L\text{-}\delta)_{th} &= (4M/L\text{-}\delta)_{exp}\end{aligned} \qquad (20)$$

By matching the experimental results (r.h.s) with the predicted curves of the theoretical model (l.h.s), the governing parameters are evaluated so as to obtain the best possible fitting, Tab. 2.

Table 2. Material parameters

FIBER CONTENT	E_c (MPa)	f_{ct} (MPa)	G_F (N/m)	w_c (mm)	w_u (mm)	k_t
0 % by volume	50000	9.4	120	0.2	0.8	60
4 % by volume	52000	14.7	9000	3	10	160

Note: 1 N = 0.225 lb., 1 m = 3.2808 ft., 1 MPa = 0.145 ksi

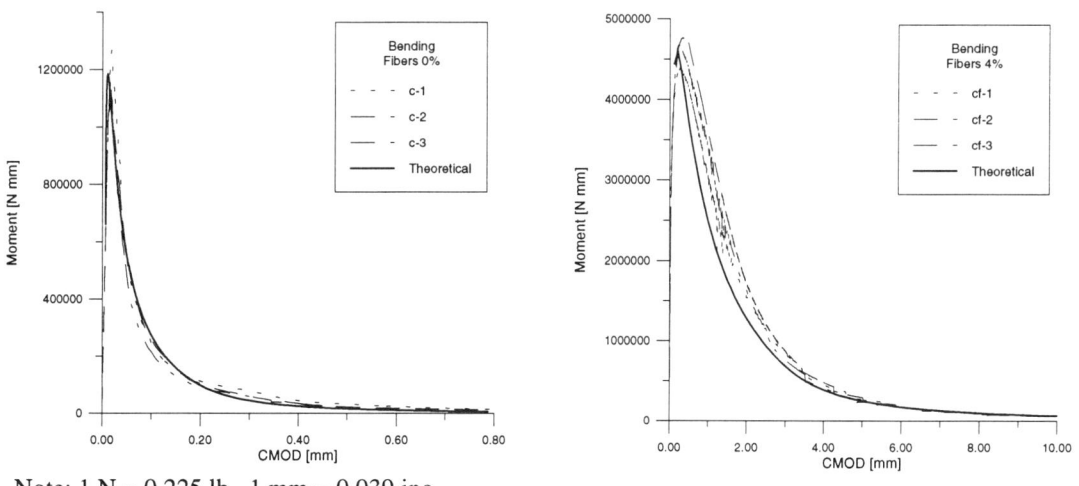

Note: 1 N = 0.225 lb., 1 mm = 0.039 inc.

Fig. 9. Experimental and theoretical curves

CONCLUDING REMARKS

The results of this research project can be summarised as follows:

- Ultra High-Strength microconcretes ($f_c \geq 130$ MPa (18.84 ksi)) with small-diameter quartzitic aggregates are very brittle, in terms of both tensile strength and toughness; however, by adding 2 to 4% of smooth metallic fibers (by volume) the material becomes stronger in tension and definitely tougher (the fracture energy increases by two orders of magnitude, from 120 to 9000 N/m (8.23 lb./ft. to 617.23 lb./ft.)).
- In the UHS microconcretes studied here, the Young's modulus is not affected significantly by fiber amount, since fiber contents as large as 2-4% by volume increase E_c by less than 10%.
- The difference between the values of the Young's modulus in tension and compression is negligible.
- An identification procedure capable of combining the curves in direct tension and those in bending is presented. This procedure allows to cleanse the stress-strain and stress-displacement curves of any undesired structural effect (lack of concrete uniformity in the notched section and rotation of press platens).
- These results are applied in a research project in progress on hollow bridge slabs. A high fiber content results in a high material cost. For this reason, the steel fiber percentage adopted in structural application (precast prestressed slabs) is close to 1% by volume. However, the first results show a sufficiently ductile response of the structure.

ACKNOWLEDGEMENTS

The authors wish to thank ITALCEMENTI and the Italian National Council for Research-CNR (Project "Special Materials for Better Structures", 1994-97) for providing the financial support for this investigation. The authors are also indebted to the staff of the laboratory for testing materials of the Politenico of Milan, whose skillfulness was instrumental in securing the success of the tests.

References

1. Bentur A. and Mindess S., "Fiber Reinforced Cementitious Composites", Elsevier Applied Science, London and New York, 1990.

2. Balaguru N. and Shah S. P., "Fiber-Reinforced Cement Composites", Mc Graw-Hill, 1992.

3. Bache H. H., "Densified Cement/Ultrafine Particle-Based Materials", Second International Conference on Superplasticizers in Concrete, June 1981, Ottawa, Canada, CBL Report No. 40, Aalborg Portland, 33 pp.

4. Van Mier J.G.M., Stang H. and Ramakrishnan V., "Practical structural applications of FRC and HPFRCC", Proceedings of the Second Internatinal Workshop on "High-Performance Fiber-Reinforced Cement Composites", Ann Arbor, USA, June 11-14, 1995, pp. 443-459.

5. Biolzi L., Guerrini G. and Rosati G., "Overall structural behaviour of high-strength concrete specimens", Construction and Building Materials, Vol. 11; No 1, 1997 pp. 57-63.

6. Naaman A. E:, Paramasivam P., Balazs G., Bayasi Z. M., Eibl J., Erdelyi L., Hassoun N. M., Krstulovic O. N., Li V. C. and Lohrmann G., "Reinforced and prestressed concrete using HPFRCC matrices", Proceedings of the Second International Workshop on "High-Performance Fiber-Reinforced Cement Composites", Ann Arbor, USA, June 11-14, 1995, pp. 291-347.

7. Neuber H., "Kerbspannungslehre", Springer-Verlag, Berlin OGH, 1937, pp. 226.

8. Rosati G. and Schumm C.,E., "Modelling of the stress intensification at a pointed-notch tip under tension and bending", Studi e Ricerche,V:13, School for the Design of R/C Structures, Milan University of Technology, Milan, Italy, 1992, pp. 241-259.

9. Rosati G. and Schumm C. E., "An identification procedure of fracture energy in concrete: mathematical modelling and experimental verification", Fracture Processes in Brittle Disordered Materials, RILEM Int. Con., Noordwijk, Holland, 1991, pp. 533-542.

10. Rosati G., Schumm C. and Ferrara G., "Evaluation of objective stress-c.o.d. relationship for cracked concrete under tension", US-Europe Workshop on Fracture and Damage in Quasibrittle Structures, E&F Spon, Prague, Czech Republic, 1994, pp. 183-190.

11. Biolzi L., Gambarova P.G., Rosati G. and Schumm C. E., "On fracture and size effect in concrete beams", EURO-C 1994 Computational Modelling of Concrete Structures, March 22-25, 1994, Innsbruck, Austria, 1994, pp. 53-62.

Proceedings of the PCI/FHWA
International Symposium on High Performance Concrete
New Orleans, Louisiana, October 20-22, 1997

CONCRETE MIXTURE OPTIMIZATION USING STATISTICAL MIXTURE DESIGN METHODS

Marcia J. Simon
Research Highway Engineer
Federal Highway Administration
McLean, Virginia

Eric S. Lagergren
Mathematical Statistician
National Institute of Standards and Technology
Gaithersburg, Maryland

Kenneth A. Snyder
Physicist
National Institute of Standards and Technology
Gaithersburg, Maryland

ABSTRACT

The optimization of mixture proportions for high-performance concretes, which contain many constituents and are often subject to several performance constraints, can be a difficult and time-consuming task. Statistical experiment design and analysis methods have been developed specifically for the purpose of optimizing mixtures, such as concrete, in which the final product properties depend on the relative *proportions* of the components rather than their absolute amounts. Although mixture methods have been used in industry to develop products such as gasoline, metal alloys, detergents and foods, they have seen little application in the concrete industry. This paper describes an experiment in which a statistical mixture experiment was used to optimize a six-component concrete mixture subject to several performance constraints. The experiment was performed in order to assess the usefulness of this technique for high performance concrete mixture proportioning in general.

INTRODUCTION

In the simplest case, portland cement concrete is a mixture of water, portland cement, fine aggregate, and coarse aggregate. Additional components, such as chemical admixtures (air entraining agents, superplasticizers) and mineral admixtures (coal fly ash, silica fume, blast furnace slag), may be added to the basic mixture to enhance certain properties of the fresh or hardened concrete. High-performance concrete mixtures, which may be required to meet several performance criteria (e.g., compressive strength, elastic modulus, rapid chloride permeability) simultaneously, typically contain at least six components.

In a recent paper, Rougeron and Aitcin[1] stated that, "The optimization of the composition of a high performance concrete (HPC) is at present more of an art than a science...." Even for conventional concrete mixes, the American Concrete Institute (ACI) guideline document for mix proportioning[2] provides a method for proportioning one mix, but it does not provide a procedure for finding the proportions which provide the best settings to meet a number of performance criteria simultaneously. The recent ACI guideline document for proportioning high-strength concrete containing fly ash[3] also does not provide a means for optimizing mixtures. The selection of mix proportions for a conventional concrete mix, with strength as the primary criterion, may not require a significant number of trial batches to find an appropriate mix. However, for a concrete mixture containing six or more components which must satisfy several performance constraints, trial and error or "one factor at a time" approaches sufficient for a conventional mix will be inefficient and costly. More importantly, they may not provide the best combination of materials at minimum cost.

In this study, a statistically designed mixture experiment was used to identify the best factor settings for optimizing properties of high performance concrete. In a mixture experiment, the total amount (mass or volume) of the mixture is fixed and the factors or component settings are proportions of the total amount. For concrete, the sum of the volume fractions is constrained to sum to one, as in the ACI mix design approach[2]. Because the volume fractions must sum to unity, the component variables in a mixture experiment are not independent.

One viable experiment design option for concrete mixtures[1,4] is the factorial design, in which the q mixture components are reduced to $q-1$ independent factors by taking the ratio of two components. There are advantages and disadvantages to both the mixture and factorial approaches. For example, the experimental region of interest is defined more naturally in the mixture experiment approach, but the analysis of such experiments is more complicated. The factorial (independent variables) approach permits the use of classsical factorial and response surface designs[5,8], but has the undesirable feature that the experimental region changes depending on how the q mixture components are reduced to $q-1$ independent factors.

Because mixture experiments have not been readily used in the concrete industry, the utility of this approach for optimizing concrete properties was investigated. An experiment was designed to find the optimum proportions for a concrete mix meeting the following conditions: 50 to 100 mm (2 to 4 in) slump for the fresh concrete, 1-day target compressive strength of 22.06 MPa (3200 psi), 28-day target compressive strength of 51.02 MPa (7400 psi), target 42-day "rapid chloride" (ASTM C1202) test (RCT) measurement less than 700 coulombs, and minimum cost (dollars per m^3). The materials (components) used included water, cement, microsilica (silica fume), high-range water reducing admixture (HRWRA), coarse aggregate, and fine aggregate.

EXPERIMENT DESIGN

Background on Mixture Experiments

As a simple (hypothetical) example of a mixture experiment, consider concrete as a mixture of three components: water (x_1), cement (x_2), and aggregate (x_3), where each x_i represents the volume fraction of a component. Assume the coarse-to-fine aggregate ratio is held fixed. The volume fractions of these components sum to one,

$$x_1 + x_2 + x_3 = 1 \tag{1}$$

and the region defined by this constraint is the regular triangle (or simplex) shown in Figure 1. The axis for each component x_i extends from the vertex it labels ($x_i = 1$) to the midpoint of the opposite side of the triangle ($x_i = 0$). The vertex represents the pure component. For example, the vertex labelled x_1 is the pure water "mixture" with $x_1 = 1$, $x_2 = 0$, and $x_3 = 0$, or (1,0,0). The coordinate where the three axes intersect is (1/3,1/3,1/3) and is called the centroid.

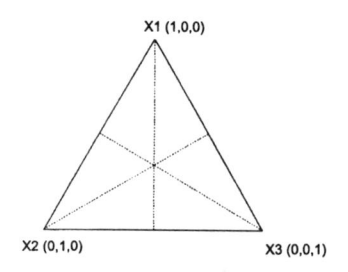

Figure 1. Experimental region for three component mixture

A good experiment design for studying properties over the entire region of a three-component mixture would be the simplex-centroid design shown in Figure 2. This example is included for illustrative purposes only, since much of this region does not represent feasible concrete mixtures. The points shown in Figure 2 represent mixtures included in the experiment. This design includes all vertices, midpoints of edges, and the overall centroid. All properties of interest would be measured for each mix in the design and modeled as a function of the components.

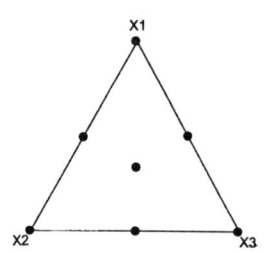

Figure 2. Layout of experiment design for three component simplex-centroid mixture

Typically, polynomial functions are used for modeling, but other functional forms can be used as well. For three components, the linear polynomial for a response y is

$$y = b_0^* + b_1^* x_1 + b_2^* x_2 + b_3^* x_3 + e \tag{2}$$

where the b_i^* are constants and e, the random error term, represents the combined effects of all variables not included in the model. This model is typically reparameterized in the form

$$y = b_1 x_1 + b_2 x_2 + b_3 x_3 + e \qquad (3)$$

using $b_0^* = b_0^*(x_1 + x_2 + x_3)$ and is called the Scheffé [9] linear mixture polynomial. Similarly the quadratic polynomial

$$y = b_0^* + b_1^* x_1 + b_2^* x_2 + b_3^* x_3 + b_{12} x_1 x_2 + b_{13} x_1 x_3 + b_{23} x_2 x_3 + b_{11} x_1^2 + b_{22} x_2^2 + b_{33} x_3^2 + e \qquad (4)$$

is reparameterized as

$$y = b_1 x_1 + b_2 x_2 + b_3 x_3 + b_{12} x_1 x_2 + b_{13} x_1 x_3 + b_{23} x_2 x_3 + e \qquad (5)$$

using $x_1^2 = x_1(1 - x_2 - x_3)$, $x_2^2 = x_2(1 - x_1 - x_3)$, $x_3^2 = x_3(1 - x_1 - x_2)$.

Since feasible concrete mixes do not exist over the entire region shown in Figure 1, a meaningful subregion of the full simplex must be defined by constraining the component proportions. An example of a possible subregion for the three component example is shown in Figure 3. It is defined by the following volume fraction constraints (x_1 = water, x_2 = cement, x_3 = aggregate):

$$0.15 \leq x_1 \leq 0.25$$
$$0.10 \leq x_2 \leq 0.20$$
$$0.60 \leq x_3 \leq 0.70$$

In this case the simplex designs are generally no longer appropriate and other designs[10] are used. These designs typically include the extreme vertices of the constrained region and a subset of the remaining centroids (e.g., centers of edges, faces, etc.).

Experiment Design for the Six-Component Study

Selection of proportions and constraints - The proportions for the six-component mixture experiment were initially selected in terms of volume fraction and converted to weights for batching. The minimum and maximum levels of each component were chosen based on typical volume fractions for non air-entrained concrete[11] with the constraint that the volume

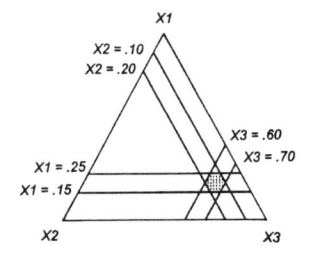

Figure 3. Example of constrained experimental region

fractions sum to unity. In addition to the individual constraints on each component, the paste fraction of the concrete (water, cement, microsilica, and HRWRA) was required to range from 25 to 35 percent by volume. Although air is incorporated into concrete during mixing, it is not an initial component and therefore was not considered to be a component of the mixture. Ignoring the air content as a mix component affects yield calculations, but these are not important for the small trial batches and can be adjusted later after a final mix is selected.

Table 1 – Mixture Components and Volume Fraction Ranges

Component	ID	Minimum volume fraction	Maximum volume fraction
Water	x_1	.16	.185
Cement	x_2	.13	.15
Microsilica	x_3	.013	.027
HRWRA	x_4	.0046	.0074
Coarse Aggregate	x_5	.40	.4424
Fine Aggregate	x_6	.25	.2924

The six components and the final ranges of their volume fractions for this experiment are shown in Table 1. The volume fractions were converted to corresponding weights using the specific gravities and percent solids (where applicable) obtained from laboratory testing or from the material supplier.

Experiment Design Details - The selection of an appropriate experiment design depends on several criteria, such as ability to estimate the underlying model, ability to provide an estimate of repeatability, and ability to check the adequacy of the fitted model. These issues are addressed below.

The "best" experiment design depends on the choice of an underlying model which will adequately explain the data. For this experiment, the following quadratic Scheffé polynomial was chosen as a reasonable model for each property as a function of the six components:

$$y = b_1 x_1 + \ldots + b_6 x_6 + b_{12} x_1 x_2 + \ldots + b_{56} x_5 x_6 + e \qquad (6)$$

This model is an extension of Equation 5 for the six component case. Since there are 21 coefficients in the model, the design must have at least 21 runs (21 distinct mixes) to estimate these coefficients. In addition to the 21 required runs, seven additional runs (distinct mixes) were included to check the adequacy of the fitted model, and five mixes were replicated to provide an estimate of repeatability allowing us to test the statistical significance of the fitted coefficients. Finally, a single mix was replicated during each week of the experiment to check statistical control of the fabrication and measurement process. In all, a total of 36 mixes were planned.

Commercially available computer software for experiment design was used for design and analysis of the experiment. The program selected thirty-six points from a list of candidate points that is known to include the best points for fitting a quadratic polynomial. A modified-distance design[6] was chosen to ensure that the design selected could estimate the quadratic mixture model while spreading points as far away as possible from one another.

Table 2 summarizes the mixes used in the experiment. The run order was randomized to reduce the effects of extraneous variables not explicitly included in the experiment. The first three mixes were repeated at the end of the program because an incorrect amount of water was used in batching them. The test results from the incorrectly batched mixes were not included in the subsequent analysis. A total of 39 batches were prepared, from which 36 sets of test results were analyzed.

SPECIMEN FABRICATION AND TESTING

The materials used in this study included a Type I/II Portland cement, tap water, #57 crushed limestone coarse aggregate, natural sand, microsilica (in slurry form), and a naphthalene-sulfonate based superplasticizer (ASTM C494 Type F/G). Thirty-nine batches of concrete, each approximately .04 m^3 (1.5 ft^3) in volume, were prepared over a four-week period. A rotating-drum mixer with a 0.17 m^3 (6.0 ft^3) capacity was used to mix the concrete.

Each batch included sufficient concrete for two slump tests, two fresh air content (ASTM C231) tests, two unit weight tests, and ten 100 mm by 200 mm (4 in by 8 in) cylinders. The cylinders were fabricated in accordance with ASTM C192. In order to obtain adequate consolidation, cylinders for concretes with slumps less than 50 mm (2 in) were vibrated on a vibrating table; otherwise, the cylinders were rodded. The cylinders were covered with plastic and left in the molds for 22 hours, after which they were stripped and placed in limewater-filled curing tanks for moist curing at 23 ± 2 °C (73 ± 3 °F).

Compressive strength tests (ASTM C39) were conducted on the cylinders at the ages of one day and 28 days. In most cases, three cylinders were tested for each age. A fourth test was performed in some cases if one result was significantly lower or higher than the others. Before testing, the cylinder ends were ground parallel to meet the ASTM C39 requirements using an end-grinding machine designed for this purpose. The three remaining cylinders from each batch were used for "rapid chloride" testing according to ASTM C1202. Three specimens (50 mm (2 in) thick slices taken from the middles of the concrete cylinders) were tested at an age of 42 days.

Table 2 - Summary of mix proportions (per cubic meter of concrete)

Design ID	Run Order	Water (kg)	Cement (kg)	Silica fume (kg)	HRWRA (l)	Coarse aggregate (dry) (kg)	Fine aggregate (dry) (kg)	w/(c+sf)
5(r)	7, 22	122.3	312.9	45.4	3.52	867.6	506.3	0.35
11(r)	6, 23	141.4	312.9	21.9	3.52	845.3	506.3	0.43
13	15	122.3	312.9	21.9	3.52	810.1	592.2	0.37
15	2*, 38	126.6	361.1	45.4	5.66	810.1	506.3	0.32
16	8	122.3	312.9	21.9	3.52	895.9	506.3	0.37
20(r)	13, 34	141.4	312.9	21.9	3.52	810.1	541.8	0.43
22	4	141.4	354.8	21.9	3.52	810.1	506.3	0.38
28	16	122.3	312.9	45.4	3.52	810.1	563.8	0.35
37	30	122.3	337.0	45.4	5.66	810.1	537.9	0.33
38(r)	3*, 26, 39	135.0	341.1	45.4	3.52	810.1	506.3	0.36
48	28	131.8	312.9	21.9	5.66	810.1	561.2	0.41
63	27	131.8	312.9	45.4	5.66	836.6	506.3	0.38
65	31	122.3	337.0	45.4	5.66	841.7	506.3	0.33
66	25	122.3	312.9	45.4	5.66	836.0	532.2	0.35
70	29	122.3	361.1	21.9	4.59	810.1	548.8	0.33
71(r)	5, 35	122.3	361.1	21.9	5.66	829.9	526.1	0.33
78	11	141.4	312.9	45.4	5.66	810.7	506.9	0.41
87	24	122.3	312.9	21.9	3.52	853.0	549.2	0.37
89	19	122.3	337.0	21.9	3.52	810.1	571.9	0.35
91	9	141.4	312.9	21.9	5.66	824.9	521.1	0.43
98	17	122.3	337.0	21.9	3.52	875.7	506.3	0.35
101	10	130.8	361.1	21.9	3.52	832.8	506.3	0.35
103	14	122.3	361.1	21.9	4.59	852.6	506.3	0.33
110	21	130.8	361.1	21.9	3.52	810.1	529.0	0.35
116	33	131.8	312.9	45.4	5.66	810.1	532.8	0.38
123	36	122.3	337.0	33.6	4.59	834.4	530.6	0.34
127(c)	1*, 12, 18, 32, 37	131.5	335.8	21.9	4.59	829.9	526.1	0.38
163	20	126.6	323.3	27.8	5.12	857.5	513.6	0.37

Notes: 1 l = 33.81 oz., 1 kg = 2.2046 lb.
(r) indicates replicated mix
(c) indicates control mix
* indicates mix which was repeated due to incorrect batching

RESULTS AND ANALYSIS

The average values for slump, 1-day strength, 28-day strength, rapid chloride test measurement (coulombs) for each batch are shown in Table 3, along with the estimated cost per cubic yard of concrete. The cost of each batch was calculated from the mix

Table 3 – Test Results and Costs

Design ID	Run	Slump (mm)	1-day str (MPa)	28-day str (MPa)	42-day RCT (coulombs)	Cost ($/ m³)
22	4	67	21.5	48.2	1278	95.18
71	5	57	27.0	55.2	862	102.22
11	6	102	16.8	48.5	1162	91.32
5	7	13	22.4	48.5	387	118.85
16	8	35	21.6	53.1	776	92.20
91	9	200	16.8	60.4	1027	96.89
101	10	22	26.6	53.6	744	96.24
78	11	127	19.2	51.7	492	123.56
127	12	99	21.5	50.2	842	96.67
20	13	118	18.2	50.9	903	91.32
103	14	64	27.4	54.6	583	99.42
13	15	57	21.8	53.2	684	92.20
28	16	29	22.2	53.6	292	118.85
98	17	32	25.3	51.9	604	94.41
127	18	92	22.3	54.1	847	96.67
89	19	38	21.8	54.3	720	94.41
163	20	95	22.1	60.8	554	103.80
110	21	51	24.7	53.2	792	96.24
5	22	25	23.4	54.1	348	118.85
11	23	114	16.5	48.0	968	91.32
87	24	67	22.9	51.0	700	92.20
66	25	76	24.7	59.8	316	124.44
38	26	29	23.0	53.2	390	120.85
63	27	124	21.7	55.2	302	123.99
48	28	171	23.0	58.1	682	97.34
70	29	51	27.5	54.5	505	99.42
37	30	35	27.3	56.0	245	126.65
65	31	32	27.2	51.1	310	126.65
127	32	121	22.4	57.2	636	96.67
116	33	114	23.9	56.2	356	123.99
20	34	127	18.6	51.6	820	91.32
71	35	108	28.8	65.3	553	102.22
123	36	99	26.6	61.0	340	110.53
127	37	102	24.2	54.6	640	96.67
15	38	51	28.8	58.1	239	128.68
38	39	25	23.6	54.5	332	120.85

Note: 1 mm = .0394 in., 1 MPa = 145 psi, 1 m³ = 1.308 yd³

proportions using approximate costs for each component material obtained from a local ready-mix concrete producer. Each of the four responses was analyzed by fitting a model, validating the model (by examining the residuals for trends and outliers), and interpreting the model graphically using contour and trace plots. The statistical analysis is described in detail for 28-day strength. The analyses for the other properties was performed in a similar manner.

Model Identification and Validation

The first step in the analysis is to identify a plausible model. Even though the design selected permits estimation of a quadratic model, a linear model may provide a better fit to the data. This is assessed using analysis of variance (ANOVA). The ANOVA results for 28-day strength are shown in Table 4[6,7]. The row with source *linear* tests whether the

Table 4 – ANOVA Table for 28-day strength

Source	Sum of Squares	DOF	Mean Square	F Value	Prob > F
Mean	1.062E+05	1	1.062E+05		
Linear	257.52	5	51.50	5.46	0.0011
Quadratic	135.19	15	9.01	0.92	0.5665
Residual	147.62	15	9.84		
Total	1.068E+05	36	2965.37		

coefficients of the linear terms are equal. In the absence of quadratic terms, this means that mixing does not affect response (i.e., any mixture would give the same response). We conclude that the coefficients differ for low values (say less than 0.05) of the *Prob > F* (also called the p-value). Since the *Prob > F* value is 0.0011, we conclude that linear terms should be included in the model. The row with source *quadratic* tests whether any quadratic coefficients differ from zero. Since the *Prob > F* value of 0.5667 exceeds 0.05, we conclude that quadratic terms should not be included in the model.

The resulting linear model for 28-day strength (y_1), fit by least squares, is

$$\hat{y}_1 = -45.22x_1 + 89.15x_2 - 3.81x_3 + 1972x_4 + 38.36x_5 + 87.19x_6 \qquad (7)$$

with residual standard deviation $s = 3.07$ MPa (445.4 psi).

The residual standard deviation s is defined as

$$s = \sqrt{\frac{\sum (y_i - \hat{y}_i)^2}{n-p}} \quad (8)$$

where the number of observations $n = 36$ and the number of parameters in the fitted model $p = 6$. A value of s near the repeatability value (replicate standard deviation) is an indication of an adequately fitting model. The repeatability value is 3.39 MPa (492.3 psi), which is close to s.

The fitted model is then validated by examining residual plots. The residuals are the deviations of the observed data from the fitted values, $y_i - \hat{y}_i$. The residual $y_i - \hat{y}_i$. estimates the error term e_i in the model. The e_i's are assumed to be random and normally distributed with mean 0 and constant standard deviation. The residuals, which estimate these errors, should exhibit similar properties. Essentially, an adequate model should capture all information in the data leaving structureless, random residuals. If structure remains in the residuals, residual plots will often suggest how to modify the model to remove the structure. In this study, a plot of residuals versus run sequence as well as a plot of the control mix data revealed a linear trend in the data for each response. However, because the run sequence was randomized, this trend had little impact on the fitted models.

Graphical Interpretation

Once a valid model is obtained, it can be interpreted graphically using response trace plots and contour plots. A response trace plot is shown in Figure 4. This figure consists of six overlaid plots, one for each component. For a given component the fitted value of the response is plotted as the component is varied from its low to high setting in the constrained region, while the other components are held in the same relative ratio as a specified reference mixture, here the centroid. The plot shows the "effect" of changing each component on 28-day strength.

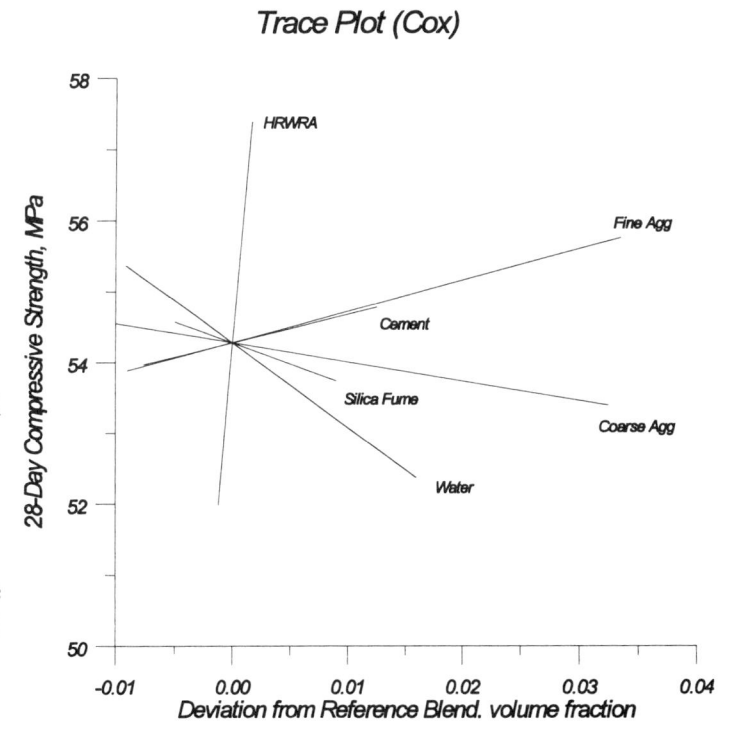

Figure 4. Trace plot for 28-day strength

As expected, increasing the amount of water decreased strength, while increasing the amount of cement increased strength. HRWRA had the largest effect with higher amounts of HRWRA yielding higher strength. This may be due to the improved dispersion of the cement and silica fume caused by higher amounts of HRWRA. Surprisingly, an increase in silica fume appears to reduce strength. This apparent reduction may not be significant when compared to the underlying experimental error.

Contour plots are used to identify conditions which give maximum (or minimum) response. Because contour plots can only show three components at a time (the others components are set at fixed conditions), several must be examined. Figure 5 is a contour plot of 28-day strength for water, cement, and HRWRA, with the other components fixed at their centroid values. The plot indicates that strength increases rapidly by increasing HRWRA, confirming the result from the response trace plot. Therefore, in subsequent

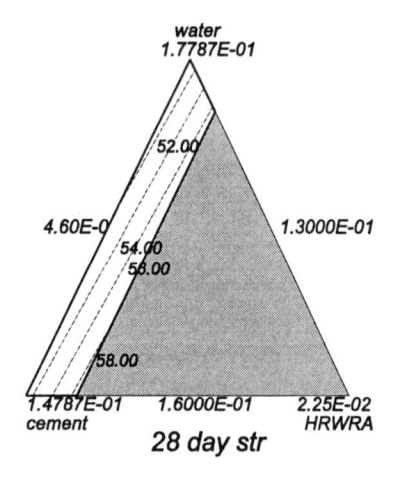

Figure 5. Contour plot for 28-day strength (MPa) in water, cement and HRWRA (microsilica = .018, CA = .410, FA = .259)

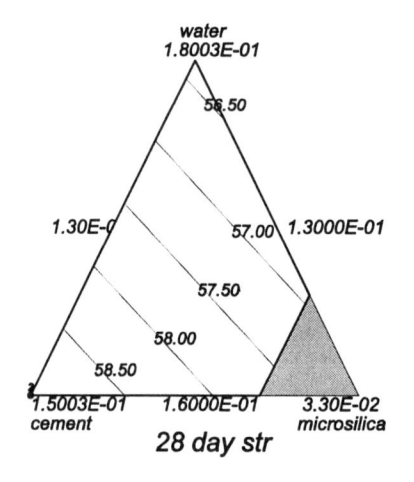

Figure 6. Contour plot for 28-day strength (MPa) in water, cement and microsilica (HRWRA = .0074, CA = .410, FA = .259)

contour plots, HRWRA will be set at its high value.

Figure 6 shows a contour plot of 28-day strength in water, cement, and microsilica, and Figure 7 shows a contour plot of 28-day strength in water, coarse aggregate, and fine aggregrate. In each case, HRWRA is fixed at its high value and the other components are fixed at the centroid settings. These plots show that strength increases for low water, high cement, low microsilica, low coarse aggregate, and high fine aggregate. The best overall settings can be found using the contour plot shown in Figure 8 for microsilica, coarse aggregate, and fine aggregate at the best settings of water, cement, and HRWRA. The best settings (expressed as volume fractions) are water = 0.16, cement = 0.15, microsilica = 0.013, HRWRA = 0.0074, coarse aggregate = 0.40, and fine aggregate = 0.27, with a predicted value of strength of 59.53 MPa (8634 psi).

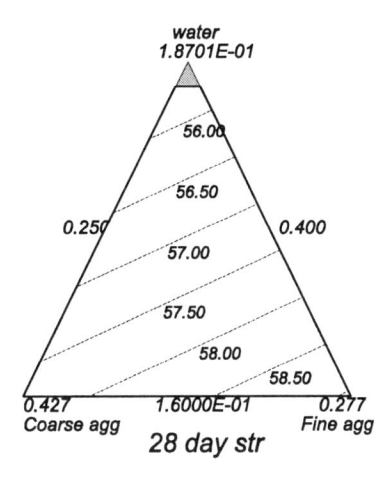

Figure 7. Contour plot for 28-day strength (MPa) in water, CA and FA (cement = .1376, microsilica = .018, HRWRA = .0074)

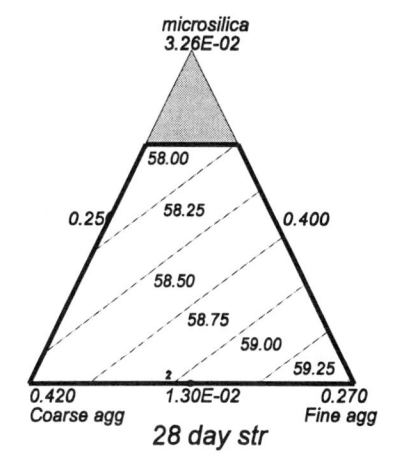

Figure 8. Contour plot for 28-day strength (MPa) in microsilica, fine aggregate, and coarse aggregate (water = .16, cement = .15, HRWRA = .0074)

Models for Other Responses

Using the same procedure described above for 28-day strength, the following models were fit to slump (y_2), 1-day strength (y_3), and 42-day RCT results (y_4):

$$\hat{y}_2 = 85.27x_1 - 94.09x_2 - 133.92x_3 + 955.63x_4 - 8.07x_5 + 6.69x_6 \tag{9}$$

$$\begin{aligned}\hat{y}_3 = &-1.752E+05x_1 + 2.573E+05x_2 - 10723x_3 - 1.732E+06x_4 + 8632x_5 - 15245x_6 \\ &+ 6.107E+05x_1x_6 - 8.118E+05x_2x_6 + 6.328E+06x_3x_4 + 6.481E+06x_4x_6\end{aligned} \tag{10}$$

$$\ln(\hat{y}_4) = 20.34x_1 - 2.99x_2 - 49.68x_3 - 29.65x_4 + 7.96x_5 + 4.15x_6 \tag{11}$$

Linear models were adequate for all responses except the 1-day strength, for which the fitted model includes four quadratic terms which were found to be significant. The natural logarithm of 42-day RCT was used for modeling, since residual plots showed that the standard deviation of 42-day RCT was proportional to the mean.

Selection of Optimum Mix

The optimum concrete mix is defined here as that mix which minimizes cost while meeting the specifications. Numerical optimization using desirability functions[8] can be used to find the optimum mix. First, a desirability function must be defined for each property. The desirability function takes on values between 0 and 1, and may be defined in several ways, as indicated in Figure 9. Minimum and maximum specifications are used for strength and RCT, respectively, resulting in desirability functions with values of 1 above the minimum or below the maximum and zero otherwise. For example, for 1-day strength the desirability value is 0 below 22.06 MPa (3200 psi) and 1 above 22.06 MPa (3200 psi). At 34.48 MPa (5000 psi) the desirability becomes 0, however this strength was chosen to be well beyond the maximum value in the observed data. Desirabilities for 28-day strength and 42-day RCT are defined similarly. For slump a range of 50 to 100 mm (2 to 4 in) was specified, but the most desirable value is the midpoint of this range, or 75 mm (3 in). Therefore, the maximum desirability is given to the target value of 75 mm (3 in), with a linear decrease in desirability to a value of zero at the lower and upper specifications (see Figure 9). Since cost is to be minimized, the desirability function for cost decreases linearly over the range of costs observed in the data (see Figure 9). It is also possible to develop more complex desirability functions (e.g., non-linear function instead of linear for cost).

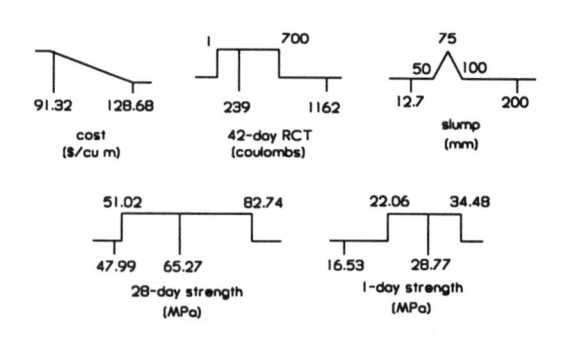

Figure 9. Desirability functions for optimization

In the numerical optimization scheme, the optimum mix maximizes the geometric mean D of the individual desirability functions d_i over the feasible region of mixtures, using the fitted models:

$$D=(d_1 d_2 d_3 d_4 d_5)^{1/5} \qquad (12)$$

Based on the experimental results, the mix which maximizes D, expressed in volume fractions, is water = 0.160, cement = 0.130, microsilica = 0.013, HRWRA = 0.00493, coarse aggregate = 0.404, and fine aggregate = 0.287, at a cost of \$92.94 per cubic meter. The response values for this mix are slump = 75 mm (3 in), 1-day strength = 22.06 MPa (3200 psi), 28-day strength = 54.62 MPa (7922 psi), and 42-day RCT value = 653 coulombs.

If the fitted functions for each property were known without error, the analysis would be complete. However, there is uncertainty in the fitted functions since they are estimated from a sample of data. For example, at the current mix the predicted 1-day target strength is 22.06 ± 0.97 MPa (3200 ± 140 psi). The uncertainty provided is for a 95% confidence interval, i.e., we are 95% confident that the interval (21.09, 23.03) contains the true 1-day target strength for this mix. So if this mix is used, it is quite possible that the true 1-day target strength would fall below 22.06 MPa (3200 psi). Therefore, each specification must be modified to account for the uncertainty in the fitted function, which depends on the location of the mix in the feasible region. The uncertainties in the properties of the current mix can be used to modify the constraints and identify a revised optimal mix for these new constraints. The revised mix must then be checked to see that the specifications are met.

The predicted values and 95% uncertainties for the remaining responses at the current best mix are slump = 75 ± 15 mm (3.0 ± 0.6 in), 28-day strength = 54.62 ± 2.99 MPa (7922 ± 434 psi), and 28-day RCT = 653 ± 81 coulombs. The modified constraints on the responses which take into account the uncertainties are 66 mm < slump < 86 mm (2.6 in < slump < 3.4 in), 1-day target strength > 23.03 MPa (3340 psi), 28-day target strength > 53.78 MPa (7800 psi), and 42-day RCT < 620 coulombs. The best mix for this new set of constraints (expressed as volume fractions) is water = 0.160, cement = 0.135, microsilica = 0.0131, HRWRA = 0.00533, coarse aggregate = 0.401, and fine aggregate = 0.285 at a cost of $72.54. The predicted values and 95% uncertainties for this mix are slump = 75 ± 15 mm (3 ± 0.6 in), 1-day strength = 23.09 ± 0.77 MPa (3349 ± 112 psi), 28-day strength = 55.48 ± 2.72 MPa (8047 ± 394 psi), and 42-day RCT = 617 ± 81 coulombs. The lower or upper bound values (as appropriate) for all responses meet the specifications.

CONCLUSIONS AND RECOMMENDATIONS

In high performance concretes consisting of many components, where several properties are of interest, it is critical to use a systematic approach for identifying optimal mixes given a set of constraints. Statistical experiment design and mixture experiments provide such an approach. They permit a thorough examination of a feasible region of interest in which to identify optimal mixes. Fitted models are obtained from the experimental data and are used to identify optimal mixes over the region.

Typically, quadratic models are assumed to provide an adequate representation of each property over the region of interest. For a six-component mixture, 21 mixes are required to fit a quadratic model, although additional runs should be included for checking the adequacy of the fitted model and estimating repeatability. A minimum of 31 runs is recommended. For the materials and conditions of this experiment, a linear model was adequate for all but one response (1-day strength). Since materials and conditions will vary by location, the quadratic model should be considered initially. However, if a linear model is found to be adequate for all responses of interest, the number of experimental runs can be halved.

Extra care is required to run a designed experiment; however, the results are well worth the additional effort. With many components and several properties of interest, trial and error methods could easily miss the optimal conditions, resulting in higher costs to producers over the long-term.

REFERENCES

1. Rougeron, P., and Aitcin, P.-C., "Optimization of the Composition of a High-Performance Concrete," *Cement, Concrete and Aggregates*, V. 16, No. 2, December, 1994, pp. 115-124.

2. ACI Committee 211, "Standard Practice for Selecting Proportions for Normal, Heavyweight, and Mass Concrete." In *ACI Manual of Concrete Practice,* Volume 1. Detroit: American Concrete Institute, 1995.

3. ACI Committee 211, "Guide for Selecting Proportions for High-Strength Concrete with Portland Cement and Fly Ash." In *ACI Manual of Concrete Practice*, Volume 1. Detroit: American Concrete Institute, 1995.

4. Luciano, J.J, Nmai, C.K., and J.R. DelGado, "A Novel Approach to Developing High-Strength Concrete," *Concrete International*, May, 1991, pp. 25-29.

5. Box, G.E.P, Hunter, J.S., & Hunter, *Statistics for Experimenters*, New York: Wiley, 1978.

6. Cornell, J.A., *Experiments with Mixtures: Designs, Models, and the Analysis of Mixture Data*, 2nd ed. New York: Wiley, 1990.

7. Neter, J., Wasserman, W., & Kutner, M.H., *Applied Linear Statistical Model*s, 3rd edition, Boston: Irwin, 1990.

8. Myers, R.H. & Montgomery, D.C., *Response Surface Methodology: Process and Product Optimization Using Designed Experiment*s, New York: Wiley, 1995.

9. Scheffé, H., "Experiments with Mixtures," *Journal of the Royal Statistical Society*, B, V. 20, 1958, pp. 344-360.

10. McLean, R.A., and Anderson, V.L., "Extreme Vertices Design of Mixture Experiments," *Technometrics*, V. 8, pp. 447-454.

11. Kosmatka, S.H. and W.C. Panarese, *Design and Control of Concrete Mixtures*, Thirteenth Edition. Skokie, Illinois: Portland Cement Association, 1988.

INFLUENCE OF SILICA FUME ON CHLORIDE RESISTANCE OF CONCRETE

R.D. Hooton (Professor, University of Toronto)
P. Pun (Graduate Student, University of Toronto)
T. Kojundic (Elkem Materials Inc.)
P. Fidjestol (Elkem Materials Inc.)

ABSTRACT

A series of concrete slabs were cast with 0, 7, and 12% by mass silica fume replacement of cement at W/CM = 0.35, 0.40, and 0.45. After either moist or steam curing, a variety of chloride resistance tests were performed. These included AASHTO T259, modified T259, T277, as well as several bulk diffusion tests and sorptivity tests.

Results indicate that the use of properly dispersed silica fume results in far more dramatic reduction in chloride penetration, by all test methods, than does reduction in W/CM from 0.45 to 0.35. In general, the AASHTO T277 rapid index test provides a good indication of the reduction in chloride diffusion coefficients.

INTRODUCTION

It is well known that chloride penetration is reduced with a reduction in water-to-cementitious materials ratio (W/CM). There are numerous references [1] to the beneficial influence of supplementary cementing materials (SCM) such as silica fume. However, precast prestressed concrete producers have typically not used SCM due to a need to obtain high early strengths. The typical practice often includes steam or heat curing to further accelerate strength. In addition, recent work sponsored by PCI has suggested that the benefits from silica fume are not so great and that the use of the AASHTO T277 Rapid Chloride Permeability Test has distorted results in favor of silica fume [2,3].

The purpose of this study was to study the influence of silica fume on strength development and chloride penetration resistance of concrete using a variety of test procedures and to examine whether silica fume could reduce the need for steam curing.

EXPERIMENTAL PROCEDURE

Materials
The Portland cement (OPC) used was Lafarge's Woodstock CSA Type 10 cement (low in C_3A and alkalis, similar to ASTM Type I or II). The silica fume tested was Microsilica Slurry 970S supplied by Elkem Materials Inc. This slurry is composed of water and silica fume (each 50% by mass). The chemical admixtures used in all the mixtures were a

naphthalene-sulfonate based superplasticizer (SP), a Type A ligin-based water reducer (WR), and Micro-Air air-entraining agent (AEA) from Master Builders Technologies. The coarse aggregate was 19 mm (0.75 in.) Manitoulin crushed limestone (absorption = 0.37%, specific gravity = 2.85) and the fine aggregate was Dufferin Sand from Mosport Pit (absorption = 0.86%, specific gravity = 2.68). Seven mix designs, each with a cementitious content of 375 kg/m^3 (630 pcy), were used. The W/CM ratios were 0.35, 0.40, and 0.45. The silica fume replacements of cement on a dry mass basis were 0, 7, and 12%.

Sample Preparation

Prior to mixing, the silica fume slurry was stirred thoroughly inside its pail with a small hand-held mixer to ensure uniformity. Concretes were mixed in a 140 L flat pan mixer. The AEA was added to and covered by the sand in the hopper. WR was mixed with water prior to adding to the dry ingredients in the mixer. Part of the SP was added to the mix right after the silica fume slurry in order to reduce the stiffness and enabled thorough mixing of the slurry. Three kinds of samples were cast for each mix: 350 x 250 x 75 mm (14 x 10 x 3 in.) slabs, 100 x 75 mm (4 x 3 in.) disks made from cylindrical molds, and 100 x 200 mm (4 x 8 in.) cylinders.

Curing Regimes

The steam-curing regime shown in Figure 1 was applied to four mixes. Including the preset period determined using ASTM C 403, the 18-hour cycle consisted of a rise from 23 °C (73 °F) to 70 °C (158 °F) in 2 hours, followed by about 7 hours at 70 °C (158 °F) and then cooling to 23 °C (73 °F) in 2 hours. Even though the relative humidity of the programmable environmental chamber was set at 100%, the samples were covered with moist burlap and plastic during the whole process. Table 1 summarises various curing regimes for each test conducted in this research. For most of the tests, 100 mm (4 in.) ± 3 mm (0.12 in.) concrete cores were obtained from 350 x 250 x 75 mm (14 x 10 x 3 in.) slabs to eliminate edge effects.

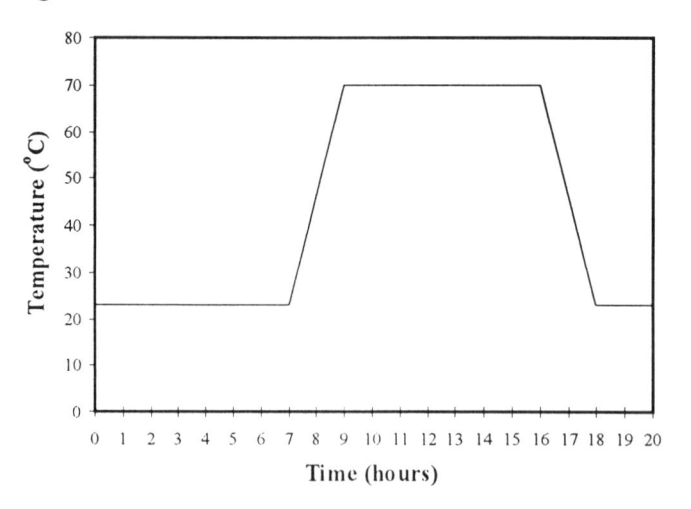

Figure 1 - 18-hour steam curing cycle

Table 1 - Summary of Curing Regimes

Test	Curing Regimes
compressive strength	ambient or steam curing for the first 24 hours + cured in lime water until test
sorptivity	ambient or steam curing for the first 24 hours + 27 days laboratory air + dry in 50 °C (122 °F) for another 7 days
RCPT (AASHTO T277)	four different curing regimes: i) 1 day moist + 27 days laboratory air ii) 18-h steam curing + 27 days laboratory air iii) 14 days moist + 14 days laboratory air iv) 28 days moist curing
bulk diffusion	ambient or steam curing for the first 24 hours + cured in lime water another 27 days
salt ponding (AASHTO T259)	ambient or steam curing for the first 24 hours + 13 days moist + 14 days laboratory air

Time of Set
The purpose of this test was to determine the optimum time for application of the steam curing cycle. The regime was applied after the initial setting of concrete (a penetration resistance of 500 psi or 3.5 MPa). The detailed procedure followed ASTM C 403.

Compressive Strength
The compressive strengths were tested at five different ages: 1, 3, 7, 28, and 56 days following ASTM C 39. For steam-cured cylinders, instead of 1-day, the 18-hour strength was determined. Two specimens were used for each test.

Surface Sorptivity
The procedure for the surface sorptivity test was based on the ASTM draft standard. Due to scheduling of the research program, cylindrical disks of 100 mm (4 in.) diameter and 75 mm (3 in.) height were cut to 50 mm (2 in.) thick on the twenty-fifth day after casting. 3 samples from each curing regime was tested. The samples were then dried in a 50 °C (122 °F) oven for 7 days. The samples were allowed to cool down to room temperature in a desiccator before testing. With the sides sealed with vinyl electrician's tape, after measuring the dry mass, the top surface was then immersed in distilled water to a depth of about 3 - 5 mm. The sample mass was repeatedly measured at times ranging from 1 to 25 minutes after wiping the surface moisture off with a damp cloth.

Rapid Chloride Permeability Test (AASHTO T277)
The same type of specimens as the surface sorptivity test was used for RCPT. The samples were epoxy-coated on the sides and vacuum-saturated before being tested with the formed face exposed to the half cell containing salt solution. The total charges at 10 minutes, 30 minutes, and 6 hours were recorded. By extrapolating the early readings to 6 hours, the heating effect in high coulomb value concrete would be minimised. This

extrapolated value is considered a better estimate of concrete's chloride penetration resistance than the standard 6 hours value. It is realised that the RCPT is for the most part a measure of a concrete's resistivity, and not permeability or diffusivity directly. However, resistivity is a useful and simple index of quality.

Bulk Diffusion Tests
Two types of bulk diffusion tests were conducted: 90 days in 1.0 mol/L NaCl at 23 °C (73 °F) and 120 days in 5.0 mol/L NaCl at 40 °C (104 °F). Two cores of 100 mm (4 in.) diameter and 50 mm (2 in.) thickness were used for each test. The sides and bottom face were epoxy-coated and the samples were also vacuum-saturated. On the twenty-eighth day after casting, the cores were immersed in chloride solutions inside plastic containers. After the tests, the samples were profile ground and the bulk diffusion coefficients (D_B) were determined by fitting Fick's second law to the profiles.

AASHTO T259 Salt Ponding Test
Two 350 x 250 x 75 mm (14 x 10 x 3 in.) slabs from each curing regime were ponded. 25 mm (1 in.) high Styrofoam strips which acted as dams, were sealed to the edges of the top face with silicone. With all the connections properly sealed with silicone, 3% (by mass) NaCl solution was ponded on top to a depth of 20 mm (0.75 in.) on the twenty-eighth day. To minimise evaporation, the open top was then covered with Saranwrap held in place with duct tape. The slabs were ponded continuously for 90 days in a 50% relative humidity chamber. After ponding, a sample of 100 mm x 50 mm (4 in. x 2 in.) was dry cut from the center. Chloride profiling was conducted and the combined "diffusion" coefficients (D_C) were determined.

Modified Salt Ponding Tests
AASHTO T259 ponding test was modified to investigate various transport mechanisms. A 3-day ponding test was used to study initial absorption, wicking action was eliminated using a 90-day non-wicking ponding test, and the test duration was extended to one year for long-term chloride ingress. 350 x 250 x 75 mm (14 x 10 x 3 in.) slabs were used for the 1-year ponding test and 100 mm diameter (4 in.) and 75 mm (3 in.) high cores were used for the other modified tests. The sides of the cores were sealed with bithuthene adhesive sheets before air curing in order to maintain an interior moisture condition. A day before commencing the non-wicking test, specimens were epoxy-coated on the bottom face. This was to isolate the bottom face of the samples from the 50% relative humidity environment. 3% (by mass) NaCl solution was also ponded on top in these tests. By fitting Fick's second law to the profiles, sorption coefficient (S_P), non-wicking coefficient (D_W), and long-term diffusion coefficient (D_L) were calculated.
It is necessary to point out that the background chloride level (700 - 800 ppm) was subtracted from the chloride profiles before Fick's second law was fitted to it.

Table 2 - Plastic Properties and Compressive Strengths Results

Material (kg/m³)	0.35-0	0.35-7	0.35-12	0.40-0	0.40-7	0.45-0	0.45-7	0.35-0s	0.35-7s	0.45-0s	0.45-7s
Portland Cement (Type 10)	380	356	334	374	353	369	350	370	357	369	347
Silica Fume	0	27	46	0	27	0	26	0	27	0	26
Coarse Aggregate	1193	1194	1174	1141	1150	1098	1108	1162	1197	1098	1101
Fine Aggregate	731	732	719	699	705	672	679	708	733	672	674
Water	133	133	132	149	152	165	169	151	131	165	167
Water Reducer (mL/100 kg)	325	325	325	325	325	325	325	325	325	325	325
Superplasticizer (mL/100 kg)	1232	1194	1555	436	1255	154	550	482	952	154	611
Air Entrainer (mL/100 kg)	74	49.9	52	45	62.8	52.1	54	52.1	52	52.1	54.2
Air (%)	5.0	4.4	5.4	6.8	5.2	8.0	6.3	5.7	4.5	8.0	7.0
Slump (mm)	100	100	180	110	165	190	125	110	80	190	80
Fresh Density (kg/m³)	2443	2449	2414	2368	2394	2307	2336	2395	2452	2307	2320
Compressive. Strength (MPa)											
1-day/18-h	11.4	24.8	26.0	10.7	23.4	12.1	10.7	28.8	41.4	19.8	37.8
3-day	39.0	47.8	44.6	31.3	42.6	22.2	24.9	29.7	50.2	20.2	40.0
7-day	50.3	55.6	64.0	38.9	54.8	24.8	37.2	32.1	57.9	21.4	42.1
28-day	58.8	82.2	76.1	46.1	62.9	31.8	51.6	37.4	58.4	29.3	46.3
56-day	69.0	85.6	--	49.6	69.6	35.7	59.1	45.9	61.0	31.1	47.1

Note: 1 kg/m³ = 1.7 pcy; 1 mm = 0.004 in.; 1 MPa = 143 psi, 1 mL/100 kg = 0.015 fl. oz./100 lb.

EXPERIMENTAL RESULTS

Plastic Properties and Compressive Strengths
The plastic properties and compressive strengths are presented in Table 2. In order to maintain similar workability, the amount of SP had to be doubled in silica fume mixes but the dosage of air-entrainer was only slightly affected. The 56-day compressive strength for the 0.35-12 concrete is not listed because, for both cylinders, the capping material failed before the cylinders. Steam curing increased the 18-hour compressive strengths but the early advantage was lost after three days. Silica fume concretes also benefited less from steam curing than the OPC ones. The use of 7% silica fume was more effective in enhancing compressive strengths than was reducing the W/CM by 0.05. A larger increase was observed between 7 and 28 days in silica fume concretes.

RCPT Results
The RCPT results shown in Table 3 are normalised to a standard diameter of 95 mm (3.75 in.) as per ASTM C 1202 and are extrapolated from the 30-minute readings by multiplying the values by 12 and the actual 6-hour readings are in brackets. The results clearly indicate that silica fume was effective in reducing the RCPT values regardless of the curing regimes applied. For each curing regime, the beneficial effect of silica fume was more profound as the W/CM ratio was increased. Moreover, silica fume enhanced chloride resistance more than reducing W/CM. This effect was confirmed by the diffusion tests shown in Table 4. The RCPT results indicate that a longer initial moist curing period was beneficial in reducing the permeability; especially for silica fume concretes. 14 days seemed to be sufficient when compared with the 28-day moist cured ones. The 18-hour steam curing cycle was found to be detrimental to OPC concretes when measured by RCPT as well as the diffusion tests. This agreed with previous research [1]; nevertheless, for silica fume concretes, the steam-cured samples performed similar to the one-day moist cured ones.

Table 3 - Rapid Chloride Permeability Test Results (Coulombs)

Mix	1-d moist + 27-d air	14-d moist + 14-d air	28-d moist	18-hr steam + 27-d air
0.35-0	2495 (3168)	2474 (2996)	2126 (2530)	3962 (5058)
0.35-7	637 (543)	377 (371)	293 (295)	607 (611)
0.35-12	272 (282)	230 (226)	198 (202)	--
0.40-0	2740 (3713)	2314 (3172)	2205 (3062)	--
0.40-7	567 (596)	452 (487)	432 (442)	--
0.45-0	4968 (5908)	3744 (4498)	2832 (3527)	6384 (7299)
0.45-7	1472 (1783)	665 (758)	648 (719)	1164 (1430)

Surface Sorptivity and 3-Day Salt Ponding Results
The surface sorptivities in Table 4 indicated that the values were decreased by steam curing. Besides, both W/CM and silica fume replacement affected the surface sorptivity of

concretes. The sorption coefficients (Sp) calculated were found to be sensitive to W/CM and silica fume. Both tests indicated that the initial sorption was decreased by steam curing in silica fume concretes but different effects were observed in OPC samples.

Bulk Diffusion Tests

The bulk diffusion coefficients (D_B) determined from tests were also presented in Table 4. Silica fume was proved to be effective in reducing chloride diffusion similar to that indicated by the RCPT results. Steam-cured concretes were found to be less resistant to chloride diffusion than the corresponding ambient-cured ones. This was different from RCPT values where silica fume concretes were found to have higher chloride resistance after steam curing. This could be due to different curing that followed the steaming cycle. As outlined in Table 1, samples for these tests were continuously moist-cured after the steam curing cycle while those for RCPT were cured in laboratory air until coring. Silica fume replacement was more beneficial than lowering the W/CM from 0.45 to 0.35.

Table 4 - Surface Sorptivity and Various Coefficients

Mix	Sorptivity (mm/min$^{1/2}$)	S_P (x 10^{-12} m^2/s)	90-d D_B (x 10^{-12} m^2/s)	120-d D_B (x 10^{-12} m^2/s)	D_C (x 10^{-12} m^2/s)	D_W (x 10^{-12} m^2/s)	D_L (x 10^{-12} m^2/s)
0.35-0	0.116	13.0	4.7	5.9	5.3	5.8	2.8
0.35-7	0.103	3.7	1.2	1.9	1.3	1.4	0.7
0.35-12	0.112	3.2	1.0	0.9	0.7	0.7	0.3
0.40-0	0.169	26.5	5.0	9.4	7.0	6.1	3.3
0.40-7	0.109	3.9	1.2	1.8	1.3	1.7	0.7
0.45-0	0.168	24.3	7.1	10.5	8.7	7.1	3.4
0.45-7	0.180	11.9	1.6	1.9	2.3	2.2	0.8
0.35-0s	0.096	18.7	10.0	8.2	6.1	--	--
0.35-7s	0.069	3.3	1.6	3.2	1.5	--	--
0.45-0s	0.122	51.7	8.5	12.1	10.7	--	--
0.45-7s	0.094	11.5	3.9	4.5	6.3	--	--

Note: 1 mm/min$^{1/2}$ = 0.004 in./min$^{1/2}$; 1 m^2/s = 10.9 ft^2/s

AASHTO T259 Salt Ponding Test and Modified Tests

   AASHTO T259 Salt Ponding Test (D_C) -- The results show that D_C was sensitive to W/CM, especially in OPC concretes. Silica fume was, again, shown to be beneficial regardless of the curing regime applied as all silica fume samples, except 0.45-7s, had a coefficient lower than that of 0.35-0. Steam curing adversely affected the chloride resistance as in previous tests. In all cases, silica fume concretes performed much better than the OPC reference concretes with a low W/CM ratio.

Non-wicking Salt Ponding Test (D_W) -- Unlike AASHTO T259, D_W of OPC and silica fume concretes were equally sensitive to the W/CM. The chloride resistance was increased by the addition of silica fume and the decrease in diffusion coefficients were comparable to those indicated by AASHTO T259.

1-year Salt Ponding Test (D_L) -- The influence of W/CM was less obvious in the 1-year data, especially in silica fume concretes. The increase in chloride resistance by silica fume was consistent with AASHTO T259 results. Silica fume was also more effective than W/CM in resisting chloride ingress in the long term.

DISCUSSION OF RESULTS

Bulk Diffusion Tests

The results from these tests are presented graphically in Figure 2. Both tests were able to demonstrate the influences of W/CM and silica fume on chloride resistance of concrete. Silica fume was shown to be more effective in enhancing the durability. Steam curing consistently decreased the resistance. Except for the steam-cured 0.35-0, the 120-day bulk diffusion coefficients calculated were higher than the 90-day ones. Chloride ingress was promoted by the higher solution temperature and salt concentration. Both tests were highly reproducible and generated a clear indication of the concrete quality.

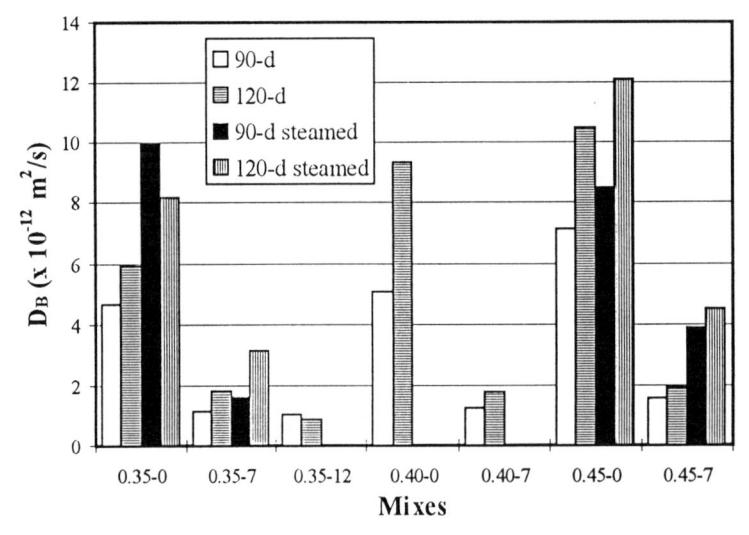

Note: 1 m^2/s = 10.9 ft^2/s

Figure 2 - Bulk diffusion coefficients (D_B) of all the mixes

AASHTO T259 and Modified Salt Ponding Tests

Figure 3 is a plot of the different coefficients calculated for each mix. Similarly, the chloride resistance was influenced more by silica fume than by W/CM. The diffusion coefficients of silica fume concretes were comparable regardless of the W/CM. For most cases, chloride ingress was increased by the wicking action, particularly in OPC concretes.

In comparing D_C with D_L, the chloride diffusion coefficient of concrete decreased by 50% between 90 days to 1 year. This percentage increased with the W/CM.

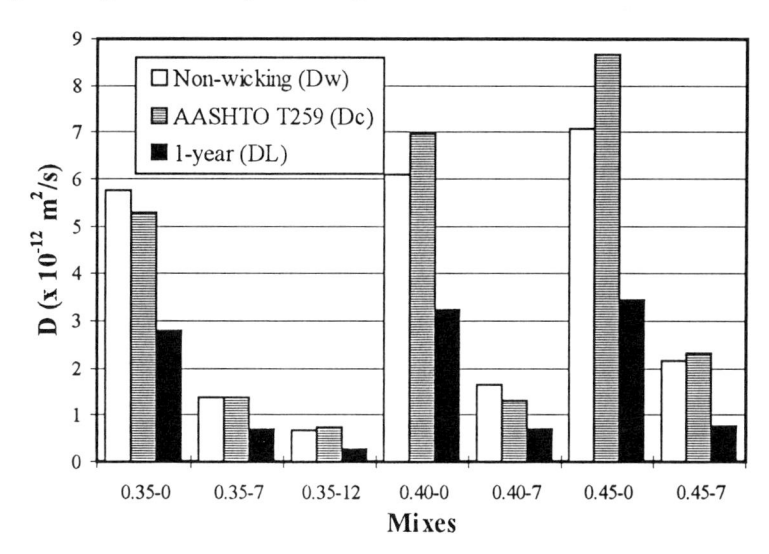

Note: 1 m^2/s = 10.9 ft^2/s

Figure 3 - Comparison of diffusion coefficients from salt ponding tests

Bulk Diffusion Tests and RCPT
In Figure 4 and Figure 5 the 90-day and 120-day bulk diffusion coefficients (D_B) are plotted against the RCPT values, respectively. All the samples were moist cured for 28

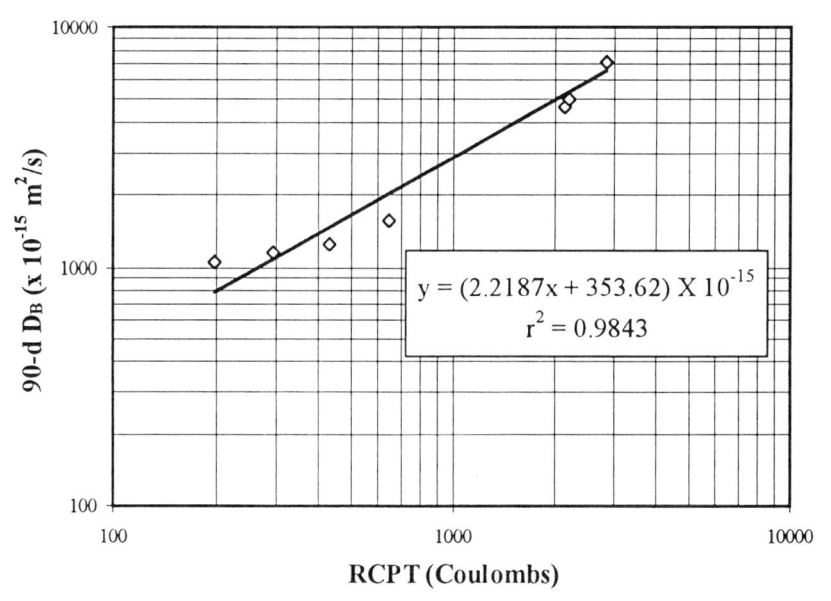

$$y = (2.2187x + 353.62) \times 10^{-15}$$
$$r^2 = 0.9843$$

Note: 1 m^2/s = 10.9 ft^2/s

Figure 4 - 90-day bulk diffusion coefficients vs. RCPT values

Figure 5 - 120-day bulk diffusion coefficients vs. RCPT values

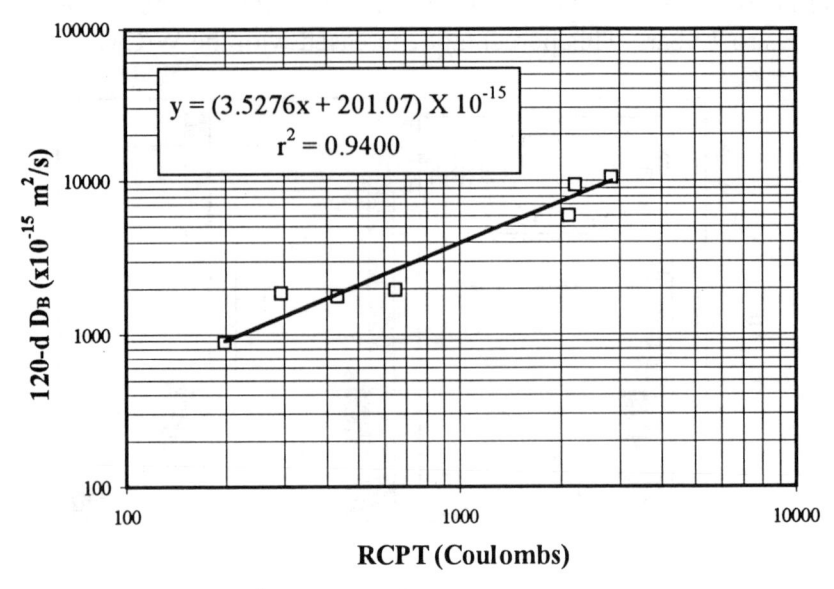

days. A linear regression was applied to both graphs. High coefficients of correlation were found from the plots; therefore, a linear function existed between the tests.

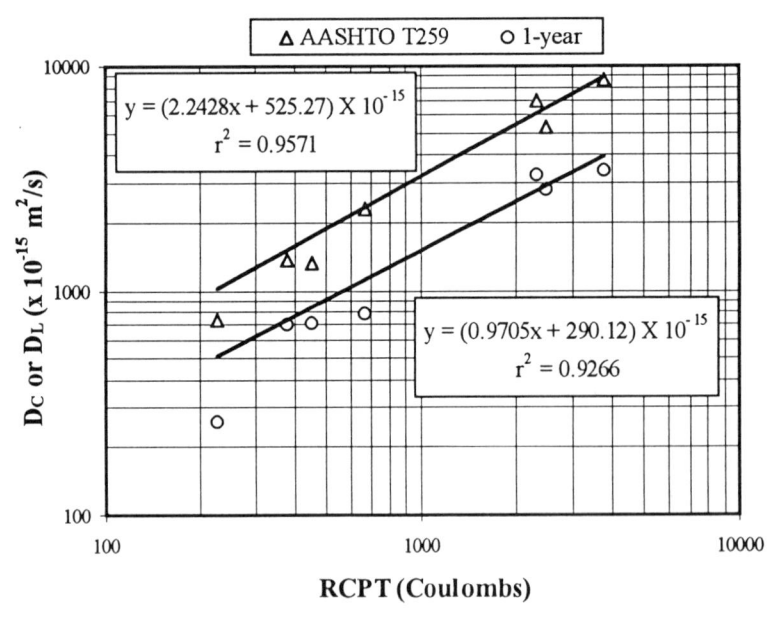

Figure 6 - Salt ponding diffusion coefficients vs. RCPT values

Salt Ponding Tests and RCPT

Both D_C and D_L are plotted against the RCPT results in Figure 6. The same curing regime was used for all the tests: 14-day moist curing then 14-day air curing. Through regression analysis, a linear relationship was found between the ponding tests and RCPT.

Effects of Silica Fume on Diffusion Coefficients

Various diffusion coefficients were normalised and the relative reduction of these coefficients is shown in Figure 7. It is clearly indicated that silica fume could enhance concrete durability by lowering the diffusion coefficients. The rate of decrease was reduced as the silica fume replacement level was beyond 7%. The diffusion coefficients were consistently lowered by 75% at this replacement level. A further 10% reduction was resulted when the silica fume replacement level was increased to 12%.

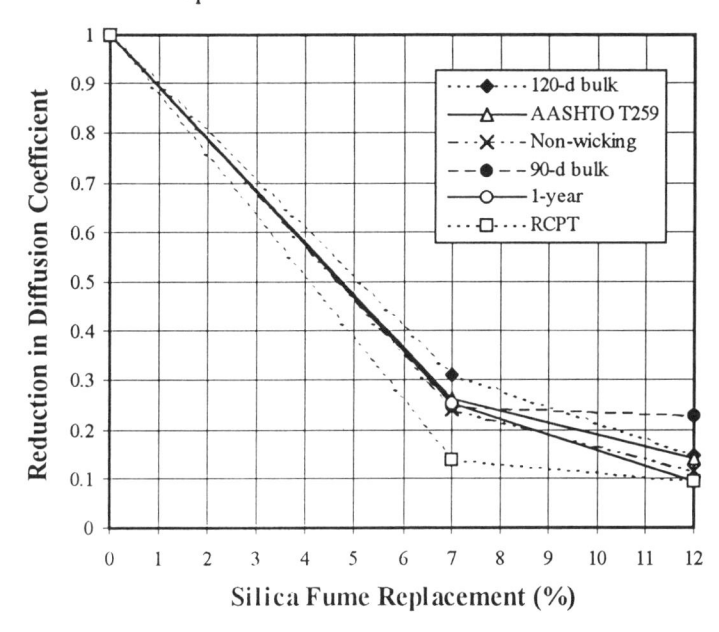

Figure 7 - Relative reduction in diffusion coefficients with silica fume (W/CM = 0.35)

CONCLUSIONS

The study has confirmed the beneficial effect of silica fume on chloride penetration resistance of concrete.
1. 7% silica fume provides a dramatic improvement in chloride penetration resistance regardless of the test procedure used.
2. The reductions exhibited by silica fume concretes in Rapid Chloride Permeability Test (AASHTO T277) are indicative of the reductions shown by both bulk diffusion tests and the AASHTO T259 90-day ponding test.
3. The AASHTO T259 test and the bulk diffusion tests indicate that steam curing is having an adverse effect on chloride penetration but the steam-cured silica fume concretes have greater resistance to chloride penetration compared to the Portland cement controls.

4. One day, ambient-cured strengths were not as high as the same mixture when steam cured. However, the 12% silica fume strength was similar to the 18-hour strength of the 0.35 W/CM Portland cement concrete. From 3 to 56 days, at the same W/CM, 7% silica fume concretes had higher strengths than the steam-cured Portland cement concretes.

REFERENCES

1. Detwiler, R.J., Fapohunda C., and Natale, J., "Use of Supplementary Cementing Materials to Increase the Resistance to Chloride Ion Penetration of Concretes Cured at Elevated Temperatures," ACI Materials Journal, V. 91, No. 1, January-February 1994, pp. 63-66.

2. Sherman, M.R., McDonald, D.B., and Pfeifer, D.W., "Durability Aspects of Precast Prestressed Concrete -- Part 1: Historical Review," PCI JOURNAL, V. 41, No. 4, July-August 1996, pp.62-74.

3. Sherman, M.R., McDonald, D.B., and Pfeifer, D.W., "Durability Aspects of Precast Prestressed Concrete -- Part 2: Chloride Permeability Study," PCI JOURNAL, V. 41, No. 4, July-August 1996, pp.76-95.

DEVELOPMENT OF HIGH PERFORMANCE GROUTS FOR CORROSION PROTECTION OF POST-TENSIONING TENDONS

A.J. Schokker
Assistant Research Engineer
Ferguson Structural Engineering Lab
The University of Texas at Austin
Austin, Texas, USA

B.D. Koester
Assistant Research Engineer
Ferguson Structural Engineering Lab
The University of Texas at Austin
Austin, Texas, USA

M.E. Kreger
Professor
Department of Civil Engineering
The University of Texas at Austin
Austin, Texas, USA

J.E. Breen
Nasser I. Al-Rashid Chair
Department of Civil Engineering
The University of Texas at Austin
Austin, Texas, USA

ABSTRACT

Post-tensioned, precast concrete bridge substructure can greatly reduce traffic interference. The substructure elements may be exposed to such aggressive environments as deicing salts, sea water, acid rain, polluted air, and sulfate-rich soils. In bonded post-tensioned construction, the cement grout acts as a "last line of defense" for preventing chlorides from reaching the steel and initiating corrosion. An optimum grout combines a high level of corrosion protection and desirable fresh properties such as fluidity and resistance to bleed. Different Portland cement grout mixes were evaluated through a series of fresh property tests, an accelerated corrosion test, and a large scale clear duct test. Variables included water to cement ratio and numerous admixtures such as superplasticizer, anti-bleed chemicals, silica fume, fly ash, and corrosion inhibitors.

INTRODUCTION

An optimum grout is fluid enough to be workable while providing corrosion protection to the prestressing strands. Bleed resistance is also important in applications requiring any change in elevation in the duct. Often admixtures used to attain these qualities improve some qualities while adversely affecting others; thus, optimum combinations must be attained. Three separate test phases were used to evaluate critical grout properties. Grout designs were first subjected to a series of fresh property tests. Favorable grouts were then tested with an accelerated corrosion testing method. Finally, selected grouts were tested in a large scale clear parabolic duct to observe grout flow and bleed lenses.

Numerous grout designs were evaluated with different combinations of admixtures. The admixtures used in testing are shown in Table 1. The standard grout mix used by the Texas Department of Transportation was also used as a comparison to grout currently in use. This standard mix has a water-cement ratio of 0.44, with 0.9% cement weight Intraplast-N (superplasticizer/expansive admixture).

Table 1: Admixtures Tested

Admixture	Brand	Manufacturer
superplasticizer	Rheobuild 1000 WRDA-19	Master Builders W.R. Grace
superplasticizer / expansive admixture	Intraplast-N	Sika Corporation
anti-bleed admixture (includes superplasticizer)	Sikament 300SC	Sika Corporation
corrosion inhibitor	DCI Rheocrete 222	W.R. Grace Master Builders
silica fume	Sikacrete 950DP	Sika Corporation
fly ash	Class C	

FRESH PROPERTY TESTS

Test Descriptions

Fluidity -- A fluidity test indicates the workability of the grout. Two variations of the flow cone test (ASTM C939, "Standard Test Method for Flow of Grout for Pre Placed-Aggregate Concrete") were used. For non-thixotropic grout, the cone was filled to the indicated level and the time for the grout to empty the cone was recorded. In the case of thixotropic grouts, the cone was filled to the top and the time for 1 liter of grout to empty from the cone was recorded. If the grout flow stopped before the end of the test, the test was null. The Post-Tensioning Institute is in the process of developing a grouting specification for post-tensioning tendons and suggests a flow cone time to exit between 10-30 seconds.

Standard bleed -- The standard bleed test indicates the bleed properties of grout under atmospheric conditions. The ASTM standard bleed test (C940, "Standard Test Method for Expansion and Bleeding of Freshly Mixed Grouts for Preplaced-Aggregate Concrete in the Laboratory") was used with a slight modification to include a 3 strand bundle that more closely simulates actual conditions. Inserting strands into the grout column tends to increase the amount of bleed. A 1000 ml graduated cylinder was filled with 800 ml of fresh grout, and volume of bleed water was measured over time. The PTI draft specification suggests a maximum bleed of 1% of grout volume.

Bleed under pressure -- Bleed under pressure tests simulate conditions where grout ducts experience a change in elevation. The grout was tested using a funnel manufactured by Gelman Sciences. The grout was placed in the funnel and pressurized with air through a stem at the top. The bottom of the funnel contained a stainless steel screen which supported a glass fiber filter that filtered out bleed water from the grout. Ten minutes after the funnel was filled with grout, the sample was pressurized in 69.0 kPa increments and held for 3 minutes at each increment up to 552 kPa where it was held for 30 minutes before pressure was released. Readings were taken at each increment and pressure at first bleed and loss of pressure were noted. At the time of testing, no established standard existed, so testing followed tests by H.R. Hamilton with a recommended maximum bleed of 2% of the grout sample volume at 345 kPa.

Results

Low water-cement ratio grouts have proven effective in providing corrosion resistance to the prestressing strand, so the majority of the grout designs in the current study had a w/c (water-cement ratio) less than or equal to 0.35. In fact, a w/c of 0.35 is the maximum allowed by the Working Party[3] in a technical report to re-establish grouted post-tensioned bridges in the United Kingdom. Results from the fresh property tests indicated that for w/c in the tested range, a superplasticizer is necessary to maintain fluidity. The addition of fly ash increases fluidity while silica fume significantly reduces fluidity. It is possible that silica fume is more applicable at higher w/c due to the large amount of superplasticizer necessary to maintain fluidity at low w/c.

Standard bleed was found to be reduced with reduced w/c and in grouts containing fly ash or silica fume, while superplasticizer had an adverse effect. The grouts tested that contained anti-bleed admixture experienced no significant bleed during the standard bleed test. Only grouts containing an anti-bleed admixture passed the pressurized bleed test.

ACCELERATED CORROSION TESTS

An accelerated corrosion testing method developed by Thompson, Lankard, and Sprinkel in a FHWA sponsored study and refined at the University of Texas at Austin by T. Hamilton[2] was used to evaluate the corrosion protection potential of several grout designs. The corrosion test used anodic polarization to accelerate corrosion by providing a potential gradient, driving negatively charged chloride ions through the grout to the steel. The test specimens consisted of a short length of prestressing strand in a grouted PVC mold casing. Prior to testing, a portion of the casing was removed to expose the grout. During testing, the specimen was immersed in a 5% NaCl solution and anodic polarization was provided using a multi-channel potentiostat. Specimens were monitored to determine the time to initiation of corrosion. A diagram and picture of one of the 12 experimental stations is shown in Figure 1. A saturated calomel electrode (SCE) was used as a reference electrode. Corrosion potential (E_{corr}) was measured from the reference electrode relative to the working electrode, and the corrosion current (i_{corr}) was found by measuring the voltage across the 100 Ω resistor in-line with the lead on the counter electrode.

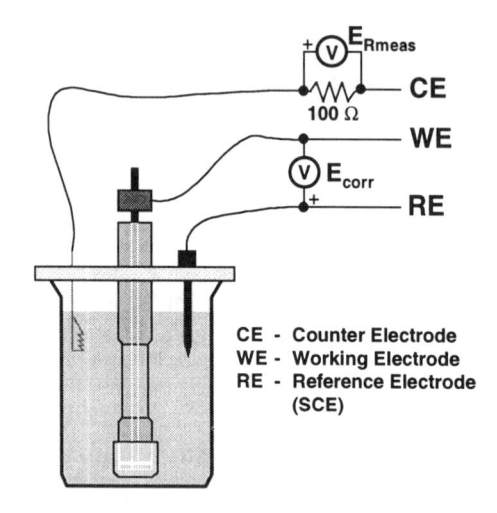

Figure 1: (a) schematic of experimental station (b) test setup

Previous Accelerated Corrosion Studies
Initial tests at the University of Texas at Austin were carried out by H.R. Hamilton. Hamilton tested various grout designs using initially cracked specimens with an applied potential of +600 mV$_{SCE}$. Hamilton tested grout mixes containing corrosion inhibitors, anti-bleed admixtures, superplasticizer, and silica fume. He found that silica fume has the potential to increase corrosion protection, while anti-bleed admixtures can reduce the effectiveness of the corrosion protection of the grout. Hamilton also found that corrosion inhibitors containing calcium nitrite can actually reduce the effectiveness of the grout in providing corrosion protection. Calcium nitrite theoretically works with the passive layer on the steel surface, but cannot work without the passive layer and may actually accelerate the corrosion process when the layer is not present. Hamilton produced anodic polarization curves for the grouted specimens in his study and found that an applied potential of +600 mV$_{SCE}$ puts the steel in the transpassive region. Further accelerated corrosion tests were carried out by B.D. Koester[5] at the University of Texas at Austin with an applied potential of +200 mV$_{SCE}$ which should be well within the passive region of the polarization curve. Other modifications to the accelerated corrosion tests included using uncracked specimens to hopefully reduce data scatter with a larger exposed grout region to accelerate the test. These modifications were also included in the present study, and the present specimen is shown in Figure 2. Koester's variables included w/c, corrosion inhibitor and tests with the current standard TxDOT mix. Koester found increased corrosion protection with lower w/c and lowered protection with the addition of Intraplast-N found in the TxDOT mix. He also found that the corrosion inhibitor calcium nitrite reduced corrosion resistance.

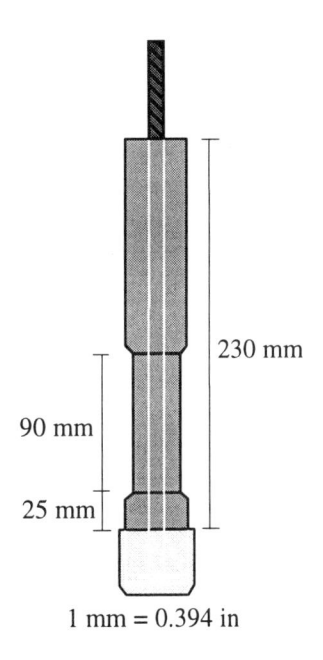

Figure 2: Specimen Dimensions

Present Accelerated Corrosion Studies

Behavior -- The initial behavior for six stations with a grout design of 0.33 water-cement ratio and 2% cement weight of anti-bleed admixture is shown in Figure 3. This initial rapid reduction in corrosion current was typical in all tests and was helpful in isolating and correcting a faulty station (such as the station showing an initial rise in corrosion current in the figure) early in testing. Once a specimen corroded, the corrosion current typically increased by orders of magnitude very rapidly as shown in Figure 4. The spike in corrosion current was immediate in many specimens, but in some cases the rise began gradually. The time to corrosion was taken at the time where the sharp rise in corrosion current began on the plot.

Results -- Present studies evaluated effects of silica fume, fly ash, anti-bleed admixture, superplasticizer, and w/c to develop an optimum corrosion resistant grout. Due to previous negative findings for grouts containing calcium nitrite, corrosion inhibitors were not included in testing. A comparison of all accelerated corrosion tests to date is shown in Figure 5. Hamilton's specimens are included for comparison even though testing conditions were different. The reduced time to corrosion for initially corroded strand is included for several grout designs. This indicates the importance of using the same strand when comparing different grout designs using the accelerated corrosion testing method. The addition of fly ash to the mix gave the highest time to corrosion of the tests to date. A mix containing anti-bleed admixture also tested well. Although anti-bleed admixture has been known to reduce time to corrosion, the very low w/c of 0.33 allowed this mix to perform favorably. This mix was also the only grout design from the

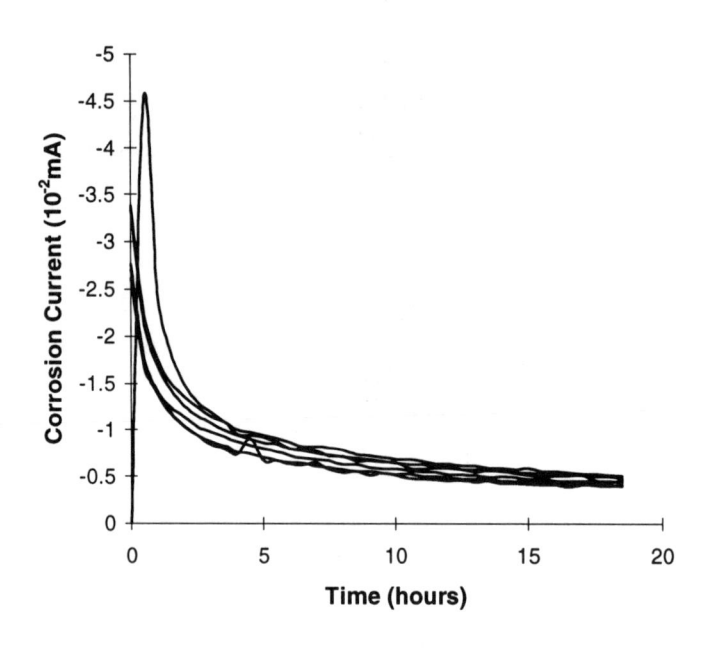

Figure 3: Initial Behavior of Corrosion Current with Time

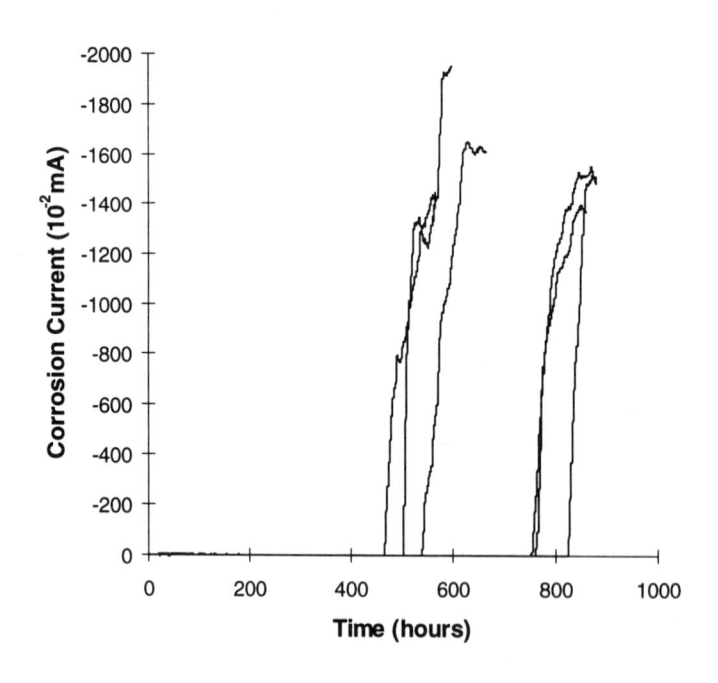

Figure 4: Behavior of Corrosion Current with Time

current tests that passed all of the fresh property tests. Surprisingly, the silica fume mix tested did not perform favorably. This was likely due to the large amount of superplasticizer needed at the 0.35 w/c for fluidity. Superplasticizer, like most chemical admixtures, has been found to reduce corrosion protection effectiveness of the grout. A grout with low permeability and with minimal chemical admixtures gives the best corrosion performance.

Mineral Admixtures	*Chemical Admixtures*	*w/c*
	2.2% Sikament, DCI	0.40
	Rheocrete	0.40
	2.2% Sikament	0.40
15 % Sikacrete		0.47
5% Sikacrete	2.2% Sikament	0.40
		0.40
	1.9 ml/kg WRDA-19	0.40
	1% Intraplast-N, DCI	0.44
	DCI-S	0.40
15% Sikacrete	16 ml/kg Rheobuild	0.35
	1% Intraplast-N	0.44
		0.50
		0.44
	2.0% Sikament	0.33
		0.40
30% fly ash	4 ml/kg Rheobuild	0.35

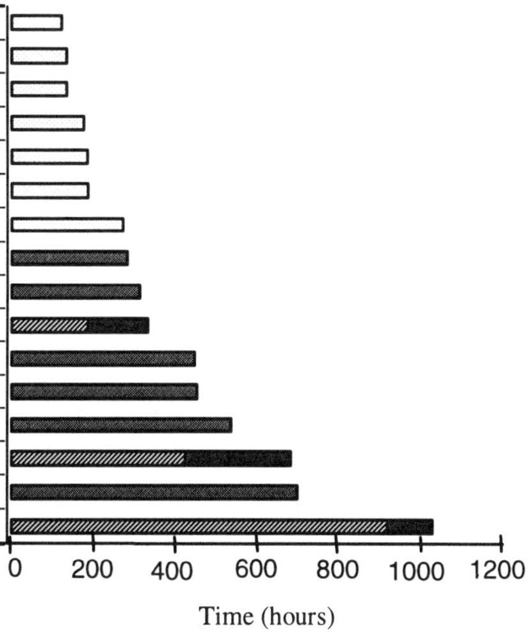

Time (hours)

Initially cracked specimens, applied potential +600 mV$_{SCE}$
Initially uncracked specimens, applied potential +200 mV$_{SCE}$
Initially uncracked specimens, applied potential +200 mV$_{SCE}$, initially corroded strand

Figure 5: Mean Time to Corrosion for Various Grout Designs

LARGE SCALE PARABOLIC DUCT TESTS

Grouts that performed favorably in the fresh property tests and the accelerated corrosion tests were then tested in a large scale clear parabolic draped duct. This test investigated grout flow along the duct and strand along with the formation of bleed water lenses due to changes in duct elevation. The workability of the grout in actual grout pumping was also observed.

Test Setup and Procedure

A clear vinyl flexible tube (inner diameter 38 mm [1 1/2"]) with a bundle of 3, 13 mm [0.5"] diameter prestressing strands was used to simulate conditions in a post-tensioning duct. The duct dimensions are shown in Figure 6 and a picture of the grouted duct is shown in Figure 7. Vents were located at the crowns and slightly downstream of the main crown. The duct was grouted from the low point on the right end. During grouting vents were closed in order from the beginning to the end of the duct with the exception that at the intermediate crest, the downstream vent was closed prior to the corresponding crown vent. Each vent was closed after a steady stream of grout was ejected with a volume greater than that of the duct between the vent and the previous vent. After grouting, the duct was sealed and allowed to set. The hardened grouted duct was autopsied by removing 50 mm [2"] slices from critical locations. The slice locations are marked with X symbols in Figure 6, and a diagram of a typical slice is shown in Figure 8. The slices were evaluated for area voids and perimeter voids as defined in the figure.

1 m = 3.281 ft

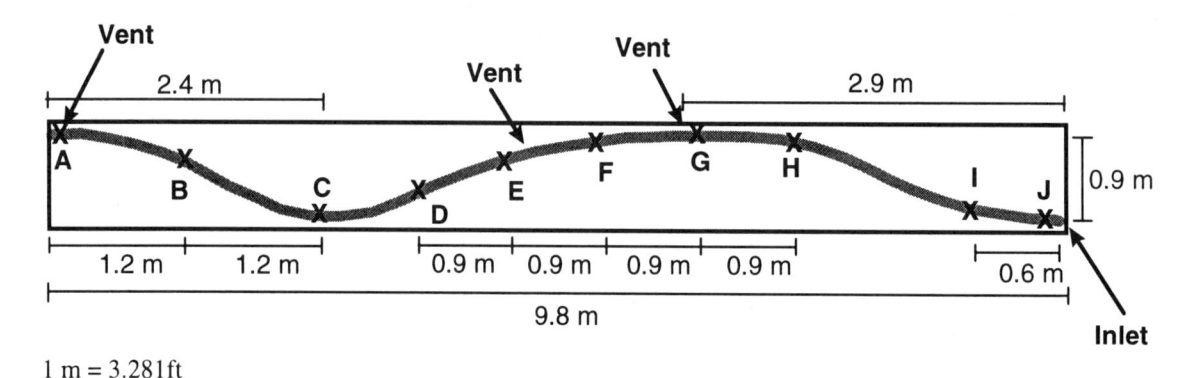

Figure 6: Duct Dimensions and Autopsy Cut Locations

Figure 7: Grouted Duct

Inner diameter = 38 mm
Outer diameter = 48 mm

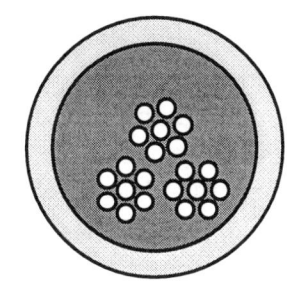

3 strand bundle
13 mm diameter,
7 wire strand

PV = perimeter voids = $\dfrac{\text{exposed strand perimeter}}{\text{total strand perimeter}}$

1 mm = 0.0394 in

Figure 8: Typical Slice

Results

A mix of w/c=0.33 with 2% Sikament (anti-bleed admixture) was tested in the duct. This mix performed exceptionally well with no voids present in any of the autopsy slices as shown by the typical slice in Figure 9. Tests with the standard TxDOT mix and with a fly ash mix will follow. Bleed lenses are more likely to be formed with these grout designs which do not contain an anti-bleed admixture, and it will be interesting to see if an anti-bleed admixture is needed in cases with vertical rises of only a few feet.

0.33 w/c
2% anti-bleed

0.35 w/c
30% fly ash

TxDOT Standard
(0.44 w/c, 0.9% Intraplast-N)

Figure 9: Comparison of Slices from Grouted Duct

CONCLUSIONS

A summary of grout performance in each phase of testing is shown in Table 2. The two top performing grouts are shown along with the TxDOT standard grout for comparison. All grouts are within the limits recommended by PTI for fluidity and standard bleed. Bleed was instantaneous (before pressure was applied) for the fly ash grout and the TxDOT standard grout in the pressurized bleed test while the anti-bleed grout performed very well. The fly ash grout demonstrated the best corrosion protection in the accelerated corrosion test phase with a mean time to corrosion more than twice as long as the TxDOT standard mix. Both the anti-bleed and fly ash mixes performed well in the large scale duct test with no noticeable voids in the autopsy slices. The TxDOT standard mix contained significant voids and exposed steel at the intermediate crest.

Table 2: Results for Each Testing Phase

Grout Mix	Fluidity (sec)	Standard Bleed (%)	Pressure Bleed (%)	Mean Time to Corrosion (hours)	Area Voids (%)	Perimeter Voids (%)
0.33 w/c 2% anti-bleed	28	0	1.9	683	0	0
0.35 w/c 30% fly ash 4 ml/kg super	17	.6	instant bleed	1028	0	0
TxDOT mix (0.44 w/c, 1% Intraplast-N)	19	.6	instant bleed	447	16	8

Development of a corrosion resistant grout is an important step in durability design of post-tensioned structures. An optimum grout combines workability with corrosion resistant properties such as low permeability and resistance to bleed lense formation. Grouts with low w/c or that contain fly ash or silica fume are less permeable and have performed favorably in accelerated corrosion tests. However, often these low permeability grouts are not workable without the addition of superplasticizer or may require an anti-bleed admixture to reduce formation of bleed water lenses. Both superplasticizer and anti-bleed admixtures have been found to reduce a grout's corrosion protection effectiveness. Corrosion inhibitors have also been found to reduce the effectiveness of corrosion protection in Portland cement grouts. At this stage in testing, the most promising grout for situations requiring a high resistance to bleed has a 0.33 w/c with 2% cement weight Sikament (anti-bleed admixture) which passed all fresh property tests, performed well in the accelerated corrosion test, and had no voids in the large scale duct test. This grout is very workable as long as it is kept agitated. The most promising grout for situations requiring a high resistance to corrosion has a 0.35 w/c with 30% cement weight replacement with fly ash and 4 ml/kg cement weight superplasticizer. The fly ash reduces permeability and slightly increases fluidity so that only minimal superplasticizer is necessary for workability. This mix is workable but is on the borderline for standard bleed. It did not pass the pressurized bleed test. The excellent performance in the large scale duct test indicated that bleed lense formation was not critical for small vertical rises. The fresh property tests coupled with accelerated corrosion testing and a large scale trial provide an excellent means for comparing grout designs for corrosion protection while assuring that workability is maintained.

REFERENCES

1. "Guide Specification for Grouting of Post-Tensioned Structures," PTI Committee on Grouting Specifications, 4th Draft, November 1996.
2. Hamilton, H.R., III, "Investigation of Corrosion Protection Systems for Bridge Stay Cables," *Ph.D. Dissertation*, The University of Texas at Austin, September 1995.
3. "Durable Bonded Post-Tensioned Concrete Bridges," *Technical Report No. 47*, the Working Party of the Concrete Society, August 1996.
4. Thompson, N.G., Lankard, D., and Sprinkel, M., "Improved Grouts for Bonded Tendons in Post-Tensioned Bridge Structures," *FHWA-RD-91-092*, Cortest Columbus Technologies, January 1992.
5. Koester, B.D., "Evaluation of Cement Grouts for Strand Protection Using Accelerated Corrosion Tests," *Masters Thesis*, The University of Texas at Austin, December 1995.

PERMEABILITY OF HIGH PERFORMANCE CONCRETE: RAPID CHLORIDE ION TEST VS. CHLORIDE PONDING TEST

John J. Myers, P.E., Graduate Research Assistant / Ph.D. Candidate
Wissam E. Touma, Graduate Research Assistant / M.S.E. Candidate
Ramon L. Carrasquillo, Ph.D., P.E., Professor of Civil Engineering
Department of Civil Engineering
The University of Texas at Austin
Austin, Texas, U.S.A.

ABSTRACT

Concrete permeability may be the most relevant concrete property affecting its durability especially under exposure to aggressive environment. As a result, permeability is an aspect of concrete performance that must be specified, designed for and monitored in the production of concrete. This is especially true in the case of a bridge deck subjected to deicing salt, a parking garage, a pavement overlay, an off-shore structure, or even a building structure where durability and life cycle costs are a critical issue. To determine the chloride permeability of concrete, many engineers have increasingly been using AASHTO T277, "Rapid Determination of the Chloride Permeability of Concrete," and specifying a coulomb rating of 1500 or less to ensure low permeability concrete. Current research studies dealing with normal to mid-range concrete strengths have questioned the validity of AASHTO T277. As a sole indicator of concrete permeability, it has been suggested that for specific concrete mixes, AASHTO T277 should not be used solely without developing initial correlation data with AASHTO T259, "Resistance of Concrete to Chloride Ion Penetration."

The paper summarizes the results of an ongoing research study which was initiated in the Fall of 1995 at the University of Texas at Austin. The main objective of the study has been to investigate the correlation between AASHTO T277 and AASHTO T259 for high performance concrete. The variables being investigated include entrained air content, fly ash content, and aggregate type. The mix designs used included a variety of mix proportions which might be used in either an exposed environment such as a bridge or in a protected environment such as inside a building. Statistical data on compressive strength, modulus of elasticity, splitting tensile strength, chloride percentage, and rapid ion permeability is presented.

INTRODUCTION

Workability and strength have long been the two major properties used for the specification and design of durable structural concrete. This was based on the tradition that strength relates to all properties of concrete and thus can also be used as a durability indicator. This general association has been seriously questioned and recently engineers have started to specify concrete for durability in addition to strength. This change can be traced in part to the rising cost of repair and maintenance of existing concrete structures many of which are experiencing serious structural distress long before their design service life is reached.

While high strength concrete, enables the design of structures with higher load capacity and smaller members, it does not automatically imply a more durable structure. It is the responsibility of the designer to specify a concrete mixture that combines both strength and durability under aggressive environments. This task is further complicated by the use of several mineral and chemical admixtures in the production of high strength concrete. Such a durable mixture is classified as a "High Performance Concrete" which includes any concrete with an enhanced property, be it strength, durability, economics, or constructability.

Specifying concrete directly on the basis of performance rather than indirectly by strength requires a suitable durability test measure that is simple and reliable. While accelerated durability tests exist, they are costly, time consuming, and limited to laboratory use. On the other hand, correlations between permeability and durability have long been justified and studied. Permeability has been shown to be a major factor affecting many types of concrete distress such as sulfate attack, frost resistance, alkali-aggregate attack, carbonation, corrosion of reinforcing steel, and other durability properties. As a result, permeability is an aspect of concrete performance that can be specified, designed for and monitored in the production of concrete.

Two tests are generally used to determine concrete's permeability to chloride ions and other aggressive solutions: The American Association of State Highway and Transportation Officials (AASHTO) test method T277, *Rapid Determination of the Chloride Permeability of Concrete,* and test method T259, *Resistance of Concrete to Chloride Ion Penetration.* The AASHTO T259 test has long been favored, among the concrete corrosion specialists, for assessing the chloride penetration of concrete. On the other hand, with a growing need for a faster and less expensive indicator of chloride permeability, AASHTO T277 is becoming more popular among owners and specifiers. Today, the rapid 6-hour test (T277) is a common requirement in construction project specifications where engineers are specifying low values between 700 and 1,000 coulombs to ensure durable concrete. The following study discusses the original correlation developed between these aforementioned tests and investigates the correlation for a series of high performance concrete mix designs developed in the laboratory.

BACKGROUND

Original Development of AASHTO T277

The process of chloride ion diffusion into concrete is a very slow one even when dealing with conventional concrete having high water to cement ratio. Thus, in 1981, the Portland Cement Association (PCA) headed a research program sponsored by the Federal Highway Administration (FHWA) to develop a test method for determining the chloride permeability of various concretes. The original test method is listed under FHWA/RD-81/119, *Rapid Determination of the Chloride Permeability of Concrete*. The American Association of State Highway and Transportation, AASHTO, adopted the same test in 1983 under AASHTO T277, *Rapid Determination of the chloride Permeability of Concrete*, and in 1991, the American Society of Testing and Materials, ASTM, designated it as ASTM C1202, *Electrical Indication of Concrete's Ability to Resist Chloride Ion Penetration*. ASTM limited the use of this test only to types of concrete where correlations to long-term chloride ponding procedures such as AASHTO T259, have been established [1].

Originally, the FHWA wanted a test that was fast, could be done in place without disturbing concrete or on cores in the laboratory, had a good correlation to data obtained from AASHTO T259, and measures the degree of impermeability. The research program, conducted in 1981, showed that the rate of chloride migration through a concrete specimen accelerates under the influence of an electric potential. Researchers also found that a good correlation between the measured amount of coulombs through the concrete specimens and the results obtained from the long term ponding tests existed. Based on these findings, the rapid chloride permeability test was developed to measure the concrete resistivity to chloride migration. It does not measure the concrete permeability nor does it measure what actually occurs to concrete in the field. It only gives an indication as to how a certain concrete mixture is relating to the original set of mixtures listed in Table 1[1].

Table 1: Charge Passed vs. Total Integral Chloride [1]

Charge Passed (Coulombs)	Chloride Permeability	Type of Concrete	Total Integral Chloride to 40.6-mm (1.6-in.) Depth After 90-day Ponding Test
>4000	High	High water-cement ratio, conventional (≥0.6) PCC *	>1.3
2000-4000	Moderate	Moderate water-cement ratio, conventional (0.4 to 0.5) PCC *	0.8 to 1.3
1000-2000	Low	Low water-cement ratio, conventional (<0.4) PCC *	0.55 to 0.8
100-1000	Very Low	Latex-modified concrete, internally sealed concrete	0.35 to 0.55
<100	Negligible	Polymer impregnated concrete, polymer concrete	<0.35

* Portland Cement Concrete

Correlation Between AASHTO T277 and AASHTO T259

While the amount of literature involving the use of AASHTO T277 and AASHTO T259 with low to mid-range strength concrete is voluminous, few research programs dealt with the use of these tests with high performance concrete. The following papers were found to be most relevant to the current research and very necessary to assess the validity of the two tests.

In the original report, *"Rapid Determination of the Chloride Permeability of Concrete" by Whiting* [1] which was published in 1981 and included the results of a research program undertaken by the FHWA, David Whiting outlined the development of a test procedure for a rapid assessment of the permeability of concrete to chloride ions (rapid 6-hours test method). He conducted a test program that included the concrete mixtures listed in Table 2 and established a correlation (Figure 1) between the 6-hour test method and AASHTO T259, the 90-day Ponding Test as shown in Table 1. The correlation was based upon tests conducted on a single core of each of the concrete mixtures. None of the concrete mixtures tested contained mineral admixtures or were high strength.

Table 2: Concrete Mixtures Used in the FHWA's 1981 study [1]

Concrete Type	Water to Cement Ratio by Weight
Conventional Portland Cement Concrete, PCC	0.60, 0.50, 0.40
Latex Modified, LMC	0.24
Internally Sealed Wax Bead, WBC (Heated and not Heated)	0.55
Polymer Impregnated Concrete, PIC	0.50
Iowa Low Slump, IOWA	0.33
Polymer Concrete, PC	N.A.

Whiting introduced the parameter of "Total Integral Chloride", *I*, which is a dimensionless value representing the integral of chloride concentration versus unit depth increments of 5.1-mm (0.2-in.) rather than actual depth. "*I*" was used to represent the results of the 90-days ponding test in only one value. In an equation form the "total integral Chloride" is represented by Equation 1.

$$I = \frac{\text{Total Area Under the Line Representing Chloride Content vs. Depth}}{5.1\text{-mm }(0.2\text{-in.})\text{ Depth Increments}} \quad \text{Equ. (1)}$$

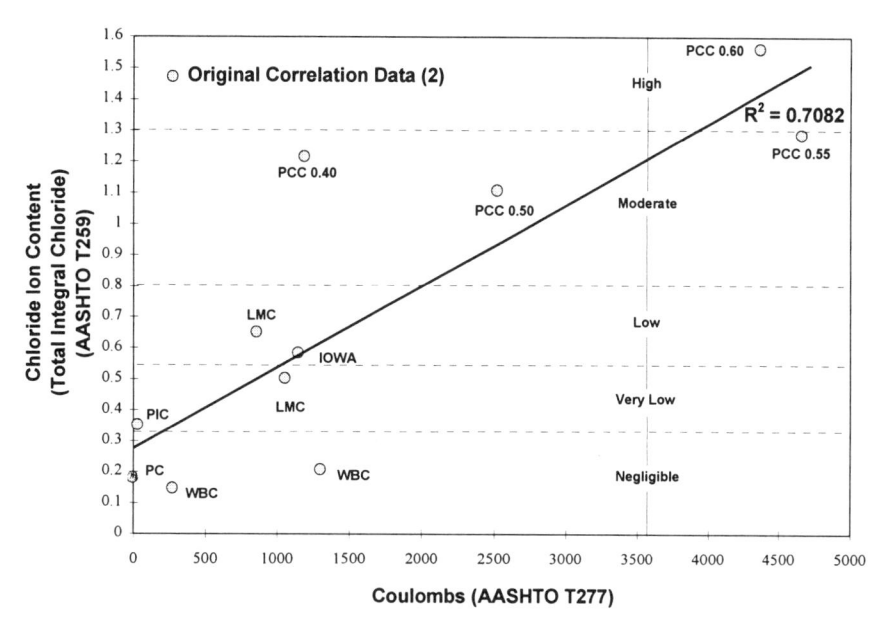

Figure 1: Correlation Data from Reference 1 of ASTM C1202 Document [2]

Using a linear regression analysis, Whiting developed a correlation between the 90-day Ponding Test and the Rapid 6-hours Permeability Test with an "R^2" coefficient of 0.708. The linear regression line of his data are reproduced in Figure 1 where it can be seen that the data show a poor correlation in the 700 to 1300 coulombs range. Whiting also concluded that polymer concrete and polymer-impregnated concrete had "negligible" chloride permeability based upon the 6-hour Test. Yet, these same concrete mixtures showed "*I*" values of 0.16 and 0.34 according to the 90-days Ponding Tests. These "*I*" values are relatively high compared with the 6-hour Test results.

Whiting and Dziedic published *"Resistance to Chloride Infiltration of Superplasticized Concrete as Compared with Currently Used Concrete Overlay Systems"* [2] in 1989. In it, they discussed the effectiveness of various materials used in bridge overlays at construction sites in Ohio, and developed a correlation between the Rapid (6-hour) and 90-day Ponding Tests. From each of the concrete mixtures listed in Table 3, two specimens were tested for AASHTO T277 and a singular specimen was tested for AASHTO T259. Concrete mixtures containing high range water reducers (HRWR) and silica fume were moist cured for 42-days and the latex-modified concrete (LMC) was moist cured for 2-days then left to air dry for 40-days. Whiting and Dziedic found a correlation coefficient, R^2, of 0.81 for data from both the Rapid and the 90-day Ponding Tests. The long term Ponding Tests on all twenty-two concrete mixtures resulted in low total integral chloride values, "*I*", ranging from 0.05 to 0.15. The corresponding coulomb values ranged from 250 to 5000. The latex-modified concretes had coulomb values between 700 and 2200, while the HRWR concretes had coulomb values between 250 and 5000.

Table 3: Concrete Mixtures Used in the 1989 Paper by Whiting and Dziedic [2]

Concrete Type	Usual w/c	Number of Mixes	Source of Concrete Mixtures
HRWR modified	0.30 to 0.32	10	2 mixtures from each of 5 sites
Latex modified	<0.39	10	2 mixtures from each of 5 sites
Silica Fume modified	0.30	2	1 site

In a paper [3] published in 1992, Mitchell and Whiting stated that results of the AASHTO T277 test are not as reliable as they were thought to be. The authors cited that the development of the original correlation did not investigate variables such as aggregate type and size, cement content and composition, density, in addition to other factors. In the authors opinion, the correlation did not represent a large data base of concrete, only single cores taken from FHWA original slabs. The authors recommended for individuals using the Rapid Chloride Permeability Test (RCPT) to establish their own correlation between charge passed and known chloride permeability for their own particular materials.

In 1994, Pfeifer, McDonald, and Krauss published *"The Rapid Chloride Permeability Test and Its Correlation to the 90-days Chloride Ponding Test"* [5] in which they seriously critiqued the use of the AASHTO T277 test. The paper included a review of the five references[1,6,7,8,9] that substantiate the use of the ASTM C1202 and several other papers concerned with the application of the Rapid Chloride Penetration Test. The authors of the paper calculated and compared total integral values, *"I"*, from all five references. They concluded that these values exhibit wide variance and no correlation with Table 1, originally devised by Whiting in 1981. Results of that comparison are presented in Table 4 which shows that there is no consistency in rating the chloride penetrability of the concrete mixtures tested.

Table 4: Comparison of the Ratings of Chloride Penetrability Based Upon Results from the Five ASTM C1202 References [5]

Coulombs	Ref. 1	Ref. 2	Ref. 3	Ref. 5
5000	High	Low	Negligible	Low
2500	Moderate	Very Low	Negligible	Very Low
1500	Low	Very Low	Negligible	Very Low
500	Very Low	Negligible	Negligible	Very Low
<100	Negligible	Negligible	Negligible	Negligible

EXPERIMENTAL PROCEDURE

Mix Designs and Materials

Throughout the study, a total of thirty-one high strength concrete mixtures were developed varying several components and ingredients. The variables being investigated include the entrained air content, fly ash content, HRWR type, and aggregate type. The mix designs used included a variety of mix proportions which might be used in either an exposed environment such as a bridge or in a protected environment such as inside a building. The proportions of these mix designs and physical properties of the constituents selected for this study may be referenced in the publication entitled *"Permeability of High Performance Concrete: The Rapid Chloride Test vs. The 90-Days Test,"* by W. Touma [16].

Mixing was conducted under laboratory conditions in a 6-cubic feet capacity rotary drum mixer using 0.01-cubic meter (3.5-cubic feet) batches. All mixtures had an initial slump, before the addition of HRWRs, between 12.7-mm to 25.4-mm (0.5-in. and 1.0-in.) and after the addition of HRWRs, between 203.2-mm to 254-mm (8-in. and 10-in.). Materials utilized were typical of

those being used by local suppliers and were approved by TxDot for use in Texas bridge structures. The following is a list of the material properties:
1. Portland Cement: Type I/II conforming with ASTM C150. Total alkalies 0.60% and 3.15 specific gravity.
2. Fly Ash: a commercially available Class C fly ash provided conforming to ASTM C618-94a
3. Coarse Aggregates: 19-mm river gravel (RG), 19-mm Burnet limestone (LS), and 19-mm trap rock (TR) (3/4-in. nominal maximum size) all conforming to ASTM C33.
4. Fine Aggregate: Natural river sands from the Colorado river conforming to ASTM C33.
5. Chemical Admixtures:

   ASTM C494, a commercially available Type B Retarder.
   ASTM C494, a naphthalene based commercially available Type A/F HRWR.
   ASTM C260, air entraining agent with a neutralized vinsol resin base.
   Additional HRWR: a water soluble acrylic graft copolymers with functional sulfonic and carboxyl groups and a molecular designed graft copolymer with functional anion group.

Test Specimens

Cylinder Specimens - Twenty one, 101.6-mm x 203.2-mm (4-in. x 8-in.), cylinders were fabricated from each batch for the purpose of testing for the rapid chloride permeability, the compressive strength, elastic modulus of elasticity, and split cylinder strength. Using plastic molds, the cylinders were made in accordance with ASTM C192, *Standard Practice for Making and Curing Concrete Test Specimens in the Laboratory*, covered with a plastic cap, and left to harden at room temperature for 24-hours. The cylinders were then demolded and subjected to the specified curing procedure (See Curing Procedure).

Ponding Block Specimens - Each batch was also used to cast four 305-mm x 305-mm x 89-mm (12-in. x 12-in. x 3.5-in.) thick ponding blocks which satisfied the requirements of AASHTO T259. Concrete was placed into the ponding block's formwork, vibrated using an electric vibrator, finished with a float, and covered with a plastic sheeting. Twenty-four hours later, the slabs were demolded and subjected to the specified curing.

Curing Procedure

Twenty-four hours after casting, specimens were placed in a moisture room at 21° C (73° F) and 100% relative humidity. All the slab specimens were moist cured for 14-days and then air dried for 42-days. Two cylinders of each mixture were cured under the same conditions as their accompanying ponding blocks. Cylinders cured under this condition were selected for the correlation study. The rest of the cylinders were cured under continuous moist curing until time of testing. Two continuous moist cured cylinders were subjected to the rapid ion permeability test to investigate any variation between the curing conditions.

Testing Procedures

Fresh Concrete Testing - Each concrete mixture was tested, in the plastic state, for slump and air content. Slump tests were performed before and after adding superplasticizers and in accordance with ASTM C143, *Standard Test Method for Slump of Portland Cement Concrete*. Air content was tested in accordance with AASHTO T152, *Air Content of Freshly Mixed Concrete by the Pressure Method*, immediately before placement of the concrete.

Compressive Strength Testing - All concrete mixtures were tested for compressive strength. Compressive strength of each mixture was determined at 7, 28, and 56 days of age in accordance with ASTM C39-93, *Standard Test Method for Compressive Strength of Cylindrical Concrete Specimens*. The compressive strength was reported as the average of three companion cylinders.

Splitting Tensile Strength - The splitting tensile strength was determined for each mixture at 28 and 56 days of age in accordance with ASTM C496-90, *Standard Test Method for Splitting Tensile Strength of Cylindrical Concrete Specimens*. The average of two companion cylinders was reported as the splitting tensile strength of the mix design. Results of the splitting tensile strength at 56 days for all the mixtures are illustrated in Figure 2. This represents the concrete properties when the 90 day ponding cycle was initiated.

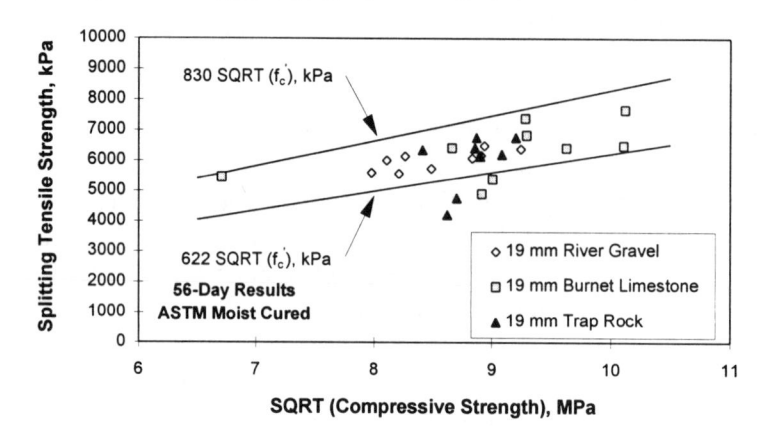

Figure 2: Splitting Tensile Strength Test Results

The data is generally concentrated in the bands between 622 to 830 Sqrt f_c', in kPa (7.5 to 10 Sqrt f_c', in psi) which exceeds the empirical equation for HPC recommended by ACI Committee 363.

Elastic Modulus - ASTM C469-94, *Standard Test Method for Static Modulus of Elasticity and Poisson's Ratio of Concrete in Compression* was used to determine the elastic moduli of 101.6-mm x 203.2-mm (4-in. x 8-in.) cylinder specimens. Three cylinders of each mixture were used to obtain an elastic modulus measurement. Figure 3 illustrates the 56 day test results for all the mixes. The elastic modulus generally satisfied the empirical equation recommended by ACI committee 318. All of the mixes exceeded the empirical equation recommended by ACI committee 363 for high strength concretes.

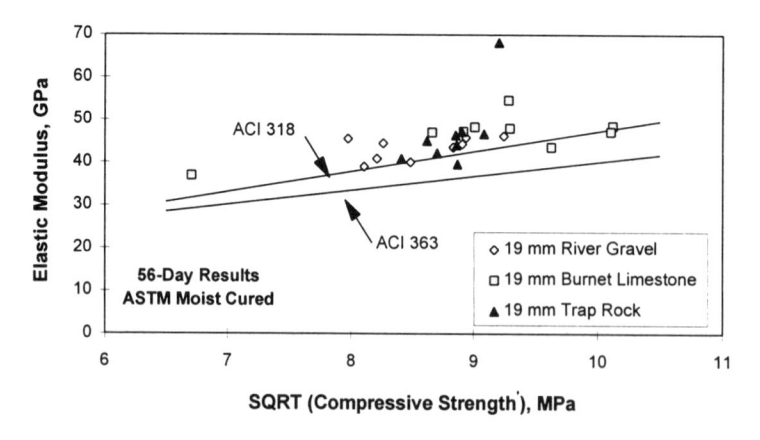

Figure 3: Elastic Modulus Development Test Results

Rapid Chloride Permeability Testing - AASHTO T277, *Standard Method of Test for Rapid Determination of the Chloride Permeability of Concrete*, was used to determine the chloride ion permeability. The permeability of each mixture was considered to be the average coulomb results of four 51-mm (2-in.) thick disks (i.e. two 101.6-mm x 203.2-mm cylinders from each mixture) which were prepared and tested following the AASHTO T277 procedure.

In this study, it was desired to investigate the correlation between AASHTO T277 and AASHTO T259. Therefore, specimens from both tests were cured identically and the Rapid Chloride Test was performed at the same time the 90-day ponding cycle for the ponding block specimens was initiated. At later ponding cycles in the study, RCPT are planned for cores taken from the ponding block specimens after being ponded for 180-days and 365 days (See Ongoing Comparison Study). This was done to ensure that specimens for both tests were subjected to identical curing conditions. The RCPT was also performed on specimens moist cured for 56-days to investigate the effect of curing on the correlation data.

Chloride Ion Penetration Testing - Diffusion of chloride ions into the slab specimens was determined using a CL-1000 chloride test system provided by James Instruments. After 90-days of continuous ponding, powdered samples were collected from each of the slabs by drilling three holes to various depths: from 1.6-mm to 12.7-mm (0.0625-in. to 0.5-in.), 12.7-mm to 25.4-mm (0.5-in. to 1.0-in), and 25.4-mm to 38.1-mm (1.0-in. to 1.5-in.). The powered concrete samples from each depth were mixed together from each slab and then divided into two 1.5-grams samples. This resulted in a total of 6 samples per mixture at each depth (three slabs per mixture). The powder was then mixed with a 15% acetic acid extraction solution which reacted with the chloride ions and produced an electrochemical reaction, the degree of which was measured using the CL-1000 electrode setting. All drilled holes were then closed using a rapid setting epoxy and the slabs were subjected to an additional 90-days period of ponding as part of the ongoing study. Measurements were converted into percent chloride ion content using a mathematical model presented in AASHTO T260, *Sampling and Testing for Chloride Ion in Concrete and Concrete Raw Materials*, Procedure C. Assuming a concrete density of 4000-pounds per cubic yard, the chloride percentage was translated into chloride content in pounds of chloride per cubic yards. For the purpose of comparing these results to literature values, the "Total Integral Chloride", *I*, factor was computed.

DISCUSSION OF TEST RESULTS

Variation Between RCPT Curing Conditions

In order to investigate the correlation between AASHTO T277 and AASHTO T259, it was important to determine the effect of the length of moist curing on the AASHTO T277 results. Therefore, specimens from all the mixtures were cured under two different conditions: 1) 56-days moist curing and 2) 14-days moist curing plus 42-days air drying. Clearly the specimens moist cured for 56-days were less permeable than those cured with the ponding blocks due to the additional cement hydration associated with the added moist curing. However, the difference between the two permeability results was not large enough to cause a shift in the permeability levels defined in AASHTO T277 (Table 1): Specimens from the same mixture but cured under the two conditions exhibited the same permeability, either "Low" or "Very Low". From this evaluation, it was concluded that for the low water to cementitious material ratio mixtures used in this study (0.25 through 0.33), the length of moist curing had a relatively minor effect on the permeability results of AASHTO T277. However, the correlation relationship was affected by the curing condition as discussed in subsequent sections of this report.

90-Day Correlation Between AASHTO T277 and AASHTO T259
 General Results - As expected from high strength concrete (HSC) mixtures with low water to cementitious material ratios, the measured coulomb values were relatively low and the 90-day chloride contents were concentrated in the uppermost 12.7-mm (0.5-in.) of all the blocks subjected to ponding; chloride content at deeper depths was negligible and in some cases was zero. The RCPT results ranged from 384 to 1956 coulomb while the corresponding integral chloride ranged from 0.250 to 1.214.
 Permeability Classifications as Defined in the FHWA-81 Original Report - Using the permeability definitions presented and discussed in the FHWA-81 report, Table 1, the mixtures were classified based upon their total integral chloride and their coulomb values. Fifty percent of the mixtures exhibited identical total integral chloride and coulomb classifications while the other fifty percent showed some discrepancies: 8 mixtures exhibited "Very Low" coulomb values while exhibiting "Low" total integral chloride values, 3 mixtures exhibited "low" coulomb values while exhibiting "Moderate" integral chloride values, and 2 mixtures exhibited "Very low" coulomb values while exhibiting "Moderate" total integral chloride values. As a result, it was concluded that for the high strength concrete mixtures tested, the two tests were not consistent in predicting the permeability levels defined in the original test development report of AASHTO T277. However, it is important to note that in the case of this study all of the AASHTO T277 results were conservative and resulted in a lower permeability classification than the 90-day Ponding Test results; All the discrepancies arose from the fact that the total integral chloride resulted in a higher classification.
 Linear Regression Analysis - The correlation between AASHTO T277 and AASHTO T259 for the HSC mixtures used in this experimental program was further investigated by performing a linear regression analysis on the two sets of data. Two cases were considered: 1) the 90-day Total Integral Chloride values, *I* values, were coupled with their corresponding

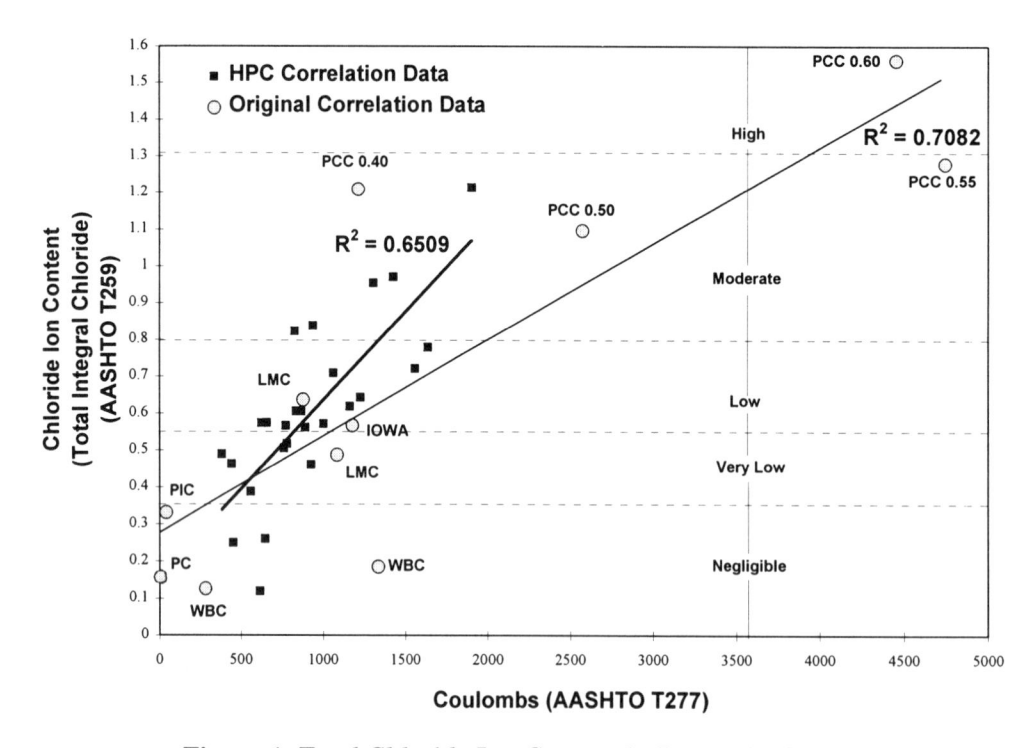

Figure 4: Total Chloride Ion Content by Integral - Correlation Data

coulomb values; and 2) the 90-day chloride content results in percent by weight of concrete in the uppermost 12.7-mm (0.5-in.) of the blocks were coupled with their corresponding coulomb values. The uppermost 12.7-mm (0.5-in.) chloride content resulted in an R^2 correlation coefficient of 0.71 and exhibited a slightly better correlation to the T277 results than the total integral chloride values which displayed an R^2 coefficient of 0.65. Results of both regression analysis are shown in Figures 4 and 5. In the original FHWA-81 report, the authors used the total integral chloride data and reported an R^2 correlation coefficient of 0.70 which is equivalent to the coefficient calculated in this study. However, the correlation was quite different due to the following: 1) 31 mixtures were investigated in the current study as compared to 11 mixtures in the original report; and 2) results obtained from the Rapid Chloride Permeability Test in this experimental program were all relatively low and did not cover a wide range of values. Figure 4 compares the Total Integral Chloride data generated in the course of the current study to data used in the development report of AASHTO T277. The following remarks pertain to Figure 4 and the Total Integral Chloride:

1. For the range of coulomb values used, between 384 and 1904 coulomb, the original data exhibited a very poor correlation while data produced during this study displayed good correlation ($R^2 = 0.65$).
2. 77 percent of the data points were higher than the original regression line, which was anticipated due to the mixture types originally investigated. However, the location of the original portland cement concrete data (w/cm ratio = 0.04 to 0.60) relates well to the water cement ratios (w/cm ratio = 0.28 to 0.33) investigated in this study.

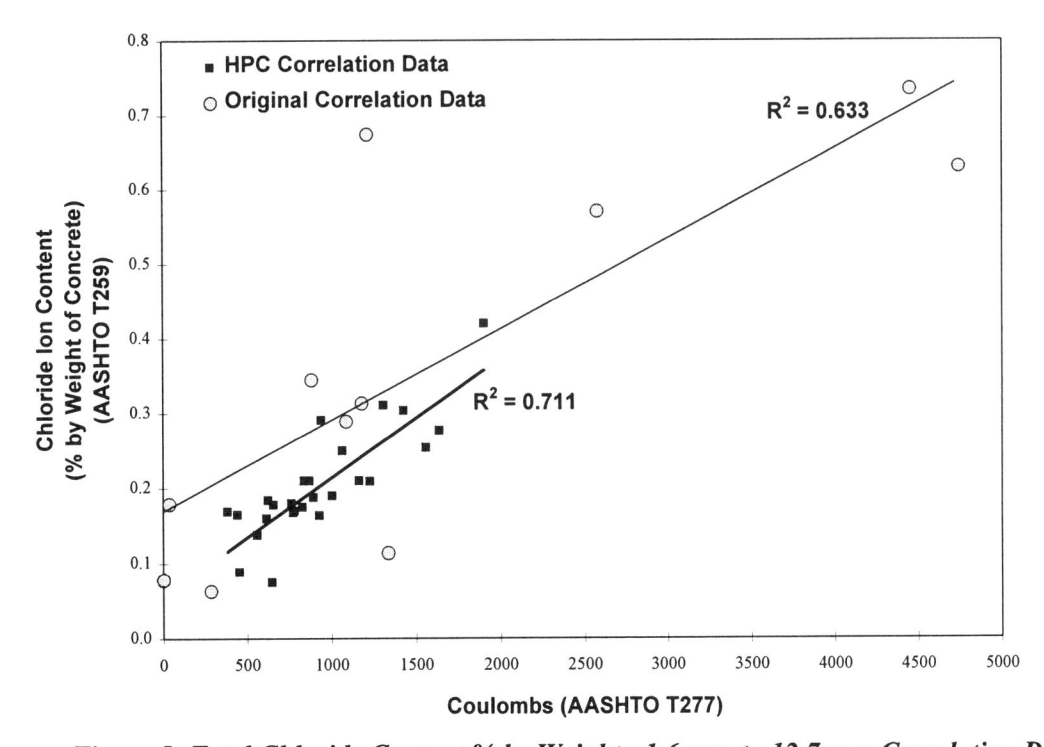

Figure 5: Total Chloride Content % by Weight - 1.6-mm to 12.7-mm Correlation Data

Figure 5 compares the Chloride Ion Content from percentage by weight of concrete to data used in the development report of AASHTO T277 [1]. In this manner, a better correlation is represented for the HSC mixes since chlorides below 19-mm (0.75-in.) depths are generally not present after the initial 90-day ponding cycle. The Chloride Integral value was averaged to the same depth (40-mm) as the original correlation study even though chlorides were not present at this depth. In order to determine how changing different constituents in the mix design affected the correlation between the two tests, mixtures were divided into 6 groups: mixtures containing River Gravel, mixtures containing Limestone, mixtures containing Trap Rock, mixtures containing fly ash, mixtures with 0% fly ash content, and mixtures with 0% entrained air content. For each group of mixtures, linear regression analysis was performed. The following sections address each of these results in detail.

Effect of Entrained Air Content on the Results of the Two Tests - The first step in investigating the correlation between AASHTO T277 and AASHTO T259 was to monitor the effect of the entrained air content on the results of the two tests. For the levels of entrained air content used in this study (between 0 and 6 percent), both tests reinforced the fact that the permeability of HSC is not affected by the addition of air entrainment. Figure 6 illustrates the T277 results of identical mixtures with the only difference being the entrained air content. The data indicated that 44 percent of the mixtures containing entrained air had higher permeability than the corresponding mixtures without air while 56 percent of these mixtures had lower permeability. The T259 results for the same mixtures also illustrated in Figure 6 where it can be seen that the addition of air entrainment caused an increase in chloride content in 69 percent of the mixtures while 31 percent of these mixtures showed a decrease. It is clear from this evaluation, that the addition of entrained air did not exhibit a definite trend on the chloride content and the coulomb values of the mixtures involved and thus the conclusion that an increase in the entrained air content had no effect on the chloride permeability of high strength concrete. Results from both tests exhibited similar behavior and conveyed the same conclusion. This is an expected behavior from HSC mixtures which can be explained by the fact that the capillary porosity is usually very low and the addition of air entrainment for HSC mixes is not expected to drastically affect the permeability.

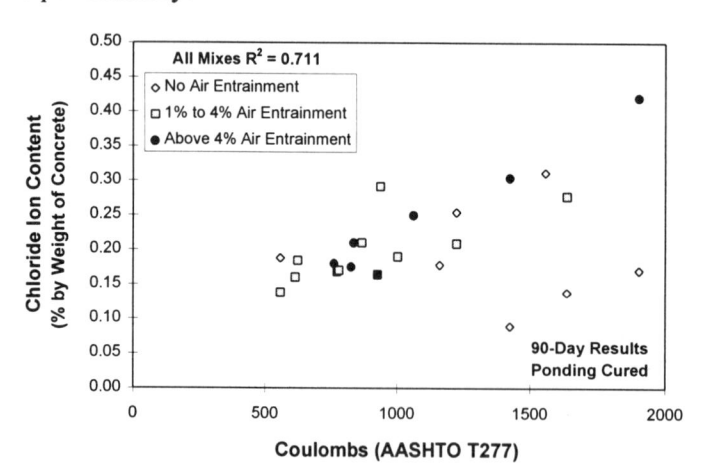

Figure 6: Air Entrainment Correlation Data

Effect of Coarse Aggregate Type on the Results of the Two Tests - The effect of three aggregate types on the permeability of HSC was also investigated using both tests (Figure 7). The aggregates were: 19-mm River Gravel, 19-mm Limestone and 19-mm Trap Rock. Based upon the AASHTO T277 results, mixtures containing Trap Rock exhibited higher permeability than mixtures containing River Gravel which in turn exhibited higher permeability than the

Limestone mixtures. The variation between the coulomb values of the River Gravel and Limestone mixtures was minimal. Based upon the AASHTO T259 results, the Trap Rock

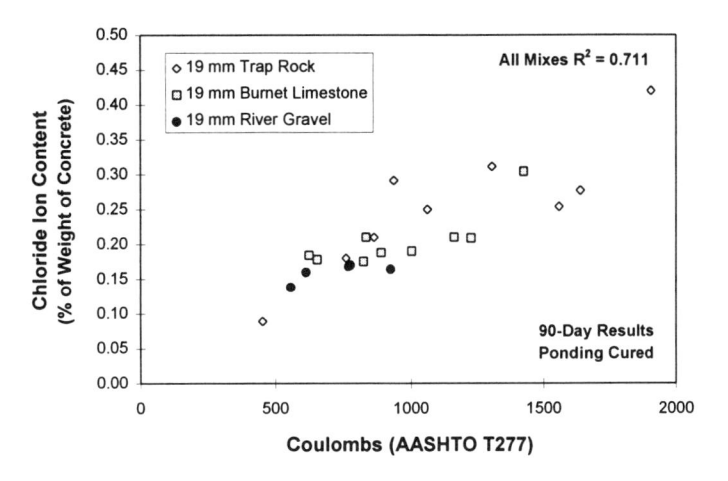

Figure 7: Aggregate Type Correlation Data

mixtures also exhibited higher permeability than the Limestone mixtures which in turn exhibited higher permeability than the River Gravel mixtures. In the same manner, total chloride integrals of the River Gravel and Limestone mixtures were comparable with minimal differences. Thus, both tests classified the Trap Rock mixtures as having the highest permeability and determined that the River Gravel and Limestone mixtures exhibited comparable permeability. The two tests had contradicting results on the classification of the River Gravel and Limestone mixtures.

Effect of Fly Ash Replacement on the Results of the Two Tests - The effect of fly ash replacement on the permeability of HSC was also investigated using both tests. Three variations of fly ash replacement were selected, 0, 25, and 35 percent respectively. Figure 8 illustrates the results of both ASSHTO T277 and T259.

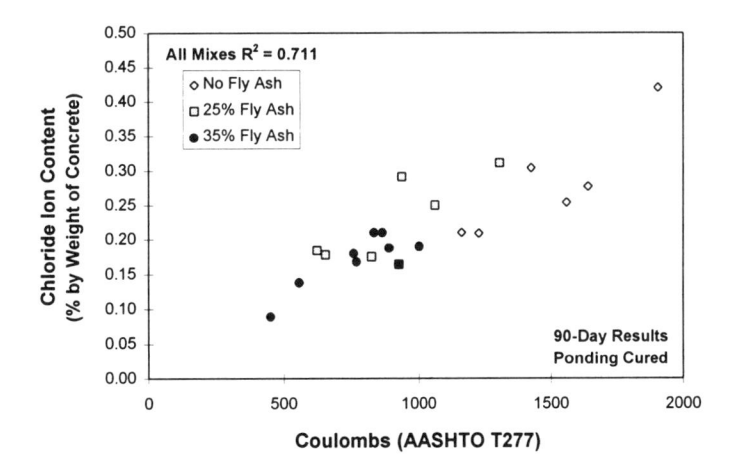

Figure 8: Fly Ash Replacement Correlation Data

Each series of replacement percentage exhibited very good grouping correlation as illustrated in Table 5. Clearly, as the Class C fly ash replacement was increased, both the chloride content and coulombs passed decreased. The results after the initial 90-day ponding cycle illustrates the reduced permeability with the use of fly ash replacement. Results based on this study at 90-days illustrated the improved permeability through the use of a Class C fly ash.

Table 5: Correlation of Mixes with and Without Fly Ash Replacement

Fly Ash Replacement (Percentage)	R^2 Correlation Value
No Fly Ash Replacement	0.8094
25 % Fly Ash Replacement	0.6445
35 % Fly Ash Replacement	0.6852

Ongoing Comparison Study - As has been mentioned previously, the correlation study is currently an ongoing research study. After the initial 90-day ponding cycle and chloride testing, the specimens were reponded for a second 90-day ponding cycle. Current plans include ponding cycles of 180, and 365-day cycles. Ongoing ponding has been scheduled to investigate the correlation at later ages due to the slow intrusion of chlorides for these high performance concrete mix designs at depths greater than 19-mm (0.75-in.). Rapid ion permeability test specimens will be core drilled from the control blocks and ponding blocks at that time to insure similar curing conditions for the specimens. Results of these test results will be reported at a future time.

SUMMARY

As designers, contractors, and producers become more familiar with its production and properties, high strength concrete is becoming more widely utilized. Strength levels which were thought to be impossible to attain 15 years ago are now commonplace. Nowadays, high strength concrete is being considered for use in a wide variety of applications and environments be it exposed such as a bridge or protected such as a building. Permeability of high strength concrete which is considered to be a good indicator of its durability and long term performance has been the topic of a large number of research studies and investigations. The main reason for this concern is that permeability has been shown to directly affect most of the durability problems of concrete.

The purpose of this study was to investigate the correlation between AASHTO T277 and AASHTO T259 for high performance concrete and to determine which constituents and percentages of materials affect its permeability. Thirty-one high strength concrete mixtures were cast which investigate varying several constituents within the mix design. The variables included the entrained air content, coarse aggregate type, and length of moist curing. The water to cementitious material ratio varied between 0.28 and 0.33 and the 28-day compressive strength varied between 48.3 and 93.1 MPa (7,000 and 13,500 psi). All mixtures were tested for compressive strength, splitting tensile strength, elastic modulus, and permeability which was evaluated using both the Rapid Chloride Penetration Test (AASHTO T277) and the 90-Days Ponding Test (AASHTO T259).

CONCLUSIONS

Based upon the testing performed during the course of this study, the following conclusions were drawn:
1. The addition of air entrainment had no appreciable effect on the permeability of high performance concrete.
2. The length of moist curing for the high performance concrete did not significantly affect the results of AASHTO T277. Mixtures moist cured for 56 days exhibited slightly lower

coulomb values than mixtures moist cured for 14 days and then air dried for 42 days. This difference was not significant and did not cause a shift between the permeability levels defined in AASHTO T277.

3. AASHTO T277 and T259 indicated that mixtures containing Trap Rock aggregate were most permeable. Based upon the T277 results, the River Gravel mixtures displayed slightly larger coulomb values than the Burnet Limestone mixtures. Based upon the T259 results, the Burnet Limestone mixtures exhibited slightly higher total integral chloride values than the River Gravel mixtures. Both tests however, indicated that the variation of the permeability due to changing the coarse aggregate type was minimal.
4. Both AASHTO T277 and T259 indicated that the use of a Class C fly ash replacement reduced the permeability of the high strength concrete mixes. Mixes which incorporated 35% fly ash replacement exhibited generally lower permeability than 25% fly ash replacement.
5. Both AASHTO T277 and T259 indicated that decreasing the w/cm ratio from 0.28 to 0.22 while keeping all other properties constant was not effective in decreasing the permeability of high strength concrete.
6. There exists a linear regression correlation between AASHTO T259 and AASHTO T277. For specimens cured under the same conditions namely 14 days of moist curing followed by 42 days of air drying, the 90-day chloride content in percent by weight of concrete in the uppermost 12.7-mm (0.5-in.) exhibited an R^2 value of 0.71 when coupled with the corresponding coulomb values. The 90-day total integral chloride and the corresponding coulomb values exhibited an R^2 value of 0.65 with a 33 percent standard error.
7. When coulomb values of specimens moist cured for 56 days are coupled with total integral chloride values of specimens moist cured for 14 days and air dried for 42 days, the R^2 correlation coefficient decreases to 0.42 indicating a poor correlation.
8. When compared against data used in the original development report, the total integral chloride results obtained during this study, displayed a better correlation in the range of coulomb values used, 400 to 2000 coulomb. The current regression line was steeper reflecting the fact that relatively large *I* values were recorded for low coulomb values.
9. When separated into groups, mixtures containing the River Gravel aggregate indicated a very poor correlation between the T277 and T259 results.
10. The original correlation between AASHTO T259 and T277 in the FHWA-81 report could not be used to accurately predict the permeability of high strength concrete. However, the RCPT can serve as a general tool to compare the relative permeability of similar mix designs with similar constituents and admixtures, in this case HPC. An initial correlation to accurately predict permeability for mix designs developed for field use is recommended.

REFERENCES

1. Whiting, D., *"Rapid Determination of the Chloride Permeability of Concrete,"* Final Report No. FHWA/RD-81/119, Federal Highway Administration, August 1981.
2. Whiting, D., Dzeidzic, W., *"Resistance to Chloride Infiltration of Superplasticized Concrete as compared with Currently Used Concrete Overlay Systems,"* Final Report No. FHWA/OH-89/009, Construction Technology Laboratories, May 1989.
3. Whiting, D. Mitchell, T.M., *"A History of the Rapid Chloride Permeability Test,"* Transportation Research Record 1335, Washington, D.C., 1992.
4. Sherman, M.R., Pfeifer, D.W., McDonald, D.B., *"Durability Aspects of Precast Prestressed Concrete. Part 1: Historical Review,"* PCI Journal, July-August 1996, pp. 62-74.

5. Pfeifer, D.W., McDonald, D.B., Krauss, P.D., *"The Rapid Chloride Permeability Test and Its Correlation to the 90-Day Chloride Ponding Test,"* PCI Journal, Vol. 39, No. 1, January-February 1994, pp. 38-47.
6. Whiting, D., *"Permeability of Selected Concretes,"* Permeability of Concrete, ACI SP-108, American Concrete Institute, Detroit, Michigan, 1988, pp. 195-222.
7. Whiting, D., Dzeidzic, W., *"Resistance to Chloride Infiltration of Superplasticized Concrete as compared with Currently Used Concrete Overlay Systems,"* Final Report No. FHWA/OH-89/009, Construction Technology Laboratories, May 1989.
8. Ozyildirim, C., Halstead, W.J., *"Use of Admixtures to Attain Low Permeability Concretes,"* Final Report No. FHWA/VA-88-R11, Virginia Transportation Research Council, February 1988.
9. Berke, N.S., Pfeifer, D.W., *"Protection Against Chloride-Induced Corrosion,"* Concrete International, Vol. 10, No. 12, December 1988. pp. 45-55.
10. Scanlon, J.M., Sherman, M.R., *"Fly Ash Concrete: An Evaluation of Chloride Penetration Testing Methods,"* Concrete International, June 1996.
11. Mosbacher, B., Mitchel, T.M., *"Laboratory Experience with the Rapid Chloride Permeability Test,"* Permeability of Concrete, ACI SP-108, American Concrete Institute, 1988, pp. 117-144.
12. ACI Committee 363, *"State-of-the-Art Report on High Strength Concrete,"* ACI 363R-92, American Concrete Institute, Detroit (1992).
13. Mindess, S., Young, J.F., *"Concrete,"* Prentice Hall, Inc. Englewood Cliffs (1990).
14. Neville, A.M., *"Properties of Concrete,"* Third Edition, Longman Scientific and Technical (1991).
15. ACI Committee 211, *"Guide for Selecting Proportions for High Strength Concrete with Portland Cement and Fly Ash,"* Committee Report, ACI Materials Journal, (May-June 1993).
16. Touma, W.E., *"Permeability of High Performance Concrete: The Rapid Chloride Test vs. The 90-Days Test,"* Master of Science Thesis, Department of Civil Engineering, The University of Texas at Austin, (May 1997).

CURING OF HIGH PERFORMANCE CONCRETE - AN OVERVIEW

Changqing (Max) Wang
Postdoctoral Fellow
Dept. of Civil Engineering
The University of Calgary
Calgary, Alberta, Canada

Walter H. Dilger
Professor of Civil Engineering
The University of Calgary
Calgary, Alberta, Canada

Wib S. Langley
Vice President
Jacques Whitford & Associates
Dartmouth, Nova Scotia, Canada

ABSTRACT

The method and duration of curing significantly affect the strength and durability of concrete as well as the construction speed and cost. It has been argued that high performance concrete (HPC) requires longer curing duration because of low water-cementitious ratio (w/cm) in the mixture and the consequent self-desiccation. However, after reviewing the basic mechanisms of moisture transport in concrete, literature information and limited experimental data by the authors, it is concluded that the lower w/cm may shorten the curing duration needed for strength and durability due to faster development of capillary "discontinuity". Further curing after that stage has very limited or no effect. That is, HPC may require shorter curing duration. Test results also confirm that the curing method and duration have no significant effects on chloride penetration into HPC containing silica fume or fly ash. Indeed, the protection against moisture loss from fresh HPC is crucial to the long-term strength and durability, as well as to the prevention of plastic shrinkage cracking.

INTRODUCTION

The development of concrete properties such as strength and impermeability depends on the hydration of cement. After placement, the only factors in cement hydration of a particular concrete are the availability of moisture and temperature. Therefore, "curing" of concrete is defined by ACI 116[1] as "the maintenance of a satisfactory moisture content and temperature in concrete during its early stages so that desired properties may develop." The implication of the definition, as well as the logical practice of concrete curing, is to protect the hardening concrete till it has reached a certain degree of hydration or certain level of maturity so that the desired long-term properties can develop in the storage or service environment without any special protection.

After an initial period of 1 to 2 days of relatively fast reaction, the cement hydration is a very slow process. Theoretically, the longer the curing duration, the better the concrete strength and impermeability due to extended hydration. However, extra long curing duration may have very limited or negligible positive effect on the long-term properties of the concrete, but a negative effect on construction speed and cost. Curing duration is a particularly significant factor in "fast track" construction of concrete structures. Also, HPC is often used in very large structures and in environments that curing is both costly and difficult[2].

It has been argued that HPC needs more or longer curing because of low w/cm[3,4]. Indeed, most of the HPC mixtures have a water content lower than the theoretical minimum for the "complete" hydration of the cementitious materials. It seems that an external water supply through wet curing is necessary for the completion of "hydration potential". On the other hand, the extremely low permeability of HPC, even at a very early age, may drastically reduce the moisture loss from young concrete when compared with normal strength concrete (NSC). As a result, HPC is more capable of protecting itself against moisture loss at early ages. That is, HPC may need shorter duration of curing for the development of strength and durability. Complete hydration of cement is rarely if ever achieved. Therefore, a determination is required as to the amount of curing for obtaining the desired long-term properties or service life for a structure. This paper reviews the practical needs of curing for HPC in order to ensure the quality of the concrete on one hand, and to avoid unnecessary conservatism for the construction industry on the other.

MEASUREMENT OF CURING REQUIREMENT

Determination of Curing Termination

When the concrete is cured under a constant temperature, the required curing duration can be measured by time. When the concrete experiences large temperature variations, maturity or mechanical properties such as strength and permeability have to be used. The maturity of concrete is a combined effect of time and temperature in a given moisture regime, and it is independent of the concrete mixture proportions. It is also commonly defined as equivalent or effective age at which a particular concrete will have reached the same degree of hydration or similar values of mechanical properties such as strength. Since the hydration of cement is accelerated at higher temperatures, at the same age concrete cured at higher temperature will have a higher maturity.

For a particular concrete mixture, the relationship between the maturity and degree of hydration (and the development of mechanical properties) is more or less a unique function which can be established by experiment. However, the relationship can be quite different for different concrete mixtures, because in addition to temperature the rate of cement hydration depends on the type of cement, w/cm, chemical and mineral admixtures. The strength and impermeability of different concretes can be significantly different at the same maturity, particularly at early ages. That is, for different concrete mixtures the required minimum maturity for curing can be different even in the same service environment.

The above problem can be avoided if the degree of hydration is used to measure the curing requirement. However, the actual determination of the degree of hydration is rather difficult because it involves complex laboratory procedures, and the accuracy of the determination is often questionable. The use of relative concrete strength to gauge the time of curing termination is probably the most convenient and reliable method for field operations. It is well established that the development of concrete strength is directly related to the degree of hydration, or the accumulation of hydration products and the reduction of void volume in the cement paste[3].

It should be cautioned that the samples for strength tests must be representative of the field concrete. This is particularly true for the cover concrete where the development of strength is usually the slowest but the durability of which is the most important. Strength determination on core samples test the interior of the concrete which is basically in a sealed condition. This strength is not necessarily relevant to the quality of the covercrete which may have experienced temperature and moisture conditions different from those in the core.

Required Curing Duration

In practice, curing is normally terminated before the concrete has reached the specified long-term properties because concrete has some potential of further development in strength and impermeability even without protection due to capillary water contained in the concrete. This potential is affected by the ambient conditions, cement type, w/cm and member size. It is evident that the required duration of external curing for a structure depends on, among other factors, the development of strength and impermeability of the concrete after the curing termination as well as the service exposure conditions, i.e., the durability requirement. For example, for normal exposure conditions, CSA-A23.1[5] specifies a minimum curing of either 3 days at a minimum temperature of 10°C (50°F) or for the time necessary to attain 35% of the specified 28-day compressive strength of the concrete. However, for severe service exposure conditions, the concrete has to be cured for an additional 4 days or for the time necessary to attain 70% of the specified 28- day compressive strength.

The use of relative strength as a curing termination gauge as included in CSA-A23.1[5] is, in the authors opinion, more rational than the use of age because it takes into account the real degree of hydration which, as mentioned before, depends on not only time, but also cement type, w/cm, temperature, admixtures etc. For example, under the same temperature history, concrete with Type III cement will need shorter time of curing than concrete with Types I or II cement to reach the same degree of hydration. As a general guideline, ACI 308[6] specifies different minimum curing requirement for concrete with different cements, namely, 14 days for Type II, 7 days for Type I and 3 days for Type III, at a minimum temperature of 10°C (50°F). Since the cement hydration rate at 20°C (68°F) is about twice as that at 10°C (50°F) according to the Arrhenius Function, the actual curing duration may be cut by half under normal temperature (20°C or 68°F) conditions. The CEB-FIP Model Code (1990)[7] also recommends different minimum curing durations for different types of cement and mixture proportions.

CURING REQUIREMENT FOR HPC

General Comments

It has been argued that HPC requires more curing than NSC because it has a very low w/cm in the mixture and an external water supply (wet curing) is needed to increase the hydration potential of the cementitious materials[4]. Indeed, the w/cm in HPC mixtures is often far below the theoretical minimum for the "complete" hydration of the cementitious materials, and self-desiccation does take place. That is, as the hydration progresses, the internal relative humidity will drop significantly, sometimes below 80%, particularly when silica fume is used[8,9]. However, by the time when significant internal relative humidity drop has taken place to hinder the further hydration of cementitious materials, the concrete has reached a very low moisture permeability. Further moisture loss or gain will be in the form of very slow process of diffusion. Longer curing may have little effect because of the difficulty in water ingress or moisture loss.

It must be pointed out that the development of both strength and durability does not depend on the complete hydration of the cement grains in the concrete. It rather depends on the filling of the spaces originally occupied by water in the fresh concrete by the hydration products[3]. HPC has much lower volume of water filled space or voids per unit volume of cementitious materials in the mixture than NSC. Therefore the filling of the voids can be completed with a much lower degree or shorter duration of hydration than for NSC. The supply of external water for further hydration may therefore not be necessary.

Curing of Fresh HPC

For fresh concrete, the moisture loss occurs in the form of evaporation from wet concrete surfaces. Water from inside the concrete continues to migrate to the surface through capillary action- an extremely fast form of moisture transport compared with diffusion. The evaporation rate is significantly affected by the temperature of the concrete, relative humidity, and wind speed of the ambient air. When the evaporation rate exceeds the bleeding rate of the fresh concrete, plastic shrinkage is likely to occur[3]. Therefore, it is true that fresh HPC needs more protection against moisture loss than NSC because of very low initial w/cm and, consequently, much less or no bleeding especially when silica fume is used. Plastic shrinkage cracking will occur when rapid drying of fresh HPC occurs. Also, any moisture loss from fresh concrete other than from the accumulated bleeding water will result in the decrease of water supply for hydration and, therefore, the decrease of long-term concrete quality. For HPC with very low w/cm in the original mixture, water loss from fresh concrete is especially harmful.

Although wet curing is often used to combat plastic shrinkage cracking problem in hot weather concreting, if applied too soon (e.g., immediately after finishing) it is harmful to the quality of the surface layer concrete which is of crucial importance to the durability of a structure. The high durability of HPC, such as very low porosity and high impermeability, is the result of very low w/cm in the mixture. If wet curing is applied to the concrete surface before final setting, the additional curing water will dilute the cement paste near the surface. As a result, the actual w/cm of the surface layer concrete may be significantly increased and the strength and durability of this critical zone seriously compromised. The damage due to wet curing which is started too soon can often be characterized by "dusty surface" or shallow surface scaling of hardened concrete, and has been confirmed by experiments[10]. In the authors' opinion, the best form of moisture protection for fresh HPC is the use of waterproof membrane such as the polyethylene sheets to prevent the evaporation. Field tests by the authors and experience have shown that curing compound is often inadequate in protecting fresh HPC from moisture loss, unless it is carefully selected and applied. Wet curing, if desired, should not start till after the final set.

Curing Duration for HPC

As mentioned before, the capillary pores in the fresh concrete are continuous, and water transport is very easy due to capillarity. As the hydration progresses, these macro pores are filled with hydration products and become discontinuous. At this stage, the migration of water takes the form of diffusion which is many orders slower than capillary action. Naturally, the moisture protection after this stage is much less important than when capillary action is predominant. If curing is terminated after this stage, the long-term properties of the concrete will not be seriously compromised.

The cement particles in HPC are placed much closer to each other than in NSC due to low w/cm, and the "discontinuity" of capillary pores can be reached much faster because smaller pores need less hydration products to fill. That is, the transition from capillary action to diffusion takes place much sooner in HPC than in NSC under the same curing conditions. The two forms of moisture migration in young concrete can be clearly demonstrated by the weight loss curve of a HPC sample (w/cm = 0.30 with 7.5% silica fume) placed inside the lab ambient air without protection (Fig.1). It can be seen from Fig. 1 that for this particular HPC, the discontinuous stage of capillary pores starts after approximately 24 hours, before which the moisture loss rate is very high and after which it is drastically reduced.

As mentioned before, when the capillary pores in concrete are discontinuous the water absorption through diffusion is very slow and external water can never reach more than a few millimetres beyond the concrete surface during any practical curing period. Test results on changes in internal relatively humidity of HPC at locations 50 mm (2") from the drying or wetting front (surface) by Persson[9] confirm the ineffectiveness of long curing period for HPC. Water just cannot penetrate the highly impermeable HPC even after more than one year of continuous submersion. The same must be true for drying, i.e., the moisture loss from hardened HPC is extremely slow. For example, test results on moisture loss on HPC with w/cm = 0.24 by de Larrard and Bostvironnois[11] show that 90 days of natural drying only affects the first 15 mm of the surface layer.

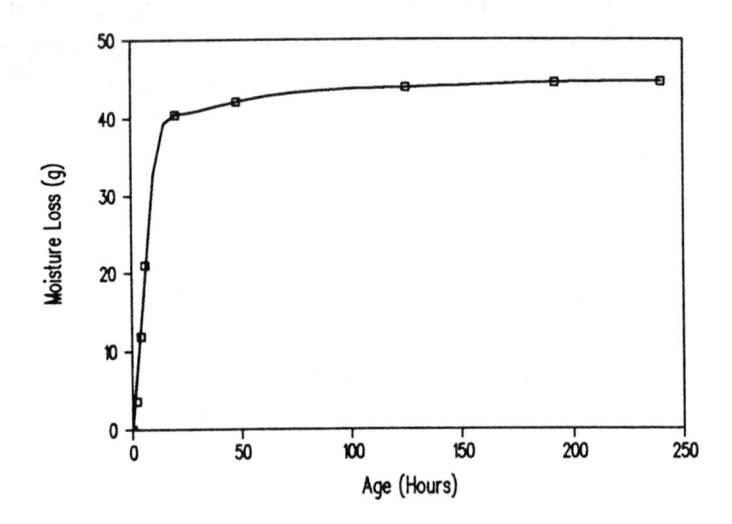

Fig. 1. Moisture loss from a fresh HPC (w/cm = 0.30, 7.5% silica fume), exposed to lab conditions (RH was around 50%).

If curing or moisture protection can be terminated when diffusion becomes the dominant form the moisture migration in the concrete, HPC will need shorter duration of curing. Furthermore, as a result of very low w/cm and the use of superplasticizers, the relative hydration rate of HPC is higher at early age, leaving less long-term hydration potential for HPC than NSC[12]. That is, longer curing is less effective for HPC than for NSC. This statement is in agreement with the recommendations of CEB-FIP (Appendix 1)[13] and the Danish Concrete Association[14] both of which indicate that low w/cm may shorten the duration of curing.

It should also be pointed out that the concrete strength tested on cylinders is affected by the moisture conditions and by micro cracking induced by drying. Some of the higher strength observed in specimens continuously wet cured till the time of testing may be attributed to the fact that these specimens are not cracked as in the dry cured ones[15]. That is, the strength increase in continuously wet cured concrete is not necessarily due to more hydration of the concrete, at least not due to it alone.

Curing of HPC for Reducing Chloride Permeability

It has been generally accepted[3] that the effect of inadequate curing on strength is greater for concrete with high w/cm. Therefore, for HPC with low w/cm the curing is more important for durability than for strength. One of the main concerns about concrete durability is the chloride-induced corrosion of reinforcement, and to achieve a low chloride permeability or diffusivity may become a key objective for the curing of HPC in chemically aggressive environment. Since the moisture permeability is extremely low even at very early ages (e.g., 1 day or later), it is expected that curing duration will have insignificant effect on the long-term chloride permeability of HPC.

The results of rapid chloride permeability tested according to ASTM C1202[16], conducted for the Confederation Bridge Project for Strait Crossing Joint Venture Inc., P.E.I., Canada, confirmed that the duration of curing was not an important factor in chloride penetration of HPC. The particular HPC mixture contains 7.5% silica fume and a w/cm of 0.3. The two variables of the tests were the in-form (sealed) curing duration varying from 1 to 5 days, and curing temperature ranging from 5°C (41°F) to 50°C (122°F). The results of rapid chloride permeability tested at the age of 3 months showed no apparent influence by curing duration. All the results were below 300 Coulombs, which is considered extremely low.

Dhir et al[17] tested chloride penetration into various grades of concrete (20 to 60 MPa, or 2.9 to 8.7 ksi) with or without fly ash, using diffusion cells without external electric current. Their results showed that drying after demoulding at 1 day significantly increases the coefficient of chloride diffusion (tested at 28 days) for low grade concrete (20 and 40 MPa, or 2.9 to 8.7 ksi) with pure cement. However, for 60 MPa (8.7 ksi) concrete without fly ash drying has virtually no effect. This confirms the hypothesis by the authors of this paper that the effect of curing duration on long-term chloride permeability is insignificant for HPC with low w/cm. It is interesting to note from the test results by Dhir et al[17] that curing has no apparent effect on coefficient of chloride diffusion of all concretes containing fly ash, regardless of the grades (strength), w/cm and cementitious content. This is contrary to the common belief that fly ash concrete requires longer curing duration for the development of durability.

A systematic experimental program on the effects of curing on long-term chloride permeability is currently underway at The University of Calgary under the auspices of Concrete Canada. Some initial results show that the duration of wet curing up to 6 days (after demoulding at 1 day) has no apparent effect on chloride permeability at 91 days of age for 80 MPa (11.6 ksi) HPC containing 10% silica fume and w/cm of 0.32 (Fig.2). It should be pointed out that the drying of these concrete specimens in the laboratory is considered very severe, because the relative humidity during the test period (winter) was only about 15%.

The above discussions seem to indicate that curing duration after the initial moisture protection (e.g., sealed in the mould for at least 1 day) has little effect on the long-term chloride permeability of HPC, at least for HPC containing silica fume or fly ash.

The current ASTM C1202[16] method of testing for rapid chloride permeability uses 50 mm (2") thick discs. However, the lack of curing affects normally thinner layer of concrete cover for HPC. It would be of interest to reduce the disc thickness to e.g., 25 mm (1"), for testing chloride permeability of the covercrete.

Effects of Curing on Shrinkage of HPC

In addition to strength and permeability, curing also affects other properties of concrete such as creep and shrinkage. Most of the time, creep and shrinkage are influenced by the same parameters in a similar manner and are not mutually exclusive[18]. Although curing does not affect the shrinkage of NSC in any significant manner[3,7], both curing method and duration have a significant effect on shrinkage of HPC with low w/cm[19]. Under sealed condition, HPC develops basic (autogenous) shrinkage which develops very fast during the first few days. Under water curing, there is basically no length change in small specimens of HPC (e.g., 100 mm or 4" diameter cylinders) for at least several months[19]. The relative volume stability of HPC under wet curing at early age is probably the result of cancellation between swelling of surface layer concrete due to water ingress and the basic shrinkage of the interior concrete due to self-desiccation. For large specimens of low water permeability and w/cm, basic shrinkage will still manifest at least at early ages[20]. Water curing up to 28 days also significantly reduce the drying shrinkage and total shrinkage of HPC (Fig.3).

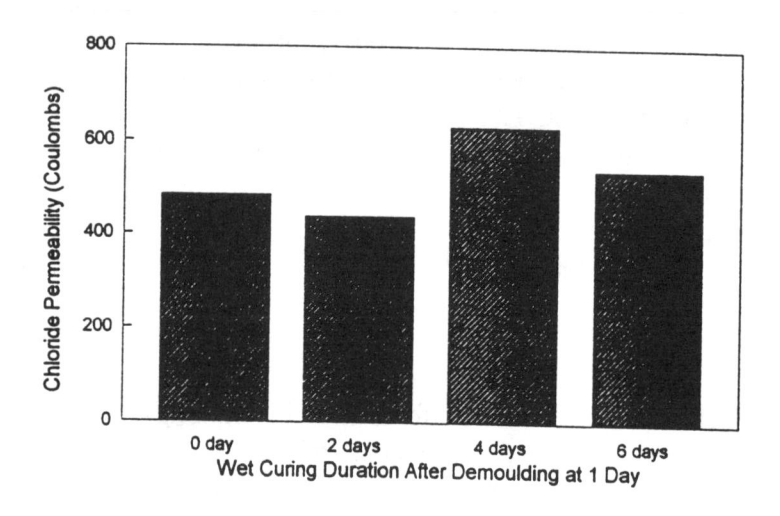

Fig.2 Chloride permeability (ASTM C1202) of a HPC (w/cm = 0.32, 10% silica fume) at the age 3 months, wet cured for different durations after demoulding at 1 day (RH of the storage environment was about 15%).

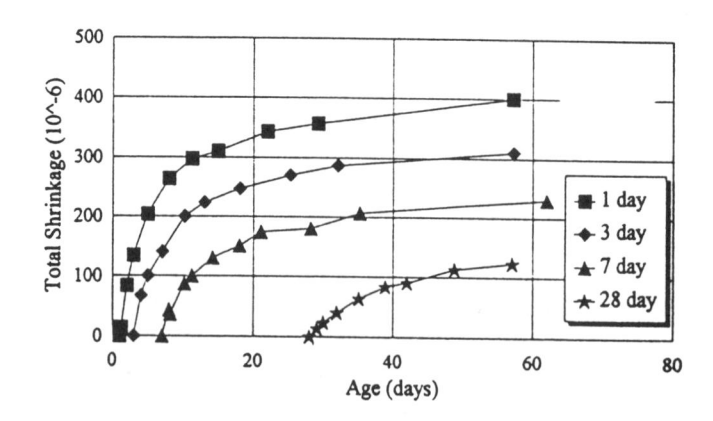

Fig.3. Effect of wet curing duration on shrinkage of HPC (w/cm = 0.23, 9% silica fume, 100 mm diameter cylinders)[19].

CONCLUSIONS AND RECOMMENDATIONS

Based on the literature information and the limited experimental data, the following preliminary conclusions and recommendations can be made:

1. The protection against moisture loss from fresh HPC is crucial to the final strength and durability of the concrete, as well as to the prevention of plastic shrinkage cracking.

2. Water curing of HPC should not be started till at least the final set to prevent weakening of the surface layer concrete.

3. HPC with low w/cm requires a shorter period of moisture protection or curing, because the time required to reach the capillary "discontinuity" stage is much shorter than for normal concrete. Further curing after this stage may have little effect on strength and durability, but may significantly increase the cost of construction.

4. It seems that after the initial moisture protection for about 1 day under normal temperature, the chloride permeability, as measured by ASTM C1202 method, is not much affected by the method and duration of curing. This is particularly true for particularly HPC containing silica fume or fly ash.

5. Water curing up to 28 days may significantly reduce drying shrinkage of HPC containing very low w/cm and silica fume.

6. Since HPC does not need longer curing than NSC, the current code recommendations on minimum curing durations for NSC can be followed for HPC, and they are believed to be more than adequate for the long-term development of both strength and durability of HPC.

7. The most practical and convenient method for field determination of curing termination time is the relative compressive strength of the concrete, as specified by CSA-A23.1-1994, which automatically includes the effect of temperature, cement type and mixture proportions. Although the degree of hydration is the ideal parameter for curing requirement measurement, it is too complicated for practical applications. The maturity method is unreliable for use on different concretes at early ages because of the effects of chemical admixtures, cement type and mixture proportions. These effects are significant but not readily determined.

8. More research is needed to confirm the effects of curing method and duration on other durability parameters of HPC such as water and gas permeability, abrasion resistance, and chloride diffusivity. It would also be of interest to reduce the disc thickness for the ASTM C1202 method to e.g. 25 mm (1") for testing the rapid chloride permeability for HPC.

ACKNOWLEDGEMENT

The financial support by Concrete Canada for this study is gratefully appreciated. The authors also acknowledge Messrs Ross Gilmour, P.Eng., Jim Turnham, P.Eng. and Paul Bragdon, P.Eng., all of Strait Crossing Joint Venture Inc. for their collection of the test data in Fig.1 during construction of the Confederation Bridge and for their discussions on the curing of HPC for the same bridge. Thanks are also due to Mr. Greg Barraclough, research assistant, for conducting the rapid chloride permeability tests at The University of Calgary.

REFERENCES

1. ACI 116, *Cement and Concrete Terminology*, American Concrete Institute, Detroit, 1990, 68pp.

2. Langley, W.S., Gilmour, R., and Tromposch, E., "The Northumberland Strait Bridge Project," *ACI Special Publication 154, Advances in Concrete Technology*, June 1995, pp.543-564.

3. Neville, A.M., *Properties of Concrete*, 4th Ed., Longman Group Ltd. England, 1995, 844pp.

4. ACI 363, *State-of-the-Art Report on High Strength Concrete*, American Concrete Institute, Detroit, 1992, 55 pp.

5. CSA-A23.1, *Concrete Materials and Methods of Concrete Construction*, Canadian Standard Association, June 1994.

6. ACI 308, *Standard Practice for Curing Concrete*, American Concrete Institute, Detroit, 1981, 11pp.

7. CEB-FIP, *Model Code for Concrete Structures, 1990*, CEB Bulletin No. 213/214, Tomas Telford, London, May 1993, 473 pp.

8. Wang, C., Dilger, W.H. and Niitani, K., "Characteristics and structural effects of creep and shrinkage of high performance concrete," *Proceedings, CSCE Annual Conference*, Edmonton, Vol.IIA, May 1996, pp.523-536.

9. Persson, B., "Hydration and strength of high performance concrete," *Advanced Cement-Based Materials*, No.3, 1996, pp.107-123.

10. Guse, U. and Hilsdorf, H.K., "Surface cracking of high strength concrete," *High Performance Concrete, Material Properties and Design*, AEDIFICATIO, Freiburg, 1995, pp.69-90.

11. de Larrard, F. and Bostvironnois, P.J., "On the long-term strength losses of silica-fume

high-strength concretes," *Magazine of Concrete Research*, Vol.35, No.155, 1991, pp. 109-119.

12. Dilger, W.H. and Wang, C., "Effects of w/cm, superplasticizers and silica fume on the development of heat of hydration and strength in HPC," *High Performance Concrete, Material Properties and Design*, AEDIFICATIO, Freiburg, pp.3-22.

13. CEB-FIP, *Durable Concrete Structures, Design Guide*, Thomas Telford, 1992, 112pp.

14. Danish Concrete Association, *Recommendations for Curing of Concrete*, Publication No.35, 1990, 27pp.

15. de Larrard, F. and Aitcin, P.C., "The strength retrogression of silica fume concrete," *ACI Materials Journal*, V.90, N.6. Nov-Dec 1993.

16. ASTM C1202, *Test for Electrical Indication of Concrete's Ability to Resist Chloride Ion Penetration*, 1994.

17. Dhir, R.K, Jones, M.R. and McCarthy, M.J., "Binder content influences on chloride ingress in concrete," *Cement and Concrete Research*, V.26, N.12, 1996, pp.1761-1766.

18. Neville, A.M., Dilger, W.H. and Brooks, J.J., *Creep of Plain and Structural Concrete*, Construction Press, New York, 1983, 361pp.

19. Dilger, W.H., Wang, C. and Niitani, K., "Experimental study on shrinkage and creep of high-performance concrete," *Proc. of 4th Inter. Sym. on Utilization of High-Strength/Hight-Performance Concrete*, Paris, June 1996, pp.311-319.

20. Tazawa, E. and Miyazawa, S., "Autogenous shrinkage of concrete and its importance in concrete," *Pros. 5th Intern. RILEM Symp. on Creep and Shrinkage of Concrete*, E & FN Spon, London, 1993, pp.159-160.

HIGH PERFORMANCE CONCRETE IN PRECAST CONCRETE TUNNEL LININGS; MEETING CHLORIDE DIFFUSION AND PERMEABILITY REQUIREMENTS

A. John R. Hart, Project Manager, CCL Joint Venture, Guelph, Ontario, Canada
John Ryell, Vice President, Trow Consulting Engineers Ltd., Brampton, Ontario, Canada
Michael D.A. Thomas, Assistant Professor, Department of Civil Engineering,
University of Toronto, Toronto, Ontario, Canada

ABSTRACT

Chloride induced corrosion of steel reinforcement is the predominant cause of premature deterioration of concrete transportation structures in North America. Recently specifications for a number of major tunnel and bridge projects in Europe, Asia and North America have required the concrete to meet chloride diffusion and permeability limits in order to ensure adequate durability in aggressive environments.

This paper discusses concrete quality in two major North American tunnel projects using precast concrete linings, i.e. the CN St. Clair River Rail Tunnel (constructed 1992-94) and the Toronto Transit Commission Sheppard Line Subway (construction commenced 1996). The concrete quality specifications and quality control programs are reviewed together with material selection, mix design trails, pre-production trials and concrete testing and evaluation.

It is demonstrated that the very demanding specification requirements covering chloride transport rates in "covercrete" concrete for precast concrete tunnel linings can be met when concrete materials and mix proportions are properly evaluated in pre-production mix design trails and the very highest standards for concrete mixing, fabrication and curing are used.

INTRODUCTION

In a number of recent major projects, owners have required designers to demonstrate that the completed structure will provide an adequate service life in its given environment. In order to achieve this it is necessary to be able to model the deterioration process and determine the relevant properties of the concrete. For structures in chloride environments, the principal process governing the service life is the penetration of chloride ion through the concrete cover to the steel reinforcement. There are a number of mathematical models available for predicting chloride ion penetration providing that the relevant transport properties of the concrete are known. For concrete tunnels exposed to chloride bearing waters and hydraulic pressure, the principle mechanisms governing chloride transport are ionic diffusion and hydraulic conductivity (water permeability).

This paper describes two major North American tunnel projects using precast concrete linings where the concrete materials specifications included requirements for chloride diffusion and water permeability.

CN ST. CLAIR RIVER RAIL TUNNEL

The precast concrete tunnel linings for the 1.8 km (1.1 mile) long, 8.4 m (27.6 ft) internal diameter rail tunnel under the St. Clair River between Sarnia, Ontario and Pt. Huron, Michigan were constructed in 1993/94 by the joint venture of the Pre-Con Company, Canada and Sehulster Tunnels, U.S.A. The contract was to manufacture 1,256 tunnel rings. Each ring consisted of six major segments and a key and weighed over 45 tons (Figure 1). An individual segment, curved in shape, measured 5 m (16 ft) long, 1.5 m (5 ft) wide, 400 mm (16 in) thick and weighed 7.5 tons. A total of 20,500 cubic meters (26,800 cubic yards) of concrete was used in the project. To manufacture the tunnel liners the contractor took possession of a vacant industrial building in Woodstock, Ontario and the installed equipment included a concrete vibration system, steam curing facilities and an epoxy powder coating line for the reinforcement. Concrete was supplied from a nearby ready mixed concrete facility equipped with a computer controlled batch plant and computer controlled truck mixers.

Figure 1. CN St. Clair Rail Tunnel, Trial Assembly of Segments (Cages of Epoxy Coated Steel Reinforcement in Background)

The heated concrete, 22°C to 30°C (72°F to 86°F) was placed in steel moulds and vibrated at a central vibration station. Following a pre-set time of approximately 3 hours the concrete segments were subjected to steam curing at 50°C (122°F) and removed from the steel moulds at approximately 9 hours. At an early stage in segment production water curing was implemented following steam curing. Water curing, using atomizer sprays, was continued for 5 days.

High groundwater chloride (4,000 ppm) and sulphate (155 ppm) levels combined with a hydrostatic head of up to 35 m (115 ft) led to concerns regarding concrete durability for the required 100 year design life[1]. The concrete quality specification for the tunnel segments focussed on achieving a dense, high quality concrete with low permeability to chloride ions.

The concrete specification included the following requirements:

- Cementitious Content 400 kg/m^3 to 550 kg/m^3 (675 lbs/yd^3 to 927 lbs/yd^3)
- Minimum Compressive Strength 60 MPa (8,700 psi) at 28 days
- Maximum Water/Cementitious Materials Ratio 0.38
- Concrete non air entrained
- Chloride Diffusion (120 days) 600 x 10^{-15} m^2/sec (6456 x 10^{-15} ft^2/sec) maximum
- Water Permeability (40 days) 25 x 10^{-15} m/sec (82 x 10^{-15} ft/sec) maximum.

Minimum proportions of fly ash or granulated slag, and maximum proportions of silica fume, were specified for the work.

It was specified that the diffusion and permeability tests would be carried out on 50 mm (2 in) thick slices cut from core samples removed from production concrete segments. Testing was required on trial mixes and segments and on up to four production segments cast during the first month of production.

TORONTO TRANSIT COMMISSION, SHEPPARD LINE SUBWAY

At a new state-of-the-art manufacturing plant near Guelph, Ontario the CCL Joint Venture of Concrete Systems, Inc., Hudson, New Hampshire, Con Cast Pipe Inc., Ontario and St. Lawrence Cement Ltd., Ontario are producing the segments for the 5.2 meter (17.1 ft) internal diameter precast concrete tunnel lining for the TTC Sheppard Subway. Originally a much larger contract that included 5,350 rings of tunnel lining for the cancelled Eglinton West Subway, the project now consists of the production of 5,550 rings for the Sheppard line only. Segment production commenced in February 1996 and is scheduled for completion in April 1997. Each ring consists of seven segments 1.4 m (4.6 ft) wide at the centre line (Figure 2). The segments are 225 mm (9 in) thick and are reinforced by wire mesh and weldable grade reinforcing steel.

Figure 2. TTC Subway Project Assembled Tunnel Ring

Manufacture of the units is highly automated and a three line carousel system with the steel moulds mounted on wheeled carriages is the heart of the high production rate technology. Concrete is produced in a 2 m^3 (2.6 yd^3) high speed central mixer and a bridge carrying a flying bucket transfers concrete to the mould filling station where the concrete is consolidated and vibrated. The heated concrete approximately 25°C to 30°C (77°F to 86°F), has a preset time of 2½ hours followed by 7½ hours of steam curing inside the kiln at a temperature of 55°C (131°F). Demoulding occurs approximately 10 hours after concrete placement followed by 5 days in moisture chambers at 100% relative humidity.

The concrete specification includes the following requirements:
- Cementitious Content 400 kg/m^3 to 550 kg/m^3 (675 lbs/yd^3 to 927 lbs/yd^3)
- Minimum Compressive Strength 60 MPa (8,700 psi) at 28 days
- Maximum Water/Cementitious Materials Ratio 0.35
- Minimum total Supplemental Cementitious Materials content 25%
- Air Entrainment: minimum percentage to meet freeze-thaw resistance requirements (ASTM C666 test method, Procedure A)
- Chloride Diffusion: Target Value (120 days) 1000 x 10^{-15} m^2/sec (10,760 x 10^{-15} ft^2/sec) maximum
  Maximum Value (120 days) 1500 x 10^{-15} m^2/sec (16,140 x 10^{-15} ft^2/sec)
- Water Permeability 100 x 10^{-15} m/sec (328 x 10^{-15} ft/sec) maximum.

The specification contains requirements for a minimum of 25% supplementary cementing materials and maximum proportions of fly ash, granulated slag and silica fume.

It is specified that the diffusion and permeability tests are to be carried out on 50 mm (2 in) thick slices cut from core samples representing the extrados of the segment. Test requirements include diffusion and permeability testing at 2 week intervals during production of the segments until consistency can be demonstrated. A concrete mix quality requirement of the specification is a program of preconstruction test mixes (10 mixes minimum) including chloride diffusion and water permeability tests.

Due to difficulties in coring the production segments without cutting reinforcing steel a separate concrete block was cast in a steel mould for core samples. The test block, of dimensions 635 mm x 285 mm x 200 mm deep (25 in x 11 in x 8 in), was stacked on the carousel production line adjacent to the segments.

DETERMINING CHLORIDE DIFFUSION COEFFICIENTS FROM PENETRATION TESTS

In the 1970's *Collepardi et al. (1970; 1972)*[2,3] demonstrated that chloride penetration from salt bearing groundwaters or seawater into saturated concrete can be approximated by Fick's second law of diffusion:

$$\frac{dC}{dt} = D_a \cdot \frac{d^2C}{dx^2} \qquad (1)$$

which can be solved for idealized boundary conditions using the following equation (adapted from *Barrer, 1951*)[4]:

$$C_{x,t} = C_s - (C_s - C_i) erf\left(\frac{x}{2\sqrt{D_a t}}\right) \qquad (2)$$

where: $C_{x,t}$ = chloride concentration at depth, x, and time, t
 C_s = chloride concentration at the surface ($x = 0$)
 C_i = background (initial) chloride concentration
 x = depth
 t = time
 D_a = 'apparent' diffusion coefficient

Eqn. 2 provides the basis for determining the diffusion coefficient of concrete experimentally using chloride penetration tests. In such tests, saturated specimens are immersed in a salt solution (e.g. NaCl) for a specific period of time (t), after which the specimen is sampled incrementally with depth from the exposed surface and a chloride concentration profile ($C_{x,t}$ vs x) established by chemical analysis. The diffusion coefficient (D_a) and surface concentration (C_s) are then found by curve fitting the profile data to Eqn. 2.

Using this approach, *Wood et al. (1989)*[5] developed a protocol for determining the diffusion coefficient of concretes with a high resistance to the penetration of chloride ions using a relatively short test duration (e.g. 40 days). The test uses an elevated temperature 40°C (104°F) and a highly concentrated salt solution (5M NaCl) to accelerate the penetration of chlorides, and utilizes precision grinding techniques to obtain depth increments of 1 mm (0.04 in) or less. A suitable profile for use with Eqn. 2 can be established even when the depth of penetration is low (e.g. <10 mm [0.4 in]). These procedures were used in the evaluation of concretes specified for both the Channel Tunnel and the Storebaelt Tunnel. A similar procedure has been standardized in Denmark as Test Method APM 207 (*Sørensen & Frederiksen, 1990*)[6].

The test procedure used for the tunnel lining segments described in this paper followed generally the procedures outlined by *Wood et al. (1989)*[5].

The chloride diffusion coefficients reported in this paper are based on computations by Mott MacDonald Consulting Engineers, Croydon, United Kingdom and Trow Consulting Engineers Ltd., Brampton, Ontario, Canada.

DETERMINING WATER PERMEABILITY USING CHLORIDE AS A TRACER

The coefficient of water permeability of concrete is typically determined by measuring fluid flow across a saturated sample under a hydraulic gradient and assuming Darcian flow. Experience has shown that steady-state flow is not readily established for concretes of low permeability (e.g. <10^{-13} m/sec) when tested in standard laboratory permeameters. This problem can be overcome by adding a suitable tracer (e.g. NaCl) to the upstream water supply in the permeameter and establishing the depth of penetration by profile grinding and analysis of the sample after a specific time period. The coefficient of permeability, k, is then calculated using a modified version of the Valenta equation (*Wood et al., 1989*)[5]:

$$k = \frac{nx_d^2}{2th}$$

where: k = coefficient of water permeability
n = porosity
x_d = depth of penetration calculated from chloride concentration profile
t = time under pressure
h = hydraulic head

The test procedure used for the tunnel lining segments described in this paper followed generally the procedures outlined by *Wood et al. (1989)*[5].

The permeability test cell applies a solution of sodium chloride (NaC1) at a concentration of 19,000 ppm at a pressure of 8 bar (116 psi) to the face of the core slice representing the trowelled surface. The test is conducted at normal room temperature (20°C [68°F]) and pressure testing in the permeability cell continues for 40 days.

The water permeability coefficients reported in this paper are based on computations by Mott MacDonald Consulting Engineers, Croydon, United Kingdom. A private communication[9] from Mott MacDonald indicated that the depth of chloride penetration (x_d) was determined as the point at which the concentration reaches the background (if this can be determined from the profile) or the point of intersection of the mathematically modelled best fit curve through the chloride profile and the background chloride concentration.

DISCUSSION OF TEST RESULTS

CN St. Clair River Rail Tunnel

The calculated diffusion coefficients are presented in Table 1. The mean strength of 381 28 day standard cylinder tests was 76.3 MPa (11,065 psi). Although not required by the contract a number of cores removed from a trial segment cast in the early stages of the project were tested for the electrical conductance of concrete in accordance with the AASHTO Designation T-277 'Rapid Determination of the Chloride Permeability of Concrete'. Core slices representing the extrados surface of the segment had coulomb values ranging from 284 to 366, samples from the segment centre had coulomb values ranging from 190 to 212; the concrete was between 29 and 33 days old at the time of test. Based on criteria contained in the AASHTO standard the concrete would be characterized as having "very low" chloride ion penetrability (100 - 1,000 coulombs). The concrete in this trial segment contained 500 kg/m^3 (843 lbs/yd^3) of cementitious material consisting of 64% normal Portland cement, 30% fly ash and 6% silica fume by mass. The standard cured 28 day concrete test cylinders are indicative of concrete of very high quality and the variation in quality, i.e. standard deviation and coefficient of variation compares favourably with data from a number of large high strength concrete projects in the Toronto area (*Ryell and Fasullo*)[7]. The specified maximum chloride diffusion coefficient of 600 x 10^{-15} m^2/sec (6,456 x 10^{-15} ft^2/sec) was only achieved on a few 60, 80 and 120 day tests and the overall mean value of 21, 120 day tests representing the segment surface and the segment centre concrete was 1,564 x 10^{-15} m^2/sec (16,829 x 10^{-15} ft^2/sec). The variability of the results, compared to the standard test cylinders, appears to be very high. This is partly due to the nature of determining mass transport measurements in concrete where values typically can range over many orders of magnitude. The high variability in the diffusion coefficient values may also be partly due to the numerous changes in concrete materials, mix proportions, supplementary cementing materials and curing procedures that occurred in the early stages of this project in an effort to achieve the very demanding concrete quality requirements of the contract and the construction needs of the contractor. It appears that concrete compressive strength is less sensitive to such changes than the diffusion coefficient.

Table 1 - Summary of Diffusion Test Data for CN St. Clair Tunnel Project

	Diffusion Coefficient (x 10^{-15} m^2/sec)					
	Segment Extrados (0-50 mm)			Segment Centre (175-225 mm)		
	40 days	80 days	120 days	40 days	80 days	120 days
No. of Tests	12	8	14	8	8	7
Minimum Value	1,047	404	302	1,833	1,282	1,110
Maximum Value	3,457	2,040	3,300	6,628	3,765	3,300
Mean Value	2,160	1,252	1,427	3,880	2,287	1,836
Standard Deviation	778	576	796	1,504	803	752
C of V %	36	46	56	39	35	41

25.4 mm = 1 in
1,000 x 10^{-15} m^2/sec = 10,760 x 10^{-15} ft^2/sec

The use of silica fume in 15 concrete mixes during the early stages of segment casting was represented by one set of chloride diffusion data from a mix containing 500 kg/m^3 (843 lbs/yd^3) of cementitious material which included 30% fly ash and 6% silica fume. Water to cementitious ratio for the silica fume mixes varied from 0.29 to 0.32 by mass. This initial diffusion data met the specification requirement at 120 days. Concerns with hairline cracking on the surface of some concrete segments and petrographic studies on segment cores that indicated a greater propensity of microcracking in concrete containing silica fume compared to non silica fume mixes, and what was considered to be "normal" microcracking in high quality structural concrete, lead to the elimination of silica fume from the concrete mix.

With one exception non silica fume concrete mixes did not meet the 120 day chloride diffusion requirement. However it is thought that the 28 day curing period prior to the 120 day immersion period diffusion test may not reflect the improved long term performance of the fly ash mixes and as indicated by *Thomas and Matthews*[8] such mixes exhibit considerable reduction in diffusivity with time.

For production segments the cementitious content of the concrete, after the early stages of casting, was in the range of 425 kg/m³ (717 lbs/yd³) to 475 kg/m³ (801 lbs/yd³). The fly ash content of the cementitious materials was between 25% and 35% by mass.

Subsequent petrographic studies of core samples representing concrete containing 25% to 35% fly ash, but no silica fume, reported a reduced number of microcracks and lower porosity in the outermost parts beneath the carbonation zone.

A 40 day water permeability test carried out on an early production segment was characterized by a permeability coefficient of 5.9×10^{-15} m/sec (19×10^{-15} ft/sec), i.e. well below the specified maximum value of 25×10^{-15} m/sec (82×10^{-15} ft/sec). This concrete contained 35% fly ash with no silica fume; the average 120 day diffusion coefficient was $1,341 \times 10^{-15}$ m²/sec ($14,429 \times 10^{-15}$ ft²/sec).

Toronto Transit Commission, Sheppard Line Subway

Typical chloride concentration profiles from the diffusion tests are characteristic of diffusion curves and Eqn. 2 can be fitted with a high degree of confidence. The profiles from permeability tests generally yielded S-shaped curves characteristic of combined convection, diffusion transport.

A substantial pre-construction laboratory concrete mix evaluation program was completed to identify materials, including admixture compatibility and proportions, that had the potential to meet the owner's design requirements and the fabrication requirements of the contractor. Chloride diffusion tests were carried out on core samples removed from test slabs cured in accordance with the profile proposed for the segment production that included steam curing at 55°C (131°F). The specified maximum diffusion value of $1,500 \times 10^{-15}$ m²/sec ($16,140 \times 10^{-15}$ ft²/sec) at 120 days was met by all of the concrete mixes containing ternary blends of cement (i.e., type 10 SF Portland cement blended with either slag or fly ash). Mixes with type 10 normal Portland cement and slag did not meet the diffusion requirement. Based on the laboratory tests full scale field trials were made using the moulds, the vibration technique and the concrete plant, the concrete materials and three selected chemical admixture systems to evaluate the performance of the concrete and its suitability to the production process. The production process has very precise requirements for the workability of the concrete, with particular needs for slump during placement and for any remaining workability when the mould pour covers are removed for trowelling the segment extrados. Concrete used initially in the field trials contained 450 kg/m³ (759 lbs/yd³) of cementitious material (68% type 10 SF silica fume Portland cement, 32% slag), a high range water reducing admixture and an air entraining agent. The water/cementitious materials ratio was approximately 0.31.

The calculated diffusion and permeability coefficients for all the completed tests at the time of writing are summarized in Table 2. In all cases, the concrete tested met the specified maximum limits of 1,500 x 10^{-15} m^2/sec (16,140 x 10^{-15} ft^2/sec) for diffusion and 100 x 10^{-15} m/sec (328 x 10^{-15} ft/sec) for permeability. The variability in test results appears to be very high, especially for the permeability coefficient.

The mean strength of 791 standard cylinder tests was 74.9 MPa (10,860 psi).

The values obtained from the diffusion tests are indicative of concrete of extremely high quality. In terms of characterizing the resistance of the concrete to the penetration of chloride ions using test methods commonly used in North America the data in Table 3 is interesting. Rapid Chloride Permeability values (AASHTO T277) were determined on core samples representing six test blocks used in earlier chloride diffusion tests. Because of the different maturity of concrete specimens in the diffusion test and the rapid chloride permeability test the data is not a very precise comparison between the two test methods, however the test data confirms the very low chloride permeability of the tunnel concrete. Coulomb values between 100 and 1000 are considered to represent "very low" chloride ion penetrability in most cases (AASHTO T277).

Table 2 - Summary of Diffusion and Permeability Test Data for TTC Subway Project

	Diffusion Coefficient (x 10^{-15} m^2/sec)			Permeability Coefficient (x 10^{-15} m/sec)
	40 days	80 days	120 days	40 days
No. of tests	21	21	19	14
Minimum Value	340	322	313	0.380
Maximum Value	1320	1110	1300	4.77
Mean Value	687	573	554	1.34
Standard Deviation	250	207	240	1.16
C of V%	36.4	36.2	43.4	86.1
Specification Requirement	N.R.	N.R.	1500 maximum	100 maximum

1,000 x 10^{-15} m^2/sec = 10,760 x 10^{-15} ft^2/sec
1 x 10^{-15} m/sec = 3.3 x 10^{-15} ft/sec

Table 3 - Summary of Chloride Diffusion Coefficients and Rapid Chloride Permeability Values (6 Tests) (ASTM 1202)

Date Cast	Mould Ref.	Chloride Diffusion $\times 10^{15}$ m^2/s			Rapid Chloride Permeability Charge Passed after 6 hours (Coulombs)*	Date Tested
		40D	80D	120D		
April 30/96	SZ 5 + 6	1,087	480	783	434	March 27/97
June 11/96	SZ 5	583	574	784	484	April 1/97
July 23/96	SZ 5	604	1,030	590	443	April 2/97
Sept. 12/96	SZ 5	1,320	1,112	517	515	April 3/97
Oct. 24/96	SZ 5	510	464	325	394	April 3/97
Dec. 5/96	SZ 5	499	429	972	91	April 4/97

* Core samples stored in moist room for initial 28 days, then in normal laboratory air until the date tested.

The permeability values obtained here are also exceptionally low. Water permeability coefficients reported in the literature generally range from 10^{-13} to 10^{-10} m/s depending on the quality of the concrete, and values below 10^{-14} are extremely rare. However, to some extent the measured coefficient is influenced by the method of measurement and the results obtained in this program cannot be directly compared with data obtained using different test methods.

CONTRACTOR'S VIEWPOINT

The production of precision precast concrete tunnel linings used with the high technology tunneling machines available today, present a challenge for the specialist, experienced manufacturing contractors. Tunnel linings typically are tapered across their width to provide the need for directional steerage of those linings during construction. These linings require tolerances of ±1mm (±1/32 inch) for individual segments to ensure that the seals between segments and adjacent rings are always in compression. In addition, the ability to mobilize construction of the tunnel with new tunneling machines within six to eight months of project award has dictated that the manufacturer of the tunnel linings mobilize and produce even faster to ensure fully cured segments are available to meet the programme.

In recent years tunnel designers have introduced an extended design life of 100 years for the structural concrete, and have introduced the methods of measurement described in this paper to determine the long term durability of the concrete. This has meant determining the level of risk to the structure from the ground conditions and acquiring knowledge of available concrete materials and their suitability to provide adequate protection required by the extended design life.

The test for chloride ion diffusion is lengthy and normally conducted after detailed analysis and tests of the available concrete materials has been made. These tests run for 120 days following the preparation of samples which are made on 28 day old concrete. This effectively means that results are not available until five months after concrete has been mixed. In addition, specifications may call for approval and submittal of concrete mix designs in advance of the durability test which are made and evaluated on concrete mixed to the design and placed and vibrated in a manner representative of the intended production technique.

Since contractors are not given five or six months to bid projects, nor is there rarely enough time after project award for the tests to be conducted before production has to commence, the contractor must assess the effect on his costs and risk. If these tests are to be adopted it follows that the contractor must be pre-qualified in his understanding of these requirements in order to provide a responsive bid. Depending on the level of durability protection required the impact on the cost of concrete materials can be as much as 25 percent. In addition, the cost of concrete mix development and quality control testing can also be significant, adding as much as a further 10 percent to the concrete costs.

Clearly the development of an accelerated test would be of benefit to the timetable for bidding and project award but it is doubtful that any new test would have much impact on the cost of providing improved concrete with a known life and durability.

CONCLUSIONS

As far as is known to the authors the two projects described in this paper represent the first major civil engineering projects in North America where chloride diffusion and water permeability requirements have formed part of the specifications and tender documents.

The experience gained on these projects leads to the following conclusions:

- Chloride diffusion coefficients in the order of 600×10^{-15} m²/sec ($6,456 \times 10^{-15}$ ft²/sec) maximum are unlikely to be consistently achieved in practice when the chloride immersion part of the test starts after only 28 days of moist curing.

- Specification requirements for chloride diffusion and water permeability characteristics must reflect the high variability of the tests which in part is due to the nature of determining mass transport measurements in concrete.

- The use of blended Portland silica fume cement, or silica fume added separately to the concrete mix, appears to be necessary when low chloride diffusion characteristics are specified.

- Well executed preconstruction concrete laboratory tests that simulate production fabrication and curing conditions are excellent indicators of the chloride diffusion characteristics that will be achieved during production.

- Both water permeability and chloride diffusion tests as described in this paper take too long to complete to be effective, practical quality control tools for typical contracts. There is a need to develop a short term test, that can be related to diffusion and permeability at the preconstruction concrete mix design stage, which can then be used for quality control purposes during production.

ACKNOWLEDGEMENTS

The authors wish to thank Canadian National Railways and the Toronto Transit Commission for permission to publish this paper.

Brian L. Garrod, P.Eng., Hatch Mott MacDonald, was a member of the Technical Review Committee for the CN St. Clair River Rail Tunnel and is Project Manager for the Toronto Transit Commission Sheppard Line Subway.

Ulrich Kuebler, Quality Assurance Manager, CCL Joint Venture, Fabio Fregonese, Laboratory Supervisor, Trow Consulting Engineers Ltd. and Michael Pratt, Manager Technical Services, Canada Building Materials Company all made valuable contributions to this paper. Their assistance is acknowledged.

References

1. Finch, A.P., "The New St. Clair River Tunnel Between Canada and the U.S.A.", Civil Engineer International, February 1997, pp. 34-44.

2. Collepardi, M., Marcialis, A. and Turriziani, R., "The Kinetics of Penetration of Chloride Ions into Concrete", Il Cemento, Vol. 4, 1970.

3. Collepardi, M., Marcialis, A. and Turriziani, R., "Penetration of Chloride Ions into Cement Pastes and Concretes", Journal of the American Ceramic Society, Vol. 55 (534), 1972, pp. 534-535.

4. Barrer, R.M., "Diffusion in and Through Solids", Cambridge University Press, London, 1952, pp. 7-12.

5. Wood, J.G.M., Wilson, J.R. and Leek, D.S., "Improved Testing for Chloride Ingress Resistance of Concretes and Relation of Results to Calculated Behaviour", Third International Conference on Deterioration and Repair of Reinforced Concrete in the Arabian Gulf, Bahrain Society of Engineers and CIRIA, 1989.

6. Sørensen, H. and Frederiksen, J.M., "Testing and Modelling of Chloride Penetration into Concrete", Nordic Concrete Research, Proceedings of Nordic Concrete Research Meeting, Trondheim, Norway, 1990, pp. 354-356.

7. Ryell, J. and Fasullo, S., "The Characteristics of Commercial High Strength Concrete in the Toronto Area", Proceedings 1993 CPCA/CSCE Structural Concrete Conference, pp. 278-293.

8. Thomas, M.D.A. and Matthews, J.D., "Chloride Penetration and Reinforcement Corrosion in Fly Ash Concrete Exposed to a Marine Environment", Third CANMET/ACI International Conference on Performance of Concrete in Marine Environment (Ed. V.M. Malhotra), ACI SP-163, American Concrete Institute, Detroit, 1996, pp. 317-338.

9. Private Communication to M.D.A. Thomas from A. Butler, Mott MacDonald Consulting Engineers, Croydon, United Kingdom, dated March 26, 1997.

DESIGN AND CONSTRUCTION DEMANDS ON CONCRETE THAT NECESSITATED A VERY LOW WATER-CEMENT RATIO AND HIGH SLUMP

Robert W. LaFraugh
Senior Consultant
Wiss, Janney, Elstner Associates, Inc.
Seattle, Washington, U.S.A.

ABSTRACT

Designers, from the Washington Department of Transportation (WSDOT), of the Lacey V. Murrow (LVM) floating bridge replacement pontoons were seeking watertight, durable concrete that would permit economical and high quality construction of these rather complex structures. Wiss, Janney, Elstner Associates (WJE) was engaged by WSDOT to develop specific material performance data and specifications. Research and development of the concrete mixture designs was followed by a full-scale test to assure constructability of the concrete in deep, heavily-reinforced walls.

BACKGROUND

Durability Requirements for Floating Bridges in Washington

The LVM replacement pontoons were necessitated by the sinking of the original 50-year-old bridge in November, 1990. Design of the new pontoons with prestressed high performance concrete was to limit or prevent cracking and leakage. Corrosion of reinforcement and prestressed steel in the fresh water of Lake Washington, site of the LVM bridge, is not considered a severe threat to structural integrity, but can occur over a long period of time with normal concrete.

Structural Design Demands on Concrete

The use of prestress in the pontoon walls and slabs, together with the need to limit the weight of the pontoons and the subsequent displacement in water, results in thin sections, congested with reinforcement and post-tensioning ducts. Minimum, but acceptable, concrete cover over reinforcement on exposed walls and slabs is dictated by concern for weight and, thus, requires concrete of low permeability. High strength is usually specified in prestressed structures to provide resistance to high compressive stresses. Generally, the post-tensioning is applied at an early age to speed construction, thus necessitating high, early strength concrete.

The concrete pontoon design incorporates some mass concrete sections to accommodate high stresses at pontoon anchors and bolted connections between pontoons. The differential thermal shrinkage and drying shrinkage created between these members and the thinner walls and slabs can create significant cracking problems. The concrete mixture design properties, needed to mitigate the thermal problem, prevent the use of high cement content (which is generally associated with high shrinkage) to produce high compressive strength. The structural design also depends on minimal creep and shrinkage to reduce prestress losses and to reduce or eliminate cracking prior to application of prestress or in non-prestressed areas.

None of the above criteria for concrete design are as demanding, however, as those for watertightness and durability. It is possible in the Seattle area to achieve high strength with relatively low cement content, because of the excellent raw materials available, especially concrete aggregates. Thus, high strength and low heat of hydration are not totally incompatible. Watertightness and durability are primarily achieved with the use of materials that produce low water-cement ratios, very dense cement paste between aggregate particles, and good paste-aggregate bond. The challenge was to make those qualities required by the pontoon design compatible with consistent, highly workable concrete that had to be placed under demanding conditions in fast track construction.

Construction Demands on Concrete
The greatest concern by WSDOT for watertightness and durability was in the outer shell of the pontoon which is directly exposed to water. This area of the pontoon requires fresh concrete properties be given special attention. Flowable concrete, with sufficient cohesiveness to prevent segregation, is needed for placing concrete in the heavily reinforced, deep walls. Workable concrete, with moderate slump and normal setting time, is required for flatwork in the slabs. While those characteristics are not necessarily conflicting, they do require a great deal of flexibility in the concrete mixture.

Contractor incentive for early completion and WSDOT's desire to reopen this vital link on the heavily-traveled Interstate 90 commuter corridor, dictated fast-track construction conditions. The design encouraged large, continuous concrete placements in walls and slabs to minimize construction joints. These factors combined to impose requirements for consistent and controllable concrete quality. Low slump loss and effective slump control were critical elements for consideration in the concrete mixture design and development.

CONCRETE MIXTURE DEVELOPMENT

Description of Test Program
The mixture design and related concrete properties were developed in three phases and described in a final report[1]. The first phase consisted of a series of trial mixtures designed to determine properties of locally available materials and combinations thereof. A review of other related literature was conducted simultaneously with the trial mixture series. In the second phase of the program, the final mixture was developed based on results from the first phase. The mixture was refined in this phase to produce properties that were anticipated to be required for construction. The final phase of testing involved

the construction of two typical walls and a slab section to evaluate the performance of the final recommended concrete mixture.

The final mixture design was selected with primary concern for watertightness of the concrete and was based on chloride permeability results, density of the concrete, and water-cementitious material ratio. Considerations for placeability of the concrete in congested, deep walls, concrete delivery methods, and quality control dictated admixture types and amounts, aggregate gradations, and slump ranges. As expected, the concrete design strength of 45 Mpa (6500 psi) for the pontoons was easily achieved with any of the trial mixtures. Thus, the mixture design was controlled by watertightness and durability criteria, and not by strength requirements.

Mixture design requirements

Parameters for the recommended final mixture design, concrete placement, and curing were selected after all results from the mockup test were analyzed. The table below compares the specified concrete proportions to the contractor's final mixture design and the WJE final test mixture.

Comparison of LVM Mixture Designs and Specifications

	WSDOT Specification	**Contractor Mixture Design**	**WJE Final Test Mixture Design**
Portland Cement, pcy	625 min.	625	640
Silica Fume, pcy	50 - 70	50	64
Fly Ash, pcy	100 min.	100	140
Water-cementitious ratio*	0.33 max.	0.33	0.33
Max. Slump, in.	9	9	8.25

Note: 1 pcy = 0.593 Kg/M^3, 1 in. = 25.4 mm

* Cementitious material includes cement, silica fume, and fly ash

Trial mixture tests and previous experience[2,3] had shown that a combination of admixtures produced optimum workability, strength, slump retention, and density. It is well known that silica fume concrete requires the use of a high range water reducer because of the extra water demand created by the extreme fineness of the admixture. The introduction of a normal range, retarding water reducer at the batch plant, with a portion or all of the high range water reducer, is standard practice in the Seattle area for high strength concrete. The retarder aids in better slump retention and reduces the total amount of high range water reducer, thereby producing better and longer lasting workability. Thus, both types of water reducers were specified.

The decision to use non-air entrained concrete for the pontoons was somewhat controversial. There is not total agreement in the concrete industry that air entrainment is essential in high strength concrete with a very low water-cement ratio. It has been demonstrated in previous research[4] that some entrained air is necessary to produce concrete that is resistant to the severe exposure of standard rapid freeze-thaw tests. However, successful experience in the mild Seattle climate, and even the severe Alaska climates, with non-air-entrained high strength concrete in piling, pier decks, and other bridge pontoons, supports the argument for omitting entrained air. The air-entraining agent, besides adding a difficult control element in high performance concrete, produces stickiness that impairs workability and placeability because of the relatively high amount of cementitious material in the mixture.

Contractor Mixture Development

Pontoon Test Section — The WSDOT Special Provisions required that the contractor build a test section mockup of a corner cell in a typical pontoon prior to start of construction (see Figs. 1 and 2). This was to demonstrate materials and techniques planned for construction. The test section was not allowed to become part of any of the pontoons as part of the final work. It was to include all the typical elements in a pontoon, such as bottom slab, cast-in-place exterior and precast interior walls, typical reinforcement, and post-tensioning ducts and anchorages. The contractor's approved LVM mixture design was to be placed in the slab and walls using the same methods intended for the final work.

Full-Scale Wall Tests — The contractor voluntarily conducted several tests of the concrete mixture and placement procedures in the deep walls on four full-scale mockups of the pontoon walls. Two of these tests were done prior to construction of the test section and two tests were conducted afterward, but prior to pontoon construction. The initial tests were performed primarily to measure form pressures, evaluate vibration methods, and measure form deformation. Load cells were installed on several form ties to measure pressure during and after placement.

The latter two wall tests were performed to measure form pressures using different concrete admixtures. The specified admixtures were thought, by the contractor, to be slow setting and produce form pressures that could cause reduction of the desired rate of placement. In fact, the form pressures induced by the "retarded" LVM mixture were equal to or less than those resulting from an accelerated concrete. Pressures measured on the bottom she-bolt form ties were about 48 to 57 kN/M^2 (1000 to 1200 psf) in both tests, while pressures in bolts higher on the forms were lower with the retarded mixture 29 to 38 kN/M^2 (600 to 800 psf) than those measured in identical locations with the accelerated mixture, 31 to 48 kN/M^2 (650 to 1000 psf). Although the forming was designed for only

600 psf, the performance was satisfactory. The forming system with external vibration was accepted based on field performance and cores taken to establish proper consolidation of the concrete and form deflection. The formwork did have to be strengthened in a few locations for Pontoons A and T as a result of excessive form deflection. The accelerated concrete was made with a proprietary, non-chloride accelerator, but otherwise identical to the design mixture. It had less workability and lower early strength in the first 48 hours than did the design mixture, apparently because of lack of compatibility with the portland cement.

Figure 1. Overall view of test section

Figure 2. Unconsolidated concrete from first placement in test section walls

These individual wall tests not only provided valuable data regarding form pressures, consolidation techniques with external vibrators, and formwork design, but also were useful to the contractor in determining the behavior of the concrete in the fresh state under simulated placement conditions. Rate of slump loss, effect of additional dosages of high range water reducer at the site, and behavior of the concrete from two different batch plants and types of concrete mixers, were all important to planning the construction quality control procedures. Concrete was placed in the various walls with slumps ranging from 140 to 240 mm (5 1/2 to 9 1/2 inches). A few isolated tests were lower or higher than this range, which was achieved by varying the dosage of high range water reducer.

No detrimental effects were observed or measured from the higher slump concrete. The contractor was allowed to proceed into construction with a deviation from the WSDOT Standard Specifications, which allowed a maximum slump of 230 mm (9 in.).

LVM CONCRETE PERFORMANCE

Adjustments to Concrete Mixture and Placement

Previous experience by WJE in the construction of an oil exploration platform built in Japan made it clear that placement of concrete in the typical deep walls was best done by using tremie systems. The tremie is lowered into the top of the wall and slowly withdrawn as the level of concrete rises. The upper lifts of concrete can be easily placed from the top of the wall without the tremie. The outer layer of vertical reinforcement is spaced to allow room for the tremie insertion. The LVM bridge contractor followed this suggestion and used a modified structural tubing attached to the concrete pump hose to place concrete in all the remaining walls and bulkheads (see Figs. 3, 4 and 5). The cross-sectional area of the tubing was approximately equal to that of the 125 mm (5 in.) diameter pump line. High slump concrete and the 9.5 mm (3/8 in.) maximum aggregate size allowed concrete to be placed as rapidly as needed with two pumps.

Figure 3. Seattle graving dock layout for concrete placement

Figure 4. Insertion of "tremie" tube in outside wall of Pontoon F

Compressive Strength Test Results

As expected, the achievement of the design compressive strength of 45 MPa (6500 psi) was never a problem during construction. The average compressive strength of all tests at 28 days was 71.7 MPa (10,390 psi). No attempt was made to establish strength beyond 28 days, but that determination was made during the LVM Mixture Design Development testing. Those tests showed a strength gain of about 15 percent from 28 to 90 days, for mixtures similar to the final LVM mixture. Thus, the 90 day strength for the pontoon construction is estimated to average 83 MPa (12,000 psi).

Figure 5. Closeup of tremie tube inserted in wall form next to form surface. Note internal vibrator on right.

Rapid Chloride Permeability Test Results

The AASHTO T-277 test for Rapid Chloride Permeability was conducted at a frequency of about one test for each 1530 cubic meters (2000 cy) of concrete placed. The rapid permeability test was developed as an evaluation tool, to replace the 90 day Chloride Ponding Test (AASHTO T-259), because of its ease and speed. Correlation of results between the two tests was established before its acceptance as a standard test method for conventional concrete, i.e. low coulomb values in the Rapid test correlated to low permeability in the Ponding test which was extended to 180 days. Acceptance of the Contractor's mixture design was based upon achieving a maximum coulomb value of 1000 as per WSDOT standards. The results for construction were informational only. They were not used as a basis for acceptance of concrete. At least two specimens from each sampling were tested at 28 days and, in many cases, other specimens from the same sample were tested at 56 days and 90 days, and a few at 7 or 14 days. A statistical summary of the tests is shown in the following table:

Rapid Chloride Permeability vs. Age

STATISTIC	28d	56d	90d
No. of Tests	109	51	22
Average Perm, Coulombs	1327	785	577
Standard Deviation	523	230	135
Range of Results	517 - 2784	368 - 1608	310 - 804

It can be seen that the permeability is well below the targeted maximum of 1000 coulombs at 56 days, thought to represent concrete with excellent resistance to chloride intrusion. There is significant reduction of permeability with age, as can be seen in the table. The high variability of results of 28 days is believed to be typical for higher performance concrete.

SUMMARY AND RECOMMENDATIONS

The risk of proceeding into construction with concrete specifications that had no history of previous performance on WSDOT projects was minimized by undertaking a rather extensive development program. Pre-construction testing further reduced the potential for major problems. However, those efforts would have been wasted had the contractor and concrete supplier not been willing to extend themselves and make these different approaches work. The successful conclusion of the pontoon construction was greatly assisted by the cooperative efforts of WSDOT, the contractor, the concrete supplier and WJE. Post-construction input from all of these parties has confirmed that the high performance concrete, external vibration, and other mitigative construction methods, were proven to be necessary.

Several recommendations for future construction of watertight structures are made that might be considered appropriate for use in ordinary transportation projects. These recommendations are listed below:

1. The LVM mixture design will work well as it is, but consideration should be given to modifying it to include some very recent knowledge. That is, silica fume content may be reduced to the 4 or 5 percent level and fly ash increased to 120 Kg or more per M^3 (200 pcy), without impairing the impermeability of the concrete. These changes will produce a more workable concrete at a given water content with better slump retention. It should be verified that this adjustment can be made without sacrificing the outstanding resistance of this concrete mixture to segregating. Further verification of low permeability may be required to satisfy the authorities.
2. External vibration is extremely valuable when placing concrete in deep walls or bulkheads (see Fig. 6). It necessitates more expensive formwork, but that is

more than offset by the improvement in concrete density and watertightness. As a minimum, the forming system must have a steel framing system which will transmit vibration to the skin of the form. The form skin should be steel plate, but it was demonstrated in this project that plywood will work. External vibrator spacing cannot be specified, as it needs to be determined by test. However, a maximum spacing of 3.7 M (12 ft) horizontally and vertically is recommended. Vibrators must impart a minimum of 7.5 kN (1700 lbs) force.

Figure 6. Example of smooth, dense inside wall surfaces (opposite side of external vibrators) on Pontoon F

3. High performance concrete, which includes silica fume and fly ash, is very cohesive and will not segregate, if properly designed. This property, enhanced with high range water reducers, allows the concrete to be placed at very high slump without loss of strength or density. Future construction with this type of mixture should not restrict the use of flowable concrete.
4. High range water reducers may be added at the jobsite to retemper concrete more than once. Research and previous experience has shown that the total amount of admixture should be within the manufacturer's recommended maximum dosage, but two or three additions are acceptable as long as the concrete will respond by regaining its plasticity.
5. 100 mm (4 in.) diameter test cylinders may be used for field tests of concrete, provided that the inspectors are familiar with the restrictions of maximum aggregate size, method of consolidation, and testing. The much reduced size and weight of the smaller specimens has indirect benefits in consistency of

results, especially when large numbers of tests are conducted. The smaller specimens become necessary for high strength concrete, if the testing machine capacity is not 50 percent greater than the ultimate load to be applied.

6. Consideration should be given to alternate methods of measuring and specifying workability limits of flowable concrete with slump greater than 175 or 200 mm (7 or 8 in.). The slump test can be misleading or meaningless, particularly when changes in the appearance of the concrete are evident.

7. Flatwork, such as decks and deck overlays, made with silica fume concrete should be fog sprayed immediately after the initial screeding or floating to prevent plastic shrinkage cracking.

8. Special provisions for curing and controlling temperature differentials are needed for thick elements, especially if they are rigidly connected to previously-cast elements. This requires further study as there is not agreement among all authorities on the limitations involved.

REFERENCES

1. Arvid Grant Associates and Wiss, Janney, Elstner Associates, Inc., "Concrete Mix Design Development, I-90 Pontoon Replacement," December, 1991.

2. Alaska Oil and Gas Association (AOGA), "Developmental Design and Testing of High-Strength Lightweight Concretes for Marine Arctic Structures," AOGA Project Report Numbers 198 and 230, May, 1983 and September, 1984.

3. LaFraugh, R.W., "Design and Placement of High Strength Lightweight and Normalweight Concrete for Glomar Beaufort Sea I," *Proceedings of the Symposium for Utilization of High Strength Concrete*, Stavenger, Norway, June 1987.

4. LaFraugh, R.W., "Feasibility Study of 14,000 psi Pretensioned Concrete for a Navigational Structure," *PCI Journal*, May/June, 1993.

USE OF QUALITY SYSTEMS TO CONSISTENTLY PRODUCE DURABLE HIGH STRENGTH CONCRETE RAILROAD TIES

Dave Millard
Asst. Vice President
CXT Incorporated, Railroad Division
Spokane, Washington, U.S.A.

ABSTRACT

The demand for precast prestressed concrete railroad ties in the United States remains steady due to excellent concrete tie performance and lower total life cycle costs compared to other types of ties. Railroad tie durability is a major concern for all tie systems, as they are subjected to severe service conditions including static wheel loads approaching 18,000 kg. (39,600 lb.), dynamic wheel loads in excess of 70,000 kg. (154,000 lb.), heavy annual tonnage, frequent freeze thaw cycles, moisture, and other environmental and operational factors. To meet these challenging performance requirements, prestressed concrete tie producers in North America have developed and implemented Quality Assurance systems that meet Precast/Prestressed Concrete Institute (PCI) and ISO 9000 guidelines. This includes use of high quality materials, qualified personnel, well developed procedures, and computer control of key operations. Prestressed concrete tie production requires reliably obtaining 12 hour compressive strengths over 31 MPa (4500 lb/in^2) and 28 day strengths in excess of 48 MPa (7,000 lb/in^2). These strengths must be achieved while minimizing risks of Alkali Silica Reactivity (ASR) or Delayed Ettringite Formation (DEF). Concrete railroad tie quality systems and product development will continue to be a focus to ensure that customers' rigid expectations are properly met.

INTRODUCTION

Prestressed concrete railroad ties have been the accepted standard in Europe, Asia, Australia, and South Africa since their introduction approximately 50 years ago. Concrete ties were not introduced in North America until the late 60's. The first substantial order for concrete ties was placed by Canadian National Railroad in the late 70's, and the first significant use in the United States did not occur until the mid 80's. This delayed acceptance in North America was due primarily to the widespread availability of good quality timber at competitive prices. Prestressed concrete railroad ties now have established a foothold in the North American market. During the past 20 years, over 15 million prestressed concrete railroad ties have been produced and installed in the United States and Canada. Although this is a significant number of ties, less than 2% of the total railroad ties in North America consist of concrete.

From the beginning, prestressed concrete railroad tie producers faced an uphill battle selling ties to traditionally "all wood" railroads in North America. Two hurdles stood in the path of concrete ties, price and a reluctance to change the standard within the railroad community. Twenty years ago, customers paid a significant premium for concrete ties. Prices for good quality timber have increased substantially, which reduces the premium customers pay to use prestressed concrete railroad ties. Prestressed concrete railroad ties are often less expensive than wood on a first cost basis for new track construction.

To convince railroad leadership to change from wood ties to concrete, producers had to prove to them that concrete ties would provide additional value compared to the wood standard, even at a cost premium. Quality was the cornerstone of acceptance of new prestressed concrete railroad ties for the initial users in North America, including Canadian National, Burlington Northern Sante Fe, and Union Pacific. These customers, among others have come to realize the additional value that concrete ties provide including longer life, reduced maintenance, fewer derailments, reduced fuel consumption, and improved ride quality.

Figure 1. Concrete Track Construction, Yakima River Valley, Washington

Figure 2. Concrete Ties in Mainline Track

DESIGN PRINCIPLES

Figure 3 shows the basic components of a concrete railroad tie system. Each concrete tie weighs between 275 - 325 kg (605 - 715 lb.) Ties have eight to twenty-four prestressing tendons that provide flexural bending strength, and four ductile iron inserts cast into the tie to hold the elastic spring clips, which in turn holds the rail in place. A pad is placed under the rail to reduce wheel impact severity. Nylon insulators are placed between the rail and the elastic spring clips to provide electrical isolation of the rail and to eliminate wear of the ductile iron inserts. The prestressed concrete tie system is a highly engineered system that requires tight quality controls to ensure optimum performance.

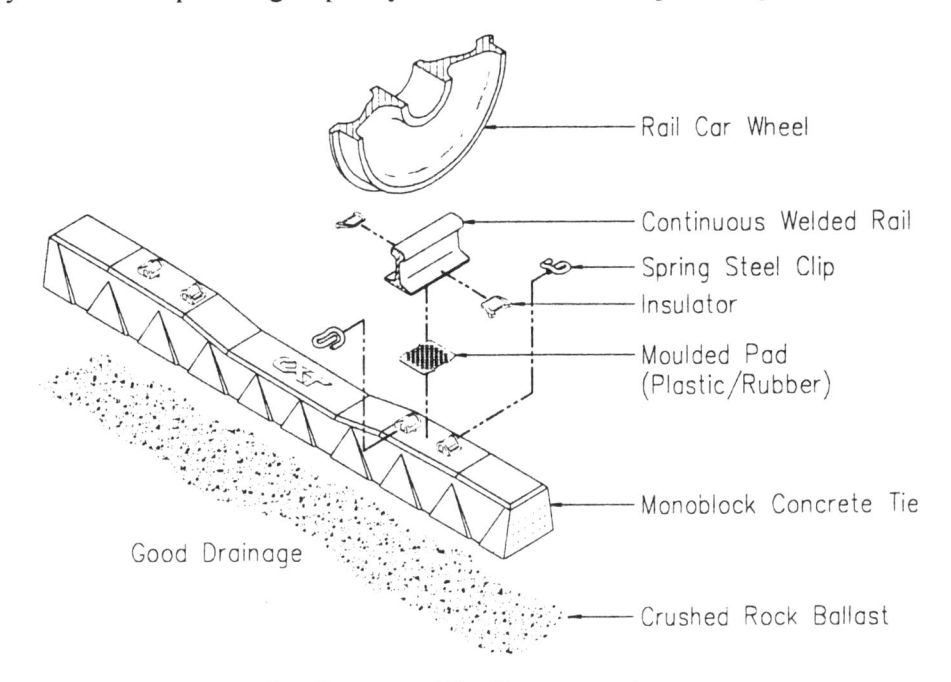

Figure 3. Concrete Tie Components

QUALITY CONTROL

To convince railroad engineers that concrete tie producers were committed to quality and performance, industry officials worked hand in hand with railroad engineers to develop detailed recommended practices. The outcome was a 54 page manual of recommended practices for concrete tie design, production, and quality control. Quality assurance and quality control are essential components of the prestressed concrete tie design and production processes. Producers are held to rigid finish product specifications. Some of these requirements include:
   1. Twenty-eight day concrete compressive strength greater than 48 MPa (7000 lb./in^2)
   2. Dimensional tolerance of 1.5 mm (0.060 in) over the length of a 2.5 m (8.25 ft.) tie
   3. No surface voids in the concrete surface greater than 10 mm (0.39 in)

CXT, Inc. has developed a simple, yet effective, quality management system to ensure consistent production of durable, high strength concrete railroad ties. The primary goal of our quality system is prevention of quality problems. US Class 1 railroads mandate that key suppliers be certified to both PCI and ISO 9000 quality standards. Like many other manufacturing businesses, the prestressed concrete tie industry was strongly encouraged by these customers to become quality certified to an ISO 9000 equivalent quality standard. CXT, Inc. became quality certified in 1992 out of business necessity. Quality certification requires a thorough examination of almost all production, engineering, quality, and customer service procedures. The certification process helps a company become more efficient and better organized to both internal and external customers. CXT, Inc. believes in the value of quality certification and plans to achieve ISO 9001 registration by September, 1998. Listed in Table 1 are the general quality system requirements for PCI and ISO 9001. ISO 9001 differs from PCI in that ISO has more extensive documentation and procedures requirements while PCI focuses on quality control, record keeping, and production processes. In addition, ISO impacts all aspects of a company including design, customer service, and installation.

Table 1. Requirements of PCI and ISO 9001

PCI	*ISO 9001*
Quality Control (15%) Prestressing (20%) Concrete (20%) Materials (5%) Production Practices (10%) Quality Control Operations (30%)	Management Responsibility Quality System Contract Review Design Control Control of Customer Supplied Product Product ID and Traceability Process Control Inspection and Testing Control of Inspection, Measuring, and Test Equipment Corrective and Preventive Action Handling, Storage, Packaging, Preservation, & Delivery Control of Quality Records Internal Quality Audits Training Inspection and Test Status Purchasing Servicing Statistical Techniques

Compliance with ISO 9001 requires procedures manuals (often referred to as a Quality Assurance Manual or Plant Procedures Manual) and documentation that show compliance with these manuals. ISO requires a detailed plan for controlling quality and proof of compliance with the plan.

A representative page of our company's QA manual is provided in Figure 4. The QA Manual describes the key quality systems that are used to manage quality. Key items required on any procedure are title, date written, author, approval, and revision number. It is essential to identify one person to lead the administration of all plant and quality procedures. The QA manager is often responsible for the manuals. The QA manager makes sure all procedures are correct, up to date, and implemented. In addition, he/she leads the annual review of the quality system and all procedures.

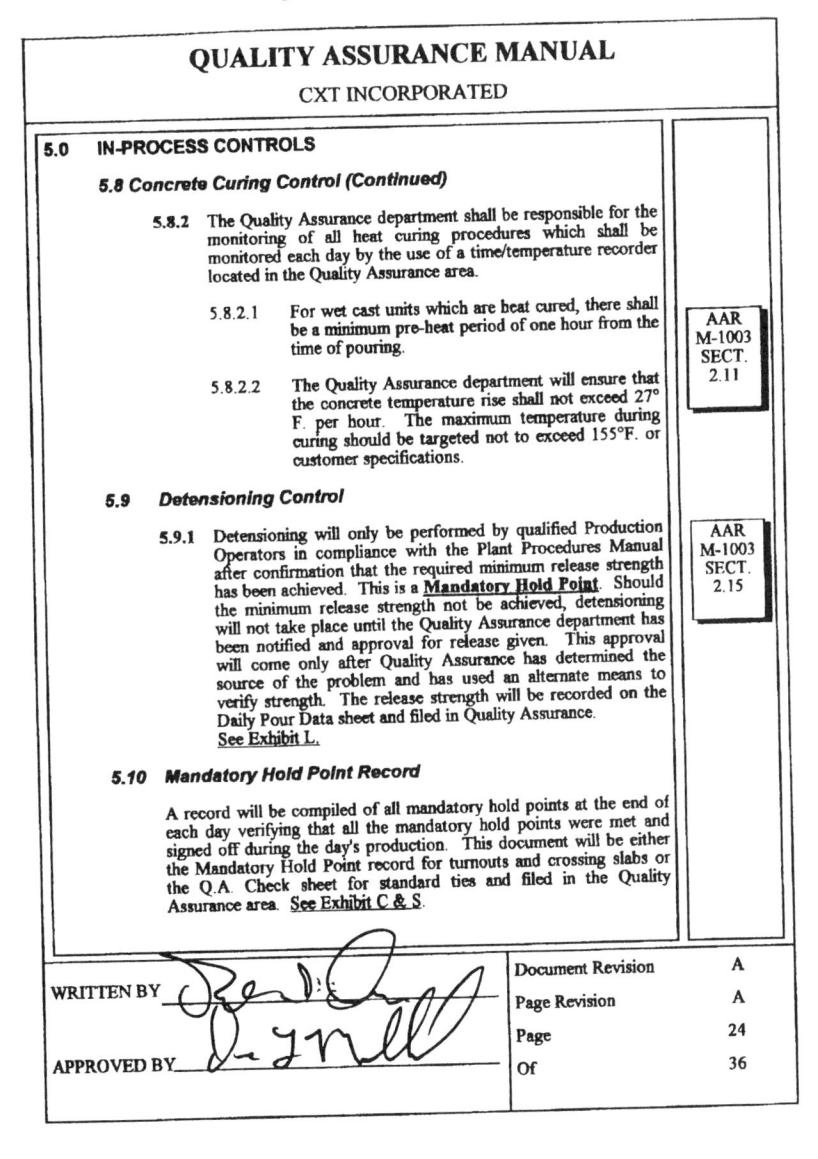

Figure 4. CXT QA Manual Page

In addition to detailed written procedures, simple flow charts which visually describe the process flow of an operation are often used. Procedures are written on every key operation that can impact quality. These include general production procedures, control of mix designs, prestressing of tendons, curing, quality control testing, material release, engineering controls, customer complaints, material shipping, and many others. Many industry related procedures from PCI, ACI, ASTM, and AREA are also followed.

Another key component of an effective quality program is the use of mandatory hold points. Mandatory hold points are defined as critical points in the production process at which key criteria must be met prior to continuing to the next step. CXT, Inc. has established five mandatory hold points which are followed to produce high strength concrete railroad ties. These are:
1. Concrete quality (slump, % air, and unit weight on 12% of the loads)
2. Final stressing of the wire or strand (per PCI)
3. Concrete compressive strength at time of detensioning must be greater than 31 MPa (4500 lb/in^2)
4. Flexural strength and bond development testing of finished concrete ties
5. Thorough visual inspection of the ties

If the criteria of one of these key points is not met, the operation is stopped and production and QA leaders jointly decide on the appropriate next step. All employees in the production facility know the five mandatory hold points.

Good quality procedures alone do not ensure that excellent finished product quality will be achieved. Excellent materials and investment in key production controls are necessary to ensure excellent finished product quality.

Extensive material testing is essential prior to using any materials. This includes testing aggregates to ASTM C33, C227 (alkali reactivity of cement/aggregates), C295 (petrographic analysis), C289 (alkali reactivity), and C1260 (mortar bar alkali reactivity). Cements are tested to ASTM C150 and C114. Admixtures must meet ASTM C494 requirements. In addition, prior to using any new source of materials, test batches are produced using a small mixer and tested for strength development and workability.

Concrete tie producers use several state of art control systems to monitor and control key processes. These include:
1. Computer controlled batch plants
2. Computer controlled curing systems
3. Computer controlled monitoring of prestressing operation

One of the most critical processes for the precast industry to manage is concrete curing. Concrete curing must be closely monitored to ensure high early strength development, while minimizing the risk of DEF and ASR. The industry has worked closely with customers and prestressed concrete experts to establish appropriate curing specifications to ensure high early concrete strength, without sacrificing durability or performance.

As an example of a quality management system, the following steps have been implemented at CXT, Inc. to manage the concrete curing process:

Concrete Curing Quality System
1. After reviewing the product specification, manufacturing and quality instructions are developed which show specific quality and production requirements
2. A specific cure cycle program based on the customer requirements is developed. This program is authorized by the QA Manager.
3. Each cure cycle is monitored every 30 minutes by both QA and production personnel to ensure that excessive temperatures are not reached, adequate strengths are achieved, and customer specifications are met.
4. QA personnel review and approve each cure cycle and sign off that strengths have been achieved.
5. If a cure cycle is out of tolerance, an exception report is written, and corrective measures are implemented. If concrete temperatures exceed the customers specification, all ties produced on that bed are rejected.

Conservative guidelines to minimize the risk of DEF are followed. A three hour delay is used prior to applying any heat to the ties. The heat rise per hour is kept below 18 °C (27°F) per hour, and the maximum temperature is kept well below 70 °C (158°F). In addition, cement and aggregate quality are monitored through evaluation of mill certificates, independent lab testing, in-house testing, and annual audits. CXT, Inc. successfully uses these procedures to manage the curing process for concrete tie production. CXT, Inc. averages over 35 MPa (5000 lb/in^2) compressive strength in as little as eight hours and over 55 MPa (8000 lb/in^2) in 28 days, while maintaining an average maximum temperature of 61°C (142°F).

The precast, prestressed concrete industry is committed to providing customers with durable, high performance concrete railroad ties that will perform for many years under severe service and environmental conditions. This can be achieved only by implementation and strict adherence to a detailed quality management system. Good quality is good business while poor quality could put the prestressed concrete tie industry out of business.

DEVELOPING HIGH PERFORMANCE CONCRETE SPECIFICATIONS FOR HIGHWAY BRIDGE CONSTRUCTION - EXPERIENCE OF THE ONTARIO MINISTRY OF TRANSPORTATION

Hannah C. Schell
Manager, Concrete Section, Engineering Materials Office
Transportation Engineering Branch

Beata Berszakiewicz
Assistant Research Engineer, Materials Research Office
Research and Development Branch

Alan K.C. Ip
Director (Acting)
Research and Development Branch

Ministry of Transportation, Downsview, Ontario, Canada

ABSTRACT

This paper describes the Ontario Ministry of Transportation's development of specifications for use of High Performance Concrete (HPC) in highway structures, and the field and laboratory testing and monitoring programs on which the Ministry's current specifications have been based. In 1995 the Ministry constructed the first full highway bridge in Ontario of HPC. Previous experience had been gained with HPC in bridge deck overlays and a bridge deck replacement. In 1996, in partnership with Concrete Canada and Canadian Highways International Constructors, Ministry staff monitored the placement of HPC in two additional highway bridges in Ontario. The intention of the Ministry is to develop specifications which maximize the benefits of HPC for the owner, which ensure the contractor has responsibility for and control over the concrete mix itself and the placement process, and which as much as possible rely on an "end-result" approach as the basis of acceptance of the product.

INTRODUCTION

Since 1992 the Ontario Ministry of Transportation (MTO) has investigated the use of high performance concretes as a means of achieving better long-term structure performance and extending service life. While the potential benefits of HPC in terms of improved concrete durability and better corrosion protection for imbedded steel are very attractive, the Ministry was also concerned with reported drawbacks of HPC and the lack of familiarity with this relatively new material in most areas of Ontario.

The intention of MTO was to develop specifications which maximized the benefits of the use of HPC for the owner, while ensuring that the contractor had responsibility for and control over the concrete mix itself and the placement process. It was also the intention of MTO to develop end-result or performance based specifications for HPC materials and construction, and to use trial projects as a means of developing and verifying the validity of such specifications.

The Ministry was approached by Concrete Canada (part of the federally-funded Network of Centres of Excellence Programme) in 1994 to discuss potential opportunities for use of HPC in Ontario highway construction. Prior to that time, MTO had utilized silica-fume concrete to rehabilitate or replace deteriorated bridge decks: in 1992 Ontario's first silica-fume concrete overlay was placed on Lynde Creek Bridge on Hwy 12, in Brooklin, Ontario; in 1993, a thin-slab bridge deck was replaced on the Montreal River Bridge on Highway 11, in Latchford, Ontario; two bridge decks were rehabilitated using silica-fume concrete overlays in 1995 on Hwy 588, over the Kaministiqua River near Thunder Bay, Ontario. The silica-fume concretes specified required a minimum cement content of 355 kg/m^3 (599 lb/yd^3), utilizing 7.5 or 8 % replacement of cement by silica fume. The 28 day compressive strengths were typically in the range of 50 to 60 MPa (7250 to 8700 psi), with excellent permeability characteristics and a well-structured air void system in the hardened concrete. In addition to field and laboratory testing of the concrete, the construction process was monitored on all structures[1]. Based on these projects, MTO was eager to make greater use of HPC in routine bridge construction contracts because of the enhanced material properties, but had concerns with several construction-related issues.

HPC DEMONSTRATION AND MONITORING PROJECTS

Highway 20 Structure (MTO Contract 95-39)

In 1995 MTO constructed their first highway structure of HPC, a new bridge replacing a corrosion-damaged structure on Highway 20 near Smithville in Southern Ontario, as a demonstration project in cooperation with Concrete Canada. The Hwy 20 structure was to be a single span, 30 m (98.4 ft) long and 13.4 m (44 ft) wide bridge with a 225 mm (9 in) slab deck on prestressed girders. It was designed with a 30 MPa (4350 psi) cast-in-place concrete deck slab, approach slabs, abutments, barrier walls and footings, and 40 MPa (5800 psi) precast, prestressed girders. The structure was to contain epoxy-coated reinforcing steel and the deck surface was to be waterproofed, as is normal MTO policy for structures exposed to deicing salts.

Once identified as a HPC demonstration project, the contract was altered such that the deck slab, approach slabs, abutments and barrier walls would be constructed using 60 MPa (8700 psi) concrete. (The design of the structure was not altered to take advantage of the higher strength of the concrete; this project was considered to be primarily a "demonstration" of the industry's ability to produce and place HPC, rather than design advantages associated with this material.) Uncoated reinforcement in the deck and barrier walls and bridge deck, and elimination of the conventional waterproofing and asphalt surface course on the deck surface, were specified to allow for monitoring of corrosion activity and concrete condition with time. Corrosion measurement probes and thermocouples would be imbedded in the deck.

Highway 407 Structures (Canadian Highway International Constructors (CHIC))

In 1996 MTO staff were involved in monitoring the construction of two HPC highway structures on Highway 407 in the Greater Toronto area. Construction of this highway, located in the Toronto area, was undertaken by Canadian Highways International Constructors which is a private industry consortium of contractors, suppliers and consultants, working in partnership with the government of Ontario. Both structures monitored are single-span, integral abutment structures. The eastbound bridge over Levi's Creek has a 34 m (111.5 ft) long and 11.7 (38.4 ft) wide span. The eastbound entrance ramp bridge span is 34 m (111.5 ft) long and 8.7 m (28.5 ft)wide. The thickness of both bridge decks was specified at 235 mm (9.3 in). As for Highway 20, ready-mixed concrete with a minimum 28 day compressive strength of 60 MPa (8700 psi) was specified for the bridge decks, barrier walls and approach slabs, and for the abutments above the level of the bearing seats. Similarly, uncoated steel was specified for the decks and barrier walls and the surface of the decks was to be exposed. (The design and construction of these structures is the subject of another paper in this symposium.)

MTO SPECIFICATIONS FOR HIGHWAY 20 HPC DEMONSTRATION STRUCTURE

Specifications for use of HPC in highway structures were developed by MTO, with input from Ontario industry and from Concrete Canada, to address the Ministry's specific needs with respect to ensuring durability of the finished product and reducing potential for difficulties or delay during the construction process. Ontario Provincial Standard Specifications for concrete materials and construction (OPSS 1350 and 904)[2] were amended as outlined below. Materials to be used were required to conform to the normal requirements for concrete-making materials, with exceptions as noted, and as is normal for most MTO structural concrete work the particular materials were to be selected and the mix designed by the contractor.

Materials and Mix Design

The minimum 28-day compressive strength of the concrete was specified to be 60 MPa (8700 psi). Acceptance of structural concrete by MTO is normally based on an end-result specification which applies payment adjustments based on the mean strength and standard deviation of the concrete strength results obtained; because of the lack of experience of the industry with high strength concretes, and the concern that high standard deviations could result in inappropriate financial penalties, the end-result specification was eliminated on this contract and alternative acceptance procedures used.

A commercially available blend of Portland cement and silica fume (Type 10SF cement)[3], consisting of Portland cement with 8% silica fume and available from three suppliers in Ontario, was specified for use. It was felt that this would avoid potential inconsistencies in the amount of silica fume in the concrete. The minimum cementing materials content was specified as 450 kg/m^3 (759 lb/yd^3). Use of a superplasticizer was specified for all 60 MPa (8700 psi) concrete, and a set retarding admixture, Type RX, was also specified for the bridge deck concrete.

Plastic Concrete Properties

The air content of the plastic mix was specified to be 6.0±1.5 %. By means of superplasticizer addition the initial specified slump of 50±20 mm (2±0.8 in) was to be increased to 180±20 m (7±0.8 in). The initial slump constraint was applied in order to ensure a low water/cement ratio in the mix, and the upper limit was placed on the superplasticized concrete in order to avoid the segregation which had been experienced in previous work with concrete slumps above 200 mm (8 in). The maximum temperature of the concrete mix at the time of discharge was limited to 25°C (81°F).

Sampling and Testing

To monitor the contractor's ability to produce on a large scale HPC having a low variability in the plastic and hardened state, the frequency of sampling for field and laboratory testing was increased substantially beyond the normal requirements. Air content, slump and temperature of each truckload of concrete was to be measured. Cylinders were to be prepared from alternate truckloads of HPC for evaluation of compressive strength (at 1, 3, 7 and 28 days of age), rapid chloride permeability and air void system parameters of the hardened concrete. Due to concern with greater variability in strength of the 60 MPa (8700 psi) concrete, compressive strength was to be determined by averaging sets of three 100 mm (4 in) diameter by 200 mm (8 in) cylinders, with a small number of 150 mm (6 in) diameter by 300 mm (12 in) cylinders taken for comparison purposes. While routine testing of concrete on MTO contracts is carried out exclusively by private laboratories, the HPC was tested in the ministry's Central Laboratory in order to provide the experience needed to develop testing requirements for HPC.

Pre-construction Trials

The specification required preparation of an initial truck-size trial batch, at least 28 days prior to the first actual concrete placement, followed by placement of a trial slab the same thickness and width as the bridge deck and 10 m (32.8 ft) in length, constructed using the same personnel, equipment and construction procedures, including curing, to be used in the structure itself. The purpose of these trials was to first ensure that the mix intended for use was in fact capable of meeting the contract requirements in terms of compressive strength (as well as target values for air void system and permeability), and to allow the concrete supplier and contractor to learn as much as possible about the practical aspects of handling the new material prior to commencing work on the structure. The trials were specified as a means of verifying the contractor's ability to produce and place HPC in conformance with the contract requirements.

Concrete Placement and Curing

Preventing the evaporation of moisture through wet curing of HPC is critical because of the potential for plastic shrinkage and cracking. Immediately after placement, fog misting of the finished concrete surface of the deck and approach slabs was specified. Two layers of wet burlap were to be applied, following the deck screed within 2 to 4 meters (6.5 to 13 ft). Use of soaker hoses was specified to provide a continuous source of moisture for finished and formed surfaces of the HPC. Wet curing was to be applied to all surfaces for seven days; on the bridge deck, immediately after removal of the wet burlap, application of two layers of curing compound was specified in order to avoid a sudden loss of moisture from the concrete surface and to extend the curing period.

Controlled permeability formwork, intended to result in a densified concrete surface that provides enhanced protection against the intrusion of water and chlorides into the concrete, was specified for the middle portion (between the deck expansion joints) of the structure barrier walls; the end portions of each wall were to be cast without textile form liners serve as control sections to evaluate the effectiveness of the liners.

Corrosion-monitoring Hardware

The installation of corrosion-monitoring probes imbedded in the concrete of the deck slab was included in the contract. Three types of imbedded probes were used to monitor concrete resistivity and steel corrosion activity. Graphite, silver-silver chloride and three element (carbon steel, stainless steel and manganese dioxide reference) probes were positioned at the level of the top mat of reinforcement in the bridge deck, at four locations, monitored remotely from a junction box mounted on the structure.

In general, the specifications for the Highway 407 HPC Demonstration Structures followed the MTO requirements for HPC in the Highway 20 Demonstration Project with some modifications. A lower total cementitious materials content of 400 kg/m^3 (674 lb/yd^3) was initially specified, and the air content was increased to 6.5 ± 1.5%. A slighter wider range of slump of 180 ± 30 mm (7±1.2 in) was specified. Sampling and testing frequencies for both plastic and hardened concrete were reduced significantly, and textile form liners were not used.

MONITORING TO VERIFY SUITABILITY OF SPECIFICATIONS

Monitoring of the construction process provided an opportunity to assess the suitability of the specifications for high-performance concrete construction.

Trial Mixes and Placements

While adding substantial cost to the work, the trial mixes and placement of HPC provided an excellent opportunity for 'first time' material suppliers and contractors to familiarize themselves with a new material and new specification requirements. Trials also serve to provide the structure owner confidence that unnecessary delays or difficulties will be avoided when work commences on the highway structure itself. By preparing a number of different trial batches on the Highway 20 project, the contractor and his concrete supplier were able to investigate a number of material combinations in order to arrive at an economical and "high-performance" concrete mix design. The trial stage also allows for verification of the behaviour of the admixtures and superplasticizer with the cementing materials used, and for identification of sequence of batching the mix components. The trial batch and slab on the Highway 20 project were also useful from the point of view of establishing what precautions would be taken to maintain the specified temperature of the mix under hot weather conditions (replacement of mixing water by ice was used as a means of maintaining the specified maximum concrete temperature).

Based on MTO experience to date, the most critical part of the HPC placement in the structure is the placing and finishing of the deck slab. Placement of a trial placement, which may be the first contact with HPC for a construction crew, providing an opportunity to learn the interaction of the material, equipment, and construction methods. The impact of the concrete discharge method (pumping, crane and bucket) on the properties of the material can be determined, the optimal range of HPC workability for mechanical finishing and the best techniques for texturing, fog misting et cetera can be established. Placing of the trial slab for the Highway 20 project is shown in Figure 1.

HPC Supply

Concrete mixes for both projects were designed by the ready-mixed concrete supplier. Table 1 shows total cement and cementitious materials (C+CM) content, water to C+CM ratio and plastic mix properties as specified and as supplied. The consistency of the concrete mix supply is characterized by the coefficient of variation of air content and initial and final slump of the mix supplied.

Two systems of batching and mixing the superplasticizer were used by the suppliers on the two projects. For the Highway 20 demonstration project, superplasticizer was added to the concrete after delivery to the jobsite (a journey of approximately 45 minutes from the plant): the initial slump and air content were measured, superplasticizer added and the concrete mixed, and the final slump and air measured to verify conformance with specifications (after satisfactory control was established, the initial air content measurement was eliminated). On the Highway 407 project, superplasticizer was added at the batching plant (approximately 15 minutes away), and slump and air content were subsequently measured at the site. Slump measurements reported below for the two projects, and illustrated in the histogram in Figure 2, would indicate that the first supplier

Figure 1. Construction of the 13 m x 10 m x 225 mm (42.6 ft x 32.8 ft x 9 in) trial slab for the Highway 20 project, using the same personnel, equipment and processes to be used on the structure.

was able to achieve a lower variability and finer control of the properties of the concrete.

Placing and Curing of HPC

Susceptibility of HPC to drying and a high potential for plastic shrinkage and cracking require that construction operations progress without interruption and that the concrete be protected at all times from loss of moisture. Based on the two HPC projects described, it was clear that fog misting of the bridge deck concrete must start as soon as possible after placing and be maintained continually until wet curing is in place. Pretesting of the fog misting rate (water pressure and fineness of the mist) in field trials appeared to be essential in order to ensure an adequate and continuous supply of water to all areas of the concrete; improper application such as intermittent fogging at a high rate can result in water collecting on the deck (Figure 3) and affect the quality of the finished surface. Although the Highway 20 deck was placed under relatively favourable conditions, drying of the deck surface was evident if fog misting was interrupted for even a few minutes. On both projects, plastic shrinkage cracks were observed in areas subject to insufficient fog misting.

Table 1. Plastic Mix Properties

HPC Cement Content and Plastic Concrete Characteristics	MTO HPC Demonstration Project - Hwy 20	HPC Demonstration Project - Hwy 407
Cement Content, kg/m3 :		
A.Specified:		
Total Cementitious Materials	450	400
B: Supplied:		
Total Cementitious Materials:	450	430
-Portland Type10 (in Blended Type10SF)	414	297
-Silica Fume (in Blended Type 10 SF)	36	28
-Slag	0	105
Nominal Water / C+CM Ratio	0.32	0.32
Plastic Concrete Properties:		
A. Specified:		
Air Content,%:	6.0 ± 1.5	6.5±1.5
Slump (before superplasticizer),mm	50 ± 20	50±30
Slump (after superplasticizer), mm	180 ± 20 (upper limit increased to 220 mm)	180±30
B. Supplied		
Air Content,%: mean	5.4	6.2
standard deviation	0.6	0.9
coeff. of variation,%	11.0	15.0
Initial Slump (before superplasticizer), mm	45	*
mean	12	*
standard deviation	27.0	*
coeff. of variation,%		
Final Slump (after superplasticizer), mm	204	154
mean	12	36
standard deviation.	6.0	23.0
coeff. of variation,%		

* Initial slump not measured
Conversion factors: 1 mm = 0.039 in, 1kg/m³ = 1.686 lb/yd³

During hot weather, rapid loss of the slump in superplasticized concrete was experienced in some instances, which contributed to difficulties in placement and finishing. Even with a finishing crew experienced with conventional bridge deck concretes, it was not possible to achieve an acceptable finish on the decks by means of mechanical finishing only. Hand-finishing of bridge deck surfaces with bullfloats was necessary on both projects to produce an adequate surface finish; this is prohibited in normal MTO practice as the quality of the finished surface is extremely vulnerable to the level of workmanship of the particular crew. Following the Highway 20 project, it was proposed that in any future MTO HPC projects representatives of the manufacturer of the finishing machine be present prior to or during the placement of the trial slab to provide the contractor's staff with training and recommendations on operation of the finishing machine for optimum handling of the selected mix.

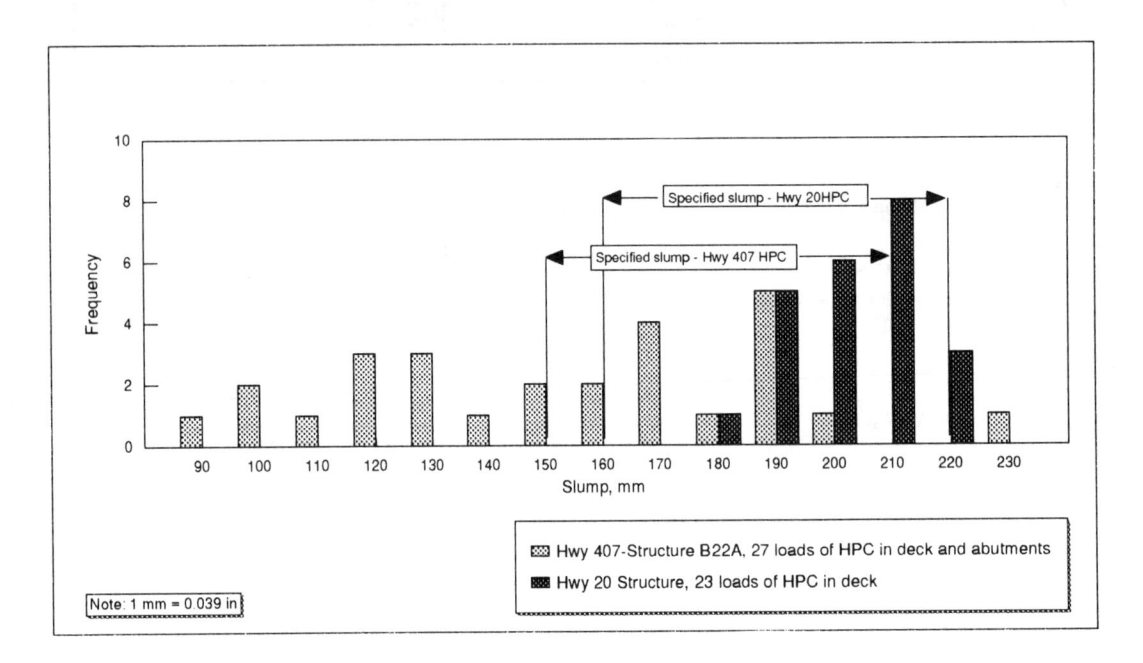

Figure 2. Distribution of the HPC slump, as delivered to the jobsite, for the two demonstration projects

In all MTO HPC applications, deck surfaces were exposed and were therefore textured to provide a skid-resistant finish. The quality of texturing has been found to be highly variable and directly dependent on the workability of the plastic concrete at the time of texturing. Rapid loss of slump during placement of a load of concrete can result in inconsistent surface texture, as illustrated in Figure 4. Generally surface textures achieved have been less than satisfactory.

Thermal Effects

Thermocouples were installed and imbedded in both the Highway 20 and Highway 407 deck slabs at depths of 25 mm (1 in) and 115 mm (4.5 in) below the surface, as well as on the slab surface to monitor the temperature of the concrete during the setting process. For both structures, the contractor was required to have insulation on site sufficient to cover the deck surface, to be put into place if temperature differentials within the slabs became excessive. The Highway 20 deck slab was cast in October, at low ambient air temperatures; high 14.4°C (58°F) to low 3.5°C (38.3°F) during the first 36 hours after placement). The Highway 407 deck slab was cast in August; ambient air temperature was 16°C (61°F) when the placement started at 4:00 a.m., rising to a high of 28°C (82.4°F) 10 hours later. The maximum temperature differential within the deck slab recorded in the HPC in both of the Highway 20 and Highway 407 structures was in the area of 12°C (21.6°F); the maximum temperature of concrete was recorded in Hwy 407 slab deck, at 52.5 °C (126.5°F). Rate of heat gain at the 115 mm (4.5 in) depth within the two deck slabs was 2.4°C/hour (4.3°F/hour) and 0.8°C/hour (1.4°F/hour) in the Highway 20 and 407 decks, respectively. Although the thermal effects in deck slabs are representative only for a thin section of the HPC structure, the temperatures recorded indicate that there is certainly potential for development and accumulation of high heat during the hydration process of HPC.

Figure 3. Water discharge rate must be controlled during fog misting to avoid ponding of water.

Guidelines specifying temperature conditions under which protective measures must be taken, based on measurement of both differential and maximum concrete temperatures, (which will be affected by factors within the contractor's control, such as combinations of materials, characteristics of the plastic mix, construction timing and procedures, as well as ambient weather conditions, etc.), will be developed for inclusion in future MTO work.

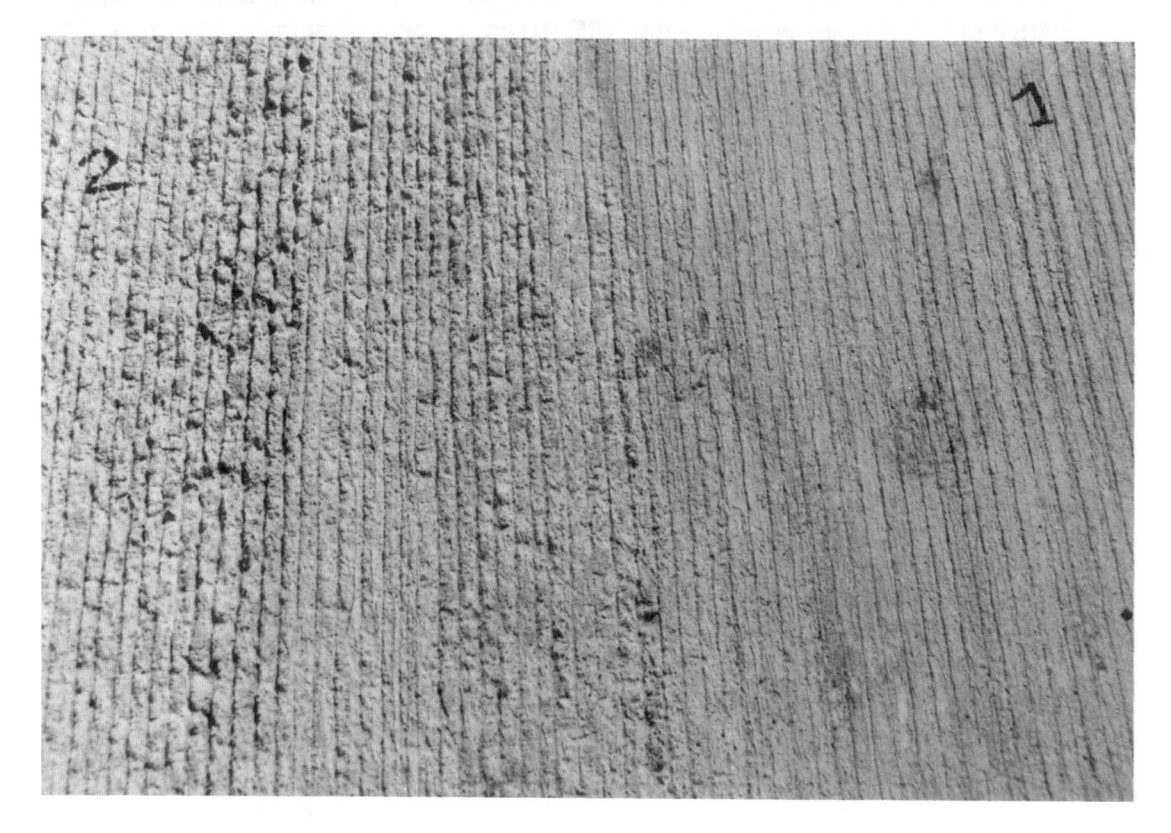

Figure 4. Rapid slump loss reduces concrete workability and impacts the quality of surface texture

PERFORMANCE-BASED EVALUATION OF HPC QUALITY

A number of means were used to evaluate the quality of the end product. Traditional, standardized laboratory test methods such as compressive strength[4], rapid chloride permeability[5], and analysis of the air void system of hardened concrete[6] were utilized to provide information about the overall quality and potential durability of the material. Table 2 summarizes results for the Highway 20 demonstration project.

Compressive strength results above are based on thirty-six 100mm (4 in) diameter cylinders. Six low breaks were experienced, which have been excluded from the average shown. All remaining cylinders significantly exceeded the contract requirement of 60 MPa (8700 psi), with the strength measurements exhibiting a standard deviation that is in the range of those measured on conventional concretes used on MTO contracts. The frequency of low breaks could not be explained by any obvious physical deficiency in the cylinders, and could only be attributed to greater sensitivity of the smaller cylinders to

Table 2. Hardened Concrete Properties

Hardened Concrete Characteristics	Hwy.20 HPC - Bridge deck and approach slabs	
28 day Compressive Strength (CSA/A.23.2-9C), MPa -mean -standard deviation -coefficient of variation (%)	80.1 4.2 5.2	
Air Void System (ASTM C 457): Air Content, % Specific Surface, mm^2/mm^3 Spacing Factor, mm	Before pumping: 4.0 40.0 0.138	After pumping: 4.5 23.8 0.212
Rapid Chloride Permeability (ASTM C 1202), Coulombs -mean -standard deviation -coefficient of variation (%)	758 136 18	

Conversion factors: 1 MPa = 145 psi, 1 mm = 0.039 inch

normal variation in preparation and handling or testing procedures. A number of additional 150 mm (6 in) cylinders were cast to compare the results; the coefficient of variation for 28-day compressive strengths was lower with the larger cylinders (4.4% versus 5.2%) although the mean strength was the same. In future HPC work, three cylinders rather than two will be averaged to produce a test result; the issue of variability is a particular concern of MTO in terms of ability to move quickly, without excess risk to the contractor or owner, to an percent-within-limits based end-result specification for strength of HPC. (On the Highway 407 projects, 28-day compressive strengths were in the area of 70 MPa (10,150 psi). As noted previously, the testing carried out was much less intensive than for the Highway 20 work so statistical data is not available.)

Air void data above is based on eight samples from bridge deck concrete, four taken from concrete at the ready-mix truck and four from concrete from the same truck after it had been pumped to its final location in the bridge deck. Samples were taken at intervals throughout the pour as the pump boom used to place the concrete was moved into an increasingly vertical position; it was found that the air void spacing factor increased as the vertical height through which the concrete moved was increased. Additional detail on these and other test results from the Highway 20 project can be found in [1]. Air void results from Highway 407, based on five samples from the two decks, averaged air content of 5.4%, with average specific surface of 23.6 mm^2/mm^3 (599 in^2/in^3) and spacing factor of 0.179 mm (0.007in).

Rapid chloride permeability measurements for both projects were indicative of high quality concrete which would be expected to be resistant to the ingress of chlorides. Results for the five samples from Highway 407 averaged 549 Coulombs.

To assess the quality of the concrete cover to steel and predict the long term performance, in-situ tests of water absorption and electrical resistivity of the concrete were performed (shown in Table 3). The rate of capillary water absorption (sorptivity) into the concrete surface was carried out on both projects using the modified vacuum attached initial water absorption apparatus (MVA-ISAT). Developed at the University of Toronto, the apparatus was further modified by MTO[7]; the test procedure is based on a British standard[8]. This new technique, illustrated in Figure 5, shows potential for use as a quality assurance tool.

Table 3. In-situ Tests of High-performance and Conventional Concrete

Concrete Characteristics	Hwy. 20 HPC	Hwy. 407 HPC	Hwy.407 Conventional (30 MPa) Concrete
Average MVA-ISAT Sorptivity of Barrier Wall Surface, 10^{-3} mm/min$^{1/2}$ (Concrete age 60-90 days, number of data point = 6)	22.7	19.2	49.6
Electrical Resistance, Imbedded 3E Probe (Carbon/Stainless Steel) Ohm x 10^3 (Concrete age 60-90 days, number of data points= 4)	40	39	5.1

Conversion factors: 1 mm = 0.039 in

Limited field monitoring of the HPC installations has been carried out to date. Over the next several years, the ministry will continue to carry out conventional condition surveys (including half-cell potential measurements, measurement of corrosion current using linear polarization, and visual examination), augmented by the use of water absorption, resistivity and air permeability measurements, to develop a base of HPC performance data.

CONCLUSIONS

Performance-based specifications for materials and construction of HPC structures are essential in order to ensure that the benefits of this promising new material are realized and that contractors and suppliers are able to be innovative in terms of utilizing available materials and techniques. The Hwy 20 and Hwy 407 HPC Demonstration Projects have shown that the Ontario road-building industry recognizes the advantages of HPC

Figure 5. Sorptivity testing was carried out using MVA-ISAT apparatus

application and is eager to learn how to work with the new material to ensure the best end results. It is the intention of the Ministry to move towards less prescriptive specifications, increasing reliance on the contractor's ability to produce the desired end-result. In 1997 and 1998 MTO is seeking to complete approximately 10 additional installations of HPC across the Province using updated specifications, establishing a sound base of laboratory and field test data that will support the development of performance-based specifications for the future.

REFERENCES

1. Schell, H.C., Ip, A.K.C., Berszakiewicz, B., " Application of High Performance Concrete for Highway Bridges - MTO Experience", 1996 Annual Conference of the Transportation Association of Canada, Charlottetown, P.E.I., October 1996

2. Ontario Provincial Standard Specifications OPSS 1350, Material Specifications for Concrete - Materials and Production, and OPSS 904, Construction Specification for Concrete Structures, Ministry of Transportation, Downsview, Ontario

3. CAN/CSA A362-93 Blended Hydraulic Cement, Canadian Standards Association, Rexdale, Ontario

4. CAN/CSA A 23.2-94 Methods For Test of Concrete - Compressive Strength of Cylindrical Concrete Specimens, Canadian Standards Association, Rexdale, Ontario

5. ASTM C 457 - Standard Test Method for Microscopical Determination of Parameters of the Air-Void System in Hardened Concrete, Annual Book of ASTM Standards, Vol. 04.02

6. ASTM C 1202 - Standard Test Method for Electrical Indication of Concrete's Ability to Resist Chloride Ion Penetration, Annual Book of ASTM Standards, Vol. 04.02

7. Hochbahn, F. X., "Modified Vacuum Attached Initial Surface Absorption Test (MVA-ISAT) - Apparatus Development", MTO R&D Work Term Report, January 1994

8. British Standard 1881: Part 5: Methods of Testing Hardened Concrete for Other than Strength, Test for Determining the Initial Surface Absorption of Concrete

HIGH PERFORMANCE SILICA FUME GROUT FOR POST-TENSIONING DUCTS

Michael Sprinkel
Research Manager
Virginia Transportation Research Council
Charlottesville, Virginia
USA

ABSTRACT

Grout containing 7% silica fume was pumped into the post-tensioning ducts in the pier caps of the Coleman Bridge. Prior to the field installation two grouting demonstrations were performed to ensure the grout could be mixed and pumped without problems. Properties of the fluid and hardened grout used in the demonstrations and the field installation were measured. The silica fume grout had a higher strength and lower permeability to chloride ions than a conventional grout without silica fume. Shrinkage and cracking were considerable for the grouts with and without silica fume thereby severely reducing the effectiveness of the grouts as protective systems for the tendons.

INTRODUCTION

The purpose of grout in a post-tensioned tendon duct is to protect the tendon from corrosion during the life of the structure and to provide a bond between the tendon and the structural concrete. Corrosion of the tendons in a post-tensioned structure could lead to a catastrophic failure. The grout is the last level of protection for a tendon. The quality of hydraulic cement concrete used in transportation structures has improved over the years with high performance concretes being used in more and more applications. Yet the grout used to fill the ducts of post-tensioned structures has stayed almost the same as in the first installations in 1966.[1]

The FHWA recently sponsored research with the objective of identifying grouts that would have improved properties and thereby provide better protection for post-tensioning tendons should chloride ions and water penetrate the surrounding structural concrete and the duct.[1,2] Conventional grouts are typically Type I and II Portland cement and water with a w/c of 0.44 to 0.53. Admixtures intended to reduce bleeding and cause expansion are sometimes specified. The study showed that adding 10% silica fume and a high range water reducing admixture and reducing the w/c to 0.37 can improve the protective properties of grout and provide a two fold increase in the time to corrosion relative to a standard grout.[1] Improvements in corrosion protection were also found for grouts containing fly ash, calcium nitrite and high range water reducing admixtures.[1] Other recent studies have also shown the benefits of adding silica fume to grout.[3-9]

OBJECTIVE

The objectives of this project were to demonstrate that a grout containing 7% silica fume could be used to fill the post-tensioning ducts in the pier caps of the Coleman Bridge, to evaluate the properties of the fluid and hardened grout, and to make recommendations with regard to improving the protective properties of grouts used to fill post-tensioning ducts.

METHODOLOGY

The project began with a review of the literature to identify potential grouts that might have improved protective properties. A grout containing 7% silica fume, a high range water reducing admixture and a maximum w/c of 0.40 was specified after consideration of the relative merits of reduced permeability to chloride ions and reduced working time at higher silica fume contents and lower w/c's. The project included fabricating and grouting a mockup segment (figure 1) of the circular pier caps of the bridge to ensure that the grout could be successfully mixed and pumped into place and to allow the properties of the fluid and hardened grout to be measured prior to moving to the field installation. A second grouting demonstration (figure 2) included the specified grout with silica fume and a conventional grout without silica fume to further ensure that the silica fume grout could fill the duct and to provide an additional opportunity to compare the properties of the grouts. The two circular and four rectangular post-tensioned pier caps of the Coleman Bridge were grouted in January and February of 1996 as part of a widening project (figure 3). The field grouting followed the two demonstrations conducted in October and December of 1995. During the two demonstrations and the field installations the fluidity of the grouts were measured after initial mixing and 30 minutes later, and specimens were fabricated. Specimens included cubes 51 mm (2 in), permeability specimens 102 mm (4 in) in diameter x 102 mm (4 in) high, and length change specimens 25 mm (1 in) x 25 mm (1 in) x 286 mm (11.25 in). Specimens were stored in a styrofoam cooler until removed from the molds at 24 to 48 hours of age and stored in the laboratory at the Research Council. The cube and permeability specimens were stored in a moist room until tested. Some of the length change specimens were stored in the moist room at 98% relative humidity until 28 days of age at which time they were moved to the laboratory and allowed to dry. Others were allowed to dry from the time they were removed from their molds. After 4 to 7 days of age, 305-mm long (12 in) sections of the segments that were grouted for the two demonstrations were saw cut and shipped to the Research Council for inspections to determine how well the grouts filled the ducts and surrounded the tendons.

Figure 1. Grouting of circular segment for demonstration 1

Figure 2. Grouting of segment for demonstration 2

Figure 3. One of the rectangular post-tensioned pier caps in the Coleman Bridge

RESULTS

Grout mixtures tested

The grouts were batched in a paddle type mixer. Approximately 5 minutes was required for proper mixing. The mixing procedure at the demonstrations consisted of adding water and a high range water reducer followed by cement and finally silica fume. Some silica fume tended to float on the surface so the procedure was changed such that the silica fume was added first for the field installations. The ingredients and proportions are shown in Table 1. The w/c for the mixtures with and without silica fume were 0.37 and 0.40, respectively. The batch sizes were 0.063 m^3 (2.25 ft^3), 0.160 m^3 (5.67 ft^3), 0.182 m^3 (6.50 ft^3), and 0.252 m^3 (9.0 ft^3), respectively, for Demo 1, Demo 2, and Field 1-6.

Table 1 - Mixture Proportions and Temperatures

Ingredient	Demo 1 10/30/95	Demo 2 12/15/95	Demo 2 12/15/95	Field 1 1/23/97	Field 2 1/26/97	Field 3 1/26/97	Field 4 2/8/96	Field 5 2/9/96	Field 6 2/14/96
Batch No.	1	2	3	4	5	6	7	8	9
Type I/II Blue Circle Portland Cement, kg (lb).	85 (188)	213 (470)	257 (565)	341 (752)	341 (752)	341 (752)	341 (752)	341 (752)	341 (752)
Elkem Micro Silica 965 (7%), kg (lb).	6 (13.2)	15 (33)	0 (0)	23 (50)	23 (50)	23 (50)	23 (50)	23 (50)	23 (50)
Sikament 10 HRWR (16 oz/bag), l (oz)	0.9 (32)	2.4 (80)	0 (0)	3.8 (128)	3.8 (128)	3.8 (128)	3.8 (128)	3.8 (128)	3.8 (128)
Water, l (gal).	34 (9)	85 (22.5)	102 (27)	136 (36)	136 (36)	136 (36)	136 (36)	136 (36)	136 (36)
W/C	0.37	0.37	0.40	0.37	0.37	0.37	0.37	0.37	0.37
Air Temperature, °C (°F)	18 (65)	16 (61)	19 (66)	-	-	-	7 (44)	11 (52)	-
Grout Temperature, °C (°F)	-	18 (64)	19 (66)	10 (50)	6 (42)	4 (40)	4 (39)	10 (50)	-

Fluidity

The flow cone efflux time (ASTM C939) was measured following the initial mixing of the grouts and again after a sample was maintained in an 18.9 L (5 gallon) pail for 30 minutes. The grout was stirred briefly with a wooden paddle before the measurements were made at 30 minutes. The specifications required an efflux time of 11 to 25 seconds. The efflux times are shown in Table 2. The grout used in ducts of the pier caps was less fluid than specified but according to the inspector it was of the correct fluidity to properly fill the ducts.

Table 2 - Efflux Time, ASTM C939, sec.

Installation	Batch	Time	Efflux Time, Sec.[a]
Demo 1	1	initial	18.2
Demo 1	1-30	after 30 minutes	23.7
Demo 2	2	initial	18
Demo 2	2-30	after 30 minutes	20
Demo 2 (no SF)	3	initial	18
Demo 2 (no SF)	3-30	after 30 minutes	30
Field 1	4	initial	25
Field 1	4-30	after 30 minutes	no flow
Field 2	5	initial	17
Field 2	5-30	after 30 minutes	no flow
Field 3	6	initial	26
Field 3	6-30	after 30 minutes	no flow
Field 4	7	initial	28
Field 4	7-30	after 30 minutes	no flow
Field 5	8	initial	23
Field 5	8-30	after 30 minutes	no flow
Field 6	9	initial	17
Field 6	9-30	after 30 minutes	no flow

a) specified time = 11 to 25 seconds

Compressive Strength

The results based on the average of 3 compressive strength tests (ASTM C109) on 51 mm (2 in.) moist cured cubes are shown in Table 3. The silica fume grout has a higher strength than the conventional grout and exceeds the minimum requirements of the specifications which were 20.7 MPa (3000 psi) at 7 days and 34.5 MPa (5000 psi) at 28 days. Strengths were less or did not increase at later ages for batches 1, 2, and 7, and the only explanation is segregation of the grout at the time the cubes were fabricated.

Table 3 - Compressive Strength MPa, (psi)

Installation	Batch	7 Day	28 Day	1 Year
Demo 1	1	37.5 (5440)	35.7 (5180)	-
Demo 1	1-30	31.6 (4585)	51.7 (7500)	-
Demo 2	2	46.8 (6795)	39.7 (5765)	-
Demo 2	2-30	46.2 (6700)	53.6 (7770)	-
Demo 2	3 (no SF)	33.2 (4820)	36.7 (5315)	-
Demo 2	3-30 (no SF)	28.4 (4125)	43.9 (6365)	-
Field 1	4	38.0 (5510)	56.0 (8120)	55.9 (8110)
Field 2	5	31.9 (4620)	50.4 (7310)	45.0 (6535)
Field 3	6	44.2 (6410)	54.9 (7960)	59.7 (8665)
Field 4	7	38.1 (5525)	36.7 (5320)	36.1 (5235)
Field 5	8	43.7 (6340)	42.9 (6220)	60.3 (8740)
Field 6	9	48.3 (7000)	47.5 (6895)	55.6 (8060)

Permeability to Chloride Ion

Although the reliability of the rapid permeability to chloride ion test (AASHTO T277) is questioned by some, the Virginia Transportation Research Council has found it to be a very reliable test for concrete and the Virginia Department of Transportation has successfully used high performance concretes specified to meet permeability test requirements. The results of tests conducted on slices 51 mm (2 in) thick from moist-cured specimens 102 mm (4 in) in diameter x 102 mm (4 in) high are shown in Table 4. The values are based on tests of one specimen for batches 1, 2, and 3 and of two specimens for batches 7, 8, and 9. Only the specimens fabricated at the first demonstration could be measured for permeability over the 6 hour test period. Other specimens were removed from the test immediately, or after ½, 1 ½ and 2 ½ hours because of over heating caused by the high permeability. Values were computed for the reduced test times for comparison. No measurements were possible for the specimens for which the values are shown as being high. In earlier work tests were done with a 30 volt power supply rather than the standard 60 volt so that values could be recorded over a 6 hour period.[1] The results show that grouts have a higher permeability than concretes because no aggregate is used in the grout, the silica fume grout is much less permeable than the conventional grout, and the permeability of the grout at 1 year is higher than at 90 days, and at 90 days higher than at 6 weeks. The permeability of the grouts differs from that of concretes in that the permeability of concretes typically decreases with age. Microcracking probably caused the higher values at 90 days and 1 year. The grout used in the first demonstration must have been of higher quality than specified. A test procedure with a 30 volt power supply is better suited to evaluate grouts.

Table 4 - Permeability to Chloride Ions, Coulombs

Batch	Time of Reading	6 Weeks	90 Days	1 Year
1	-	7186	4297	-
1-30	-	7226	4405	-
2	½ hour	644	775	>high
2-30	½ hour	611	690	>high
3	-	>high	>high	>high
3-30	-	>high	>high	>high
7	½ hour	-	306	431
	1 ½ hours	-	1165	1656
	2 ½ hours	-	2341	-
8	½ hour	-	301	370
	1 ½ hours	-	1142	1370
	2 ½ hours	-	2278	-
9	½ hour	-	324	428
	1 ½ hours	-	1272	1670
	2 ½ hours	-	2606	-

Shrinkage

The changes in length of specimens moist cured for the first 28 days, fabricated, and tested in accordance with ASTM C157 are shown in Figure 4. The values are based on measurements of two specimens with the exception that some data are based on one specimen. One of the two specimens representing 5 Moist and 6 Moist failed at 42 days. Like concrete, the grouts increase in length during the moist curing period. Although the early shrinkage is greater for the conventional grout used in Demo 2, the later age shrinkage is less than that of any specimens containing silica fume that were moist cured. The shrinkage of the moist cured silica fume specimens made at the bridge averaged 0.32% at 36 weeks.

The change in length of specimens air cured is shown in Figure 5. One of the specimens representing 8 Air failed at 16 weeks. One of the specimens representing 9 Air was faulty and unsuitable for use. At 36 weeks of age, shrinkage averaged 0.29% for the silica fume mixtures and 0.27% for the conventional grout. The shrinkage of the air-cured silica fume specimens from Demo 2 was 0.24% at 32 weeks as compared to 0.25% for the control specimens. The shrinkage of the silica fume specimens made at the bridge that were not moist cured was the greatest and averaged 0.47% at 32 weeks.

The change in length of the specimens made at Demo 2 and not moist cured was similar to those that were moist cured. The data indicated that shrinkage of grouts is typically 3 to 7 times greater than concrete mixtures with the same w/c and that mixtures with silica fume have the same or greater long-term shrinkage than mixtures without silica fume that have a higher w/c.

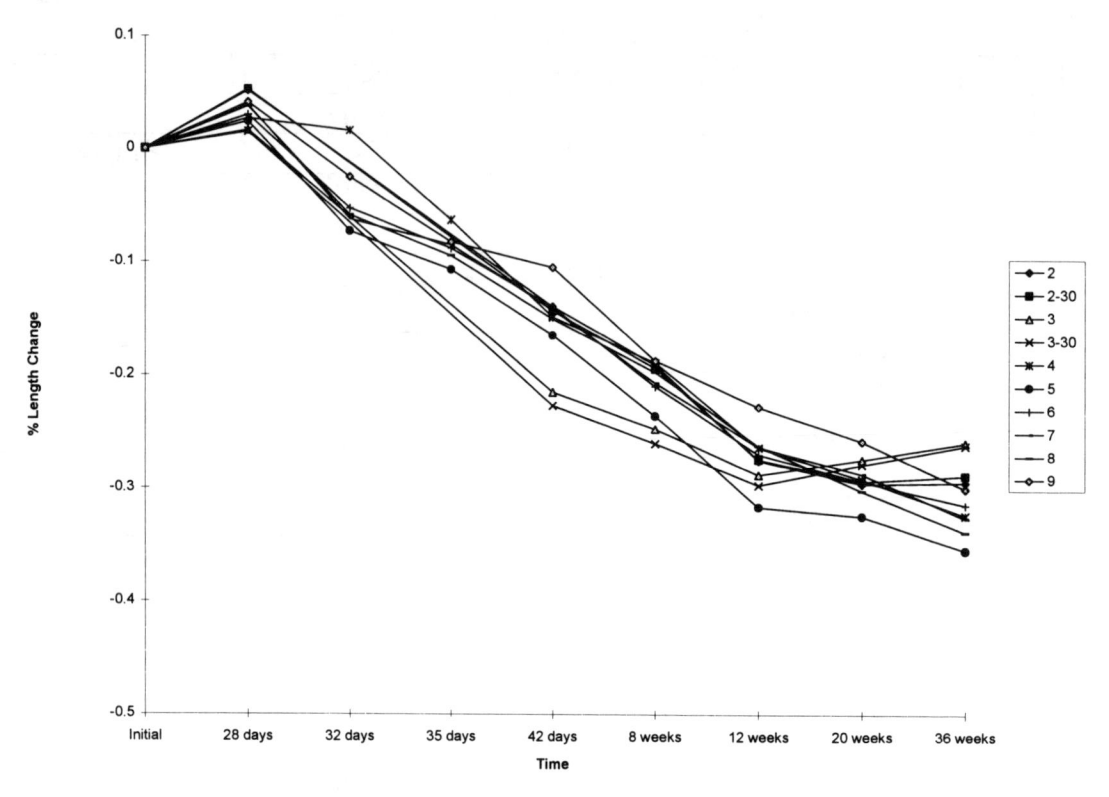

Figure 4. Length Change of Moist Cured Grouts

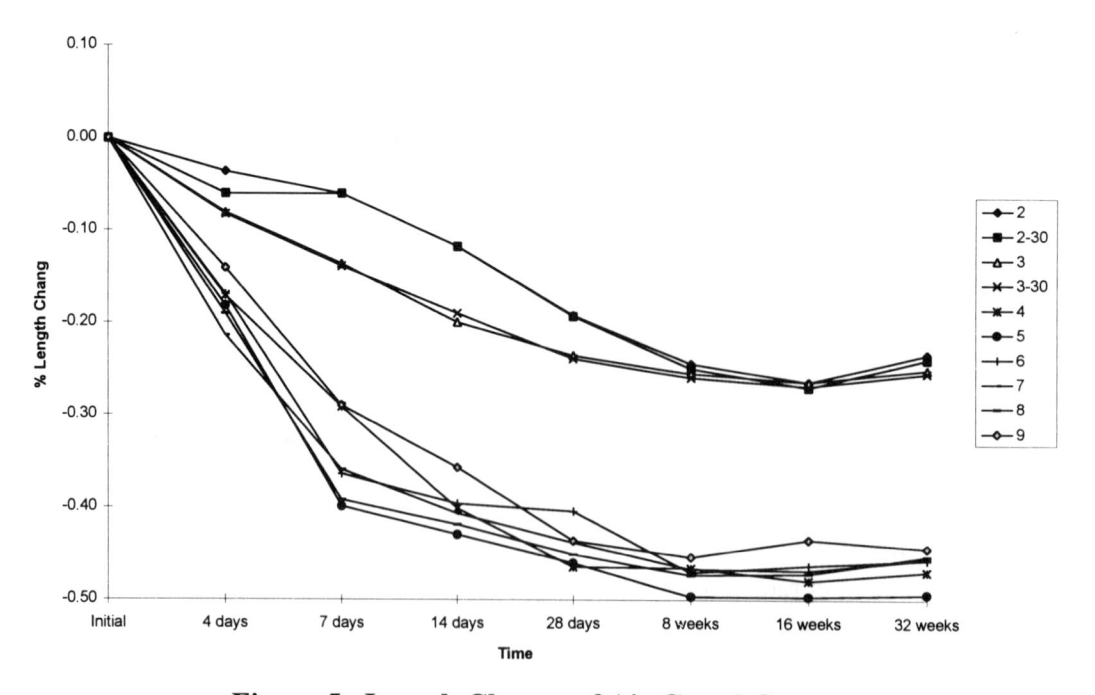

Figure 5. Length Change of Air Cured Grouts

Figure 6 shows cracks in typical specimens with and without silica fume prepared during Demo 2 and during the grouting of pier caps 1 (field 1, batch 4) and 5 (field 5, batch 8). All specimens showed shrinkage cracks with time, and the measurement of some specimens had to be discontinued because they broke into two or more pieces due to cracking during storage. Because of the high level of shrinkage and the resulting cracks, grouts offer little protection to post-tensioning tendons, and the mixtures containing silica fume provide little or no improvement.

Confinement of Strands and Duct Filling, Demonstrations 1 and 2
An inspection of sections cut at the ¼, ½, and ¾ points of the circular segment that was fabricated for demonstration 1 revealed that the duct was not full and the strands were not encapsulated as shown in figure 7. Although the contractor claimed to have followed normal grouting procedures the results were discouraging and a second demonstration was scheduled. The cracks in figure 7 are believed to have been partially caused by the release of the prestress force when the sections were saw cut.

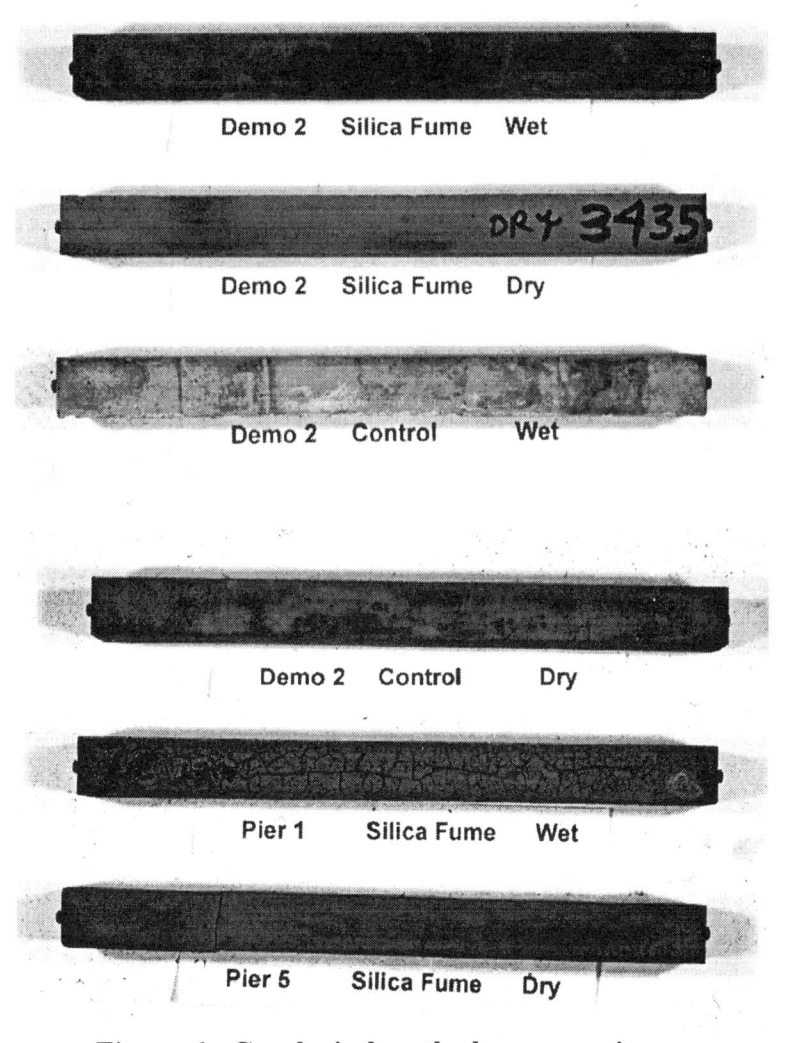

Figure 6. Cracks in length change specimens

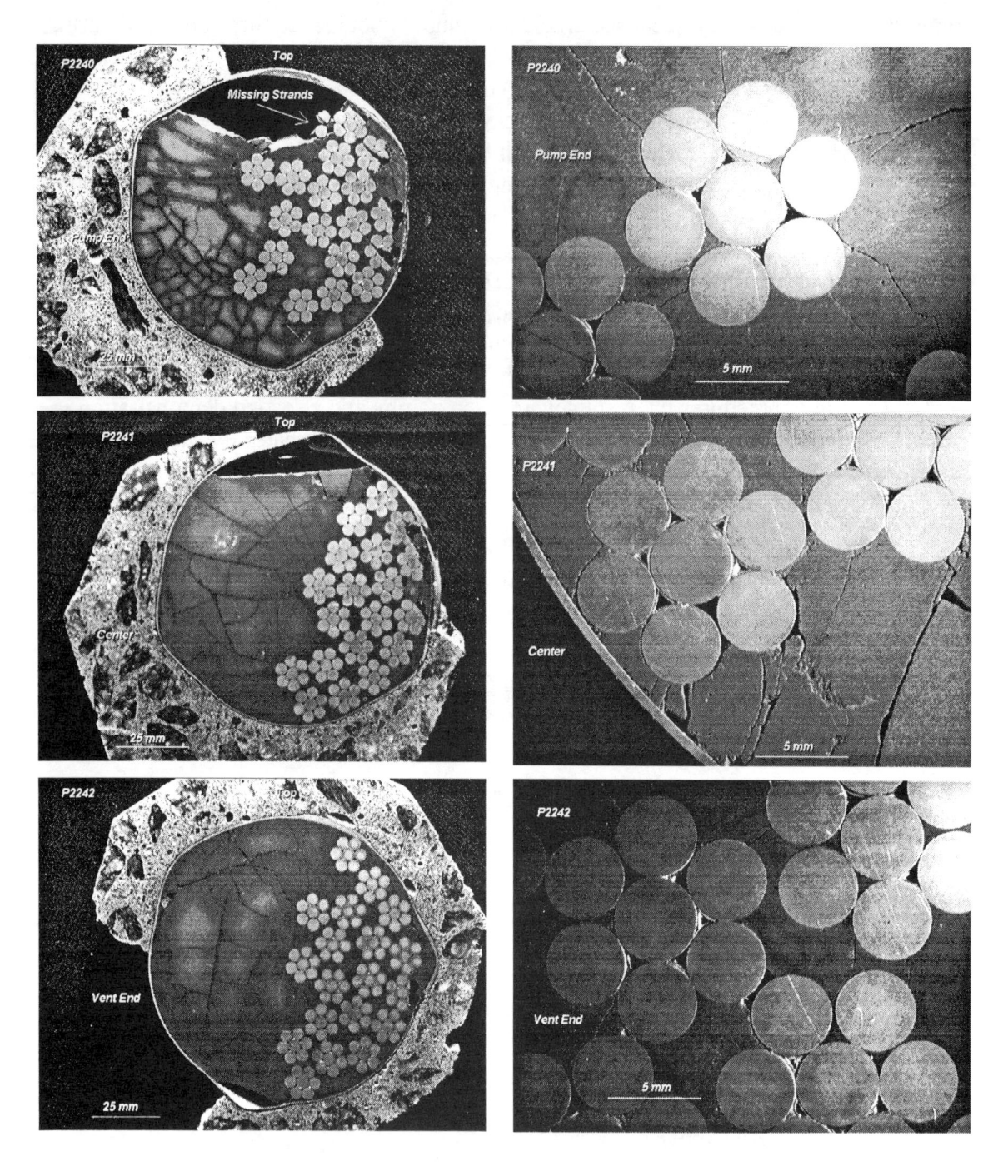

Figure 7. Sections cut from segments taken @ quarter points showing voids in the duct and unprotected strands

For demonstration 2, we placed two ducts in a straight unreinforced segment. Because the piers were being readied for grouting and time was of the essence the segment was not reinforced and no post tensioning strands were placed in the ducts. One duct was filled with grout containing silica fume and one with a conventional grout. To ensure that the duct was full the pump pressure was increased to 345 kPa (50 psi) and the duct ruptured allowing the silica fume grout to leak out. Once again the duct with silica fume grout was not full. The duct with the conventional grout was full when sections cut at 7 days of age were inspected. As a result of the problems with filling the ducts during the demonstrations, we required that an inspector experienced with grouting post-tensioning strands be present during the grouting of the ducts in the six pier caps of the Coleman Bridge.

CONCLUSIONS

1. The protection provided by grouts is compromised because of high shrinkage and the cracking that results. The addition of silica fume did not reduce shrinkage or cracking and may have made it worse.
2. Grouts have more than a very high permeability to chloride ions and the addition of silica fume provides some reduction.
3. The lower w/c of the silica fume grouts provided for higher compressive strengths.
4. Problems with filling the ducts during the two demonstrations suggest that qualified inspectors need to be present during grouting operations to increase the chances that protection of the strands is not compromised due to a failure to fill the ducts.
5. If grouts are expected to protect post-tensioning strands from chloride ion and water considerable work needs to be done to improve the grouts.
6. The silica fume grouts evaluated in this study did not improve the overall protection provided by grouts.

RECOMMENDATIONS

1. Additional research needs to be done to improve the quality of grouts.

REFERENCES

1. Thompson, N. G. D. Landark, M. Sprinkel, "Improved Grouts for Bonded Tendons in Post-Tensioned Bridge Structures," FHWA-RD-91-092, Federal Highway Administration, McLean, VA, October, 1991.
2. Lankard, D. R. et al, "Grouts for Bonded Post-Tensioned Concrete Construction: Protecting Prestressing Steel from Corrosion," <u>ACI Materials Journal</u>, Detroit, Vol. 90, No. 5, Sept. - Oct. 1993, pp. 406-414.
3. Hope, B. B., and A. K. C. Ip, "Improvements in Grouts and Grouting Techniques for Cable Ducts," Ontario Ministry of Transportation and Communications, Publication No. ME-87-04.
4. Ranisch, E. H., et al., "Properties of Cement Grouts with Silica Fume Addition for the Injection of Post-Tensioning Ducts," <u>Fly Ash, Silica Fume, Slag and Natural Pozzolans in Concrete</u>, Proceedings of the Third International Conference, Sp-114, Vol. 2, American Concrete Institute, Detroit, Michigan, 1989, pp. 1159-1171.
5. Diederichs, U. and K. Schutt, "Silica Fume- Modified Grouts for Corrosion Protection of Post-Tensioning Tendons," <u>Fly Ash, Silica Fume, Slag and Natural Pozzolans in Concrete</u>, Proceedings of the Third International Conference, Sp-114, Vol. 2, American Concrete Institute, Detroit, Michigan, 1989, pp. 1173-1195.
6. Gautefall, O., and J. Haudahi, "Effect of Condensed Silica Fume on the Mechanism of Chloride Diffusion in Hardened Cement Paste," <u>Fly Ash, Silica Fume, Slag and Natural Pozzolans in Concrete</u>, Proceedings of the Third International Conference, Sp-114, Vol. 2, American Concrete Institute, Detroit, Michigan, 1989, pp. 849-860.
7. Paillere, A. M. et al., "Use of Silica Fume and Super Plasticizers in Cement Grouts for Injection of Fine Cracks," <u>Fly Ash, Silica Fume, Slag and Natural Pozzolans in Concrete</u>, Proceedings of the Third International Conference, Sp-114, Vol. 2, American Concrete Institute, Detroit, Michigan, 1989, pp. 1131-1157.
8. Donnone, P. L. and S. B. Tank, " Use of Condensed Silica Fume in Portland Cement Grouts," <u>Fly Ash, Silica Fume, Slag and Natural Pozzolans in Concrete</u>, Proceedings of the Third International Conference, Sp-114, Vol. 2, American Concrete Institute, Detroit, Michigan, 1989, pp. 1231-1260.
9. Aitcin, P. C., et al., "Use of Condensed Silica Fume in Grouts," <u>Innovative Cement Grouting</u>, American Concrete Institute, Detroit, 1984, pp. 1-18.

HIGH-PERFORMANCE CONCRETE BRIDGES:
THE CANADIAN EXPERIENCE

Denis Mitchell, Professor of Civil Engineering,
McGill University, Montreal, Quebec, Canada

Pierre-Claude Aïtcin, Professor of Civil Engineering,
University of Sherbrooke, Sherbrooke, Quebec, Canada.

John A. Bickley, Implementation Manager,
Concrete Canada, Toronto, Ontario, Canada.

ABSTRACT

Since 1990 an expanding acceptance of High-Performance Concrete (HPC) for use in bridges has led to the construction of about 100 bridges in Canada. The program started in Quebec, and bridges have now been completed in Ontario and New Brunswick. To be completed in May 1997 is the 14 km (11 mile), precast concrete PEI Link between New Brunswick and Prince Edward Island. The first HPC bridge in Nova Scotia will be completed this year. A wide variety of designs has been involved, including precast and cast-in-place elements, with normal reinforcement and with prestressing. In some cases lower first costs have been achieved. Additionally HPC has been used in the rehabilitation of bridges in Quebec and British Columbia. On rehabilitation contracts quicker construction schedules have resulted in lower social and construction costs.

QUEBEC

A special program called "Project on New Directions for Concrete" involving participants from the Canadian Portland Cement Association, the Quebec Ministry of Transportation, cement producers, precasters, structural designers and researchers from the Centre of Excellence on High-Performance Concrete resulted in the construction of 3 HPC bridges in Quebec in 1992. These innovative projects, some of which are described below, demonstrated the economic feasibility of using HPC in the construction of bridges in Canada.

Pedestrian Bridge in Laval (1992)
Figure 1 shows the cross section of the 35 m (115 ft) span replacement pedestrian bridge made with high-performance concrete[1] in Laval.

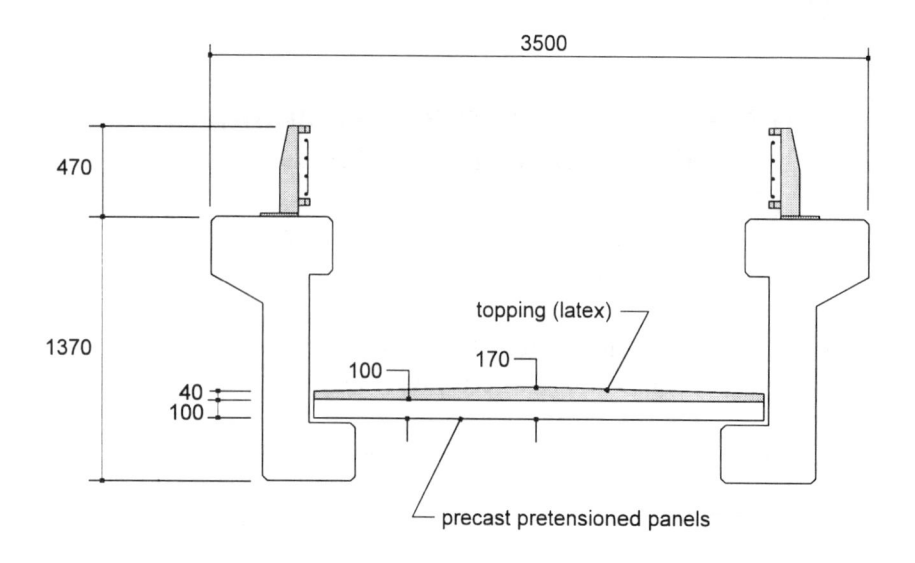

Figure 1. Cross Section of HPC pedestrian bridge in Laval. Dimensions in mm (1 in. = 25.4 mm).

The cross section consists of two Z-shaped precast girders with the shape providing a bottom ledge for supporting the precast panels for the deck slab. Each of the girders was pretensioned with 40 - 15 mm (0.6 in.) diameter pretensioned strands. The HPC precast girders were erected in one night limiting the disruption of traffic. The reinforcement in the bridge girders and precast pretensioned panels was all uncoated. The specified 28-day compressive strength of the concrete for the girders and the panels was 70 MPa (10,000 psi). The water/cement ratio was 0.30 and the specifications called for an air content of 5% ± 1%. The use of HPC for this bridge there had the following advantages:

i) There were smaller long-term prestress losses due to the larger modulus of elasticity, the smaller creep strains and the smaller long-term shrinkage strains,
ii) The HSC resulted in larger permissible stresses in the concrete,
iii) The smaller resulting section resulted in a practical precast concrete solution,
iv) HPC resulted in smaller deflections due to the larger modulus of elasticity, and
v) The improved durability of HPC will significantly extend the service life.

Portneuf Bridge near Quebec City (1992)

Figure 2 shows the cross section of the bridge which has a clear span of 24.8 m (81.4 ft). The superstructure consists of 5 precast, post-tensioned rectangular beams which were positioned on their supports and a temporary support was placed at midspan during construction. Each beam was post-tensioned with 4 tendons, each containing 6 - 15 mm (0.6 in.) diameter strands. After casting the 170 mm (6.7 in.) thick deck slab the bridge was post-tensioned longitudinally with external tendons located between the 5 main beams. The external tendons passed under a saddle formed with a transverse cast-in-place diaphragm at midspan. The precast beams were made

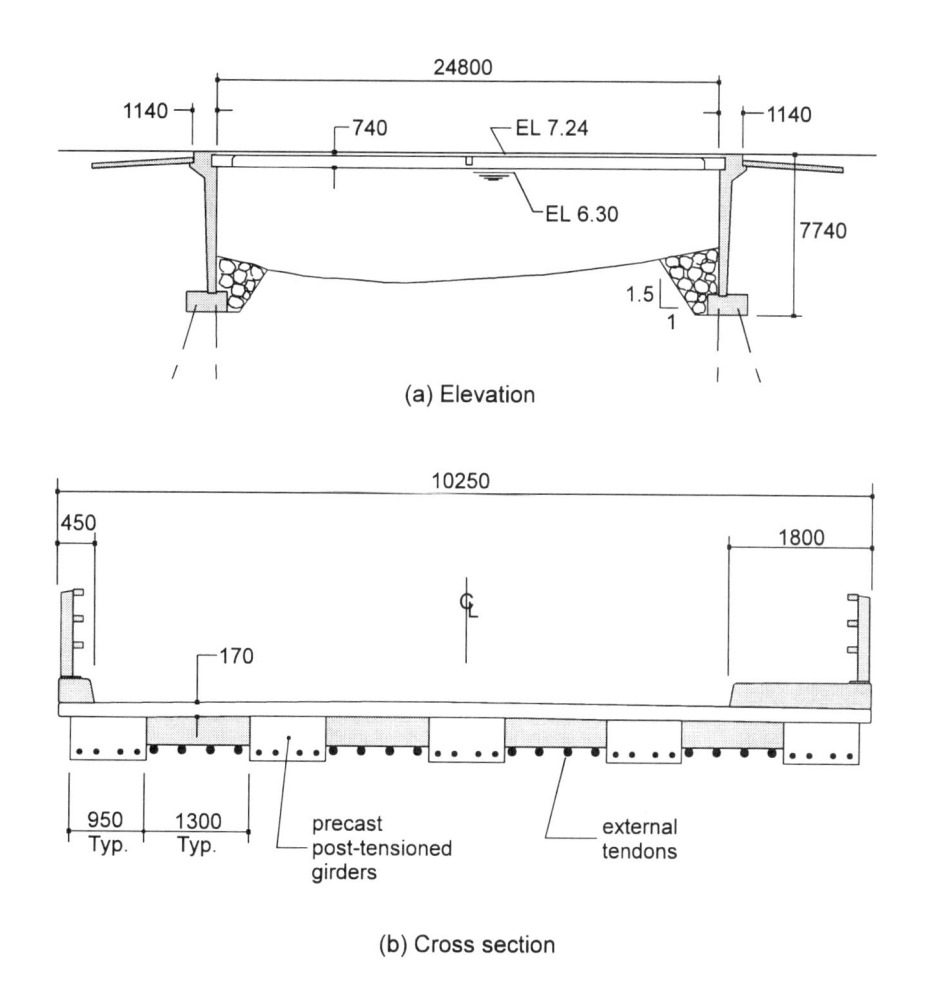

Figure 2. Portneuf bridge in Québec. Dimensions in mm (1 in. = 25.4 mm).

composite with the deck slab. The superstructure was transversely post-tensioned at both ends. The precast, post-tensioned beams, diaphragms and deck slab were all constructed with HPC having a minimum specified concrete compressive strength of 60 MPa (8700 psi). An air content of 6% resulted in spacing factors less than 200 µm^2. The blended cement contained 7-8% silica fume and a water cement ratio of 0.30 was used. Since the deck slab was cast in October, insulating blankets were used to protect the concrete from freezing during the evenings. The air content of the concrete delivered to the site varied from 5% to 7.5%. The average 28-day compressive strength was 75 MPa (10,900 psi) with a coefficient of variation of 4.3%. More details on the material characteristics is given by Aïtcin et al.[2] and other construction details are given by Coulombe and Ouellet[3].

The Saint-Rémi Bridge crossing Autoroute 50 in Mirabel

Figure 3 compares the normal-strength and high-strength concrete design alternatives for the Saint-Rémi bridge crossing the Autoroute 50 north of Montreal[3]. The continuous bridge, with 2 - 41 m (134.5 ft) spans, was built with cast-in-place post-tensioned construction. The normal-strength alternative was designed with 35 MPa (5100 psi) concrete and used post-tensioned tendons with parabolic tendon profiles, which were cast in the main girders. The high-strength (60 MPa, 8700 psi) concrete solution utilized external post-tensioned tendons, resulting in narrower main beams than the normal-strength concrete alternative. The combination of the high-strength concrete and the use of external tendons led to a 21% reduction in the volume of concrete in the superstructure. The Ministry of Transportation of Quebec estimated that the high-strength concrete solution with external tendons resulted in construction cost savings of about 5%[3].

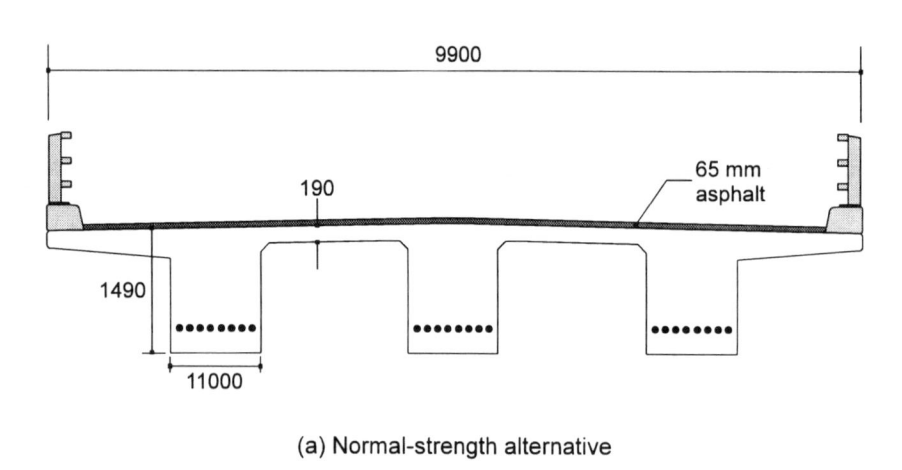

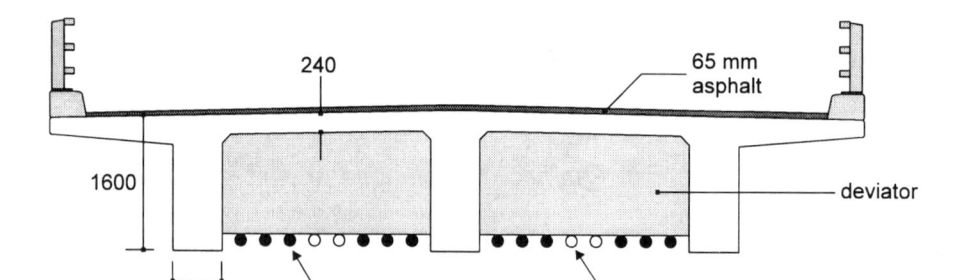

Figure 3. The normal-strength and HPC alternatives for the Saint-Rémi Bridge.

NEW BRUNSWICK[4]

By the end of 1996 the Province of New Brunswick had built about 85 HPC bridges. The adoption of HPC was to improve the durability and hence the service life and life-cycle costs of these bridges. The inevitably higher strengths achieved were not taken into account in the designs. The following is an abstract from the specification requirements.

Cementitious materials: Type 10 low alkali SF cement or
                                 Type 10 low alkali cement plus SF added as
                                 an admixture.
SF content: 7.5% to 10% of total cementitious
Maximum water to cementitious ratio: 0.40
Air content of fresh concrete: Cast-in-place   7.5±1.5%
                                 Pre-cast   6.0±1.0%
Specified strength: 30 MPa (4350 psi)
Maximum slump: Cast-in-place   N.A.
                                 Precast   180 mm
Admixtures to: CSA A23.1 - 94
Superplasticizer: Not more than 50% added at plant.
Minimum cement content: Cast-in-place   N.A.
                                 Precast   410 kg/m^3 (690 lb/yd^3)

Air void system of hardened concrete to meet CSA A23.1-94 requirements. Calcium nitrite in all precast concrete and in cast-in-place concrete above the elevation of the bridge seats. Fresh concrete to be fog cured immediately after finishing until wet burlap applied. Burlap to remain in place for seven days. It is understood that practice now is to use an evaporation reducer instead of fog curing. Recently the chloride permeability of the HPC bridges was compared to that of bridges made with concrete typical of normal highway specifications. Data collected from 25 HPC projects resulted in an average coulomb reading of 730, compared with an average reading of 4024 from four projects using normal strength concrete (Type 10 cement).

A problem reported for these bridges which is common to the use of typical high slump HPC mixes, is the loss of air during placing and the resulting degradation of the air void system. It was found that the loss of air content was 2% to 6% for pumping and 1% to 2% for placement with a bucket. Higher doses of air-entraining agent were needed due to the low water-cement ratios used in HPC mixes.

ONTARIO[4]

In Ontario three bridges have been completed to the end of 1996. The first was the Highway 20 bridge for the Ministry of Transportation in 1995. A specified strength of 60 MPa (8700 psi) was adopted as this is the highest strength allowed by the Ontario Bridge code unless special permission is sought.

Pre-bid Meeting
It was a condition of bidding that all those intending to bid had to attend a pre-bid meeting. This was held on May 23, 1995. At this meeting the prospective bidders were informed of the specific differences in this contract resulting from the use of HPC. These were presented under the following headings.
- Concrete materials and mix design
- Concrete handling

- Construction features
- Placement of trial slabs
- Curing requirements
- Sampling and testing
- Concerns

Special Provisions
a) Trial Mixes - A full scale trial of mixes proposed for consideration was required to be carried out at least 28 days in advance of concreting the deck.
b) Trial Slab - It was considered very important that the ability to place and finish HPC be demonstrated by the contractor.

Control of Thermal Effects
The bridge incorporated uncoated normal reinforcing bars and is not waterproofed or paved. Consequently the cover concrete is the sole corrosion protection for the reinforcement. It is vital that after the deck has been cast and cured that it should not contain cracks wide enough to allow corrosion of the reinforcement to occur at crack locations. The maximum temperature of the concrete on discharge was specified as 25°C. A maximum of 20°C would have been preferable. With the anticipated summertime casting of the deck, the lower delivery temperature was not considered to be practical. To meet concerns the thermal gradients or thermal shock could cause cracking that could adversely affect the durability of the slab CIMS technology was used to characterise the thermal properties of the trial mixes. This data together with the trial slab thermal data was used to provide guidance criteria to be used after the deck had been placed. The contract specified that insulation should be available, and the CIMS data was available to determine if it was needed, taking into account the ambient conditions at the time the slab was cast and from then until the differential between the slab and ambient temperatures was less than 20°C.

Embedded Instrumentation[5]
Three types of probes were installed and embedded in the deck to monitor the corrosion activity of the reinforcing steel and the corrosion resistance. The probe types are:
- graphite electrode, 40x40x150 mm (1.6 x 1.6 x 6 in.)
- silver/silver chloride electrode, ⌀ 20mm (0.8 in.)
- three element electrode (carbon steel / stainless steel / manganese dioxide

Construction
Finishing problems were encountered in casting the deck, and hand finishing with bull floats had to be used. On future contracts a pre-construction meeting should be held at which the correct finishing procedures are reviewed, and the contractor confirms that these will be used. The casting of the trial slab proved to be a valuable exercise for all concerned, and the lessons learned contributed to the successful completion of the deck.

Test Data

The quality control exercised by the supplier was effective, meeting the ACI 214 limits. Test results at age 28 days are given in Table 1.

Table 1 - Strength and Air Void Test Data

	All Test Results	**Deck Test Results Only**
Strength		
Number of Results	23	12
Mean Strength	77.5 MPa (11,240 psi)	80.1 MPa (11,620 psi)
Standard Deviation	4.77 MPa (690 psi)	4.2 MPa (610 psi)
Coefficient of Variation	6.15 %	5.7 %
Rapid Chloride Permeability		
Number of Results	28	12
Mean Strength	823 coulombs	758 coulombs
Standard Deviation	191 coulombs	135 coulombs
Coefficient of Variation	23.2 %	17.8 %
Air Voids, Before Pumping		
Air Content	4.3 %	4.0 %
Specific Surface	37.4 mm^{-1} (960 in.$^{-1}$)	40.0 mm^{-1} (1,025 in.$^{-1}$)
Spacing Factor	0.143 mm (0.006 in.)	0.138 mm (0.005 in.)
Air Voids, After Pumping		
Air Content	4.0 %	4.5 %
Specific Surface	26.0 mm^{-1} (666 in.$^{-1}$)	23.8 mm^{-1} (610 in.$^{-1}$)
Spacing Factor	0.206 mm (0.008 in.)	0.212 mm (0.008 in.)

Conclusions and Recommendations

There is concern that hand finishing had to be used on the deck. On future contracts the correct operation of finishing equipment should be determined prior to concreting the deck. A trial slab is a useful tool on contracts where there is no prior experience of HPC and until the industry becomes used to placing and finishing slabs made with this type of concrete. At the age of 28 days bridge deck concrete was surveyed to detect any surface defects, measure concrete cover and half cell potential of the embedded steel. At that time no visible signs of surface concrete distress were observed. The steel was found to be passive. No cracks were seen in the deck. Cover to top reinforcing bars was found to average 72.6 mm (2.9 in) with a standard deviation of 7.2 mm (0.3 in). The RCP and, air void system, test data and initial observations on cover and lack of cracking suggest that the prognosis for durability is good. Annual monitoring will confirm the actual performance of the deck.

HIGHWAY 407

Highway 407 is a 68 km (42 mile) toll highway north of Toronto. It comprises 600 lane kilometers (375 lane miles) of exposed concrete pavement and 128 bridges. An innovation committee determined that there could be first cost as well as life-cycle cost savings in building bridges in HPC. Two were built in 1996 as a demonstration project and others will be built in 1997.

Construction
Both bridges were integral abutment designs with cast-in-place decks and barrier walls and pre-tensioned pre-cast girders. Both bridges had an HPC deck and barrier walls. The second bridge had girders upgraded to 60 MPa (8700 psi).

Economics
In advance of construction it was determined that the elimination of waterproofing and paving and the use of uncoated steel and the increased cost of HPC would result in a net saving of $15/m^2 ($1.50/sq.ft.) of deck. The upgrading of the girders on the second bridge permitted the reduction from four girders to three, a saving of $30,000.

Conclusion
The test data show, which will form the subject of a later report, that the quality of the two bridges is satisfactory and it is expected that durability will be excellent. Once the initial experience is gained it appears that significant savings in first cost can be achieved, particularly by upgrading girders and reducing their number.

NOVA SCOTIA

The East River Bridge in Nova Scotia
The bridge has two 34.7 m (114 ft) spans which are to be made continuous for live load. Figures 4a and 4b show a comparison of the design alternatives using normal (35 MPa, 5100 psi) and high-strength (65 MPa, 9400 psi) concrete. The normal-strength alternative required 4 - 1900 mm (75 in.) deep bulb tee girders, each containing 52 - 13 mm (0.5 in.) diameter strands. The high-strength concrete solution utilized 15 mm diameter strands and the girder and deck slab concrete had 28-day compressive strengths of 65 and 60 MPa, respectively. The high-strength concrete solution required only 3 girders, each containing 42 - 15 mm diameter strands. Cost comparisons of the normal and high-strength alternatives indicated cost savings of about 4% for the high-strength solution. The elimination of both the epoxy coating on the deck slab reinforcement and the water proofing membrane for the high-performance concrete solution resulted in a total cost reduction of 8% below the normal-strength alternative.

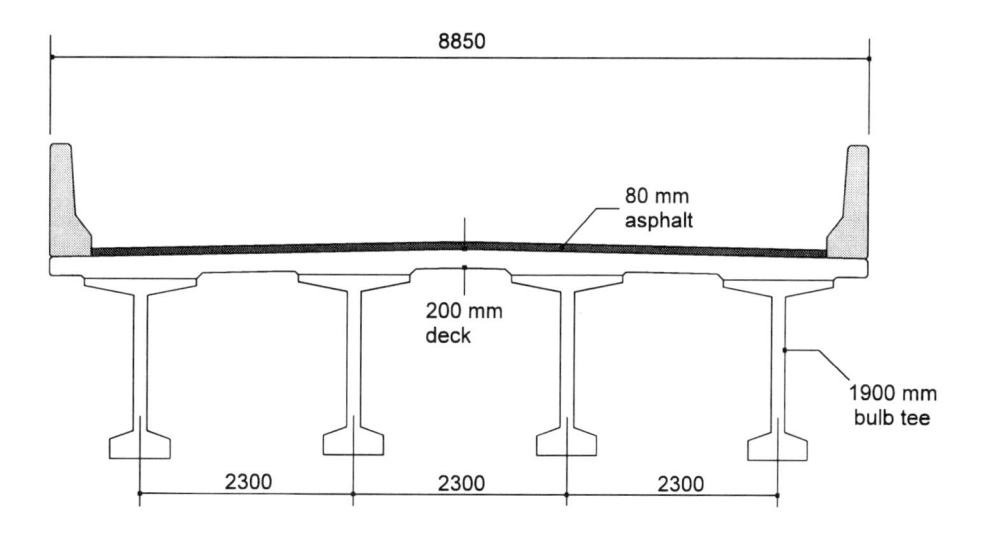

Figure 4. Normal and high-strength concrete designs for the East River Bridge.

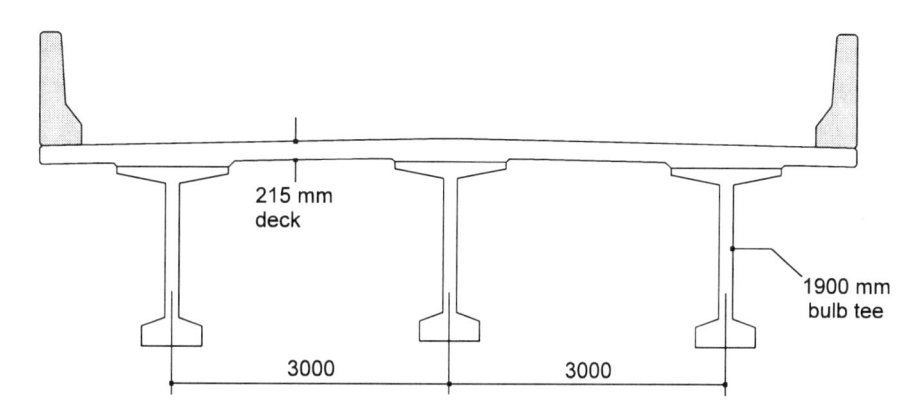

FRASER RIVER BRIDGE, HOPE, B.C.[5]

The Ministry of Highways, B.C. undertook to replace the deck on this 70 year old heritage steel structure. The original design was to use a membrane overlay system. Consideration of bridge closure, construction speed and cost led to the alternate design with HPC as a stand-alone riding surface. Epoxy top rebar was retained although it might not be critical in this system.

Monitoring Installation
Specially designed embedded probe assemblies, providing electrodes at three depths of cover, were installed in the HPC repair to monitor future corrosion activity. The location of the probes was at areas near expansion joints and on the downslope of the drainage system such that it should represent the more corrosive exposures. The depth of probes was varied at three elevations in order to obtain early indications at the onset of corrosion and subsequent direct indications of the corrosion state at the

cover of the rebar itself. Included in the probes are epoxy bars completely coated and some containing large holidays created by the use of a hacksaw blade.

Problems experienced
Maintaining the plastic life of the concrete (i.e. placement window) was a challenge - rapid slump loss from traffic delays resulted in numerous pump blockages; site retempering with superplasticizer is difficult to achieve since the effectiveness of a particular dosage is markedly reduced and it is difficult to gauge the balance between further delays and the need for increased workability. Some problems with stability of the air contents were experienced; in part, this related to the placement window problem above.

Positive features
When the HPC is properly prepared at the mixing plant, it is an extremely worker-friendly concrete. Others have experienced significant problems in finishing of surfaces of bridge decks in the form of tearing and open texture; in this situation, none of the above was experienced and a simple finishing procedure immediately behind the concrete placement resulted in an excellent riding surface.

Summary
As a result of Concrete Canada research, it has been possible to install advanced technology corrosion monitoring probes in this and a number of infrastructure projects including both new construction and rehabilitation. The installations are relatively straightforward and do not impact on the construction system. A typical cost of an installation is in the order of $ 5,000.00 This cost includes the initial (baseline) readings. As the process becomes refined, it is expected that this can be reduced. It is far too early to anticipate the effectiveness of the corrosion rate readings in estimating the times required for maintenance of the structure. Such information will take at least a decade to develop. However, initial readings suggest that the monitoring systems are responsive and durable and Concrete Canada is, therefore, optimistic that such installation can become a part of many infrastructures and thus achieve the Owner's objectives.

REHABILITATION OF THE JACQUES-CARTIER BRIDGE[6]

After forty-one years of service, the deck of Sherbrooke's Jacques-Cartier Bridge needed to be replaced. The bridge deck is 378 m (1150 ft.) long and resting on a structural steel frame.

Approximately 1,800 m^3 (2350 yd^3) of air-entrained, high-performance concrete having a specified compressive strength of 60 MPa (8700 psi), was used in the reconstruction which started in the August 5, 1995 and was completed at the end of November 26, 1995. During August and September, the concrete placement was done almost exclusively at night and during the weekends using crushed ice in order to ensure that the temperature of the concrete was approximately 18°C (64°F) upon delivery. During the late October and November, the concrete was delivered during the day, and the water was heated, whenever necessary, in order to ensure that the temperature of the concrete at the time of delivery was 22°C (72°F). Because a highly efficient quality control program was initiated, the concrete producer was able to complete more than 300 concrete deliveries with only three of them being rejected. Because of its low chloride-ion permeability and near absence of cracks, the new bridge is expected to have a very long service life.

Socio-economic considerations
HPC has a high early strength so that the needed 24 MPa (3480 psi) to resume work is reached after 24 or 36 hours. This early strength development accelerated the construction process and enabled the completion of the bridge deck replacement in only four months. If the bridge deck had been repaired using ordinary concrete, it would have been necessary to wait 7 days after each concrete placement before resuming the work. Since the 22,000 people who drive across the bridge every day had to make an average 3 kilometer (1.9 mile) detour, this represents a socio-economic cost of $22,000 per day, if the cost of each kilometer is calculated at $0.33 ($0.53 per mile). Hence, the use of HPC resulted in socio-economic gains of $150,000 during each concrete placement.

Mixing Sequence
Problems in generating an adequate network of air bubbles and, more importantly, air bubble spacing factor conforming to the specifications of the Transport Ministry of Quebec led to a reconsideration of the conventional mixing sequence of a concrete at the batch plant. A mixing sequence was formulated and adapted to the conditions of the concrete producer's plant. The air-void system was fairly stable during the transportation of the concrete to the site and placement. The mixing sequence was as follows:

- 1s: Introduction of the coarse aggregate, water, and ice
- 5s: Introduction of the sand
- 20s: Development of the air bubble network
- 80s: Introduction of the cement
- 140s: Initial plasticification of the concrete (introduction of the water reducer)
- 200s: Final plasticification of concrete to 180 ±40 mm (introduction of superplasticizer)
- 310s: End of mixing

Measured Characteristics -Three concrete pours were sampled by the team of the Concrete Canada-Sherbrooke during the replacement of the bridge deck. In total, 56 trucks were sampled, which is almost 20% of the total production. Two of the samplings took place in August, while the third took place in October. The average characteristics of the concrete taken from the sampled concrete are presented in the following Table.

Table 2 - Characteristics of the high-performance concrete used for the Jacques-Cartier Bridge repair

	Mean Value	**Standard Deviation**
Slump	190 mm	25 mm
Air content	5.4%	0.7%
Unit weight	2350 kg/m^3	20 kg/m^3
Temperature	18°C (64°F)	1.5°C (2.7°F)
Compressive strength at 28 days	76.3 MPa (11070 psi)	4.1 MPa (590 psi)

Conclusion

The reconstruction of Sherbrooke's Jacques-Cartier Bridge necessitated the placement of 1,800 m^3 (2350 yd^3) of HPC having a specified compressive strength of 60 MPa (8700 psi). A strict, but realistic specification enabled a closely-knit team to produce a high-quality structure to the benefit of all of the contributors, including the local community.

ACKNOWLEDGEMENTS

The authors are grateful to all of the participants in all of the projects reported in this paper. Special thanks are extended to the Ministries of Transportation of Québec, Ontario, British Columbia, New Brunswick and Nova Scotia and to the Canadian Portland Cement Association. This paper was compiled from reports and papers produced by Principal Investigators of Concrete Canada. Concrete Canada is a Network of centres of excellence on High Performance Concrete financed by the Canadian Federal Government.

REFERENCES

1. Mitchell, D., Pigeon, M., Zaki, A.R. and Coulombe, L.-G., "Experimental Use of High-Performance Concrete in Bridges in Quebec", Proceedings, 1993 CPCA/CSCE Structural Concrete Conference, Toronto, May 1993, pp.63-75.

2. Aïtcin, P-C., Ballivy, G., Mitchell, D., Pigeon, M. and Coulombe, L-G., "The Use of Air-Entrained HPC for the Construction of the Portneuf Bridge", ACI SP-140, Nov.1993, pp. 53-72.

3. Coulombe, L.-G. and Ouellet, C., "Construction of Two Experimental Bridges Using High-Performance Air-Entrained Concrete", Proceedings, Transportation Research Board, paper 95-1060, Jan., 1995, 26p.

4. Bickley, J.A., High Performance Concrete In Transportation, Technology Transfer Day on High Performance Concrete, Moncton, August 1996, Proceedings, pp. 37-60.

5. Seabrook, P.T. and Hansson, C.M., Applications of In-Situ Monitoring in HPC Structures, Technology Transfer Day on High Performance Concrete, Moncton, August 1996, Proceedings, pp. 97-120.

6. Blais, F.A., Dallaire,E., Lessard, M. and Aïtcin, P-C., The Reconstruction Of The Bridge Deck Of The Jacques-Cartier Bridge In Sherbrooke (Quebec) Using High Performance Concrete, Proceedings C.S.C.E. 1st Structural Specialty Conference, Edmonton, June 1996, pp. 501-507.

QUALITY CONTROL & QUALITY ASSURANCE PROGRAM FOR PRECAST PLANT PRODUCED HIGH PERFORMANCE CONCRETE U-BEAMS

John J. Myers, P.E., Graduate Research Assistant / Ph.D. Candidate
Ramon L. Carrasquillo, Ph.D., P.E., Professor of Civil Engineering
Department of Civil Engineering
The University of Texas at Austin
Austin, Texas, U.S.A.

ABSTRACT

High performance concrete (HPC) with its improved service under load and improved resistance to environmental conditions, represents a promising material to assist with the rehabilitation of the crumbling infrastructures. Although HPC has found widespread application in buildings in certain pockets of the country, its incorporation into transportation structures has been very recent. To demonstrate the suitability of HPC for use in highway structures, the Federal Highway Administration (FHWA) initiated a series of projects that include the complete incorporation of HPC from design to long-term monitoring of the bridges in service. The design and construction of Louetta Road Overpass in Houston, Texas, one of these projects, was conducted as a joint effort by the University of Texas at Austin and the Texas Department of Transportation (TxDOT). The Louetta Road Overpass project included the use of pretensioned precast U-Beams. These beams required very high initial prestressing forces to take full advantage of the high performance concrete in addition to higher than usual elastic and flexural strength properties of the concrete. This was accomplished by using larger 15.2 mm (0.6 in.) prestressing strands upon FHWA approval. The high initial prestressing forces required high early release strengths of 63.4 MPa (9,200 psi) and 56 day design strengths of 91.0 MPa (13,200 psi). The designers (TxDOT) also required a high initial modulus of elasticity at release and long-term of 41.3 kPa (6,000 ksi) to meet serviceability requirements for the beams. The mix design was optimized in the laboratory using locally available materials to produce the most economical mix design. The following paper discusses the evolution and optimization of the mix design and it's subsequent use in the field. The quality control and quality assurance program will also be discussed with the statistical evaluation of the mechanical properties of the HPC used in actual construction. Among others, statistical data on compressive strength, modulus of elasticity, flexural strength, splitting tensile strength, and rapid ion permeability is presented.

BACKGROUND

Description of The Louetta Road Overpass Bridge Project

The Louetta Road Overpass Bridge Project in Houston, Texas, shown in Figure 1, was the first High Performance Concrete project in the United States to fully take advantage of HPC from design of the substructure to the superstructure. The project involved the incorporation of a non-standard U-shaped precast beam member in lieu of traditional AASHTO I-shaped precast members. The U-shaped member, commonly referred to as the U-Beam by TxDOT personnel,

Figure 1: Erected Precast / Prestressed U-Beams at Louetta Road Overpass Bridge Site in Houston, Texas

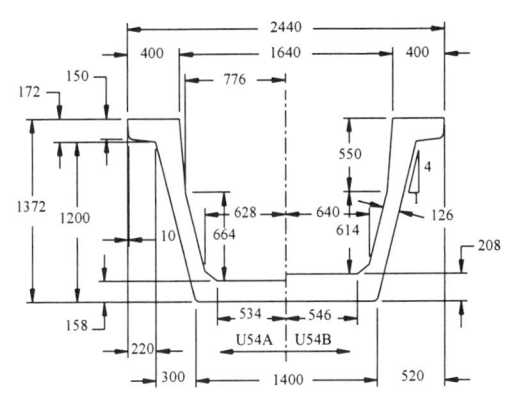

Figure 2: Cross Sectional Dimensions of Precast U-Beam Section (in mm)

was conceived in the late 1980's by TxDOT.

While aesthetics was one of the driving forces behind the conception of the U-beam, it was not the only design consideration. Economy, durability, function, and safety were also very important. Advantages of the U-shaped member over more traditional I-shaped members were recognized early on. These included reduced production requirements at the precast plant, fewer members to erect and transport to the site, and members that provide greater stability during erection and shipping. Figure 2 illustrates the cross sectional dimensions of both the U54A and U54B. Two bottom flange widths were developed for various span conditions. The U54A (158 mm flange thickness) allows for 2 vertical layers of prestressing strands in the bottom flange, while the U54B (208 mm flange thickness) allows for 3 vertical layers of prestressing strands. The HPC U-Beams utilized the U54B section detailed above.

Definition of High Performance Concrete (HPC)

To date, there is no unique definition of HPC. In broad terms, it can be defined as concrete that meets special requirements that cannot be obtained using conventional concrete. Special requirements could be an enhanced property over others such as strength, elastic modulus, durability, permeability, economics, special construction practices, or a combination of them. For the purpose of this project, HPC is defined as concrete with the following criteria:

1. a minimum compressive strength of 90.3 MPa (13,100 psi) at 56 days.
2. a minimum compressive strength of 60.6 MPa (8,800 psi) at one day.
3. a minimum elastic modulus of 41,300 MPa (6,000 ksi) at 1 day.
4. a flexural strength of 830 Sqrt (f_c'), in kPa (10 Sqrt (f_c'), in psi).
5. flowing concrete, a minimum slump requirement of approximately 229 mm (9 in.) at the time of casting.

MIX DESIGN DEVELOPMENT AND OPTIMIZATION

Optimization of the Mix Design

In an effort to develop a HPC mix which was economical, the use of locally available materials was emphasized. Only if the required material properties could not be attained using readily available materials in Texas would outside materials sources be considered. A variety of materials and sources were initially selected to investigate the interaction of the various constituents. A laboratory trial phase was initiated at the Construction Materials Research Group at the University of Texas in Austin to develop and optimize the materials selected for the most promising mix designs [3]. The following sections detail the various materials which were selected for this initial trial phase.

Coarse Aggregate - Five types of coarse aggregate were used in the overall study, namely crushed gravel (CG, Nominal Maximum Size of Aggregate = 19 mm), trap rock (TR, NMSA = 19 mm), dolomitic limestone (LS and LL, NMSA = 13 and 19 mm), and calcitic limestone (CL, NMSA = 19 mm).

Sand - Local concrete sand from the Colorado River was used in the overall study. It had a specific gravity of 2.63 and a fineness modulus of 2.45.

Portland Cement - Commercially available Type I cements designated as BI/II and CI, and Type III cements designated as AIII, CIII and LIII were used. They conformed to ASTM C150-94, "Standard Specification for Portland Cement".

Fly Ash - Class C fly ash from three different sources (J, D and L) and Class F fly ash (R) were used in varying amounts. They conformed to ASTM C618-94a, "Standard Specification for Fly Ash and Raw and Calcined Natural Pozzolan for Use as a Mineral Admixture in Portland Cement Concrete".

Chemical Admixtures - Two combinations of chemical admixtures, each containing a retarder and a superplasticizer, were used. One combination contained a naphthalene based superplasticizer conforming to ASTM C494-92 for Type A/F, and a retarder conforming to ASTM C494-92 for Type B. The other combination contained a naphthalene based superplasticizer conforming to ASTM C494-92 for Type A, a retarder conforming to ASTM C494-92 for Type B.

LABORATORY MIX DESIGNS

In the summer of 1993, more than 75 trial batches were made using locally available materials at the laboratory of the Construction Materials Research Group at the University of Texas in Austin. As an initial starting point, guidelines provided by the ACI Committee 211 were followed to come up with an initial set of mix proportions for high performance concrete. The mix designs were later adjusted through laboratory trial mixes designed to optimize the performance of the constituent materials used. Cylinder specimens were cast in 102 mm x 203 mm (4 in. x 8 in.) plastic cylinder molds and moist-cured in accordance with ASTM C192-90a "Standard Practice for Making and Curing Concrete Test Specimens in the Laboratory" until the test age. The specimens were tested in compression at the ages of 1, 3, 7 and 28 days. The average of three tests were reported as the compressive strength.

Elastic modulus was determined using 102 mm x 203 mm (4 in. x 8 in.) cylinder specimens from a limited number of mixes. Elastic modulus was investigated at early and later ages on standard and accelerated heat cured cylinders.

Using a temperature-matched curing system, two of the laboratory trial mixes were subjected to accelerated heat curing. The curing regime applied to these trial mixes consisted of an approximate four to five hour preset time followed by an eight hour soaking time at 88 degC (190 degF). This was designed to investigate the effects of higher curing temperatures on both the compressive strength and elastic modulus of the concrete. It was anticipated that higher curing temperatures would be encountered at the U-Beam end block locations which would impact the concrete properties. It was important to closely simulate the concrete in the member to verify that the project specifications and guidelines were being met. Higher end block temperatures within the U-Beam members were noted during casting of the beams and are discussed in further detail later in this report.

FIELD TRIAL MIXES

Following the successful development and optimization of the laboratory mix design, a field trial mix phase was initiated to batch the mix design at Texas Concrete Precast Plant in Victoria, Texas where the beams would later be fabricated. The purpose of this phase was two-fold. First, to investigate the strength loss of the mix design which typically occurs when a laboratory mix design is utilized in the field. Secondly, the field trial mixes were performed at the precast plant for plant personnel to gain added experience with HRWR's and the use of mineral admixtures. Table 1 illustrates the final mix design used in production of the U-Beams.

Table 1: High Performance Concrete Mix Design utilized for the Precast / Prestressed U-beams

COMPONENT	QUANTITY		TYPE
Superplasticizer	1.1-1.4 liter/100 kg cem.	(17.0-21.6 oz/cwt)	ASTM C494 Type F
Retarder	0.168 liter/100 kg cem.	(2.57 oz/cwt)	ASTM C494 Type B
Water	147 kg/m^3	(247 pcy)	Potable
Cement	398 kg/m^3	(671 pcy)	ASTM C150 Type III
Fly Ash	187 kg/m^3	(316 pcy)	ASTM C618 Class C
Fine Aggregate	610 kg/m^3	(1029 pcy)	Natural River Sand
Coarse Aggregate	1138 kg/m^3	(1918 pcy)	Crushed Dolomitic Limestone, 1/2" max, ASTM GR 7

QC / QA PROGRAM

The QC / QA program was developed to monitor and document both early and later age concrete properties which were important in the production of these precast / prestressed concrete U-Beams. In addition, since this project was the first high performance concrete bridge project in the United States which used 15.2 mm (0.6 in.) prestressing stands and HPC throughout, The Texas Department of Transportation and The Federal Highway Administration wanted to create a data base which investigated and monitored various high performance concrete properties. The early age aspects which were closely monitored included compressive strength, modulus of elasticity, and flexural strength at release of the strands. As previously documented, the specifications addressed several requirements at release including compressive strength 63.4 MPa (9,200 psi), modulus of elasticity 41.3 kPa (6,000 ksi), and flexural strength 830 Sqrt (f_{ci}'), kPa (10 Sqrt. (f_{ci}'), psi). These specifications were set for strength and serviceability requirements for movement and shipping of the U-Beams from the beds. Release of the prestressing strands typically occurred at 22 to 28 hours after concrete placement. The production of these precast U-Beams did not differ from the standard production of any other precast elements at the plant which included more "traditional" concrete mix designs. Turnover of the U-beams were important to the precaster to free up the precasting beds and thereby increase production. Additional performance related testing incorporated within the QC / QA program included splitting tensile strength, and permeability. Additional concrete durability testing and monitoring such as abrasion resistance, freeze-thaw, or scaling deicing were not warranted for the precast members due to the mild Houston climate. Generally, these concrete durability items are deemed to be more important for superstructure elements such as the cast-in-place deck where durability is considered more critical.

The QC / QA program also investigated various concrete properties under a variety of curing conditions. These included ASTM moist cured cylinders, TxDOT member cured cylinders, and match cured cylinders. Because of the high required strength and MOE at release and 56 days, it was very important to determine the concrete properties of the member as accurately as possible. This meant trying to duplicate the curing conditions of the test cylinders with respect to the actual curing condition of the U-beams. A match curing system was selected to accomplish this. The specifics of this are discussed in the following sections.

Sampling Frequency

Guidelines for sampling frequency were developed for fresh and hardened concrete properties. The fresh concrete performance related tests which were conducted on the precast / prestressed concrete U-Beams included air content, slump (before and after addition of HRWR), unit weight, and concrete temperature. The hardened concrete performance related tests which were conducted on the precast / prestressed concrete U-Beams included compressive strength gain with time, elastic modulus, flexural strength, splitting tensile strength, rapid chloride permeability, and concrete temperature rise. Specimens tested included ASTM Moist Cured, Member Cured, and Match Cured. Table 2 details the testing at which the hardened concrete performance related tests were conducted. The fresh and hardened performance related testing was conducted on each placement for the U-Beams at the precast plant in Victoria, Texas.

Table 2: Performance Related Testing Schedule

PERFORMANCE RELATED TEST	TESTING DATES
Compressive Strength	at Release, 28 days, 56 Days
Modulus of Elasticity	at Release, 28 days, 56 Days
Flexural Strength	at Release, 28 days, 56 Days
Splitting Tensile Strength	at Release, 28 days, 56 Days
Rapid Ion Permeability	at 56 Days

Utilization of the Match Cure System

A commercially available match curing system was selected to produce the match cured cylinders. The system incorporates steel molds that include internal coils that cure the cylinders at the same temperature profile as the thermocouple location in the member. The match curing steel molds were located in a room where the temperature of the environment could be controlled. The room was maintained at a cool temperature since the match curing system could not cool the specimens, only heat the specimens. This system was selected to more closely simulate the actual concrete temperature profile in various locations of the member.

QC /QA PROGRAM RESULTS

Laboratory Trial Mixes

Fly Ash Effects - The use of a Class C fly ash was beneficial in several aspects. The fly ash replacement from an economic standpoint reduced the cost of the cemetitious material by approximately 20 percent. The beneficial fresh concrete properties of the fly ash included improved workability and finishability. The beneficial hardened concrete properties of the fly ash included improved long term strength and reduced permeability. Figure 3 and 4 illustrate permeability results of HPC laboratory mixes incorporating 0, 25, and 35 percent fly ash replacement. As illustrated in Figure 3, the mix designs incorporating 35 percent fly ash replacement exhibited the lowest permeability. Figure 4 illustrates the permeability versus the water to binder ratio or the water to cementitious material ratio (w/cm). Of particular note is that today with the use of high range water reducers and mineral admixtures, the w/cm ratio is no longer an acceptable means to insure a low permeable concrete. Some construction specifications today dictate a w/cm ratio range as a means to address permeability and durability. This may have been an effective means in the days prior to the widespread use of mineral admixtures, but it is no longer applicable today, particularly for high performance concrete. It may be noted that in the laboratory trial mixes for a w/cm ratio of around 0.30, the resulting permeability ranged between 500 and 2,500 coulombs passed. This supports the notion that today's HPC's are an engineered concrete. Recently, several research studies and publications have questioned the correlation validity of the Rapid Chloride Permeability (RCPT) Test (AASHTO T-277) to the Chloride Ponding Test (AASHTO T-259). This research topic is part of an ongoing research study at the Construction Materials Research Group at The University of Texas in Austin and may be referenced as part of these proceedings [1]. However, the current RCPT test does serve as a comparison tool for similar mix designs under investigation.

In addition to the fly ash benefits discussed, the fly ash replacement reduced the initial hydration temperature of the concrete during placement. This also resulted in a lower thermal cracking potential for the in-place concrete. Furthermore, the lower initial hydration temperature of the concrete benefited the long term strength gain as developed in the temperature effects section below.

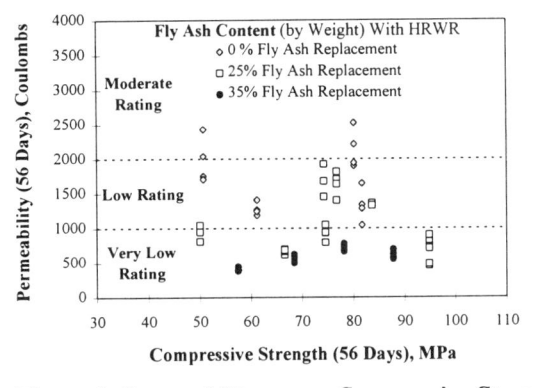

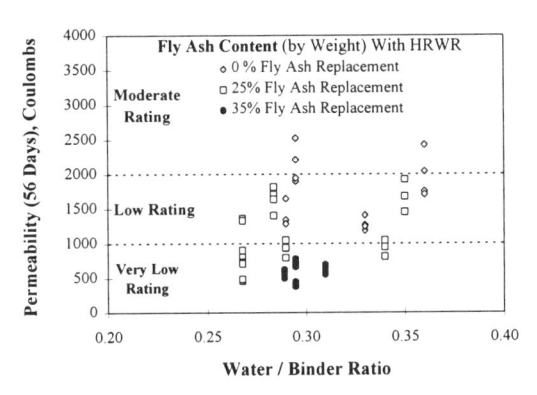

Figure 3: Permeability verses Compressive Strength for the Laboratory Trial Mixes

Figure 4: Permeability verses Water to Binder Ratio for the Laboratory Trial Mixes

Coarse Aggregate Effects - Three different aggregate contents 36, 40, and 44 percent and five aggregate types were investigated within this study. The coarse aggregate type and gradation played an important role in the compressive strength development of the concrete mix designs investigated. The trap rock (TR), which is the hardest and densest aggregate selected for the development of the mix design, resulted in a lower compressive strength when compared to the dolomitic limestone (DL) for the same mix proportions and materials. The difference between the stiffnesses of the mortar and trap rock caused stress concentrations at higher stress levels resulting in failure of the cylinders at a lower compressive value. Cracks generally formed in the cement paste around the aggregate rather than through it. The dense trap rock acted as stress riser under high stress levels. The crushed river gravel mix designs investigated also tended to develop lower compressive strengths. The crushed river gravel tended to fail in the range of 75 to 83 MPa. Fracture of the aggregate was clearly visible during compression testing with no strength gain noted beyond 83 MPa (12,000 psi) at long-term testing ages. The crushed river gravel (CG) also provided poor bond characteristics which are highly desirable for HPC due to their partially smooth surfaces. The calcitic limestone (CL) provided good compressive strength development, however the elastic modulus development was not suitable for the U-Beams requirements.

Concretes with smaller maximum size aggregate gradation of 13 mm yielded slightly higher compressive strength test results than the larger maximum size aggregate gradation of 19 mm. The smaller aggregate size, and consequently higher surface area for a given aggregate content, results in a lower bond stress at a given load level and consequently higher strength. The concrete tends to act more like a homogenous material.

For all aggregate types, increasing the coarse aggregate content beyond 40 percent did not appear to benefit the compressive strength development. However, the elastic modulus appeared to be a function of the coarse aggregate content and type. The elastic modulus development for different aggregates are illustrated in Figures 5, 6, and 7 for 36, 40, and 44 percent aggregate contents from the laboratory trial mixes. The only coarse aggregate which exhibited an elastic modulus below the empirical equation recommended by ACI Committee 318 was the calcitic limestone. The elastic modulus of the calcitic limestone exceed the empirical equation recommended by ACI Committee 363 for high strength concrete. The lower elastic modulus values may be attributed to the abundance of calcite which is a soft mineral.

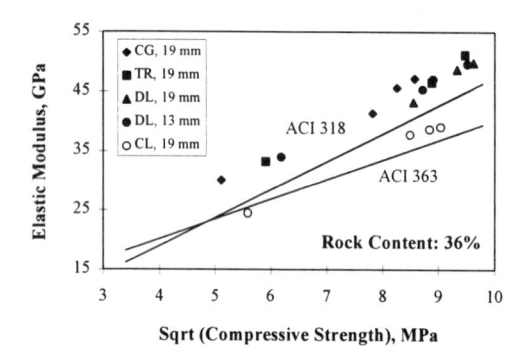

Figure 5: Elastic Modulus Development for Concretes with 36% Aggregate Content

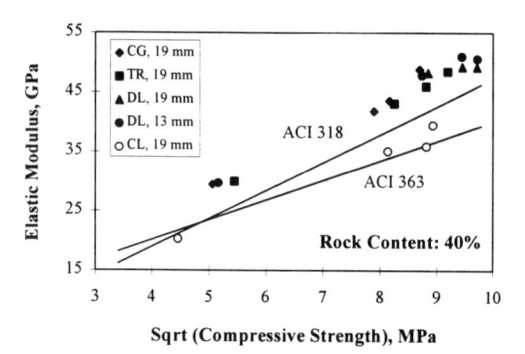

Figure 6: Elastic Modulus Development for Concretes with 40% Aggregate Content

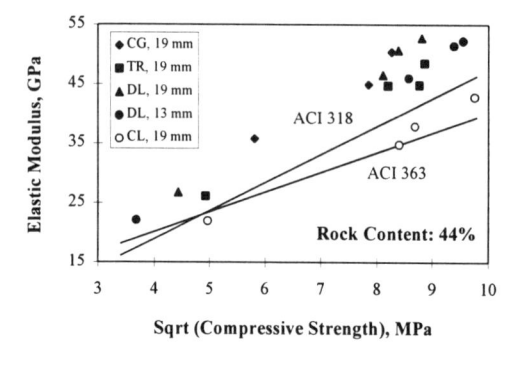

Figure 7: Elastic Modulus Development for Concretes with 44% Aggregate Content

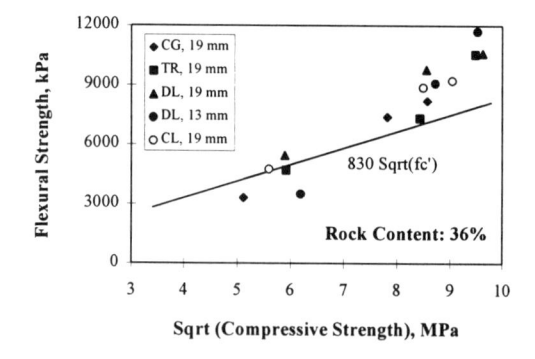

Figure 8: Flexural Strength Verses Compressive Strength for Concretes with 36% Aggregate Content

In addition to the elastic modulus, the flexural strength of the trial mix designs were monitored. The flexural strength for different aggregates are illustrated in Figures 8, 9, and 10 for 36, 40, and 44 percent aggregate contents from the laboratory trial mixes. Increasing the coarse aggregate content appeared to result in a slight reduction in the flexural strength. Increasing the aggregate

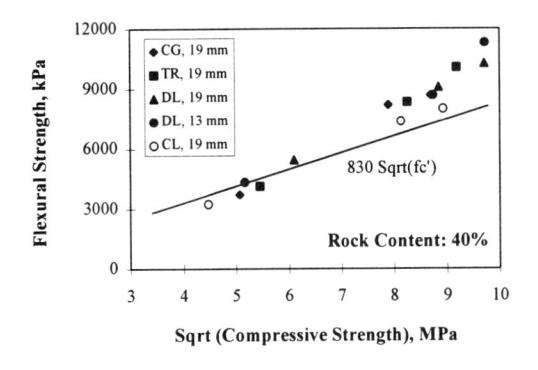

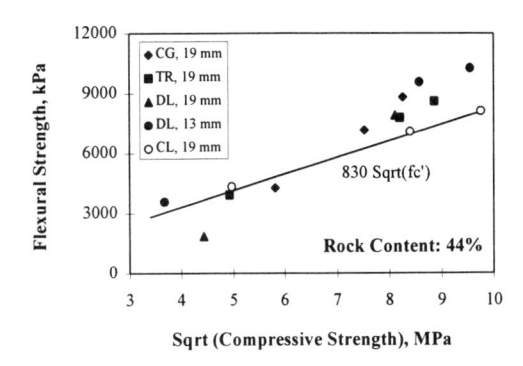

Figure 9: Flexural Strength Verses Compressive Strength for Concretes with 40% Aggregate Content

Figure 10: Flexural Strength Verses Compressive Strength for Concretes with 44% Aggregate Content

content without a change in the aggregate size results in an increased interface area. This results in a potentially weaker zone than either mortar or aggregate resulting in a reduction in the flexural strength. Based on the results of the trial mixes, the 13 mm dolomitic limestone from central Texas resulted in the aggregate most suitable to meet the required compressive strength and modulus of elasticity specified in the project specifications. The dolomitic limestone with its angular shape and compatible elastic modulus to the mortar provided an aggregate with excellent bond characteristics and homogeneous performance. The aggregate was also locally available for the precast plant subsequently resulting in an economical mix design.

Temperature Effects - Two laboratory trial mix designs were subjected to accelerated heat curing to investigate any variation in compressive strength and elastic modulus due to temperature effects. Five curing temperatures were selected. Accelerated temperatures of 43, 66, and 88 degC were selected. Furthermore, cylinders were also standard cured (no accelerated heat curing) and cured under a hydration chamber condition. The hydration chamber involved the curing condition of cylinders with no external ambient influences. One of the two mix designs incorporated 30 percent fly ash replacement.

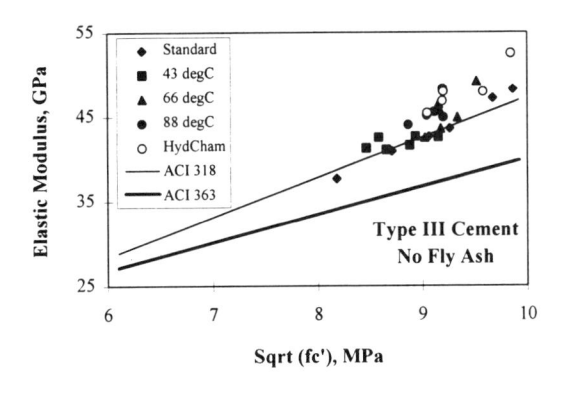

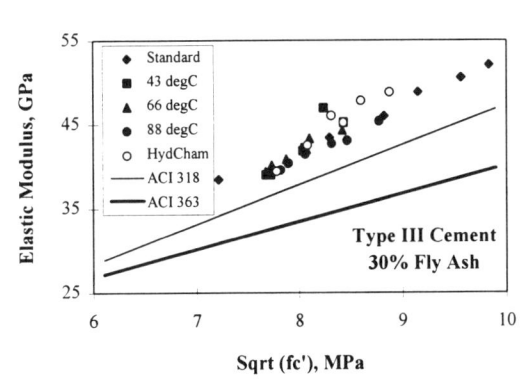

Figure 11: Elastic Modulus Development for Concretes without Fly Ash Replacement

Figure 12: Elastic Modulus Development for Concretes with 30% Fly Ash Replacement

Figures 11 and 12 illustrate the compressive strength and elastic modulus results. While the results varied between the two mix designs, the elastic modulus results appeared independent of the accelerated heat curing and primarily a function of the compressive strength of the concrete.

The mix design incorporating fly ash replacement resulted in higher compressive strengths and elastic modulus.

Precast Plant Produced U-Beam Mixes

Fabrication of the U-Beams occurred from November 1993 to March 1995. These included a total of nine casting dates. Table 3 summarizes the fresh concrete properties monitored during casting of the high performance concrete U-beams. The slump of the concrete ranged from 200 to 250 mm after the addition of the high range water reducer. Little if any slump loss was noted since the concrete was immediately placed upon completion of batching. The temperature of the concrete mix varied from 24 to 35 degC.

Table 3: Summary of Fresh Concrete Properties Monitored during Casting of the U-Beams

PROPERTY	VALUE
Water / Binder (C+FA) Ratio	0.25
Fly Ash Replacement	32% by weight
Slump (after HRWR)	200 to 250 mm (8 to 10 in.)
Concrete Temperature	24° to 35° C (76° to 95° F)
Ambient Temperature	7° to 32° C (45° to 90° F)
Unit Weight	2467 kg/m^3 (154 pcf)

The temperature of the concrete at placement was influenced mainly by the ambient temperature at the time of casting. The unit weight of the mix design was 2,467 kg/m^3. The mix incorporated a high coarse aggregate content for two reasons. First, to meet the elastic modulus requirements set in the project specifications. The elastic modulus was concluded to be a function of the coarse aggregate and type. Secondly, a high coarse aggregate content was selected for placement reasons. The fabricator selected to use a continuous inner steel form which required the concrete to flow across the bottom flange and up the adjacent web to avoid concerns of possible air voids in the bottom flange. The natural tendency for concrete is to flow in the path of least resistance which would naturally be down the strands rather than across the strands which would lead to air voids in the bottom flange. It was critical to develop a mix design which would create a bulkhead during placement of the concrete. Sandy mixes or mixes with lower coarse aggregate contents would not be able to create a bulkhead. The final mix design developed in the field trial mixes used a slightly higher aggregate content to accomplish this. Concrete was placed from the top web, vibrated down the web and across the bottom flange. Form and hand held vibrators were used to consolidate the concrete and avoid any voids or honeycombing. This casting sequence is illustrated in Figure 13. Hardened concrete properties and performance related test results of the field mixes are discussed in below.

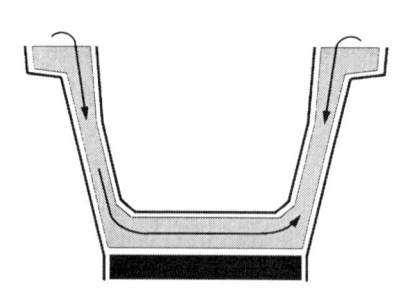

Figure 13: Casting Sequence

Compressive Strength - The required compressive strength for the U-beams at release ranged from 47.6 to 63.4 MPa (6,900 to 9,200 psi). Release strengths were attained at 24 to 27 hours from casting. The required design compressive strength for the U-beams ranged from 67.6 to 90.3 MPa (9,800 to 13,100 psi) at 56 days from casting. Each of these casting dates were monitored for compressive strength at release, 28 days, and 56 days. As previously noted, the monitoring involved various curing conditions. These include match curing, ASTM moist curing, and TxDOT member curing. The 28 day strengths varied from approximately 82.7 to 103.4 MPa (12,000 to 15,000 psi) at 28 days. The ASTM moist cured cylinders displayed the highest strengths at 28 days typically as discussed in the following section. In all cases, the required strengths at release and 56 days were met by the precaster. Figure 14 illustrates the running standard deviation on the sequential pours.

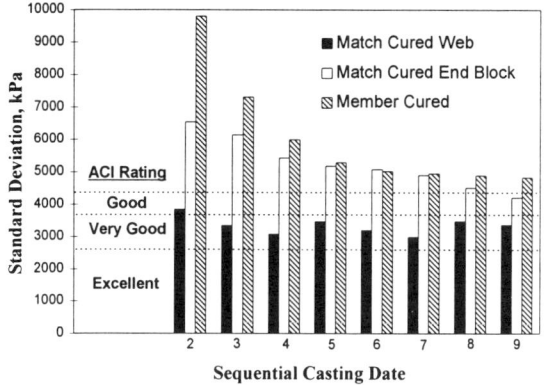

Figure 14: Running Standard Deviation for Sequential U-Beam Pour Dates

Figure 15: Typical Skewed End Block Region for a Precast / Prestressed U-Beam

The data indicates a standard deviation of 2,760 to 4,140 kPa (400 to 600 psi) for the match cured concrete. These match cured cylinders best represent the concrete in the U-Beam members. As the precast plant gains more and more experience producing the high performance concrete, the standard deviation has continued to indicate a declining trend in the good to very good range as defined by ACI 214.

Temperature Effects - The temperature development within the U-Beam precast members dramatically influenced the long term strength gain of the concrete within the member. These temperatures were recorded with thermocouples which were placed at various location within the member. The peak temperatures were recorded in the skewed end blocks of the beams which consistently attained the highest curing temperatures due to the mass of concrete. Figure 15 shows a typical skewed end block. Two of these precast beams were fabricated in November when the precaster chose to steam cure the members to offset cooler nighttime temperatures. The steam curing did not exceed 66 degC (150 degF), the limit specified by TxDOT. The steam curing was not intended to provide any accelerated curing, rather it was provided to serve as protection against subjecting the early curing of the member to an excessive thermal gradient.

Figure 16: U-Beam AA-23 Curing Temperatures

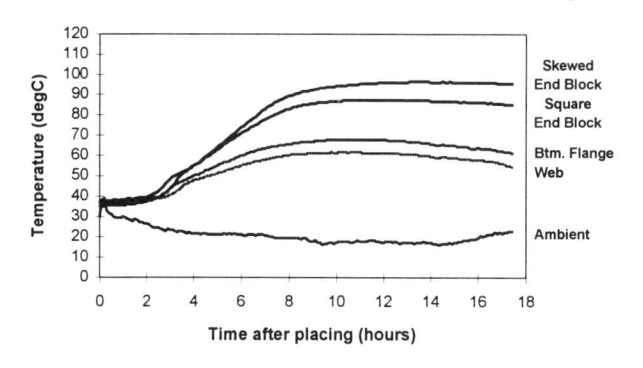

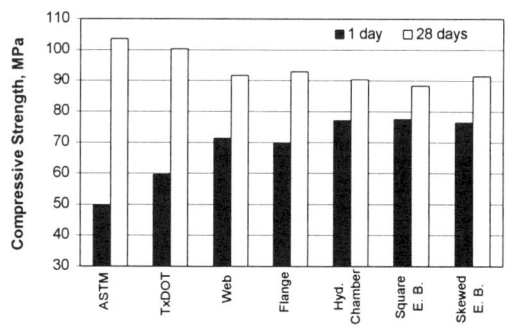

Figure 17: U-Beam AA-23 Compressive Strength at Release & 28 days Under Varied Curing Conditions

Figure 16 illustrates the curing temperatures for Beam AA-23 cast on November 10, 1994 which was representative of all castings. It may be noted that the end blocks undergo the highest curing temperatures followed by the flange and web locations. The TxDOT cylinders are cylinders which are placed next to the formwork prior to release of the strands. These cylinders are used to determine when the prestressing stands may be released. ASTM cylinders are cylinders placed in a 21 degC (70 degF) environment for a 24 hour period followed by standard ASTM moist curing. Figure 17 illustrates the curing type or location versus the compression strength at release (24 hours) and 28 days. Clearly, the curing condition which attained the highest early temperature displayed the highest early strength, but the lowest long term (28 days) strength. Therefore, the strength of the member is typically underestimated at release, but overestimated long term if ASTM most cured or TxDOT cured cylinders are used for design strength verification. For members with greater surface area to volume ratios than this U-Beam such as an AASHTO I-shaped beam, the variation in the early and long term compressive strengths between the various curing conditions would not be as dramatic as illustrated in this U-Beam section. This is mainly due to the fact that the members' curing temperature would not be as great.

Modulus of Elasticity - The modulus of elasticity of the precast plant produced concrete which was sampled from the beam exceeded the empirical equation recommended by ACI Committee 363 for high strength concrete. The match cured cylinders generally exhibited slightly lower elastic modulus than the ASTM moist cured and member cured cylinders as illustrated in Figure 18. The member cured cylinders developed the majority of it's elastic modulus at early ages with a limited amount of gain at later ages. This was due to the high initial curing temperatures. The elastic modulus development of the member cured cylinders was similar to the compressive strength development where initial values exceeded ASTM moist cured and member cured cylinders. The ASTM moist cured cylinders generally exhibited the highest elastic modulus at 56 days when these cylinders attained the highest compressive strength.

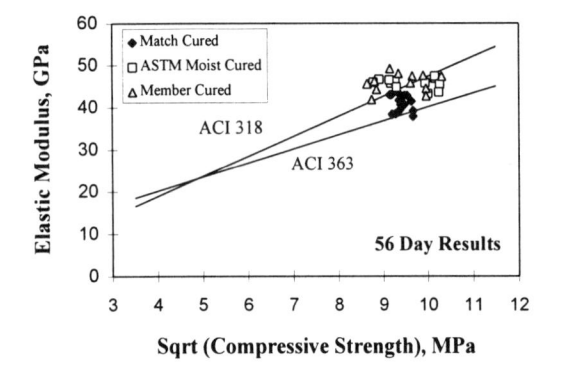

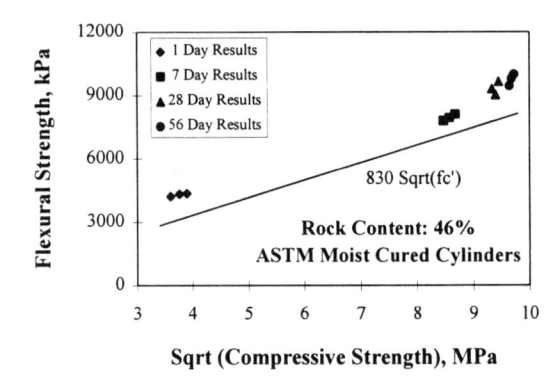

Figure 18: Elastic Modulus Development for Conc. Sampled from the Precast U-Beam Members

Figure 19: Flexural Strength Results for Concrete Sampled from the Precast U-Beam Members

Flexural Strength - The flexural strength of the precast plant produced concrete exceeded the flexural strength project requirement of 830 Sqrt (f_c'), kPa. Figure 19 illustrates the results at 1, 7, 28, and 56 days. Note that the final mix design used at the precast plant incorporated a 46 percent aggregate content. Since the match cured system was not applicable for beam molds nor could the U-Beam forms accommodate beam molds, only ASTM moist cured beams were investigated at the precast plant.

Splitting Tensile Strength - The splitting tensile strength of the precast plant produced concrete generally tested below the flexural strength project requirement of 830 Sqrt (f_c'), kPa. Match cured cylinders exhibited the greatest splitting tensile strength as illustrated in Figure 20. The higher initial curing temperature appears to slightly improve the splitting tensile strength of the concrete.

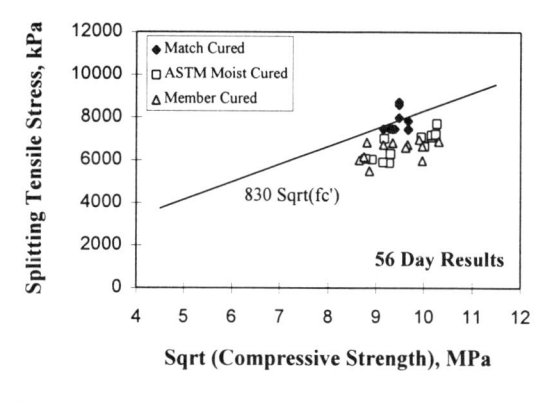

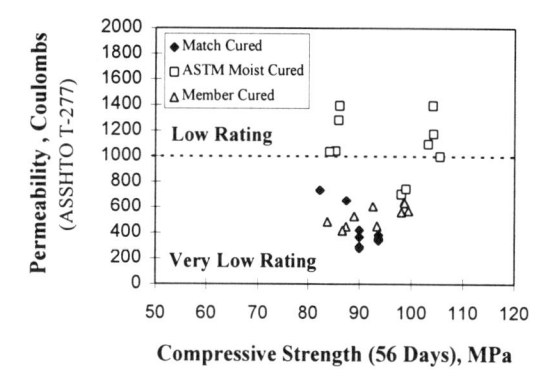

Figure 20: Splitting Tensile Strength for Concrete Sampled from the Precast U-Beam Members

Figure 21: Rapid Ion Permeability for Concrete Sampled from the Precast U-Beam Members

Rapid Ion Permeability - The project objectives provided by the Texas Department of Transportation and the Federal Highway Administration at the start of this project was to develop a performance concrete mix design with a permeability rating of 1500 coulombs passed or less. It was determined that this would result in a more durable structure with longer service life and lower maintenance costs. Rapid ion permeability testing (AASHTO T-277) was performed on precast plant produced concrete cylinders under various curing conditions as illustrated in Figure 21. Over the years, precast plant produced members have traditionally had a reputation for being very durable, impermeable concrete members. This is illustrated in the match cured cylinders which exhibited the lowest permeability. Previous research studies have generally found that ASTM moist cured cylinders yield lower permeability over site or field cured cylinders from the added cement hydration which takes place due to the moist curing. It is interesting to note the influence on the higher curing temperatures on both the match cured and member cured cylinders.

Creep & Shrinkage - Creep and shrinkage data of the mix design was monitored for four months. The specimens were member cured with the beam until release. The peak curing temperature of these creep and shrinkage specimens was 63 degC (145 degF). The curing temperatures of the specimens was similar to temperatures found in the bottom flange and side webs of the U-Beams. DEMAC points were placed on unloaded shrinkage specimens at 200 mm (8 in.) gauge lengths. Strains were measured at three levels (top, middle, and bottom) on three sides of the specimen. The nine measurements from one specimen are then averaged with two other specimens for an average shrinkage value. Creep and shrinkage specimens were monitored in a similar matter except that the specimens were loaded to 20.7 MPa (3,000 psi) at 1 day. This value corresponds to approximately 40 percent of f_c' at release. The difference between the loaded and the unloaded specimens is defined as the creep of the concrete. Figure 22 outlines the measured shrinkage strain and current ACI 209 prediction for shrinkage. The fitted curve quantifies the measured results for the HPC mix design and is defined by Equation 1 as follows:

$$\varepsilon_{sh} = (0.000510)\frac{t}{35+t} \quad \textit{(Eqn. 1)}$$

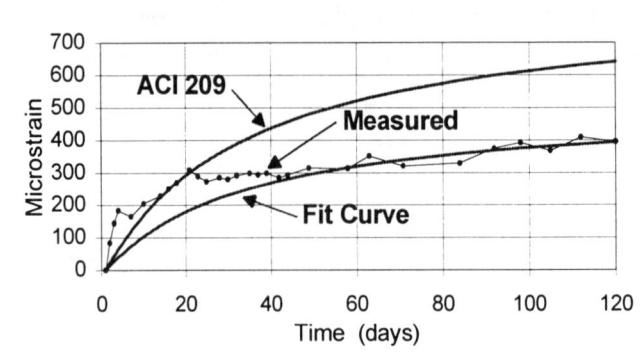

Figure 22: Average Shrinkage Strain for Concrete Sampled from the Precast U-Beam Members [2]

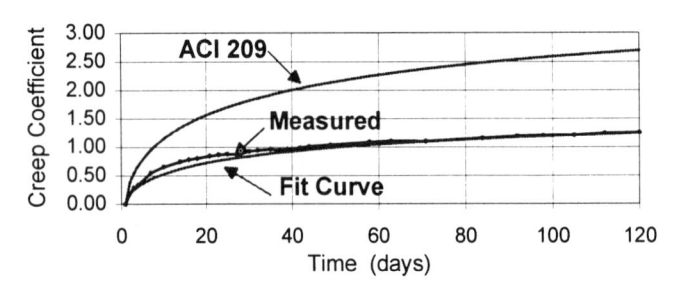

Figure 23: Creep Coefficient for Concrete Sampled from the Precast U-Beam Members [2]

Figure 23 outlines the measured creep coefficient and current ACI 209 prediction for the creep coefficient. The fitted curve quantifies the measured results for the creep coefficient for the HPC mix design and is defined by Equation 2 as follows:

$$C_{ct} = (1.95)\frac{t^{0.6}}{10+t^{0.6}} \quad \textit{(Eqn. 2)}$$

The following characteristics were noted of the HPC mix related to creep and shrinkage:
1. Less ultimate creep and shrinkage
2. 20 % lower creep rate for loading at 28 days.
3. No significant effects due to the high curing temperatures.

The high performance concrete exhibited less creep and shrinkage than comparable normal strength concrete mixes. The impact that this plays on the serviceability and structural aspects of the U-Beams are discussed in greater detail by Burns, Gross, and Byle as part of these proceedings [2].

SUMMARY AND CONCLUSIONS

The precast / prestressed U-Beam members produced for the Louetta Road Overpass Bridge Project in Houston, Texas has illustrated the suitability of HPC as a means of improving the quality and durability of bridge components within our infrastructure. The concrete mix selected for the production of these beams after a series of laboratory and field trial mixes incorporated locally available materials including, cement, fly ash, natural river sand, and dolomitic limestone from central Texas. With the optimization of HPC, increased demands and performance requirements may be anticipated. The U-Beams with their longer spans, high number of 15.2 mm prestressing strands, and thinner elements, has demonstrated this point clearly. The project has also demonstrated how an engineered HPC mix design was developed to satisfy several project requirements including compressive strength, elastic modulus, and flexural strength. While the use of HPC generally dictates increased demands and performance requirements, the benefits are clearly evident. The increased durability benefits of longer life structures with lower maintenance costs will benefit the infrastructure for decades to come.

Based upon the quality control / quality assurance performance related testing, the following conclusions were drawn on various aspects of the HPC U-Beams:

Laboratory Trial Mix Results
1. Concretes with smaller maximum size aggregate gradation (13 mm) yielded slightly higher (4.5 % on average) compressive strength results than the larger maximum aggregate gradation (19 mm) at a given aggregate content level.
2. Concretes incorporating crushed gravel exhibited low compression strengths generally caused by poor mechanical bond due to its shape and surface texture. Aggregates which were significantly harder than the mortar, such as the trap rock, are likely to cause stress concentrations at higher stress levels introducing microcracking in the transition zones thus resulting in lower strengths. The mineralogical characteristics of coarse aggregates appear to be an important factor influencing the mechanical properties of concrete.
3. Low elastic modulus and flexural strength test values resulted for concretes incorporating a high abundance of calcite as illustrated by the calcitic limestone test results.
4. Increasing the coarse aggregate content appears to result in a reduction in flexural strength for a given aggregate size. The increased aggregate content translates into increased interfacial area that is potentially weaker in tension than mortar or aggregate. The elastic modulus appears to be independent of aggregate size for a given aggregate content based on the results of this study. This may be attributed to the fact that bond strength is not critical at lower loading levels where the elastic modulus is determined.
5. Fly ash replacement was incorporated to aid in maintaining lower initial curing temperatures which resulted in a lower thermal cracking potential for the in-place concrete. The fly ash also improved the concrete durability by lowering the concrete permeability and benefiting the long term strength gain of the mix design. The fly ash reduced the required amount of cement resulting in an economical mix design.

Precast Plant Produced U-Beam Results
1. Member cured cylinders used to determine release of the prestressing strands underestimated the compressive strength by as much as 27.9% based on the results of the match cured cylinders. In this study the match cured cylinders exceeded the compressive strength of the member cured cylinders by an average of 19.1% at release. Using the match curing system would allow the precaster to release the prestressing strands earlier and thereby increase plant productivity.
2. ASTM moist cured cylinders used to verify the design strength at 56 days overestimated the compressive strength by as much as 15.9% based on the results of the match cured cylinders. In this study the ASTM moist cured cylinders exceeded the compressive strength of the match cured cylinders by an average of 9.8% at 56 days.

3. The majority of the U-Beams' elastic modulus is developed at early ages with a limited amount of gain at later ages. The elastic modulus of the concrete is clearly a function of the compressive strength development within the member.
4. For the precast plant produced concrete sampled, the match cured cylinders clearly displayed the lowest permeability at 56 days under the curing regimes investigated. The high initial curing temperatures appear to significantly influence the permeability and matrix of the concrete.
5. Less ultimate creep (20% at 28 days) and shrinkage was noted compared to normal strength concrete for this high performance concrete mix design with no significant effects due to high curing temperatures.
6. For all the concrete specimens sampled in this study, ACI Committee 363 empirical equations for high performance concrete were satisfied.
7. In addition to satisfying concrete property specifications, allowances for unexpected occurrences during production may be anticipated due to the nature of most high performance mix designs and placement requirements. This begins with closely monitoring of a well developed QC/QA program. Allowances for a reduction in mechanical properties should be accounted for when a laboratory mix design is used in the field. A recommended value for high performance concrete is a minimum 15% overdesign strength based on the standard deviations noted on the precast plant produced concrete on these projects.

ACKNOWLEDGMENTS

The authors wish to thank the joint sponsors of this research project, The Federal Highway Administration and The Texas Department of Transportation, for their support and encouragement. In addition, the authors would like to thank the fabricator, Texas Concrete Company, for their assistance, interest, and involvement in this research study.

REFERENCES

1. Myers, J.J., Touma, W.E., Carrasquillo, R.L., *Permeability of High Performance Concrete: Rapid Chloride Ion Test vs. Chloride Ponding Test,* PCI Annual Meeting Proceedings, New Orleans (1997).
2. Burns, N.H., Gross, S.P., Byle, K., *Instramentation and Measurements - Behavior of Long Span Prestress High Performance Concrete Bridges,* PCI Annual Meeting Proceedings, New Orleans (1997).
3. Cetin, A., *Effect of Accelerated Heat Curing and Mix Characteristics on the Heat Development and Mechanical Properties of High Performance Concrete,* Dissertation, The University of Texas at Austin (1995).
4. Carlton, M.P., Carrasquillo, R.L., *Quality Control of High Performance Concrete For Highway Bridges,* Research Report 580-3, Center for Transportation Research, The University of Texas at Austin (1995).
5. ACI Committee 363, *State-of-the-Art Report on High Strength Concrete,* ACI 363R-92, American Concrete Institute, Detroit (1992).
6. Peterman, M.B., Carrasquillo, R.L., *Production of High Strength Concrete,* Research report 315-1F, Center for Transportation Research, The University of Texas at Austin (1993).
7. ACI Committee 211, *Guide for Selecting Proportions for High Strength Concrete with Portland Cement and Fly Ash,* Committee Report, ACI Materials Journal, (May-June 1993).

Proceedings of the PCI/FHWA
International Symposium on High Performance Concrete
New Orleans, Louisiana, October 20-22, 1997

SEISMIC BEHAVIOR OF HIGH STRENGTH CONCRETE FILLED TUBE (CFT) COLUMNS

A. El-Remaily
Ph.D. Candidate
University of Nebraska-Lincoln
Lincoln, NE, U.S.A.

A. Azizinamini, Ph.D., P.E.
Associate Professor
University of Nebraska-Lincoln
Lincoln, NE, U.S.A.

M. Zaki
Former Graduate Student
University of Nebraska-Lincoln
Lincoln, NE, U.S.A.

F. Filippou, Ph.D.
Associate Professor
University of California-Berkeley
Berkeley, CA, U.S.A.

ABSTRACT

An investigation of the behavior of concrete filled tube columns under lateral seismic load is being conducted at the University of Nebraska-Lincoln. As part of the project, four concrete filled tube columns utilizing concrete with compressive strength exceeding 70 MPa (10,000 psi) were tested to define the parameters that control the behavior and study their effect. The tested columns were simply supported at both ends and subjected to constant axial and cyclic lateral loads. The test columns showed high ductility and maintained their strength up to the end of the test. Failure was due to tensile cracking in the steel shell portion of the column.

INTRODUCTION

The concrete filled tube (CFT) as a structural member has many advantages over conventional sections. The concrete core provides considerable axial load capacity and prolongs local buckling of the steel tube wall. The steel tube contributes to the axial load capacity and provides confinement to the concrete core which significantly increases the concrete compressive strength. Having the steel at the extremities of the cross section provides the most effective contribution of the steel to the section moment of inertia and hence the lateral load capacity. The significant increase of the section capacity and stiffness over the conventional sections allows the use of smaller cross sections which means a larger free building space and lighter construction. The steel tube also provides formwork for the concrete core which reduces the construction cost.

The behavior of these columns has not been investigated thoroughly due to the number of factors involved in the behavior and the uncertainty regarding the extent of the contribution of each of these factors.[1] Although there has been extensive research to describe the behavior and establish design guidelines for monotonic loading[2], very few studies have dealt with the behavior of these columns under seismic conditions.[3] These studies have not fully provided a sound base for the equations used in building codes and many of the design criteria found in these building codes are believed to be over conservative. The objective of the current study is mainly to define the behavior of the concrete filled tube, the basic failure mode and failure criteria. This paper presents partial results of tests conducted on CFT columns.

EXPERIMENTAL STUDY

Test Specimen
Four high strength concrete filled tube specimens were tested. The specimens were all 305 mm (12.75 in.) in external diameter with steel tube thickness of 6.4 mm (0.25 in.) or 9.5 mm (0.375 in.). The corresponding diameter to thickness ratios (D/t) are 48 and 32 which satisfy the NEHRP[4] limitation given by the equation

$$\frac{D}{t} < \sqrt{\frac{5E_s}{F_y}} \qquad (1)$$

where E_s is the steel Young's modulus and F_y is the steel yield stress.

The supported span of the specimen is 2185 mm (7 ft). The axial load level (P) ranged between 0.2 and 0.4 from the nominal axial load capacity (P_o) of the section. The yield stress of the steel tubes is 373 MPa (54 ksi). High strength concrete with compressive strength (f_c') larger than 70 MPa (10,000 psi) was used. Table 1 shows the parameters for different specimens in the testing plan.

Test Setup
Each column was tested under a constant axial load and a cyclic lateral load. The specimen is confined in the middle portion by a rigid stub made of a steel box filled with concrete. The stub provides high confinement along the middle 356 mm (14 in.) of the specimen. This effect simulates the effect of a rigid floor system intersecting the beam column at the floor level. The axial reaction is supported using conventional pin supports. The lateral force reactions are supported using specially manufactured reaction fixtures

that allow full rotation of the specimen in the plane of loading. Figure 1 shows an overall view of the test setup.

Table 1- Properties of test specimens

Specimen No.	D/t	t (in.)	P/P_o	f'_c psi
CFT1	48	0.25	0.32	15000
CFT2	32	0.375	0.20	15000
CFT3	32	0.375	0.40	15000
CFT6	48	0.25	0.32	10000

Note: 1in.=25.4 mm; 1ksi=6.9 MPa

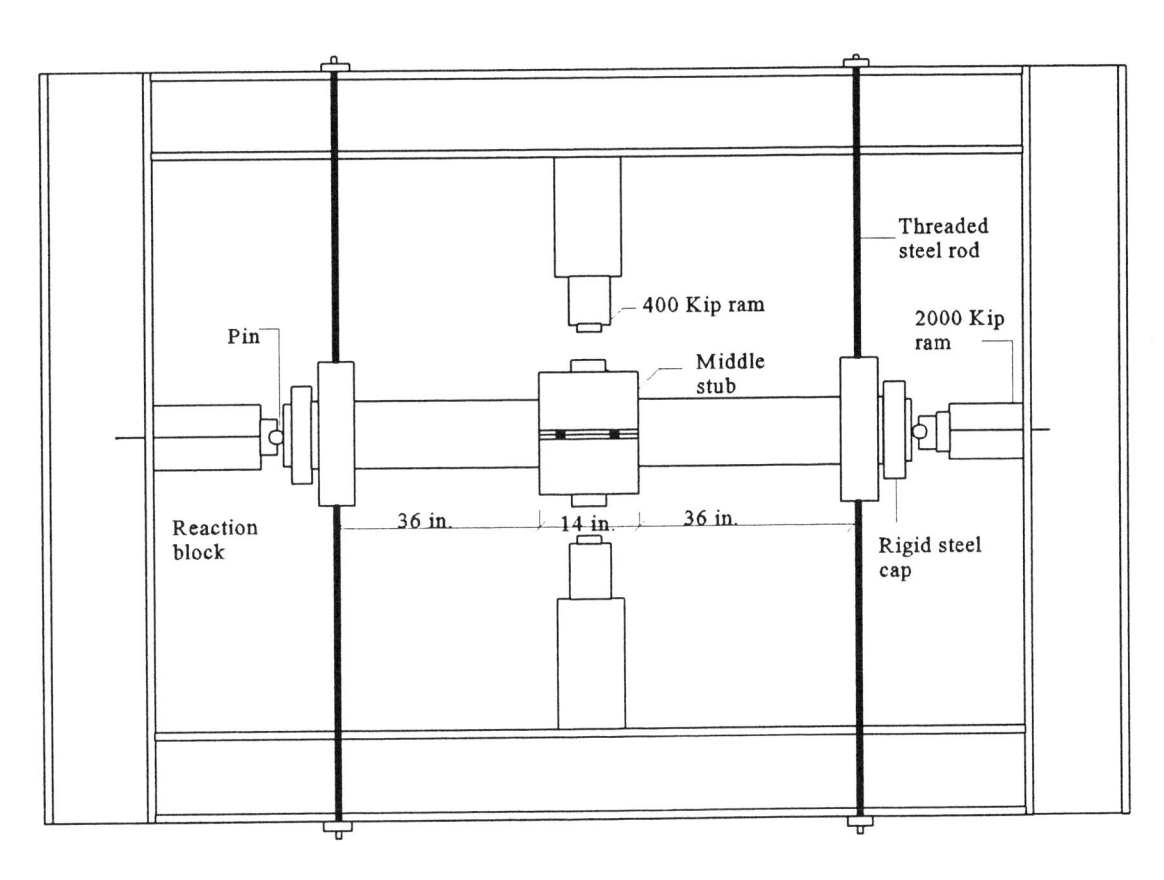

Note: 1in.=25.4 mm; 1kip=4.448 kN

Figure 1. Test Setup

Test Procedure
The specimen is loaded with a permanent axial load equal to 0.2, 0.32 or 0.4 of the nominal axial capacity of the beam column as shown in Table 1. The nominal axial capacity (P_o) of the column is calculated by summing the ultimate axial capacity of both the steel and concrete and is given by the following equation.

$$P_o = A_s F_y + A_c f_c' \qquad (2)$$

The test is displacement controlled in the lateral direction. That is by applying loads until the specimen reaches predefined levels of displacement ductility in the lateral direction at midheight. The ductility level is defined by the ratio of the specimen lateral displacement during the test (Δ) to the first yield displacement at the same location (Δ_y). The first yield displacement is determined from the moment displacement curve for the first loading cycle. It was considered as the displacement at the point of intersection of the initial slope and final tangents to that curve. Each specimen was cycled for two cycles at 1, 2, 3, 4, 6, 8 and 10 times the first yield displacement.

TEST RESULTS AND OBSERVATIONS

Failure Mode
The specimens performed in a very ductile manner and maintained their moment capacity up to a high ductility ratio of 10. Failure was due to fracture of the steel shell. As the number of cycles increased, the permanent deformation of the specimen increased noticeably. The first yield took place at location next to the middle stub. After a few cycles and approximately at $4\Delta y$ a small bulge in the steel tube started to pop out at a location next to the stub. The bulge grew during the test to form a complete ring around the specimen on each side of the stub. No drop in the moment capacity of the section was encountered with increasing ductility level.

Lateral Load against Deflection
The net deflection is obtained by correcting the deflection measured at the middle stub for support movement. Figures 2 to 5 represent the net deflection at the middle stub on the horizontal axis against the lateral load on the vertical axis. It is indicated from the curves that the specimens produce large hysteresis loops and that they can sustain their load up to ductility levels equal to 10 . It can also be detected that the behavior of the specimens is almost perfectly plastic after the elastic stage.

Axial Shortening against Lateral Load
Figures 6 to 8 present the axial shortening of the specimen through progressive loading cycles. Axial deformation was not measured for specimen CFT1. Since the axial load remains constant during the test, the major cause of the axial shortening is the formation of the bulge. The axial shortening of the specimens ranged between 43 to 78 mm (1.7 to 3.1 in.) which is a relatively high range. The growth of the bulge can be observed by the

rate of increase of axial shortening. Figure 7 shows a sudden change in the rate of axial shortening of the high strength concrete specimen CFT3 after the second cycle at $4\Delta_y$.

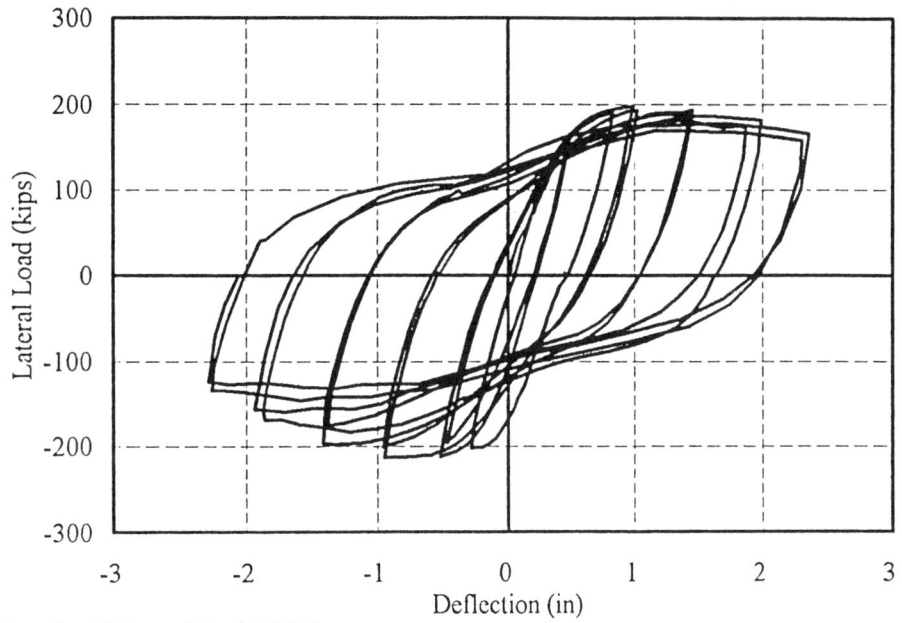

Figure 2. Lateral Load vs Deflection
CFT1, t=0.25 in. P/Po=0.32 f'c=15 ksi

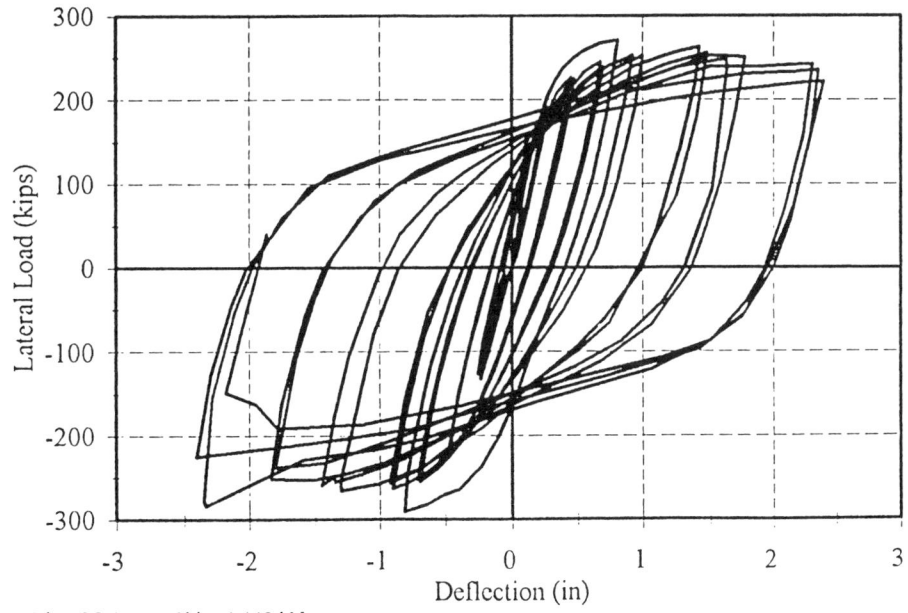

Figure 3. Lateral Load vs Deflection
CFT2, t=0.375 in. P/Po=0.2 f'c=15 ksi

Note: 1in.=25.4 mm; 1kip=4.448 kN

Figure 4. Lateral Load vs Deflection
CFT3, t=0.375 in. P/Po=0.4 f'c=15 ksi

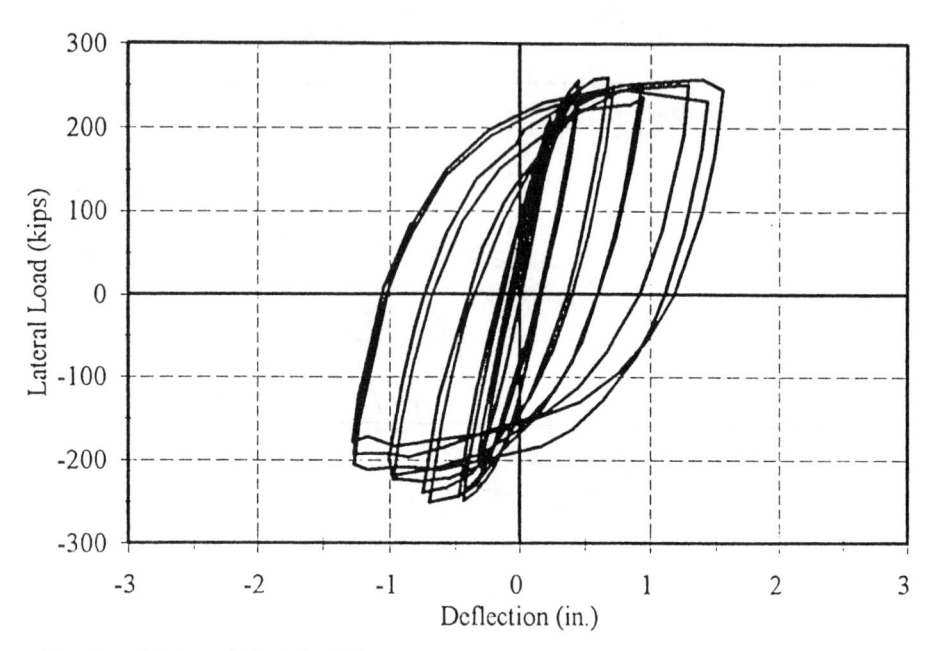

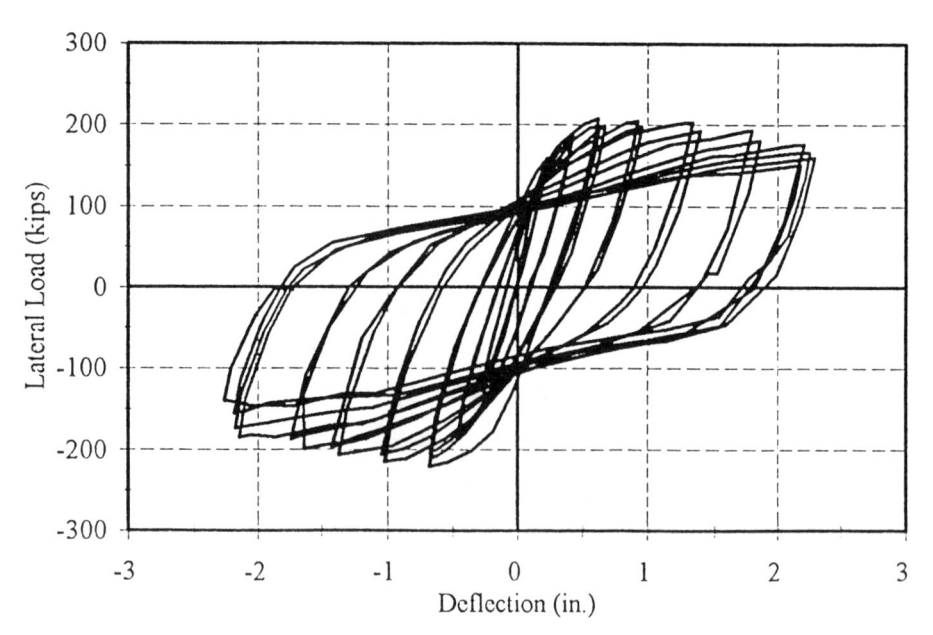

Note: 1in.=25.4 mm; 1kip=4.448 kN

Figure 5. Lateral Load vs Deflection
CFT6, t=0.25 in. P/Po=0.32 f'c=10 ksi

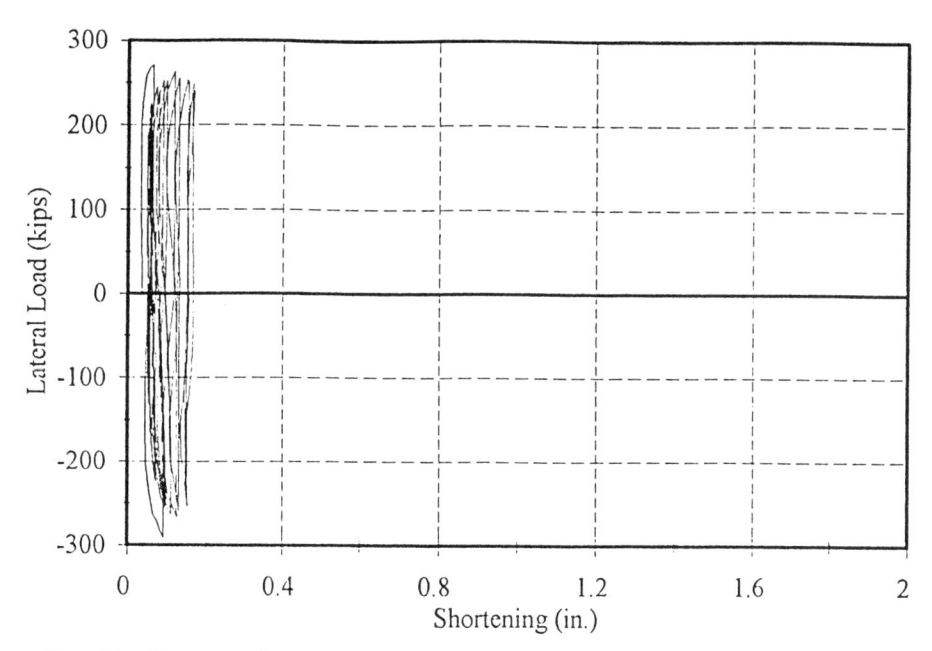

Figure 6. Lateral Load vs Shortening
CFT2, t=0.375 in. P/Po=0.2 f'c=15 ksi

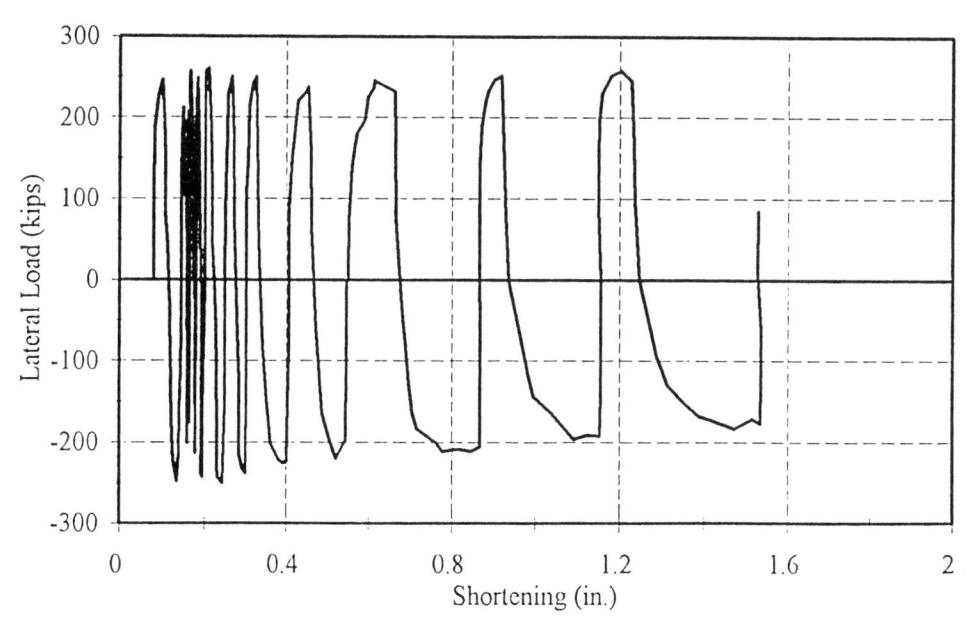

Figure 7. Lateral Load vs Shortening
CFT3, t=0.375 in. P/Po=0.4 f'c=15 ksi

Figure 8. Lateral Load vs Shortening
CFT6, t=0.25 in. P/Po=0.32 f'c=10 ksi

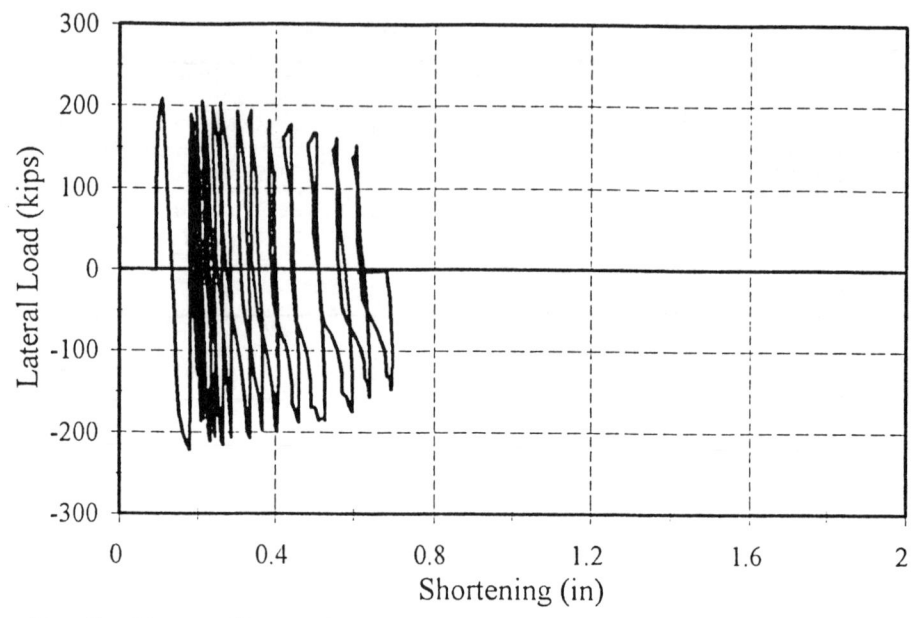

IMPLEMENTATION OF TEST RESULTS

Maximum Test Moments, Plastic Moments and AISC LRFD Design Moments
Table 2 shows the values of the maximum moments obtained from the tests for each specimen, the calculated plastic moments and the AISC LRFD design moments. The plastic moments are calculated from a section moment curvature analysis and taking strain hardening of the steel section into consideration. The last column in the table shows that the ratio of the test moment to the AISC LRFD design moment. The values in this column indicate the AISC LRFD design procedure severely underestimates the strength of these beam columns by values as high as 4.5 times. This is mainly because the AISC LRFD design procedure calculates the moment capacity of the section based on the plastic modulus of the steel section alone. The ratio between the test moment and the plastic moments for different specimens is in the range of 1.1 to 1.2. This indicates that the specimen capacity can be adequately described by the plastic capacity of its section assuming full composite action. The 10% to 20% excess can be contributed to the effect of confinement.

Stiffness Degradation
Figure 9 shows the ratio between the column stiffness as obtained from the test data and the transformed section stiffness at different cycles of the loading. The transformed section

stiffness is calculated by adding the stiffnesses of the steel and concrete assuming no interaction is taking place. It can be seen from the graph that the initial stiffness for the specimens lies between 0.7 and 0.8 of the transformed section stiffness. Only CFT3 gave an initial stiffness approximately equal to the transformed section stiffness. The degradation of the stiffness with progress of the cycles is mild and most of the specimens kept more than 65% of their initial stiffness till the end of the tests.

Table 2 Bending Moment Values from Tests, Plastic Analysis, and AISC LRFD

Specimen	Maximum Test Moment kip-in.	Plastic Moment kip-in.	LRFD Design Moment kip-in.	Ratio of Test Moment to Plastic Moment	Ratio of Test Moment to LRFD Moment
CFT1	5000	4432	1100	1.128	4.54
CFT2	5400	4910	2100	1.099	2.57
CFT3	5900	5074	1320	1.162	4.47
CFT6	4400	3562	1100	1.230	4.00

Note: 1kip-in.=0.113 kN-m

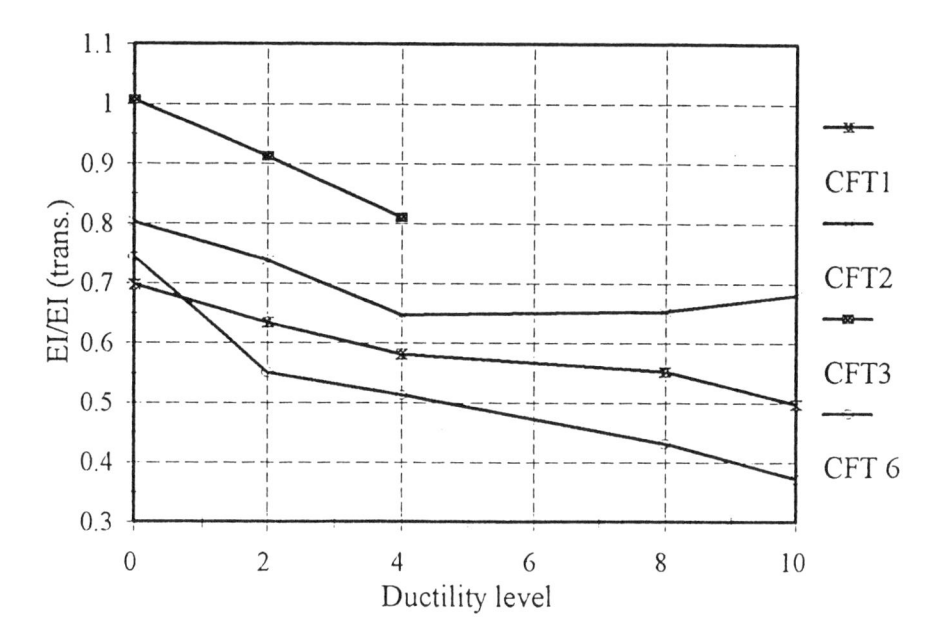

Figure 9. Stiffness Degradation

ONGOING WORK

The ongoing work includes calibration of a finite element model of composite element developed by Ayoub et al.[5], to predict behavior of CFT columns. Figure 10 shows the composite model, which consists of two beam-column elements and a bond element. The lower and upper elements represent the steel and concrete portion of the composite structure respectively. The beam-column portion of the model is capable of representing the hysteretic behavior of line element under combinations of cyclic biaxial bending and axial load. The bond element describes the interaction between steel and concrete portion of the composite structure. Even though the springs are lumped at the end nodes, the interface shear is distributed over the element length. This model has been successfully applied to predict behavior of composite beams.[5] Using this model the ongoing work will include analysis of composite structures under seismic loads.

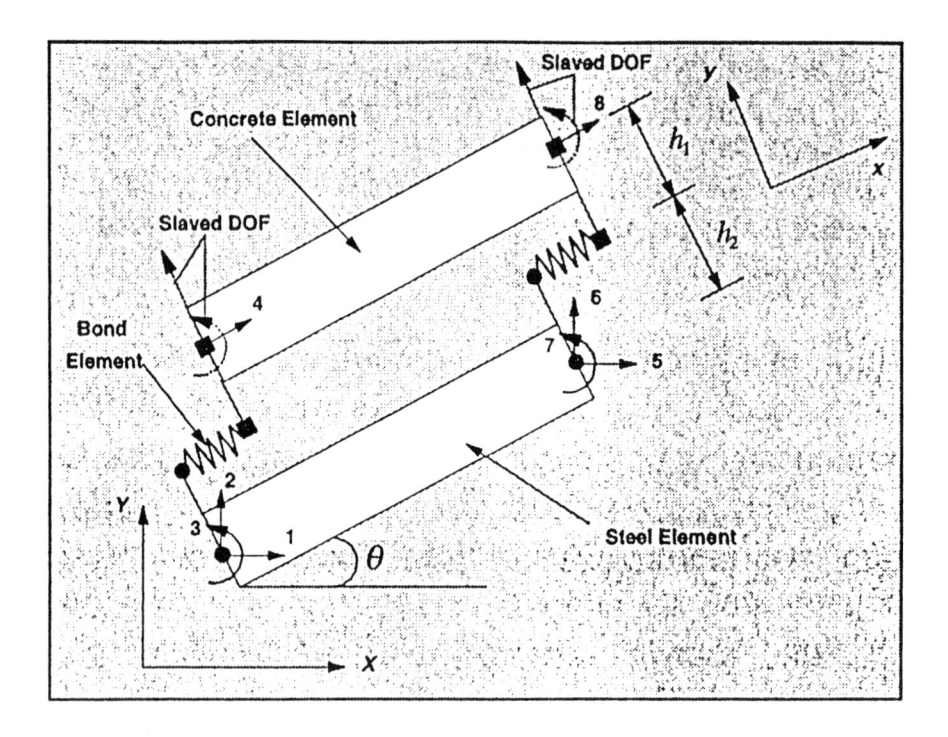

Figure 10. Composite Beam-Column Element

CONCLUSIONS

An experimental testing program was conducted for seismically loaded concrete filled tube columns. The results and observations obtained from the tests point to the following:

- The columns have very high energy dissipation capabilities and produce large hysteresis loops at ductility levels up to 10. These columns are suited for bridge columns located in seismic regions.
- The behavior of the column depends to a large extent on the critical section where the tube section plastically bulges.

- The envelope to the load displacement relationship of the column can be idealized by an elastic-perfectly plastic curve.
- The moment capacity of the specimen is in the range of 1.1 to 1.2 times the calculated plastic moment of its section.
- The initial stiffness of the column is in the range of 0.7 to 0.8 of the transformed section stiffness and stiffness degradation due to cyclic loading is mild.

ACKNOWLEDGMENT

The presented study in this paper is part of the U.S.-Japan Cooperative Research Program on Composite/Hybrid Structures, which is funded by National Science Foundation, with Dr. S. C. Liu as program director. Authors are grateful for this support. Authors would also like to acknowledge contribution of Valmont Industry, who are providing all test specimens. Results presented in this paper are the opinion of the authors and do not necessarily reflect the opinion of the sponsors.

REFERENCES

1. Gourley, B., Hajjar, J., and Schiller, P., "A Synopsis of Studies of the Monotonic and Cyclic Behavior of Concrete-Filled Steel Tube Beam-Column", *Structural Engineering Report No. ST-93-5.2*, Department of Civil Engineering and Mineral Engineering, University of Minnesota, Minneapolis, Minnesota, 1995.

2. Grauers, M., "Composite Columns of Hollow Steel Sections Filled with High Strength Concrete", *Ph.D. dissertation*, Chalmers University of Technology, Sweden, 1993.

3. Sugano S and Nagashima T and Kei T 1992, "Seismic Behavior of Concrete Filled Tubular Steel Columns", *Proceedings of the ASCE Tenth Structures Congress*, San Antonio, Texas, April 1992.

4. "NEHRP Recommended Provisions for Seismic Regulations for New Buildings", Building Seismic Safety Council, Washington, D.C. 1994.

5. Ayoub, A. and Filippou, C. F., " A Model for Composite Steel-Concrete Girders under Cyclic Loading", *Proceedings of Structures Congress*, Portland Oregon, PP 721-725, 1997.

TESTS OF TWO HIGH PERFORMANCE CONCRETE PRESTRESSED BRIDGE GIRDERS

Catherine French, Professor
Carol Shield, Associate Professor
Department of Civil Engineering
University of Minnesota
Minneapolis, Minnesota, U.S.A.

Theresa Ahlborn, Assistant Professor
Department of Civil Engineering
Michigan Technological University
Houghton, Michigan, U.S.A.

ABSTRACT

Two long-span prestressed bridge girders were fabricated using high performance concrete (HPC) and large diameter strand. Information including transfer lengths, prestress losses over time, cracking loads, behavior under repeated loads, and ultimate flexural and shear strength was obtained. Transfer lengths of large diameter strand in HPC were conservatively predicted by current design practice. There was no stiffness degradation observed through three million cycles of service loads. The flexural strength of the composite section was controlled ultimately by the compressive strength of the deck fabricated with normal strength concrete. Ultimate shear strengths were predicted conservatively using current design procedures.

INTRODUCTION

This paper summarizes the results of a study conducted at the University of Minnesota regarding the application of high performance concrete (HPC) to prestressed bridge girders. Advantages of using HPC include increasing bridge span lengths and spacings of bridge girders. The increased span lengths enable greater underpass clearance widths (fewer supports, bridge piers). In addition more slender, shallower long span members can be used to replace deeper members. This facilitates the transportation and placement of the members (lower loads on equipment transporting and erecting the girders). Using shallower members also enables reduced embankment heights or increased underpass clearance heights. The latter option can be used for retrofitting bridges with deficient underpass clearances by replacing deeper members with shallower members. Increasing the bridge girder spacings results in economic savings through requiring fewer girders per bridge. Savings are realized in material reductions and reduced transportation and erection costs involved with fewer required girders. Another primary benefit of high performance concrete is the improved durability and reduced long term maintenance to the bridge

The Department of Civil Engineering at the University of Minnesota research project comprised a comprehensive material test program and structural tests of two full-scale long span prestressed bridge girders. The combination of large diameter prestressing strands and HPC was used to push the envelope of the precast bridge girder span lengths to an excess of 30% longer than currently achieved in practice. Because current design codes were developed based on empirical results of tests on members made with lower strength concretes ($f'_c < 40$ MPa (6 ksi)), one of the purposes of the tests was to investigate whether present design standards are adequate for HPC.

MATERIAL TEST PROGRAM

The objective of this phase of the research was to document the effects of mix composition and proportions, curing, age, and test procedures on the mechanical properties and freeze-thaw durability of HPC[1]. Nearly 7000 specimens were tested from 142 HPC mixes. The mixes had water to cementitious materials ratios between 0.28 and 0.32.

The investigation showed that the aggregate type was the dominant variable. For high strength concretes, the strength of the paste and paste aggregate interface are sufficiently increased such that the strength becomes limited by failure of the aggregate. Consequently, further reductions in the water to cementitious materials ratio may not increase strength, and may cause problems by reducing the workability of the mix.

Based on the results of the material tests, it was decided to include mix design as a variable in the prestressed bridge girder investigation. Two types of mixes were considered for this purpose: a limestone mix because of its high strength without the need for addition of more expensive materials such as micro-silica, and a glacial gravel mix with micro-silica. The latter mix was considered because the use of round river gravel was typical for the local precast industry, and micro-silica was required in the mix to bring it to the desired compressive strength.

DESIGN, FABRICATION, AND TESTING OF
TWO LONG-SPAN HPC PRESTRESSED BRIDGE GIRDERS

To investigate the structural performance of prestressed bridge girders fabricated with HPC, two full size long-span composite girders were fabricated and tested. The preliminary design of the girders was based on results of a parametric study to investigate the viability and design implications of using HPC for Minnesota Department of Transportation (MnDOT) prestressed bridge girders.

Parametric Study
The parametric study was based on the assumption that the girders were members of a hypothetical bridge with a total width of 15.9 m (52 ft.) loaded with an AASHTO HS25

vehicle[2]. As the girder depth, spacing, and material strengths were varied, the maximum achievable span lengths were plotted with respect to the required number of strands.

An example of the results is shown in Figure1 for a given girder section, MnDOT 45M (Figure 2). Figure 1 shows the maximum achievable span lengths and required 15.3 mm (0.6 in.) diameter 1860 MPa (270 ksi) strands as a function of the girder spacing (1.2, 2.1 and 3.1 m (4, 7 and 10 ft.)) and concrete strength (48, 69, 83, 103 MPa (7, 10, 12 and 15 ksi)). The current MnDOT standard is to use 12.7 mm Gr. 1860 MPa (0.5 in. Gr. 270 ksi) strands with a maximum 28-day concrete compressive strength of 48 MPa (7000 psi).

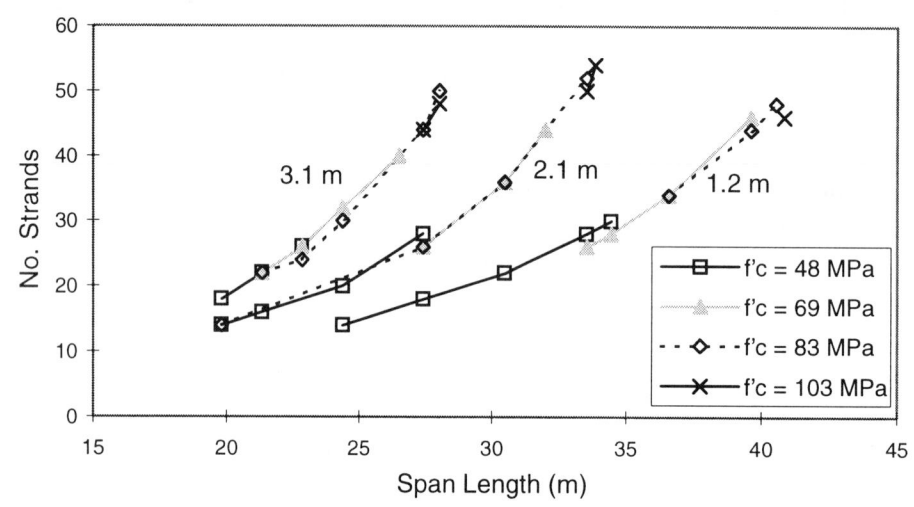

Figure 1. Parametric Study

The parametric study indicated that increases in maximum span lengths on the order of 20-35% can be achieved for concrete compressive strengths on the order of 69 MPa (10,000 psi). The use of higher concrete strengths resulted in smaller percentage gains in length, due to the limitation of the number of strands (15.3 mm, 1860 MPa (0.6 in., 270ksi)) that can be placed in the section; concrete strengths above 83 MPa (12 ksi) did not indicate any beneficial effect. As the concrete strength enables increases in span length, more and more strands are required. The larger number of strands results in decreased strand eccentricity and consequently reduced effectiveness of additional strands. Larger diameter (15.3 mm (0.6 in.)), higher grade (2070 MPa (Gr. 300 ksi)) strands can be used to increase the reinforcement effectiveness by providing a larger force at a larger eccentricity than smaller diameter, lower strength strands. Because of difficulties experienced in the fabrication of 2070 MPa (300 ksi) strand, 1860 MPa (270 ksi) strand was used for the experimental study.

Figure 1 also shows the correlation between increased span lengths and reduced girder spacings. It was decided to fabricate the experimental girders assuming a close girder spacing (1.2 m (4 ft.)) for which case later age strengths tend to control. As the girder spacing decreases, the self weight of the girder becomes a larger portion of the total load the girder must resist. As a consequence, later age strengths tend to control these cases. In the case of widely spaced girders, a large amount of prestressing force must be "stored" in the girders at release which will be required to carry the larger proportion of the loads at service. In this case the release strengths tend to control, and it may be more economical to allow the girder to cure for a longer period of time than to fabricate it with higher strength concrete that may not be required at later ages.

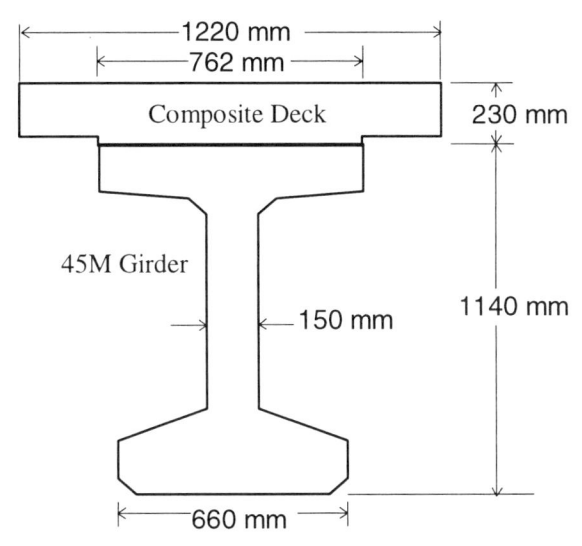

Figure 2. Composite Cross Section

Description of the Two HPC Test Girders

The two high-strength girders chosen for study were MnDOT 45M sections, 1140mm (45 in.) deep, reinforced with forty-six 15.3 mm (0.6 in.) diameter 1860 MPa (270 ksi) low-relaxation prestressing strands at 50.8 mm (2 in.) on center. A nominal 28-day concrete strength of 72 MPa (10.5 ksi) was required to maximize the span length to 40.5 m (132.75 ft.) for the assumed 1.2 m (4 ft.) girder spacing. The corresponding nominal release strength was 61 MPa (8.9 ksi). The actual target 28-day concrete compressive strength at the fabrication plant was designed to exceed 83 MPa (12 ksi) to insure that all of the test cylinders would exceed the nominal required strength. The noncomposite span-to-depth ratio of the girders was 35. Composite concrete decks were added to each individual girder using unshored construction, giving a composite span-to-depth ratio of 29. Figure 2 depicts the cross section of the composite members.

The girders were designed identically with the exception of three variations: the mix design (Girder I - limestone aggregate mix; Girder II - glacial gravel with silica fume mix); end strand patterns (Girder I - 4 draped / 8 debonded on each end; Girder II - 12 draped strands on one end and 4 draped / 8 debonded on the other end); and stirrup anchorage details (modified U stirrups with leg extensions on all ends except End A of Girder I which used standard U stirrups).

Fabrication of the HPC Test Girders

The two prestressed bridge girders were cast on the same bed in an outdoor precasting yard in August 1993. No modifications were made to standard construction techniques. Both girder mixes showed good workability and consolidation during placement. The girders were heat-cured under tarps using their own heat of hydration. Although both girders were cast on the same bed, Girder I was cast approximately 1.5 hours prior to the casting of Girder II. Girder II, with the silica fume mix, achieved its nominal release strength at an age of 14 hours and form removal began at 17 hours. The forms on Girder I were removed at an age of 22 hours, although Girder I did not achieve its release strength until a half hour later.

A total of 5.5 hours elapsed between the time the forms were removed from Girder II and release (2 hours for Girder I). During this time, cracks were observed to develop in Girder II. A total of fifteen vertical cracks were observed along the length of the girder, concentrated within the middle 50% of the span length. The cracks extended from the top flange towards the bottom flange. Eleven of the cracks extended approximately 864 mm (34 in.) deep (nearly to the bottom flange). Four of the cracks were less than 152 mm (6 in.) deep. Prestressing strands were flame cut to release them from the prestressing bed. Upon strand release the vertical cracks in the girder closed completely. If not for the lines drawn on Girder II to identify the initial crack locations, there was no indication the girder had experienced cracking prior to release.

Using unshored construction, composite decks were cast on the individual girders at an age of 200 days. Nominal and measured material properties for the girders and deck are given in Table 1.

Table 1. Material Properties

	Nominal	Girder I Measured	Girder II Measured
Girder Concrete			
Release Strength, f'_{ci} (MPa)	62	64	72
28-Day Strength, f'_c (MPa)	72	83	77
Release Modulus of Elasticity, E_{ci} (MPa)	41[†]	30	33
28 Day Modulus of Elasticity, E_c (MPa)	45[†]	33	33
Deck Concrete			
28-Day Strength (MPa)	28	40	40

[†]Based on $E_c = 0.043 w^{1.5} \sqrt{f'_c}$ with $w = 2480 \text{ kg/m}^3$

Test Program

A test program was developed to monitor girder transfer lengths, cambers, and prestress losses. The girders were each subjected to a series of static and cyclic load tests to investigate serviceability. The girders were then loaded with monotonically increasing loads to investigate the ultimate flexural strength. Following the flexural tests, each of the girder ends was loaded separately to investigate the shear strengths.

TRANSFER LENGTHS, PRESTRESS LOSSES, AND CAMBER OF TWO-LONG SPAN HPC BRIDGE GIRDERS

A summary of the transfer lengths, prestress losses, and initial cambers are described in this section. More detailed information can be found in Reference 3.

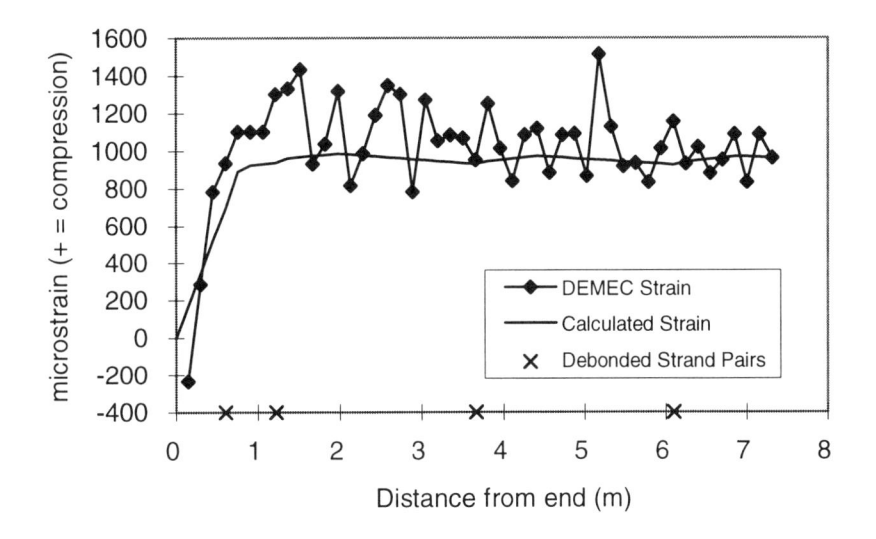

Transfer Length

The strain measured at release in the transfer region of Girder I (draped/debonded End A) is shown in Figure 3. Superimposed on the figure is the predicted strain distribution assuming the AASHTO[2] transfer length of 50 d_b (strand diameter). The shallow dip in the calculated strains indicates the effect of gravity load causing a decrease in the concrete compression strain at the level of the strands. The slight increases are caused by the initiation of bonding pairs of debonded strands along the length of the girder. The increases appear minor due to the small percentage of debonded strands in the cross section.

Figure 3. Transfer Length

Transfer lengths were determined graphically from the measured surface strains using the "95% Average Maximum Strain Method," and "Final Average Method[4]." These methods yielded transfer lengths in the range of 565 to 725 mm (22.2 to 28.5 in.) which were all less than those predicted using the AASHTO relationship of 50d_b (780 mm (30.7 in.)). This indicates that the AASHTO estimated transfer length relationship was a conservative predictor for these high strength concrete girders with the arrangement of 15.3 mm (0.6 in.) diameter strands on 50.8 mm (2 in.) centers.

Prestress Losses

Prestress losses occur instantaneously due to elastic shortening at release and over time due to steel relaxation, and creep and shrinkage of concrete. The components generating the prestress losses are interdependent, leading to the complex nature of predicting prestress losses and the state of stress in a member at any given time. The force in the prestressing strands continuously decreases until such time when the losses stabilize; the majority of the losses occur within the first 6 to 12 months of the member life.

Data were recorded from vibrating wire gages installed at the center of gravity of the strands (cgs) to monitor the change in concrete strain with time. Gages were installed at each of the following locations: 0.45L, 0.50L, and 0.55L. The change in concrete strain at the cgs was assumed equal to the change in strand strain. Multiplying this change in strain by the elastic modulus of the strand gave the change in strand stress. This method of experimentally determining losses does not account for strand relaxation losses, but this component is typically small for members with low-relaxation strands (typically less than 2%) and was superimposed on the measured data. In addition, the strands were tensioned four days prior to release. Consequently, a large proportion of the total strand relaxation took place prior to release.

The measured and predicted prestress losses are given in Table2 at release, 28-days, deck casting (200 days), time of the crack tests (Girder I - 598 days; Girder II - 727 days), time of ultimate flexural testing (850 days) and final service-life. The predicted losses have the same values for Girders I and II because they are based on nominal material properties (e.g. f'_c of 72 MPa (10,500 psi)).

Table 2. Predicted and Measured Prestress Losses (%)

Time	Girder I Measured	Girder II Measured	Predicted PCI[5]	Predicted AASHTO[2]
Release	14.5	12.8	10.5	10.6
28 days	20.7	16.4	19.6	---
Deck casting (200 days)	22.7	17.7	26.1	---
Cracking	25.6	20.2	28.9[†]	---
Flexural testing	26.3	20.7	29.0	---
Final Service	---	---	33.2	33.1

[†]Predicted losses were the same for both girders at the time of cracking tests (598 and 727 days)

At release, the measured losses were greater than predicted, especially in the case of Girder I (14.5 vs. 10.5%); whereas at later ages the predicted losses exceeded the measured losses. The early age losses tend to be underpredicted due to an overestimate in the girder stiffness. Conventional modulus of elasticity relations overpredict the stiffness of high strength concrete which causes an underestimation of the girder elastic shortening at release. At later ages, conventional concrete creep relations overpredict the concrete creep which results in an overprediction of the later losses.

Camber
Camber was monitored for both girders since the time of strand release. A predicted initial camber of 148 mm (5.82 in.) was computed for Girder I using measured material

properties and self weight based on the PCI Design Handbook Method[6]. Girder I had a measured (on-bed) camber of 121 mm (4.76 in.) which increased to 139 mm (5.47 in.) after lifting the girder from the bed and immediately setting it back down. The "actual" initial camber lies between these values. Friction between the precasting bed and the girder tends to reduce the initial on-bed camber and increase the lift/set camber measurement. A predicted initial camber of 132 mm (5.19 in.) was computed for Girder II using measured material properties and self weight. Girder II had an initial (on-bed) camber of 98 mm (3.86 in.) and a lift/set camber of 103 mm (4.06 in.).

FLEXURAL LOADING RESPONSE

Static and cyclic loads were applied to each composite girder using servo-controlled hydraulic actuators located 0.4L from the supports to simulate the moment induced by an AASHTO HS25 truck. The load was proportioned assuming a girder spacing of 1.2 m (4 ft.) center to center. Overload testing at 125% HS25 was also performed.

No stiffness degradation was observed after undergoing 1.0 million cycles in the uncracked state. An additional 2 million cycles (1 million at HS25, 1 million at overload) were applied to the cracked girders. During these cycles, no appreciable stiffness degradation was observed for either Girder I or Girder II.

The static load - deflection responses of Girders I and II are shown in Figure 4. During the flexural cracking tests, the girders were monitored for acoustic emissions (AE) which indicated the onset of microcracking in the concrete[7]. The y-axis represents the percent of HS25 truck loading applied. The AE monitoring equipment indicated that crack initiation began to occur at 178% HS25, and the first visual crack appeared at 217% HS25. These observed cracking loads were below the predicted cracking loads but well above the intended service capacity of the bridge girder. Girder II had a higher predicted cracking load due to the lower measured losses relative to Girder I, but the AE equipment indicated crack initiation at 136% HS25 and visual cracks were observed at 159% HS25.

Even though the predicted cracking loads were higher for Girder II, cracks were observed at much lower loads in Girder II than those observed in Girder I, although still above the intended service capacity. It was suspected that the pre-release cracks, which had only been observed in Girder II prior to strand release, could account for part of this discrepancy. To close the cracks upon release, geometric compatibility requires bending of the beam about the tip of the cracks. This deformation shortens the top fiber length of the girder by the sum of the lengths of the pre-release crack top openings and elongates the bottom fiber length of the girder. Consequently, concrete strains below the crack tips become less compressive than expected, thereby causing flexural cracking to initiate at a lower than expected load. The net effect of the pre-release crack closure is a reduction in the girder camber and flexural cracking load.

The pre-release cracks did not have a significant impact on the girders because the live load was a relatively small portion of the total load. The stress ranges due to the live load were less than 35 MPa (5 ksi). These cracks would have a more significant impact on wider spaced girders where live load is a greater proportion of the girder load; also wider spaced girders may be more susceptible to such cracks if they remain on the bed longer to achieve required release strengths.

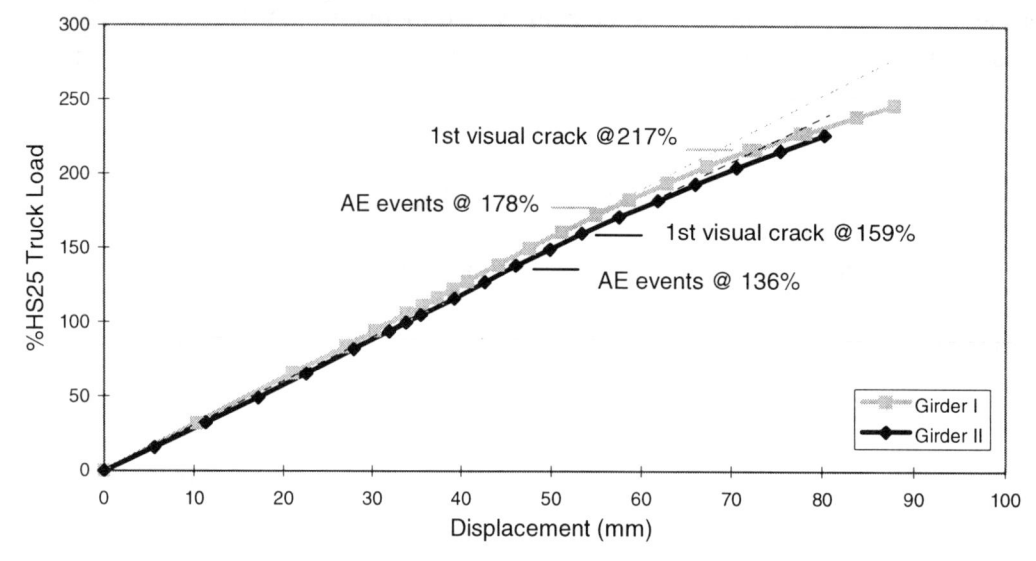

Figure 4. Load vs. Deflection Measured During Initial Flexural Crack Tests

Ultimate Flexural Strength

Following conclusion of the static and cyclic flexural tests, the two girders were loaded to failure. Both girders showed a substantial amount of ductility and strength (Figure5). Eventually the girders exhibited an explosive failure due to the anticipated compression failure of the normal strength concrete deck. Girder I carried a peak applied moment of 685% HS25 (9480 kNm (83,900 in-k)) with a corresponding displacement of over 760 mm (30 in.) at failure. Girder II had a peak load of 670% HS25 (9300 kNm (82,300 in-k)) with a corresponding centerline displacement of 870 mm (35 in.) at failure.

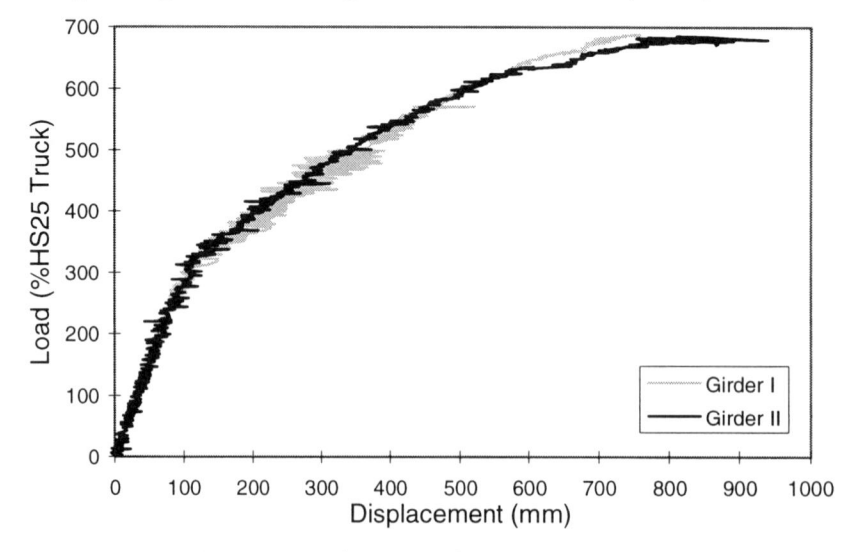

Figure 5. Ultimate Flexural Behavior

SHEAR LOADING RESPONSE

To investigate the shear capacity of the end regions of the girders, the girder ends were tested individually with a concentrated load applied approximately 4.2 m (13.9 ft.) from the end of an approximately 12.2 m (40 ft.) section of the girder. To prevent a shear failure from occurring at the severed end of the girder (damaged during the flexural tests), sheets of 0°/90° fiberglass were epoxied to the girder web for strengthening. The fiberglass sheets extended from the severed end of the girder to a distance of approximately 6 m (20 ft.) along the length of the girder.

The design shear strength for the girders was determined using the nominal section properties (e.g. f'_c of 10,500 psi). Based on AASHTO[2] and the ACI 318-95 Code[8], the girder ends were expected to carry loads of 1560 kN (350 kips) in the case of the ends with the debonded strands and 1740 kN (390 kips) for Girder End II-D which contained the 12 draped strands at the end. Measured results are shown in Figure 6.

The limestone girder (Girder I, Ends A and B) actually achieved strengths from 2230 to 2310 kN (500 to 520 kips). The limestone girder ends developed diagonal shear cracks which eventually extended into horizontal cracks at the interface between the web and the bottom flange accompanied by a drop in load capacity. The enhanced anchorage provided by the large amount of prestressing reinforcement in the bottom flange, limited the variation in results among the end reinforced with standard U-shaped stirrups (End I-A) and the end with modified U-shaped stirrups (End I-B). Both types of stirrups were observed to yield during the tests.

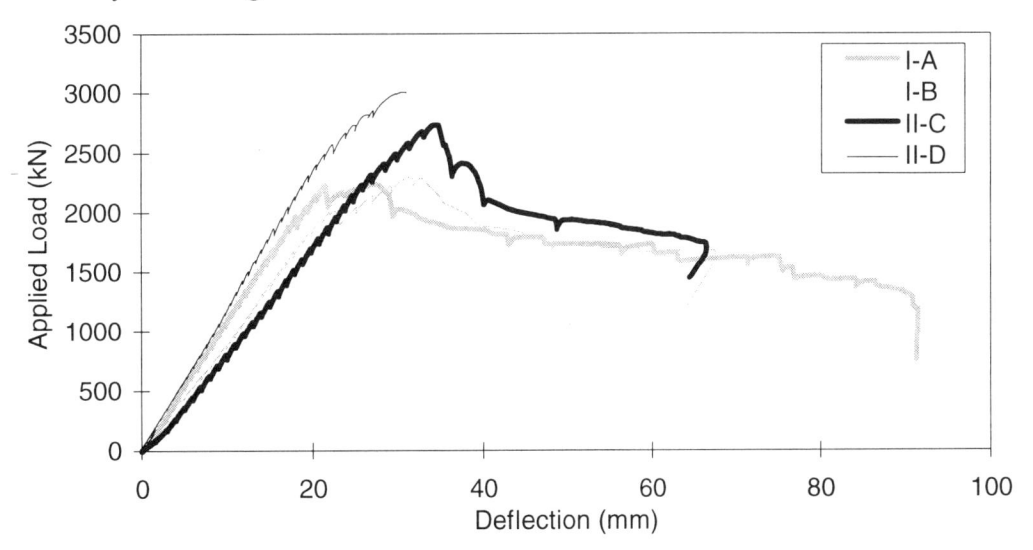

Figure 6. Shear Behavior

The glacial gravel girders incorporating micro-silica exhibited even greater strengths. The girder end with the modified U-stirrups and debonded strands, Girder End II-C, (similar in all respects to Girder End I-B, with the exception of concrete mix) carried a load of 2730 kN (614 kips) before failure. Girder End II-D was similar to Girder End II-C in all respects, with the exception that Girder End II-D did not have any debonded strands (12 draped strands rather than 4 draped / 8 debonded strands). The investigators were unable to develop a shear failure in Girder End II-D with the Universal Testing Machine in the Structural Engineering Laboratory at the University of Minnesota. The girder exhibited nearly elastic behavior even following the peak load level of 3010 kN (676 kips). Girder II eventually developed diagonal cracks which extended towards the support reaction.

The phenomenon of the cracks extending along the top of the bottom flange in the case of Girder I and diagonally to the support at the very end of the girder, in the case of Girder II, was attributed to the failure surface developed for the respective mix designs. The failure of the limestone mix tended to develop relatively smooth planes of failure, whereas the glacial gravel girder tended to have a more uneven failure plane due to the fracture passing around some of the aggregate rather than through all of the aggregate particles. This behavior would tend to increase the aggregate interlock in the case of the glacial gravel girders. It is believed that the compression stresses generated by the large numbers of strand in the bottom flange helped to confine/anchor each other, thereby preventing strand anchorage failure.

CONCLUSIONS

Two full size prestressed concrete bridge girders were monitored over time and load tested to investigate behavior. Results indicated that transfer lengths in both girders were approximately 80 percent of those predicted by AASHTO. AASHTO relationships were conservative for these larger diameter strands when used in high strength concrete girders and placed on 50.8 mm (2 in.) centers.

The static response of the girders to design truck loading was favorable. Serviceability testing under HS25 truck and overload (125% HS25) testing showed no stiffness degradation in either girder after being subjected to 1 million cycles in the uncracked state. Following flexural cracking, the girders were subjected to an additional 2 million cycles of loading. There was no change in girder stiffness observed during these load cycles at the HS25 and overload levels. The normal strength concrete of the composite decks eventually controlled the peak flexural loads as expected.

The shear strengths of the girders were more affected by the HPC because the shear failures were generated in the HPC precast portions of the girders. The observed shear capacities were much greater than anticipated. The increased capacities were attributed to the large amount of prestressing strands in the bottom flange which confined each other and the stirrups. The aggregate type was found to have a significant effect on the shear capacity (attributed to aggregate interlock). In the case of the limestone concrete mix, the crack plane tended to be much smoother than the case of the crack plane in the glacial gravel mix.

ACKNOWLEDGMENTS

This project has been collectively sponsored by the Minnesota Department of Transportation, Minnesota Prestress Association, University of Minnesota Center for Transportation Studies, Precast/Prestressed Concrete Institute, and the National Science Foundation Grant No. NSF/GER-9023596-02. The authors also wish to acknowledge the generous donations of materials, equipment and technical support by Elk River Concrete Products, Truck/Crane Services, Union Wire & Rope, Simcote, Inc., Lefebvre & Sons Trucking, Golden Valley Rigging, Atlas Foundation, Borg Adjustable Joist Hanger Co., United Technologies, Lehigh Cement Company, Holnam, Inc., National Minerals Corporation, J.L. Shiely Company, Edward Kraemer & Sons, Inc., Meridian Aggregates, W. R. Grace & Co., and Cormix Construction Chemicals Appreciation is also expressed for the assistance of graduate students Ali Mokhtarzadeh, Roxanne Kriesel, Dave Cumming, Jeffrey Kielb, Jeffrey Kannel, and Douglass Woolf. The views expressed herein are those of the authors and do not necessarily reflect the views of the sponsors.

REFERENCES

1. Mokhtarzadeh, A., Kriesel, R., French, C., and Snyder, M., "Mechanical Properties and Durability of High-Strength Concrete for Prestressed Bridge Girders," *Transportation Research Record No. 1478*, Washington, D.C., 1995, pp. 20-29.

2. American Association of State Highway Transportation Officials (AASHTO), "Standard Specifications for Highway Bridges" 15th Edition, Washington DC, 1993.

3. Ahlborn, T., Shield, C., and French, C., "Behavior of Two High Strength Prestressed Bridge Girders," *Worldwide Advances in Structural Concrete and Masonry, Proceedings of the CCMS Symposium/Structures Congress XIV*, 1996, pp. 141-152.

4. Ahlborn, T., French, C., and Leon. R., "Applications of High-Strength Concrete to Long-Span Prestressed Bridge Girders," *Transportation Research Record No. 1476*, Washington, D.C., 1995, pp. 22-30.

5. PCI Committee on Prestress Losses, "Recommendations for Estimating Prestress Losses," *PCI Journal*, Vol. 20, No. 4, 1975, pp. 44-75.

6. PCI, "PCI Design Handbook - Precast and Prestressed Concrete," 4th Edition, Chicago, IL, 1992.

7. Hearn, S., and Shield, C., "Acoustic Emission Monitoring as a Nondestructive Testing Technique in Reinforced Concrete," *ACI Materials Journal*, in Press 1997.

8. ACI Committee 318 "Building Code Requirements for Concrete, 318-95". American Concrete Institute, Detroit, MI, 1992.

PRESTRESSED I-GIRDER DESIGN USING HIGH PERFORMANCE CONCRETE AND THE NEW AASHTO LRFD SPECIFICATIONS

M. Myint Lwin, P.E.
Bijan Khaleghi, P.E.
Jen-Chi Hsieh, P.E.
Washington State Department of Transportation
Olympia, Washington, U.S.A.

ABSTRACT

Washington State Department of Transportation (WSDOT) is involved in the design, fabrication, and construction of prestressed concrete bridges using high performance concrete (HPC) and the AASHTO LRFD Bridge Design Specifications. A design example is used to illustrate the WSDOT design practice and the application of the LRFD Specifications to predict the time-dependent prestress losses by the Time-Step Method and the Modified Rate of Creep Method.

INTRODUCTION

In Washington State, the use of prestressed I-girders started in the 1950s. At that time, construction of highways and freeways was greatly accelerated under the new Interstate Highway Program. There was a challenge to quickly and cost-effectively build grade separations at highway crossings. The economy, quality in fabrication, and ease in construction of prestressed I-girder bridges met the challenge. By the late-1950s, WSDOT had developed standard I-girder sections to facilitate design and construction, and to save cost. Today, over 80% of the state highway bridges in Washington State are prestressed I-girder bridges.

The design of I-girders in WSDOT has been based on the provisions of the AASHTO Standard Specifications for Highway Bridges, except as modified and supplemented by WSDOT office practices. The prestressed I-girders are designed using allowable stress design for service conditions at all stages of loading and checked for ultimate strength capacity in accordance with load factor design. The concrete used in the prestressed I-girders generally has a minimum 28-day compressive strength in the range of 41 MPa (6 ksi) to 48 MPa (7 ksi).

AASHTO adopted and published the first edition of the new LRFD Bridge Design Specifications[1] in June 1994. At about the same time, Federal Highway Administration (FHWA) was encouraging state DOTs to use high performance concrete in bridge construction. FHWA was sponsoring demonstration projects to acquire information on the design, fabrication and construction of concrete bridges using HPC. WSDOT seized the opportunity to design a prestressed I-girder superstructure using the new AASHTO LRFD Bridge Design Specifications and HPC.

It is the purpose of this paper to share WSDOT's experience in using HPC and the LRFD Specifications in the design of prestressed I-girders.

HIGH PERFORMANCE CONCRETE

High performance concrete is one of many products researched and selected under the Strategic Highway Research Program (SHRP) as having higher quality for use in highway structures. HPC has enhanced durability characteristics and strength parameters not normally attainable by using conventional ingredients, normal mixing procedures, and normal curing practices. The enhanced durability characteristics improve resistance to thermal freeze/thaw cycles and to salts and other chemicals. The enhanced strength parameters include reduced creep and shrinkage, increased abrasion resistance and higher strength concrete[2]. It may be noted that high strength is only one of many attributes of HPC. The bridge engineers have the flexibility to specify the durability characteristics and strength parameters to meet the short and long term demands of a project with the goal of reducing construction and maintenance costs[7].

Use of HPC has the potential to:
1. Reduce weight and construction cost through using fewer lines of prestressed I-girders. The design example in this paper shows that it is feasible to reduce from 7 lines of girders to 5 lines;
2. Solve vertical clearance problem by using shallower girders;
3. Increase the applicability of prestressed I-girders by making longer lengths;
4. Minimize maintenance cost through increased durability.

Research efforts in recent years have demonstrated that HPC is constructable and can be used advantageously in highway bridge construction. Many state DOTs are beginning to use HPC in the construction of highway bridges. Washington State is one of six lead states tasked by the AASHTO SHRP Implementation Task Force to acquire and share knowledge in proper use of HPC in highway construction. The mission of the lead state team is to promote implementation of HPC technology for use in highway structures and share knowledge, benefits and challenges with other states.

This paper and several other papers on HPC in this conference are aimed at accomplishing the mission of the HPC Lead State Team.

DESIGN EXAMPLES

Design Conditions
This design example illustrates design of a typical interior girder for a simple span bridge.

Simple span of 40.56 m (133.1 ft) with 11.58 m (38 ft) roadway width and a 40.37° skew angle.
HL-93 live load - 3 lanes.
Use WSDOT's W74MG[3] girders at 2.44 m (8 ft) spacing.
Consider composite construction with 190 mm (7.5 in.) deck slab, which includes a 10 mm (0.4 in.) integral wearing surface. The typical section of the bridge is shown in Figure 1.

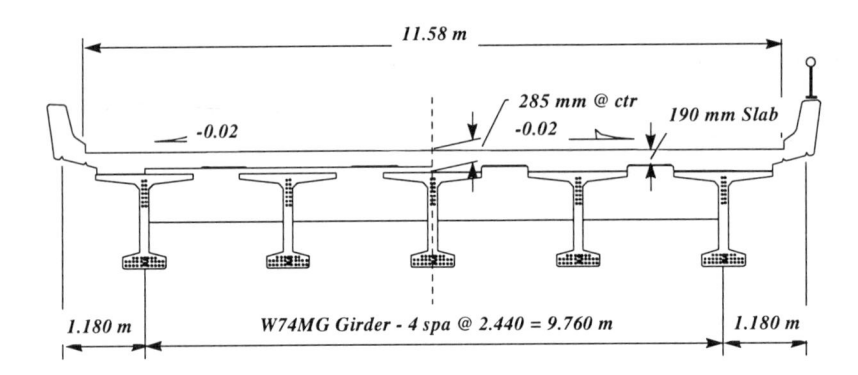

Figure 1 - Bridge Typical Section

Materials
Precast concrete: normal weight, 56 day's f'$_c$ = 69 MPa (10 ksi), f'$_{ci}$ = 51 MPa (7.4 ksi) (note: it can be shown f'$_c$ can be reduced to 55 MPa (8 ksi)).
Slab concrete: normal weight, 28 day's f'$_c$ = 28 MPa (4 ksi).
Density of concrete: for computing E$_c$, y$_c$ = 2480 kg/m^3; for weight, y$_c$ = 2560 kg/m^3.
Prestressing steel: AASHTO M-203M, uncoated 15 mm (0.6 in.) diameter, 7 wire, 1860 MPa (270 ksi) low-relaxation strands. Strand area = 140 mm^2 (0.217 in^2). E$_p$ = 197000 MPa (28500 ksi).
Reinforcing Steel: AASHTO M-31M, Grade 400 (60 ksi), E$_s$ = 200000 MPa (29000 ksi).

Section Properties
Use modular ratio, n = $\sqrt{f'_c}$(girder) / $\sqrt{f'_c}$(slab) = 1.57 ; section properties are shown in Table 1.

Limit States
Service I - 1.0 DC + 1.0 (LL+IM)
  Compression in prestressed components is investigated using this load combination.
Service III - 1.0 DC + 0.8 (LL+IM)
  Load combination relating only to tension in prestressed concrete structures with the objective of crack control.

Vehicular Live Load

Design live load, HL-93, shall be taken as:
(Design truck (HS20) or tandem) (1 + IM) + Lane

Dynamic load allowance, IM = 33%
The distribution factor can be shown to be D.F. = 0.63.

Table 1 - Section Properties

	Girder	Composite
depth, mm	1 865	2 045
Area, mm^2	485 300	765 100
I, mm^4	227.5 x 10^9	400.4 x 10^9
y$_b$, mm	970	1 330
S$_b$, mm^3	234.4 x 10^6	301.0 x 10^6
y$_t$ girder, mm	895	535
S$_t$ girder, mm^3	254.3 x 10^6	748.8 x 10^6
y$_t$ slab, mm	-	715
S$_t$ slab, mm^3	-	560.2 x 10^6

Note: 1 mm = 0.0394 in.

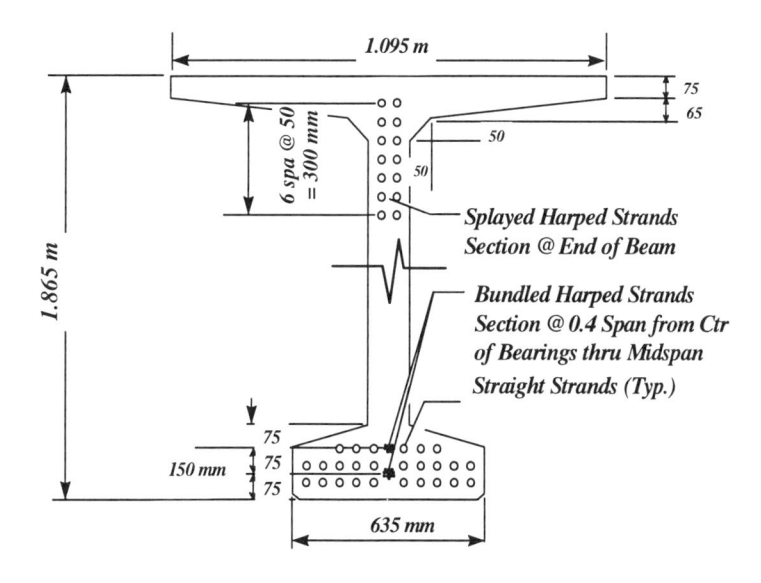

Figure 2 - Typical W74MG Prestressed Girder and Strand Pattern

Allowable Concrete Stresses at Service Limit State

Current WSDOT[3] design practice does not allow any tension at the bottom of prestressed girder at Service III limit state. The zero tension criteria practiced by WSDOT provides some reserve to mitigate construction, operational and fatigue related problems. It is usually achieved by adding one or two strands to the total number of strands per girder. The other allowable concrete stresses follow the LRFD Specifications.

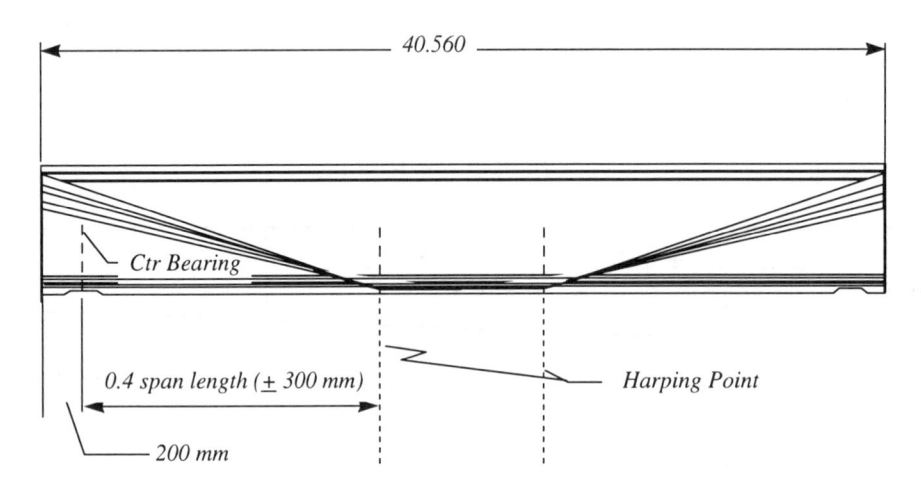

Figure 3 - Prestressed Girder Elevation

Determination of Prestressing Forces

A trial and error method was used to determine the proper number of prestressing strands. The jacking stress at transfer, $0.75 f_{pu}$, is used as allowed in the LRFD code.

Table 2 - Summary of Stresses at Service

	At mid-span			At harping point		
Stress (MPa)	f_b	f_t(girder)	f_t(slab)	f_b	f_t(girder)	f_t(slab)
Girder	10.15	-9.35	-	9.74	-8.98	-
Slab + haunch	12.26	-11.30	-	11.77	-10.85	-
Diaphragm	1.41	-1.30	-	1.27	-1.17	-
Traffic barrier	1.85	-0.74	-0.99	1.77	-0.71	-0.95
ΣDL	25.67	-22.69	-0.99	24.55	-21.71	-0.95
LL - Service I	-	-4.73	-6.33	-	-4.58	-6.12
LL - Service III	9.42	-	-	9.11	-	-
Prestressing	-35.98	8.66	-	-35.98	8.66	-
Stresses under permanent load	-	-14.03	-0.99	-	-13.04	-0.95
Allowables	-	-31.05	-12.60	-	-31.05	-12.60
Stresses under all loads	-0.89	-18.76	-7.32	-2.32	-17.62	-7.07
Allowables	0.00	-41.40	-16.80	0.00	-41.40	-16.80

Note: Tension (+); 1 MPa = 0.145 ksi

For the final design, 14 harped and 26 straight strands are used, as shown in Figures 2 and 3. The total prestress loss at service is calculated to be 283.5 MPa (42 ksi) using the Modified Rate of Creep Method (see next section). The final stresses at service at the mid-span and the harping point are summarized in Table 2. It can also be shown that concrete stresses at transfer satisfy the code requirement.

PRESTRESS LOSSES IN PRESTRESSED GIRDERS

Despite wide use of prestressing, there is no simple practical method for predicting accurately the time-dependent losses of prestressed concrete structures. This is partly because of the difficulty in predicting the time-dependent properties of concrete and prestressing steel, and the uncertain environmental conditions in which the structure will be subjected to after prestressing[5]. For HPC, the difficulty also comes from the limited experience and data available.

TIME-DEPENDENT PRESTRESS LOSSES

Prestress Loss Due to Creep

Creep loss due to permanent loads occurs immediately after initial prestressing. The stress in the concrete at the level of prestressing steel at transfer is the elastic response, while the creep effect will occur over a long period of time under a sustained load. For composite girders, part of initial compressive strain induced in the concrete immediately after transfer is reduced by the tensile stress resulting from superimposed permanent loads. Loss of prestress due to creep is proportional to the net compressive stress in the concrete. In accordance with the AASHTO LRFD Specifications, the creep loss is given by:

$$\Delta f_{PCR} = 12.0\, f_{cgp} - 7.0\, \Delta f_{cdp} \quad > 0.0 \tag{1}$$

where:
f_{cgp} = sum of the concrete stresses due to prestressing and weight of the girder at the center of gravity of prestressing strands at mid-span;
Δf_{cdp} = change in concrete stress at the level of prestressing strands due to applied dead loads of concrete deck slab and diaphragms.

The term $7.0\, \Delta f_{cdp}$ in the above equation is the approximate estimate of prestress gain due to dead load of slab and diaphragm.

In the Modified Rate of Creep Method, prestress loss due to creep may be considered in two stages:
Stage 1 is the creep loss between time of transfer and slab casting which may be expressed as:

$$\Delta f_{PCR1} = n\, f_{cgp}\, \Psi_{t,tisc}\, (1 - \Delta F_{SC}/2F_o) \tag{2}$$

Stage 2 is the creep loss for any time after slab casting which may be expressed as:

$$\Delta f_{PCR2} = n\, f_{cgp}\, (\Psi_{t,ti} - \Psi_{t,tisc})\, (1 - (\Delta F_{SC} + \Delta F_t)/2F_o)\, I_g/I_c \qquad (3)$$

where:

$\Psi_{t,ti}$ = creep coefficient of girder at any time;
$\Psi_{t,tisc}$ = creep coefficient of girder at the time of slab casting;
ΔF_{SC} = total loss at the time of slab casting minus initial elastic shortening loss;
F_o = prestressing force at transfer after elastic losses;
ΔF_t = total prestressing loss at any time minus initial elastic shortening loss;
I_g/I_c = ratio of moment of inertia of prestressed girder to composite girder.

In above equations, the terms $\Psi_{t,tisc}(1 - \Delta F_{SC}/2F_o)$ and $(\Psi_{t,ti} - \Psi_{t,tisc})(1 - (\Delta F_{SC} + \Delta F_t)/2F_o)$ represent the effect of variable stress history from the time of transfer to the time of slab casting and from slab casting to final conditions, respectively. The term I_g/I_c represents the effect of composite section properties after slab casting.

Prestress Loss Due to Shrinkage

Shrinkage of concrete can vary over a wide range depending on the material properties and surface drying conditions. Prestress loss due to shrinkage of concrete for prestressed members according to AASHTO LRFD Specifications is given by:

$$\Delta f_{PSR} = (117.0 - 1.035\, H) \qquad (4)$$

where:

H = relative humidity;
Δf_{PSR} = shrinkage loss in MPa.

Prestress Loss Due to Relaxation of Prestressing Strands

Relaxation of prestressing strands depends upon the stress level in the strands. However, because of the other prestressing losses, there is a continued reduction of the strand stress, thus causing a reduction in prestress. The reduction in strand stress due to elastic shortening of concrete occurs instantaneously. On the other hand, the reduction in strand stress due to creep and shrinkage takes place for a long period of time.

The loss due to relaxation of prestressing strands given by AASHTO LRFD Specifications is considered in two stages. Relaxation loss at transfer and relaxation loss after transfer of prestress force. In a prestressed member, with low-relaxation strands, the relaxation loss in prestressing steel, initially stressed in excess of $0.5 f_{pu}$, is given by:

$$\Delta f_{PR1} = \log(24t)/40.0\, [f_{pj}/f_{py} - 0.55]\, f_{pj} \qquad \text{at transfer} \qquad (5)$$

$$\Delta f_{PR2} = 0.30 \, [\, 138 - 0.4 \, \Delta f_{PES} - 0.2(\Delta f_{PSR} + \Delta f_{PCR})] \quad \text{after transfer} \quad (6)$$

In prestressed members, the part of the loss due to relaxation which occurs before transfer may be deducted from the total relaxation loss by an adjustment in initial prestress force.

Prestress Gain Due to Slab Casting

The AASHTO LRFD Specifications recognizes the prestress gain by the term $7.0 \, \Delta f_{cdp}$ in the creep equation. In the Modified Rate of Creep Method, the creep effect of slab and diaphragms dead load may be considered as a prestress gain. Part of initial compressive strain induced in the concrete immediately after transfer is reduced by the tensile strain resulting from permanent loads. The prestress gain due to slab dead load consists of two parts. The first part is due to instantaneous elastic prestress gain. The second part is time-dependent creep effect. Prestress gain due to elastic and creep effect of slab casting is given as[6]:

$$\Delta f_{EG} = n_{SC} \, f_{S+D} \qquad \text{elastic prestress gain} \qquad (7)$$

$$\Delta f_{CRG} = n_{SC} \, f_{S+D} \, (\Psi_{t,ti} - \Psi_{t,tisc}) \, I_g/I_c \qquad \text{creep effect prestress gain} \qquad (8)$$

where:

n_{SC} = modular ratio at the time of slab casting;
f_{S+D} = stress in concrete at the level of prestressing strands due to dead load of slab and diaphragms.

Prestress Gain Due to Differential Shrinkage

In composite prestressed girders bridges, the concrete in the girder is steam-cured while the concrete in slab is usually cast-in-place and moist-cured. Slab concrete is also cast at a later time when girders are in place. Due to the difference in the quality of concrete, curing process and time of casting, a prestress gain due to differential shrinkage may be considered as[6]

$$\Delta f_{DS} = n_{SC} \, f_{CD} \qquad (9)$$

where:

$f_{CD} = [\Delta \varepsilon_{S\text{-}G} \, A_S \, E_S \, / \, (1 + \Psi_{t,ti})] \, (y_{CS} \, e_c \, / \, I_c)$ = concrete stress at the level of prestressing strands

$\Delta \varepsilon_{S\text{-}G}$ = differential shrinkage strain;
A_S = area of concrete deck slab;
E_S = modulus of elasticity of slab;
y_{CS} = distance between the c.g. of composite section to the c.g. of slab;
e_c = eccentricity of prestressing strands in composite section;
I_c = moment of inertia of composite section.

The denominator $(1 + \Psi_{t,ti})$ approximates the long term creep effect.

METHODS OF DETERMINATION OF TIME-DEPENDENT LOSSES

Approximate Lump Sum Estimate of Time-dependent Prestress Losses

According to AASHTO LRFD Specifications, the approximate time-dependent prestress losses, for I-shaped girders, prestressed with low-relaxation strands, and ultimate strength of 1860 MPa (270 ksi), without partial prestressing and strength of concrete above 41 MPa (6 ksi), is given by:

$$\Delta f_{PT} = 230 \left[1 - 0.15 (f'_c - 41) / 41\right] \tag{10}$$

For specified concrete strength above 35 MPa (5 ksi), methods such as, AASHTO-LRFD Specifications, the Time-Step Method and Modified Rate of Creep Method may be used.

Refined Estimates of Time-Dependent Losses

The refined estimate of time-dependent prestress losses, according to AASHTO LRFD Specifications, is given as the summation of losses due to shrinkage and creep of concrete and relaxation of prestressing strands.

$$\Delta f_{PT} = \Delta f_{PSH} + \Delta f_{PCR} + \Delta f_{PR} \tag{11}$$

Time-Step Method

The Time-Step Method is a numerical step by step procedure based on having constant values over short time intervals. During each time interval, all the factors having an influence on the long-term deformations and prestress losses are taken into account.

Prestress force in a prestressed concrete member continuously decreases and at any time may be estimated based on the effective prestressing force in the prestressing strands. Effective prestressing force at any time is considered as the initial applied force at the time of transfer minus the total estimated loss at that time. Several cycles of loss calculation are needed in order to converge to an acceptable range of variation[1].

Modified Rate of Creep Method

The Modified Rate of Creep Method takes into account the instantaneous and time-dependent effect of slab casting, and the transition from non-composite to composite section properties.

Prestress losses due to each time-dependent source, may be evaluated in detail by considering the effect of creep and shrinkage on different construction stages. Casting of deck slab in composite girders affects prestressing force in the member by the influence on the creep coefficient, shrinkage strain of concrete and steel relaxation. When an added load is applied sometimes after initial prestressing, and sustained on the structure up to time t_i, its contribution to the prestress force may be considered by including the elastic and creep effect gain due to dead load of slab, using appropriate creep factors before and after slab casting. The effect of differential shrinkage between concrete deck slab and prestressed girder on prestress losses shall be included in time-dependent losses. In detailed evaluation of prestress losses due to each time-dependent parameter, the appropriate section properties of the stage under consideration shall be used. Time-dependent prestress losses using the Modified Rate of Creep may be given as:

$$\Delta f_{PT} = \Delta f_{PSH} + \Delta f_{PCR1} + \Delta f_{PCR2} + \Delta f_{PR} - \Delta f_{EG} - \Delta f_{CRG} - \Delta f_{DSH} \qquad (12)$$

For non-composite girders, the time-dependent prestress losses may be taken as:

$$\Delta f_{PT} = \Delta f_{PSH} + \Delta f_{PCR1} + \Delta f_{PR} \qquad (13)$$

Comparison of Prestress Losses

Table 3 summarizes time-dependent prestress losses obtained from different methods discussed in this paper. Losses from the Time-Step Method are lower than the ones obtained from the AASHTO LRFD refined estimate analysis, because the Time-Step Method is based on effective prestress force rather than the initial force at transfer. Prestress losses computed from the Modified Rate of Creep Method are lower than the ones obtained from other methods, because this method performs an in-depth analysis of the effects of slab casting. It includes, as mentioned previously, the elastic and creep effect of slab casting as well as the prestress gain obtained from differential shrinkage, using appropriate creep coefficient and shrinkage strain of concrete before and after slab casting.

Table 3 - Comparison of Prestress Losses in MPa

	AASHTO-LRFD Approx. Method	AASHTO-LRFD Refined Method	Time-step Analysis	Modified Rate of Creep Method
Transfer	159.2	159.2	159.2	159.2
Before Slab Cast.	-	238.5	227.3	197.3
After Slab Cast.	-	232.5	214.2	175.8
Final	348.2	376.1	327.5	283.5

1 MPa = 0.145 ksi

DEFLECTION OF PRESTRESSED GIRDER

Prestressed girder deflection may be evaluated in detail by considering the effect of creep and shrinkage of concrete on different construction stages. Prestress force produces moment and axial force in the girder tending to bow the girder upward. Girder dead load resists this upward deflection, but is overpowered by the deflection due to prestressing and the girder continues to deflect upward due to creep effect. The result for a non-composite girder prior to slab casting is a net upward deflection. Casting of slab results in a downward deflection, it is also accompanied by continuing downward deflection due to creep and differential shrinkage between precast girder and slab which are cast in different times and cured in different process. The long-term downward deflections after slab casting, in the most part, is compensated for by the long-term upward deflection due to prestressing. Many measurements of actual structure deflections have shown, that once slab is cast, the girder tends to acts as it is locked in position. The final deflection of a prestressed girder according to the Modified Rate of Creep Method may be given as[6]:

$$\Delta t = \Delta_{DLg} + \Delta_{DLgbsCR} + \Delta_{P/S} + \Delta_{P/SbsCR} + \Delta_{P/SasCR} + \Delta_{DLs} + \Delta_{DLgasCR} + \Delta_{DLsCR} + \Delta_{SH} \quad (14)$$

where:

Δ_{DLg} is the deflection due to girder dead load and may be taken as:

$$\Delta_{DLg} = 5 w_g L^4 / 384 E_g I_g \quad (15)$$

$\Delta_{DLgbsCR}$ is the deflection due to creep effect of girder dead load and may be taken as:

$$\Delta_{DLgbsCR} = \Psi_{t,ti} \Delta_{DLg} \quad (16)$$

$\Delta_{DLgasCR}$ is the creep effect deflection of composite girder due to slab load after slab casting and may be taken as:

$$\Delta_{DLgasCR} = \Delta_{DLg} (\Psi_{t,ti} - \Psi_{t,tisc}) I_g / I_c \quad (17)$$

Δ_{DLs} is the deflection due to slab plus diaphragm dead and may be taken as:

$$\Delta_{DLs} = 5 w_s L^4 / 384 E_g I_g + 19 P L^3 / 384 E_g I_g \quad (18)$$

Δ_{DLsCR} is the creep effect deflection of composite girder after slab casting and may be taken as:

$$\Delta_{DLsCR} = \Psi_{t,ti} \Delta_{DLs} I_g / I_c \quad (19)$$

$\Delta_{P/S}$ is the deflection due to straight and harped prestressing strands and may be taken as:

$$\Delta_{P/S} = P_s e_s L^2 / (8 E_g I_g) + P_h / E_g I_g (L^2/8 - a^2/6)(e_s - e_h) \quad (20)$$

$\Delta_{P/SbsCR}$ is the creep effect deflection of prestressing before slab casting and may be taken as:

$$\Delta_{P/SbsCR} = [(1-\Delta F_{SC}/2F_o)\Psi_{t,ti} - \Delta F_{SC}/F_o]\Delta_{P/S} \quad (21)$$

$\Delta_{P/SasCR}$ is the creep effect deflection of prestressing of composite girder after slab casting and may be taken as:

$$\Delta_{P/SasCR} = [(1-(\Delta F_{SC}+\Delta F_t)/2F_o)(\Psi_{t,ti}-\Psi_{t,tisc}) - (\Delta F_t-\Delta F_{SC})/F_o]\Delta_{P/S} I_g/I_c \quad (22)$$

Δ_{SH} is the deflection due to differential shrinkage and may be taken as:

$$\Delta_{SH} = [\Delta\varepsilon_{S-G} A_S E_S / (1 + \Psi_{t,ti})](y_{CS} L^2 / 8E_g I_c) \quad (23)$$

where:
- w_g = unit weight of girder;
- E_g = modulus of elasticity of girder;
- w_s = unit weight of slab and pad;
- P = weight of diaphragm;
- P_s = prestressing force in straight strands;
- P_h = prestressing force in harped strands;
- a = distance from the centerline of bearing to the harping point;
- e_s = eccentricity of straight strands from neutral axis;
- e_h = eccentricity of harped strands from neutral axis.

The denominator $(1 + \Psi_{t,ti})$ approximates the long term creep effect.

In above equations, the terms $\Psi_{t,ti}(1-\Delta F_{SC}/2F_o)$ and $(\Psi_{t,ti}-\Psi_{t,tisc})(1-(\Delta F_{SC}+\Delta F_t)/2F_o)$ represent the effect of variable stress history from the time of transfer to the time of slab casting and from slab casting to final conditions, respectively. The term I_g/I_c represents the effect of composite section properties after slab casting.

Prestressed girder deflection according to Modified Rate of Creep Method is presented in Table 4.

Table 4 - Prestressed Girder Deflection

	Total Deflection in mm
At Transfer	-80
Before Slab Casting	-107
After Slab Casting	-56
Final	-44

Note: Upward deflection (-); 1 mm = 0.0394 in.

CONCLUDING REMARKS

The new LRFD Specifications are logical and relatively straightforward for the design of prestressed I-girders. The LRFD Specifications provide reasonable methods for predicting the prestress losses due to creep and shrinkage. However, the Modified Rate of Creep Method produces more accurate prestress losses. The LRFD design yields similar results as the LFD design using HS-25 live loading.

High performance concrete is constructable and can be used advantageously in highway construction. The enhanced strength parameters result in lower creep and shrinkage values and make it possible to increase span length, build shallower girders and reduce the number of lines of girders.

REFERENCES

1. AASHTO-LRFD Bridge Design Specifications, First Edition, 1994 and Interim Specifications through 1996.

2. Roy, D. M., "Superior Microstructure of High Performance Concrete for Long-term Durability," Report No. 1478, Material Research Laboratory, The Pennsylvania State University, University Park, PA 16802, 1995.

3. Bridge Design Manual, Publication No. M23-50, Washington State Department of Transportation, Bridge and Structures Office, Olympia, Washington, 1997.

4. AASHTO Standard Specifications for Highway Bridges, Sixteenth Edition, 1996.

5. Hernandez H.D. and Gamble W.L., "Time-Dependent Prestressed Losses in Pretensioned Concrete Construction," Report NO. UILU - ENG - 75 - 2005, Department of Civil Engineering, University of Illinois at Urbana Champaign, Illinois, May 1975.

6. Branson D.E. and Panarayanan K.M, "Loss of Prestress of Non-composite and Composite Prestressed Concrete Structures," PCI Journal, September-October 1971.

7. Goodspeed, C. et al., "HPC Defined for Highway Structures," ACI Concrete International, Feb. 1996.

TIME-DEPENDENT EFFECTS
IN HIGH PERFORMANCE CONCRETE BRIDGE MEMBERS

XIAOMING HUO
Ph.D. Candidate
University of Nebraska
Omaha, Nebraska, USA

MAHER K. TADROS
Cheryl Prewett Professor
University of Nebraska
Omaha, Nebraska, USA

ABSTRACT

This paper shows how to calculate creep and shrinkage deformations in precast pretensioned bridge I-girders with cast-in-place composite deck. It further illustrate how reduced creep and shrinkage properties of high performance concrete (HPC) affect the prediction of prestress losses and deflections at various stages of construction and bridge service. Analysis of differential creep and shrinkage between the precast girder and the cast-in-place deck is given. Three examples are presented to illustrate the theory and the significance of accurate analysis compared to standard multipliers.

INTRODUCTION

Properties of HPC, which include creep, shrinkage and modulus of elasticity, are important for assessment of time-dependent stress redistribution and member deformation. The interaction between precast girder concrete and cast-in-place deck concrete causes time-dependent changes in stresses and deformations in the member which should be accurately accounted for. This paper shows the basic assumption used in the calculation of time-dependent effects in precast pretensioned girders acting compositely with CIP decks. If the member is simply supported and if the time-dependent interaction between steel and concrete is ignored, it is possible to determine the effects by hand. Two examples are given to illustrate how to conduct the analysis and to show the impact of the time-dependent interaction.

If the member reinforcement is taken into account, if loading is applied in several stages and if the member is made continuous over the piers; the problem becomes too detailed to do calculation by hand. Example 3 is intended to show how a computer analysis, using the program CREEP3 and the theory explained for Examples 1 and 2, could be utilized to complete the analysis. It further shows that stress and deflection changes could be significantly different from those resulting from the PCI Design Handbook multipliers. Finally the results of a parameteric study are given to show the impact of using reduced creep and shrinkage parameters corresponding to HPC.

TIME-DEPENDENT MATERIAL PROPERTIES

Creep and Shrinkage of Concrete

The creep and shrinkage of conventional concrete may be predicted using the procedures specified in ACI Committee 209.[1]

The elastic strain due to a constant stress $f_c(t_1)$ applied at time t_1, in days, can be expressed as

$$\varepsilon_c(t_1) = \frac{f_c(t_1)}{E_c(t_1)} \tag{1}$$

When a constant stress f_c is applied at time t_1 and sustained until time t_2, total strain at time t_2, $\varepsilon_c(t_2)$ can be expressed as

$$\varepsilon_c(t_2) = \varepsilon_c(t_1)[1 + C(t_2, t_1)] + \varepsilon_{sh}(t_2, t_1) \tag{2}$$

where $C(t_2, t_1)$ is a creep coefficient that represents the ratio of creep to instantaneous strain. It is a function of concrete age at loading t_1 and duration for which creep is calculated. $C(t_2, t_1)$ in ACI Committee 209 Report is presented as follows:

$$C(t_2, t_1) = \frac{(t_2 - t_1)^{0.60}}{10 + (t_2 - t_1)^{0.60}} C_u \tag{3}$$

C_u is the ultimate creep coefficient for concrete loaded at 1 to 3 days for steam cured concrete, or at 7 days for moist cured concrete, with the load sustained until time infinity. The "standard" value of C_u is 2.35 for members exposed to a relative humidity (RH) of 40%. A correction factor

$$\gamma_{RH} = 1.27 - 0.0067\,(RH) \qquad \text{for RH} > 40 \tag{4}$$

is used for variation in relative humidity. Other correction factors for member thickness, volume-to-surface ratio, temperature, and concrete composition are available in Ref. 1. However these correction factors have little impact on the $C(t_2, t_1)$ value.

Equations for predicting unrestrained shrinkage strain at any time are recommended in ACI Committee 209 Report. It is expressed as

$$\varepsilon_{sh} = 780 \times 10^{-6} \frac{t_2 - t_1}{a + (t_2 - t_1)} \tag{5}$$

where a = 35 days for concrete moist cured for 7 days and a = 55 days for concrete steam cured for 1-3 days. 780×10^{-6} is the ultimate value of unrestrained shrinkage strain under standard conditions.

For ambient relative humidity greater than 40 percent, use the correction factor in Eq.(6)

$$\gamma_{RH} = b - c \, (RH) \tag{6}$$

where RH is relative humidity in percent, b = 1.40, c = 0.010, for $40 \leq RH \leq 80$ and b = 3.00, c = 0.030, for $80 > RH \leq 100$.

The main factors influencing shrinkage are similar to those of creep: ambient relative humidity, thickness of members, water-cement ratio, aggregate type, and method of curing. The equations to calculate shrinkage are given in the ACI Committee Report. Research is underway at the University of Nebraska to establish correction factors for creep and shrinkage to allow use of the above prediction formulas for concrete strength up to 83 MPa (12,000 psi). The research confirms that both creep and shrinkage of HPC develop at a relatively fast rate in the first several weeks. Then, the time development considerably slows down, with ultimate values being as low as 50 percent of those of conventional concrete. The results of this research should be available for presentation at the Symposium.

Relaxation of Prestressed Steel
Relaxation is defined as the loss of stress in prestressed steel as it maintains constant length. The amount of the relaxation is dependent mainly upon initial prestress force and time. The equation of steel relaxation at time t_2 for low relaxation steel is based on a PCI Committee Report, and reported in Reference 2.

$$f_{ro} = \frac{1}{45} f_{pi} \left(\frac{f_{pi}}{f_{py}} - 0.55 \right) log \left(\frac{24 t_2 + 1}{24 t_1 + 1} \right) \tag{7}$$

f_{pi} = initial stress at time t_1 in days, and f_{py} = the yield strength which is taken = 0.9 f_{pu} for low-relaxation strands.

EXAMPLE 1 - TIME-DEPENDENT EFFECTS IN A NON-COMPOSITE MEMBER

This example illustrates the time-dependent stresses and deformations of a non-composite simply supported beam. A 12.2 m (40 ft) long beam with no reinforcement is shown in Fig. 1. The modulus of elasticity of concrete at the time of loading is $E_c(t_1)$ = 30,000 MPa (4,300 ksi). The creep coefficient C = 2.0 and the shrinkage strain ε_{sh} = 400 $\times 10^{-6}$.

w = 0.3 k/ft

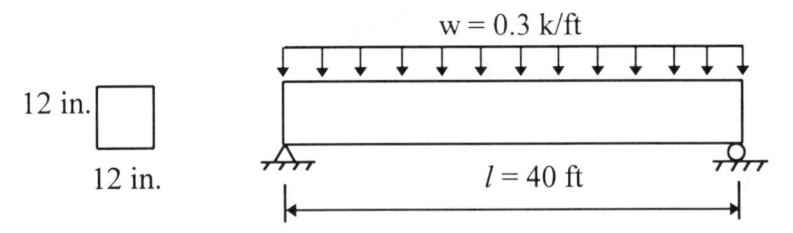

12 in.
12 in.

l = 40 ft

Note: 1 in = 25.4 mm, 1 ft = 0.3048 m, 1 k/ft = 14.59 kN/m
Figure 1. A Simply Supported Prism Beam in Example 1

Initial stress and deflection of the beam are calculated by using elastic analysis theory. The initial shortening of the beam is zero because no axial load is involved. At time infinity, the final stress is the same as the initial stress because time-dependent deformations do not cause any additional stresses. The final deflection is the sum of the initial deflection and the deflection due to creep. Since C= 2.0, the final deflection is (1+2) times the initial deflection. Shrinkage causes shortening of the beam equal to ε_{sh} times the beam length. The calculated results are given in Table 1.

STIFFNESS ANALYSIS FOR TIME-DEPENDENT EFFECTS IN COMPOSITE MEMBERS

The stiffness method of structural analysis is the most convenient approach to use for composite members, especially when analysis is done by computer. The steps used in analysis are as follows: (1) separate the section components (girder & deck) and allow each component to deform with time; (2) force the components to have zero deformation (restore compatibility); (3) reconnect the components and apply equal and opposite forces to the values in Step 2 (restore equilibrium); (4) conduct structural analysis on the composite member using the forces in Step (3) and an equivalent "age-adjusted" effective modulus of elasticity of the component materials for calculation of section properties; and (5) the sum of stresses and deformations obtained in Steps (1) through (4) produces the final values.

EXAMPLE 2 - TIME-DEPENDENT EFFECTS IN A COMPOSITE MEMBER

This example illustrates the impact of creep and shrinkage on time dependent stresses and deformations in a composite member containing the same precast concrete beam given in Example 1 plus a 152 mm × 610 mm (6 in. × 24 in.) cast-in-place deck. The deck is assumed to act compositely with the girder after the girder has undergone some time-dependent deformation. The assumed properties are: ultimate creep = 2.0 and shrinkage = 400 × 10^{-6}. Creep of girder due to deck weight = 1.0, shrinkage of girder after composite action = 200 × 10^{-6}. Modulus of elasticity of girder concrete = 30,000 MPa (4,300 ksi) at deck loading. Modulus of elasticity of deck at start of composite action = 26,000 MPa (3,800 ksi).

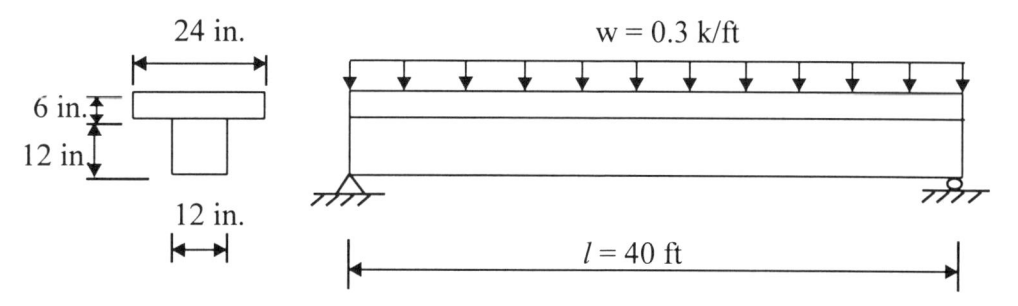

Note: 1 in = 25.4 mm, 1 ft = 0.3048 m, 1 k/ft = 14.59 kN/m

Figure 2. A Simply Supported Composite Beam in Example 2

The stiffness method was used to determine the time-dependent stresses and deformations of the composite member. Calculation results are summarized in Table 1. The table shows, for each example, the values with time-dependent effects ignored, and with time-dependent effects considered.

Table 1 - Summary of Results in Examples 1 & 2

	Initial Values	Final Values			
		Example 1		Example 2	
		T.D. Ignored	T.D. Considered	T.D. Ignored	T.D. Considered
$f_{top,\ deck}$ (ksi)	0	-	-	0	-0.24
$f_{bottom,\ deck}$ (ksi)	0	-	-	0	-0.01
$f_{top,\ girder}$ (ksi)	-2.50	-2.50	-2.50	-2.50	-1.31
$f_{bottom,\ girder}$ (ksi)	2.50	2.50	2.50	2.50	1.71
Deflection (in.)	2.33	2.33	6.99	2.33	7.36
Shortening (in.)	0	0	0.19	0	0.26

Note: 1 in. = 25.4 mm, 1 ksi = 6.8948 MPa

The analysis results in Table 1 show that ignoring the time-dependent effects could produce a significant error. The stresses in the beam and deck are redistributed and deformation of the member increases due to the creep and shrinkage developed with time. It is necessary to consider the time-dependent effects in order to accurately predict the stresses and deformation of members. Note that because the beam of Example 1 is non-composite, i.e. the concrete does not interact with another type of concrete or with reinforcement, the initial elastic analysis stresses remain unchanged with time. The deflection, however, increases with time due to creep. Also member shortening develops with time due to concrete shrinkage. The same beam combined with a concrete topping, in Example 2, undergoes continuous changes in stresses due to the interaction between the two concretes. The top beam fibers continue to lose compression and to transmit that compression to the deck.

COMPUTER ANALYSIS OF COMPOSITE MEMBERS

CREEP3 is a structural analysis program which considers the time-dependent effects of creep and shrinkage of concrete, and relaxation of steel[2,3]. Time dependent analysis in the step-by-step method is an incremental linear analysis of strains and stress of the member. The stress at the end of each interval is calculated in terms of stress increments that have occurred in preceding intervals. Linear creep law is adopted, and the stress distribution of a cross section is always linear. During interval i an increment axial strain $\Delta\varepsilon_c(i)$ occurs.

$$\Delta\varepsilon_c(i) = \frac{\Delta N_c(i)}{A_c E_{ce}(i)} + \Delta\varepsilon_c'(i) \tag{8}$$

where $\Delta\varepsilon_c(i)$ is the strain increment in concrete during interval i, $\Delta\varepsilon_c'(i)$ is defined by

$$\Delta\varepsilon_c'(i) = \sum_{j=1}^{i-1} \frac{\Delta N_c(j)}{A_c E_c(j)} \left[C\left(i+\frac{1}{2},j\right) - C\left(i-\frac{1}{2},j\right) \right] + \Delta\varepsilon_{sh}(i) \tag{9}$$

where j refers to the time at the middle of the jth intervals, and $(i-1/2)$ and $(i+1/2)$ are the time at the beginning and end of the ith interval, respectively; $\Delta\varepsilon_{sh}(i)$ is free shrinkage strain during interval i, $\Delta N_c(i)$ and $E_c(j)$ are the axial force increment and the modulus of elasticity of concrete at the middle of interval j, respectively, and A_c is the cross-sectional area of concrete.

The program is applicable for multi-stage construction, such as stages of pretensioning and post-tensioning. Using the computer program, stresses in concrete and steel can be evaluated and the deflection at various stages of construction can be determined.

EXAMPLE 3 - COMPUTER ANALYSIS OF A PRESTRESSED CONCRETE BRIDGE WITH COMPOSITE SECTION

A simple span bridge with Bulb-Tee 1829 mm (72 in.) girders is analyzed with consideration of time-dependent effects [4]. The span length is 36.27 m (119 ft). The girder spacing is 3.66 m (12 ft). The elevation and typical cross section of the bridge are shown in the Fig. 3. 44 - 13 mm (0.5 in.) φ low relaxation strands are used to prestress the girder. The concrete strength of girders at 28 days is 48 MPa (7,000 psi) and at release is 34 MPa (5,000 psi). The concrete strength of the cast-in-place deck is 27 MPa (4,000 psi). The construction stages are as follows: (1) release of prestress and application of girder self weight at 1 day; (2) placement of cast-in-place deck at 35 days; and (3) application of superimposed dead load at 120 days.

The computer program CREEP3 was used to determine the time-dependent stresses and deformation in each stage of construction. For the purpose of comparison, structural analysis ignoring the time-dependent effects is also conducted for this example. The analysis results are shown in the following figures.

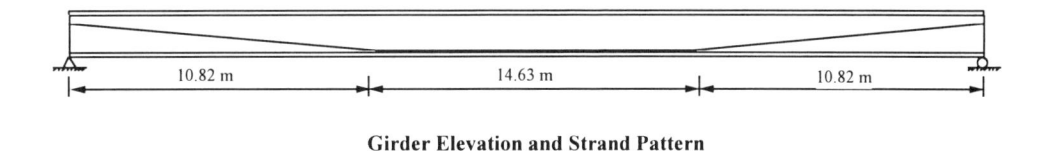

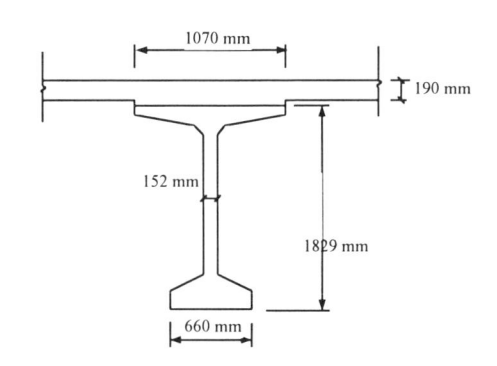

Girder Elevation and Strand Pattern

1829 mm Bulb Tee Girder Section

Figure 3. Bridge elevation and typical cross section

Deflection

Bridge deflections at midspan of the girder from computer analysis and hand-calculation are shown in Fig. 4. In the figure, positive deflection is downward.

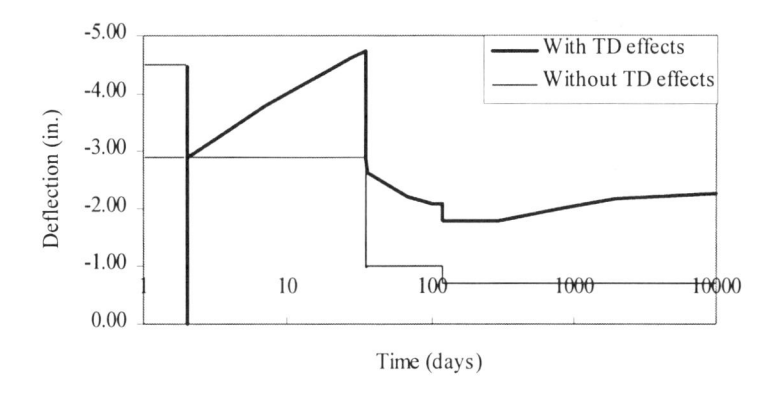

Note: 1 in. = 25.4 mm

Figure 4. Effect of Creep, Shrinkage, and Relaxation on deflections

The immediate camber due to prestress is 113 mm (4.45 in). Girder self weight reduces the camber of the girder to 73 mm (2.88 in.). The computer analysis shows that the deformation of the girder changes with time. The time-dependent deformation by computer analysis is 57 mm (2.26 in.) at time infinity, while the deformation from hand-calculation changes only at the time of load application. Ignoring time-dependent effects causes about 40 mm (1.56 in.) difference in final deformation at about 10,000 days, which corresponds to about 70% error.

Table 2 shows the individual immediate deflections due to prestress, girder self weight, CIP deck weight, and superimposed dead load. The long term deflection determined by the PCI multipliers and by the computer analysis are also presented in Table 2.

Table 2 - Comparison of Deflection by PCI Multipliers and Computer Analysis

Cause	Elastic deflection in.	35 days PCI multiplier	35 days PCI Δ in.	35 days Computer Δ in.	Final PCI multiplier	Final PCI Δ in.	Final Computer Δ in.
Prestress	4.45 ↑	1.8	8.01 ↑		2.2	9.79 ↑	
Girder Self Weight	1.57 ↓	1.85	2.90 ↓		2.4	3.77 ↓	
Deck Weight	1.88 ↓	1.0	1.88 ↓		2.3	4.32 ↓	
Superimposed Dead Load	0.30 ↓				3.0	0.9 ↓	
Total			3.23 ↑	2.86 ↑		0.80 ↑	2.26 ↑
Percent Difference			13 %			65 %	

Note: 1 in. = 25.4 mm

The results show that, from the prestress release to the time of casting the CIP deck (35 days), the deflection determined by PCI multiplier and computer analysis are within 13% of each other. But at time infinity, there is a large difference between the two methods. The PCI multiplier method does not correctly predict the long term deflection of girders due to its inability to adequately account for differential creep and shrinkage between the girder and the deck.

Stresses in Prestressed Girder

The analysis results shown in Figs. 5 and 6 are the top and bottom fiber stresses at midspan section of the bridge with the time-dependent effects fully ignored and fully considered. The deck stresses are also shown in Fig. 5. The analysis results show that the compressive stress at the top fiber of girder decreased with time, while the compressive stress at the top fiber of deck increased. This indicates that creep of girder top was more dominant and shrinkage of the deck bottom. After the girder self weight is applied, both the top and bottom fibers of the girder are under compression. Without considering the time-dependent effects the stresses in the girder are changed only when new loads are applied to the girder, such as casting the deck or applying the superimposed dead load. The computer analysis shows the stress change with time.

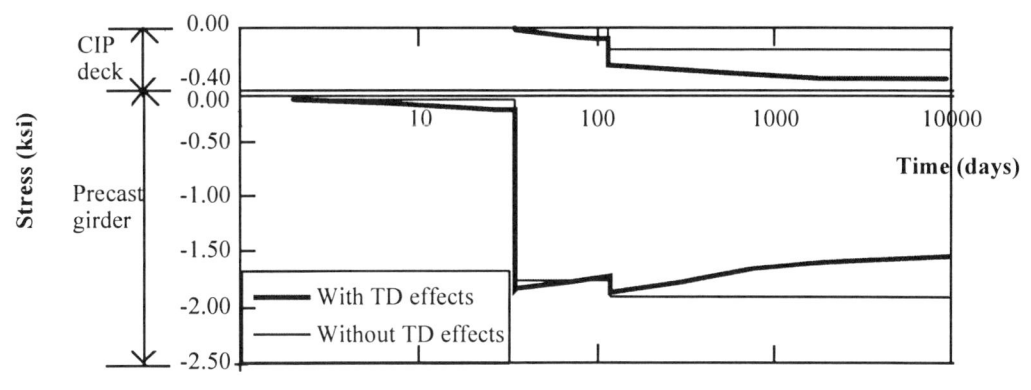

Figure 5. Effect of Creep, Shrinkage and Relaxation on Top Fiber Stresses

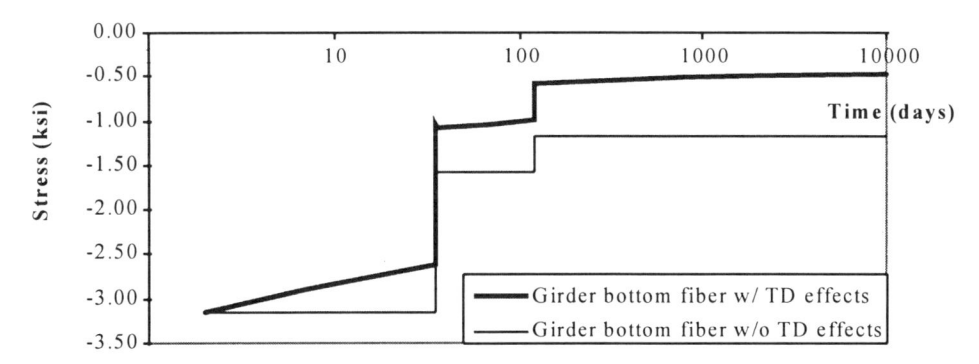

Figure 6. Effect of Creep, Shrinkage and Relaxation on Bottom Fiber Stresses

At time infinity, the actual compressive stress at the top fiber of girder is 10.57 MPa (1.533 ksi) while the compressive stress from elastic analysis is 13.03 MPa (1.891 ksi). The difference between the two methods is about 20%. At the bottom fiber of the girder, stress without time-dependent effect is 8.07 MPa (1.171 ksi) in compression while the final compressive stress with time-dependent effects is decreased to 3.34 MPa (0.485 ksi). The difference between the two methods was about 35%. The stresses in the girder are redistributed to the deck with time. It should be noted that the difference in stress in the bottom fiber of girder could result in unsafe conditions. The increased tension stress may cause the girder to crack.

Stress in Prestressing Strands
Fig. 7 shows the change of stress in the prestressing strands with time by computer analysis. The initial prestress of strands is $f_i = 0.75 f_p$ = 1396 MPa (202.5 ksi). The final stress in prestressed strands is 1051.5 MPa (152.5 ksi). According to the computer analysis, the initial prestress loss is 143.4 MPa (20.8 ksi) and the final prestress loss is 344.7 MPa (50 ksi), which is about 10.3% and 24.7% of initial prestress force, respectively.

Note: 1 ksi = 6.8948 MPa

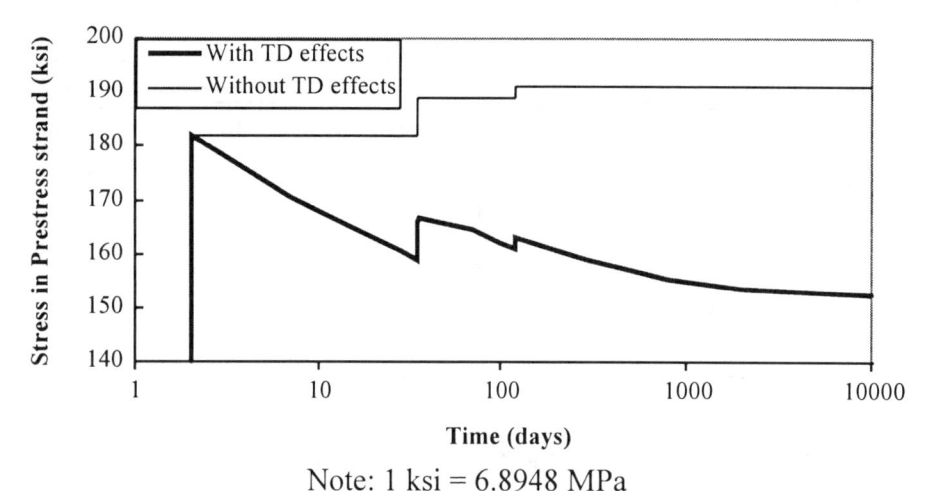

Figure 7. Effect of Creep, Shrinkage and Relaxation on Prestress Strands

For the purpose of comparison, the results from predication formulas in the AASHTO Specifications are also presented here. The total prestress losses can be expressed as:

$$\text{Total losses} = SH + ES + CR_c + CR_s \tag{10}$$

where SH = loss of prestress due to concrete shrinkage
ES = loss of prestress due to elastic shortening
CR_c = loss of prestress due to creep of concrete
CR_s = loss of prestress due to relaxation of prestressing steel

Using Eq. 10 the initial prestress losses, ES, are 143.4 MPa (20.8 ksi) and the final prestress losses are 344.7 MPa (50 ksi). Details of calculation of the prestress losses are shown in the Ref. 4. The difference between the results from the computer analysis and the AASHTO formula is small for the case of conventional concrete with relatively large creep and shrinkage coefficients.

TIME-DEPENDENT EFFECTS IN HPC BRIDGE

For the same bridge in Example 3, two more cases were considered using the computer program to study the effects of HPC. In the second case of analysis, 44 - 13 mm (0.5 in.) φ strands were used with HPC. In the third case, 44 - 15 mm (0.6 in.) φ strands were used with HPC. For both cases, the concrete strength of girder was 90 MPa (13,000 psi) at service and 59 MPa (8,500 psi) at release. The concrete strength of deck was 34 MPa (5,000 psi). The creep coefficient used in the analysis is 1.41, instead of the 1.88 for the first case, in order to account for lower creep deformation of HPC. Ongoing research of the University of Nebraska will be presented at the Symposium to show how to predict creep for HPC.

Table 3 lists the analysis results of prestress losses and deflection from the three case studied. The difference between Case 1 and Case 2 is that HPC was used in Case 2 while conventional concrete was used in Case 1. The results in Table 3 show that the initial prestress losses in these two cases were close. Because of using HPC and the lower creep coefficient and shrinkage strain, the final losses in Case 2 are much smaller than in Case 1. The difference of final losses between the two cases is about 40%. The initial deflection in Case 1 is larger than that in Case 2. The difference of deflections between the two cases is about 28% at the time of prestress release. The difference increases to 34% at the time of casting the CIP deck due to different time-dependent effects. The small final deflection in Case 2 shows less impact of time-dependent effects than in Case 1.

Table 3 - prestress losses and deflection of conventional concrete and HPC bridges

	Case 1 44 - 0.5 in. ϕ strands $f'_{c, girder}$ = 7,000 psi $f'_{c, deck}$ = 5,000 psi	Case 2 44 - 0.5 in. ϕ strands $f'_{c, girder}$ = 13,000 psi $f'_{c, deck}$ = 5,000 psi	Case 3 44 - 0.6 in. ϕ strands $f'_{c, girder}$ = 13,000 psi $f'_{c, deck}$ = 5,000 psi
Elastic losses (ksi)	20.8 (10.3 % of initial prestress)	19.2 (9.5 % of initial prestress)	23.3 (11.5 % of initial prestress)
Final losses (ksi)	50.0 (24.7 % of initial prestress)	33.5 (15.0 % of initial prestress)	49.9 (21.2 % of initial prestress)
Initial deflection (in.)	-2.88	-2.08	-3.47
Deflection at erection (in.)	-2.86	-1.88	-4.02
Final deflection (in.)	-2.26	-2.05	-4.91

Note: 1 ksi = 6.9 MPa; 1 in. = 25.4 mm

Both Case 2 and Case 3 used HPC and related properties in the analysis. The only difference between the two cases is that the girder in Case 2 had 44 - 12.7 mm (0.5 in.) ϕ strands while the one in Case 3 had 44 - 15 mm (0.6 in.) ϕ strands. The initial prestress force in Case 3 is larger than the one in Case 2. The analysis results show that both the initial and final prestress losses are larger than in Case 2. This illustrates that the higher initial prestress force in the section, caused by the availability of higher concrete compression capacity, results in higher prestress losses. The deflection at midspan in Case 3 is 88 mm (3.47 in.) upward at release. This deflection is higher than the one in Case 2 because of higher initial prestress force. The difference of deflection between two cases is 35 mm (1.39 in.) at initial and 73 mm (2.86 in.) at final. The higher prestress force in Case 3 is the main reason for the large difference in time-dependent deflection. It could also be seen that the higher the prestress force, the higher the time-dependent deflection could be. This fact is important to realize: HPC allows for higher amounts of prestress than conventional concrete, which in turn results in higher initial and long term cambers.

The lower creep coefficient and shrinkage could cause lower prestress losses and time-dependent deflection in the girder. But higher initial prestress force, could cause significantly larger time-dependent effect. One should be aware that the time-dependent effects could cause tremendous difference in the camber of bridge girders, especially when large prestress forces are used. These large cambers must be accounted for in determining cast-in-place deck elevations during construction.

CONCLUSIONS

The analysis results from the first two examples show that the effect of creep and shrinkage causes significant differences in stress and deflection when the beam is composite with a CIP deck. Creep and shrinkage will always increase the deflection of the member regardless of whether the section is composite or non-composite.

The bridge examples with the composite section also show the impact of time-dependent effects when the girder is under multi-stage loading. Considering the time-dependent effects can cause large differences in stresses and deflection of bridges. The analysis results also show that the deflection determined by PCI deflection multipliers is reasonable up to the time of casting of the deck. After that, the section becomes composite, and the PCI deflection multipliers become invalid. Significant stress redistribution takes place in the girder and the deck due to the time-dependent effects.

Three case studies of conventional concrete bridges and HPC bridges are presented to show the impact of the time-dependent performance of bridges. When using HPC, bridges are expected to have less time-dependent effects, such as lower prestress losses of strands and smaller time-dependent deflection, only if the same levels of prestress are used as for conventional concrete. However, higher concrete strength would likely be associated with higher prestress forces. This would result in a situation where the initial and long-term cambers can be significantly higher than for conventional concrete.

REFERENCES

1. "Prediction of Creep, Shrinkage, and Temperature Effects in Concrete Structures," ACI Committee 209 Report, 1992.
2. CREEP3, Computer software developed by Professor M.K. Tadros, University of Nebraska-Lincoln, Omaha, Nebraska.
3. Tadros, M.K., Ghali, A. and Dilger, W.H., "Time-Dependent Analysis of Composite Frames," ASCE, Journal of the Structural Division, April 1977.
4. *PCI Bridge Design Manual*, Precast/Prestressed Concrete Institute, Chicago, IL, 1997(expected).
5. "Standard specifications for highway bridges", American Association of State Highway and Transportation Officials, Inc., Washington D.C.

SHEAR LIMIT OF HPC BRIDGE I-GIRDERS

Zhongguo (John) Ma	Maher K. Tadros	Mantu Baishya
Ph.D. Candidate	Cheryl Prewett Professor	Research Assistant Professor
University of Nebraska	University of Nebraska	University of Nebraska
Omaha, Nebraska, USA	Omaha, Nebraska, USA	Omaha, Nebraska, USA

ABSTRACT

Due to recent optimization of I-girder shapes for maximum flexural efficiency and increasing use of high performance concrete (HPC), shear strength controlled member design is possible. AASHTO Standard Specifications require that V_s not exceed $8\sqrt{f_c'}b_w d$ (f_c' in psi), otherwise the section web must be widened or the depth increased. On the other hand, LRFD Specifications require a maximum shear limit of $0.25f_c'b_w d_v$. The difference between the two limits may be as high as 100 percent. The purpose of this research is to theoretically and experimentally establish a limit on shear capacity before member size must be increased. Recent experiments have shown that a key factor is anchorage of the prestressing steel at the member end.

INTRODUCTION

Shear strength of prestressed concrete beams is too complex to establish exclusively with theory. Available design methods attempt to offer semi-empirical procedures of varying theoretical rigor, which attempt to account for the parameters that influence design. Such parameters include material properties, level of prestress, shear span/depth ratio, amount of reinforcement, moment/shear ratio at cross section in question, impact of support conditions, etc. Most of the complexity is in determining the concrete contribution, V_c, to the overall shear resistance. A recent study by Ma, Saleh, and Tadros (1997)[1] compared the design of a two-span I-girder and a single span box girder bridges using various shear design methods. The purpose of that study was to determine the significance of varying the methods of shear design on the overall cost of the precast girder. The results of the study showed that, although V_c varied very significantly from one method to the other, the overall cost of the girder caused by the variation of the total amount of shear reinforcement was affected a maximum of 1.1 percent.

Another important issue besides refining the value of V_c is the maximum allowed shear reinforcement. The AASHTO Standard Specifications'(AASHTO-STD) limit on $V_s/b_w d$ is $8\sqrt{f_c'}$ (f_c' in psi) which is 716 psi (4.9 MPa) for 8,000 psi (55.2 MPa). The AASHTO LRFD (AASHTO-LRFD) limit on $V_n/b_w d_v$ is $0.25f_c'$ which is 2,000 psi (13.8 MPa) for 8,000 psi (55.2 MPa) concrete. The difference between the two limits is very significant. A maximum limit that is too low could result in unnecessary increase in member depth, web width or concrete strength. It may, in an extreme situation, result in the designer or owner abandoning a precast concrete alternate in favor of structural steel. This situation is becoming possible recently due to the economies achieved in using high strength concrete shallow bridge I-girders for relatively long spans at relatively wide girder spacings.

The shear limit ($8\sqrt{f_c'}$ on the V_s/b_wd term) of the AASHTO-STD was based on 166 tests of relatively short, deep reinforced concrete beams (149 rectangular beams and 17 T-beams) with the concrete strength ranging from 1,500 psi to 7,000 psi (10.4 to 48.3 MPa) and with vertical stirrups as reported by ACI-ASCE Committee 326[2]. No tests have been done to check its suitability to prestressed concrete bridge girders except for PCA's 13 tests of half-scale models of AASHTO-PCI Type III prestressed bridge girders[3]. PCA's research (1961) appears to be the source of the $8\sqrt{f_c'}$ limit on V_s/b_wd. That research was conducted on specimens whose shear reinforcement corresponded to a V_s/b_wd not exceeding $7\sqrt{f_c'}$. That maximum experimental value appears to have been part of the cause for the code limitation of $8\sqrt{f_c'}$.

In the 1960s, web crushing of flanged beams was brought into research focus in America. As a result of the research, the CEB-FIP limits the shear force in vertical thin webs to $0.2f_c' b_wd$ in beams with vertical stirrups and $0.25f_c' b_wd$ in beams with 45 degree stirrups. These limits are close to the AASHTO-LRFD shear limit of $0.25f_c' b_wd_v$. However, no effort has been undertaken to verify this limit experimentally for the I-sections used in bridge construction which consist of significant top and bottom flanges, especially when high strength concrete is used. The objective of the University of Nebraska research reported here is to determine an acceptable shear limit for thin webbed prestressed HPC bridge I-girders. This paper presents the initial testing results. The testing program is expected to be completed in the summer of 1997. Results and final recommendations will be ready for reporting at the PCI/FHWA Symposium.

EXPERIMENTAL PROGRAM

Four NU 1100 girders were fabricated for this research program, and are shown in Figs 1 to 4. Two of these girders were designed for 8,000 psi (55.2 MPa) concrete strength. They are referred to as specimens "A" and "B". The other two girders were designed for 12,000 psi (82.8 MPa) concrete. They are referred to as specimens "C" and "D". Other major variables besides concrete strength are draped strands versus shielded strands (as shown in Fig. 1), different types of shear reinforcement with different steel grades, including conventional double-legged stirrups (as shown in Fig. 2), "vertical" welded wire fabric (WWF) (Figs. 2 and 4), and orthogonal WWF (as shown in Fig. 4). Please note that "vertical" welded wire fabric designates a fabric where the main reinforcement is the vertical wires; top and bottom longitudinal wires are used for anchorage only, Figs. 2 and 4. Fig. 5 shows the heaviest orthogonal WWF used, consisting of D31 wires at 4

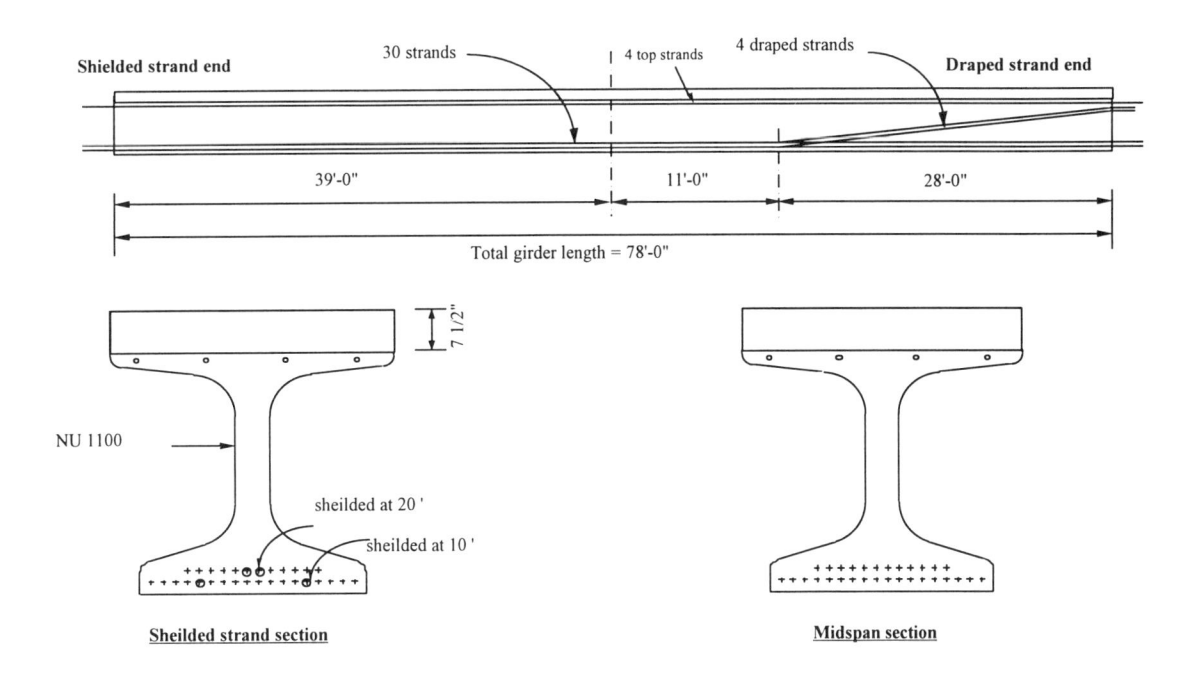

Figure 1. Elevation and Cross Section of Specimens A and C

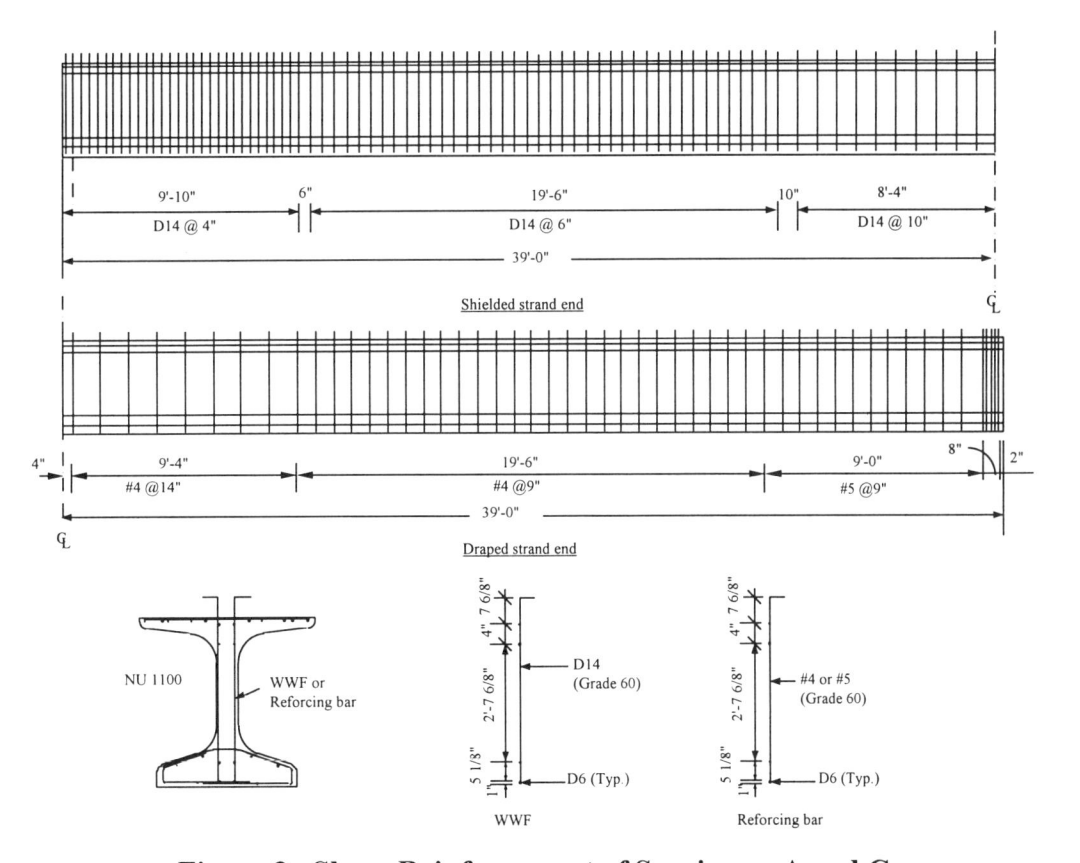

Figure 2. Shear Reinforcement of Specimens A and C

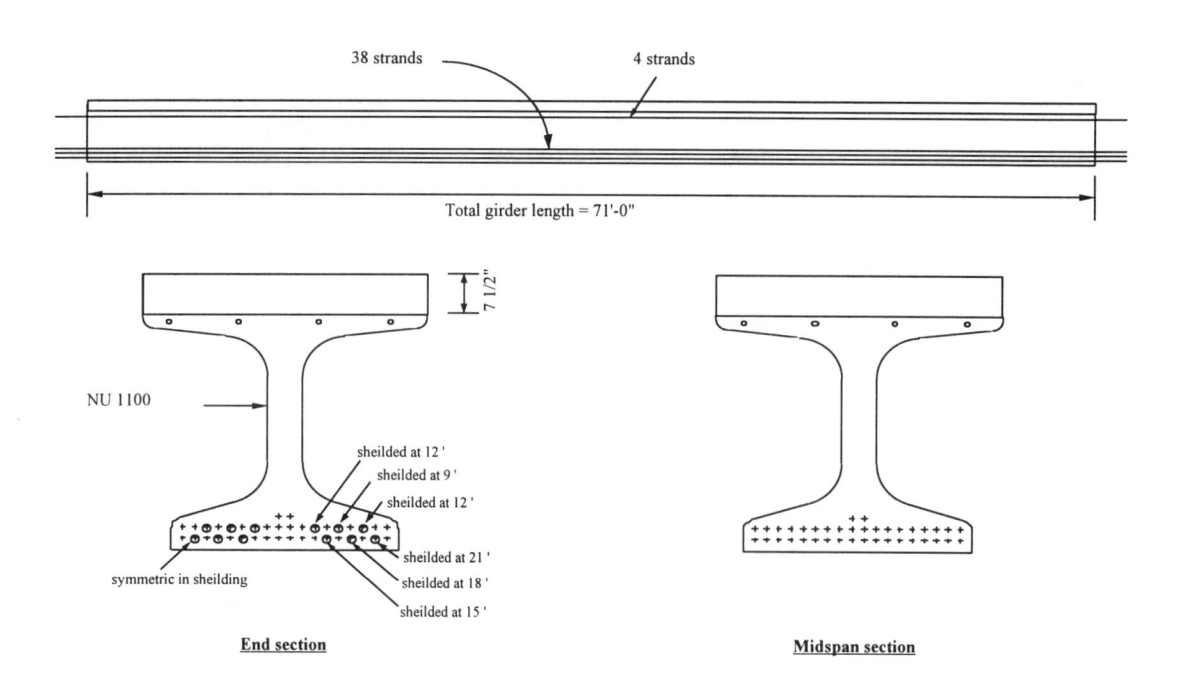

Figure 3. Elevation and Cross Section of Specimens B and D

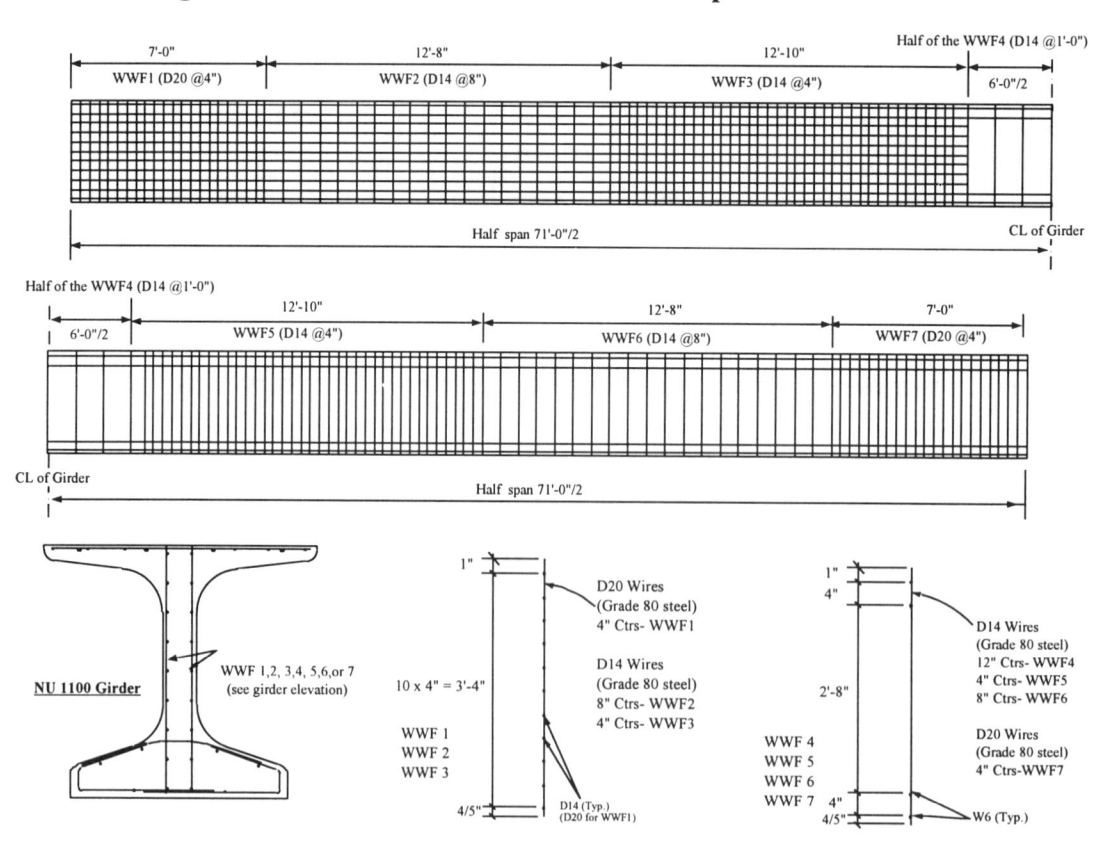

Figure 4. Shear Reinforcement of Specimens B and D

in. (101.6 mm) spacing both horizontally and vertically. Each specimen was planned to have at least two shear tests (both ends). And each test is labeled according to the specimen designation (A, B, ...etc), the web reinforcement and concrete strength of the tested region. For example, "DOW31412X" is interpreted as follows: D = specimen "D", OW = orthogonal WWF, the first two numbers "31" = D31 wire, the third number "4" = 4 in.(101.6 mm) of wire spacing, the last two numbers "12" = 12,000 psi (82.8 MPa) concrete strength, and X = strand anchorage is provided by the end block. Other labels used are: VW = vertical WWF, R = reinforcing bar, and Y = strand anchorage is not provided.

Specimen Design

A bridge design utilizing NU 1100 girders was developed for the purpose of developing a representative girder details for the test specimens. Draped prestressing strands at one end and shielded strands at the other end were used for specimens A and C. Shielded strands at both ends were used for specimens B and D. A 28-day concrete compressive strength of 8,000 psi (55.2 MPa) and 12,000 psi (55.2 MPa) and a release compressive strength of 6000 psi (41.4 MPa) were assumed in the design of the girders. The girder spacing in the bridge model varied from 8 ft (2.4 m) to 12 ft (3.6 m). To be conservative and to save deck forming costs, the 7 1/2 in. (190 mm) thick deck slab width was kept the same as the girder top flange width. The cross section of the girder specimen with the added deck slab is shown in Figs. 1 and 3. Please note that Figs. 2 and 4 do not show the deck slab or the strands. Deck slab reinforcement was intentionally omitted. It was decided that the slab reinforcement would not have any significant effect on the measured shear behavior of the girder specimen. The specified concrete compressive strength for the deck slab was 5000 psi (34.5 MPa).

Specimen Fabrication and Materials

The specimens were fabricated in the Bellevue plant of Wilson Concrete Company. The welded wire fabric used was manufactured by Ivy Steel & Wire, Texas. The shear reinforcement for specimens A and C had a specified yield strength of 60 ksi (414 MPa), and shear reinforcement for other two specimens had a specified yield strength of 80 ksi (552 MPa). During the placement of reinforcement, it was found that the placement of WWF as girder reinforcement was much simpler and resulted in better location accuracy than the use of conventional reinforcing bars. Figs. 2 and 4 show the details of the shear reinforcement.

The specimens were prestressed using 1/2 in. (12.7 mm) diameter, Grade 270 (1863 MPa), low-relaxation 7-wire strands. For specimens A and C, four of the 30 strands were draped at one end and four were shielded at the other end. For specimens B and D, all 38 strands were straight and with 12 strands shielded at the ends. The bottom strands were pretensioned to a stress level of 202.5 ksi (1397 MPa). The four top strands were only pulled to a stress level of 13.1 ksi (90 MPa).

Figure 5. Othogonal WWF Shear Reinforcement

Cylinders were prepared in accordance to ASTM C 31-87. Cylinder compressive test results are listed in Table 1. Also included in the table are test results of two cores taken from each specimen web and the results of deck concrete strength.

Table 1 - Measured Concrete Strength

Specimen	Strength at release (psi) Specified	Actual	Strength at time of testing (psi) Specified	Actual	Strength of web cores (psi)	Deck strength at time of testing (psi) Specified	Actual
A	6000	6180	8000	8100	8490	5000	7200
B	6000	6530	8000	10780		5000	5100
C	6000	6180	12000			5000	
D	6000	8400	12000			5000	

Note: 1 ksi = 6.9 MPa

Instrumentation

Internal strain gauges were attached to vertical and horizontal (for orthogonal WWF) stirrups at both girder ends. For vertical stirrups, the gauges were positioned on a leg of the stirrup so that when the stirrup was placed in the girder the gage would be at approximately mid-depth of the web. During testing, deflection and strain readings were measured. Deflection was measured at the one-quarter point and the midspan of the girder specimen using the position transducer. Concrete strains in the web were measured using 8 in. Demec gauges. The Demec gauge measures the relative displacement of gauge points with an accuracy of 4.0×10^{-6} strains.

Test Setup and Procedure

The specimen was tested in three different test setups. The first setup was to test the specimen in flexure up to 80% of the calculated theoretical ultimate flexural capacity (flexural testing results will be reported elsewhere). After the flexural testing, the loading frames were moved to one of the girder ends for the shear testing. After the shear failure, the testing frames were moved to the other end for another shear testing. Fig. 6 shows a schematic of the method of load application.

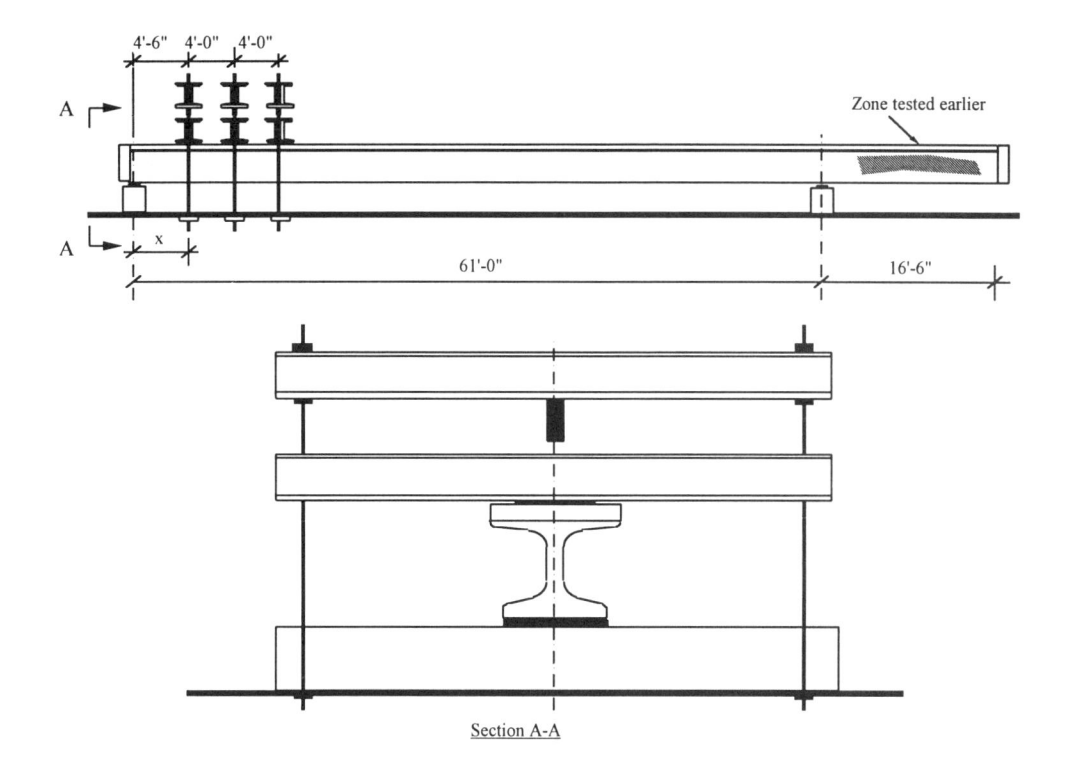

Figure 6. Shear Test Set-up for Member Left End
(Note that right end had been tested earlier)

TEST RESULTS AND DISCUSSION

The discussion of this section is limited to specimens A and B which have already been tested. The maximum shear force at diagonal tension (web-shear) cracking ($V_{cr-test}$) and at ultimate strength (V_{u-test}) are summarized in Table 2.

Table 2 - Summary of Test Results

Tested Region	r_v	r_h (%)	f_y (ksi)	b_w (in.)	d (in.)	d_v (in.)	$V_{cr-test}$ (kips)	V_{u-test} (kips)	Mode of failure
AR05908X	1.168	0	60	5.9	43.8	42.0	243.02	629.47	web crushing
AVW14408X	1.186	0	60	5.9	48.0	46.2	134.20	593.01	web crushing
BVW20408X	1.695	0	80	5.9	47.6	45.4	195.33	589.8	web crushing
BOW20408X	1.695	1.695	80	5.9	47.6	45.4	250.00	>820.3	

Note: 1 in. = 25.4 mm; 1 ksi = 6.9 MPa; 1 kip = 4.45 KN. And r_v = vertical shear reinforcement percentage = $A_v/b_w s_v$%, r_h = horizontal shear reinforcement percentage = $A_h/b_w s_h$%.

Stirrup Strain
Fig. 7 shows the variation of strain in a typical stirrup within the shear span. It can be seen that until the web-shear cracks opened the stirrup was only lightly stressed, and that this stress could be either tension or compression. It can also seen that the stirrup yielded almost immediately on formation of the diagonal tension cracks. In general, the strains of greatest significance occurred in the stirrup crossing diagonal tension cracks in the region of maximum shear, that is, between the support and the first applied load. In all cases, the strain in these stirrups at ultimate girder strength exceeded the yield point strain for the stirrup reinforcement. The pattern of behavior was similar for all tested regions.

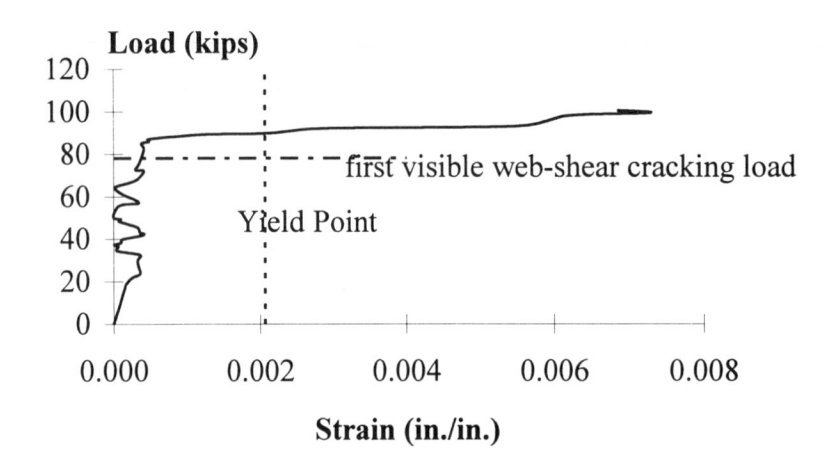

Figure 7. Stirrup Strain

Cracking and Failure Mode
In general, shear testing of specimen ends except BOW20408X behaved in a similar manner. When the cracking load was reached, a cracking sound was heard and the first visible web-shear crack was found between the support and the first applied load (as shown in Fig. 8). The cracking angle relative to girder span direction was about 35 degrees. The width of the crack was about 0.0039 in. (0.1 mm). With incremental loading, the first crack extended from both crack-tips. At the same time the second crack appeared about 12 inches (304.8 mm) away toward the applied load direction. Additional loading caused additional web-shear cracks with a little flattened angle (as depicted in Fig. 8). In the late development of cracking, the flexural cracks near the third applied load region (i.e., the load closer to the midspan) were extended into flexural-shear cracks. However, different cracking behavior was observed by comparing BVW20408X and BOW20408X. Because of the horizontal shear reinforcement in the orthogonal WWF, the first web crack appeared at a larger shear force. And the width and spacing of the cracks are smaller.

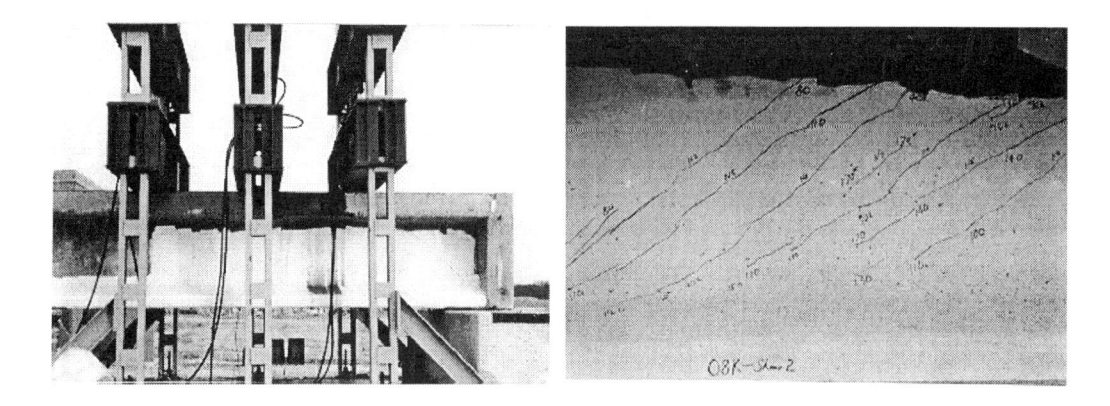

Figure 8. Web-Shear Cracking (AVW14408X)

Failure was finally due to diagonal compression crushing of the concrete in the webs of the girder. Compression spalling of the surface of the web was first visible in the middle region of the first web-shear crack. As the last increment of load was applied, a diagonal compression failure of the web commenced in this same region. Example of the appearance of the specimen which failed in shear is shown in Fig. 9. For BOW20408X, however, only the sign of the web concrete spalling was observed at the maximum loading capacity of the testing frame.

Figure 9. Web Crushing Shear Failure (AR05908X)

Discussion of Test Results
In three cases of shear failure, the final collapse of the girder ends resulted from a diagonal compression failure of the concrete in the web. This mode of failure differs from the "shear compression" type of failure usually observed in rectangular section girders, in which diagonal tension cracks reduce the depth of the flexural compression zone and finally bring about flexural failure at a reduced ultimate moment. It also differs

from the "shear-bond"[4] or "shear-tension"[5] failure experienced by recent shear tests, where the strands of the girder have no anchorage to develop their tension tie capacity.

Recent shear tests [4,5] and brittle shear failures of many parking structures during the Northridge Earthquake [6] have raised the issue of the premature "shear-bond"[4] or "shear-tension"[5] failure of prestressed concrete members. However, it has been found that this premature shear failure can be avoided by emphasizing the anchorage of the bottom flange tension tie – one of the very important truss components in shear resistance. The traditional and simple 45° truss model clearly and correctly shows that the stresses in the longitudinal tensile reinforcement in the shear span are larger than those predicted from beam theory. If the longitudinal tensile reinforcement is not well anchored in the girder support region, premature shear failure is unavoidable. For bridge I-girders advantage can be taken of the existing girder end diaphragm, where the girder strands can be anchored. To test this concept, end-blocks were cast at both ends of the specimen as shown in Fig. 10. It can be seen that the strands were well anchored to develop the web crushing failure.

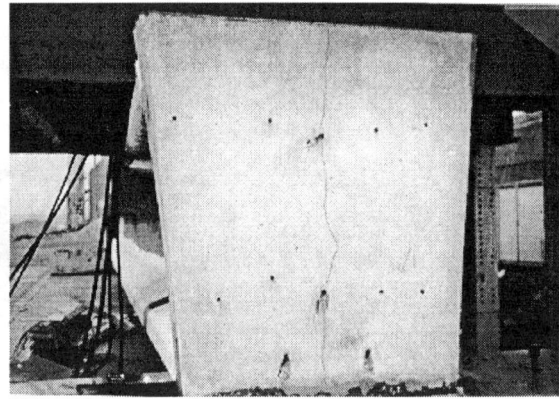

Figure 10. Strands Anchored into End Block

Comparison with AASHTO Specifications

Bridge design in the U.S. is mostly done in accordance with AASHTO-STD. However, the new AASHTO-LRFD is gaining acceptance and is expected to eventually replace AASHTO-STD. Table 3 provides a summary of the formulas used to predict the concrete contribution, V_c, the steel contribution V_s and also the maximum shear reinforcement requirement using both methods. Please note that f_c' in the formulas given in Table 3 is the pound per square inch. This is a modification of the AASHTO-LRFD formulas which require that f_c' be in kips per square inch. The effective depth is defined somewhat differently between AASHTO-STD (d) and AASHTO-LRFD (d_v). In AASHTO-LRFD, it is the distance between the resultants of tensile and compressive forces due to flexure but not less than the greater of 0.9d or 0.72h.

Table 3 - AASHTO Shear Design Methods

Methods	V_c	V_s	Limit on maximum A_v
AASHTO Standard	min (V_{ci}, V_{cw}), where: $V_{ci} = 0.6\sqrt{f'_c}\,b_w d + V_d + \dfrac{V_i M_{cr}}{M_{max}} \geq 1.7\sqrt{f'_c}\,b_w d$ $V_{cw} = (3.5\sqrt{f'_c} + 0.3 f_{pc})b_w d + V_p$	$\dfrac{A_v f_y d}{s}$	$V_s \leq 8\sqrt{f'_c}\,b_w d$
AASHTO LRFD	$\beta\sqrt{f'_c}\,b_w d_v$	$\dfrac{A_v f_y d_v}{s}\left[\dfrac{\sin\alpha}{\tan\theta} + \cos\alpha\right]$	$V_n \leq 0.25 f'_c\,b_w d_v + V_p$

Note: f_{pc} = compressive stress in concrete at centriod of section resisting externally applied loads or at junction of web and flange when the centroid lies within the flange, M_{cr} = moment causing flexural cracking at section due to externally applied loads, M_{max} = maximum factored moment at section due to externally applied loads, s = longitudinal spacing of the web reinforcement, V_d = shear force at section due to unfactored dead load, V_i = factored shear force at section due to externally applied loads occurring simultaneously with M_{max}, V_p = vertical component of effective prestress force at section, α = angle of inclination of transverse reinforcement to longitudinal axis, β = factor relating effect of longitudinal strain on the shear capacity of concrete, as indicated by the ability of diagonally cracked concrete to transmit tension, θ = angle of inclination of diagonal compressive stresses.

For specimen A, the shear reinforcement is #5 at 9 in. (228.6 mm) spacing at the draped end (AR05908X) and D14 at 4 in. (101.6 mm) at the shielded end (AVW14408X). This reinforcement corresponds to approximately 0.84 in²/ft (1.778 mm²/mm). This value is equal to the maximum reinforcement allowed by AASHTO-STD. The shear reinforcement for specimen B is 1.20 in²/ft (2.540 mm²/mm). This value is almost twice of the maximum reinforcement of 0.63 in²/ft (1.334 mm²/mm) allowed by AASHTO-STD. A proportionate increase in shear strength is observed in Fig.11, which indicates that the AASHTO-STD limits may be too restrictive. The experimental shear at failure in all cases exceeds the $0.25 f'_c\,b_w d_v$ limit. Extremely good performance is found when orthogonal WWF is used.

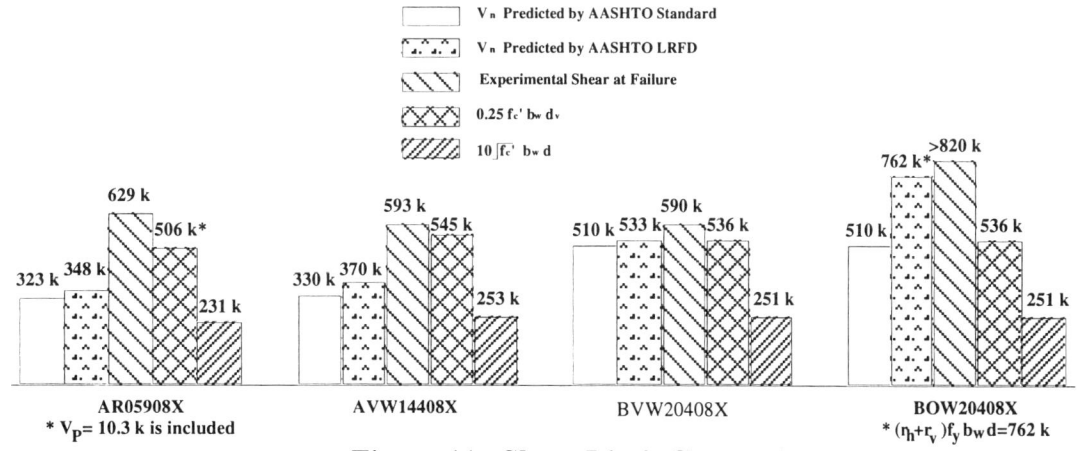

Figure 11. Shear Limit Comparison

CONCLUSION

The shear reinforcement in all four tested regions either corresponds to or exceeds the maximum reinforcement allowed according to AASHTO-STD. Test results show the proportionate increase in shear strength because of the increase of the shear reinforcement. The maximum shear capacity V_n at all tested regions exceeds the AASHTO-LRFD limit of $0.25f_c' b_w d_v$. The test results clearly indicate that the procedures given in the AASHTO-STD are too conservative for the details used in this testing. It is believed at this time that the outstanding performance of the first two specimens was a result of a number of factors. Most significantly, the strands were fully anchored into an end block, forming a strong tie of the flexural reinforcement. This detail resulted in "pure" shear or web crushing, rather than bond or flexural failure combined with the shear failure, as reported in most shear testing programs. Secondly, it is possible that the contribution of the member flanges makes the assumption of b_w equal to the web width conservative.

REFERENCES

1. Ma, Zhongguo (John), Saleh, Mohsen, and Tadros, Maher K., "Shear design of stemmed bridge members – How complex should it be?" *PCI JOURNAL*, 1997 (to be published).
2. ACI-ASCE Committee 326, "Shear and diagonal tension," Proceedings, ACI, 1962, Vol. 59, January, Febuary, and March, p.1-30, p.277-334, and p.353-396.
3. Mattock, Alan H. and Kaar, Paul H., "Precast-prestressed concrete bridges - 4. Shear tests of continuous girders," *Journal of the PCA Research Development Laboratories*, January 1961, p. 19-47.
4. Shahawy, M. A. and Batchelor, B., "Shear behavior of full-scale prestressed concrete girders: Comparison between AASHTO specifications and LRFD code," *PCI Journal*, 1996, Vol. 41, No. 3, p.48-62.
5. Kaufman, M. K. and Ramirez, J. A., "Re-evaluation of the ultimate shear behavior of high-strength concrete prestressed I-beams," *ACI Strucutral Journal*, 1989, Vol. 86, No. 4, p. 376-382.
6. Englekirk, Robert E. and Beres, Attila, "The need to develop shear ductility in prestressed members," *Concrete International*, October, 1994, p. 49-56.

SHEAR BEHAVIOR OF PRETENSIONED I-SHAPED GIRDERS MADE WITH HIGH STRENGTH CONCRETE

Dr. Bruce W. Russell, Ph.D., P.E.	Mr. James H. Allen, III
Assistant Professor	Design Engineer-in-Training
The University of Oklahoma	Coreslab Structures, Inc.
Norman, Oklahoma USA	Oklahoma City, Oklahoma USA

ABSTRACT

Tests were conducted on I-shaped composite bridge girders to examine their shear behavior when relatively small amounts of horizontal shear reinforcement are cast within the webs. Four 3/8 scale Type IV girders with a composite deck slab were tested. Girder strengths approached 70 Mpa (10,000 psi). The results demonstrate that horizontal shear reinforcement, placed within the webs of the beams, improves the shear ductility of the I-beams two to three times beyond the ductility measured on companion beams where the horizontal shear reinforcement was not included.

INTRODUCTION AND BACKGROUND

The use of high strength concrete in the construction of pretensioned girder bridges can dramatically increase the span limits for a given girder section; or alternatively, the number of girders required for a given design can be significantly reduced. In either scenario, designs that incorporate high strength concrete can dramatically increase the shear forces applied to each individual girder compared to the shear forces experienced by girders in normal strength designs. For example, the State of Oklahoma is considering the construction of a multi-span pretensioned girder bridge with clear spans of 50 m (165 ft) coupled with girder spacings of 3.7 m (12 ft). HS25 loadings will require a shear capacity of approximately 2000 kN (450 kips) for each girder. Additionally, shear forces from service loads will approach or exceed the shear that theoretically will cause web shear cracking in the I-beams. Therefore, it appears likely that these beams will develop web shear cracks during their service life and it becomes critical to provide reinforcement detailing that prevents excessive propagation of these cracks.

The potential problems caused by web shear cracking were highlighted by development length and shear tests on I-shaped beams.[1,2,3,4] These tests revealed that the propagation of web shear cracking through the bottom flange of pretensioned beams disrupts the anchorages of pretensioned strands subsequently causing strand anchorage failures. In turn, the sudden failure and collapse of the member can be caused by the disruption of pretensioned strand anchorage by web shear cracking. Typically, these failures have been labeled "shear/bond failures" because of the interdependence between strand anchorage and shear capacity.

A typical shear/bond failure is shown in the photograph in Fig. 1. This beam specimen was loaded until its sudden and violent failure. These tests demonstrated that strand anchorages failed as web shear cracks propagated through the bottom flange of the I-beam. At failure, concrete chunks burst from the beam as crack propagation disrupted strand anchorage and large shear deformations caused delamination below the strands.[2]

Figure 1. Typical Shear/Bond Failure

The load vs. deflection curve for this specimen is shown in Fig. 2. Significantly, strand slip was initiated by web shear cracking that propagated through strand anchorage zones. From this diagram, the beam's sudden loss of shear resistance is mirrored with the sudden increase in strand slippage. Disruption of the strand anchorage zone resulted to shear/bond failure.[2]

Shahawy and Batchelor tested AASHTO Type II girders with composite slabs for development length of prestressing strands.[4] These composite girders were loaded monotonically until failure. All of the girders tested in this series failed by the formation of web shear cracks and their propagation through the transfer (anchorage) zones of pretensioned strands. Interestingly, by increasing the amount of vertical shear reinforcement well beyond the requirements of AASHTO, the beams achieved slightly greater capacity than specimens without the additional shear reinforcement. However, and significantly, the overall behavior of the beams was not changed by the additional vertical shear reinforcement, and little improvement in ductility was noted. Instead, all of the beams in these tests failed by shear/bond where web shear cracking disrupted the anchorage zones of the pretensioned strands.

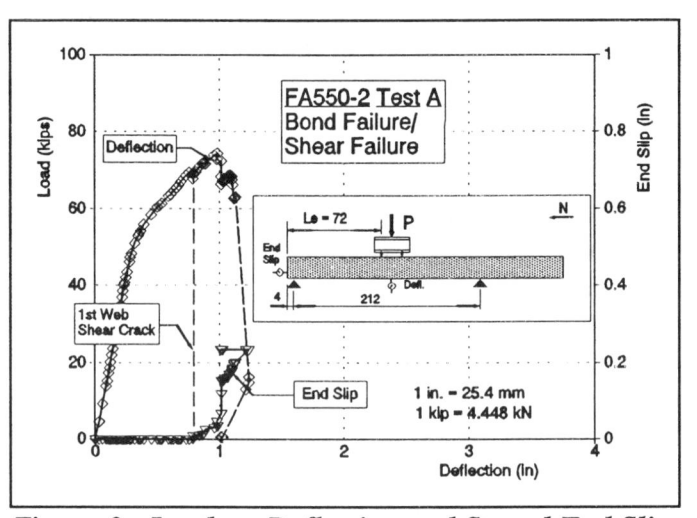

Figure 2. Load vs. Deflection and Strand End Slips for Typical Shear/Bond Failure (1 in. = 25.4 mm; 1 kip = 4.448 kN).

The potential for problems was highlighted when the Northridge Earthquake struck Southern California in January of 1994. Double-tee and inverted-tee beams in parking garages failed and collapsed when shear cracking disrupted the anchorage zones of pretensioned strands. These failures emphasize the interaction between strand anchorage and web shear cracking plus the inherent lack of shear ductility in end regions of pretensioned I-beams, double tees and inverted tees.[5]

Therefore, research was undertaken at the University of Oklahoma to examine the shear behavior affected by the inclusion of relatively small amounts of horizontal shear steel contained within the webs of pretensioned I-beams. Results demonstrate that the addition of horizontal mild shear reinforcement in the end regions of pretensioned I-beams dramatically improves beam behavior. High strength concrete girders containing horizontal web reinforcement doubled and tripled the beams' deformation capacities beyond that of companion specimens where horizontal reinforcement was omitted. Because the positive effects of these improved reinforcement details are relatively inexpensive, they should be easily accepted and incorporated into standard practice.

TESTING PROGRAM

Scope
Four pretensioned concrete I-shaped girders were fabricated at the Fears Structural Engineering Laboratory (FSEL) at the University of Oklahoma (OU). The design concrete strength for each of the girders was 70 MPa (10,000 psi). Each of the four beams were made composite with the addition of a deck slab. Each of the four beams was tested at each end for a total of eight tests. The beams were loaded monotonically until failure. Beam loads and loading geometry were designed to ensure web shear cracking occurred before flexural capacity was achieved, and failures that resulted were consistently shear or shear/bond failures.

Beam Design and End Reinforcement Details
The beam dimensions are selected based on a 3/8 scale AASHTO Type IV girder. Design details are shown in Fig. 3. Six 12.7 mm (0.5 in.) diameter strands are contained within the cross section; each were initially stressed to 75 percent of f_{pu}. The beams' flexural and shear capacities, and resistance to web cracking were designed to illicit the cracking response and capacities necessary to conduct the tests.

The "South" end of each beam contained four #3 bars placed horizontally and cast within the web. Each of the horizontal bars measured 2.4 m (8 ft) long with standard 90° hooks for anchorage at the beam ends. The "North" end of each beam contained identical vertical shear reinforcement to the South end, but the North end did not contain the horizontal shear reinforcement. The photograph in Fig. 4 depicts the shear reinforcement details for a beam during fabrication. The "South" end of each beam contained four #3 bars placed horizontally within the web of the I-beam. In contrast, horizontal shear reinforcement was omitted from the "North" ends of the beams. Stirrup spacings were held constant over the length of each beam at the spacing indicated in Table 1.

Vertical shear reinforcement consisted of single legged #3 bars, extending from the bottom of the tensile reinforcement into the composite slab. Standard 90° hooks were provided at each end of the stirrups to ensure proper anchorage.

Prestressing reinforcement consisted of 12.7 mm (0.5 in.) Grade 270, low relaxation strand conforming to ASTM A 416 and was donated by Shinko America, Inc. The strand possessed particularly good bonding properties as attested by the relatively short transfer lengths that were measured. The measured transfer lengths were shorter than ACI[6] and AASHTO[7] representations of 635 mm (25 in.) for 12.7 mm (0.5 in.) strand. Furthermore, the strand's pull-out capacity exceeded 169 kN (38,000 lbs) as measured by the Moustafa Test method.[8]

To simulate strand with transfer lengths in the "normal" range of 635 to 762 mm (25 to 30 in.), patterned debonding was applied to the strands through first 1.02 m (40 in.) of bond length, effectively lengthening the pretensioned transfer zones. For Beams #1, #2, and #4, debonding was placed on the pretensioned strands to lengthen the transfer zone. For control and comparison, strands in Beam #3 remained bare without debonding. Table 1 also reports end reinforcement and debonding details for each end of each specimen, and the transfer lengths measured at release of prestressing.

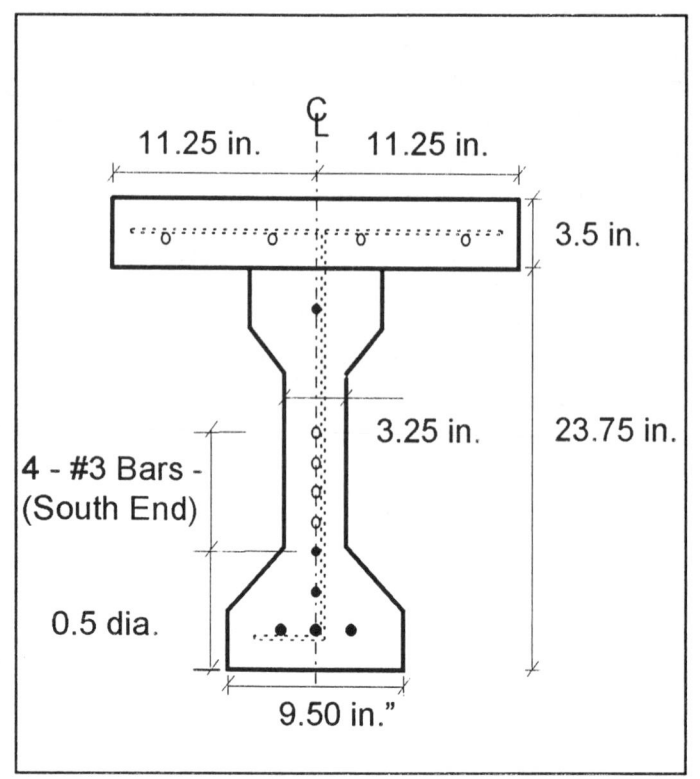

Figure 3. Design Details for the Beam Cross Section (1 in. = 25.4 mm).

Figure 4. Typical Shear Reinforcement at South End

Table 1. Summary of End Reinforcement Details for Each Beam Specimen

Test Specimen	Horizontal Reinforcement	Stirrup Spacing (mm)	Debonding (D) or Not (N)	Transfer Lengths (mm)
Beam #1 - N	none	229	D	610
Beam #1 - S	(4) #3's in web	229	D	610
Beam #2 - N	none	152	D	762
Beam #2 - S	(4) #3's in web	152	D	762
Beam #3 - N	none	229	N	457
Beam #3 - S	(4) #3's in web	229	N	559
Beam #4 - N	none	152	D	610
Beam #4 - S	(4) #3's in web	152	D	711
Note: Transfer lengths were calculated from the average of the end slips measured on all six strands, measured on the day of testing. (1.0 in. = 25.4 mm).				

Beam #2 did not achieve the required release strengths because of the inadvertent addition of water reducers in lieu of HRWR. Table 1 indicates the longer transfer lengths for Beam #2 which are indicative of lower strength concrete. The beam was tested in an attempt to acquire additional data; however, results from Beam #2 should be skeptically evaluated.

Testing Apparatus and Setup

Loading Geometry - Figure 5 illustrates the loading geometry, and the shear and bending moment diagrams for the beam tests. Load was applied through an hydraulic actuator to a spreader beam. The spreader beam distributed load to two load points and created a region of constant moment 0.61 m (2 ft) in length. The strand embedment length measured 1.68 m (66 in.) with a 152 mm (6 in.) overhang at the support.

The calculated nominal flexural capacity, M_n for the cross section is 467 kN-m (4133 kip-in.). If flexural failure occurs with the shear span of 1.52 m (60 in.), then the beam must resist maximum shear of 307 kN (68.9 kips), which exceeds slightly the calculated nominal shear capacity, V_n of 304 kN (68.3 kips). More importantly, because the applied shear will exceed the web cracking shear, V_{cw} of 220 kN (49.4 kips), the loading arrangement ensures that web shear cracking will occur before the flexural (or shear) capacity of the beam is achieved. This allows the test to examine the effects of the horizontal shear reinforcement on limiting the widening and propagation of web shear cracking, and the subsequent effects on the shear and flexural behaviors of the beam.

Instrumentation - The applied load, beam deflections and strand end slips of prestressing strands were measured and recorded electronically through a 445 kN (100 k) load cell, wire potentiometers and linear voltage displacement transducers

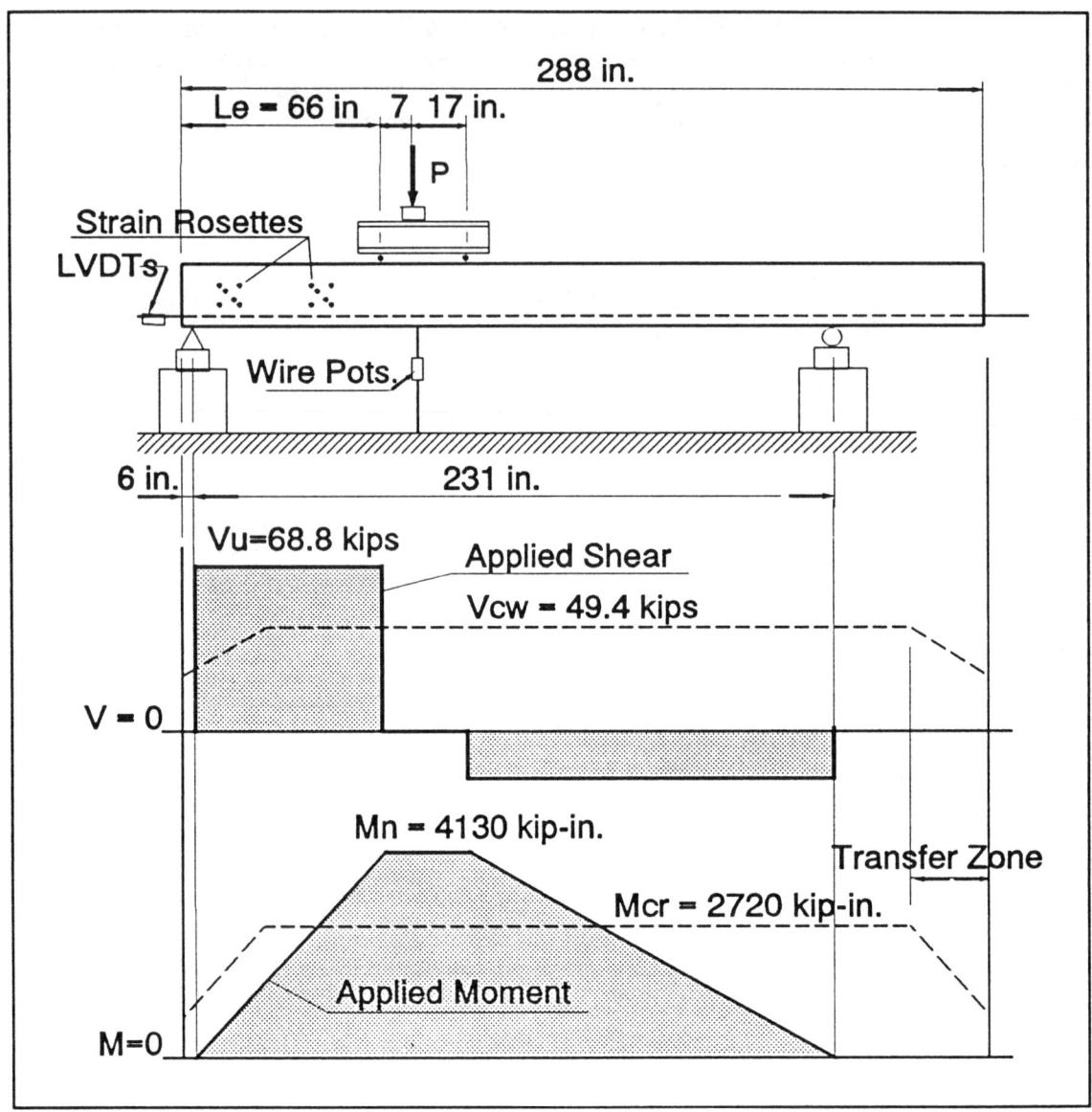

Figure 5. Loading Geometry for Beam Tests with Corresponding Shear and Moment Diagrams (1 in. = 25.4 mm; 1 kip = 4.448 kN; 1 kip-in. = 0.1130 kN-m).

(LVDTs). Shearing strains were measured mechanically at four different strain rosettes, two on each side of the beam, using a detachable mechanical strain (DEMEC) gage. Rosettes #1 and Rosettes #2 were located at mid-height on the web and 0.46 m (18 in.) and 1.22 m (48 in.) from the end of the beam. Shear strains were recorded at regular intervals throughout testing.

Testing Procedure
Each beam test proceeded with monotonic increases in load. At the beginning of each test, all measurements were initialized before external load was applied to the beam. Load was applied in approximate 89 kN (20 k) increments until cracking occurred and shear strain readings were taken at these intervals. After cracking, shear strains were

measured and collected at approximately 20 kN (4.5 k) increments in load, or at 63 mm (0.25 in.) increases in deflection. Electronic data was collected more often than manual data. Load vs. deflection was plotted and recorded in real time.

RESULTS FROM TESTING AND DISCUSSION

Effects of Horizontal Shear Reinforcement

Figures 6 and 7 plot shear force vs. deflection from tests on North and South ends, respectively, of Beam #1. Although both ends of the Beam #1 failed in shear with significant end slips, the differences between the test results from the two ends demonstrate the improvement in behavior resulting from the horizontal web reinforcement. In the test on the North end of Beam #1 (Fig. 6) where horizontal mild reinforcement was omitted, a shear capacity of about 231 kN (52 k) was achieved. However and more importantly, ductility was not achieved in the North end. The shear vs. deflection plot illustrates that the virtual collapse of the beam occurred at a deflection of 10 mm (0.40 in.). Beyond that deformation, load could not be sustained.

On the other hand, the South end of the Beam #1 (Fig. 7) demonstrated the ability to sustain the shear capacity of approximately 223 kN (50 k) through much larger deformations. The South end did not fail until deflections exceeded 30 mm (1.1 in.). The comparison of the opposite ends of the same beam clearly indicates that the inclusion of the horizontal reinforcement helped improve the shear ductility of the member, even though shear capacity was not improved.

The comparison of shear vs. shear strains also provides evidence that shear ductility is improved by inclusion of horizontal shear reinforcement. In Fig. 8, shear force is plotted vs. shear strain for both the North and South ends of Beam #1. The North end failed at a shear strain of only 0.0025 rads whereas the South end failed at a shear strain of 0.0115 rads, nearly 5 times greater than the shear strain at the North end. The plot clearly shows the improved shear ductility of the South end over the North end of the beam.

Effects of Varying Transfer Length

When comparing the behavior of Beam #3 to the behavior of Beam #1, the beneficial effects of shorter transfer lengths become apparent. Beam #3 contained strand with "good quality" bond (no debonding) and possessed an average transfer length 508 mm (20 in.) whereas Beam #1 had bond intentionally worsened by debonding and possessed transfer lengths that averaged of 610 mm (24 in.). In Figs. 9 and 10, the shear vs. deflection plots for Beam #3 are illustrated. The figures illustrate that the response to load was nearly identical for both the North end and the South end. Both ends failed in shear at a capacity of about 280 kN (63 k) and both ends achieved a deflection exceeding 30 mm (1.2 in.) before failure. Unlike Beam #1, the North end of Beam #3 was able to achieve adequate shear ductility despite not containing the horizontal web steel. Sudden collapse would not be expected from large gravity loads. Also, the measured strand end slips in Beam #3 are significantly smaller than

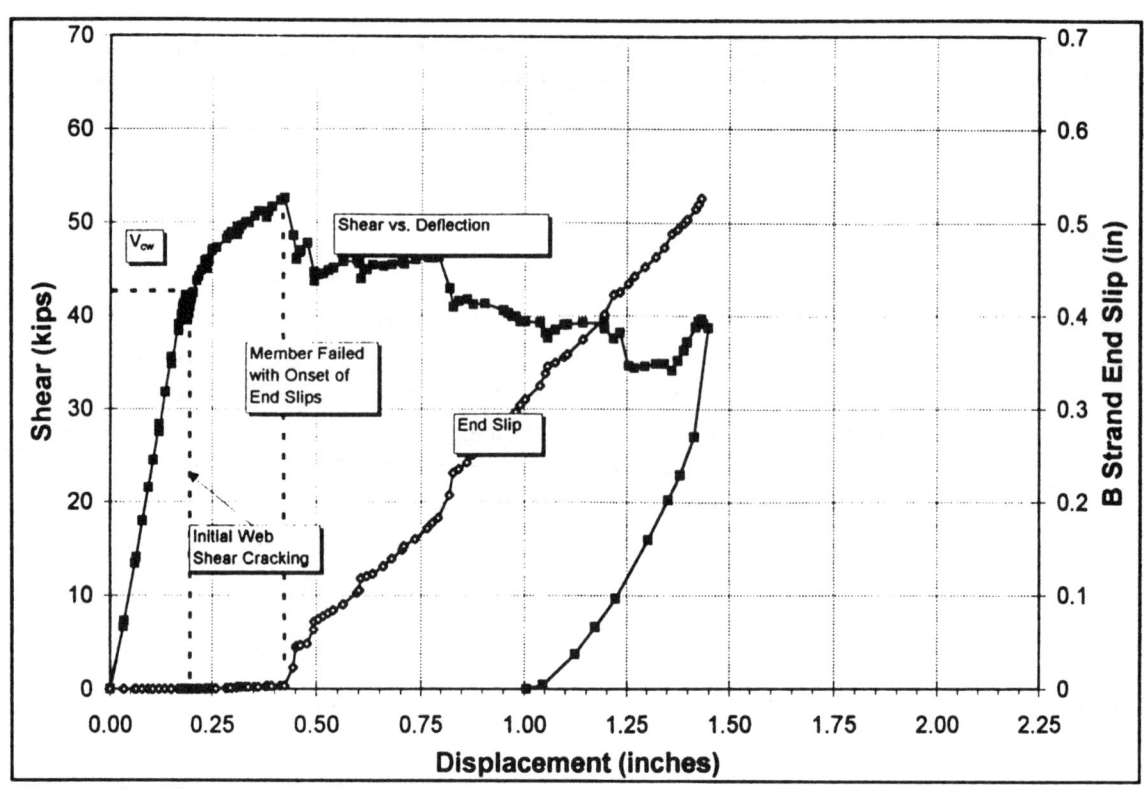

Figure 6. Shear vs. Deflection; Beam #1 - North End (1 in. = 25.4 mm; 1 kip = 4.448 kN).

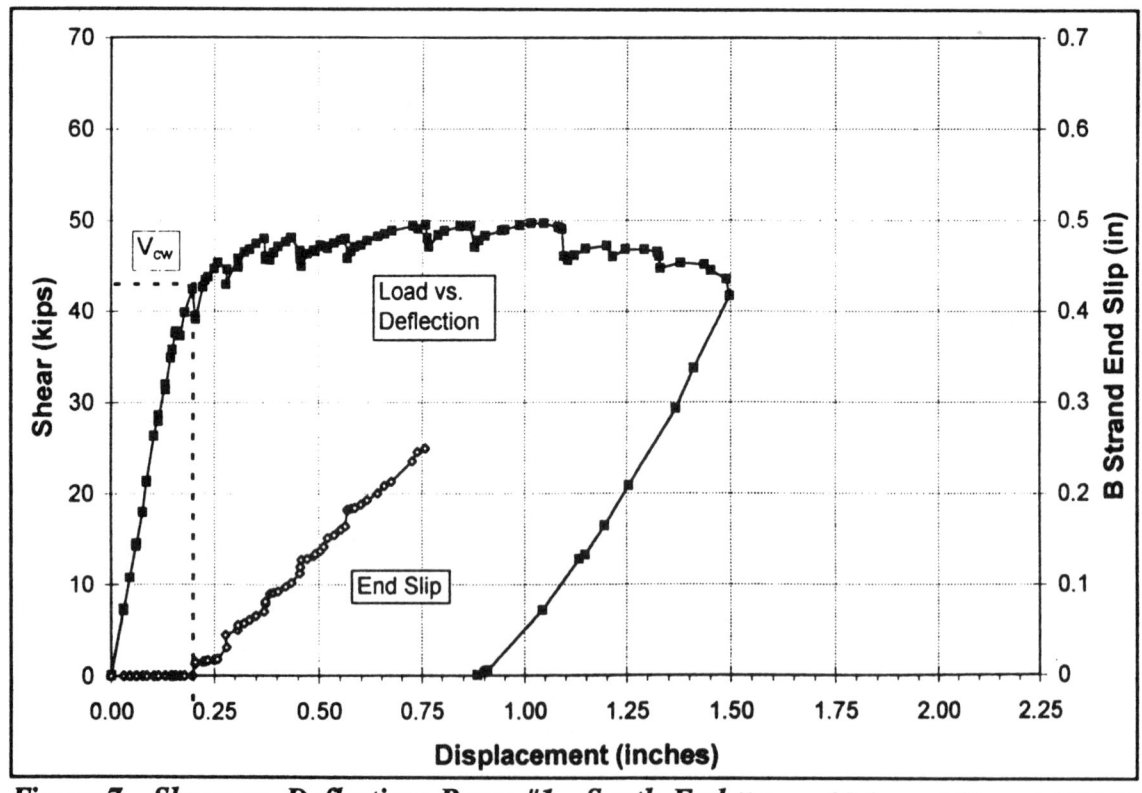

Figure 7. Shear vs. Deflection; Beam #1 - South End (1 in. = 25.4 mm; 1 kip = 4.448 kN).

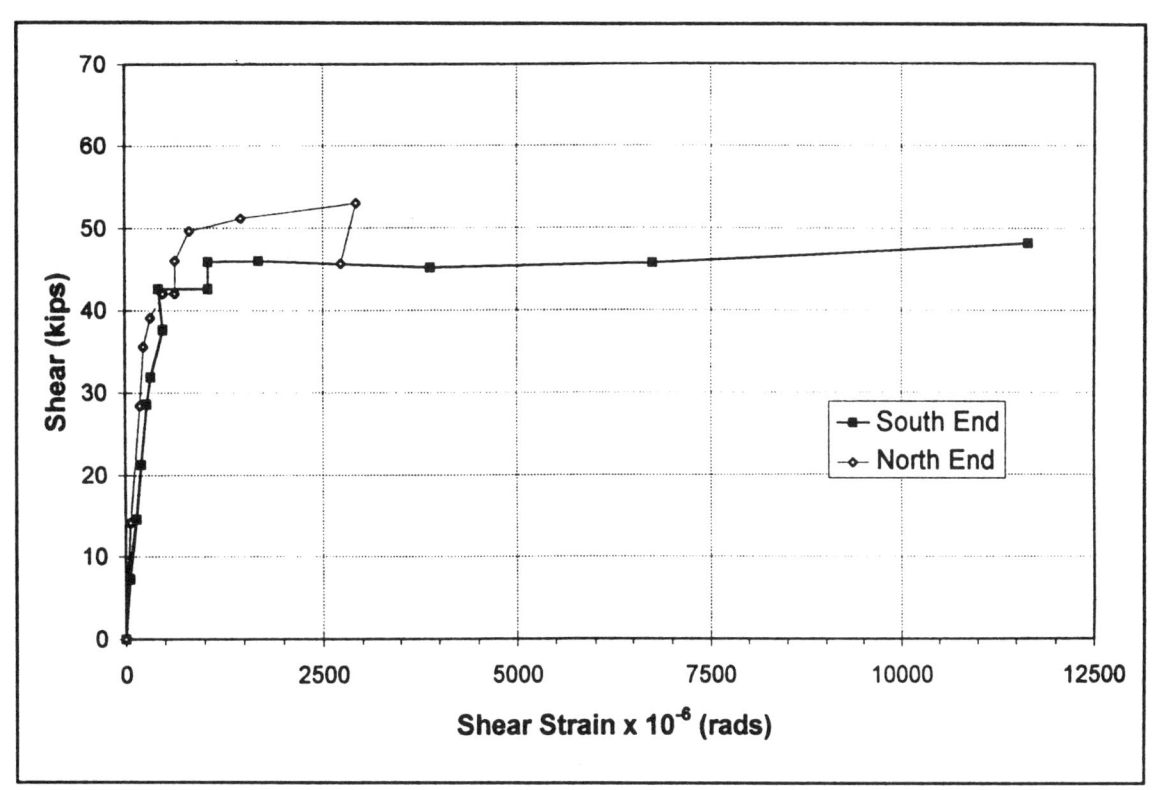

Figure 8. Shear vs. Shear Strain; Beam #1 - North End vs. South End (1 kip = 4.448 kN).

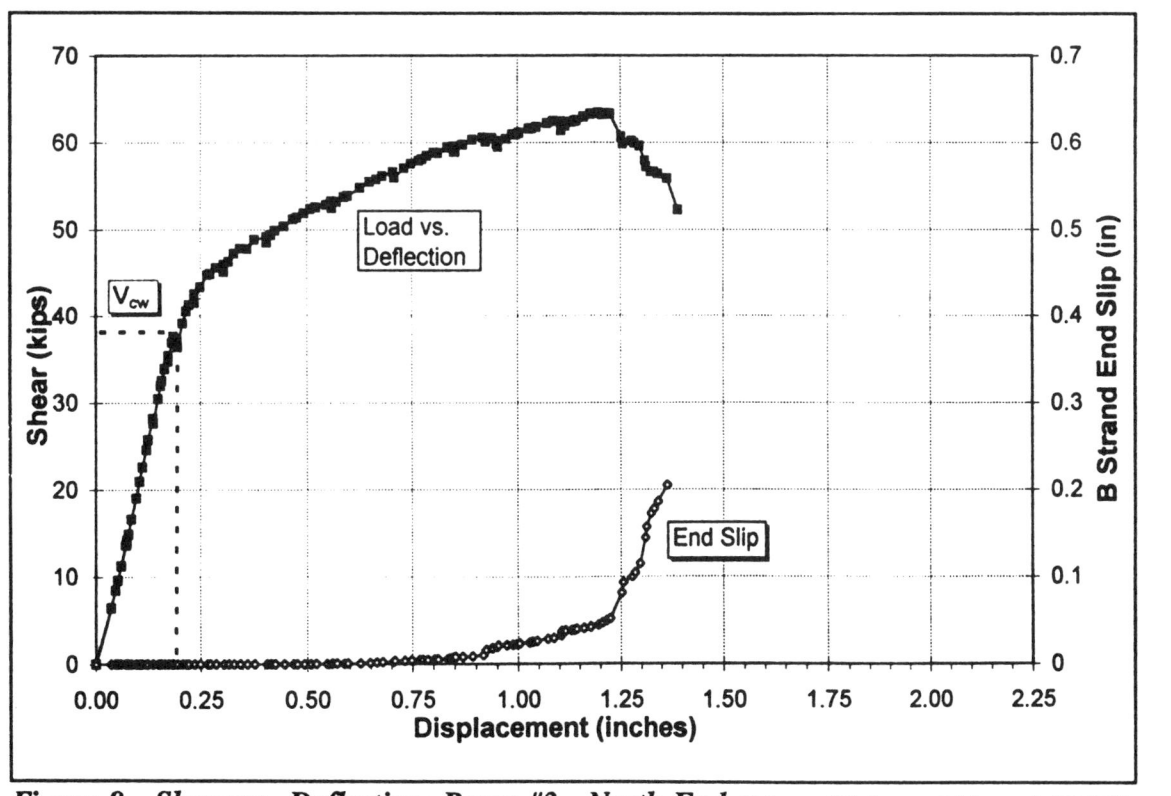

Figure 9. Shear vs. Deflection; Beam #3 - North End (1 in. = 25.4 mm; 1 kip = 4.448 kN).

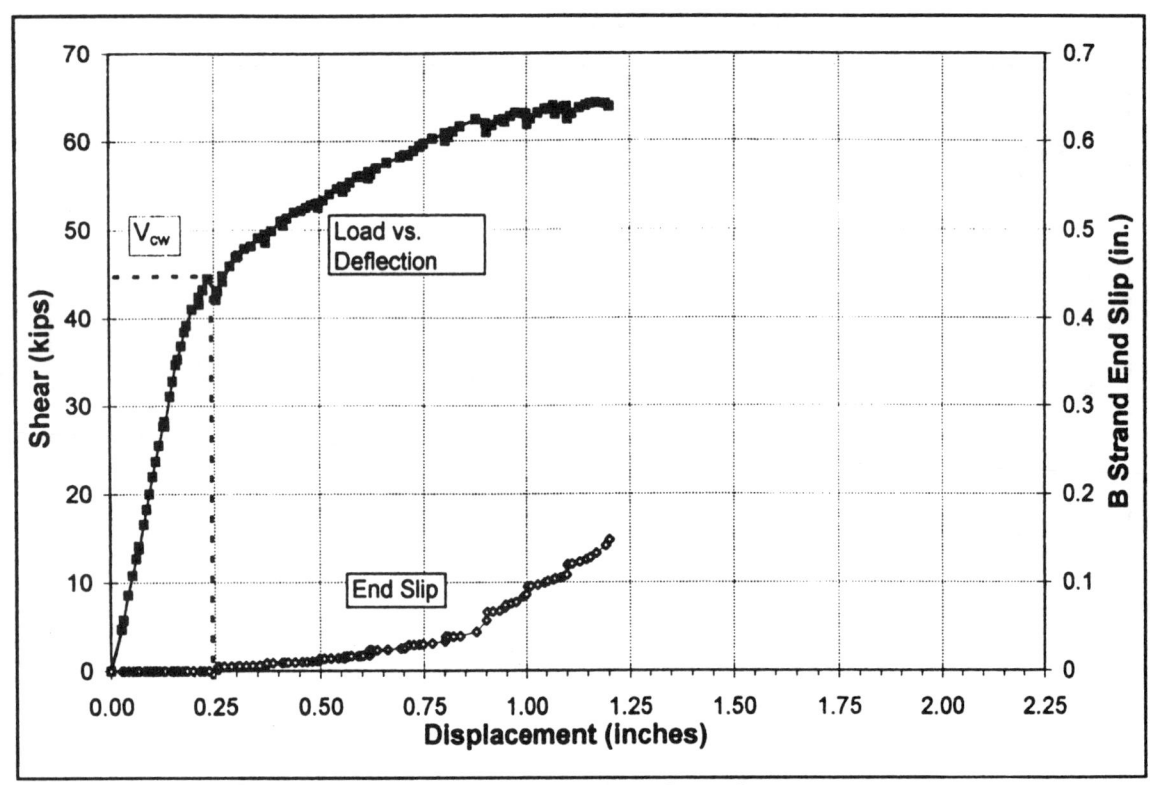

Figure 10. Shear vs. Deflection; Beam #3 - South End (1 in = 25.4 mm; 1 kip = 4.448 kN).

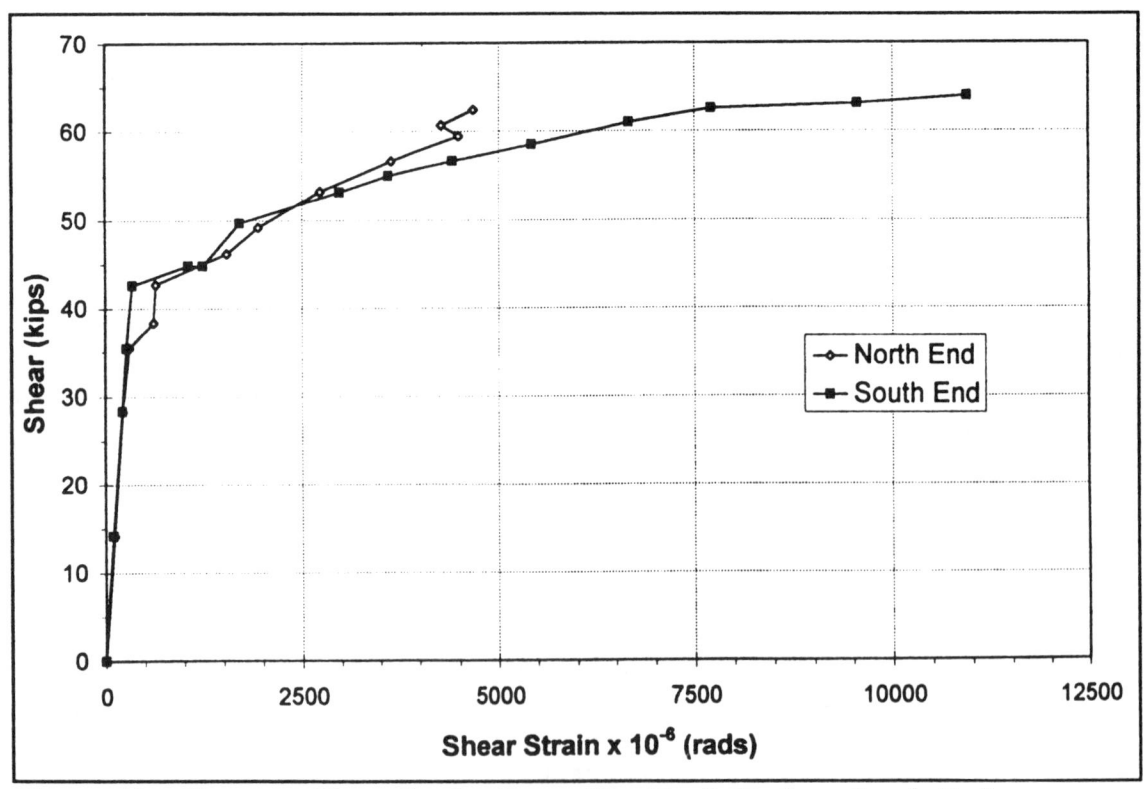

Figure 11. Shear vs. Shear Strain; Beam #3 - North End vs. South End (1 kip = 4.448 kN).

the strand end slips measured from Beam #1, indicating that the strands in Beam #3 had greater ability to resist additional strand tension required by the cracked concrete.

These facts point towards the strands' ability to develop the additional tension required to resist shear forces in Beam #3 whereas Beam #1 had significantly less shear capacity. Therefore, Beam #3 provides evidence that "good quality" strand bond can offset some of the need for horizontal shear reinforcement.

On the other hand, the South end of Beam #3 exhibited greater capacity for shear deformations than the North end. In Fig. 11, the shear vs. shear strains for both ends of Beam #3 are plotted. Like Beam #1, the South end was able to develop shear ductility well beyond that of the North end, indicating that the horizontal reinforcement continues to positively impact shear behavior, even with "good quality" bond on the prestressing strands.

Effects of Closer Stirrup Spacing

Beam #4 contained stirrups spaced at 152 mm (6 in.) on center. The shear force vs. deflection plots for this beam (not shown), like the results from Beam #3, demonstrate very little difference in load vs. deflection response between South and North ends. However, like Beam #3, the South end of Beam #4 demonstrated larger capacity for shearing strains than the beam's North end, providing further evidence that the horizontal reinforcement acts to improve shear ductility if not the shear capacity.

Comparison of Shear Ductility by Energy Methods

For each of the tests, the plot of shear force vs. the shear strain was integrated from the beginning of the beam test until failure. The resulting calculation provides a measure of shear strain energy density contained and absorbed beam from the beginning of the test until failure. Some of the energy is elastic; however, the majority of the energy represents inelastic, unrecoverable energy that was absorbed by the beams after web shear cracking had occurred. The energy density is also a measure of shear ductility. In the case of Beam #1, energy density in the South end (containing the horizontal shear reinforcement) was improved 276 percent over that of the North end. Likewise in Beams #3 and #4, significant improvement in shear ductility was noted. As measured by shear strain energy density, the shear ductility of the South ends improved 170 and 78 percent in Beams #3 and #4, respectively, over the shear ductility of the North ends.

CONCLUSIONS

1. The inclusion of horizontal shear reinforcement, contained within the webs of the I-shaped beams, significantly improved shear ductility.

2. Shear ductility can be improved through the use of strand with "good quality bond," as evidenced by shorter transfer lengths. However, the inclusion of horizontal shear reinforcement in the web provides even more improvement in shear ductility and energy absorbing capacity.

3. Shear ductility can be improved through the use of tighter stirrup spacings. However, the inclusion of horizontal shear reinforcement in the web provides even more improvement in shear ductility and energy absorbing capacity.

4. The horizontal shear reinforcement did not affect shear capacity. In every beam, the South end containing the horizontal shear was matched in capacity by the North end which did not contain the horizontal steel.

REFERENCES

1. Ramirez, J.A., and Breen, J.E., *Experimental Verification of Design Procedures for Shear and Torsion in Reinforced and Prestressed Concrete,* Research Report 248-3, Center for Transportation Research, The University of Texas at Austin, November 1983.

2. Russell (1991), B.W., and Burns, N.H., *Development Length and Flexural Bond Behavior of AASHTO-Type Girders with Fully Bonded and Blanketed Strands,* Technical Memo 1210-2, Center for Transportation Research, The University of Texas at Austin, March 1991.

3. Deatherage (1991), H.J., and Burdette, E.G., *Development Length and Lateral Spacing Requirements of Prestressing Strand for Prestressed Concrete Bridge Products,* Transportation Center, The University of Tennessee, Knoxville, February, 1991.

4. Shahawy (1991), M., and Batchelor, B., *Bond and Shear Behavior of Prestressed AASHTO Type II Beams,* Progress Report No. 1, Structural Research Center, Florida Department of Transportation, February, 1991.

5. Englekirk, R.E., and Beres, Attila, "The Need to Develop Shear Ductility In Prestressed Members," *Concrete International,* Vol. 16, No. 4, October 1994, pp. 49-56.

6. ACI Committee 318, *Building Code Requirements and Commentary for Structural Concrete (ACI 318-95 and 318-95R),* American Concrete Institute, Farmington Hills, MI, 1995.

7. AASTHO, *Standard Specifications for Highway Bridges,* 15th Edition, American Association of State Highway and Transportation Officials, Washington, D.C., 1992.

8. Logan, D.R., "Acceptance Criteria for Bond Quality of Strand for Pretensioned Prestressed Concrete Applications," *PCI JOURNAL,* Vol. 42, No. 2, March-April 1997, pp. 52-90.

THE INFLUENCE OF TENSILE STRENGTH OF HIGH STRENGTH CONCRETE ON THE BEHAVIOUR OF DEFLECTED REINFORCED CONCRETE BEAMS IN SERVICEABILITY CONDITIONS

Salvatore Russo, Technical Researcher, Venice Institute University, Laboratory of Strength of Materials, Tolentini n.191, 30135, Venice, Italy

ABSTRACT

The topic considered in this study mainly concerns the influence of effective tensile strength of high strength concrete, measured in presence of steel reinforcement, on the behaviour of a reinforced concrete deflected beam. The study consideres the behaviour of deflected beams made from 30 MPa and 70 MPa concrete class with a constant percentage of reinforcement, equal to 1.22%, and specifically measured, during the test, the effective tensile strength of concrete, f_{cte}. In addition, to ensure a broad evaluation of concrete tensile strength, and its reduction in the presence of reinforcement, the value of f_{ct} was also measured for the material concrete, without reinforcement, using conventional laboratory tests on concrete coming from the same casting as was used for preparing the beams.

INTRODUCTION

Currently, is difficult to ignore the difficulty calculating the the effective tensile strength of a concrete, i.e. the value of tensile strength measured in presence of steel reinforcement, for the purposes of accurate evaluation of structural behavior of reinforced concrete elements[1].
In fact, with specific reference to a possible utilization of tensile strength of concrete in concrete design, it is necessary to know the effective value of concrete tensile strength, , f_{cte}, especially in the case of high strength concretes of which the contribution of tensile strength is far from negligible,[2, 3, 4-5]. It is common knowledge that the value of tensile strength in plain concrete, f_{ct}, is higher than that measured in reinforced concrete elements, f_{cte}. This is causing by phenomena such as shrinkage and creep, and also by the percentage of the reinforcement involved [6-7].
In addition, today, for an effective evaluation of the structural element's behavior in a serviceability state, reference is generally made to a conventional tensile strength of concrete (deduced from its corresponding concrete class), since it would be extremely difficult to establish its exact contribution case by case. Moreover, the adoption of a conventional parameter offers a good balance between the higher values obtained from laboratory tests on the material concrete and the lower results obtained on site, which are considerably influenced by interaction between the steel and concrete, and consequently by the bond[8].On the other hand, the priority attributed these days to

the evaluation of cracking phenomena and durability[9], makes it necessary to verify the extent of the f_{cte}.

As for as the behavior of the material concrete under tensile stress, is concerned recent experimental trials have suggested deformation values under maximum loads that are virtually comparable for all compressive strength classes from 30 MPa(4351 psi) to 90 MPa(13053 psi) [10]. The increment in the strength class does have a considerable effect, however, on the peak load, with a marked stiffening of the initial branch of the curve. Instead, we have little information to evaluate the effective tensile strength of concretes. Said parameter can be established considering the degree of hardness of the mixture, the geometric parameter and the the mean tensile strength 28 days after casting[11-12].

On the basis of the above considerations and the results already obtained[13-14], this study compare the experimental value of f_{cte} deduced at the first crack in reinforced concrete deflected beam with C30 and C70 concrete class, with the value of f_{ct} deduced by splitting test in the same concrete using to cast the beams.

MATERIALS

The 30 MPa (4351 psi) concrete was prepared with type 32.5R cement, at a dosage of 300 kg/m³ (0.6 kip/40in.³), with a water-cement ratio of 0.58, and a mean compressive strength measured on cylindrical samples at 28 days of 34.2 MPa (4960 psi).

The 70 MPa (10152 psi) class concrete was prepared with cement type 52.5R, at a dosage of 500 kg/m³ (1.1kip/40in.³) with a water-cement ratio of 0.32, adding 30 kg/m³ (0.06kip/40in.³) of Silica Fume and a superplasticizing admixture (RH 5000) in a proportion of 1.5% of the cement. The mean compressive strength established on cylindrical samples at 28 days was 72.5 MPa (10515).

Table 1 - Concrete

Concrete Class f_c	Cement (kg/m³)	Portland type	w/c	Admixture (%)	Silica Fume MS610 kg/m³
30 MPa	300	32.5	0.58	/	/
70 MPa	433	52.5	0.32	RH 5000(1.5%)	30

(1 ksi=6.9 MPa)

The steel bars for the preparation of the RC elements were in diameters of 6, 8 and 12 mm (0.23, 0.31 and 0.47 in.), belonging to italian class FeB44K - comparable to B500 european steel - with a mean yield strength f_y of 519 MPa (75ksi), an ultimate tensile strength f_t of 638 MPa (92.5 ksi), a stress hardening ratio f_t/f_y of 1.23, and $\varepsilon_u = 10\%$.

EXPERIMENTAL MODEL AND TEST PROGRAM

The tests were performed on beams that were 100 x 200 x 1300 mm (3.94x7.87x51.2 in.) in size, with a constant geometric percentage of reinforcement amounting to

1.22%, and varying only the concrete's strength class from 30 MPa (4351 psi) to 70 MPa (10152 psi), in order to bring to light any differences in behavior caused by the type of concrete mixture involved. The beams were simply supported and a concentrated load was applied at the middle span, Figure 1.

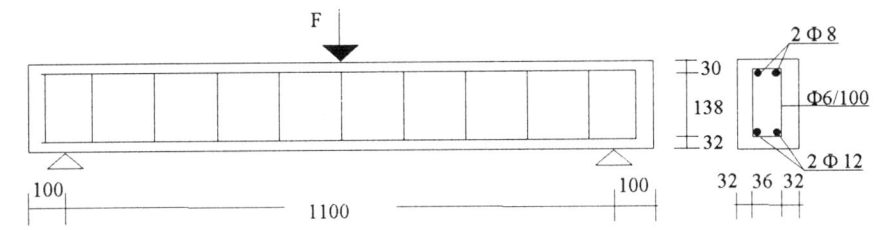

Figure 1: Beam Cross Section and Loading Scheme (dimensions in mm; in.=25.4 mm)

The tests were carried out under controlled displacement conditions at a constant rate of 0.01 mm/s (0.0004 in./s) up until the cracking phenomena were completed; thereafter, up to the failure of the beam, the test was performed at a rate ten times faster, i.e. 0.1 mm/s (0.004 in./s).
The data acquisition system enables the beam's reaction to the deflecting load to be constantly monitored. To determine f_{cte} value we utilize also the experimental value of deformation of the concrete under tensile strength, ε_{ct} (up to the opening of the first crack), were taken on the surface of the concrete itself by means of fixed bases attached to the bottom fiber of the beam and a precision dial gauge with a reading scale of 100 mm, as know in figure 2. Readings were taken continuously as the imposed displacement increased, with a pre-set increment of 0.1 mm/s (0.004 in./s), during the phase in which the concrete was still intact, i.e. up until the limit of its effective tensile strength, f_{cte}. During the tests, the beam's deflection at the mid-span was also measured by means of a precision electronic transducer. The tests were performed on a total of 10 beams, 5 made with 30 MPa (4351 psi) concrete and 5 with 70 MPa, (10152 psi). During the casting of each beam, and using the same mixture, 6 test cylinders were prepared for the evaluation of the concrete's compressive strength, and 2 cylinders for the indirect tensile stress test. This provided a comparison between the behavior of plain concrete and that of reinforced concrete.

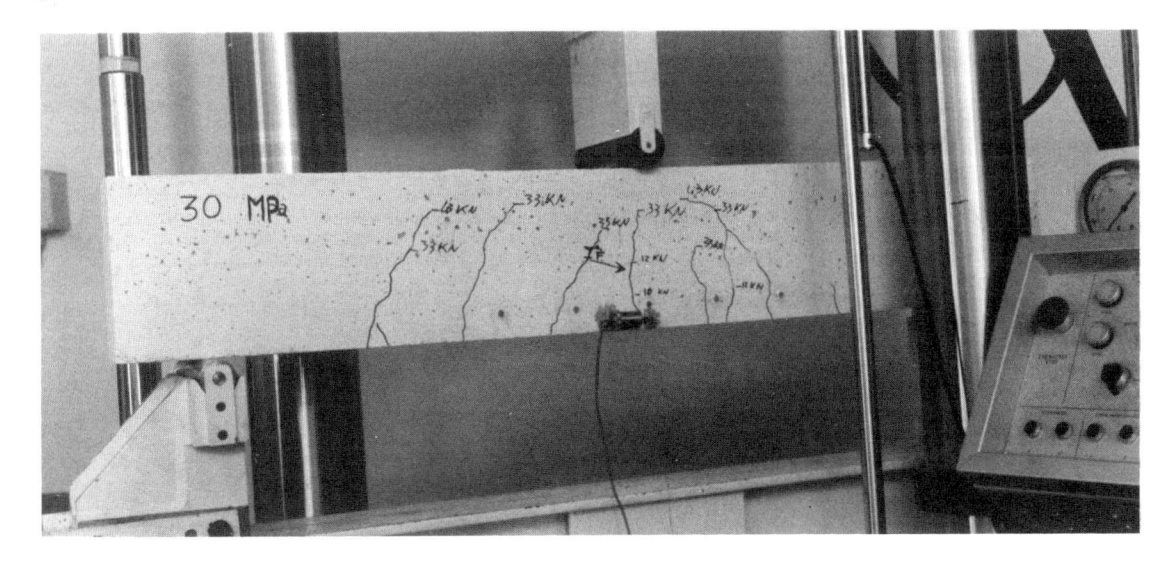

Figure 2: Details of Test Setup and Instrumentation

TEST RESULTS

It is common knowledge that there are two different methods for calculating tensile strength of concrete, i.e. by laboratory tests - using indirect (cylinder-splitting and bending) or direct tests, and by means of a calculation directly from the compressive strength of the material[15].

The latter method is the most often used because it is easier to implement and, albeit without experimental proof, it enables a more or less reliable evaluation of f_{ct}; it has recently been used even for the 100 MPa (14503 psi) concrete strength class[2].

Table 2 lists the values of f_{ct} deduced from the concrete class, according to Internationals Codes [2,3,5].

Table 3 shows the values of direct tensile strength of concrete, fct_{dir}, deduced by experimental evaluation of indirect tensile strength and adopting the conversion coefficient equal to 0.9.

The value f_{ceff} is the effective compressive strength class determinated by experimetal test, and f_c coincide with the nominal compressive strength class.

Table 2 - Tensile Strength Values by Concrete Class

f_c	f_{ct} A.C.I.	f_{ct} C.E.B.	f_{ct} EC2
30 MPa	2.7 MPa	2.8 MPa	2.9 MPa
70 MPa	4.2 MPa	4.4 MPa	/

(1ksi=6.9MPa)

Table 3 - directe tensile strength deduced by splitting tests

f_c (MPa)	f_{ceff} (MPa)	f_{ctdir} (MPa)
30	34.2	2.72
70	72.5	3.43

(1ksi=6.9 MPa)

Regarding the evaluation of f_{cte} value in reinforced concrete beams, it was assumed that this value could be calculated using the formula (1):

$$f_{cte} = \varepsilon_{ct} E_c \qquad (1)$$

where ε_{ct} is the deformation of the concrete measured during the test in the tension zone of beam, when the crack is going to open, and E_c is the modulus of elasticity for the concrete under compressive strain.

The value of f_{cte} is also established by means of an equation of equilibrium to the rotation between the middle span of the still intact beam and the experimental moment $M_{cr(exp)}$ at the time of the opening of the first crack (Figure 3).

M_cr: Moment of cracking

d₁: Distance between compressed concrete and reinforcement steel

d_ct: Distance between resultants of concrete areas being loaded

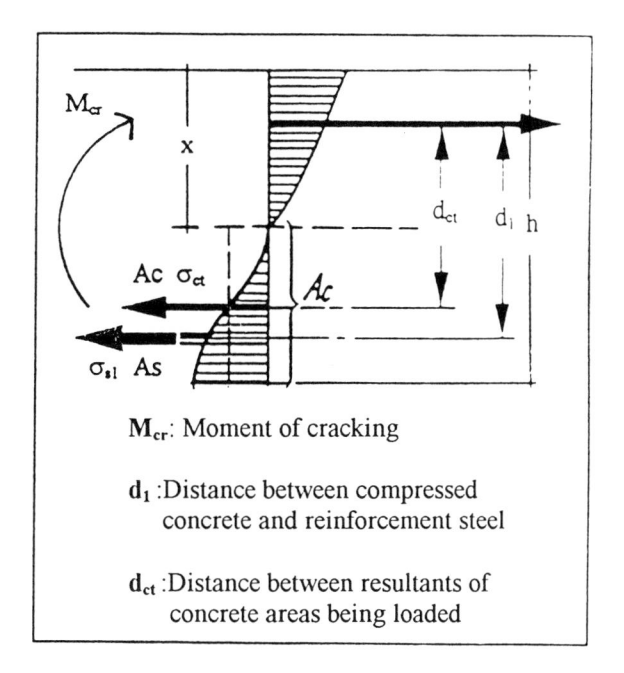

Figure 3: Stresses Diagram for Uncracked Section of Deflected Beam

Hence the following formula:

$$M_{cr(exp)} = \sigma_{s1} A_s d_1 + A_c \sigma_{ct} d_{ct} \qquad (2)$$

where σ_{s1} is assumed to equate to the stress in the reinforcement coming under the tensile load coinciding with the opening of the first crack, calculated on the basis of a perfect bond between the steel and concrete, with $\varepsilon_s = \varepsilon_c$, so that:

$$\sigma_{s1} = \varepsilon_{ct} E_s \qquad (3)$$

and d_1 is the distance between the centroid of the reinforcement coming under tensile load and the resultant of the area of compressed concrete; d_{ct} is the distance with respect to the resultant of the area of the concrete being loaded.

Assuming also that:

$$\sigma_{ct} = f_{cte(Mcr)} \qquad (4)$$

then:

$$f_{cte(Mcr)} = (M_{cr(exp)} - \sigma_{s1} A_s d_1) / A_c d_{ct} \qquad (5)$$

hence:

$$f_{cte(Mcr)} = M_{cr(exp)}/A_c d_{ct} - \rho (d_1/d_{ct}) \sigma_{s1} \qquad (6)$$

and, after simplification:

$$f_{cte(Mcr)} = 1/d_{ct}(M_{cr(exp)}/A_c - \rho\, d_1 \sigma_{s1}) \quad (7)$$

The moment of cracking, called $M_{cr(exp)}$, was calculated experimentally in every beam and coincided with the opening of the first crack.

The results relating to the values of f_{cte} and of $f_{cte(Mcr)}$, are given in Table 4, for both the concrete strength classes considered and for all the beams tested.

Table 4 - f_{cte} values

f_c	Number of beams	$f_{cte(\varepsilon)}$ (MPa)	$f_{cte(Mcr)}$ (MPa)
30 MPa	1	2.24	1.10
	2	2.32	1.06
	3	2.70	1.12
	4	2.90	1.15
	5	2.15	1.04
70 MPa	1	5.25	1.10
	2	1.75	1.60
	3	4.20	1.48
	4	3.28	1.14
	5	2.8	1.32

(1ksi=6.9 MPa)

The data in Table 4 indicate a marked difference between the effective tensile strength values obtained by means of the formula n.1 and n.7, and the values deduced experimentally, regardless of the type of concrete involved.
In other words, the values established on site, both as a function of the load applied to the element and considering the deformation of the concrete, seem to be much lower than f_{ct}, confirming the incidence of the presence of the reinforcement.
As for the comparison between the value of f_{ct} deduced directly from the concrete strength class and the value established by the laboratory tests, the diagram in Figure 4 shows that there is a fair consistency in the results for the 30 MPa (4351 psi) class of concrete, whereas for the 70 MPa (10152 psi) class of concrete the experimental values are considerably lower than the outcome of direct calculation. The difference between the value of f_{ct} measured on the material concrete - for which the mean values are given - and in association with the reinforcement, f_{cte} measured on each beam by means of formula 1, is also clearly evident, Figure 5.

Figures 6 and 7 shows the crack width for C30 and C70 beams and the corresponding load, this referring only the fixed bases 1 to 4 in the midspan, figure 2. Table 5 show the value of experimental cracking moment $Mcr_{(exp)}$ deduced by cracking load (at the first crack). Instead, the $Mcr_{(t)}$ value represents the theoretical value of cracking deduced by model[16], using the fct_{dir} value.

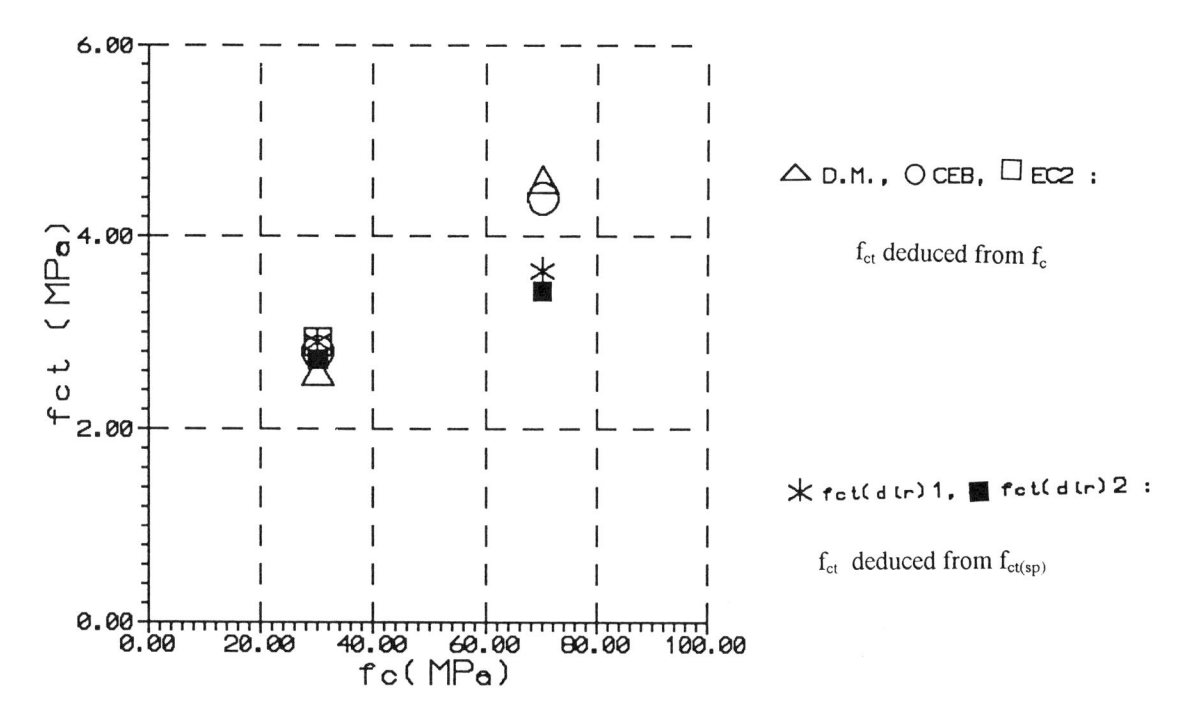

Figure 4. f_{ct} values deduced by compressive strength class and from laboratory tests

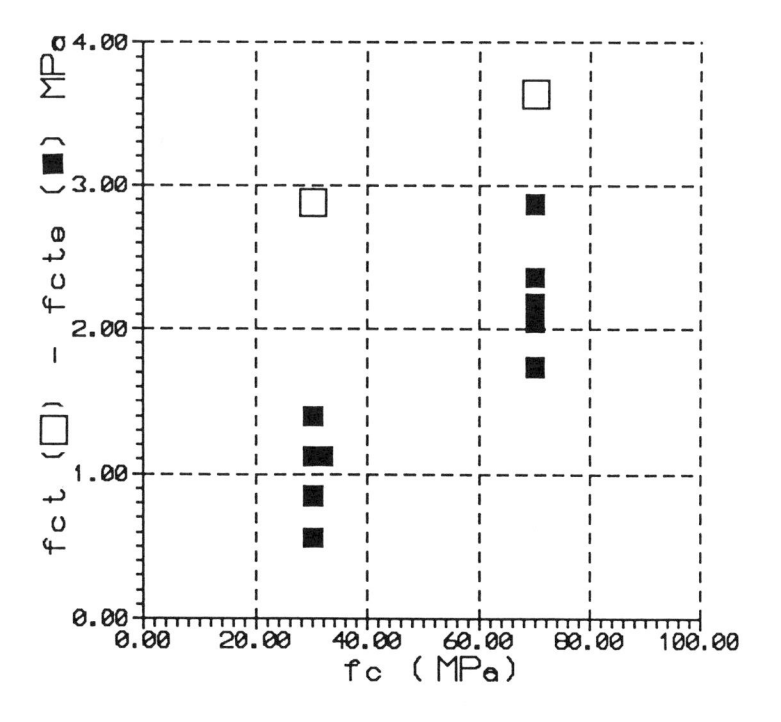

Figure 5. f_{ct} for plain concrete and corresponding f_{cte} value

Tab.5: Experimental value of bending moment

fc	Number of beams	$M_{cr(exp)}$ (kNmm)	$M_{cr(t)}$ (kNmm)
30 MPa	1	2200	
	2	2475	
	3	2337	4510
	4	2475	
	5	2502	
70 MPa	1	2282	
	2	2502	
	3	3437	5280
	4	2750	
	5	3025	

(1ksi=6.9 MPa)

kN	kip
10	2.25
20	4.5
30	6.75
40	9.0
50	11.25

mm	in.
0.05	0.002
0.10	0.004
0.15	0.006
0.20	0.008
0.25	0.01
0.30	0.012
0.40	0.016

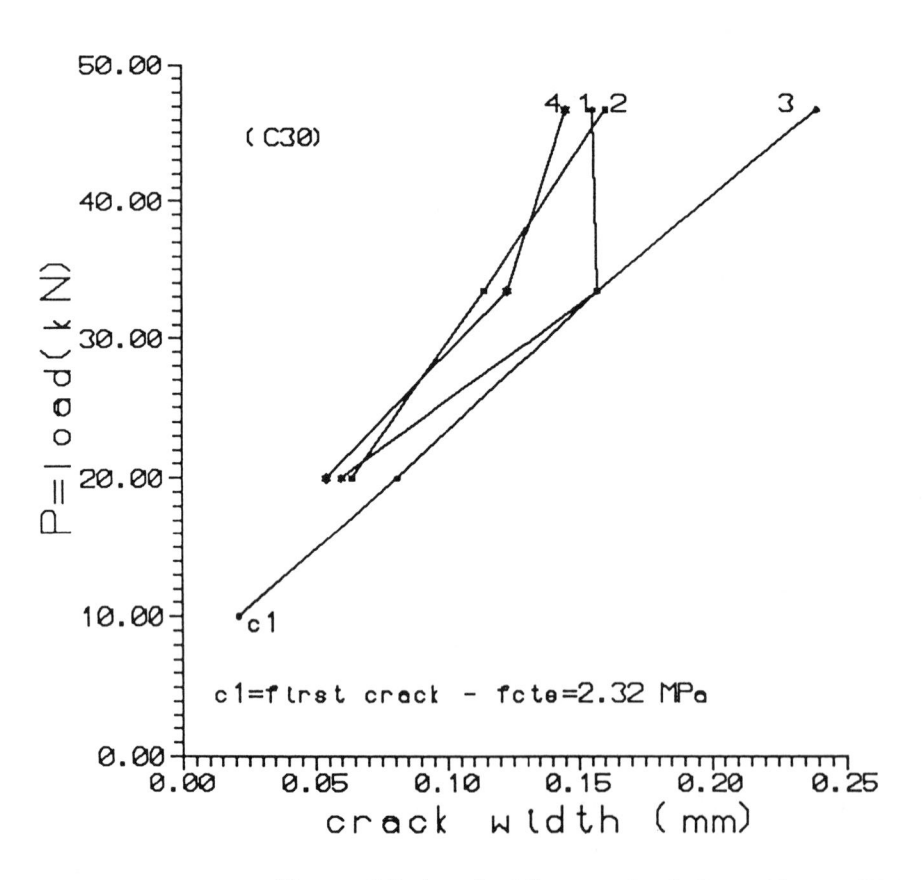

Figure 6: Point cI of first crack; deflected beam C30

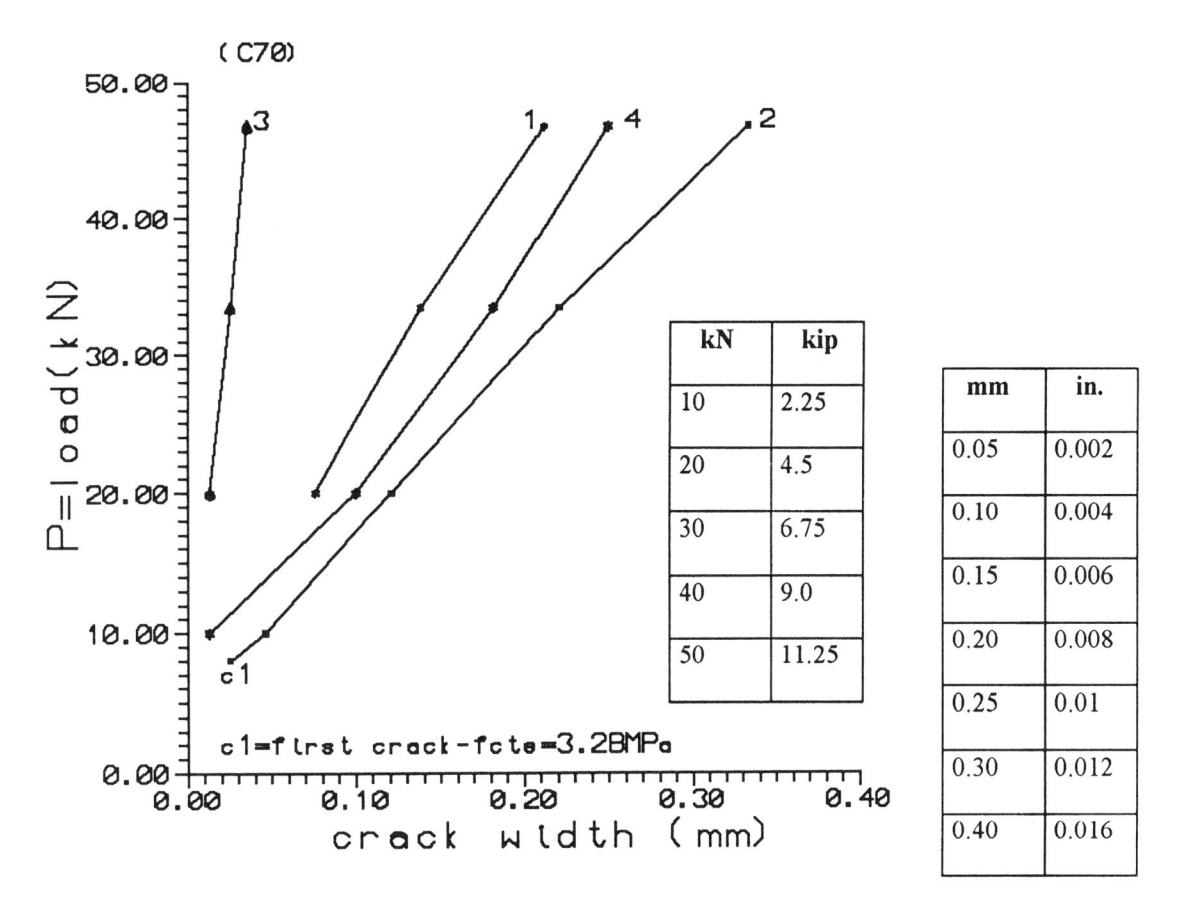

Figure 7: Point cI of first crack; deflected beam C70

CONCLUSIONS

This study proposes a calculation of the effective tensile strength of concrete f_{cts}, reinforced with steel reinforcement. In the preliminary phase described here, even before considering the size effect, the study was confined to the use of a simple experimental model, three point bending test, in order to verify the difference between f_{cte} (calculated by means of the traditional formulas valid in the elastic field) and f_{ct}, calculated in plain concrete by traditionals tests, for various classes of concrete.
The study showed that:
1. The tensile strength in plain concrete are considerably higher than for the tensile strength concrete measured in presence of steel reinforcement, even for high performance concrete mixtures;
2. The influence of concrete compressive strength on the calculation of effective tensile strength of concerte is not negligible;
3. The effective tensile strength concrete determined for mixtures higher than C30 could be used for design purposes, to control the cracking phenomena under the serviceability conditions.
4. The theoretical value of cracking moment is higher than the corresponding value deduced by experimental tests. This is probably caused by a significant difference between conventional and effective tensile strength concrete.

The study is currently underway and the next stage will consider the parameters relating to the mechanical features of the fracture behaviour and the size effect, and a comparison with other formulas presented in the field.

REFERENCES

1-Di Marco, R., Russo, S., Siviero, E. (1996), Minimum reinforcement areas in high strength concrete, 15° Congress IABSE, Copenhagen, June 1996.

2-CEB-FIP(1995), High performance concrete. Recommended Extension to the Model Code 90, Bulletin d'Information No. 228, July

3 - Eurocode 2, "Design of concrete structures - Part 1-1: General rules and rules for buildings", UNI ENV, 1992.

4-Farra, B.(1995), Influence de la résistance du béton et de son adhérence avec l'armature sur la fissuration, EPFL, Thèse n.1359, Losanna.

5-ACI, State of the art report on high strength concrete, reported by Committee 363, ACI 363R-84.

6 -Favre, R., Jaccoud, J.P., Koprna, M., Radojicic, A.,(1994), Progettare in calcestruzzo armato, HOEPLI.

7-Ghali, A., Favre, R. (1994), Concrete structures, stresses and deformation, 2nd Edition E-Fn Spon, Chapman & Hall.

8-Creazza, G., Di Marco, R., Russo, S., Siviero, E.(1996), Tension stiffening effect in high strength concrete, 4th International Symposium on Utilization of high strength/high performance concrete, Paris, May 1996.

9-Siviero, E., Cantoni, R., Forin, M.(1995), Durabilità delle opere in calcestruzzo, Franco Angeli Editore-Edilizia.

10 - Gerard, B., Marchand, J., Breysse, D., Ammouche, A.(1996), Constitutive law of high performance concrete under tensile strain, 4th Int. Sym. on Utilization of HSC, May, Paris.

11- Rostasy, F. S. (1985), Risse infolge Zwang und Eigenspannungen. Vortrag am Deutschen Betontag, Deutscher Beton-Verein.

12-Schiessl, P.(1985), Mindestbewehrung zur Vermeidung klaffender Risse, Institut für Betonstahl und Stahlbetonbau e.V., 284/85.

13- Russo, S., Siviero, E.(1994), Comportamento a trazione di elementi in c.a. ad elevata resistenza, Congresso CTE Milano.

14-Russo, S., Siviero, E.(1994), Tensile behaviour of high strength concrete. Preliminary results of an experimental investigation, Workshop on 'Development of EN 1992 in relation to new research results and to the CEB-FIP M.C. '90, Prague, October.

15 - Phillips, D. V., Binsheng, Z.,(1993), Direct tension tests on notched and un-notched plain concrete specimens, Magazine of Concrete Research, 45.

16- Arduini, a., Russo, S., Siviero, E., "The influence of high strength concrete in ductility of RC deflected beams", next publication in Bulletin CEB 'Ductility', 1997

Symbols

f_{cte}: effective tensile strength of concrete, measured in presence of reinforcement steel

f_{ct}: tensile strength of plain concrete

w/c: ratio water-cement

f_y: yield strength of reinforcement steel

f_t: tensile strength of reinforcement steel

f_t/f_y: stress hardening ratio in reinforcement steel

ε_u: uniform elongation of reinforcement steel, in correspondence of maximum load

ε_{ct}: deformation of concrete (in presence of steel reinforcement) under tensile strength, before the first crack

f_{ctdir}: directe tensile strength of concrete deduced by splitting test and adopting the conversion coefficient equal to 0.9

f_c: nominal compressive strength class of concrete

f_{ceff}: compressive strength class of concrete deduced by experimental test

E_c: modulus of elasticity of concrete

$M_{cr(exp)}$: cracking moment deduced by experimental test

σ_{s1}: stress in the reinforcement steel at the first crack

A_s: reinforcement steel area

d_1: distance between compressed concrete and reinforcement steel

A_c: concrete area

σ_{ct}: stress of concrete in tension

d_{ct}: distance between resultants of concrete areas being loaded

ε_c: deformation of concrete

$f_{cte(Mcr)}$: effective tensile strength of concrete in presence of reinforcement steel, evaluated in function of cracking moment

$M_{cr(t)}$: Theoretical value of cracking moment

$f_{ct(sp)}$: tensile strength of concrete deduced by splitting test

D.M.: Italian Code

CEB: Comité Euro-International du Béton

EC2: Eurocode n.2

Proceedings of the PCI/FHWA
International Symposium on High Performance Concrete
New Orleans, Louisiana, October 20-22, 1997

IMPLEMENTATION OF HIGH STRENGTH CONCRETE RESEARCH FOR PRESTRESSED GIRDERS - A DOT'S PERSPECTIVE

Daniel Dorgan, Assistant Bridge Engineer-Planning
Minnesota Department of Transportation
Roseville, Minnesota, USA

ABSTRACT

Since Minnesota constructed its first prestressed bridge in 1957, steady improvements have been made in prestressed design and fabrication to achieve longer and shallower spans. The results of research jointly funded by industry and the Minnesota Department of Transportation (Mn/DOT) will be highlighted which will further increase Minnesota design strengths and girder spans.

HISTORICAL DEVELOPMENT OF PRESTRESS BRIDGES IN MINNESOTA

Minnesota entered the prestressed concrete world in 1957 with the construction of its first bridge utilizing prestressed beams. Bridge Number 9053 was a four-span structure carrying 94th Street in Bloomington, Minnesota over Interstate 35W. The required concrete strength for the 18.3 m (60 ft.) long 915 mm (36 in.) deep prestressed I-girders was a lofty 34 MPa (5000 psi). That year Minnesota began construction of three other prestressed beam bridges embarking on a new era in bridge design and construction. Today, those four bridges continue in service as pioneers of a major change in bridge technology for Minnesota.

Since 1957, over 1,800 prestressed bridges have been built in Minnesota on the state and local road systems. A variety of prestressed sections have been developed and modified over the years to address varying span length and superstructure depth needs. Required concrete strengths, steel stresses, and beam lengths have steadily increased as improvements in materials and fabrication allowed Mn/DOT to push span lengths to dimensions that were not anticipated in 1957. The 34 MPa (5000 psi) design concrete strength of 1957 was increased to 41 MPa (6000 psi) in 1968 and 48 MPa (7000 psi) in 1986. For a limited number of projects with special needs, designers have been allowed to specify concrete strengths up to 59 MPa (8500 psi). The longest prestressed girder bridge span in Minnesota was built in 1994 and incorporated 2060 mm (81 in.) deep girders 47.1 m (154'-6") in length. This bridge is located on Trunk Highway 101 over the Mississippi River.

In February 1997, based on the research described in this report, Mn/DOT jointly decided with local prestress fabricators to again increase standard concrete design strengths. Designers can now routinely specify concrete strengths up to 48 MPa (7000 psi) at release and 59 MPa (8500 psi) final.

Since the introduction of the 915 mm (36 in.) section in 1957, we have added 710 mm (28 in.), 1015 mm (40 in.), 1140 mm (45 in.), 1370 mm (54 in.), 1600 mm (63 in.), 1830 mm (72 in.) and 2060 mm (81 in.) deep I girders. Additionally, Double T, Quad T, and Voided Slab girders have been used for shorter span bridges. Many of these sections have been modified over the years to increase the structural efficiency and span of the girders. Often these changes have been driven by the desire to take full advantage of higher prestress force. The amount of available prestressing force has increased with larger strand diameters, low relaxation strand, and greater yield strengths. In response, the top flange concrete area for 1140 mm (45 in.) and 1370 mm (54 in.) sections was increased in 1986 to withstand greater compressive forces. These modifications have increased the span length of the girders. Figure 1 shows the various prestress sections currently used by Mn/DOT.

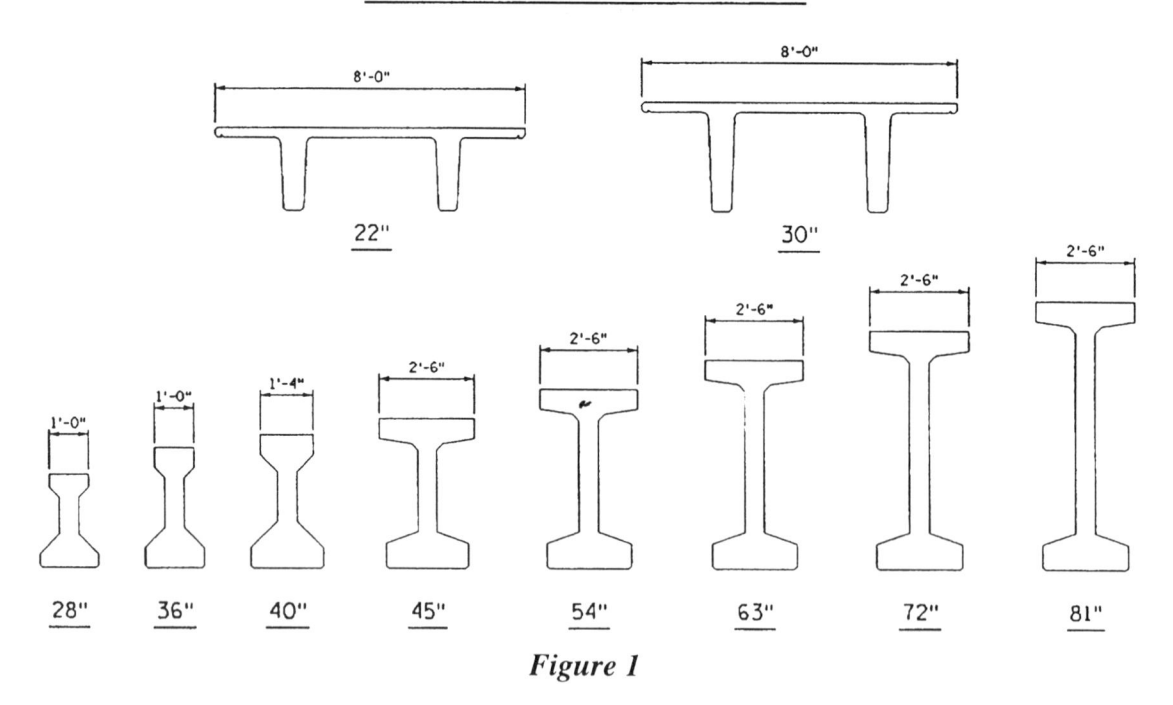

Figure 1

Note: 1 inch = 25.4 mm

RESEARCH PARTNERSHIP

Throughout the development and application of prestressing bridge technology during the last 40 years, Mn/DOT has benefitted from an excellent working relationship with local prestress fabricators. Industry and Mn/DOT have jointly worked to apply new technological improvements as they became available. Mn/DOT has also consistently approved fabricator requests to modify designs to achieve production efficiencies when project requirements were maintained. In 1991, the Minnesota Department of Transportation, the Center For Transportation Studies (CTS) at the University of Minnesota, and the Minnesota Prestress Association (Mn/PA) initiated a research project with the University of Minnesota to investigate the use of higher strength concrete mix designs for prestressed girders. Mn/PA is an industry association comprised of Minnesota prestressed concrete fabricators. Utilizing funding from the three partners, the research was expanded to include durability testing, camber, and full-size girder testing. In addition to the research funding provided by Mn/PA, Mn/DOT and CTS, a large number of local contractors and suppliers donated materials and services along with a Daniel P. Jenny Research Fellowship from the Precast/Prestressed Concrete Institute. Due to the partnership involving designers, academics, and industry, we believe the research was well focused on addressing the concerns of all parties and is leading to results that will have immediate practical application.

RESEARCH PROGRAM

Materials Testing

A variety of trial mix designs were developed using several locally available aggregates. From these studies, two mixes were selected that achieved the target 28-day compressive strength of 72 MPa (10,500 psi). A crushed limestone aggregate mix and a rounded glacial gravel aggregate with microsilica were selected from the trial designs. The addition of microsilica improved the performance of the glacial gravel mix but did not significantly benefit the limestone compressive strength. Two full-size test girders would eventually be cast, each utilizing one of these mixes.

Durability testing was conducted for a variety of mixes and local aggregates using a freeze-thaw chamber. Concrete durability did vary between aggregates and curing methods. One of the local limestone aggregates consistently outperformed the glacial gravel, granite and other limestone aggregates. Additionally, moist-cured specimens generally had better freeze-thaw durability than their heat-cured counterparts.

Girder Design and Fabrication

The showcase of the research was the fabrication and testing of two full-size test girders. To develop confidence in the use of high strength concrete, the girders were tested to evaluate prestress losses, fatigue behavior, prestress transfer lengths, camber, and ultimate flexural strength. The 1140 mm (45 in.) deep section was selected with a girder length of 40.5 m (132.75 ft.). Figure 2 shows a cross section of the girder and the composite slab along with design information. The typical maximum compressive strength currently used by Mn/DOT was 48 MPa (7000 psi) at the time of the research. Design analyses were performed and it was found that increasing the compressive strength from 48 MPa (7000 psi) to 69 MPa (10,000 psi) yielded a 12% to 18% increase in span length for a given girder spacing. For the 1140 mm (45 in.) section tested, it was also found that the use of higher strength concretes was limited by the amount of prestressing that could be effectively placed in the cross section. Tension in the bottom fiber under service loads limited the girder length to 40.5 m (132.75 ft.). Little benefit was realized from increasing the compressive strength above 69 MPa (10,000 psi).

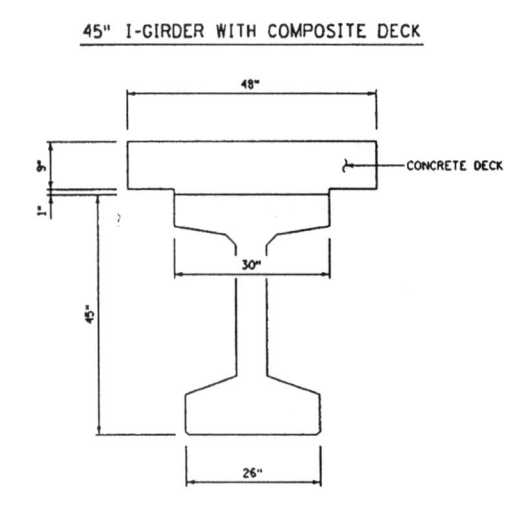

Figure 2

Note: 1 inch = 25.4 mm
1 ksi = 6.9 mPa

The test girders were fabricated in August – 1993 at Elk River Concrete Products in Elk River, Minnesota. During fabrication, the girders were extensively instrumented to measure losses, transfer length, creep and shrinkage, temperature effects, reinforcing steel and prestress strand strains, and strain in the concrete and steel due to flexure and shear. The instrumentation was monitored throughout fabrication and eventual girder testing. The strands were released 24 hours after casting. The compressive strength at that time was 64 MPa (9300 psi) for Girder I (limestone) and 72 MPa (10,400 psi) for Girder II (glacial gravel). Normal construction tarps were placed over the girders during curing, and no additional methods were employed to enhance curing. Girder I had an initial camber of 122 mm (4.8 in.) that grew to 203 mm (8.0 in.) after two months. Girder II's initial camber was 97 mm (3.8 in.) with a maximum of 146 mm (5.75 in.) in two months. After 28 days, Girder I achieved a compressive strength of 83 MPa (12,100 psi) and Girder II reached 77 MPa (11,100 psi).

Girder Testing

The girders were transported by truck to a Mn/DOT warehouse for the testing program. The hauling company regularly transports prestressed girders for Minnesota bridge projects and used the same equipment they typically employ. Prior to the trip, there was concern regarding how the section would behave during transport since 1140 mm (45 in.) sections of this length had never been shipped. However, the driver negotiated the 48 km (30 mi.) trip, complete with freeway cloverleafs, without any problems at normal hauling speeds. In fact, the skilled driver backed the girders with their 760 mm (30 in.)-wide top flange through a 860 mm (34 in.)-wide door to deliver them into the testing building.

Following the installation of the girders on their support blocks, a 230 mm (9 in.)-thick reinforced concrete deck 1220 mm (48 in.) in width was cast on each of the girders to provide a composite section. The girders were designed for an assumed girder spacing of 1.2 m (4.0 ft.) on center. Twin-load frames were erected at the 4/10 point and 6/10 point of the span as depicted in Figure 3. The frames were anchored in reaction piles that were augured into the soil. Hydraulic load actuators were installed to apply the loading for fatigue and failure testing.

An extensive fatigue testing program was conducted on both girders. The girders were loaded dynamically with a HS25 design load for 1 million cycles. No change was observed in the stiffness of the girders due to this fatigue loading. The girders were then cracked and an additional 1 million cycles at HS25 and 1 million cycles at 125% of HS25 were applied. Neither girder exhibited an appreciable reduction in stiffness due to these loadings. The load of 125% of HS25 was chosen as representative of routine overloads on our highway system. With over 3 million cycles applied to each test girder, we were satisfied with the fatigue performance of the girders.

LOAD POINTS FOR FATIGUE LOADING

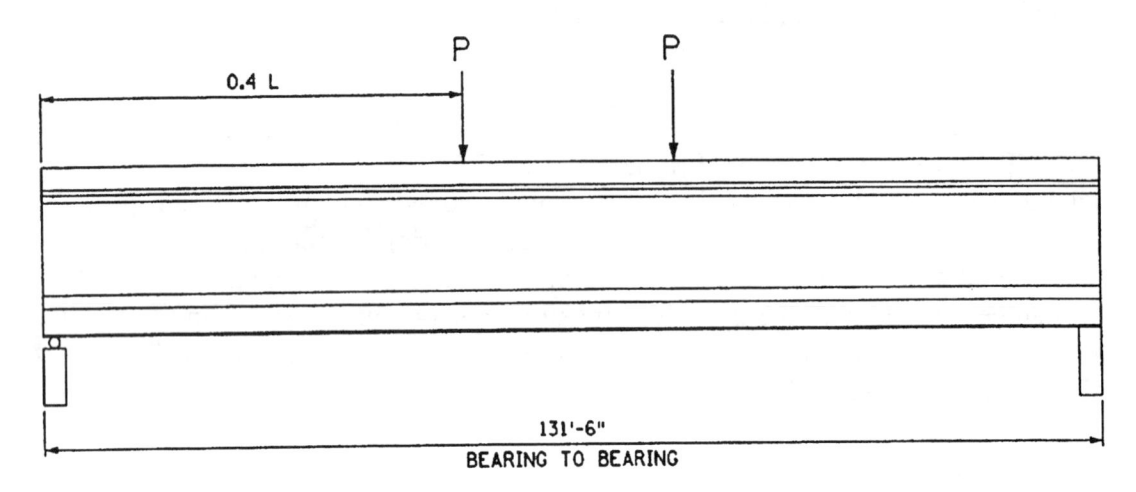

Figure 3

Note: 1 ft. = .3048 m

The prestress losses were monitored in the girders from fabrication through testing. The losses measured at the time of release were higher than predicted by AASHTO equations and other methods. The measured losses at release ranged from 11.8 to 13.4% of the prestress force versus predicted losses of about 10.4%. This underprediction of losses was attributed to an overestimate of the girder stiffness. Conventional modulus of elasticity equations appear to overestimate the stiffness of high strength concrete causing an underestimation of the elastic shortening at release. However, this was offset by the fact that the final total losses measured were consistently well below the losses predicted by AASHTO and other loss calculations. The measured final losses ranged from 16 to 21% which was below the predicted values of 25 to 31% by various methods. This was a particularly important issue to Mn/DOT since we wanted to determine if the current loss equations could be extended for use with concrete strengths of 69 MPa (10,000 psi).

The girders were loaded to failure in late 1995. At the time of failure, the deflection at mid-span was 890 mm (35 in.). A compression failure occurred in the composite deck near one of the loading points. The loading on the beams was 600 kN (135 k) and 645 kN (145 k) at the two load points at failure. In terms of design loading, this is approximately equivalent to 670% of an HS25 load.

Following the ultimate flexural load tests, the girder ends were tested in shear. Based on AASHTO and ACI codes the girders were expected to carry loads of 1560 kN (350 k) to

1730 kN (390 k), depending on the strand configuration at each end. The actual loads at failure ranged from 2220 kN (500 k) to 2730 kN (614 k). One section, which did not employ debonded strands and had a greater number of draped strands withstood a peak load of 3010 kN (676 k). This section did not fail in shear since its capacity exceeded the testing equipment load capabilities.

RESEARCH IMPLEMENTATION

In February 1997 Mn/DOT personnel with the four fabricators that supply prestress bridge girders for Minnesota projects. The fabricators included Wells Concrete Products of Wells, Minnesota, County Prestress Corporation of Osseo, Minnesota, Andrews Prestressed Concrete of Clear Lake, Iowa, and Elk River Concrete Products of Maple Grove, Minnesota. The purpose of the meeting was to discuss implementation of the research and also to discuss issues and concerns the fabricators may have regarding our designs, specifications, or inspection procedures. Mn/DOT was seeking the fabricators input and concurrence as they considered increasing Mn/DOT allowable concrete strengths for routine prestressed design projects. Since 1986 Mn/DOT had been limiting our release strengths to 45 MPa (6500 psi) and final strengths to 48 MPa (7000 psi).

As Mn/DOT discussed implementation of the research a variety of factors were considered. For the most efficient and economical operation of fabricating plants, suppliers strive to attain release strengths in about 18 hours. This allows them to cast a beam a day on their beds. It was important to the fabricators to limit the release strengths to a value they could achieve while maintaining their one day bed turnover. The ability to consistently achieve high strengths during cold weather pours is an additional limiting consideration. The fabricators wished to increase concrete strengths incrementally and obtain experience working with these strengths at full production prior to proceeding to higher compressive strengths. Additionally, experience in transporting and erecting these sections which generally will be longer, heavier, and more limber will also be valuable. At that meeting, Mn/DOT agreed with the fabricators to implement an increase to 48 MPa (7000 psi) at release and 59 MPa (8500 psi) for the final compressive strengths. These would be the limits used for routine design work. For unique projects, those limits might be exceeded on a case by case basis. For these unique projects, the practice has generally been to first consult with fabricators prior to designing for higher strengths.

During the implementation meeting with the fabricators Mn/DOT also discussed the use of 15.2 mm (0.60 in.) diameter strand at 51 mm (2 in.) spacing. Although conventional 12.7 mm (½ in.) strand is our standard, Mn/DOT has used 15.2 mm (0.60 in.) diameter or 12.7 mm (½ in.) oversized on a few projects and the fabricators had experience with the larger strand. With the current I girder sections we are not able to add enough prestressing force with 12.7 mm (½ in.) strand to utilize strengths above the 59-62 MPa

(8500-9000 psi) range. The fabricators related some of the difficulties they had working with the larger strand. It was more difficult to work with since it was stiffer and during colder weather they have problems tensioning the larger strand due to the limitations of their jacks. Given their concerns, it was decided to continue using conventional 12.7 mm (½ in.) strand as our standard. Larger strand diameters will only be used on a limited basis for unique designs.

Mn/DOT agreed with the fabricators to implement the 48 MPa (7000 psi) release and 59 MPa (8500 psi) final strengths at this time and continue using 12.7 mm (½ in.) strand as the standard. After gaining experience at this level for several production seasons, it was agreed to revisit these limits with our fabricators. To receive further benefits from high strength concrete, the amount of prestressing force that can place in the section will have to be increased. From a DOT perspective, the open discussion and input from our fabricators was very much appreciated..

ACKNOWLEDGMENTS

This research was jointly funded by the Minnesota Prestress Association, the University of Minnesota Center for Transportation Studies, and the Minnesota Department of Transportation. The cooperation and willingness of industry, academics, and government in working together on this issue is greatly appreciated and will lead directly to savings in our bridge construction costs. In particular, the conscientious efforts of the University's research team which made the project a success are greatly appreciated. This effort included Dr. Catherine French, Dr. Roberto Leon, Dr. Carol Shield, and their graduate assistants Theresa Ahlborn, Alireza Mokhtarzadeh, Jeffrey Kielb, and Roxanne Kriesel. The Daniel P. Jenny Research Fellowship awarded by the Precast/Prestressed Concrete Institute to support this research effort was also appreciated. With the help of all of these individuals and partners, we will take another step forward in prestress design.

References

1. Ahlborn, T.M., Shield, C.K., French, C.W., "Behavior of Two Long-Span High Strength Concrete Prestressed Bridge Girders," submitted to the ASCE Structures Congress 1996.

2. Kielb, J.A., "Instrumentation and Fabrication of Two High Strength Concrete Prestressed Bridge Girders", University of Minnesota Department of Civil and Mineral Engineering, 1994.

APPLICATION OF HIGH PERFORMANCE CONCRETE IN TWO BRIDGES IN NEW HAMPSHIRE

Michelle L. Juliano, P. E.
Design Engineer
New Hampshire Department of Transportation
Concord, New Hampshire, USA

Christopher M. Waszczuk, P. E.
Project Engineer
New Hampshire Department of Transportation
Concord, New Hampshire, USA

ABSTRACT

High performance concrete, with its enhanced durability and strength characteristics, may be one means to provide economical bridges with longer life expectancies. Two such bridges are scheduled for construction in New Hampshire. Research being conducted on the high quality mix includes the development of the mix design and study of procedures for placement and curing. In-situ performance evaluations of the structural members will offer a comparison between measured and theoretical values of durability and strength.

INTRODUCTION

Background

With nearly 31% of the bridges across the nation rated as substandard,[1] the need to replace and maintain bridges in a cost-effective manner has become a paramount issue. One promising solution is the use of a new product technology in bridge improvement applications: high performance concrete.

High performance concrete (HPC) is generally defined as a concrete which meets increased strength and enhanced durability characteristics. Increased flexural strength allows for the design of longer span lengths, wider girder spacing and shallower members. The need for fewer girders and piers can lower initial bridge costs. In addition, the use of smaller and fewer girders can result in lighter superstructures which, in turn, can decrease the substructure cost. The enhanced durability characteristics provide HPC with a greater resistance to deterioration caused by climate effects, exposure to de-icing chemicals and stresses due to loading conditions.[2] Increased concrete durability can be expected to provide the nation's bridges with longer life and less required maintenance, thus reducing long-term costs.

To promote the use of HPC, the Federal Highway Administration (FHWA) is entering into agreements with state highway agencies to design and construct bridges using this new product technology. The bridges are designed, built, evaluated and documented and the experience gained is shared through the development and presentation of workshops. The New Hampshire Department of Transportation (NHDOT), in coordination with the University of New Hampshire (UNH), Department of Civil Engineering, has entered into such an agreement to build two HPC bridges in Bristol, NH. In October 1996, the construction of the first bridge was completed. The construction of the second HPC bridge is scheduled to begin in the Spring of 1999 and will finish in the Fall of the same year.

Literature Review
HPC began as high strength concrete. In 1949, the Walnut Lane Bridge in Philadelphia, PA was constructed with 37 MPa (5400 psi) concrete. In the successive decades, the use of higher strength concrete was evident especially in the tall building industry which required stiffer and stronger compression members. In 1988, the Two Union Square Building in Seattle, WA used 131 MPa (19,000 psi) concrete with a modulus of elasticity of 50 GPa (7.2×10^6 psi). In the 1990s, as more bridges throughout the world are being constructed with higher strength concrete, the focus is gradually shifting from high strength concrete to high performance concrete in which long term performance is additionally addressed.[3]

A Strategic Highway Research Program study [4] defined HPC as having: 1) a maximum water-cement ratio (w/c) of 0.35; 2) a minimum durability factor of 80%; and 3) a minimum strength of either: a) 21 MPa (3000 psi) within 4 hours of placement (very early strength); b) 34 MPa (5,000 psi) within 24 hours (high early strength); or c) 69 MPa (10,000 psi) within 28 days (very high strength). The FHWA has recently developed a new definition in which HPC is categorized into 4 performance grades.[2] Each grade represents a different performance level of the four durability and four strength parameters. The durability parameters are: freeze-thaw, scaling, abrasion and chloride penetration. The strength parameters are: compressive strength, elasticity, shrinkage and creep. See Table 1.

One strength parameter, compressive strength, will have a dramatic influence on section capacity. Increased sectional capacity will allow for the design of longer span lengths of standard girder sections. Designs can include wider girder spacing, resulting in fewer girders, and shallower members for use in areas where vertical underclearance is a concern. Increasing the initial prestressing force by using larger diameter prestressing strands, such as 15 mm (0.6 in.), will further increase sectional capacity. However, the benefits of higher concrete strength are limited. Often the additional strands needed to increase capacity must be placed higher in the web of a standard girder I-section, decreasing the effectiveness of the strands.[5] Lateral stability becomes a concern during transportation and erection of longer girders.

Table 1. FHWA Grades of HPC performance characteristics[2]

Performance Characteristic	Grade 1	Grade 2	Grade 3	Grade 4
Freeze-thaw durability (after 300 cycles)	60% to 80%	≥ 80%		
Scaling resistance (after 50 cycles)	4.5	2.3	0.1	
Abrasion resistance (avg. wear depth, mm)	1.0 to 2.0	0.5 to 1.0	≤ 0.5	
Chloride penetration (coulombs)	2000 to 3000	800 to 2000	≤ 800	
Strength $f'c$	41 to 55 MPa (6 to 8 ksi)	55 to 69 MPa (8 to 10 ksi)	69 to 97 MPa (10 to 14 ksi)	≥ 97 MPa (≥ 14 ksi)
Elasticity E	28 to 40 GPa (4 to 6x10^6 psi)	40 to 50 GPa (6 to 7.5x10^6 psi)	≥ 50 GPa (≥ 7.5x10^6 psi)	
Shrinkage (microstrain)	600 to 800	400 to 600	≤ 400	
Creep (microstrain/pressure unit)	60 to 75/MPa (0.41 to 0.52/psi)	45 to 60/MPa (0.31 to 0.41/psi)	30 to 45/MPa (0.21 to 0.31/psi)	≤ 30/MPa (≤ 0.21/psi)

Low permeability is one characteristic which improves the durability of HPC. Low permeability reduces the rate of chloride ion diffusion into the concrete slowing the corrosion of reinforcing steel and ultimately concrete spalling. HPC's resistance to deterioration caused by abrasion, scaling and freeze-thaw cycles is influenced by curing procedures, w/c and air content. Greater resistance to deterioration will improve the concrete's performance thereby requiring less maintenance and extending its life.[2]

Increased testing and tighter quality control is essential in achieving the desired strength and durability characteristics of HPC. Development of trial batches, for example, is an important step in optimizing a mix design. Several trial batches, simulating actual in-field placement and curing procedures, should take place until all specifications are met. Equally important are the methods of finishing and curing. Finishing of the surface should be limited to only what is absolutely necessary. Curing should begin directly after finishing is completed and proper moisture conditions should be maintained in order to prevent plastic shrinkage cracking.[6]

Objectives and Scope

The objectives of this case study are to become familiar with the performance characteristics of HPC, how to achieve these characteristics, and how to benefit from them. This information will then be disseminated to bridge engineers, contractors, fabricators and concrete producers in the northeast United States. The scope of the case study is as follows:

First HPC Bridge - Choose concrete design strengths which are readily attainable in air-entrained concrete by local concrete producers for the precast/prestressed girders and the cast-in-place bridge deck. Select an HPC mix design for the deck following the testing of several mixes. Encourage competitive bidding and lower costs by allowing the girder fabricator the flexibility of developing the girder concrete mix, provided all required specifications are met. Design the most economical bridge superstructure by taking advantage of the increased concrete strengths. Evaluate the materials used in the HPC mixes and monitor the in-situ bridge performance.

Second HPC Bridge - Design and construct a second HPC bridge employing the knowledge gained from the development of the first.

Showcase - Develop and host a 2-day workshop showcasing HPC through presentations from those associated with the research, design, fabrication and construction of the HPC bridges in New Hampshire and other regions.

METHODOLOGY

Site Location
The site chosen for the first HPC bridge is on NH Route 104 over the Newfound River in Bristol, NH. The town of Bristol is a rural community situated in the central portion of the state. The second HPC bridge site is located on NH Route 3A also over the Newfound River in Bristol, NH. The second site is approximately 2.4 km (1.5 mi.) northwest of the first HPC structure. The two sites have similar characteristics: the span lengths are nearly identical, both are located at river crossings and carry similar volumes and types of traffic. The similarities between the two sites provide the opportunity for an accurate comparison of the long-term performance of each bridge.

Concrete Mix Designs
The initial phase of the HPC research included a UNH recommendation of a HPC mix design for use in the two bridge decks. This mix was chosen as a result of an in-situ evaluation of three different mix designs. Test slabs representing each mix were installed at a UNH bridge deck testing facility located in Rochester, NH. The slabs were subjected to heavy truck traffic for a period of six months during a winter season. The mix design exhibiting the best performance under these conditions was selected for use in the HPC bridges.

A girder mix design was not specified. The girder fabricator was required to develop a mix which would meet specific mix requirements (e.g., slump, air content) as well as performance characteristics. It was believed that if the fabricator selected the mix ingredients, products that were cost-effective, commonly available and familiar would be chosen, thereby, potentially reducing the overall cost of the girders.

Structural Design

Each bridge design optimized the capacity of the prestressed girder section and achieved the maximum girder spacing possible with a section shallow enough to provide adequate clearance over the 100-year flood elevation. The design of the first HPC bridge was then compared to a normal performance concrete (NPC) equivalent design for an estimated initial cost comparison.

Materials Testing and Monitoring of the First HPC Bridge

Tests performed on the fresh girder and deck concrete included slump, unit weight, air content and w/c. Tests of the hardened girder concrete included compressive strength, modulus of elasticity, freeze-thaw durability and chloride ion permeability. Some of the cylinders used for testing were cured in the conventional manner in accordance with ASTM C31 while others were match-cured for data comparison. Match curing ensures that the cylinder concrete is cured at the same temperature as the girder concrete. Strength determinations were also made from cores taken from the girders. Tests on the hardened deck concrete included compressive strength, modulus of elasticity, freeze-thaw durability and scaling. Rapid chloride permeability tests were performed on cores taken from the deck. Strain and temperature gages were placed in two of the girders and temperature gages were embedded into the deck. Data from these probes were collected and analyzed to compare measured strains due to creep, shrinkage and load deflections to theoretical values. Ambient temperature fluctuations were also compared with those inside the concrete.

STUDY FINDINGS

Preliminary Research

The first phase of the UNH research portion of the project included an in-situ evaluation of three concrete mix designs.[7] The required specifications for the three mixes were: design strength of 41.4 MPa (6000 psi); 28-day cylinder strength of 49.6 MPa (7200 psi); and a maximum chloride ion permeability of 1000 coulombs at 56 days.

Table 2. Preliminary Mix Designs[7]

Mix	Design w/c	Slag	Silica Fume	Air
1	0.35	50%	---	6%
2	0.35	50%	2%	6%
3	0.35	---	8%	6%

Six slabs were installed; two slabs representing each mix. Each 1.2 m x 4.6 m x 203 mm (4'x15'x8") slab was simply supported on two steel I-beams with a clear span of 3.2 m (10.5 ft.). The three preliminary mix designs are shown in Table 2.

Table 3. Mix Performance Results[7]

Mix	Scaling 0 = none 5 = severe	Slump	Chloride Permeability	28-Day Strength
1	3.7	64 mm (2.5 in.)	1016 C	61.4 MPa (8910 psi)
2	2.9	112 mm (4.4 in.)	625 C	49.8 MPa (7220 psi)
3	2.1	76 mm (3.0 in.)	641 C	66.9 MPa (9700 psi)

Each had a total cementitious content of 386 kg/m^3 (650 lbs/yd^3) and the percentages of pozzolans shown are the percent of cement replaced.

The performances of the three mix designs were compared using scaling (ASTM C672), slump (ASTM C143), chloride ion permeability (ASTM C1202) and compressive strength (ASTM C39). The test results are shown in Table 3. Additionally, test slabs were inspected for surface wear and total crack lengths and then loaded to failure. Mix #3 exhibited a superior overall performance compared with the others and was used as a basis in the development of the eventual HPC bridge deck mixes.

Structural Design of the First HPC Bridge

The first HPC bridge carries NH Route 104 over the Newfound River and is located in close proximity to a signalized intersection. It is a 19.8 m (65 ft.) single span structure with a total width of 17.5 m (57.5 ft.) and was designed to accommodate three lanes of traffic and a sidewalk. The bridge is on a 12° horizontal curve and is skewed 8°. The superstructure consists of a cast-in-place concrete deck in composite action with five precast/prestressed concrete girders. The abutments are founded on spread footings and the approach slabs are at grade. Clearance over the 100-year flood needed to be maintained.

This bridge was designed for the NHDOT standard AASHTO MS22 (HS25) truck loading. The girder 28-day concrete strength was set at 55.2 MPa (8000 psi) with a release strength of 44.8 MPa (6500 psi). Thirteen millimeter (0.5 in.) diameter, low-relaxation strands were specified. The required deck 28-day concrete strength was 41.4 MPa (6000 psi) and the reinforcing steel was epoxy coated.

AASHTO Type III girders were chosen and spaced at 3.8 m (12.5 ft.). By using HPC, the girder section accommodated significantly more strands, greatly increasing the capacity and allowing for the wide girder spacing. A draped strand pattern was required to satisfy the release and final stresses. Tension of $0.25\sqrt{f'c}$ ($3\sqrt{f'c}$) was allowed in the bottom of the girder for the final condition under service loads. The depth of the Type III girder allowed a small clearance above the 100-year flood elevation. A section of the girder near the end is shown in Figure 1. Strands

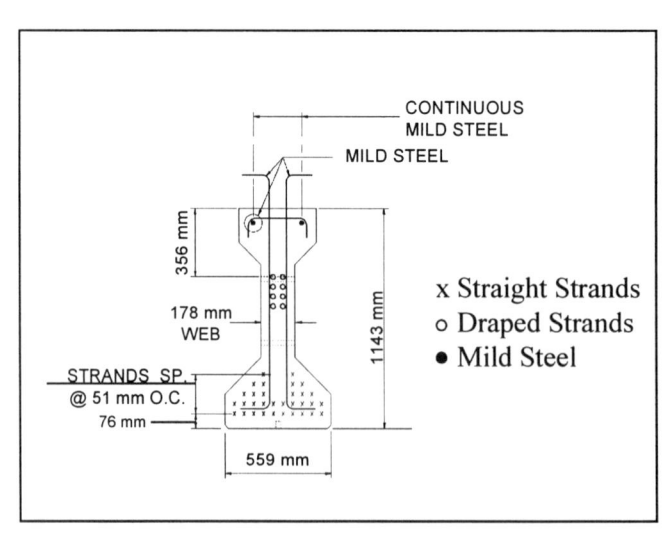

Figure 1. AASHTO Type III Girder Section

were spaced at 51 mm (2 in.). Forty strands were required per design, eight of which were draped. Mild steel was placed in the top flange to resist tensile stresses at release.

The reinforced concrete deck was 229 mm (9 in.) thick. Additional clear cover, 76 mm (3 in.) as compared with the NH standard of 64 mm (2.5 in.), was provided for the top reinforcement due to the absence of a wearing course and allowance for the saw-cut grooves.

Table 4. Superstructure Comparison

Design Requirements	HPC	NPC
Girder design strength, f'c	55.2 MPa (8000 psi)	34.5 MPa (5000 psi)
Deck design strength, f'c	41.4 MPa (6000 psi)	27.6 MPa (4000 psi)
Number of girders and spacing required	5 @ 3.8 m (5 @ 12.5 ft.)	7 @ 2.2 m (7 @ 8.3 ft.)
Number of 13 mm diameter (0.5 in.) strands required	40	24
Deck thickness	229 mm (9 in.)	216 mm (8.5 in.)

A comparison of the proposed HPC superstructure design and a NPC equivalent design using the same site conditions is shown in Table 4. Two fewer girders were required with the HPC design at the expense of a slightly thicker deck. More strands were necessary to achieve the required capacity of the HPC girders.

A cost comparison between the HPC bridge and several recently constructed NPC bridges in NH is shown in Figure 2. The HPC bridge has an approximate initial superstructure cost of $635/m² ($59/sf). Compared to the average superstructure cost of the NPC bridges, approximately $517/m² ($48/sf), the HPC cost is 23% higher. There are several possible explanations for this outcome. One possibility is that

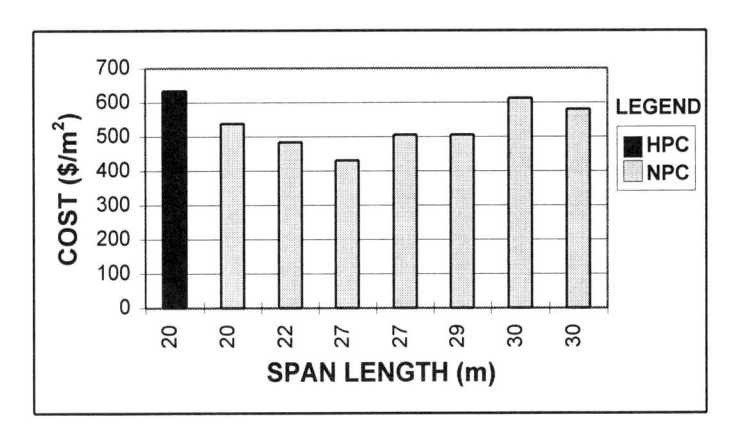

Figure 2. Superstructure Cost Comparison

HPC is a new concept in the region and because of its unknown nature, contractors assume a higher risk level and subsequently submit higher bids. Another possibility is HPC requires additional mix development and testing which increases cost. Since these costs were distributed over a small volume of concrete (the bridge had five girders and a relatively short span length) this could have inflated the concrete cost. Another possible factor contributing to the higher HPC cost was the winning low bidder was a road contractor. The bridge work had to be sublet which would inflate the road contractor's bid. The second low bidder was a bridge contractor who bid 10% lower superstructure costs than the winning bidder. Lastly, two of the girders were instrumented for testing purposes. Although the instrumentation was provided by FHWA and installed by UNH, the contractor could have included additional costs associated with potential coordination delays during the girder fabrication and bridge construction.

Construction

Girder Fabrication - Prior to the final design of the bridge, several precasters in the northeast region were solicited by NHDOT to submit HPC mix designs which included 5-7% air entrainment and met strength requirements of 55.2 MPa (8000 psi) at 28 days and 44.8 MPa (6500 psi) at release. After receiving the cylinder break results, NHDOT felt that these strengths were attainable in air-entrained concrete and could be used as the design strengths of the prestressed girders. A 28-day strength of 64.8 MPa (9400 psi) based on the average of three cylinder breaks was subsequently required in the girder concrete specifications. This higher cylinder strength was chosen to ensure the attainment of the design strength in the member. Since there would be no history of the mix design used, a standard deviation between the cylinder break results and the actual strength in the member would not be established. NHDOT believed the permeability requirement of less than 1000 coulombs could be reached with a low w/c and the use of a pozzolanic material (e.g., silica fume), both of which would also be needed to satisfy the strength requirement. The specifications developed for the girder concrete are shown in Table 5. The girder concrete mix provides FHWA Grade 2 strength ($\geq$ 55.2 MPa) and chloride ion permeability ($\leq$1000 C).

Table 5. Girder Concrete Specifications

Cement:	Type II or III
Slump:	127 to 178 mm (5 to 7 in.)
Air Content:	5 to 8%
28-Day Cylinder Strength, f'cr:	64.8 MPa (9400 psi)
Chloride Ion Permeability:	1000 C (max.)
Corrosion Inhibitor:	19.8 l/m3 (4 gal/cy)
Curing Procedure:	Steam

Table 6. Girder Concrete Test Results

Test	Results
Slump	127 to 178 mm (5 to 7 in.)
Unit Weight	66.2 to 67.8 kg (146.0 to 149.5 lbs.)
Air Content	5.2 to 7.4%
w/c	0.357 to 0.416
* Modulus of Elasticity, E	32.0 to 37.5 GPa (4.6 to 5.4x10^6 psi)
28-Day Cylinder Strength, f'cr	47.3 to 60.5 MPa (6862 to 8780 psi)
Freeze-Thaw Durability	104 to 110%
Chloride Ion Permeability	1280 to 1855 C

* Results determined by UNH Graduate Student, Cheryl R. Wilson.

The initial mix design submitted by the fabricator for NHDOT approval included Type II cement and met all fresh concrete specifications. The test results also submitted exceeded 68.9 MPa (10,000 psi) at 28 days but did not achieve the release strength until 4-5 days. The fabricator contended that with steam curing, the release strength could be attained within 24 hours without hindering the 28-day strength.

When difficulty achieving the release strength within 48 hours arose, the fabricator requested that the mix design be modified to include a Type III cement which would provide a higher early strength. Time constraints imposed by the construction schedule prevented 28-day cylinder strength tests of this modified mix design and the fabrication of the five girders proceeded. In addition to Type III cement, the final mix design developed by the precaster for use in the girders included 6% silica fume and w/c of 0.35. A considerable amount of superplasticizer was used to achieve the desired slump for workability.

Concrete test specimens were prepared at the precast plant in order to evaluate freeze-thaw durability, chloride ion permeability, modulus of elasticity and compressive strength of the concrete. The NHDOT test results are shown in Table 6. None of the five girders achieved the 64.8 MPa (9400 psi) requirement and two of the girders had an average cylinder break of less than the 55.2 MPa (8000 psi) design strength at 28 days. Consequently, three of the girder webs were cored to verify the in-place concrete strength. Two of the girders revealed average core strengths greater than the design strength and were therefore accepted. The average core strength of the third girder fell below the design strength. A subsequent analysis determined that the lower concrete strength provided the girder with the adequate structural capacity for its intended use. Ultimately, all five girders were accepted for use in the HPC bridge in Bristol.

The modulus of elasticity results were comparable with theoretical values. The 56-day permeability tests revealed slightly higher values than the maximum specified, however, the freeze-thaw durability tests revealed excellent results. The dynamic modulus of the samples actually increased (>100%) after 300 cycles.

Deck Installation - The concrete producer performed several trial batches to refine the deck mix design. Once the mix was approved by NHDOT and prior to pouring the bridge deck, a required 3.8 m^3 (5 cy) trial pour simulating actual pouring, finishing and curing conditions exposed a few minor problems which were corrected for the deck pour.

The criteria for the deck concrete mix were specified based on UNH Mix #3, as previously discussed, and listed in Table 7. The mix design provides a FHWA Grade 1 strength ($f'c \geq 41$ MPa) and Grade 2 chloride ion permeability (≤ 1000 C). A corrosion inhibitor was required due to the deviation from NHDOT's standard practice of protecting concrete decks with barrier membrane and an asphalt overlay. The deck was finished with a self-propelled finishing machine pulling a steel pan to strike off the surface. A textured finish was simultaneously applied with an attached burlap drag following the finishing pan. Within a few minutes of finishing, dry cotton mats were placed, wetted down and kept wet for a duration of four days. Requirements for water evaporation rate based on climate conditions at time of pouring were strictly enforced. For adequate friction resistance to tire skidding, the hardened finish was transversely saw-cut every 38 mm (1.5 in.) and approximately 6 mm (0.25 in.) deep. Random spacing of grooves to reduce noise was not considered due to the low traffic speed at the intersection.

There was some difficulty maintaining the required air content, however, all other specifications were met. The average 28-day cylinder strength exceeded the 49.6 MPa (7200 psi) requirement and the results from the 56-day rapid chloride permeability tests performed on deck cores were within the specified requirements. A visual inspection of the deck was performed before the bridge was opened to traffic and cracks were not evident in either the deck, sidewalk or brush curb. The NHDOT concrete test results are listed in Table 8.

Table 7. Deck Concrete Specifications

Cement:	Type II
Silica Fume:	7.50%
w/c:	0.38 (max.)
Air Content	6 to 9%
28-Day Cylinder Strength, f'cr:	49.6 MPa (7200 psi)
Chloride Ion Permeability:	1000 Coulombs (max.)
Corrosion Inhibitor:	19.8 l/m3 (4 gal/cy)
Curing Procedure:	w/ Cotton Mats

Table 8. Deck Concrete Test Results

Test	Results
Slump	76 to 127 mm (3 to 5 in.)
Unit Weight	65.3 to 66.5 kg (144.0 to 146.7 lbs.)
Air Content	4.0 to 5.8%
w/c	0.39
28-Day Cylinder Strength, f'cr	56.3 to 66.3 MPa (8163 to 9611 psi)
* Modulus of Elasticity, E	29.0 to 30.0 GPa (4.2 to 4.3x10^6 psi)
Chloride Ion Permeability	609 to 896 C
Freeze-Thaw Durability	96 to 99%
* Scaling	0 to 1

* Results determined by UNH Graduate Student, Cheryl R. Wilson.

Bridge Monitoring - The exterior girder beneath the sidewalk and the adjacent interior girder were instrumented by UNH for monitoring purposes.[8] Strain gages were located in the bottom flange of the two girders and thermistors (temperature-measuring device) were located throughout the girder depth. Girder strain measurements were taken at the release of the prestressing strands, prior to transportation to site, after erection and will be taken continually for one year after deck placement. The girder concrete creep and shrinkage, as well as the relaxation of the prestressing strands, will be determined by examining the data collected from the instrumentation. The measured values will be correlated with time, temperature and humidity and then compared to theoretical values.

Thermistors were also placed in the deck directly above the two instrumented girders and at the center of the bay between the girders. Measurements were collected over the winter season and are expected to indicate the correlation between the ambient freeze-thaw cycles and the cycles in the deck concrete. Deck temperature fluctuations will also be compared to that of the girders. If the girders exhibit fewer freeze-thaw cycles, consideration will be given to lowering the future air-entrainment requirements for the girder concrete.

The bridge deflection was monitored during the girder non-composite and composite stages.[8] Stainless steel inserts were located at the quarter points along the bridge span. The inserts were placed on the underside of the bottom flange of all the girders and underneath the deck at the center of the bays. The deflection measurements were taken at the insert locations. The girders were surveyed for deflection at the time of erection, deck placement and removal of the deck falsework. A load test was conducted just prior to opening the bridge to traffic. A 390 kN (88 k) truck was positioned in several locations on the bridge and a survey of girder and deck deflections was taken for each truck position. The results of these and all previous measurements will not only determine the actual deflections under different loading conditions, but also the composite behavior of the bridge under service loads.

Second HPC Bridge

The second HPC bridge in Bristol is currently in the design phase and is expected to begin construction in the Spring of 1999. This bridge will be an 18.3 m (60 ft.) simple span structure with a deck width of 12.0 m (39.5 ft.) carrying two lanes of traffic and one sidewalk. The superstructure will consist of four New England Bulb Tee (NE1000) girders composed of 55.2 MPa (8000 psi) concrete and 15 mm (0.6 in.) diameter prestressing strands. The girders will be spaced at 3.5 m (11.5 ft.) and will support a 41.4 MPa (6000 psi) deck. The deck will be constructed of 102 mm (4 in.) precast/prestressed panels and a 140 mm (5.5 in.) cast-in-place overlay. The deck will not have an asphalt wearing course, so the surface will be saw-cut to provide adequate friction for skid resistance. Random saw-cut spacing will be used to reduce noise.

The same procedures for the development and testing of the girder and deck concrete for the first HPC bridge will apply to the second structure. Similar instrumentation will be placed in the superstructure and the performance will be monitored in the same fashion.

CONCLUSIONS and RECOMMENDATIONS

UNH Mix #3 outperformed the other mixes in nearly all strength and durability tests. As a result, the NHDOT deck concrete specifications for the two HPC bridges were based on the design of Mix #3, with a few modifications. The amount of silica fume was decreased from 8 to 7.5% to reduce the "stickiness" of the mix and a 127 mm to 178 mm (5 to 7 in.) slump was required for improved workability.

The higher strength of HPC provided efficient use of the standard AASHTO Type III section. The maximum girder capacity possible using a 55.2 MPa (8000 psi) concrete design strength and 13 mm (0.5 in.) diameter prestressing strands was nearly reached. More efficient use of the section could only have been achieved with even higher concrete strengths in conjunction with 15 mm (0.6 in.) diameter strands. Greater girder capacity made a wider spacing possible, reducing the number of required girders. It was therefore expected that the initial superstructure cost would be reduced, even when considering the higher raw material costs of HPC, additional prestressing strands necessary and the thicker deck required. However, a cost comparison with several NPC bridges revealed a slightly higher HPC bridge cost. There are several possible reasons contributing to the higher cost. The main reason is thought to be HPC is a new concept in the region and is therefore, associated with a higher risk cost. As contractors and fabricators become more familiar with HPC, a decline in bid prices and a lower initial bridge cost should ensue.

Additional testing and a trial pour simulating the actual pouring and curing conditions should have been performed on the girder concrete mix prior to granting approval. An adequate time frame (possibly a three month period prior to casting) should be allotted to provide ample opportunity to develop, modify and complete the testing and trial pour. Final cylinder strength information verifying the release strength and 28-day strength requirements should be received prior to authorization to proceed.

Girder concrete tests indicated the strength characteristics varied between FHWA Grade 1 and Grade 2 performance levels and the durability characteristics were Grade 2. Although the chloride ion permeability slightly exceeded the maximum specified, the permeability is still considered to be quite low. The freeze-thaw tests revealed excellent results, thus consideration for specifying a lower air entrainment requirement for the girders may be warranted and is being investigated. It was believed that the high air entrainment requirement may have contributed to the difficulty in attaining the design strength in the girder concrete.

The trial batches and trial pour played an integral role in optimizing the development and placement of the HPC deck. Modifications were made throughout the pre-pour process ensuring the desired end result. The strength characteristics of the deck concrete fell within the FHWA Grade 2 performance level. Although the 28-day strength exceeded specifications, the modulus of elasticity was lower than expected. Durability characteristics all fell within the Grade 2 performance level. The freeze-thaw durability, chloride ion permeability and scaling test results revealed a concrete highly resistant to deterioration and the absence of shrinkage cracks should only further increase its resistance.

Performance monitoring of the bridge is continuing and conclusions have not been drawn as of the date of this publication. The results will determine the accuracy of current theoretical equations used in calculating the effects of concrete stress due to creep, shrinkage and relaxation of prestressing strands. Evaluation of freeze-thaw cycles in the girders will determine if a lower air entrainment requirement is warranted. Deflection measurements will be compared with theoretical values and the composite behavior of the bridge will be investigated.

A more efficient girder section and larger diameter strands will be used in the design of the second HPC structure to better utilize the benefits of HPC. The same deck HPC mix will be used in the second structure and the girder HPC mix will again be developed by the fabricator. However, the specifications will require cylinder strength results at release and 28 days and a trial pour for the girders as well as the deck.

To promote the use of HPC, NHDOT has scheduled a showcase in September of 1997. Information learned as a result of designing, constructing, and monitoring the first HPC bridge in Bristol and preliminary information of the second HPC bridge will be presented.

REFERENCES

[1] Federal Highway Administration database of structurally deficient and functionally obsolete bridges, Federal Aid and Off System combined, August 1996

[2] Goodspeed, Charles H.; Vanikar, Suneel; and Cook, Raymond A., "High Performance Concrete Defined for Highway Structures". Concrete International, February 1996, Vol. 18, No. 2

[3] Russell, Henry; "HPC Developmental History", SHRP High Performance Concrete Bridge Showcase Proceedings Notebook, March 1996, Houston, Texas

[4] Zia, P.; Leming, M. L.; and Ahmad, S. H., "High Performance Concretes: A State-of-the-Art Report". SHRP-C/FR-91-103, January 1991, Washington, DC

[5] Castrodale, R. W.; Kreger, M. E.; and Burns, N. H., "A Study of Pretensioned High-Strength Concrete Girders in Composite Highway Bridges: Design Considerations". Research Report 381-4F, Center for Transportation Research, University of Texas at Austin, January 1988, Austin, Texas

[6] PCI Committee on Durability, "Guide to using Silica Fume in Precast/Prestressed Concrete Products". PCI Journal, Vol. 39, No. 5, September/October 1994, Chicago, Illinois

[7] Fratzel, Todd M., "Evaluation of High Performance Concrete Slabs Including In-Situ Testing at a Bridge Deck Testing Facility". Graduate Thesis, University of New Hampshire, May 1996, Durham, NH

[8] Wilson, Cheryl R.; and Cook, Raymond A., "Measuring Performance: Preliminary Use of High Performance Concrete in New Hampshire Bridges". New Hampshire Journal of Civil Engineers, Vol. 1, No. 2, Autumn 1996

Fabrication of Prestressed Concrete Beams
for
Two High Performance Concrete Bridge Projects in Texas

Burson Patton
Vice- President
Texas Concrete Co,
Victoria, Texas USA

ABSTRACT

The fabrication of the prestressed concrete beams for the two high performance concrete bridges projects, Louetta Road overpass in Houston, Texas and the North Concho River, U. S. 87 and S. O. Railroad Overpass in San Angelo, Texas will be focus of this address. Topics which be addressed are the high performance concrete, prestressing techniques, delivery to the job site and erection of the prestressed concrete members.

INTRODUCTION

The first high performance concrete project in Texas was let for contract in February 1994. This project included twelve normal-strength concrete U-Beams bridges and the two High Performance Concrete bridges, which were in the Louetta Road Overpass (See Figure 1). The total project had approximately 31,000 linear feet of U-Beams of which 4,200 linear feet were High Performance Concrete U-Beams. There were approximately 1,300 cubic yards of High Performance Concrete. These High Performance Concrete beams required a concrete that would give a release strength of 8,800 psi and a fifty-six day strength of 13,100 psi. The beams were designed to use 0.6 inch strands with a maximum of eighty-seven strands per beam..

Figure 1 - Louetta Road Overpass

In bidding this job, Texas Concrete Company (TCC), was concerned with stressing the eighty-seven 0.6 inch strands which gave a total load of 3,824 kips. TCC did not have a single line capable of stressing this large force. TCC did have a bed designed for 4,000 kips, however it was being used to pour two lines of AASHTO Type IV beams with a maximum of 2,000 kips per line. If this bed was going to be used, the stressing abutments and stressing plates had to be redesigned to be able to pull eighty-seven 0.6 inch strands. In Texas, the releasing of strands must be done gradually and simultaneously. The redesigning of the abutment and stressing plates would cost TCC approximately $100,000. (See Figure 2). The 0.6 inch strand chucks had a diameter of 50mm and the spacing of the stands were also at an absolute 50mm. The diameter and spacing were going to create problems at the back of the stressing plates. TCC had to use 1 ½ round scheduled 80 pipe sleeves alternately to keep the 0.6 inch chuck at 50mm. (See Figure 3)

Figure 2- Stressing Plates

Figure 3- Strand Spacing

The coarse aggregate for the HPC was determined by the researchers at the University of Texas. TCC had all other materials on hand as they were needed in the producing of other projects. The course aggregate was from Burnett, Texas and was a ½ inch crushed dolomitic limestone. (See Table 1). The concrete had to obtain its strength properties and also had to flow down one side and across the bottom flange and up approximately twelve inches in the opposite leg of the U-Beam. After the placement of the bottom, the vertical legs of the U-Beams were then placed. (See Figure 4). Texas Concrete Co. had been pouring U-Beams on other projects for about a year. These pourings have been using 5/8 inch pea gravel at a slump of six to seven inches and have not had any problems in placing. During the first two pours, the placing of the concrete went without any difficulties. These beams had a total of sixty-eight stands and a design strength at fifty-six days of 11,600 psi. The third pour had eighty-seven stands and a design strength of 13,100 psi. By the addition of nineteen 0.6 inch stands in the bottom flange the six to seven inch slump had difficulties in moving in and around the strands which cause honeycomb area centered under the void form in two out of three beams poured. The honeycomb area was near the end and was acceptable to repair. On the remaining pours the slump was increased from about six to seven inches to eight to nine inches and this eliminated the problems with the honeycomb areas. (See Figure 5).

Table 1- Mix Design

COMPONENT	QUANTITY	TYPE
Coarse Aggregate	1918 pcy	Crushed dolomitic limestone, 1/2" max ASTM GR 7
Fine Aggregate	1029 pcy	Sand
Water	247 pcy	Potable
Cement	671 pcy	Type III
Fly Ash	316 psy	ASTM Class C
Retarder	27 oz/cy	ASTM Type B
Superplasticizer	178-227 oz/cy	ASTM Type F

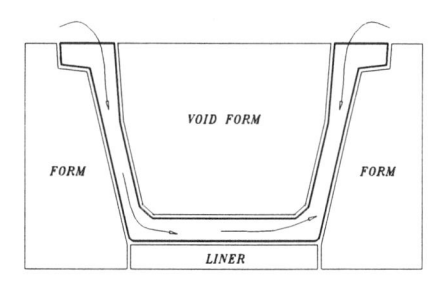

CONCRETE PLACEMENT U-BEAMS

Figure 4 -Concrete Placement

Figure 5 - U-Beams

Due to the skew angles on some of the U-Beams, a large quantity of concrete was required on the end sections. (See Figure 6). This caused curing temperatures to approach 200° F.

Figure 6 - End Sections

The unit weight of HPC was 155 pounds per cubic foot and the longest beams was one hundred thirty-seven feet long. This gave TCC a total weight of 196^k Four fork lift trucks were required to lift this beam. A straddle beam was built so two fork lift truck could be used at each end. (See Figure 7). Special equipment had to be built by our transporting trucker in order to transport these beams. (See Figure 8). All of the beams were fabricated and shipped to the job site without any difficulties.

Figure 7 - Straddle Beam

Figure 8 - Special Transporting Equipment

The next HPC project that was let by TxDot was the North Concho River in San Angelo. (See Figure 9). The project was put out for bids in June 1995. This project was composed of two bridges. The eastbound bridge was to be HPC and the west bound

493

bridge was to be normal strength concrete. The job had approximately 12,500 linear feet of AASHTO Type IV beams of which there were 5,800 linear feet of HPC. Approximately 3,600 linear feet of beams required 0.6 inch diameter strands with eighty-four strands required for some of the beams. The maximum concrete strength required at fifty-six days was 14,700 psi.

Figure 9 - North Concho River

These beams were also going to exceed Texas Concrete Company's stressing capability. If TCC was going to prestress these beams, our Bed "E" would have to be modified to cast one line of beams in lieu of two lines. This would also require additional stressing hardware. The hardware that was used to pour the U-Beams for the Louetta project was not designed to cast "I" sections. The estimated cost for these new stressing plates was $100,000. The deflection of the strands caused TCC some concerns. Some of these beams would require deflecting thirty-four 0.6 inch stands. (See Figure 10). Texas Concrete Company had not had any experience deflecting 0.6" stands and felt that from a safety aspect, this should not be done at this particular time. The extreme modifications to the bed to pour these beams was a great concern for TCC. There always existed the possibility that the beams could become damaged in transporting to and erection at the job site. The cost involved, if the bed had to be reset to pour one beam that had become damaged, was not acceptable.

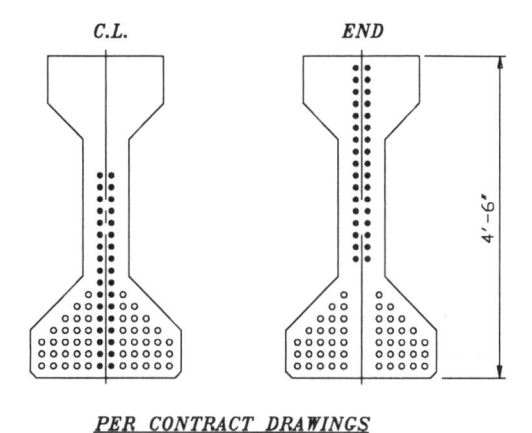

Figure 10 - Deflection of Strands

Texas Concrete Company decided to bid the job using a combination of pretension and post-tension. (See Figure 10). The beams would have pretension up to our maximum bed capacity and the remaining would be post-tension. TCC maximum bed capacity was 2,500 kips which would allow us to stress fifty-six 0.6" strands at 43,950 lbs. each. TCC decided that six strands would be in the top flange to stabilize the beam during the transporting and erecting at the job site. This was calculated by using Robert Mast's Report in PCI Journal, Volume.38 No.7 1993.

After erection at the job site, the strands would then be cut. The strands would be bonded 20 feet from each end and de-bonded the remainder. This would also keep the tension at the end of the beam under control until the post-tension was completed. The post-tension was set up to be performed from seven to twenty-one days after removal from the bed.

At Texas Concrete Company's suggestion, it was decided during the pre-construction meeting that one beam would be fabricated and shipped to the job site to verify design and hauling requirements. Researchers from the University of Texas were to monitor the beam with strain gages and deflection. TCC proceeded to set up fabrication of one beam. (See Figure 11). Special slings also had to be fabricated to move the beams from the stressing bed. (See Figure 12). After the post-tension, the beam was transported to the job site. (See Figure 13 and 14) After being transported to the job site the beam was off-loaded and placed in storage to await erection. (See Figure 15) Continuous monitoring of this beam will be done at the job site by the researchers.

Figure 11 - Fabrication of Beam #1

Figure 12 - Special Slings for Transporting

Figure 13 - Transporting Beams

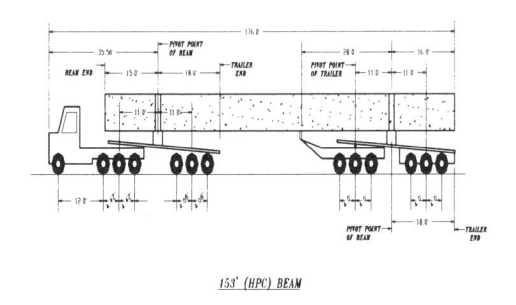

Figure 14 - Transporting Beams

Figure 15- Off-Loading of Beam

After reviewing the performance of this beam, it was decided that the calculated camber was not going to be obtained. The remaining beams would need additional prestressing to achieve the necessary camber to compensate for the dead load deflection. After the redesigning, the remaining twenty-four (24) beams were cast. These beams had approximately 1200 psi compression in the bottom flange after all loads were applied.

There were eleven (11) pours to complete the HPC beams on the San Angelo project. During the third to the last pour one of the beams encountered a blockage in one of the post tension ducts. The blockage was approximately 73 feet from one end of the beam and about 10 inches long. Various methods to remove the blockage were tried. The first effort was a hydro-blasting unit at 20,00 psi. This method was not successful. Next, a oil field bit made by one of the local suppliers was tried. Carbide particles were braised on a three legged bit. (See Figure 16) Sucker rods were screwed into the bit which was then turned by a 1 hp hand-held drill. (See Figure 17) This was successful in removing the blockage, however it did take about three hours to complete the task.

Figure 16- Drill Bit

Figure 17 - Drilling

After all the beams had obtained their concrete design breaks, the beams were shipped to the job site using the same route as the test beam. The first beam was erected without any problems. The test beam, which had been stored on the job site, was to be the second beam erected. The crane operator neglected to use the inside lifting loops for erection which resulted in the beam beginning to roll. (See Figure 18) The beam was at almost a 45° lean, but because of the six strands in the top flange, the beam did not fail. This resulted in only small tension cracks on the top flange on the tension side. The beam was then up-righted and erected. (See Figure 19) All the beams were erected and braced. (See Figure 20) TCC installed weld plates for bracing in the top flange for the contractor to use during erecting until the steel diaphragms could be applied. These steel diaphragms were only used to stabilize the beams during erecting and casting of the slab. Then diaphragms were removed by the contractor.

Figure 18- Leaning Beam

Figure 19- Erecting Beams

Figure 20- Bracing of Beams

These two projects resulted in the obtainment of much data. Our average release break was 9280 psi, the 28 day break was 14,129 psi and the 56 day break was 15,264 psi. Descriptive data are illustrated in Table II.

Texas Concrete Company always takes pride in each of their projects. We are especially proud to be a part of these two projects and the fact that we were able to meet each of the challenges that the projects presented. These two projects helped us to add much to our repertoire about High Performance Concrete. This information has enabled us to increase our standard concrete mix design from 1,500 to 2,000 psi and still utilize materials available locally.

We would like to thank Drs. Ned Burns and Ramon Carrasquillo and Mary Lou Ralls, who provided help and assistance throughout the projects. Without you, three, it would have been more difficult. These two projects also provided TCC with the opportunity to meet many other people in the industry and it is always a pleasure to internet with each of you.

PT-BB   TEXAS CONCRETE COMPANY

BATCH DESIGN NO. PT-SWR-BB (H.P.C.)
CEMENT TYPE III   BURNET STONE 1/2"
SACKS OF CEMENT 11.0   CYLINDER SIZE 4 X 8
ADMIX 1000 OZ./CWT 24   FLY ASH 35%

	SLUMP	HRS	REL 1	REL 2	AVG REL 1	28 DAY 1	28 DAY 2	28 DAY 3	AVG 28D	56 DAY 1	56 DAY 2	56 DAY 3	AVG 56D 1
MAX>	9.00	45	11690	11570	11630	15080	15320	15400	15240	16760	16560	16950	16757
MIN>	7.00	16	7120	6870	6995	11940	12420	12260	12427	13960	13660	13560	13810
NUMBER	25.00	25	25	25	25	25	25	24	25	19	19	15	19
MEAN>	9	23	9291	9270	9280	13991	14168	14278	14129	15307	15243	15333	15264
STD DEV	0.64	7	1102	1140	1112	885	712	656	694	605	683	763	654
COV(%)>					12				5				4

DATE	BED NO.	SLUMP	HRS	REL 1	REL 2	AVG REL	28 DAY 1	28 DAY 2	28 DAY 3	AVG 28D	56 DAY 1	56 DAY 2	56 DAY 3	AVG 56D
SEP 23 9	6	8.00	16.00	8400	8490	8445	12600	12420	12260	12427	14320	14360	13560	14080
SEP 30 9	6	8.00	18.00	8690	8720	8705	14440	14210	14680	14443	15000	15520		15260
OCT 07 9	6	8.00	18.00	9280	8480	8880	14170	13930	13920	14007	15400	15400		15400
OCT 21 9	6	9.00	19.00	7980	8240	8110	12720	14500	14080	13767	15920	15200	15400	15507
OCT 28 9	6	9.00	20.00	9400	8970	9185	13770	14160	14030	13987	15440	14400	15040	14960
NOV 03 9	6	8.50	21.00	9840	9520	9680	14120	14370	14480	14323				0
NOV 10 9	6	9.00	21.00	8650	8640	8645	14840	14320	14480	14547				0
FEB 05 9	1	7.00	21.00	7120	6870	6995	12560	13520		13040	15080	15120		15100
FEB 15 9	1	8.00	20.00	8550	8550	8550	14240	14640	14440	14440	13960	13660		13810
FEB 26 9	1	8.00	21.00	9120	9760	9440	11940	13560	13990	13163				0
MAR 08	1	7.75	19.00	7840	7620	7730	12760	12420	12980	12720				0
MAR 15	1	8.00	19.00	8800	8840	8820	14200	14480	14240	14307				0
APR 01 9	5	7.00	21.00	9600	9440	9520	14840	14760	14640	14747	15920	16000	15960	15960
FEB 19 9	5	9.00	19.00	8520	8710	8615	15080	15320	14960	15120	15600	15720	15880	15733
FEB 25 9	5	9.00	36.00	10720	10920	10820	14890	15120	14760	14923	15560	15360	15760	15560
MAR 03	5	9.00	19.00	8470	8950	8710	14070	13870	13990	13977	14760	15040	14880	14893
MAR 08	5	9.00	36.00	11690	11570	11630	14030	14110	13990	14043	14920	14720	14800	14813
MAR 15	5	9.00	33.00	10520	10320	10420	15080	15240	15400	15240	15440	15200	15640	15427
MAR 22	5	9.00	30.00	11090	11410	11250	14480	14270	14110	14287	15600	15520	15400	15507
MAR 29	5	9.00	30.00	10970	10890	10930	14640	14190	14520	14450	15080	15280	14640	15000
APR 07 9	5	9.00	19.00	8990	8730	8860	14270	14150	14560	14327	15200	14840	14720	14920
APR 12 9	5	9.00	45.00	10240	10240	10240	13470	13150	13630	13417				0
APR 18 9	5	9.00	21.00	8310	8230	8270	14840	14400	14560	14600	15040	15200	15280	15173
APR 28 9	5	9.00	21.00	9920	10160	10040	14720	14480	14760	14653	15840	16520	16080	16147
APR 29 9	5	9.00	16.00	9560	9480	9520	13000	14600	15200	14267	16760	16560	16950	16757

Table II

Proceedings of the PCI/FHWA
International Symposium on High Performance Concrete
New Orleans, Louisiana, October 20-22, 1997

DESIGN AND ANALYSIS OF PRETENSIONED/POST-TENSIONED LONG-SPAN HIGH PERFORMANCE CONCRETE I-BEAMS

Lisa Carter Powell, P.E.
Principal
P.E. Structural Consultants
Austin, Texas, U.S.A.

ABSTRACT

The North Concho River, US 87 & S. O. Railroad Overpass, a high performance concrete bridge project sponsored by the Federal Highway Administration, includes five spans of precast high performance concrete Type IV beams. The beams, as originally designed by TxDOT, required up to eighty-four 15 mm (0.6 in) diameter depressed and straight strands and concrete strengths of approximately 96 MPa (14,000 psi) to span a record length of over 46 m (152 feet). Beams were redesigned utilizing a combination of straight pretensioning and draped post-tensioning to facilitate fabrication. One of the beams from the bridge was fabricated early as a test beam. This beam is being monitored by University of Texas researchers in order to provide data to substantiate and refine analysis procedures.

Details of design processes and analysis procedures will be given, including discussions of time dependent behavior, stress and deflection criteria, and comparison of analysis results with test beam data.

INTRODUCTION

Background
The North Concho River, US 87 & South Orient Railroad Overpass in San Angelo is the second HPC bridge project in Texas and is part of a research project funded by FHWA, TxDOT, and ten HPC Pooled Fund States, in cooperation with the Center for Transportation Research at The University of Texas at Austin. The structure was originally designed by Bridge Design Section engineers at TxDOT's Design Division in Austin. The 8-span HPC bridge is 290 m (950 ft) long, and carries the Eastbound Mainlanes of US 67. It is adjacent to the 9-span Westbound Mainlanes, which was designed as a normal strength concrete bridge. The bridges span the North Concho River, US 87 and a railroad. The eastbound river crossing, which required a beam span of 46 m (152 ft), mandated the use of HPC; that span length is well beyond the practical limit for conventional AASHTO Type IV girders. This presented an ideal opportunity to test the extended limits of prestressed beams utilizing HPC.

TxDOT's original design for the Eastbound structure required high concrete strengths and the use of 15mm diameter strands for Spans 1 through 5. Required strengths ranged from 61 to 74 MPa (8900 to 10,800 psi) initial, and 75 to 101 MPa (10,900 to 14,700 psi) final, with required prestressing ranging between sixty and eighty-four 15mm (0.6in) diameter depressed strands. Due to concerns about bed capacity and scheduling, harping of the larger diameter strands, and stability during hauling and erection, the fabricator, Texas Concrete Company (TCC) in Victoria, Texas, decided to bid the job using a pretensioned/post-tensioned option for these spans. TCC also suggested that one of the longest beams be fabricated early and shipped to the jobsite in order to check actual cambers and prestressing losses, as well as the hauling procedures. Fabrication of this "Test Beam" presented a unique opportunity to gain feedback for design and analysis prior to the fabrication of the remaining beams. This paper presents the methodology used for the analysis of the record length Test Beam, comparisons of predicted and actual behavior, and the subsequent refinement of design procedures which resulted.

Design Considerations for Long Term Behavior of HPC Beams

Several aspects of these long span, high strength beams warrant special considerations in the design process. The increase in expected service life and in allowable stresses for HPC beams compound the designer's responsibility to ensure that allowable service criteria are met. As span to depth ratio increases, the predictability of deflections becomes more critical, as design gravitates from stress-controlled to deflection-controlled. The combination of prestressing methods introduces a challenge in calculating losses due to time-dependent effects, which directly affects the ability to calculate short and long term deflections. More sophisticated analytical tools become necessary to meet these challenges.

AASHTO[1] Serviceability Requirements for Allowable Stress and Deflection – As for normal strength prestressed concrete, the value used for concrete strength directly impacts design. Wherever f'c appears in AASHTO Specifications, a larger value directly increases computed allowable stress. Allowable flexural tensile stresses, which are based on the modulus of rupture, are a function of a multiplier and the square root of concrete strength. Materials research for HPC[2] indicates that the AASHTO multipliers may be increased for HPC; allowable tension at transfer increases from $7.5\sqrt{(f'ci)}$ to $10\sqrt{(f'ci)}$, while final allowable tension at service conditions can be increased from $6\sqrt{(f'ci)}$ to $8\sqrt{(f'ci)}$. These increased limits have been accepted by TxDOT as standard design practice.

Regarding deflections for prestressed members, AASHTO specifies that instantaneous deflections due to live load shall be less than Span/800 (8.9.3.1), but does not specifically set limits for dead load deflections of prestressed beams. For normal strength prestressed concrete construction, long-term deflection generally is not critical; enough initial camber is inherent in typical beam designs to ensure that substantial reserve camber remains under long-term loading conditions. Any variation between the initial camber that is predicted and the camber actually attained is easily accommodated in the construction process. However, camber predictions for long span prestressed girders have historically overestimated the camber actually attained. With the even longer spans attainable with HPC, this aspect of behavior warrants special attention.

Creep and Shrinkage – For typical prestressed beam design, the effects of creep and shrinkage on prestress losses (and on the calculation of resulting deflections due to prestress) are computed according to AASHTO methods, using computer programs such as

TxDOT's PRSTR14[2]. A more exact prediction creep and shrinkage affects may be obtained using ACI 209R-92[3]. For the composite slab, which is composed of either all cast-in-place concrete or a combination of precast panels and c-i-p concrete, typical shrinkage and creep coefficients were used, with corrections for ambient relative humidity, average thickness, mild steel content, cement content and slump. For the prestressed beam, experimental data from long term studies with the specific mix developed for the TxDOT HPC projects was provided by the UT Researchers, and corrections were made for difference in ambient relative humidity and volume-to-surface ratios between the study test specimens and the Test Beam.

Analytical Tools – In order to accurately predict the behavior of the Test Beam, the analytical tools used must account for the following: composite construction with varying construction histories, the combination of grouted post-tensioning tendons and partially debonded pretensioned strands, shrinkage and creep of concrete (with the ability to model the shrinkage and creep parameters unique to the HPC mix), the aging of concrete (increase of strength and modulus of elasticity with time), and the loss of prestress force due to relaxation and strain in the concrete. The tool chosen for this task is a commercially available computer program called ADAPT-ABI[5].

Test Beam

The "Test Beam" is an outside beam (Beam 1 or 6) from Span 2, with a center-to-center of bearing span of 46.4 m (152.17 ft). This beam was to be instrumented and monitored by UT researchers to provide deformation, prestress and other data from fabrication through storage, and on into the service life of the beam[6]. It was expected that data obtained from the first few months of monitoring would be characteristic of the beam's long term deformation behavior. The Test Beam analysis and design proceeded on the basis of initial assumptions regarding effective prestress at transfer, concrete material parameters, and anticipated construction history. Then computed time dependent deflections from release through storage at the job site would be compared with measured deflections. Comparative results would then enable a reassessment of initial assumptions and refinement of design procedures if required.

TEST BEAM REDESIGN AND ANALYSIS

Original Design

TxDOT's original design of the designated Test Beam, using the PRSTR14 program, required 70 15mm (0.6in) diameter strands, 20 of which were depressed (see Figure 1). PSTRS14 predicted a maximum camber of 206 mm (8.1 in) and dead load deflection of 126 mm (4.9 in) at midspan based on a modulus of elasticity of 41,400 MPa (6000ksi). Prestressing losses were calculated to be 11 percent at release and 37 percent final. The design provided an ultimate moment capacity at midspan of 11,355 kN•m (15,396 k-ft) as an under-reinforced flanged section.

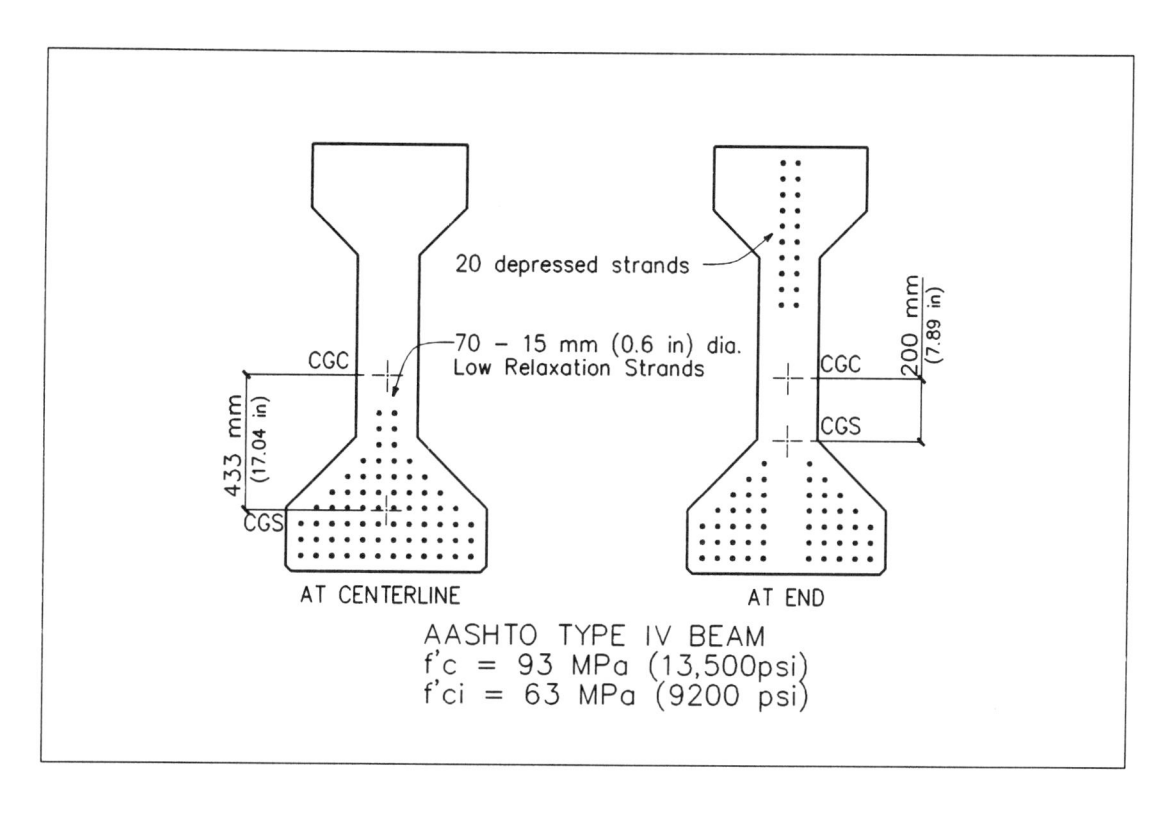

Figure 1. TxDOT Contract Design for Span 2, Beams 1 and 6

Test Beam Redesign Parameters
While the redesign was driven in large part by the fabrication and handling issues, one of TxDOT's primary concerns was that the redesigned beams meet their serviceability requirements.

Fabrication – Based on a preliminary analysis by others, and on the fabricator's stability analysis for handling, the cross-section of the Test Beam was developed (see Figure 2) and materials were ordered on that basis. Forty-six straight pretensioned strands were located in the bottom flange, with various degrees of debonding to control top fiber stresses at the ends of the beam. Six strands located in the top flange were to be unbonded throughout the center portion of the beam, with a 6 m (20 ft) length of strand bonded at each end. The unbonded portion of these six strands were to be detensioned after erection of the beam at the site (blockouts in the top flange provided access to cut strands). These strands provided stability during hauling and erection, and helped to control top fiber tensile stresses in the end regions as well. Two multi-strand parabolic post-tensioning tendons provided an additional 20 strands. The addition of a solid endblock accommodated the post-tensioning anchorages and anchorage zone reinforcing. Since analysis indicated that total shortening of the 47 m (153 ft) beam would be at least 50 mm (2 in) by the end of the construction period, the cast length was increased accordingly. The standard correction made on the basis of field experience for normal strength and length Type IV girders is on the order of 25 mm (1 in).

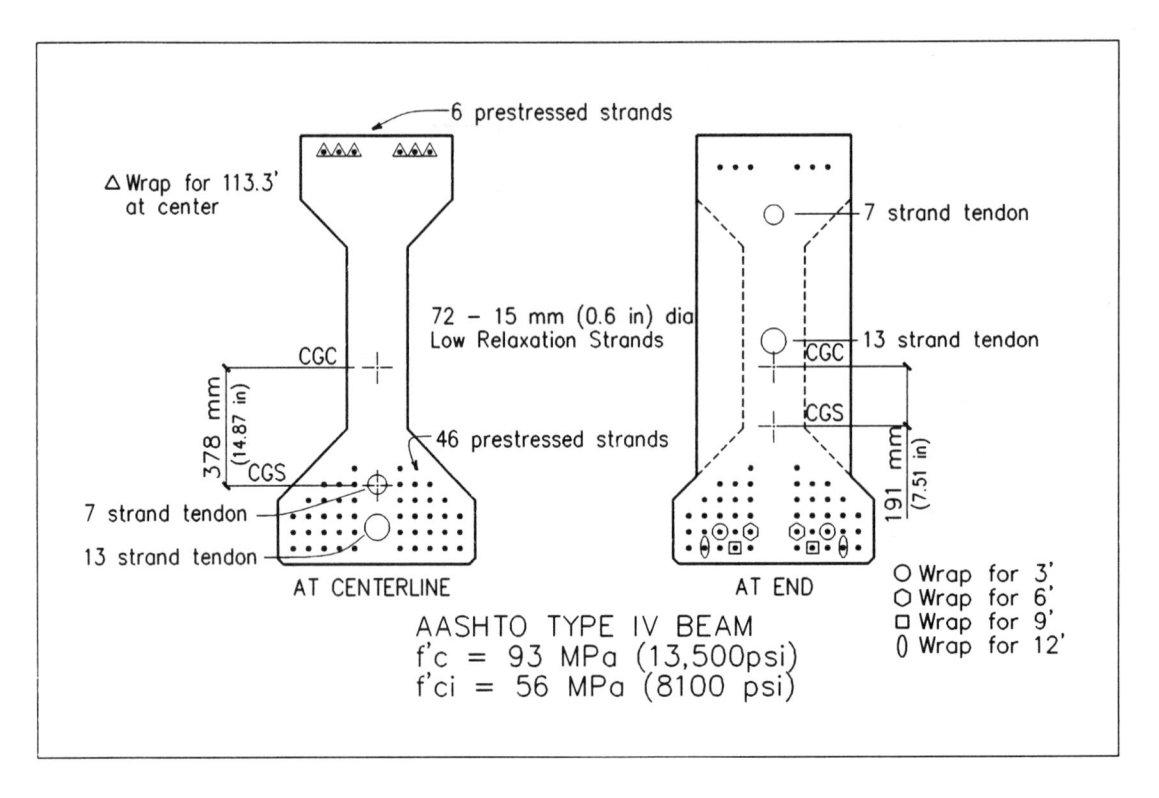

Figure 2. Test Beam Cross-section

Serviceability and Ultimate Criteria – The AASHTO live load plus impact instantaneous deflection limitation of Span/800 is basically a user comfort factor, and even for the long span length of the Test Beam, this criteria was easily met. Computed live load plus impact deflection for the Test Beam was 41 mm (1.6 in), or Span/1100. TxDOT's stated criteria for long-term deflection under sustained load was simply that they wanted no "saggy" beams. The analysis described below predicted a slight sag of less than 25 mm (1 in), and consideration was given to increasing the number of post-tensioning strands in the top duct in order to increase camber. Rather than delaying fabrication to add strands in order to eliminate this slight sag, the decision was made to construct the Test Beam as shown in Figure 2. Minimum initial and final concrete strengths were set so that computed beam stresses would fall within allowable limits. Long-term prestress losses predicted by the ADAPT analysis averaged approximately 17 to 18 percent for the prestressing strands and 5.5 percent for the post-tensioned tendons. Ultimate capacity was computed to be 11,336 kN•m (15,370 k-ft), well above required ultimate strength of 7,268 kN•m (9,854 k-ft).

Analysis Methodology

Sensitivity of Creep and Shrinkage Characteristics – Since the prediction of time dependent behavior of concrete is closely tied to assumed histories for water content, temperature and loading, a solution based on one set of assumptions can be misleading.

Various parameters can be identified which affect creep and shrinkage response, such as mix design, ambient moisture (RH), volume to surface ratios, age of loading, mild steel content, and curing conditions. For each class of concrete used in this project (ie. the 14 ksi Class H (HPC) beam concrete and the 6 ksi Class K (HPC) concrete for composite slab), some of these parameters can be accounted for with reasonable certainty while others can vary within a significant range. For example, the mix design for the prestressed beam has been refined and studied extensively; the ultimate creep and shrinkage coefficients for test cylinders should represent the beam concrete with reasonable accuracy, once corrected for the difference between the volume-to-surface ratio and curing conditions of the beam to those of the test cylinders. On the other hand, the age of the beam at various loading stages could vary considerably, depending on the fabrication and construction schedule, with loading at earlier ages leading to higher creep and shrinkage. Likewise, the composite slab could be entirely cast-in-place, or it could be constructed with one half of its thickness composed of prestressed panels, which are less susceptible to creep and shrinkage effects than freshly cast concrete. Therefore, one set of assumptions and conditions which would lead to increased creep and shrinkage effects were established and dubbed the "Upper Bound" analysis. Conversely, conditions leading to decreased shrinkage and creep were included for a "Lower Bound" analysis. Corrections for each of these sets of assumptions were then applied to the ultimate creep and shrinkage coefficients provided by researchers for the beam, and to the ACI creep and shrinkage coefficients recommended for standard conditions for the slab.

Modeling – For analysis using the ADAPT-ABI program, the beam was discretized into a series of beam elements and offset composite slab elements. The cross-sectional properties of the composite slab were computed on the basis of AASHTO effective flange width criteria. Beam elements modeled both end block and typical beam sections. Element nodes included debonding points for each prestress strand. Upper Bound and Lower Bound models included input describing material properties, creep and shrinkage parameters, and construction histories as described above. It should be noted that the 28-day concrete strengths used in the computer analysis were 96,525 MPa (14 ksi) for the beam elements and 41,368 MPa (6 ksi) for the composite slab elements. Strength gain with time, and the corresponding increase in modulus of elasticity, is computed by ADAPT according to strength gain function coefficients input by the user. Actual concrete strengths required for fabrication were then adjusted based on maximum beam stresses computed by the program.

ANALYSIS VS. REALITY

Predicted Behavior Vs. Camber Data

A primary purpose for fabricating and monitoring the Test Beam was to provide an indication of the predictability of the system deflections due to prestressing, self weight and time dependent response of the high strength concrete. It can be said that if actual conditions match closely with the expected response, then nothing much is learned. The test beam offered may lessons with regard to the prediction of HPC beam deflections.

Figure 3 compares Test Beam deflections versus time for four cases; the Upper Bound prediction, the Lower Bound prediction, the measured camber at various stages to date, and results for a "corrected" analysis, which is discussed below.

These comparisons show that the Test Beam did not attain the camber predicted by the Upper and Lower Bound analyses; in fact measured camber was essentially zero at release. A brainstorming session which included UT researchers, the fabricator, and design engineers from TxDOT and P.E. Structural Consultants led to a re-evaluation of the Test Beam analysis; and the formulation of various corrections based on the following observations and conclusions Note that the results for the "Corrected" analysis, which incorporates these corrections and actual time history of construction (see note in figure), correlate closely with observed behavior to date.

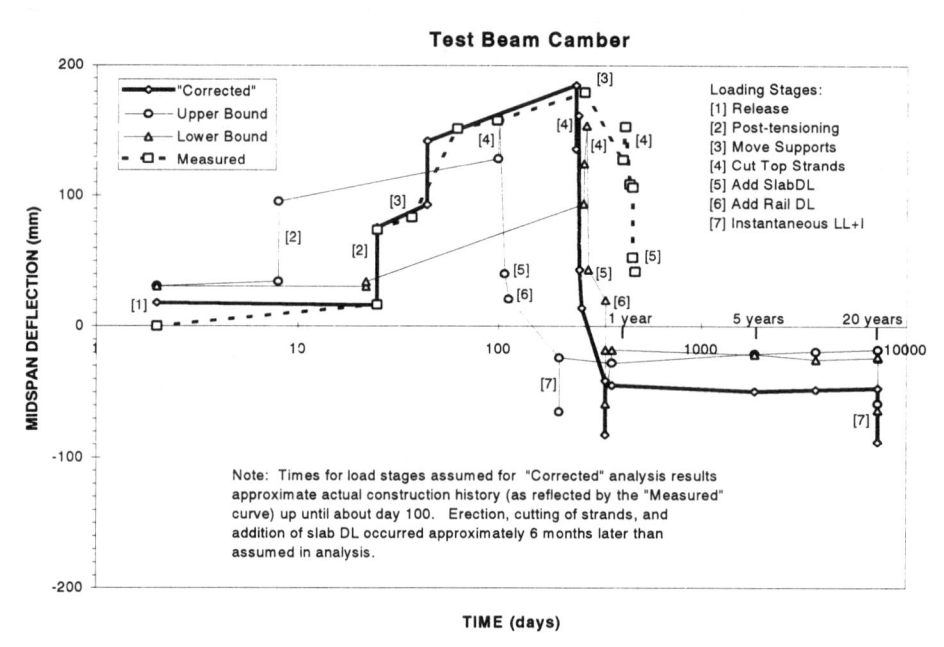

Figure 3. Test Beam Deflection vs. Time

Sensitivity of Structure Due to Length
It was concluded that small variations (ie., less than 5 percent) in prestressing force and dead load for such a long and highly stressed beam can result in considerable variations in predicted behavior. The result of this sensitivity is that sources of error which may typically be neglected, for normal ranges of strength and span length, need to be considered and accounted for in the design of long-span HPC beams..

Fabrication Conditions: Unaccounted-for Effects
Prestress Losses -- The lack of initial camber, combined with strain measurements of the concrete, led to the conclusion that the prestress force actually transferred to the beam was less than had been anticipated. Thermal affects likely are one contributing factor. The

ambient temperature during stressing over a total bed length of 132 m (435 ft) was approximately 50 degrees F. The ambient temperature of the strand during casting rose to approximately 70 degrees F over the 86m (283 ft) length of bed not occupied by the beam, and at least 120 degrees F over the 46 m (152 ft) beam length due to the high temperatures at which the high performance concrete cures prior to bonding with the strand. Accounting for these conditions results in a calculated reduction in the stress of the strand of 43 MPa (6.3 ksi). In addition, the prestressing took place 5 days prior to casting. Relaxation losses for the lo-lax strand were calculated to be 19 MPa (2.7 ksi). The total loss of prestress from thermal effects and relaxation was estimated at 62 MPa (9 ksi), or about 4.4 percent of the initial prestress force. This small adjustment in prestressing force would not typically be considered for normal construction, but it is significant for long-span HPC construction.

Estimation of Dead Load – The unit weight of the high performance mix for the beam has been measured to be 9.55 kg/m^3 (153 pcf), which was the value input into the ADAPT program for calculation of self weight. A calculation of the actual weight of the volume of prestressing and mild steel embedded in the beam resulted in an effective increase in unit weight of roughly 4 percent, or 9.93 kg/m^3 (159.1 pcf). Again, an adjustment of this magnitude would typically be neglected for normal construction, but it is significant for long-span HPC construction.

Differential Shrinkage and Unintended Restraint-- Under ideal conditions, a beam with zero initial camber and subjected to no additional loads would be expected to maintain its shape over time, since the opposing effects of prestressing and dead load have theoretically balanced each other to produce zero net deflection. However, camber was observed to grow from zero to approximately 10 mm (3/8 in) during the first 2 weeks of storage in the fabrication yard. This observed growth may be attributed to the recovery of differential shrinkage between the bottom of the beam, which was less exposed to drying, and the top surface of the beam which had more severe exposure during initial storage. When the beam was moved to a different location at about 3 weeks after casting, an instantaneous increase in camber of approximately 10 mm (3/8 in) was observed. This increase can be attributed to the release of horizontal restraint which had been provided by the temporary blocking. While these effects contributed to the difference between predicted and observed initial camber, no correction was made to the analysis, since the effects appear to be recoverable.

Upper Bound Vs. Lower Bound

Since the opposing effects of prestressing force and sustained loading are so closely balanced for the Test Beam, the difference in behavior due to time dependent effects is minimal, as evidenced by the similar results over the service life for the Upper and Lower bound curves shown in Figure 3.

REFINEMENTS FOR REMAINING DESIGNS

Lessons learned from the fabrication and monitoring of the Test Beam allowed refinement of analysis and design procedures for the remaining beams in Spans 1 through 5.

Worst Case Conditions for Deflections and Stresses

Selection of prestress which would be required to produce the desired long-term deflections for Spans 1 through 5 was based on a set of worst case conditions. These conditions assumed maximum possible prestress loss and maximum self weight. Determination of the required concrete strength at critical stages to meet allowable stress limitations was based on an additional analysis which assumed full prestress and minimum dead load. Investigation of Upper and Lower Bound conditions was continued for a basis of comparison for different span lengths and prestress levels.

Reduction in Effective Prestress

For the production of the Test Beam, only a portion of the total bed length was utilized, subjecting only about one third of the total strand length to a large temperature differential. Under typical fabrication conditions for the remaining beams, the bed could be fully utilized, and the ambient temperature at stressing would be closer to 80 degrees F, resulting in a differential of 40 degrees F over 132 m (434 ft). Relaxation losses over the period between stressing and casting were assumed to be the same as for the Test Beam, with a total possible reduction in initial prestress from 1396 MPa (202.5 ksi) to 1331 MPa (193 ksi).

Steel Volume Allowance

As with the test beam, an increase of 4 percent in unit weight for the beam was assumed for computing worst case deflections.

Efficiency of Fabrication

To increase the fabricator's flexibility for scheduling and utilization of his casting beds, pretensioning was standardized for all beams. In order to increase attainable camber for longer spans, additional straight strands were added to fill all available strand locations within the 2 inch grid, providing 50 strands in the bottom flanges of all beams. Then either 7 strand or 13 strand draped post-tensioning tendons were added as required to obtain the desired long-term deflections. The additional straight strands necessitated increased debonding, above the traditional limits set by TxDOT for Type IV beams, to control top fiber tension at release. A debonding pattern was developed based on joint discussions with Mary Lou Ralls, P.E. of TxDOT and Dr. Ned Burns, P.E. of the University of Texas. In that pattern, 60 percent of the strands in the bottom row and 40 percent in the next highest row were debonded to various lengths.

IMPLEMENTATION AND RECOMMENDATIONS

As the limits of material performance are stretched beyond traditional values to enable us to design lighter, longer, and more durable structures, so must our design and analysis methodologies reach beyond standard convention. The sensitivity of longer spans warrants special attention to sources of error in assumptions used for design and analysis. Relatively small variances that could comfortably be neglected in a conventional design may result in pronounced differences between predicted and actual behavior. More sophisticated analytical tools may be required to perform analyses with reasonable accuracy.

Allowance for Fabrication Conditions
The use of High Performance Concrete for the construction of precast beams of record length led to the development of an alternate design which has several attractive features from both a design and fabrication standpoint. The use of pretensioned strands combined with post-tensioned tendons allowed for increased efficiency in the fabrication process; the requirements for bed capacity are reduced and the standardization of pretensioning for a wide range of beam designs offers flexibility in the utilization of casting beds. The ability to prestress in stages allows more flexibility to the designer to control stresses and deflections at various stages of construction. The use of partially unbonded strands in the top flange helps to ensure stability during handling and to control stress levels in the end regions. The effect of high curing temperatures for HPC should be taken into account when calculating the initial prestress force to be transferred to the section. Large temperature differentials between the time of stressing and bonding of the strands to the concrete may reduce initial prestress force, and consequentially reduce reserve camber below desirable levels. Inadvertent relaxation losses and increases in dead load should also be accounted for.

Satisfying Serviceability Criteria
Checks should be performed assuming unique sets of worst case conditions that will result in either increased maximum stress or increased deflection. The inherent material properties of high performance concrete offer a bonus over conventional strength concrete when design is controlled by tensile stresses

Creep and Shrinkage Parameters
The use of high performance concrete structures which are deflection sensitive may warrant the use of more refined methods for predicting time dependent behavior due to creep and shrinkage effects. ACI 209R-92 can be used to compute corrections to the standard ultimate shrinkage strains and creep coefficients to account for special concrete mixes; or to experimental data for ultimate shrinkage and creep to account for the difference between test specimens and actual structures. The effect of concrete properties, moisture and loading history become more important for the longer, more highly stressed beam designs attainable with HPC; the variability of these parameters should be investigated.

Requirements for Computer Analysis
Computer programs are commercially available which enable the design engineer to more accurately model the various parameters affecting the design and analysis of long-span High Performance Concrete bridges. As always, comparison of analytical results with approximate hand methods, with other computer analyses, and with observed behavior will increase confidence in the ability to predict behavior. Continued monitoring of existing and future HPC structures will be of particular value in this process.

Acknowledgments

The author wishes to acknowledge Mr. Burson Patton, of the Texas Concrete Company, Dr. Ned Burns, P.E. and Mr. Shawn Gross of The University of Texas, Dr. Bijan Aalami of the Adapt Corporation, and Ms. Mary Lou Ralls, P.E. of TxDOT for their cooperation and contributions to this paper.

References

1. AASHTO, Standard Specifications for Highway Bridges, 15th Edition with Revisions, 1994.
2. Carrasquillo, R.L., Slate, F.O., and Nilson, A.H., "Properties of High Strength Concrete Subject to Short-Term Loading," *Journal of the American Concrete Institute*, Vol. 78, No. 3, May-June, 1981, pp. 171-179.
3. PSTRS14, "Prestressed Concrete Beam Design/Analysis Program, Version 3.20, December 1991, Texas Department of Transportation.
4. ACI 209R-92, "Prediction of Creep, Shrinkage, and Temperature Effects in Concrete Structures", American Concrete Institute, Detroit, 1992.
5. ADAPT-ABI, "ADAPT-Bridge Incremental Structural Concrete Software System", Version 2.00, ADAPT Corporation, Redwood City, California, January 1996.
6. Burns, Ned H., et al., "Instrumentation and Measurements - Behavior of Long-Span Pre-tensioned High-Performance Concrete Bridges", Concurrent Symposium Paper.

COVINGTON HIGH PERFORMANCE CONCRETE BRIDGE

Chuck Prussack
Engineering Manager
Central Pre-Mix Prestress Co.
Spokane, WA   USA

ABSTRACT

This paper provides information on a High Performance Concrete (HPC) Project for a state that has a well developed relationship between the State DOT and the prestressed concrete industry in the state. This was the first designated HPC Project the State, academia, and industry would be working together on. Since it was the first project of this type all parties went through a learning process. The steps and processes that were taken, and the fabrication of the HPC project itself is described, from the viewpoint of the prestressed concrete manufacturer. Much was learned by the precaster doing this project, such as what to expect when working with researchers doing instrumentation, and the difference between standard concrete mixes and this HPC mix.

INTRODUCTION

Washington State Department of Transportation (WSDOT), in its annual meetings with the Pacific Northwest PCI group of producers, had been expressing interest in working with the FHWA to undertake a HPC project. A project was identified near Kent, Washington, that was a three-span prestressed concrete girder bridge. By utilizing HPC with its higher strengths and using 15mm (.6") diameter prestressing strand, girder lines could be reduced from 7 to 5, and the WSDOT could gather data for subsequent HPC work. An initial project meeting was held with WSDOT, academia, and industry to discuss project goals. Girder strength requirement was 51 MPa (7,400 psi) at release and 68.9 MPa (10,000 psi) at 56 days. Conventional WSDOT and industry practice to this point was 44.8 Mpa (6,500 psi) maximum at release and 48.3 Mpa (7,000 psi) at 28 days. Although the precasters in the area had little experience working with these elevated strengths, they felt this goal was achievable. In addition to the strength requirement, the girders also had a coulomb limitation based on the AASHTO T277 rapid chloride permeability test, and a freeze thaw requirement using the AASHTO T161 test. The project had a test girder requirement that called for fabricating a 6.1m (20') long girder using 40-15mm (.6") diameter strand to give the precaster, WSDOT, and the researchers practice and background data for instrumentation. Project schedule was discussed, which showed bidding summer of 1996, test girder fabrication fall 1996, project girder casting winter 1997, and bridge construction spring 1997. Researchers selected were a team from the University of Washington headed by Dr. John Stanton and Dr. Marc Eberhard.

PROJECT

The project was bid July 2, 1996. Low general contractor was Mowat Construction of Kirkland, Washington, and girder supplier low bid was from Central Pre-Mix Prestress Co. (CPPC) of Spokane, Washington. This came as some surprise to the precaster, since Spokane is about 483 km (300 miles) from the jobsite and another precaster is about 32 km (20 miles) from the jobsite.

When CPPC was notified they had the job, the first step undertaken was to begin trial batch work. CPPC had just completed a 600 panel bridge redecking in Montana with a 55.2 Mpa (8,000 psi) strength and a small county bridge with a 68.9 Mpa (10,000 psi) strength requirement. CPPC hoped to build on the experience learned from these two jobs plus tap into what was known in the industry to generate a mix design.

Unlike some other states, WSDOT had no material lab expertise to pass on, nor did the University of Washington researchers have any experience with HPC mix designs. So CPPC began their own trial batch program. The other HPC work done by CPPC used Type F fly ash, so it was felt fly ash should be a constituent. The previous 68.9 Mpa (10,000 psi) job used fly ash plus silica fume, so it was felt silica fume should be a constituent. Also, with the AASHTO T277 coulomb limitation, these pozzolans would help achieve the less than 1,000 coulombs required.

The trial batch process was not smooth. CPPC had its largest sales year in its history, so plant personnel were busy. Furthermore, even though this author had been to Texas' HPC showcase and had some HPC exposure, the plant personnel doing the trial batch work had none, and didn't understand the interactions that occur between cement, superplasticizer, fly ash, and silica fume. Months were lost experimenting with trying to find the right combinations--consistent strengths were not achieved. CPPC had difficulty finding someone in house with enough time to put trial designs together in a methodical manner. Holding all variables constant and letting only one material vary at a time is a time consuming process when plant personnel are busy doing production related activities. Especially troublesome were the low strength gains from the first trial batches. Strength rise from release through 28 day was not great enough to meet project needs. At the 2 and 3 week points where strengths needed to be approaching project requirements, strength gains were rising very slowly. Clearly, some things needed to be done differently and some good advice was needed. Texas Dep't. of Transportation was very helpful, along with another prestress plant material engineer from Minnesota, and some admixture suppliers. A switch to Type C from Type F fly ash was suggested and implemented. The batch plant needed modifications to better control the consistency of trial batches. To make HPC requires that the batch plant have the ability to match slumps consistently. To do this requires an accurate system of measuring the moisture content of the aggregates so that the amount of mixing water to add is known. Amounts of dry constituents going into the mixer needed to be done more precisely than had typically been done. Improvements were made to the batch plant to enable this to happen.

The other variable that was being dealt with concurrent with the mix proportions was determining the optimum curing cycle. Unlike a typical girder project where the objective is to get release strength in a short time so the bed can be turned on a daily cycle and the 28 strength nearly always follows, this project had a large rise in strength from release through 56 days. A concern was that if too much heat was used early for release strengths, the subsequent rise in strength would be too flat. This was borne out by testing. With the 51 Mpa (7,400 psi) release strength, it was felt a 2 day curing cycle would be needed, with the concrete temperature not to exceed about 56 C (120 F). The objective at the time of release was to be just a little over the required release strength. That would give the steepest curve after release and hence greatest strength gain at 56 days. CPPC uses a match curing "Sure Cure" system, and the WSDOT accepts cylinders from that as the acceptance cylinders. The use of this system, where you are correlating girder curing conditions to that of the test cylinder, was a key component in making HPC. After considerable tests, plant QC personnel iterated to the optimum computer settings to provide just the right amount of heat into the "envelope" so that the girders heat of hydration coupled with the heat externally supplied would not exceed the sought after maximum temperature.

Trying to sort out all the data and philosophies about how to make HPC is confusing and frustrating, as it seems each materials engineer takes a different approach. The use of fly ash and/or silica fume, and in what proportion diverges widely. As the time for casting drew nearer, and strength rise in trial batches became acceptable, CPPC iterated to the mix design shown in Table 1. As of this writing, even within CPPC, there is not a consensus that the mix chosen is the best one for making HPC concrete.

The test girder was cast on December 11, 1996. CPPC and University of Washington were anxious that the pour would be successful, that the instrumentation would perform, and that the concrete strengths would be acceptable. Fortunately, all these things did happen, girder release strength was 56.7 Mpa (8,230 psi) in 27 hours. The test girder gave CPPC crews an idea of the time requirements University of Washington would have for installing instrumentation and taking readings during prestress release. Considerable time was spent doing these tasks, which is not conducive to a typical production cycle. After the reinforcement bar is placed, University of Washington needed about 5 hours to tie in their instruments, prior to release about 2 hours were needed to affix gauges, and during release after each increment of strands were detensioned, time was required for gauge readings.

From this experience, project girders were scheduled so that University of Washington would have access to the girders at night after the steel was tied. A very early pour was scheduled for the next day, so that the release strength would be achieved the following day as early as possible. That way, the time to strip the girder, pull strand, stress, and tie reinforcing bar could be complete for the University of Washington to do their work the next night. CPPC crews were not accustomed to having third parties to work around and to accommodate, and at times it was difficult making everyone's needs work. Credit goes to both CPPC production crews and University of Washington researchers for working together to best meet each others objectives. Some long hours and creative crew schedules were used to achieve this.

Project girder scheduling was carefully coordinated with University of Washington so the girders with large amounts of instrumentation could be cast at times they were available to stay in Spokane. With the University schedule of finals, spring break , etc., meshing with a production schedule, and meeting contractor requirements, it was like threading a needle.

Girders were cast on the dates shown in Table 2, which shows as well times prior to release and strengths at release, 28 and 56 days. As can be seen, all strengths were in excess of project requirements. The emphasis placed on careful batching and curing by procedure paid off.

Tests done by WJE Engineers showed permeability average by the AASHTO T277 Test to be 1010 coulombs, and the freeze thaw resistance average by the AASHTO T161 Test was 100%.

SUMMARY

CPPC learned much about manufacturing HPC concrete with this project. The girders were shipped to the contractor on schedule, all having achieved their needed strengths, which to a precaster is the mark of a successful project. CPPC made a profit on the job, and looks forward to bidding on subsequent HPC work. Material cost increment for HPC concrete was about $26.00/ cubic meter ($20.00/ cubic yard), which seems to be a modest premium for the enhanced properties of HPC. In a non-instrumented application, where release strengths are in the range of 48 Mpa (7,000 psi), girders could be cast on a normal daily cycle, which would give normal labor costs, so cost increment above non-HPC work would be only slightly higher if HPC became the standard.

TABLE 1

CONCRETE MIX DESIGN CERTIFICATE

Date: 11/6/96 (Revision)

Project: Covington HPC Bridge　　　　CPM Mix No. 399F

Contractor: Mowat Construction　　　Slump: 6" +/-

Mix Design: 7.75 SK + Fly Ash + Silica Fume　Entrained Air: None

Cement Type: Holnam III ASTM　　　Water/Cement: 265/1000 = .265

Strength Required: 10,000 psi @ 56 days

	SIZE	BULK SPECIFIC GRAVITY
Gravel	1/2" to #4	2.66
Sand	Blended	2.63

Source of Aggregates: Central Pre-Mix Concrete Co.

	DESCRIPTION	ABSOLUTE VOLUME (CU.FT.)	SATURATED SURFACE DRY WEIGHTS (LBS.)
MB Rheomac SF100 Silica Fume	.05 Sack	.37	50
Fly Ash	2.3 Sack	1.32	222
Cement	7.75 Sack	4.09	728
Water	31.81 gal	4.25	265
Gravel	1/2" to #4	11.28	1870
Sand	Blended	5.42	890
Air	1.0% (entrapped)	.27	
TOTAL		27.00	4025

Admixtures: MB RheoBuild 1000 Superplasticizer per manufacturer's recommendations, MBL 82 Water reducing admixture.

Remarks:

Signature: *[signed]*
Chuck Prussack, P.E.
Title: Engineering Manager

TABLE 2

CENTRAL PRE-MIX PRESTRESS CO.
CONCRETE STRENGTH REPORT

Project	Covington Way Bridge	Job. No.: 98110
Bed	NW I Line	
Contractor	Mowat Construction	
56 Day Strength	10,000	
Destress Strength	7,400; 5,100	
Concrete Mix Nos.	399C	
Cement Type	Holnam III	
Commments	HPC Mix	

CASTING DATE	SLUMP (in.)	HRS TO DEST	DESTRESS STRENGTH	MARK #	28 DAY (PSI)	AVE 28 DAY (PSI)	56 DAY (PSI)	AVE 56 DAY (PSI)
3/6/97	3 1/2 / 5 1/4	24	8400 / 7600	20-1A	11,040/11,125	11,082	12,170 / 11,140	11,655
3/10	2 3/4 / 5 1/2	24	7680 / 7460	20-1	10,760/11,840	11,300	12,360 / 12,110	12,235
3/12	6 / 4	24	7280 / 7710	20-1B	10,990/11,150	11,070	12,720 / 12,190	12,455
3/14	4 3/4 / 4 3/4	60	8380 / 8060	20-1C	11,525/10,200	10,862	12,910 / 12,560	12,735
3/18	6 / 3 1/4	22	8060 / 8110	20-1D	11,180/11,210	11,195	11,600 / 11,410	11,505
3/24	5 / 6	20	7490 / 7160	10-1C/10-2C	12,951/10,761	11,856	12,880/12,100	12,490
3/26	5 1/2 / 5 1/2	20	8250 / 7780	10-1D/10-2D	11,620/11,849	11,734	12,300/11,460	11,880
3/28	4 / 5	60	9940 / 9850	10-1/10-2B	12,503/11,595	12,049	12,570/11,530	12,050
4/2	4 1/2 / 5	18	7920 / 7670	10-1A/10-1B	11,289/11,275	11,282	12,460/11,940	12,200
4/4	4 1/2 / 4 1/2	56	9020 / 9260	10-2/10-2A	11,216/11,325	11,270	13,030/12,940	12,985

APPLICATIONS OF A NEW HIGH PERFORMANCE POLYOLEFIN FIBER REINFORCED CONCRETE IN TRANSPORTATION STRUCTURES

V. Ramakrishnan
Distinguished Professor
Dept. of Civil & Environmental Engineering
South Dakota School of Mines & Technology
501 E. St. Joseph Street
Rapid City, SD 57701 USA

ABSTRACT

This paper presents the construction of a bridge deck and jersey barriers with the newly developed polyolefin fiber reinforced concrete. This is the first time this synthetic fiber-reinforced concrete was used in the construction of reinforced concrete structural elements such as bridge deck and jersey barriers. The mixture proportions used, the procedure used for mixing, transporting, placing, consolidating, finishing and curing are described. This new polyolefin fiber-reinforced concrete with enhanced fatigue, impact resistance, modulus of rupture, ductility, and toughness properties is particularly suitable for the construction of durable highway structures.

INTRODUCTION

Due to a decaying infrastructure and tightening budget constraints, transportation engineers are challenged to rehabilitate existing facilities economically with an increase in performance. However, simultaneous improvements in cost and performance are unlikely unless new material technology can be exploited. The recently developed polyolefin fiber-reinforced concrete is one material that promises to provide many advantages, providing a practical approach to enhanced durability and cost-effectiveness in concrete compositions. It eliminates problems such as staining, inherent corrosion and potentially harmful protrusions. (1, 2) It has been shown in earlier research and publications by the author (6 to 9) that fiber reinforced concrete (FRC) with its enhanced properties beneficial in structural applications is a highly suitable material for the construction and/or rebuilding bridges and other transportation structures.

Polyolefin fiber-reinforced concrete incorporates 50 mm by 0.64 mm (2" by 0.025") fibers into the concrete mix. These fibers are longer and stronger than plastic fibers previously used to reinforce concrete, and a proprietary packaging technology enables rapid and uniform mixing into the concrete matrix at quantities up to 2% by volume. These volumes of fiber significantly alter the concrete's physical properties, especially toughness, impact resistance, fatigue strength, ductility, and resistance to shrinkage cracking. (1, 2).

The South Dakota Department of Transportation has sponsored research to investigate the properties and practicality of polyolefin fiber reinforced concrete. Through laboratory tests at the South Dakota School of Mines and Technology and construction of a segment of pavement, a bridge deck overlay, concrete barrier replacement, and a thin unbonded overlay of asphalt bridge approaches, the material proved to be workable and significantly more resistant to early cracking than ordinary concrete. The research results demonstrated increased fatigue capacity of 150%, crack width reductions below American Concrete Institute (ACI Committee 224 Report on Cracking) recommendations (10) for chloride intrusion, and skid resistant surface texture. The favorable research results warrant more widespread use of polyolefin fiber-reinforced concrete in other applications, including the construction of new bridge decks and barriers and the rehabilitation of deteriorating bridge structures. Both prestressed and reinforced concrete structural elements built with FRC, would have significantly increased toughness and ductility and would better resist earthquake forces and suddenly applied loads (3 to 5).

APPLICATIONS

The construction projects undertaken to evaluate the non-metallic polyolefin fiber-reinforced concrete were a part of repair, rehabilitation and construction of the following structures:

1. Thin bridge-deck bonded overlay on the bridge at Vivian (the bridge on the U.S. 83 number 43-026-195, over I-90 south of Pierre, SD).
2. Reinforced concrete Jersey barrier on the above referred bridge.
3. A total replacement of bridge-deck slab on the bridge at Spearfish, SD (bridge over I-90, exit 10).

The research activities involved were the development of mixture proportions, quality control testing, and advice on the construction, monitoring and evaluation of the above structures. The research activities also involved periodic condition surveys to evaluate the performance of the constructed bridge-deck overlays, barriers, and the bridge-deck slab.

RESULTS OF CONTROL TESTS AND DISCUSSION

Bridge-Deck Overlay Concrete Properties
The same mixture proportions were used for the plain and fiber reinforced concrete as shown in Table B1. Fresh concretes were tested for slump (ASTM C143), air content (ASTM C231), fresh concrete unit weight and yield (ASTM C138). The fresh properties and the concrete temperature are given Table B2. The table also includes the results of

Table :B1 **Mixture Propertions Used For Bridge Deck Overlay Concrete**

Mix Type	Mixture Proportions Kg/m³ (lbs /cu.yd)					AEA
	Cement	Coarse Aggregate	Fine Aggregate	Fibers	Water	mL/m³ (ounce/cu.yd)
PLAIN B1	488 (823)	827 (1394)	827 (1394)	0	160 (270)	657.6 (17)
FRC B2	488 (823)	827 (1394)	827 (1394)	11.83(20*)	160 (270)	657.6 (17)
PLAIN B3	488 (823)	827 (1394)	827 (1394)	0	160 (270)	657.6 (17)
FRC B4	488 (823)	827 (1394)	827 (1394)	14.86(25*)	160 (270)	657.6 (17)

AEA - Air Entraining Agent mL/m³ (ounces/cu.yd.).
* - Polyolefin fibers 50.8 mm (2 in) long and 0.635 mm (0.025 in) diameter.

Table B2: **Properties Of Fresh Concrete** *(Bridge Deck)*

Mix Type	Test #	Time a.m.	Slump (mm)	Air Content (%)	Concrete Temp (°C)	Unit Weight (Kg/m³)	Yield (m)³	Fiber Content (Kg/m³)
PLAIN B1	1*	6.50	12.70	6.2	20.0	2303.7	---	0
	2*	8.50	25.40	7.0	24.4	2322.9	---	0
	3*	10.40	19.05	6.3	27.7	2297.3	---	0
FRC B2	1*	6.45	6.35	4.1	18.9	---	---	0
	2*	6.50	12.70	5.4	18.9	---	---	0
	3	7.00	3.17	5.4	20.0	2356.2	3.00	11.68
	4	7.45	6.35	6.0	21.1	---	---	---
	5	8.30	38.10	9.0	21.1	2310.1	3.06	11.28
	6*	8.45	19.05	6.0	23.3	---	---	---
PLAIN B3	1	6.45	57.15	5.6	23.3	2376.1	---	0
FRC B4	1*	10.45	6.35	6.2	26.6	2376.1	3.73	15.37
	2	11.30	12.70	6.2	26.6	2376.1	3.73	16.04
	3	12.15	19.05	5.6	28.9	---	---	---

Notes * Tests conducted by SDDOT personnel. Conversions :
--- No tests were done. 1 lb/cu.yd. = 0.593 Kg/m³ 1 (yd)³ = 0.765 m³
1 lb / (ft)³ = 16.02 Kg/m³ °C = 0.56 (°F - 32)

Table B3: **Hardened Concrete Properties** *(Bridge Deck)*

MIX TYPE	E* Mpa (10⁶ psi)	fc'* MPa (psi)	IMPACT STRENGTH** Number of blows First Crack / Failure		JCI Equivalent*** Flexural Strength Mpa (psi)
PLAIN B1	36846 (5.34)	44.298 (6420)	12	15	-
FRC B2	36777 (5.33)	42.504 (6160)	37	200	2.704 (392)
PLAIN B3	36639 (5.31)	42.263 (6125)	21	24	-
FRC B4	29394 (4.26)	37.571 (5445)	31	231	2.863 (415)

NOTE : E = Static Modulus   * Average of three specimens
fc'= Compressive Strength   ** Average of five specimens
                            *** Average of four specimens

tests conducted by the DOT personnel. The slumps and air contents measured were satisfactory and they were within the range specified by DOT. The concrete temperatures varied between 25.6°C to 28.9°C (68°F to 84°F). The actual measured fiber content in the samples taken from the field concrete were close to the specified amounts.
The hardened concrete properties such as the compression strength, modulus of elasticity, and impact strengths are given in Table B3. The compressive strengths varied from 37.6 Mpa to 44.3 Mpa (5445 psi to 6420 psi) which is a tolerable variation in the field concrete. The concrete was tested for impact strength by the drop weight test method (ACI Committee 544) (11). There was a high impact strength due to the addition of polyolefin fibers in concrete. The number of blows for ultimate failure in fiber concrete was above 200 whereas it was between 20 to 25 for plain concrete. The first crack strength and modulus of rupture values are compared in Fig. B1. There was not a significant variation in the modulus of rupture (flexural strength) and first crack strength for different batches and it was about 5.5 MPa (800 psi). The flexural strength of fiber concrete at 28 days was about 4% higher than that of plain concrete. The toughness indices calculated according to the ASTM C1018 standard procedures are shown in Fig. B2 and the residual strengths factor (R-values) are compared in Fig. B3. The first crack toughness is compared in Fig. B4. These comparisons showed positively that the

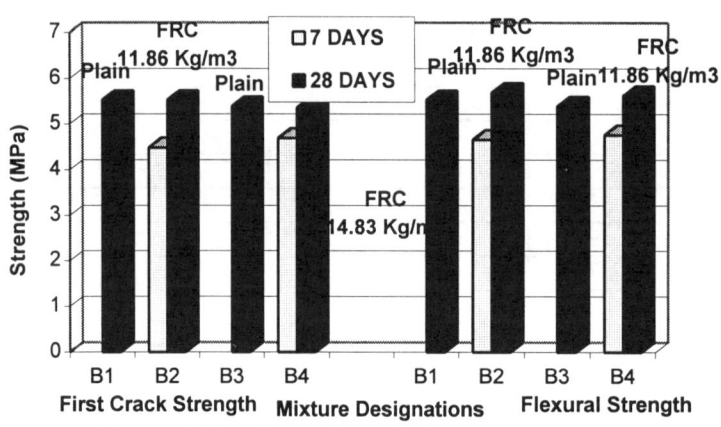

FIG. B1: COMPARISON OF FIRST CRACK AND FLEXURAL STRENGTH FOR DIFFERENT MIXTURES

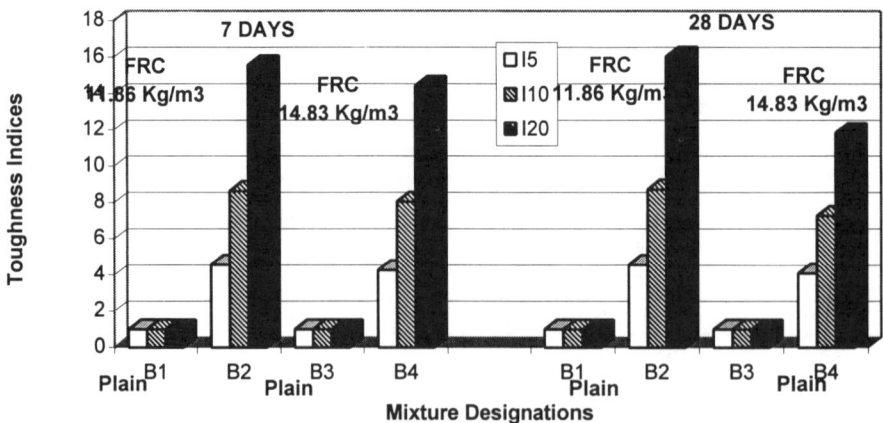

FIG. B2: COMPARISON OF TOUGHNESS INDICES, I5, I10, I20 FOR DIFFERENT MIXTURES

addition of polyolefin fibers had increased the toughness of the concrete. The R-values indicated a ductile behavior. The Japanese Standard (JCI) flexural toughness factors are compared in Fig. B5. This comparison also confirms the increase in toughness and ductility of the concrete due to the addition of polyolefin fibers.

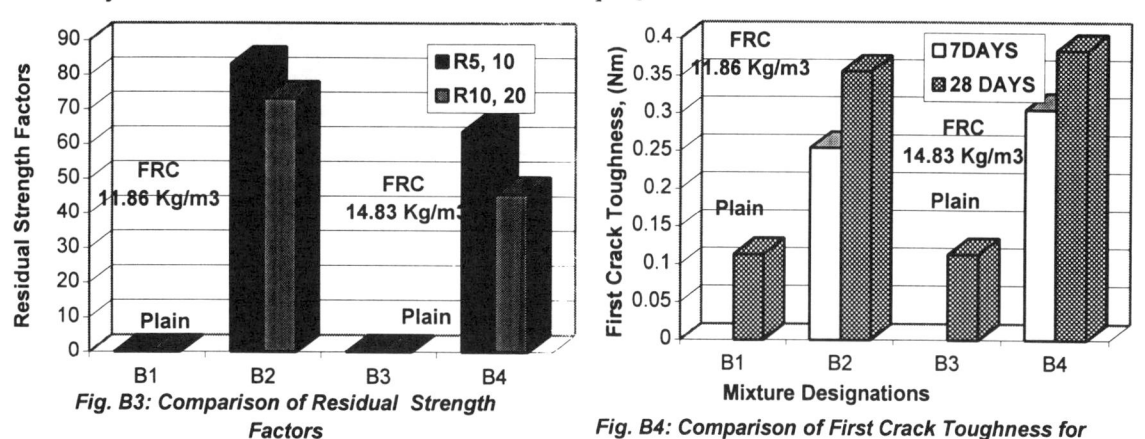

Fig. B3: Comparison of Residual Strength Factors

Fig. B4: Comparison of First Crack Toughness for Different Mixtures

Jersey Barrier Concrete Properties

The standard class A45 concrete, as specified in the SD DOT Standard Specifications for Roads and Bridges Section 460 was used for the construction of the barriers. Polyolefin fibers were added to the basic mixture proportions as given in Table J1. The same quality control tests for fresh and hardened concrete were done as described above. The fresh concrete properties are given in Table J2. As anticipated, the addition of fibers reduced the slump of the concrete. There was not a significant difference in the air contents of plain and fiber concretes and the unit weights were nearly the same. The hardened concrete properties such as compressive strength, modulus of elasticity and impact strength are given in Table J3. There was not much difference in these values for the three mixes except the impact strength which was significantly higher for the fiber concretes. The average number of blows for ultimate failure were 128 and 232 respectively for polyolefin fiber concretes with 11.9 Kg/m^3(20 lbs./cu.yd.) and 14.8 Kg/m^3(25 lbs./cu.yd.).

The first crack strength and modulus of rupture values are plotted in Fig. J1. There was not a significant difference between the cracking strength and the ultimate strength due to the addition of fibers. The modulus of rupture was about 4.5 MPa (650 psi). The first crack toughness and the toughness indices calculated from the load-deflection curves according to ASTM C1018 Procedures are given respectively in Fig. J1 and J2. The residual strength factors (R-values) are plotted in Fig. J3, the comparison of first crack toughness is plotted in Fig. J4, and the flexural toughness factors calculated according to Japanese Standard (JCI) are plotted in Fig. J5. As anticipated, the addition of polyolefin fibers greatly increased the toughness of the concrete. The ASTM toughness indices I5, I10, and I20 are approximately 4, 8, and 15 times higher than that of plain concrete.

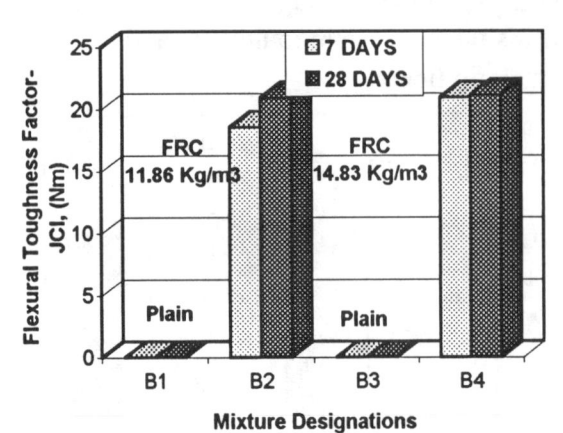

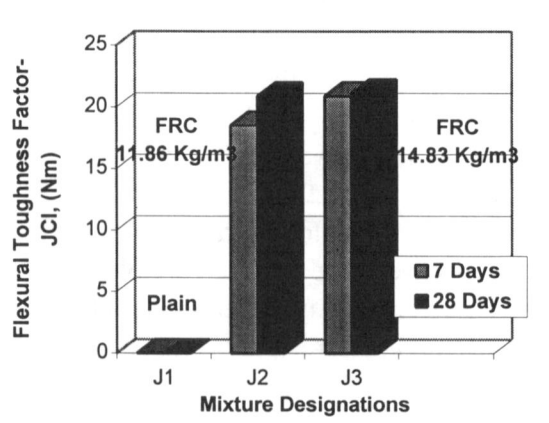

FIG. B5: Comparison of Flexural Toughness Factor (JCI) for Different Mixtures

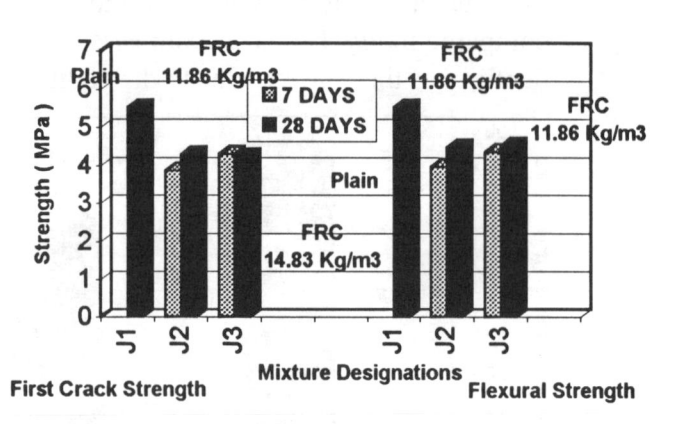

Fig. J1: Comparison of first Crack and Flexural Strength for Different Mixtures

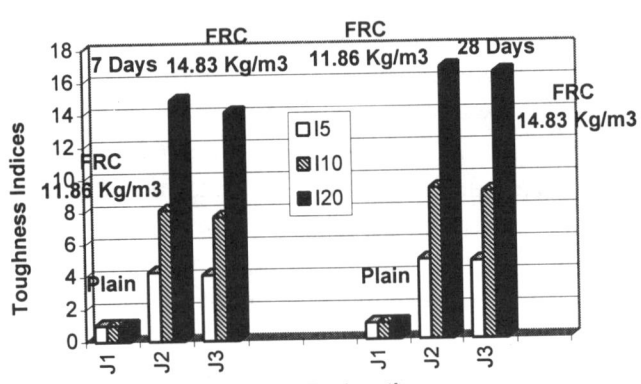

Fig. J2: Comparison of Toughness Indices, I5, I10, I20 for Different Mixtures

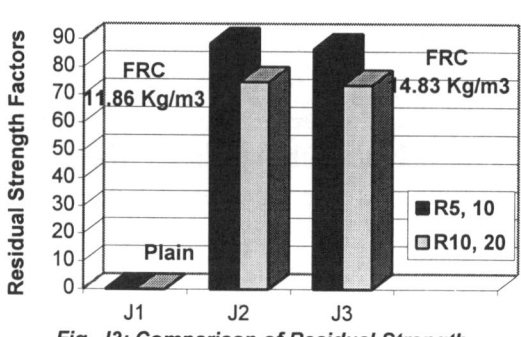

Fig. J3: Comparison of Residual Strength Factors

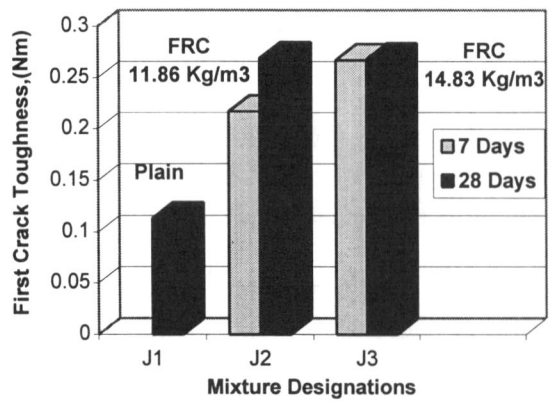

Fig. J4: Comparison of First Crack Toughness for Different Mixtures

Fig. J5: Comparison of Flexural Toughness Factor (JCI) for Different Mixtures

Table :J1 *Mixture Proportions Used For Jersey Barrier Concrete*

Mix Type	Mixture Proportions Kg/m³ (lbs/cu.yd)					AEA
	Cement	Coarse Aggregate	Fine Aggregate	Fibers	Water	mL/m³ (ounce/cu.yd)
PLAIN J1	397 (670)	1025 (1728)	705 (1189)	0	161.3 (272)	464.2 (12)
FRC J2	397 (670)	1025 (1728)	705 (1189)	11.83(20*)	161.3 (272)	464.2 (12)
FRC J3	397 (670)	1025 (1728)	705 (1189)	14.86(25*)	161.3 (272)	386.8 (10)

AEA - Air Entraining Agent mL/m³ (ounces/cu.yd.).
* - Polyolefin fibers 50.8 mm (2 in) long and 0.635 mm (0.025 in) diameter.

Table J2: *Properties Of Fresh Concrete* *(Jersey Barrier)*

Mix Type	Test #	Time	Slump (mm)	Air Content (%)	Concrete Temp (°C)	Unit Weight (Kg/m³)	Yield (m)³	Fiber Content (Kg/m³)
PLAIN J1	1*	3.15p.m	120.6	8.0	29.4	---	---	0
	2*	3.15p.m	127.0	8.5	28.9	---	---	0
FRC J2	1*	3.25p.m	44.45	7.6	31.5	2296.8	4.59	10.887
	2*	3.30p.m	60.32	8.0	29.4	---	---	---
	3*	3.50p.m	44.25	6.0	30.0	---	---	---
	4	9.00a.m	31.75	7.0	26.1	2310.1	3.87	---
	5*	9.10a.m	44.25	6.2	26.1	---	---	---
FRC J3	1*	8.25a.m	63.50	9.2	24.4	2237.5	4.74	17.339
	2	8.25a.m	73.02	10.0	24.4	---	---	---
	3*	8.50a.m	57.15	7.8	24.4	---	---	---
	4	9.00a.m	57.15	8.0	25.5	2257.2	4.00	---
	5*	9.00a.m	38.10	6.6	25.5	---	---	---

Notes * Tests conducted by SDDOT personnel.   Conversions :
--- No tests were done.   1 (yd)³ = 0.765 m³   1 lb/cu.yd. = 0.593 Kg/m³
1 lb / (ft)³ = 16.02 Kg/m³   °C = 0.56 (°F - 32)

Table J3: *Hardened Concrete Properties* *(Jersey Barrier)*

MIX TYPE	E* Mpa (10⁶ psi)	fc'* Mpa (psi)	IMPACT STRENGTH** Number of blows First Crack Failure		JCI Equivalent*** Flexural Strength MPa (psi)
PLAIN J1	---	---	--	--	--
FRC J2	31050 (4.50)	30.670 (4445)	31	128	2.74 (397)
FRC J3	32085 (4.65)	32.906 (4769)	38	232	2.73 (395)

NOTE :   E = Static Modulus   * Average of three specimens
fc'= Compressive Strength   ** Average of five specimens
*** Average of four specimens

CONSTRUCTION DETAILS AND PERFORMANCE

Construction of Bridge-Deck Overlay

A thin bridge-deck bonded overlay was constructed on the bridge at Vivian (on the U.S. 83 number 43-026-195, over I-90 South of Pierre, SD). One side of the bridge deck overlay was constructed with the polyolefin FRC and the other side with normal portland cement concrete (PCC). Different quantities of fibers 11.9 and 14.8 kg/m^3 (20 lbs./cu.yd. and 25 lbs./cu.yd.) were used. The concrete was batched at a central mixing plant and transported to the site in a truck mixer. The fibers were then added at site and mixed for additional time. The same mixture proportion used for the plain, low slump dense concrete, as specified in the SDDOT Standard Specifications for Roads and Bridges, was used for the polyolefin FRC concrete. The slump was specified as 25 mm (1.0 inch) maximum.

The same paving machine and the same procedure for placing, consolidating and finishing were used for the polyolefin FRC concrete also. There were no difficulties encountered in the placing, consolidating and finishing operations. The curing procedure used was as per the SD DOT specifications. The same procedures were used for both mobile mixed low slump concrete and ready-mixed polyolefin FRC concrete.

Inspection of the bridge-deck overlay showed no shrinkage cracks. The bonding appeared to be good. The appearance of the east side and west side of the bridge deck was almost the same, except for a few fibers visible on the west side. The tining appeared to be the same. There was a good bond between the plain low slump and the polyolefin FRC concrete overlays along the centerline. The part of the deck slab that was chipped off showed a uniform distribution of the fibers over the entire depth of the overlay. The fibers were well bonded to the concrete matrix. There was no distress, popouts, scaling, sign of debonding or delamination on the deck slab.

Construction of Jersey Barrier

The standard class A45 concrete, as specified in the SD DOT Standard Specifications for Roads and Bridges Section 460 was used for the construction of the jersey barrier. The fine to coarse aggregate ratio was 40.8/59.2 and the cement content was 397 kg/m^3 (670 lbs./.cu.yd.) and the water to cement ratio was 0.4. The polyolefin FRC barrier on the west side was constructed in three days. The north half of the barrier on the west side of the bridge was constructed using polyolefin FRC with fiber equal to 11.9 kg/m^3 (20 lbs./cu.yd.). It was decided to use the same mixture proportions for the construction of the Jersey Barrier on both sides of the bridge using plain concrete and polyolefin FRC concrete.

The barriers were carefully inspected as soon as the form work was removed and also after a week. During the inspection, it was observed that in both barriers with and without fiber- reinforced concrete, there were numerous shrinkage cracks. These were drying shrinkage cracks caused by the reinforcement restraint and no expansion, contraction, or construction joints over a long length. Initially when the side forms were removed, there were no cracks. Hence these cracks are not considered as plastic shrinkage cracks. The length and width of these cracks were accurately measured using a crack comparator which can measure the width accurate to 0.05 mm (0.002 in.). These widths were also verified by a crack measuring, hand held, microscope with graduated cross wires. In longer cracks, the widths were measured at 3 to 5 locations and the average was taken.

The ages of the concretes were the same (about 10 days) for both east and west barriers when the crack measurements were made. In the barrier with the fiber reinforced concrete, 88 cracks were observed and these were uniformly distributed. The widths of these cracks were very small, less than 0.18 mm (0.007 in.), except a few. In the barrier with the plain concrete, cracks could not be measured for a small part (about 10 ft.) due to an obstruction. In the remaining section of the barrier, 50 cracks were observed. Most of these cracks were wider than 0.18 mm (0.007in.). The American Concrete Institute Committee 224 on cracking has recommended the maximum permissible crack widths for different conditions of exposure (10). If the cracks are narrower than recommended widths, then it can be assumed that the possibility of corrosion of the reinforcement due to moisture penetration is negligible and the concrete is durable. The maximum crack width that is permissible under the environmental conditions at the bridge (exposed surface subjected to deicing chemicals) is 0.18mm (0.007 in.). Calculations had shown that only 7 percent of the cracks were wider than 0.18 mm (0.007 in.) in the barrier with the fiber-reinforced concrete. Whereas in the case of the plain concrete barrier, 85 percent of the cracks were wider than 0.18 mm (0.007 in.). It is known that any reinforcement, including randomly oriented fiber reinforcement, can not prevent cracking of concrete. The addition of fibers would prevent widening of the cracks once they were formed and the cracking would be more uniform. Therefore the observations made here were anticipated. The crack size distribution is shown in Fig. J6, and the average crack widths for barriers with and without fibers are given below.

	Number of Cracks	Crack Width in mm (inch) Total	Average
w/ Fibers (NMFRC)	88	8.05 (0.317)	.09 (0.0036)
w/o Fibers (NF)	50	15.67 (0.617)	.31 (0.0123)
NF / NMFRC	57%	195%	343%

New Deck Slab and Barrier
Inspection of the bridge deck top surface indicated no cracks. There was no spalling, no scaling or any other distress. Fibers were visible at the surface; however they were well bonded to the concrete. It is expected that these exposed fibers would wear out in the course of time due to the traffic. They did not cause any durability problems.

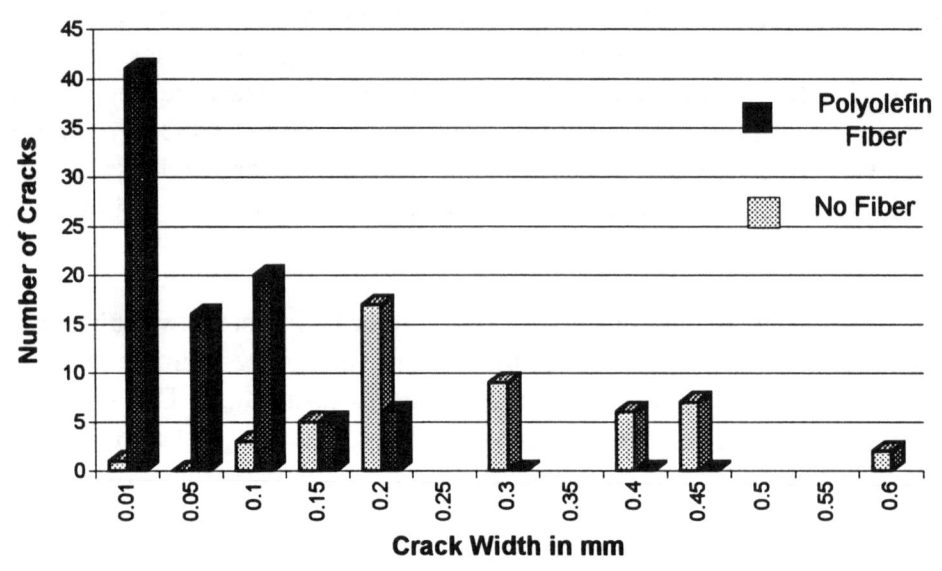

Fig. J6: Crack Size Distribution

Inspection of the barrier indicated two hair line cracks less than 0.1 mm (0.0039 inch) wide in the already constructed portion of the barrier. Subsequent inspections revealed a few additional hairline cracks. They were not easily visible. After about 10 weeks, a detailed inspection was made using magnifying lenses. The crack widths were also measured. There were three hair line cracks on the east side barrier. One crack was 0.1 mm (0.0039 inch) wide and the other two were 0.08 mm (0.003 inch) wide. All these cracks were shrinkage cracks. The widths of these cracks were less than 0.18 mm (0.007 inch), the ACI Committee 224 recommended maximum tolerable crack width under the environmental conditions at the bridge.

The deck slab bottom side was inspected using "snoopers" to look closely at the bottom side of the deck slab. A detailed and careful inspection of the entire deck slab indicated that there were totally 8 cracks. The length and width of all these cracks were measured. There were a total of six cracks on the cantilevered part of the slab on the east side of the deck. Three of the cracks were wider than 0.18 mm (0.007 inch) and the other three were less than 0.18 mm (0.007 inch) in width. There were only three cracks in the main slab, two of them just above the two end pylon supports. The third one was a small crack about the middle of the bridge 38.1 mm (15 inch) long and 0.2 mm (0.0079 inch) wide. No crack extended over the entire width of the slab. The south end crack was 2 meters (6 ft. 8 in.) long and 0.47 mm (0.0183) wide and the north end crack was 2.56 m (8 ft. 5 in.) long and only 0.13 mm (0.005 inch) wide.

There was white efflorescence in all these cracks which enhanced the visibility of these cracks and made these cracks look wider and longer. All the cracks were perpendicular to the traffic direction and parallel to the main reinforcement. Therefore the cracks were not induced due to any bending action. These cracks were due to restrained shrinkage.

CONCLUSIONS

The purpose of adding fibers was not to increase the flexural strength (the plain concrete had adequate flexural strength), but primarily to reduce the initial plastic shrinkage cracking and to reduce the overall cracking in the structures due to drying shrinkage and thermal changes during the lifetime. The minimum and maximum temperatures may vary approximately from -29°C (-20°F) to 43° C (110°F). The fibers were also added to enhance the fatigue strength, the impact strength, toughness and ductility of the concrete. The wearing quality would also be improved. The fiber dosage was selected to achieve the above stated objectives. The properties of the field concretes determined from control tests and the three years observed performance of the structures had shown that the objectives were achieved.

The newly constructed bridge deck with FRC had shown considerable less longitudinal cracking that normally occurs at the bottom side of the slab in such newly constructed bridges. A comparison of the visually observed cracks in the identical bridge built without the addition of fibers in the concrete at the same I-90 exit 10 over the east bound lanes has confirmed the above observation. There had been practically no shrinkage cracking on the top surface and the barriers.

Addition of polyolefin fibers at 11.9 kg/m^3 or 14.8 kg/m^3 (20 lbs./cu.yd. or 25 lbs./cu.yd.) enhanced the structural properties of concrete. There was a slight increase in flexural strength, and a considerable increase in toughness, impact, fatigue, endurance limit, and post-crack load-carrying capacity. There was no difficulty or problem encountered during the mixing, transporting, and placing. The same construction techniques and construction equipment without any modification could be used in the construction of bridge deck overlays and barriers using polyolefin FRC.

In the jersey barrier, the addition of fibers reduced the shrinkage crack width to a level permissible in the exposed surfaces (less than 0.18mm (0.007 inch) and a more desirable crack distribution (more number of uniformly distributed thinner cracks than a fewer number of wide cracks) was obtained. In the plain concrete jersey barrier, almost 85 percent of the cracks were wider than 0.18 mm (0.007 inch) whereas in polyolefin FRC barrier, there were only a few cracks (7 percent) that were wider than 0.18 mm (0.007 inch). In addition to reduction of crack width, the addition of fibers had increased the impact strength and toughness of the concrete.

ACKNOWLEDGMENT

The author gratefully acknowledges funding received from SD DOT and expresses gratitude to David Huft and the Technical Panel of the SD DOT for supporting and encouraging this research. The support and encouragement received for this research from the 3M Company is also gratefully acknowledged.

REFERENCES

1. Ramakrishnan, V., "Performance Characteristics of 3M Polyolefin Fiber Reinforced Concrete", *Report submitted to the 3M Company*, St. Paul, MN, 1993.
2. Ramakrishnan, V., "Evaluation of Non-Metallic Fiber Reinforced Concrete in PCC Pavements and Structures", *Report No. SD94-04-I, South Dakota Department of Transportation*, Pierre, SD, 1995, 319 pages.
3. Ramakrishnan, V., "Concrete Fiber Composites for the Twenty-First Century", *Real World Concrete,* Editor: G. Singh, Elsevier Science Ltd., U.K., 1995, pp. 111- 144.
4. Ramakrishnan, V., "Recent Advancements in Concrete Fibre Composites", Concrete Lecture - 1993, Published by American Concrete Institute, Singapore Chapter, Singapore, 1993, 28 pages.
5. ACI Committee 544, "State-of-the-Art on Fiber Reinforced Concrete", Report ACI 544 1R-82, *Concrete International Design and Construction,* May, 1982.
6. Balaguru, P., and Ramakrishnan, V., "Mechanical Properties of Superplasticized Fiber Reinforced Concrete Developed for Bridge Decks and Highway Pavement", *American Concrete Institute, Special Publication SP-93, Concrete in Transportation,* ACI, Detroit, 1986, pp. 563-584.
7. Ramakrishnan, V., "Superplasticized Fiber Reinforced Concrete for the Rehabilitation of Bridges and Pavements", *Transportation Research Record 1003,* Transportation Research Board, National Research Council, Washington, DC, 1985, pp. 4-12.
8. Ramakrishnan, V., and Coyle, W.V., "Steel Fiber Reinforced Superplasticized Concrete for Rehabilitation of Bridge Decks and Highway Pavements", *Report DOT/RSPA/DMA-50-84-2, Office of University Research, US Department of Transportation,* p. 410 (Available from the National Technical Information Service, Springfield, Virginia-22161), 1983.
9. ACI Committee 224, "Control of Cracking in Concrete Structures, 224-84" *American Concrete Institute,* Detroit, 1989.
10. ACI Committee 544, "Measurement of Properties of Fibre Reinforced Concrete", ACI 544.2R.78, A*CI Manual of Concrete Practice,* Part 5, 1982.

DESIGN AND CONSTRUCTION OF HIGH PERFORMANCE CONCRETE BRIDGES ON 407 EXPRESS TOLL ROUTE

Hari K. Jagasia, P.Eng.
Manager, Structures
Ontario Transportation Capital Corporation
Thornhill, Ontario, Canada

ABSTRACT

Last year, deck slabs and barrier walls of two bridges on Ontario's All-Electronic Highway 407 Express Toll Route were constructed using high performance concrete. Precast prestressed concrete girders in one of the structures also utilized high performance concrete technology.

The paper describes significant features including a number of design and construction details of the project. In addition, the durability and cost-effectiveness aspects of the designs are discussed and recommendations regarding an alternative approach to the current Ontario practice for design and construction of bridge structures are presented.

HIGHWAY 407ETR – EXPRESS TOLL ROUTE

This summer's opening of the first phase of Highway 407 ETR – Express Toll Route signifies the introduction of a new generation of transportation infrastructure projects.

407 ETR's central section, a 69km (42.9 miles) multi-lane urban toll expressway is located across the northern part of Metropolitan Toronto. The initial section of this all-electronic highway, a stretch of 36km (22.4 miles), is now operational. Although it was officially opened to the travelling public on June 7, 1997, a vast majority of design/build engineering activities were completed by December 1996, just 2½ years after the start of construction. Work on the balance of project is ahead of schedule and the second phase of the facility will be ready for traffic next year.

Above east-west alternative route is intended to improve transportation access in Greater Toronto Area. It will also alleviate prevailing chronic congestion for by-passing traffic on existing highways estimated to be costing some $2 billion annually to Ontario businesses and industries in lost time and productivity and contributing to air pollution and energy waste.

At a cost of over $1 billion, the 407 ETR facility is the largest single civil engineering contract tendered in Canadian history. Highway 407 is the first open-road all-electronic toll highway in the world. It is also one of the first major infrastructure projects in Canada developed through innovative approaches including a "public - private partnership" between the Ontario Transportation Capital Corporation (OTCC), a crown agency of the provincial government and Canadian Highways International Corporation (CHIC).

The above undertakings and proposed east and west extensions to 154 km (95.7 miles) corridor are being implemented in three stages:

        Central   69 km (42.9 miles)
        East   61 km (37.9 miles)
        West   24 km (14.9 miles)

As mentioned previously, construction of the Highway 407 central section is currently well underway. Further expansions of the facility–east and west–would be pursued through future contracts.

Completion of the highway 407 central section required the design and construction of 127 bridges, 38 culverts and a number of other structures such as retaining walls, hydraulic structures, high mast pole foundations, 256 tolling gantries and 128 overhead and bridge mounted sign support frames.

A variety of arrangements and types of structures span over and function as flyovers, overpasses, grade separations, interchanges, water course crossings, widenings and railroad crossings in the 407 ETR corridor. As indicated in Table 1, the structures consist of:

- cast-in-place reinforced concrete;
- precast concrete girders;
- structural steel;
- post-tensioned concrete;
- rigid frames.

Table 1 - 407 ETR STRUCTURAL INVENTORY

STRUCTURE TYPE	PHASE I	PHASE II
Precast/Prestressed Concrete Girders → Integral Abutments → Conventional Designs	14 38	22 20
Prestressed/Post-Tensioned Concrete	9	6
Rigid Frames	9	2
Structural Steel	3	2
Cast-In-Place Concrete Decks	2	0
Cast-In-Place Concrete Culverts	26	12

HIGH PERFORMANCE CONCRETE PROJECT

As part of an innovation applications program, a number of structures on the Highway 407 were selected for the implementation of new technologies, processes or products. The objectives were to explore the opportunities for reducing initial and life cycle costs and improve durability and other qualities of structures. Furthermore, it was anticipated that these projects would result in or promote the development of innovative designs and establish new construction standards or trends for bridge structures in Ontario and elsewhere.

One such application involved the use of high performance concrete (HPC) in the structures. The deck slabs and barrier walls of two Highway 407 bridges were constructed using this technology. The precast prestressed concrete girders in one of the structures also utilized HPC.

HPC project was developed through an Innovative Applications Committee. It was composed of representatives from OTCC, CHIC and the Ontario Ministry of Transportation (MTO) and chaired by the Implementation Manager, Concrete Canada. Experts from other organizations assisted on an as-needed basis during the development of the project and as construction activities were progressing.

The design and construction aspects of these HPC bridges are discussed below.

PROJECT GOALS

Epoxy-coated reinforcement is widely employed by most transportation agencies in Ontario as the primary means of corrosion protection in concrete bridge components exposed to de-icing chemicals. In addition, as a standard practice, the riding surfaces of decks are protected by applying waterproofing membranes and two layers of asphaltic concrete.

The experience relative to the use of epoxy - coated reinforcing steel to date, however, indicates that it may not be providing satisfactory long-term level of corrosion protection for the designed service life of the structures exposed to a salt-induced corrosive environment. Also, there is a premium associated with epoxy-coated reinforcement which adds significantly to the overall costs of structures.

The purpose of HPC undertakings was to incorporate alternative materials in selected structures that would alleviate the foregoing concerns and to evaluate their effectiveness or feasibility. It was anticipated that the experience gained would result in improved structural durability which, in turn, would lead to economical and functional bridge designs.

STRUCTURE DETAILS

Highway 407 EBL Over Levi's Creek (Structure B22A)

- Span   34.0m (111' - 6")
- Width   11.7m (38' - 5")
- 5-CPCI 1900 (6' - 3") Girders @ 2.35m (7' - 8½") c/c
- Deck Slab   235mm (9¼")
- Class of Concrete:
  - Deck Slab and Barrier Walls   60MPa (8,700 psi)
  - Precast Prestressed Concrete Girders   40 MPa (5,800 psi)
  - Remaining Structural Components   30 MPa (4,350 psi)
- Concrete Covers:
  - Top/Deck   80 ± 20mm (3⅛ ± ¾")
  - Bottom/Deck   40 ± 10mm (1½ ± ⅜")
  - Remainder   70 ± 20mm (2¾ ± ¾")
- Integral Abutment Design
- No Epoxy-Coated Reinforcement
- Exposed Concrete Riding Surface

Ramp Mississauga Road S - 407E Over Levi's Creek (Structure B22C)

- Span   34.0m (111' - 6")
- Width   8.71m (28' - 7")
- 3-CPCI 1900 (7" - 8½") Girders @ 3.0m (9' - 10") c/c
- Deck Slab   235mm (9¼")
- Class of Concrete:
  - Deck Slab and Barrier Walls   60MPa (8,700 psi)
  - Precast Prestressed Concrete Girders   60 MPa* (8,700 psi)
  - Remaining Structural Components   30 MPa (4,350 psi)
- Concrete Covers:
  - Top/Deck   80 ± 20mm (3⅛ ± ¾")
  - Bottom/Deck   40 ± 10mm (1½ ± ⅜")
  - Precast Girders   25 + 5/-3mm (1 + 3/16/ - ⅛")
  - Remainder   70 ± 20mm (2¾ ± ¾")
  - Prestressing Strands (½" special)   32 Straight
      18 Deflected
- Integral Abutment Design
- No Epoxy-Coated Reinforcement
- Exposed Concrete Riding Surface

*Transfer Strength   45MPa (6,530 psi)

HPC MIX DESIGNS

Decks and Barrier Walls (Structures B22A and B22C)

- Cement Type 10SF   325 kg/m³ (20.3 #/ft³)
- Slag   105 kg/m³ (6.6 #/ft³)
- Aggregates
  - Coarse: 19mm (¾")   1065kg/m³ (66.5 #/ft³)
  - Fine   735kg/m³   (45.9 #/ft³)
- Water   140 kg/m³ (8.7 #/ft³)
- Air Content, %   6.5 ± 0.5
- Slump   180 ± 30mm (7.1 ± 1.2 in)
- Admixtures (Euclid)
  - Water Reducer   Eucon DX
  - Air Entrainment   AirExtra
  - Superplasticizer   Eucon 37
- Concrete Mix Temperature < 25°C (77°F)

Precast Concrete Girders (Structure B22C)

Mix design adopted for the project is proprietary information. It generally consisted of the following admixtures:

- Type 30 Cement
- Silica Fume   Rheomac
- Water Reducing Agent   WRA Type RX
- Air Entrainment   AEA Micro-Air
- Accelerator/Plasticizer   Rheobuild 2500
- Water/14mm (9/16") Limestone/Sand
- Standard Accelerated Steam Curing

CONSTRUCTION ISSUES

Decks and Barrier Walls

Mix designs, curing, finishing, surface texturing and trial placement operations were refined, readjusted and optimized as the design and construction of the project progressed.

The final mix proportions of the HPC for deck and barrier walls are indicated above.

It should be noted that the contract documents called for total cementitious material, including Type 10SF cement, fly-ash, slag and silica fume, as a minimum of 400 kg/m³ (25#/ft³), whereas on a previous MTO bridge project the minimum cementing content of 450 kg/m³ (28.1 #/ft³) was specified for the HPC mix. The revised mix as proposed by the supplier was tested in field trials. Based on the strengths achieved and finishing suitability, it was decided that this mix design, with cementing material content of 325 kg/m³ (20.3#/ft³), be adopted for the project.

The number of test specimens and frequency of testing were reduced as compared to previous HPC projects while maintaining an acceptable level of quality control. As a result, concreting operations were carried out efficiently and with reduced frequency of interference.

HPC was moist cured by continuous fog-mist prior to placing burlap and wet cured thereafter with two layers of burlap covered with a moisture vapour barrier. The deck surface was maintained in wet condition for seven days. After this period a curing compound was applied to the concrete surface.

Test results indicated that the compressive strengths of concrete at 28 days for both structures varied between 70 to 80 MPa (10,150 to 11,600 psi).

Precast Concrete Girders

HPC girders for structure B22C were fabricated in the Armbro Precasting Plant on May 10, 13 and 14, 1996. The mix design consisting of admixtures detailed above was developed jointly by Canada Building Materials (CBM), a division of St. Marys Cement and Master Builders Technology.

In order to make pretensioning more efficient "½" special" seven-wire low-relaxation grade 270 strands in the girders were used. This provided an area of 0.167in²/strand in lieu of 0.153in² for the regular ½" strands commonly used in Ontario. As a result of designing with a 9.2% larger strand area, approximately 10% less strands in the girders were required.

Ready mix concrete was delivered from CBM's plant located approximately 20 minutes away from the precasting yard. The slumps of supplied concrete ranged from 0 to 15mm (0.6"). Adequate quantity of Master Builders Rheobuild 2500 was therefore added to achieve slumps of 200 to 220mm (7.9 to 8.7"). The concrete temperatures at delivery times ranged between 17 - 21°C (63 - 70°F) which contributed favourably to the success of concrete placement and handling operations.

Table 2 documents the delivery temperatures, air content, initial and final slumps and compressive strengths at 16¼ hours, 18 hours, 7 days and 28 days for HPC precast girders. It shows that in all cases the transfer and 28-day concrete strengths exceeded the specified design requirements.

Table 2 -Mixing/Handling/Test Results for HPC Girders

Concrete Delivery Temperature	17-21°C (63-70°F)
Air	6.2%
Slump → Initial → Final	0-15 mm (0-0.6") 120mm (4.7")
Concrete Strengths → 16¼ hrs → 16¼ hrs → 18 hrs → 18 hrs → 7-day → 7-day → 7-day → 28-day → 28-day	 61.5 MPa(8,920 psi) 60.9 (8,830 psi) 63.8 (9,250 psi) 61.3 (8,890 psi) 66.7 (9,670 psi) 72.4 (10,500 psi) 71.6 (10,390 psi) 79.8 (11,570 psi) 77.3 (11,210 psi)

MONITORING AND EVALUATION PROGRAM

The quality assurance staff carried out visual inspections, monitored construction and conducted field and laboratory tests related to the use of HPC in the structures.

Thermal curing conditions on the deck and girders were monitored by installing temperature probes which were placed strategically at various locations. Temperature measurements were recorded at different intervals after the concrete was placed in the structural components.

Tests and other data indicated the following results:
1. Satisfactory compressive strengths.

2. Excellent air void systems.

3. Low Rapid Chloride Permeability Coulomb counts.

4. Resistivity greater than for 30 MPa (4,350 psi) concrete.

5. Good and consistent concrete covers.

6. Proper curing and no signs of distress caused by plastic shrinkage.

7. Results obtained from a thermal analysis of hardening concrete in decks and barrier walls and during the early life of precast girders, and the stresses due to cooling, were satisfactory.

There is also an elaborate monitoring program in place to evaluate the long-term performance of HPC bridges in the 407 ETR facility.

To assess closely the overall performance and durability of HPC, instrumentation consisting of corrosion monitoring probes and thermocouples was installed in the deck of structure B22A. Performance and durability criteria are to be evaluated against a nearby conventional structure, designated as a control, in the Highway 407 corridor with similar environmental exposure and loading patterns.

A preliminary condition survey on the structure B22A was performed approximately two months after the deck and barrier walls were cast. Testing was done to ensure that instrumentation and wiring are performing as intended and to establish baseline data.

The initial corrosion readings will be recorded this year after the reinforcement has sufficiently passivated.

MTO has agreed to monitor these installations annually or as considered essential and distribute the data relating to the long-term performance of the structures to participating agencies.

INITIAL AND LIFE CYCLE COSTS

Construction cost of HPC structure B22A, based on actual material and labour prices, was approximately 5% greater as compared to a conventional structure. The savings as a result of eliminating epoxy coated reinforcing steel, waterproofing and asphalt paving on the decks are reflected in this estimate.

An initial increase in cost was mainly due to the higher material prices and first-time developmental premiums that were added on to the two relatively small structures.

There was no overall additional initial cost involved with the structure B22C which utilized HPC in precast concrete girders. The primary contributing factors were:

- number of girders for the structure was reduced from 4 to 3
- 56 "normal ½" strands/girder substituted with 50 "special ½" strands.

Substantial overall net savings from the girders neutralized the additional HPC costs related to the deck and barrier walls of the structure.

It is anticipated that initial costs associated with HPC will diminish significantly with an increased frequency of these applications since developmental, testing and trial premiums would spread proportionally. A competitive bidding process in lieu of an add-on order for awarding HPC contracts would also help. In addition, considerable savings are expected to be realized now that the mix designs are established and related parameters influencing economy are optimized.

Life cycle costings were investigated by considering early age characteristics of the concrete and projected long-term performance, including the associated maintenance and rehabilitation expenditure criteria, of HPC bridges. These costs were evaluated against conventional structures with similar functional characteristics. Given improved durability characteristics, including lower permeability and reduced cracking potential, and other beneficial impacts of HPC, it was determined that on the basis of life cycle costs, HPC bridges are feasible and in many instances may offer a preferred option.

FUTURE PROJECTS

To enhance or gain further experience with the technology, two more HPC bridges on 407 ETR, with variations to the mix criteria and placement procedures, are scheduled to be constructed this year. There may be additional opportunities to incorporate HPC in bridge decks of the structures designed for the second phase of this facility.

CONCLUSIONS

Based on the experience with the completed project and field performance and effectiveness of the installations to date, significant conclusions emerge as follows:

1. HPC can play a key role and represents a viable strategy, to mitigate the factors affecting corrosion of reinforcing steel, which would enhance the long-term durability of bridge structures.

2. The project was developed through the OTCC/CHIC Innovation Applications Committee. The discussions between all parties at the pre-construction meetings and throughout the implementation stage contributed to the success of these projects.

3. The design and constructibility requirements for HPC installations including mix proportions, air content, slump, placement, finishing, texturing and curing operations conformed to the specifications or manufacturers/supplier's recommendations and were satisfactorily achieved.

4. Lower chloride permeability, higher strength and the improved air void system are beneficial characteristics of HPC placements. These would result in reduced chloride access to reinforcing steel, greater resistance to cracking and extended design service life of structures.

5. The initial incremental costs associated with the construction of HPC deck slabs and barrier walls will be offset by the longer lasting structural components and hence a lower life-cycle cost.

6. By incorporating HPC girders and optimizing the design (reducing the number of girders and utilizing more efficient pretensioning with ½" special strands), significant savings in the construction costs were realized.

RECOMMENDATIONS

1. Promote HPC as a competitive and preferable product and an effective technology in transportation infrastructures.

2. Where feasible, incorporate HPC bridge girders in conjunction with other structural components in contracts.

3. Revise/update specifications and other contractual requirements related to HPC, and ensure that the experiences of various participants involved with the project are utilized or reflected in future applications.

4. Optimize structural design features (maximum spans and reduced sections) by utilizing the high strength of concrete in HPC and adopting shallower systems. Also incorporate other measures to achieve further economies.

5. Undertake additional laboratory and field work for evaluating alternative materials and construction processes to determine if HPC operations could be further improved.

ACKNOWLEDGEMENTS

The author wishes to recognize the contributions of the participants associated with following groups in the above projects:

- Canadian Highways International Corporation
- Armbro Construction Limited
- Concrete Canada
- Agra Earth and Environmental
- Ministry of Transportation, R&D Branch
- St. Lawrence Cement

The generous assistance and cooperation of the suppliers and industry are also greatly appreciated.

Any observations, opinions, conclusions or recommendations expressed or implied in this paper are those of the author and are not to be attributed to OTCC or the parties listed above.

DESIGN OF TENSION LAP SPLICES IN HIGH STRENGTH CONCRETE

Atorod Azizinamini, Ph.D., P.E.
Associate Professor
Civil Engineering Department
University of Nebraska-Lincoln
Lincoln, Nebraska, U.S.A.

ABSTRACT

Safety concerns and a lack of test data on bond capacity of deformed reinforcing bars embedded in high-strength concrete have been reasons for the ACI 318 building code imposing an arbitrary limitation of 10,000 psi (69 MPa) in calculating the tension development and splice lengths.

In an attempt to evaluate the impact of this limitation and develop provisions for its removal, an investigation was carried out at the University of Nebraska-Lincoln, partial result of which will be presented in this paper. Results of the investigation are used to discuss the differences that exist between normal and high strength concrete, develop hypotheses to explain these observed differences, and suggest alternatives for removal of the current concrete compressive limitations existing in the ACI 318 building code for calculating tension development and splice lengths.

INTRODUCTION

Due to a lack of test data, the ACI 318-95 building code requirements include an arbitrary limitation of 10,000 psi (69 MPa) on specified compressive strength of concrete, f'_c, that may be used in calculating tension development and tension splice lengths. This limitation is stated in section 12.1.2 of the ACI 318-95 building code. As a result, an investigation was conducted to evaluate the bond performance of reinforcing bars embedded in high strength concrete. Results of the investigation provided a basis to develop a behavioral model in the form of a failure hypothesis to explain the observed differences between reinforcing bars embedded in Normal and high strength concrete. The major conclusion is that increasing the splice length is not an efficient approach for improving the bond performance of reinforcing bars embedded in high strength concrete and that placing some minimum amount of transverse reinforcement over the splice regions is an efficient and safe approach.

EXPERIMENTAL WORK

A typical test specimen is shown in figure 1. Each specimen consisted of 2 or 3 reinforcing

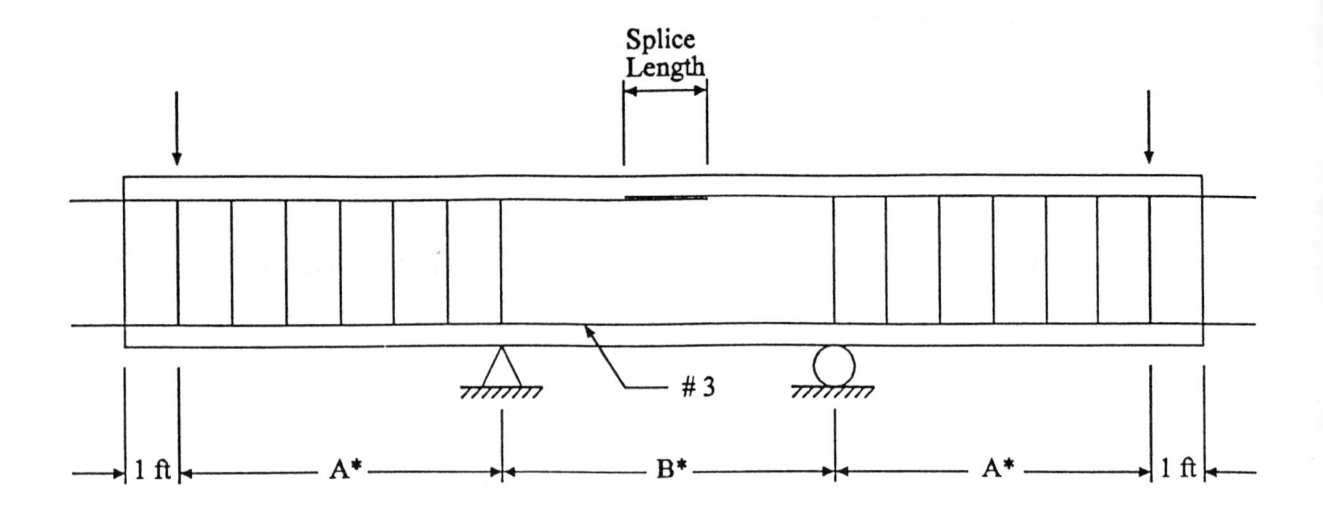

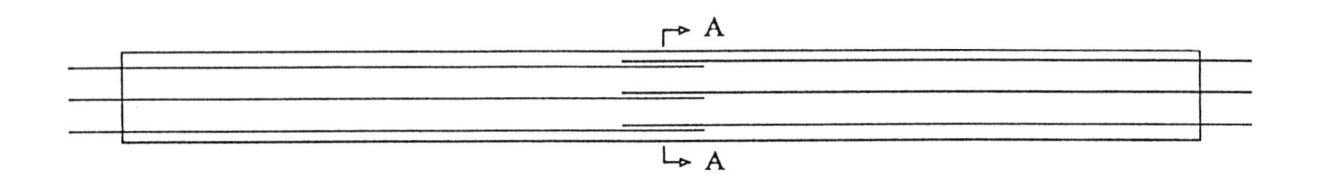

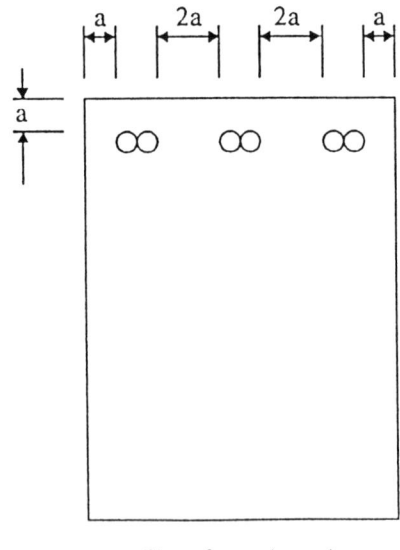

Figure 1: Typical Test Specimen.

bars spliced at midspan. The longitudinal reinforcement was either ASTM A615 Grade 60 (specified minimum yield strength of 414 MPa), No. 11 (36 mm diameter) or No. 8 (25.4 mm diameter) steel reinforcing. The transverse reinforcement used in some of the specimens over splice regions consisted of ASTM A615 grade 60, No. 3 (9.5 mm diameter) steel reinforcing. The thickness of side or top concrete cover was equal to one or two times the bar diameter. The clear spacing between reinforcing bars at midspan of each specimen was approximately two times the side cover. Compressive strength of the concrete at the time of testing were approximately 15000 psi (103 Mpa), which was obtained by testing 4 by 8-in. (102 by 204 mm) diameter cylinders cast at the same time as the specimen and cured alongside the specimens. Two different deformation types were used for longitudinal reinforcement. Majority of test specimens were bottom cast.

The high strength concrete mix used in construction of test specimens included Type I cement, Class C fly ash, silica fume and superplasticizer. The maximum aggregate size was ½ in (12.7 mm). Water-cementatious material ratios ranged from 0.21 to 0.27.

Concrete was provided by local ready-mix suppliers. Construction of each specimen consisted of pouring the concrete at one end of the wood form and proceeding to the other end. The concrete slump for the high strength concrete test specimens was about 9 in. (229 mm). Little effort was required to finish the top surface of the beams. Immediately following casting, the specimens were covered with a plastic sheet. After approximately five days the forms were removed and the specimens were covered with wet burlap and plastic sheets until generally two days before testing. The casting of each specimen lasted approximately 30 minutes.

TEST SET-UP AND TESTING PROCEDURES

The test set-up and loading arrangements for each test are shown schematically in Fig. 2. The test set-up consisted of beam specimens placed on two roller type supports and loaded equally at each end using two hydraulic rams and spreader beams. The applied load and resulting deflections at each beam end and midspan and strains from longitudinal bars and stirrups were monitored and the data stored in a computer.

The test was begun by applying equal loads at each end of the beam. The load at each end was applied in increments ranging from 0.5 (2.22 KN) to 2 (8.89 KN) kips, depending on the estimated strength of the beam specimen. Displacement control was used, for specimens with stirrups, following yielding of the longitudinal bars. The load was held constant for approximately 5 minutes after each load or displacement increment, during which time cracks were mapped and test observations recorded. Load or displacement increments continued until the specimen failed.

FAILURE HYPOTHESIS

A behavioral model in the form of failure hypothesis was developed using the results of

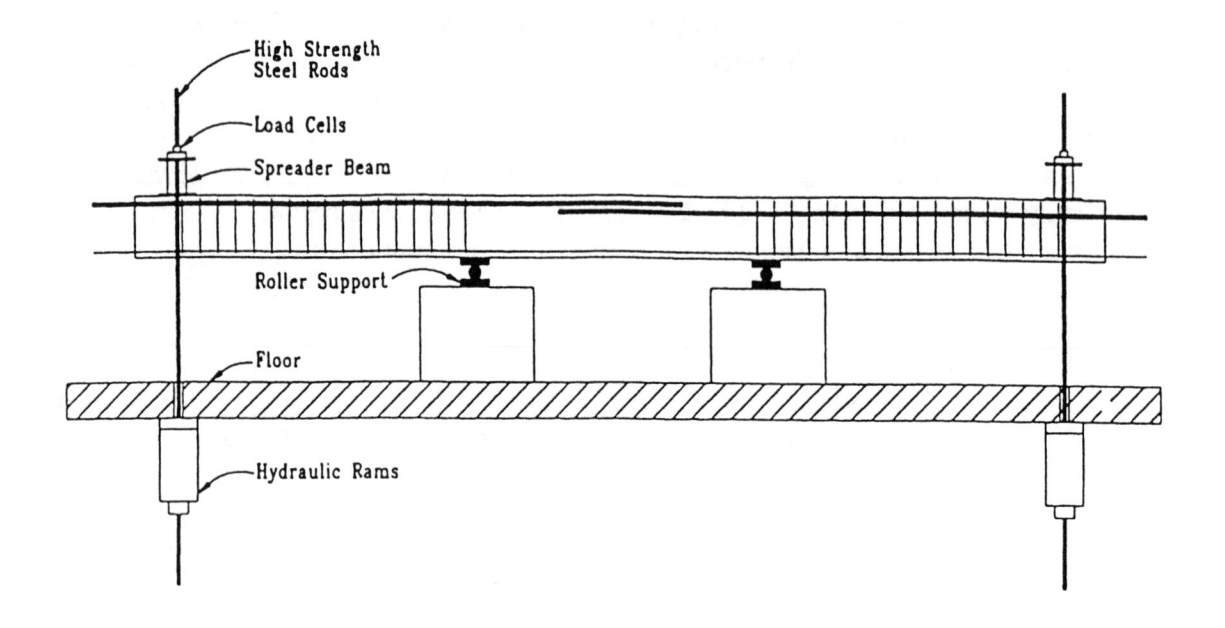

Figure 2: Typical Test Setup.

test data. A detailed discussion and analyses of data culminating in development of this behavioral model is provided in Ref. 1. Using this model, an attempt was made to describe the observed differences between bond performance of deformed reinforcing bars embedded in normal and high strength concrete. A brief description of the above noted behavioral model is given below.

For the sake of simplicity and to briefly outline the general concept, Fig. 3 shows a segment of a deformed bar embedded in concrete and subjected to different levels of axial tensile forces. This figure shows the free body diagram of the reinforcing bar at several load stages. At low axial load levels (see Fig. 3a)), the outermost lugs (i.e. those closest to the loading point) come in contact with concrete. Consequently, these lugs exert a bearing force on the concrete. The horizontal component of this force produces bond stress. (The horizontal component of the friction force is not shown in this figure but also adds to the bond strength.) Figure 3 also shows the corresponding bond stress distribution. As load increases, this bearing force causes crushing of concrete in the vicinity of the lug. This action allows the next adjacent lug to come in contact with concrete and participate in resisting the applied axial tension (see Fig. 3). The ACI building code assumes that at ultimate the bond stress distribution is uniform, which implies that all the lugs bear against concrete at the ultimate stage (see Fig. 3c) and help in resisting the applied axial force. This is a reasonable assumption to make for NSC and has been shown to be valid by experimental testing. In the investigation reported in Ref. 1, experimental evidence did not indicate the same behavior for HSC. This observation could be explained as follows.

Referring to Fig. 3, when the first lug comes in contact with concrete, a bearing force acting against the lug is created. The horizontal component of this bearing force results in what is referred to as bond stress. The vertical component of the bearing force creates a radial force which is responsible for splitting the surrounding concrete. Note also that the bearing capacity of concrete is related to f'c, whereas the tensile capacity is related to $\sqrt{f'_c}$. Therefore, as an example, assuming that the bearing capacity and tensile capacity of concrete are given by $0.85f'_c$ and $5\sqrt{f'_c}$, respectively, the ratio of the bearing capacity of 15,000 psi (104 MPa) concrete over that of 5000 psi (35 MPa) concrete would be 3, whereas the ratio of tensile capacity of 15,000 psi (104 MPa) concrete over that of 5000 psi (35 MPa) concrete would be 1.73. In other words, increasing the compressive strength from 5000 psi to 15000 psi would result in a bearing capacity three times as large, while the tensile capacity increases only 1.73 times.

In the case of HSC, the higher bearing capacity of the concrete will prevent crushing of the concrete in the vicinity of each lug to the extent that would otherwise take place in normal strength concrete. This implies that, at ultimate, all lugs may not participate in resisting applied axial forces, and demands that the first few lugs contribute the most. With the first few lugs being more active, and considering the fact that in HSC tensile capacity does not increase at the same rate as bearing capacity, it could be concluded that in the case of HSC failure could be by splitting of concrete prior to achieving uniform load distribution.

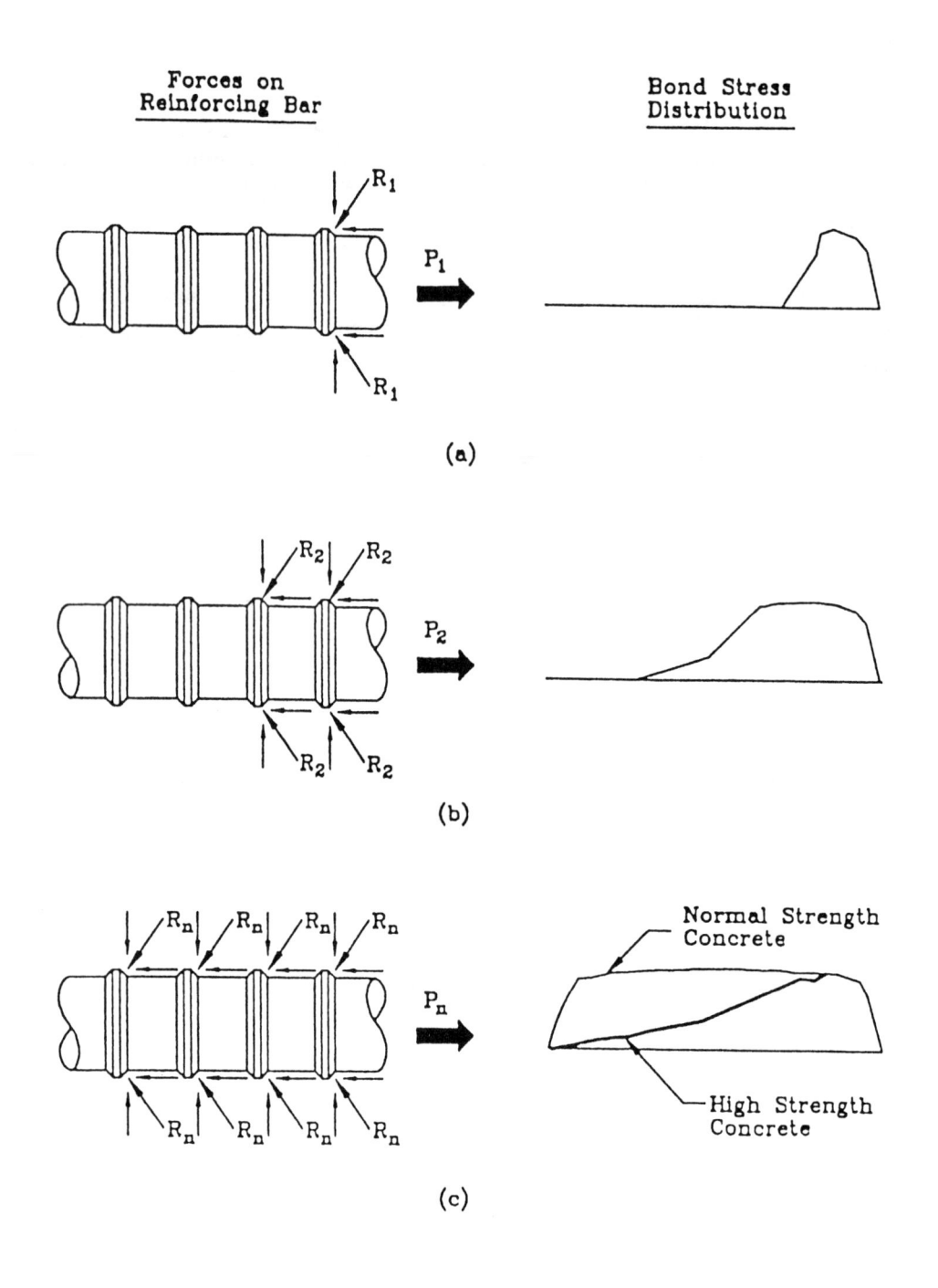

Figure 3: Freebody Diagram of a Reinforcing Bar Embedded in Concrete and Subjected to Tension.

The experimental evidence and use of the above behavioral model [1] led to the following major conclusions.

(a) In the case of HSC, especially in the presence of small cover, increasing the splice length is not an efficient approach for increasing bond capacity and, in fact, may be a waste of material. A mechanism which could delay splitting of the concrete over development or splice length would be more effective in increasing the bond capacity of deformed reinforcing bars embedded in HSC. This mechanism could be provided by requiring some minimum amount of stirrups over development or splice lengths.

(b) In the case of HSC, the assumption of uniform bond stress over the development length may not be valid.

(C) In the case of HSC, top cast bars exhibit slightly higher bond strength than bottom cast bars, which is the opposite of that generally reported for the case of NSC. Using the behavioral model described above, Ref. 1 provides a possible explanation for this observation.

(d) In the past, researchers have used the ratio of bond stress obtained from tests, U_{TEST}, over bond stress implied by ACI codes, U_{ACI}, as an index when investigating the bond capacity of reinforcing bars embedded in NSC.

U_{TEST} is defined by the following equation:

$$U_{TEST} = f_s d_b / 4 l_s$$

where f_s is the maximum bar stress in the reinforcement obtained from test, d_b is the diameter of the reinforcement, and l_s is the splice length.

When the value of this index exceeds unity it is assumed that the bond capacity is adequate. However, in the case of HSC, it was concluded that resorting only to the U_{TEST}/U_{ACI} ratio exceeding one in assessing the bond capacity of reinforcing bars is a criterion that is necessary but not sufficient. It was shown that this criterion does not guarantee that members with tension splices will fail in a ductile manner.

SELECTION OF AN INDEX TO ASSESS BOND DATA

Results of this investigation indicate that use of U_{TEST}/U_{ACI} ratio as a criteria to study the safety of reinforcing bars embedded in high strength concrete is not adequate. For instance, for two tests carried out in this investigation, which we will refer them, hereafter, as tests 29 and 30, the splice lengths were 80 in. (2.03 m) and 57.5 in. (1.46 m), respectively. The splice lengths for Tests 29 and 30 were calculated based on ACI 318-89 (ignoring the f'c limitation) and ACI 318-83 building code requirements. The ACI 318-95 code requirement for the same condition (No. 11 reinforcing bars, concrete cover of one times the bar diameter and f'$_c$ of 103 MPa) would require approximately 45 in. (1.14 m) splice length (ignoring the f'c limitation). The U_{TEST}/U_{ACI} ratio for specimens 29 and 30 were 1.17 and 1.63, respectively, which is greater than 1. However, both specimens failed in a very brittle and violent manner without exhibiting ductility. Figure 4 gives end load versus the midspan displacement for these two tests. As will be shown later, providing some minimum amount of transverse reinforcement results in a significant increase in ductility. In this study, the displacement ductility ratio, as an index, was selected to assess the test results. Figure 5 gives the definition of the displacement ductility ratio. For most tests, when incorporating transverse reinforcement over the splice region, the plot of applied end load versus mid span displacement exhibited a relationship of the type shown in Fig. 5. The displacement ductility ratio is defined as the ratio of the maximum midspan displacement over the first yield displacement. The first yield displacement is defined as the intersection of the two dashed lines, approximating the load displacement curve, as indicated in Fig. 5. A displacement ductility ratio of greater than 1 signifies firstly, that longitudinal bars are capable of developing at least their actual yield stress and, secondly, specimens will exhibit some level of ductility. Therefore, the use of the displacement ductility ratio ensures the dual criteria of strength and ductility, whereas the bond stress ratio is only a strength criteria.

BRIEF DISCUSSION OF EFFECT OF STIRRUPS OVER THE SPLICE REGION

The observed effect of transverse reinforcement over the splice region on bond performance of reinforcing bars for test specimens with #8 or #11 bars and concrete cover thicknesses of one or two times the cover concrete were similar, in general. In this paper, only discussion of the effect of stirrups is provided for specimens utilizing #11 bars and having one times the bar diameter as concrete cover thickness.

As discussed earlier, specimens 29 and 30 had 77% and 27% more splice length than required by ACI 318-95 provisions, if the f'$_c$ limitation were ignored. However, both specimens failed in a very brittle and violent manner. Figure 6 shows the midspan displacement versus total applied end load curves for tests specimens with same splice length as that of 30; however, with different amounts of transverse reinforcement over the splice region. Figures 6a, 6b, 6c and 6d give load displacement response of specimens having 57.5 inch (1.46 m) splice length and varying amount of stirrups over the splice. Each figure also gives the details of the specimens, in terms of number of longitudinal bars spliced, splice lengths and spacing of stirrups over the splice length. Figure 6 indicate that incorporating some transverse reinforcement over the splice region results in a significant increase in ductility.

CONCLUSIONS

By reviewing the information generated during the investigation, following main conclusions were drawn:

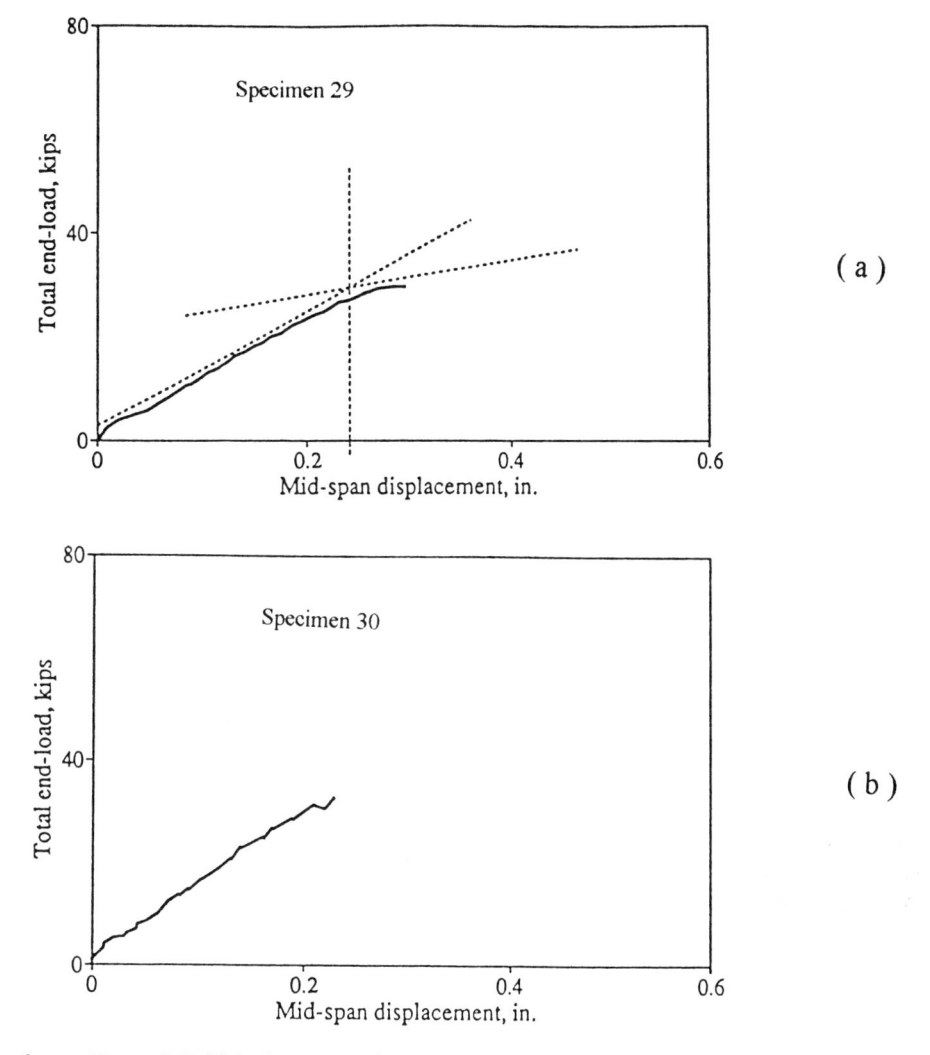

Figure 4: Load Mid-Span Displacement Response of Specimens 29 and 30.

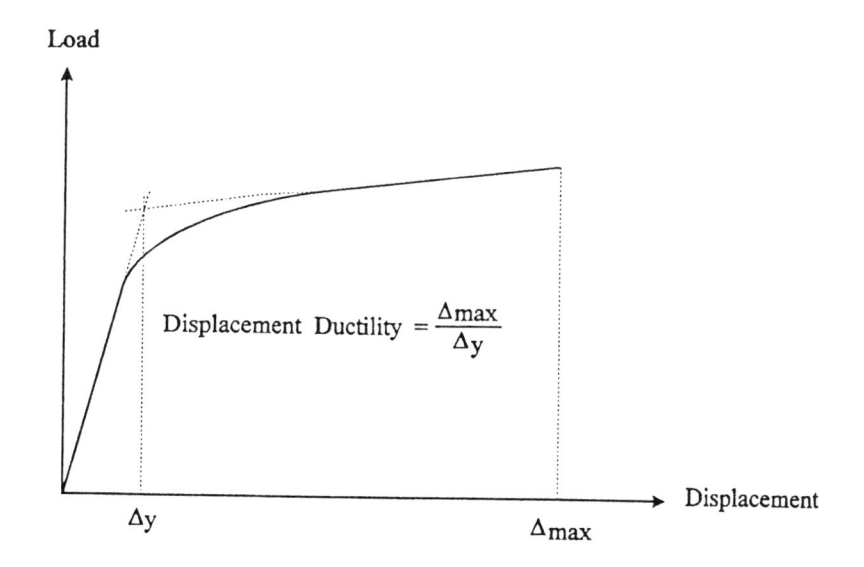

$$\text{Displacement Ductility} = \frac{\Delta_{max}}{\Delta_y}$$

Figure 5: Definition of Ductility Displacement Ratio.

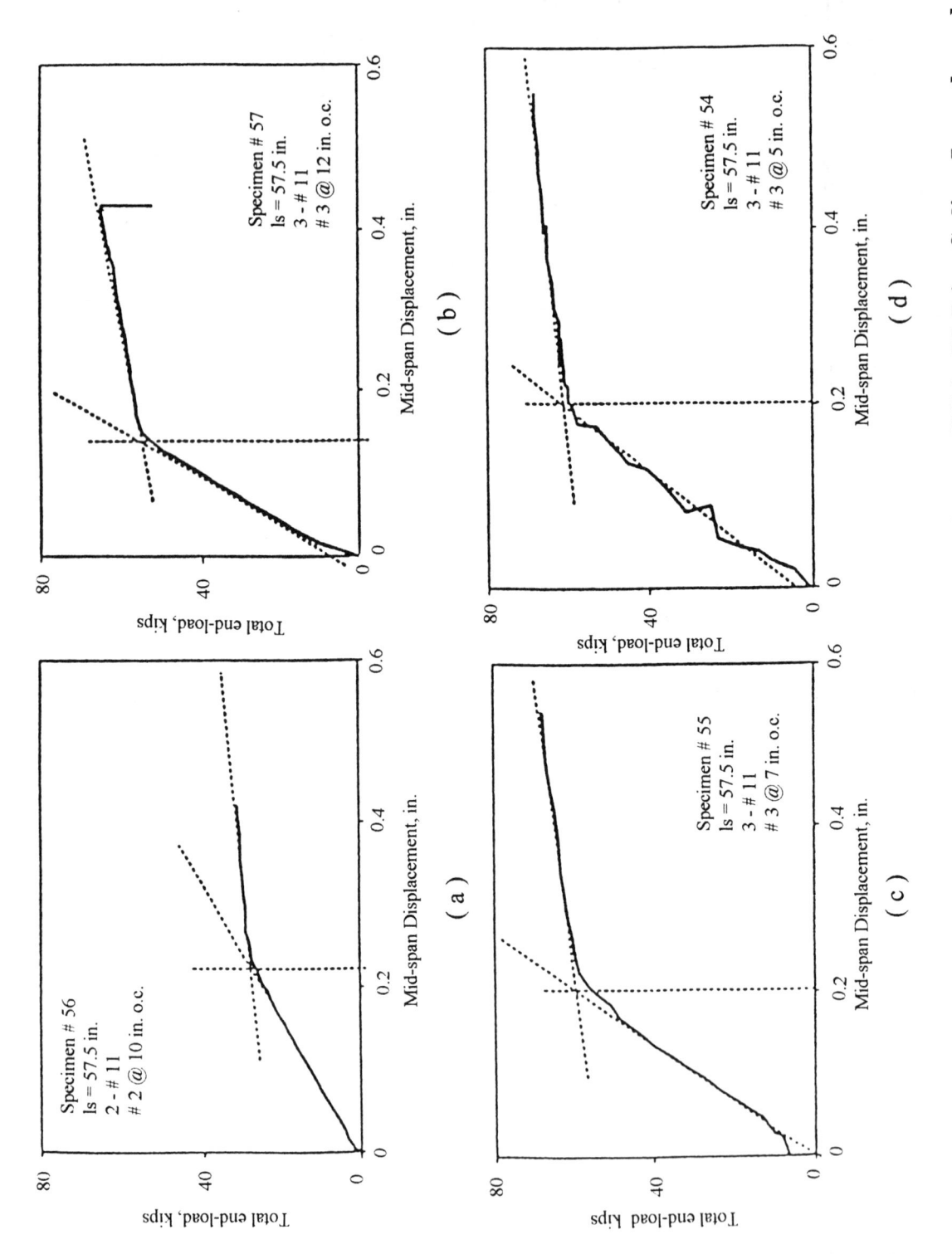

Figure 6: Load Mid-Span Displacement Response of Specimens With 57.7 in. Splice Length and Various Amount of Stirrups.

(a) In the case of HSC, especially in the presence of small cover, increasing the splice length is not an efficient approach for increasing bond capacity. A mechanism which could delay splitting of the concrete over tension development or tension splice length would be more effective in increasing the bond capacity of deformed reinforcing bars embedded in HSC. This mechanism could be provided by requiring some minimum amount of stirrups over development or splice lengths.

(b) In the past, researchers have used the ratio of bond stress obtained from tests, U_{TEST}, over bond stress implied by ACI codes, U_{ACI}, as an index when investigating the bond capacity of reinforcing bars embedded in NSC. When the value of this index exceeds unity it is assumed that the bond capacity is adequate. However, in the case of HSC, it was concluded that resorting only to the U_{TEST}/U_{ACI} ratio exceeding one in assessing the bond capacity of reinforcing bars is a criterion that is necessary but not sufficient. It was shown that this criterion does not guarantee that members with tension splices will fail in a ductile manner.

c) By studying the behavior of specimens having # 8 and # 11 reinforcing bars and varying amount of concrete cover, it was concluded that even in the case of larger concrete cover thicknesses, one will have to provide some minimum amount of stirrups over the tension development or tension splice lengths. However, as concrete cover thicknesses increases, the minimum stirrup requirements should be decreasing.

ACKNOWLEDGMENTS

This project was carried out in two phases. Phase II of the project, which was the major part of the investigation, was supported by the National Science Foundation. The authors are very grateful for this support and to Dr. Ken Chong, program director at NSF. Phase I of the project was funded by Portland Cement Foundation. This support is gratefully acknowledged. Partial support for this project was also provided by the Center for Infrastructure Research at the University of Nebraska-Lincoln, for which the authors are greatly appreciative.

The contents of this paper reflect the views of the authors, who are responsible for the facts and accuracy of the data presented, and do not necessarily represent the views of the sponsors.

REFERENCES

1- Azizinamini, A., Stark, M., Roller, J.J. and Ghosh, S.K., "Bond performance of reinforcing bars embedded in high strength concrete", ACI Structural Journal, Sept.-Oct. 1993, PP 554-561.

INVESTIGATION OF ALLOWABLE COMPRESSIVE STRESSES FOR HIGH STRENGTH, PRESTRESSED CONCRETE

Bruce W. Russell, Ph.D., P.E.
Assistant Professor
The University of Oklahoma
Norman, Oklahoma   USA

Joo Pin Pang
Graduate Research Assistant
The University of Oklahoma
Norman, Oklahoma   USA

ABSTRACT

Current ACI and AASHTO code provisions for prestressed concrete limit allowable compressive stresses to 60% of the concrete's breaking strength at release. Design cases that incorporate high strength concretes can be limited by this design requirement. Therefore, an experimental research program was conducted to investigate the adverse effects that relatively large, sustained compressive stresses may have on concrete strength. Compressive loads were applied to concrete test cylinders and sustained. After the prescribed load duration, breaking strengths of the test cylinders were measured and compared to strength of control cylinders that had not been subjected to sustained loads. The research results indicate that the allowable compressive stress at release could be increased from $0.6\, f'_{ci}$ to $0.7\, f'_{ci}$.

INTRODUCTION AND BACKGROUND

Current code provisions contained within the *Building Code Requirements and Commentary for Structural Concrete (ACI 318-95)*[1] and the *Standard Specifications for Highway Bridges (AASHTO)*[2] limit allowable compressive stresses at release to 60% of the concrete's breaking strength at release. This limitation has not often been a factor in the past because the allowable tension stress at service conditions usually controlled cross section size and prestress requirements. However, the use of high strength concrete can permit longer span bridges with effectively shallower depths. Frequently, these design cases, which necessarily incorporate high strength concrete, are limited by the allowable compressive stress at release.

In prestressed concrete, the concrete is precompressed by restraining tension in prestressing steel. Future tensile stresses from external loads are counteracted by precompression stresses built into the concrete. In this manner, prestressed concrete is an active composite material where larger precompression stresses can increase a structure's resistance to cracking from external loads. By using higher strength concrete, even larger precompression stresses can be imposed on the concrete which can further increase the member's resistance to external loads before cracking, thereby improving the prestressed member's resistance to applied loads.

The demand for larger precompression stresses is restricted because the maximum prestressing force is usually imposed at an early age when the concrete material has gained only a portion of its 28 day design strength. This factor is especially critical for pretensioned concrete where prestress transfer is usually accomplished within 24 hours after casting the concrete. Therefore, the concrete compressive strength at early ages and its corresponding allowable stress of 0.6 f'_{ci} become critical elements in the design and fabrication of precast/prestressed concrete structures.

Therefore, an experimental research project was conducted to investigate the effects that relatively large, sustained compressive stresses may have on concrete strength. In the testing program, compressive stresses at and exceeding 0.6 f'_{ci} (0.6 f'_{ci}, 0.7 f'_{ci} or 0.8 f'_{ci}) were imposed on concrete test cylinders and maintained for a prescribed load duration. Concrete "test" cylinders were loaded at one day of age and compressive stresses were sustained for the prescribed load duration of 7 days, 28 days, 63 days, 90 days or 180 days. At the end of the prescribed loading duration, breaking strengths of the "test" cylinders were measured and compared to those of the "control" cylinders that were not subjected to sustained loads.

TESTING PROGRAM

Scope
Altogether, twelve (12) concrete batches were cast and tested. Of the twelve batches, seven (7) batches were made with Concrete Mixture A, possessing a nominal 1 day (release) strength of about 35 MPa (5000 psi), and five (5) batches were made with Concrete Mixture B, possessing nominal 1 day (release) strength of 50 MPa (7000 psi). The concrete materials and mixture proportions are reported in Table 1.

In addition to concrete strength, other research variables included: 1) the age at loading (1 day or 28 days after casting); 2) the magnitude of applied compressive stress (0.6 f'_{ci}, 0.7 f'_{ci}, or 0.8 f'_{ci}), and; 3) the duration of sustained loading (7 days, 28 days, 63 days, 90 days, or 180 days). For each of the twelve casts, thirty-six (36) 100 x 200 mm (4 x 8 in.) cylinders were made. Of these, eighteen (18) "test" cylinders were subjected to sustained compressive stress for the entire length of prescribed load duration. Of the eighteen (18) "test" cylinders, six (6) cylinders were loaded to a compressive stress of 0.60 f'_{ci}, six (6) cylinders were loaded to 0.70 f'_{ci} and six (6) cylinders were loaded to 80% f'_{ci}. The "control" cylinders were not subjected to sustained loads. At the end of each prescribed load duration, three (3) "test" cylinders and three (3) "control" cylinders were tested for compression strength. The compressive strengths of "test" cylinders were then compared to the compressive strengths of "control" cylinders. The "control" cylinders are cast from the same concrete batch, cured in the same manner and stored in nearly identical environmental conditions as their companion "test" cylinders.

Casting
Casting procedures followed the guidelines of ASTM C 192.[3] Conventional water reducing admixture was added to the mixture with the initial mixing water. The high range water reducing admixture (HRWR) was added to the mixture immediately before the final two minute mixing period. The concrete mixtures were prepared in a rotating drum mixer. Concrete slumps were measured before and after addition of

Table 1. Concrete Mixture Proportions

Material	Quantity (per m³)	Quantity (per yd³)
Concrete Mixture A (Nominal 50 MPa)		
Type III Cement	386 kg	650 lbs
Water	154 kg	260 lbs
Crushed Limestone, ASTM #67	1053 kg	1775 lbs
Fine Aggregate, "Dover Sand"	795 kg	1340 lbs
Water Reducer: Daratard 17 (W.R. Grace)	774 mL (2.0 mL/kg)	20 fl oz (3 oz/cwt)
High Range Water Reducer: Daracem 19 (W.R. Grace)	3020 mL (7.8 mL/kg)	78 fl oz (12 oz/cwt)
Concrete Mixture B (Nominal 60 MPa)		
Type III Cement	460 kg	775 lbs
Water	157 kg	265 lbs
Crushed Limestone, "3/8 chips"	1009 kg	1700 lbs
Fine Aggregate, "Dover Sand"	753 kg	1270 lbs
Water Reducer: Daratard 17 (W.R. Grace)	890 mL (1.9 mL/kg)	23 fl oz (3 oz/cwt)
High Range Water Reducer: Daracem 19 (W.R. Grace)	4180 mL (9.1 mL/kg)	108 fl oz (14 oz/cwt)

HRWR in conformance with ASTM C 143.[4] Slumps averaged 25 mm (1 in.) before and 200 mm (8 in.) after addition of HRWR. Concrete cylinders were cast as quickly as possible after the concrete batch had been prepared, and in conformance with ASTM C 192 except that the cylinders were consolidated using external vibration.

Curing and Storage of the Concrete Cylinders

Because the testing program was intended to emulate precast concrete production, the fresh concrete cylinders were transferred to a water bath with an elevated temperature maintained at 38 °C (100 °F). The cylinders were stored in the water bath for 24 hours. Both the "test" specimens and the "control" cylinders were subjected to elevated curing. Precautions were taken so that the water bath would not affect the fresh concrete. A submersible pump circulated continuously the hot water within the batch to sustain constant curing temperature for the concrete specimens.

After the initial 24 hr. period, the concrete cylinders were removed from the elevated curing tank and the plastic molds were stripped from the concrete. As quickly as possible, the "test" cylinders were placed in the loading frame and loaded with the correct level of compressive stress, and the "control" cylinders were stored adjacent to the loading frame so that the "control" cylinders would experience the same environmental conditions as the "test" cylinders.

One should note that one cylinder from Cast #4 and and one cylinder from Cast #7 failed in compression as a direct result of sustained loading to 0.80 f'_{ci}. In cast #7, the cylinder failed after only one day of sustained loading. In cast #4, one cylinder failed after 50 days of sustained loading. A typical compression failure, caused by sustained loading, is pictured in Fig. 3. The photograph clearly illustrates a failure from pure compression without apparent influence of eccentric or flexural loading. The failure of these two cylinders indicates that significant damage to concrete can be caused by sustained compressive stresses at 80 percent of the one day breaking strength. Conversely, if the "test" cylinders survived the sustained loading through the prescribed period, then little or no impairment in compressive strength is observed.

Figure 3. Compression Failure of Cylinder Under Sustained Load of $0.80 f'_{ci}$.

Table 5 contains the compressive strength data for "test" cylinders that were loaded at 28 days of age. These specimens were loaded to compressive stress levels corresponding to 60, 70 and 80 percent of their 28 day strengths (0.60 f'_c, 0.70 f'_c, and 0.80 f'_c). In both casts, cylinders loaded to 80 percent of their 28 day strengths failed as a direct result of sustained compressive stresses. Significant reductions in compressive strength were also exhibited by cylinders loaded at 28 days and subjected to sustained stress levels of 0.6 f'_c and 0.7 f'_c.

DISCUSSION OF TEST RESULTS

The data contained in Tables 3 and 4 indicate that, with a few exceptions, concrete breaking strengths are not reduced as a result of the sustained compressive stresses. This fact is supported by the nearly identical strengths of the "test" cylinders that were subjected to sustained loading when compared to the strengths of their companion "control" cylinders that remained unloaded throughout the test duration. However, for cylinders loaded to relatively large compressive stresses at 28 days of age, significant strength reductions were noted. The data presented in Table 5 clearly

The concrete release strengths (f'_{ci}) were established by the compressive strength measured from one set of three (3) one day old "control" cylinders. From the one day breaking strengths, the load levels corresponding to 0.6 f'_{ci}, 0.7 f'_{ci}, and 0.8 f'_{ci} were calculated. Concrete strengths and corresponding sustained compressive stresses are listed in Table 2.

For example, concrete from Cast No. 1 developed a 24 hr. compressive strength of 37.3 MPa (5410 psi). For the cylinders loaded to 0.60 f'_{ci}, the compressive stress was 0.60 times 37.3 MPa, or 22.4 MPa (3250 psi). For cylinders loaded to 0.70 f'_{ci}, the sustained compressive stress was 0.70 times 37.3 MPa, or 26.1 MPa (3790 psi). And finally for cylinders loaded to 0.80 times f'_{ci}, the sustained compressive stress was 0.80 times 37.3 MPa, or 29.8 MPa (4330 psi).

Table 2. Concrete Strengths and Sustained Compressive Stresses

Cast No.	Concrete Mixture (A or B)[c]	f'_{ci} (@ 1 d) Concrete Strengths (psi)[a]	f'_c (@ 28 d)	Age at Loading (days)	0.6 f'_{ci} Sustained Compressive Stress (psi)[b]	0.7 f'_{ci}	0.8 f'_{ci}
1	A	5410	7680	1	3250	3790	4330
2	A	5750	7750	1	3450	4030	4600
3	A	6720	8530	28	5120	5970	6820
4	B	6780	9200	1	4070	4750	5420
5	B	7210	8830	1	4330	5050	5770
6	B	6960	8130	28	4880	5690	6500
7	A	5400	6680	1	3240	3780	4320
8	B	7390	9070	1	4430	5170	5910
9	A	5790	7320	1	3470	4050	4630
10	A	4900	6430	1	2940	3430	3920
11	B	6720	8300	1	4030	4700	5380
12	A	6620	7550	1	3970	4630	5300

Notes:
a. Reported concrete strengths are the average of three cylinder breaks.
b. Sustained compressive stresses for Cast Nos. 3 and 6 were calculated from the 28 days breaking strength. All other casts used the one day breaking strength to calculate the sustained compressive stresses.
c. Average concrete strengths for Concrete "A:" f'_{ci} = 5800 psi (40.0 MPa); f'_c = 7420 psi (51.2 MPa). Average concrete strengths for Concrete "B:" f'_{ci} = 7010 psi (48.3 MPa); f'_c = 8710 psi (60.0 MPa).
d. 1000 psi = 6.895 MPa.

Tests for Compressive Strengths

Compressive stresses were maintained on the concrete test cylinders from one day of age (or 28 days of age for Cast Nos. 3 and 6) through the prescribed loading duration. At the conclusion of the prescribed sustained load duration, the "test" cylinders were removed from the loading frame and tested for breaking strength. On the same day, three (3) companion "control" cylinders were also tested for compressive strength. The compressive strengths from the "test" cylinders and the "control" cylinders were then compared. All concrete compressive strength tests were conducted in conformance with ASTM C 39.[5]

TEST RESULTS

Compressive Strength Test Results

Table 3 reports the compressive strengths of Cast Nos. 1, 2, 7, 9, 10 and 12, encompassing all data from concrete that was loaded at one day of age and made from Concrete Mixture "A." For example, compressive strengths were measured on concrete from Cast No. 1 at 7 days and 28 days of age. For concrete cylinders that had been subjected sustained compressive stresses of 0.60 f'_{ci}, the compressive strengths were measured as 47.4 MPa (6870 psi) at 7 days and 53.8 MPa (7800 psi) at 28 days. These strengths are approximately the same strength as the control cylinders, 47.4 MPa (6880 psi) at 7 days and 53.0 MPa (7680 psi) at 28 days, which had remained unloaded throughout the durational loading period.

For concrete from Cast No. 2, compressive strengths were measured on "test" cylinders after 90 days and 180 days of sustained compression loading. Concrete cylinders that had been stressed to 80 percent of f'_{ci} exhibited strengths of 56.6 MPa (8210 psi) at 90 days and 50.9 MPa (7380 psi) at 180 days. Again these strengths are approximately the same strength as the companion "control" cylinders with strengths of 56.5 MPa (8200 psi) at 90 days and 49.8 MPa (7220 psi) at 180 days.

The data contained in Table 3 consistently demonstrates that concrete compressive strengths were not diminished by the effects of sustained loading. In other words, all of the compressive strengths measured on the "test" cylinders matched, for practical purposes, the compressive strengths of the "control" cylinders.

Concrete strengths from Cast Nos. 4, 5 8 and 11 are reported in Table 4. Concrete strengths reported in Table 4 were made from Concrete Mixture "B" and were loaded at one day of age. Again, like results reported in Table 3, the compressive strengths of the "test" cylinders matched the compressive strengths of their companion "control" cylinders. This fact was observed for both Concrete Mixtures "A" and "B" and for all lengths of sustained loading. Additionally, Tables 3 and 4 also report the ratio of "test" cylinder strength to the strength of the "control" cylinders. In all cases, the ratio of "test" cylinder strength to "control" cylinder strength are 1.00 or larger with only a few exceptions.

Table 3. Compressive Strengths of Specimens Loaded at One Day of Age; Concrete Mixture "A"

Cast No.	Level of Sustained Compressive Stress (% f'_{ci})	Concrete Strengths (psi)[a] and Ratio of "Test" to "Control" Strengths				
		7 days	28 days	63 days	90 days	180 days
1	"Control"	6880 (1.00)	7680 (1.00)	N.A.	7580	7080
	60%	6870 (1.00)	7800 (1.02)		-	-
	70%	6980 (1.01)	7720 (1.01)		-	-
	80%	7260 (1.06)	8350 (1.09)		-	-
2	"Control"	7060	7750	N.A.	8200 (1.00)	7220 (1.00)
	60%	-	-		7810 (0.95)	7670 (1.06)
	70%	-	-		7740 (0.94)	7680 (1.06)
	80%	-	-		8210 (1.00)	7380 (1.02)
7	"Control"	6400	6680 (1.00)	N.A.	6510 (1.00)	N.A.
	60%	-	6820 (1.02)		6880 (1.06)	
	70%	-	6920 (1.04)		6970 (1.07)	
	80%	-	--[b]		--[b]	
9	"Control"	7240	7320 (1.00)	N.A.	7070 (1.00)	N.A.
	60%	-	7310 (1.00)		7390 (1.05)	
	70%	-	7400 (1.01)		7310 (1.03)	
	80%	-	7410 (1.01)		7030 (0.99)	
10	"Control"	6410	6430	6480 (1.00)	N.A.	N.A.
	60%	-	-	--[c]		
	70%	-	-	6940 (1.07)		
	80%	-	-	6605 (1.02)		
12	"Control"	7110	7550	7340 (1.00)	N.A.	N.A.
	60%	-	-	7450 (1.01)		
	70%	-	-	7520 (1.02)		
	80%	-	-	7530 (1.03)		

Notes:
a. Reported concrete strengths are the average of three cylinder tests.
b. One cylinder from Cast #7, loaded to 0.80 f'_{ci}, failed after 1 day of sustained loading and compressive strengths were not attained at 28 or 90 days.
c. No tests were performed at the 0.6 f'_{ci} stress level.
d. N.A. - Not applicable.
e. 1000 psi = 6.895 MPa.

Table 4. Compressive Strengths of Specimens Loaded at One Day of Age; Concrete Mixture "B"

Cast No.	Level of Sustained Compressive Stress (% f'_{ci})	Concrete Strengths (psi)[a] and Ratio of "Test" to "Control" Strengths				
		7 days	28 days	63 days	90 days	180 days
4	"Control" 60% 70% 80%	8100 - - -	9200 - - -	N.A.	8710 (1.00) 9000 (1.03) 8670 (1.00) --[b]	8640 (1.00) 8290 (0.96) 8200 (0.95) --[b]
5	"Control" 60% 70% 80%	8380 (1.00) 8500 (1.01) 8520 (1.02) 8560 (1.02)	8830 (1.00) 9300 (1.05) 8560 (0.97) 8430 (0.95)	N.A.	8700 - - -	8380 - - -
8	"Control" 60% 70% 80%	8560 - - -	9070 (1.00) 9270 (1.02) 9120 (1.01) 9090 (1.00)	N.A.	8530 (1.00) 8980 (1.05) 8680 (1.02) 8420 (0.99)	N.A.
11	"Control" 60% 70% 80%	8270 - - -	8300 - - -	8670 (1.00) 8610 (0.99) 8730 (1.01) 8810 (1.02)	N.A.	N.A.

Notes:
a. Reported concrete strengths are the average of three cylinder tests.
b. One cylinder from Cast #4, loaded to 0.80 f'_{ci}, failed after 50 days of sustained loading and compressive strength tests were not attained at 90 or 180 days.
c. N.A. - Not applicable.
d. 1000 psi = 6.895 MPa.

An environmental chamber was constructed to encapsulate the testing frame using plastic sheeting as a vapor barrier. Inside the plastic sheeting, relative humidity was maintained between 40 and 60 percent and temperature was controlled between 18 and 27 °C (65 and 80 °F). Both the "test" cylinders and "control" cylinders experienced identical storage conditions within the environmental chamber. Figure 1 pictures the test frame enclosed in the plastic sheeting.

Figure 1. Loading Frame Encased in Plastic Sheeting

Sustained Loading

The loading frame contained test apparatus for twelve (12) columns with six (6) test cylinders in each column. The twelve columns were organized into four (4) separate loading systems. Each of the four loading systems included an hydraulic pressure manifold, an hydraulic pressure indicator, an hydraulic accumulator, and three hydraulic cylinders. The

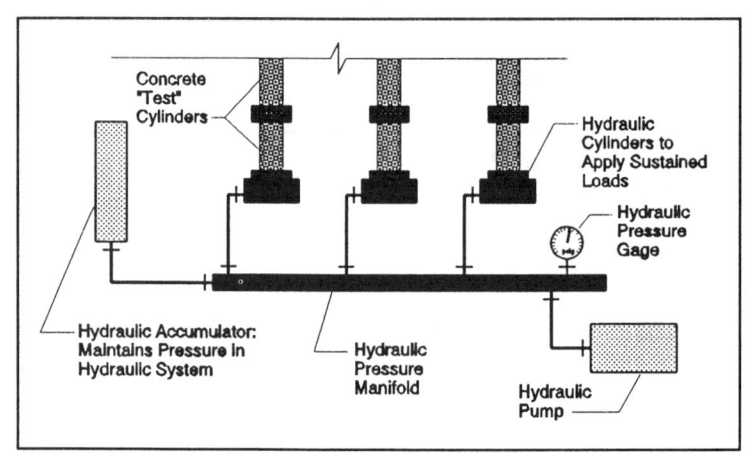

Figure 2. Schematic of Hydraulic System for Sustained Loads on Concrete "Test" Cylinders

hydraulic cylinders were made from steel tubing with inside diameters of 152, 165, and 178 mm (6, 6.5 and 7 in.) successively, thus providing the desired ratio of hydraulic area, and thus the hydraulic force, of 6:7:8. These ratios correspond to the loading requirements of 0.6 f'_{ci}, 0.7 f'_{ci}, and 0.8 f'_{ci} for each casting of concrete. Each 2 L (0.5 gal) accumulator functioned as a hydraulic "spring" which sustained hydraulic pressure on the concrete test cylinders while overcoming creep and shrinkage deformation of the cylinders. A photograph of one loading system is shown in Fig. 2.

demonstrates that sustained compressive stresses equal to and in excess of 60 percent of the 28 day breaking strength causes significant strength reductions in concrete.

In Fig. 4, concrete strengths measured at 28 days are normalized with respect to "control" cylinders. The bar heights represent the average strength of each set of three (3) concrete cylinders. The figure graphically illustrates the similarity in compressive strengths between the "control" cylinders and the "test" cylinders. Also shown are barred lines representing the 90 percent confidence intervals calculated from the strength data. The 90 percent confidence interval represents the 90 percent probability that the true average strength is contained within the interval, if an infinite set of data were available for testing. Overlapping confidence intervals designates two sets of data as statistically similar. In these cases, the concrete strengths of the "test" cylinders are essentially the same as the concrete strengths of the "control" cylinders.

Table 5. Concrete Strengths of Specimens Loaded at 28 Days of Age

Cast No.	Level of Sustained Compressive Stress (% f'_{ci})	Concrete Strengths (psi)[a] and Ratio of "Test" to "Control" Strengths			
		7 days	28 days	90 days	180 days
3	"Control"	7760	8530	8460	7580
	60%	-	-	7940	-[d]
	70%	-	-	6830	-[d]
	80%	-	-	7860[b]	-[d]
6	"Control"	8100	8130	8040	8010
	60%	-	-	8350	-[d]
	70%	-	-	7650	-[d]
	80%	-	-	8270[c]	-[d]

Notes:
a. Reported concrete strengths are the average of three cylinder tests.
b. One cylinder from Cast #3, loaded to 0.80 f'_c, failed after 10 days of sustained loading. Compression strength tests were performed on remaining cylinders at 53 days of age.
c. One cylinder from Cast #6, loaded to 0.80 f'_c, failed after 4 days of sustained loading. Compression strength tests were performed on remaining cylinders at 47 days of age.
d. Test specimens were damaged after only 90 days of testing and sustained load tests were discontinued.
e. 1000 psi = 6.895 MPa.

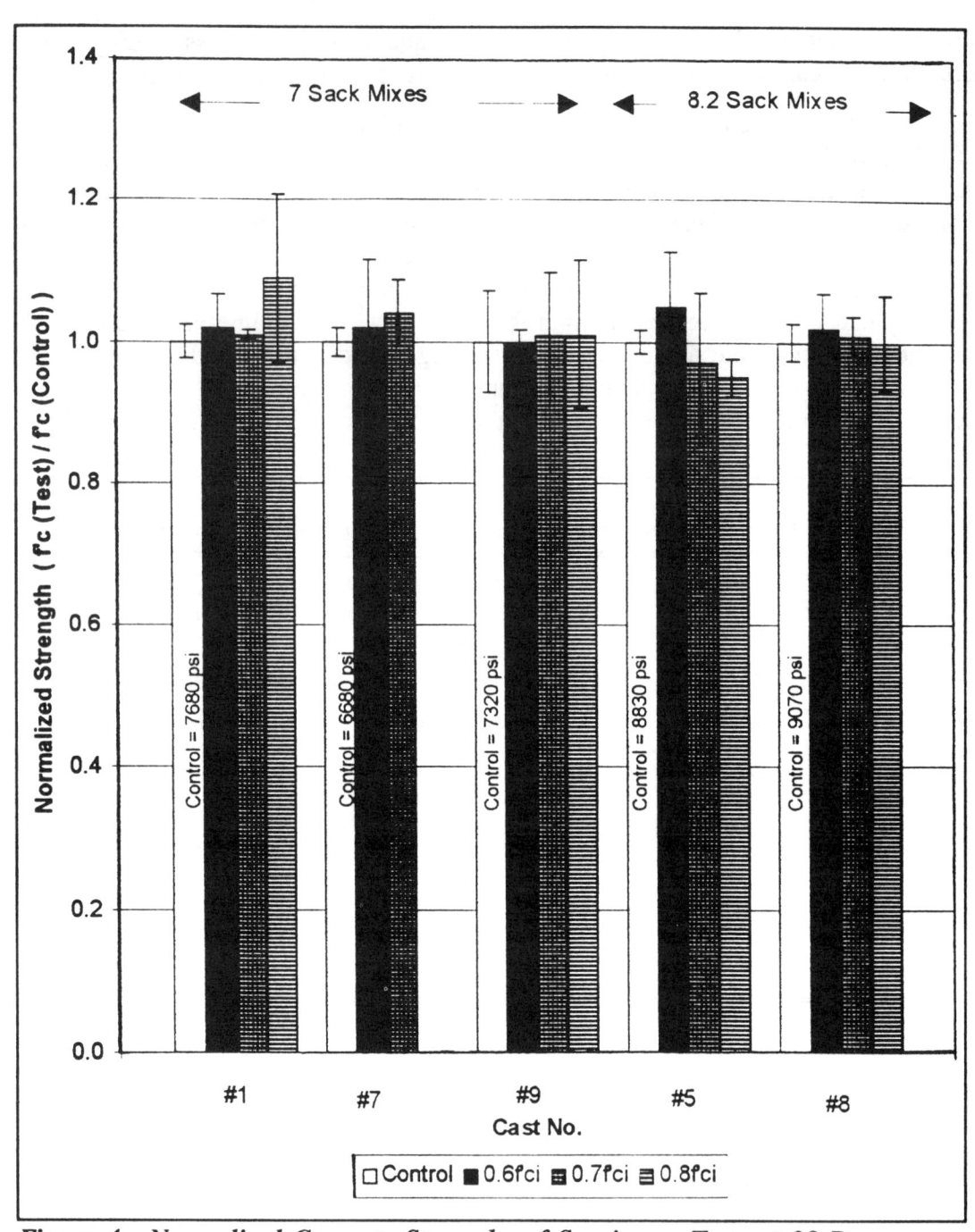

Figure 4. Normalized Concrete Strengths of Specimens Tests at 28 Days (1000 psi = 6.895 MPa).

CONCLUSIONS

1. When examining the data from concrete that was subjected to sustained loads equivalent to 0.60 f'_{ci} and 0.70 f'_{ci}, no evidence is found that suggests that the allowable compressive stress at release could not be raised to 0.70 f'_{ci}. In other words, the compressive strengths of the "test" cylinders that were subjected to sustained loading matched the compressive strengths of "control" cylinders that were not subjected to sustained loadings. The research signals a possibility of raising the current allowable stress limits in current building and bridge codes.

2. In two casts, concrete loaded at one day of age failed from the result of sustained compressive stresses with magnitudes equivalent to 0.80 f'_{ci}. Therefore, an increase in allowable stress at release to 0.80 f'_{ci} is not recommended.

3. For concrete loaded at 28 days of age, and loaded to stress levels that corresponds to 0.60 f'_c, 0.70 f'_c, and 0.80 f'_c, significant strength reductions and failures were noted. Therefore, these data do not support an increase in the allowable compressive stresses for sustained loads.

RECOMMENDATIONS

This research suggests the possibility that allowable compressive stress requirements at prestressed release can be relaxed to 0.7 f'_{ci} from 0.6 f'_{ci}. However, additional experimental research be is recommended before the allowable compressive stress limits are relaxed from the 1995 edition of the ACI Building Code. More comprehensive studies are required to develop the knowledge of allowable stress design and the impact of increasing the limit on allowable stress.

REFERENCES

1. ACI Committee 318, *Building Code Requirements and Commentary for Structural Concrete (ACI318-95 and 318-95R)*, Detroit, MI, 1995.

2. *Standard Specifications for Highway Bridges*, 15th Edition, American Association of State Highway and Transportation Officials, Washington, D.C., 1992.

3. "ASTM Standard Practice for Making and Curing Concrete Test Specimens in the Laboratory (C 192-90)," *Annual Book of ASTM Standards,* V. 04-02, Philadelphia, 1995, 116-122.

4. "ASTM Standard Test Method for Slump of Hydraulic Cement Concrete (C 143-90)," *Annual Book of ASTM Standards,* V. 04-02, Philadelphia, 1995, 88-90.

5. "ASTM Standard Test Method for Compressive Strength of Cylindrical Concrete Specimens (C 39-94)," *Annual Book of ASTM Standards,* V. 04-02, Philadelphia, 1995, 17-23.

6. Pang, Joo Pin, "Allowable Compressive Stresses for Prestressed Concrete," M.S. Thesis, Graduate College, The University of Oklahoma, May 1997.

INSTRUMENTATION AND MEASUREMENTS - BEHAVIOR OF LONG-SPAN PRESTRESSED HIGH PERFORMANCE CONCRETE BRIDGES

Ned H. Burns, Ph.D., P.E.
Zarrow Centennial Professor in Engineering
Department of Civil Engineering
The University of Texas at Austin
Austin, Texas, U.S.A.

Shawn P. Gross
Graduate Research Assistant / Ph.D. Candidate
Department of Civil Engineering
The University of Texas at Austin
Austin, Texas, U.S.A.

Kenneth A. Byle
Graduate Research Assistant / M.S.E. Candidate
Department of Civil Engineering
The University of Texas at Austin
Austin, Texas, U.S.A.

ABSTRACT

Two long-span high performance concrete (HPC) bridges in Texas have been provided with instrumentation to monitor concrete strains and temperatures. In these projects, the utilization of HPC with compressive strengths of up to 96.5 MPa (14,000 psi) allowed the design of spans as long as 47.9 m (157 ft.) using 1372 mm (54 in.) deep girders. This paper presents a brief discussion of the instrumentation plan along with results from measurements collected over the past two and a half years. Emphasis is placed on measurement of prestress loss at release and over time, as well as measurements of hydration temperatures and thermal gradients.

INTRODUCTION

The beneficial strength properties of high performance concrete (HPC), including high compressive strength and modulus of elasticity, are already having a significant impact on the design of prestressed concrete bridges. The modern engineer is able to accommodate higher prestress forces in the same girder cross-sections, allowing for more flexible and efficient designs. Longer span lengths, wider girder spacing, and thinner bridge decks are possible using these materials.

However, several questions must be answered to make such designs both safe and efficient. How does the use of these materials affect loss of prestress? What is the structural response of bridges with such long span lengths and or large girder spacing? Are there structural consequences to the high hydration temperatures generated by these materials? The current research discussed in this paper gives some insight into the answers of these important questions.

EXPERIMENTAL PROGRAM

Two HPC bridges in Texas are being studied as part of research projects sponsored jointly by the Federal Highway Administration (FHWA) and the Texas Department of Transportation (TxDOT). Embedded instrumentation has been installed in various components of each bridge during construction to monitor concrete strains and temperatures. Camber and deflection of the prestressed beams is also being monitored at several stages, and is discussed in detail in a separate paper by the authors[1]. More information about all aspects of the overall research projects is presented by Ralls[2].

Project Descriptions

Louetta Road Overpass (Houston, Texas) -- The Louetta Road Overpass on S.H. 249 in Houston was let in February 1994 and consists of two three-span highway structures. All components, including precast beams, precast piers, precast deck panels, and cast-in-place decks were constructed with HPC. The Texas U54 beam, a 1372 mm (54 in.) deep open-top U-shaped cross-section was used for all girders in the two structures. Girder design concrete strengths were as high as 90.3 MPa (13,100 psi) and most girders incorporated the use of 15 mm (0.6 in.) diameter prestressing strand. Span lengths range from 37.0 to 41.3 m (121.4 to 135.5 ft.) and girder spacing from 3.57 to 4.82 m (11.7 to 15.8 ft.). Construction was completed in May 1997.

North Concho River/U.S. 87/S.O.R.R. Overpass (San Angelo, Texas) -- The San Angelo HPC project, let in June 1995, consists of a 290 m (951 ft.) long, 8-span HPC bridge adjacent to a 292 m (958 ft.) long, 9-span bridge designed using normal concrete. Spans 1 through 5 of the Eastbound HPC bridge range from 39.9 to 47.9 m (131 to 157 ft.) in length and were designed using 1372 mm (54 in.) deep AASHTO Type IV girders. Due to the extremely high prestress forces required to accommodate these span lengths, a two-stage (pre-tension / post-tension) process was used by the fabricator. These HPC girders utilized 15 mm (0.6 in.) diameter strands and design concrete strengths up to 96.5 MPa (14,000 psi). The design and fabrication of these girders are discussed in detail by Powell[3] and Patton[4], respectively. The normal concrete Westbound bridge was also designed using AASHTO Type IV girders, with span lengths up to 42.7 m (140 ft.). Design concrete strengths of up to 61 MPa (8,900 psi) and 13 mm (0.5 in) diameter strands were used in these girders. Construction at the jobsite is ongoing and expected to be completed in late 1997.

Instrumentation Plan

Several girders from each HPC project were instrumented with embedded gauges during fabrication. Embedded gauges included vibrating wire gauges and bonded resistance strain gauges to measure concrete strain, and thermocouples to measure concrete temperatures. All gauges were scanned using one of several portable data acquisition systems constructed by the researchers. Measurements were recorded at several stages of construction and service, including release of prestress, storage at the precast plant, erection of the girders, erection of the precast deck panels, and casting of the bridge deck. Readings continue to be recorded several times per day in the completed bridges.

Gauges were typically placed at midspan of the girder, at six locations through the depth, as shown in the cross-sections in Figure 1. The locations correspond to 50 mm (2 in.) from the bottom surface, the c.g. (center of gravity) of pretensioned strands, the c.g. of the girder section, the c.g. of the composite section, the top of the web, and 50 mm (2 in.) from the top surface. In general, both strain and temperature measurements were recorded at these locations.

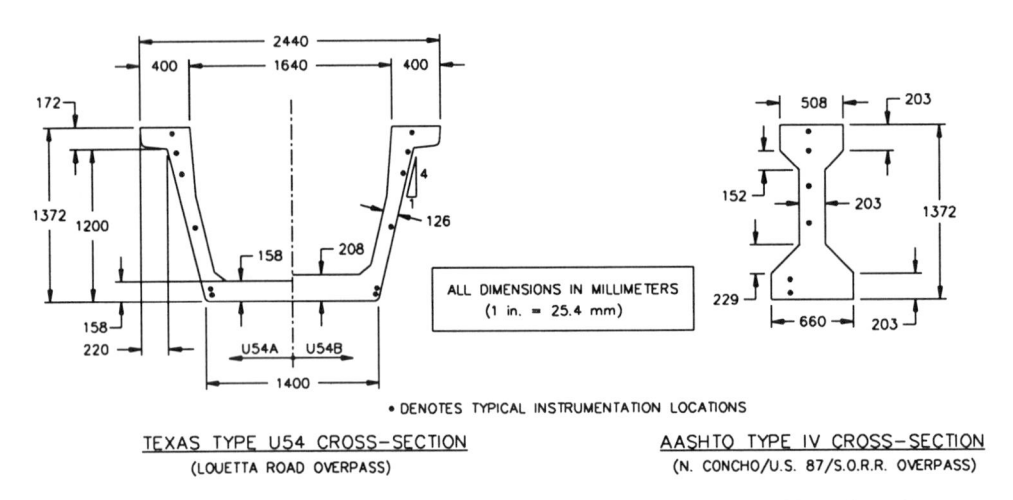

Figure 1. Beam cross-sections for the Louetta Road and N. Concho/U.S. 87/S.O.R.R. Overpasses

A total of twelve U54 girders were instrumented in the Louetta Road Overpass, six in the Northbound bridge and six in the Southbound bridge. Instrumented Louetta beams are designated in Figure 2. Fourteen AASHTO Type IV girders were instrumented in the San Angelo project, including ten HPC Eastbound beams. Since Span 1 of the two bridges were designed as comparison spans (same span length but different girder spacing), four normal concrete beams from Span 1 of the Westbound bridge were also instrumented. The instrumented beam layout for the San Angelo project is shown in Figure 3.

Girders were selected for instrumentation on the basis of span length, concrete strength, and location. Summaries of instrumented girder properties are presented in Tables 1 and 2 for the Louetta and San Angelo projects, respectively.

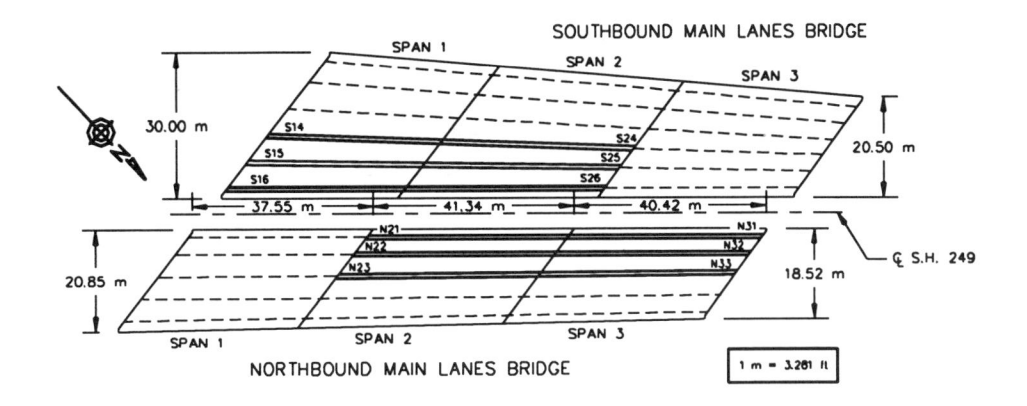

*Figure 2. Plan of the Louetta Road Overpass
(with instrumented beams labeled)*

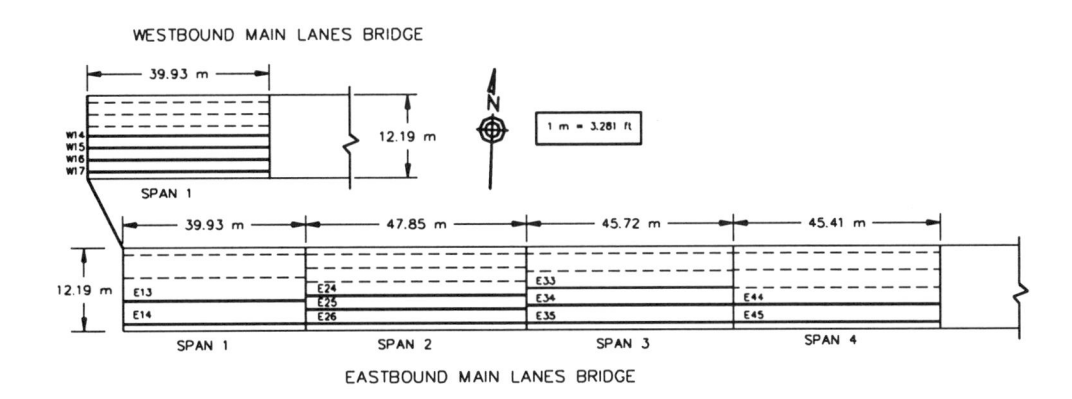

*Figure 3. Plan view of the N. Concho/U.S. 87/S.O.R.R. Overpass
(with instrumented beams labeled)*

Table 1. Summary of instrumented beams in the Louetta Road Overpass

Beam(s)	Casting Date	Section Type	Number of Strands	Design Length(s) (m)	Spacing (m)	Release Strength (MPa)	Design Strength (MPa)
N23	9-23-94	U54B	68	41.58	3.73	53.1	80.0
S16, N22	9-30-94	U54B	68	36.89/41.40	4.94/3.73	53.1	80.0
S26	10-7-94	U54B	87	41.18	4.30	60.7	90.3
N21, N31	10-28-94	U54B	87/83	41.23/40.74	3.73/3.54	60.7	90.3
S24, S25	11-10-94	U54B	68	40.16/40.66	4.30	53.1	80.0
N32, N33	2-15-96	U54A	64	40.90/41.06	3.54	53.1	80.0
S14, S15	2-26-96	U54A	64	35.94/36.41	4.94	53.1	80.0
(1 m = 3.281 ft; 1 MPa = 6.895 ksi)							

Table 2. Summary of instrumented beams in the N. Concho/U.S. 87 Overpass

Beam(s)	Casting Date	Section Type	Number of Strands	Design Length(s) (m)	Spacing (m)	Release Strength (MPa)	Design Strength (MPa)
W14-W17	3-7-96　3-12-96	Type IV	52[1]	39.30	1.73	40.5	55.2
E25	4-1-96	Type IV	72[2]	46.74	2.01	55.8	93.1
E13, E14	2-19-97	Type IV	76[3]	39.30	3.35	55.8	89.6
E24, E26	3-8-97	Type IV	76[3]	46.74	2.01	60.7	90.3
E33, E34	3-22-97	Type IV	76[3]	44.60	2.51	55.2	95.1
E35	3-29-97	Type IV	76[3]	44.60	2.51	55.2	95.1
E44, E45	4-12-97	Type IV	76[3]	44.40/44.48	2.51	55.2	94.5

(1 m = 3.281 ft; 1 MPa = 6.895 ksi)
1. Strand diameter was 13 mm for these beams.
2. Includes 50 straight pretensioned strands in the bottom flange, 6 straight pretensioned strands in the top flange, and 16 draped post-tensioned strands. Strand diameter was 15 mm for all strands in these beams.
3. Includes 50 straight pretensioned strands in the bottom flange, 6 straight pretensioned strands in the top flange, and 20 draped post-tensioned strands. Strand diameter was 15 mm for all strands in these beams.

Additional instrumentation was placed in selected precast deck panels corresponding to spans with instrumented beams. Embedded gauges were also placed in the cast-in-place decks of these spans prior to casting. These strain and temperature measurements will be of specific interest during live load testing and in determination of thermal gradients in the composite bridge.

MEASURED MATERIAL PROPERTIES

Concrete material properties, including compressive strength and modulus of elasticity, were measured on all HPC and selected normal strength components of the two projects as part of an extensive quality control/quality assurance program. An extensive discussion of the QC/QA program, including results, is presented by Carrasquillo and Myers[5]. Average measured strength and modulus values for the instrumented beams are summarized in Table 3. An extensive study of the creep and drying shrinkage properties of the Louetta beam mix was performed by Farrington[6]. Measured parameters from this mix are presented in Table 4. In general, creep and drying shrinkage were both observed to be substantially less than predicted by ACI 209[7] procedures. Creep and drying shrinkage were also observed to occur much faster than predicted by ACI 209, which is based on the work of Branson and Kripanarayanan[8], and others, using data from normal concrete mixes. Creep and shrinkage properties are also being investigated for all other beam, panel, and deck mixes. Those tests are ongoing and not yet complete.

The coefficient of thermal expansion is also being measured on all mixes, and is necessary for adequate prediction of thermal effects and temperature corrections to measured data. The coefficient of thermal expansion for the Louetta beam mix was measured to be $11.0 \times 10^{-6}/°C$ ($6.1 \times 10^{-6}/°F$).

Table 3. Summary of strength and modulus data for instrumented beams pours

	Louetta HPC U-Beams (Both Bridges)	San Angelo HPC Type IV Beams (Eastbound)	San Angelo Normal Type IV Beams (Westbound)
Number of Instrumented Beam Pours	7	6	2
Compressive Strength (MPa)			
Release	62.8[1]	69.4[2]	57.3[1]
28 days	88.7	93.7	70.3
56 days	98.3	96.9	76.8
Modulus of Elasticity (GPa)			
Release	41.1[1]	41.2[2]	40.7[1]
56 days	46.3	43.9	44.5

1 MPa = 145 psi; 1 GPa = 145 ksi

Notes:
1. Release typically 24 hours after casting.
2. Release typically 48 hours after casting.

Table 4. Creep and drying shrinkage properties for the HPC mix used for the Louetta U-beams

Creep[1]		Drying Shrinkage[2]	
28 days after loading		28 days after stripping	
Specific Creep (10^{-6}/MPa)	25		
Creep Coefficient	1.06	Shrinkage Strain	301×10^{-6}
180 days after loading		180 days after stripping	
Specific Creep (10^{-6}/MPa)	37		
Creep Coefficient	1.53	Shrinkage Strain	399×10^{-6}
Regression Curve Fit to Data	$C_{ct} = \dfrac{t^{0.6}}{7 + t^{0.6}}(1.81)$	Regression Curve Fit to Data	$\varepsilon_{sh} = \dfrac{t^{0.6}}{3 + t^{0.6}}(433 \times 10^{-6})$

Notes: Specimens are 100 mm x 600 mm cylinders cast from Louetta HPC Beam mix. All data corrected to "standard" conditions of 40% relative humidity and volume-to-surface ratio of 1.5 in accordance with ACI 209 recommendations.

1. Average of three specimens loaded at 48 hours to loads of 6.9, 24.7, and 34.5 MPa respectively.
2. Average of six unloaded specimens stripped approximately 24 hours after casting.

TEMPERATURE MEASUREMENTS

Hydration Temperatures

Temperatures were monitored during casting through the depth of the cross-section for several instrumented beams. A typical set of hydration curves are shown in Figure 4, with temperatures measured at six depths through the section. As hydration takes place, the top portions of the section get significantly hotter than the lower portions. The presence of this temperature gradient as the beam is forming may have a significant impact on beam camber. This is discussed in more detail in a separate paper by the authors[1]. Though the peak hydration temperature and magnitude of the gradient are generally a function of section geometry and material properties, all beams monitored during casting exhibited similar behavior.

One consequence of the high hydration temperatures in the beams is the presence of thermal cracking prior to release. On most pours, vertical cracks were observed at several locations along the length of the beams. Cracks extended through the width of the section, whether U54 or Type IV, from the top surface to within approximately 250 mm

(12 in.) of the bottom surface. These cracks were caused by tension induced by the restraint of the bed against shortening of the beam as it cools. Typically on the order of 0.2 to 0.8 mm (0.01 to 0.03 in.) in width at the top surface, all of these cracks closed upon release of prestress and are not likely to pose any long-term structural problem.

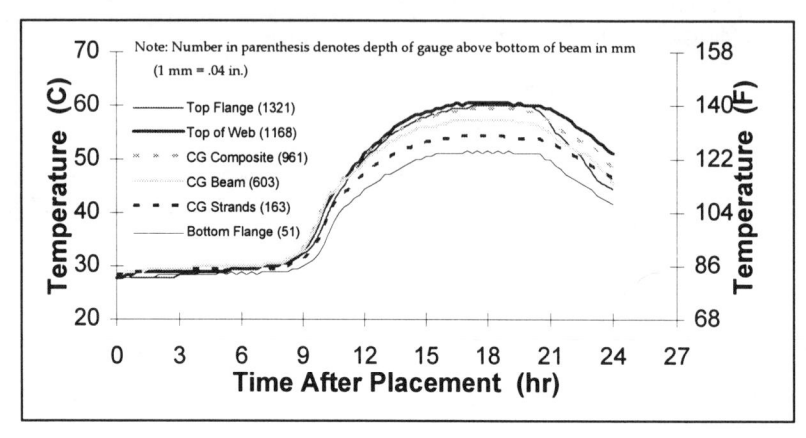

Figure 4. Typical hydration curve for the HPC beam mix

Temperature Gradients in the Completed Bridge

Temperatures in the composite bridge were measured to determine variations in temperature gradients throughout the day and the year, and to determine the maximum gradient that could be expected in the composite section. Temperature gradients, which are nonuniform vertical distributions of temperature, develop in the composite beam due to uneven heating and cooling. Heat energy is provided to the bridge superstructure by means of solar radiation. Additional heat may be gained or lost due to convection to or from the surrounding atmosphere. The amount of temperature change created by these heat sources depends upon wind speed, ambient temperature, relative humidity, weather conditions (clear or cloudy), material properties of the bridge, surface characteristics, the time of day, and the time of year[9].

The large nonlinear temperature gradients that were measured in exterior and interior composite U-beam sections on March 10, 1997 are shown in Figure 5. Temperature distributions early in the morning were uniform or slightly negative with the deck cooler than the bottom flange. By mid-afternoon, the temperature in the deck above both beams had increased significantly, while most of the beam remained much cooler. In addition, the shape of the afternoon temperature gradients of the two beams differed because the web of the exterior beam was exposed to the sun. Temperatures in the web of the interior beam, which is completely shaded, were lower and more uniform than in the exterior beam. Similar temperature gradients were measured in composite Type IV beams in the North Concho River Overpass.

Measured temperature gradients for both types of composite sections exhibit similar magnitudes and shapes. Between November of 1996 and April of 1997, the largest temperature gradient observed in either bridge was 12 °C (22 °F) and the maximum increase in deck temperature throughout a single day was 21 °C (37 °F). During that time period the number of freeze-thaw cycles observed in the Louetta Road

Overpass and the N. Concho River/U.S. 87/S.O.R.R. Overpass were 5 and 13, respectively.

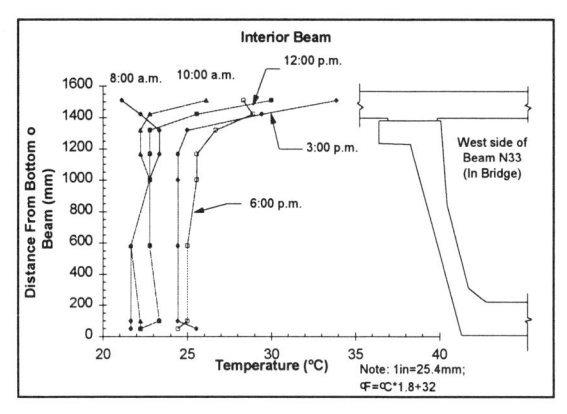

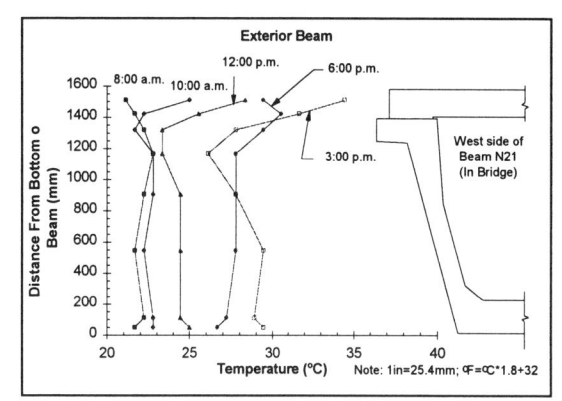

Figure 5. Temperature gradients composite U- beams on March 10, 1997

PRESTRESS LOSSES

Losses Prior to Release of Prestress

Relaxation, temperature changes, and concrete drying shrinkage contribute to prestress losses prior to release. In an attempt to estimate the magnitude of these losses, load cells were placed on selected strands at the bulkhead of the prestressing bed for several of the San Angelo HPC beam pours. Only the loss prior to casting could be measured directly; losses after placement were estimated using an analytical procedure that considered the measured change in force at the bulkhead, strand temperatures, and lengths of free and bonded strand.

Measured losses before casting ranged from 1 to 3 percent, depending on the difference in stressing and placement temperatures. Total losses prior to release were estimated between 4 and 6 percent. A large component of the total loss was due to the high heat of hydration observed in all beams. Total losses were also significantly influenced by the fact that many beams remained on the bed for two days after casting prior to release, resulting in substantial loss due to drying shrinkage. In general, losses before release were slightly higher than expected.

Measured Prestress Losses After Release

Prestress losses during and after release were measured using the strain compatibility principle. Embedded gauges measured changes in concrete strain at the c.g. of the prestressing strands, which is assumed to be equal to the changes in strand strain (at the c.g. of the strands). Multiplying by the modulus of elasticity of the strand gives a measured stress change, or prestress loss. Loss due to strand relaxation must be added analytically since it does not cause changes in strain. Losses before the baseline strain measurement, which was always recorded just before release, must also be added to determine the total loss at any time. Measured losses are shown graphically for all Louetta and San Angelo beams in Figure 6.

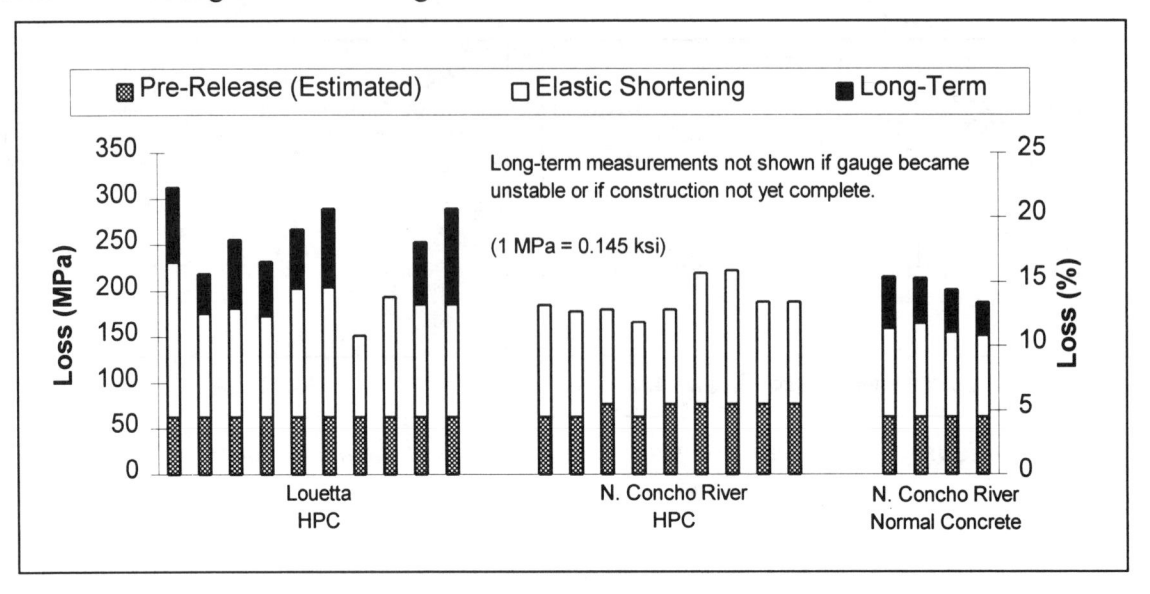

Figure 6. Measured prestress losses

Elastic Shortening Loss at Release

Elastic shortening losses were determined for each beam using the measured strain immediately after release. These measurements were taken while the beams were still on the prestressing bed, prior to their removal to storage. Measured and predicted elastic shortening losses are compared in Table 5. Predicted losses are computed based on known material and transformed section properties, using a standard superposition of stresses method. Since the procedure is recursive (the computed change in stress is a function of the assumed prestress force), several iterations were performed until convergence was observed.

Table 5. Comparison of measured and predicted elastic shortening loss

	Measured ES Loss % of 0.75 f_{pu}	Difference (Measured - Predicted) % of 0.75 f_{pu}	Ratio (Measured / Predicted)
Louetta HPC U-Beams			
Mean	9.02	+0.89	1.11
Range	6.36 to 12.10	-0.74 to +3.66	0.90 to 1.43
Standard Deviation	1.64	1.23	0.15
San Angelo Normal Type IV Beams			
Mean	6.78	+1.45	1.27
Range	6.36 to 7.27	+1.03 to +1.94	1.19 to 1.36
Standard Deviation	0.40	0.40	0.07
San Angelo HPC Type IV Beams			
Mean	8.63	+1.93	1.29
Range	7.37 to 10.44	+0.64 to +3.84	1.08 to 1.58
Standard Deviation	1.27	1.28	0.20
1 MPa = 0.145 ksi			

Table 6. Comparison of measured and predicted long-term (total) prestress losses

	Measured Long-Term Loss % of 0.75 f_{pu}	Difference (Measured - Predicted) % of 0.75 f_{pu}	Ratio (Measured / Predicted)
Louetta HPC U-Beams			
Mean	19.00	-1.03	0.95
Range	15.70 to 22.35	-3.63 to +2.76	0.81 to 1.14
Standard Deviation	2.28	2.08	0.10
San Angelo Normal Type IV Beams			
Mean	14.67	-0.60	0.96
Range	13.44 to 15.45	-1.69 to +0.14	0.89 to 1.01
Standard Deviation	0.94	0.86	0.06
1 MPa = 0.145 ksi			
Note: Long-term losses are not yet available for the San Angelo HPC Type IV Beams			

Measured elastic shortening losses were higher than predicted for all except two beams. A possible source of these higher losses is a lower than expected modulus of elasticity. However, this would be inconsistent with other behavior at release, including low observed camber[1]. A more likely cause is the restraint provided by the prestressing bed against shortening of the member (due to thermal cooling and drying shrinkage) prior to release. The presence and orientation of several cracks, as described previously, implies that this restraint was significant at the lower levels of the beam. Recall that the baseline strain readings were taken immediately prior to release, when this restraint would still be present. All subsequent readings would include the effect of the release of this restraint, adding an apparent compressive strain to the readings. Unfortunately, this apparent strain cannot be directly measured, and therefore cannot be corrected for.

Typical values for measured elastic shortening losses were in the range of 6 to 12 percent. Elastic shortening losses were also significantly higher for HPC beams than for the normal concrete beams. This difference was expected since the HPC beams utilized a higher prestress force at release. Recall that the modulus of elasticity for the normal concrete mix was comparable to that of the HPC mixes.

Long-Term Prestress Losses

Measurement of long-term prestress losses is possible through the use of embedded vibrating wire strain gauges, which are durable and typically remain stable for several years. Measurements of this type have been made for several months after casting of the decks in the Louetta and normal concrete San Angelo bridges. Measurements are summarized and compared to predicted values of long-term losses in Table 6. Predictions were calculated with a time-step procedure using a personal computer spreadsheet program developed by the authors. The known material properties, construction sequence, and transformed section properties were used in these analyses.

Measured and predicted long-term losses agreed reasonably well, with measured losses typically slightly lower than predicted. Since almost all elastic shortening losses were higher than predicted, the prediction models must have overestimated the net time-dependent losses after release. It is likely that the differences come largely from errors in estimating the material properties of the members, including modulus of elasticity and creep and drying shrinkage parameters.

Total losses in the completed structures ranged from 13 to 22 percent. Total losses were higher in the Louetta HPC bridge than in the San Angelo normal strength bridge, which is consistent with the higher measured elastic shortening loss discussed earlier. Losses in the Louetta HPC bridge, however, are clearly not excessive, and are typical of the general range of total losses found in most highway bridge structures.

SUMMARY

The following observations have been made in this research program:

1. High hydration temperatures developed in the monitored beams. Significant cracking occurred prior to release because the prestressing bed prevented the beams from shortening freely as they cooled from these high temperatures of hydration.
2. Nonlinear temperature gradients as large as 12°C (22°F) were observed in interior U- and I-beams of the composite bridges during the winter months of 1997.
3. Prestress losses before casting were observed to be in the range of 1 to 3 percent. Total losses prior to release are estimated to be in the range of 4 to 6 percent.
4. Measured elastic shortening losses were significantly higher than predicted. However, it is probable that the difference is caused by an unavoidable measurement error due to restraint from the prestressing bed.
5. Long-term prestress losses were measured to be in the range of 13 to 22 percent, but were significantly lower for the normal concrete Type IV beams than for the HPC U54 beams. Measured long-term losses for all beams were reasonably close to predicted values.

ACKNOWLEDGMENTS

The authors wish to thank the joint sponsors of these research projects, The Federal Highway Administration and The Texas Department of Transportation for their support and encouragement. In addition, the authors would like to thank the fabricators and contractors (Williams Brothers Construction Co., Jascon Inc., Texas Concrete Company, and Bexar Concrete Works) involved in the many phases of these projects for their assistance, patience, and interest.

REFERENCES

1. Gross, S. P., Byle, K. A., and Burns, N. H., "Deformation Behavior of Long-Span Prestressed High Performance Concrete Bridge Girders," PCI/FHWA International Symposium on High Performance Concrete, Oct. 20-22, 1997.

2. Ralls, M. L., "High Performance Concrete Bridge Construction in Texas," PCI/FHWA International Symposium on High Performance Concrete, Oct. 20-22, 1997.

3. Powell, L. C., "Design and Analysis of Pretensioned/Post-Tensioned Long-Span High Performance Concrete I-Beams," PCI/FHWA International Symposium on High Performance Concrete, Oct. 20-22, 1997.

4. Patton, B., "A Fabricator's Challenges," PCI/FHWA International Symposium on High Performance Concrete, Oct. 20-22, 1997.

5. Carrasquillo, R. L., and Myers, J. J., "Quality Control and Quality Assurance Program for Precast Plant Produced High Performance Concrete U-Beams," PCI/FHWA International Symposium on High Performance Concrete, Oct. 20-22, 1997.

6. Farrington, E. W., Burns, N. H., and Carrasquillo, R. L., "Creep and Shrinkage of High Performance Concrete," Research Report 580-5 (Preliminary Review Copy), Center for Transportation Research, The University of Texas at Austin, February 1996, 92 pp.

7. ACI Committee 209, "Prediction of Creep, Shrinkage, and Temperature Effects in Concrete Structures," *Designing for Creep and Shrinkage in Concrete Structures*, SP-76, American Concrete Institute, Detroit, 1982, pp. 193-300.

8. Branson, D. E. and Kripanarayanan, K. M., "Loss of Prestress, Camber and Deflection of Non-composite and Composite Prestressed Concrete Structures," PCI JOURNAL, V. 16, No. 5, Sept.-Oct. 1971, pp. 22-52.

9. Radolli, M., and Green, R., "Thermal Stresses in Concrete Bridge Superstructures Under Summer Conditions," *Transportation Research Board*, Transportation Research Record, No. 547, 1975, pp. 23-36.

TREATMENT OF HIGH-STRENGTH CONCRETE IN U. S. CODES

S. K. Ghosh
Director, Engineering Services, Codes and Standards
Portland Cement Association
Skokie, Illinois, U.S.A.

ABSTRACT

The latest edition of ACI 318 *Building Code Requirements for Structural Concrete* contains several provisions related specifically to high-strength concrete, that mostly reflect insufficient knowledge concerning the shear (diagonal tension) strength of concrete and the bond between reinforcing steel and concrete in the ranges of very high concrete strength. This paper details and discusses these provisions. It then explores areas where changes in ACI 318 provisions may be needed to ensure continued satisfactory usage of high-strength concrete in non-seismic applications. Seismic applications are essentially outside the scope of this paper. The treatment of high-strength concrete in AASHTO Standard Specifications and LRFD Bridge Design Specifications are also briefly discussed.

DEFINITION OF HIGH-STRENGTH CONCRETE

The definition of high-strength concrete has changed over the years, and should not be considered static. The precise strength defining high-strength concrete also tends to vary by geographic location. In a 1984 American Concrete Institute (ACI) committee report, revised and reissued in 1992[1], 6,000 psi (41 MPa) was selected as a lower limit for high-strength concrete. According to that report, although 6,000 psi (41 MPa) was selected as the lower limit, it was not intended to imply that there is a drastic change in material properties or in production techniques that occur at this compressive strength. In reality, all changes that take place above 6,000 psi (41 MPa) represent a process which starts with the lower-strength concretes and continues into high-strength concretes.

Concrete with a strength in excess of 6,000 psi (41 MPa) is more sensitive to quality control procedures than lower-strength concrete. Close attention to each facet of concrete production becomes more important.

UPPER LIMIT ON THE STRENGTH OF CONCRETE

Neither ACI 318-95[2] nor any of the three model codes in the United States (the Uniform Building Code[3], the BOCA/National Building Code[4], and the Standard Building Code[5]) imposes an upper limit on the strength of normal-weight concrete that can be used in construction, even in regions of high seismicity. Only the City of Los Angeles has so far informally imposed a limit of 6,000 psi (41 MPa) on the specified compressive strength (f'_c) of concrete used in special (meaning specially detailed) moment frames.

ACI 318-95 requires that the specified compressive strength of lightweight-aggregate concrete used in the design of members of special moment frames in any seismic zone[3] or seismic performance category [4-5], and of members of seismic resisting systems in seismic zones 3, 4[3] or seismic performance categories D and E, [4-5] shall not exceed 4,000 psi (28 MPa). Lightweight-aggregate concrete with higher design compressive strength may be used if demonstrated by experimental evidence that structural members made with that lightweight-aggregate concrete provide strength and toughness equal to or exceeding those of comparable members made with normal weight aggregate concrete of the same strength. The Uniform Building Code imposes an absolute upper limit of 6,000 psi (41 MPa) on the specified compressive strength of lightweight concrete.

MODULUS OF ELASTICITY

According to ACI 318-95 Section 8.5.1, the modulus of elasticity E_c for concrete may be taken as $w_c^{1.5} 33\sqrt{f'_c}$ (in psi) or $w_c^{1.5} 0.043\sqrt{f'_c}$ (in MPa) for values of w_c (unit weight of concrete) between 90 and 155 pounds per cubic foot or 1500 and 2500 kg/m^3. This formula does not make any distinction between high-strength and lower-strength concretes.

In view of test results from Cornell University[6] indicating that the moduli of elasticity of concretes with very high levels of strength might be lower than values given by the ACI 318 provision, the following formula proposed by Cornell researchers was endorsed by ACI Committee 363[1]:

$$E_c = 40,000\sqrt{f'_c} + 1,000,000 \text{ psi} \qquad (1)$$
$$\text{for } 3,000 \text{ psi} < f'_c < 12,000 \text{ psi}$$
$$(E_c = 3320\sqrt{f'_c} + 6900 \text{ MPa for } 21 \text{ MPa} < f'_c < 83 \text{ MPa})$$

It should be noted that ACI 318-95 has not adopted the ACI 363 formula. Recent evidence[7] suggests that this was probably a prudent decision.

MODULUS OF RUPTURE

According to ACI 318-95 Section 9.5.2.3, the modulus of rupture of normal weight concrete is:

$$f_r = 7.5\sqrt{f'_c} \qquad (2)$$

$$(f_r = 0.7\sqrt{f'_c})$$

Again, this formula does not make any distinction between high-strength and normal-strength concretes, although there are indications from Cornell research[8] that the constant relating f_r and f'_c might be higher for concretes with very high-strength levels.

LONG-TERM DEFLECTION MULTIPLIER

ACI 318-95 Section 9.5.2.5 requires that unless values are obtained by a more comprehensive analysis, additional long-term deflection resulting from creep and shrinkage of flexural members (normal-weight or lightweight concrete) shall be determined by multiplying the immediate deflection, caused by the sustained load considered, by the factor:

$$\lambda = \frac{T}{1 + 50\rho'} \qquad (3)$$

where the compression reinforcement ratio ρ' shall be the value at midspan for simple and continuous spans, and at support for cantilevers. The time-dependent factor T for sustained loads may be taken equal to:

2.0	for a loading duration of	5 years or more
1.4	for a loading duration of	12 months
1.2	for a loading duration of	6 months
1.0	for a loading duration of	3 months

Shrinkage values for normal-strength and high-strength concretes are basically comparable. Creep per unit stress (specific creep), however, decreases significantly as concrete strength increases.[1] This fact is not reflected in the long-term deflection multiplier (T) given by the Code. It is believed that the multiplier might be in the right range for high-strength concrete members, and might be unconservative for moderate-to low-strength concrete members.

STRENGTH DESIGN OF MEMBERS FOR FLEXURE AND AXIAL LOADS

The strength design of members subject to flexure and axial loads is usually based on the equivalent rectangular concrete stress distribution defined in ACI 318. According to Section 10.2.7.1, a concrete stress of $0.85 f'_c$ shall be assumed uniformly distributed over an equivalent compression zone bounded by edges of the cross section and a straight line located parallel to the neutral axis at a distance $a = \beta_1 c$ from the fiber of maximum compressive strain (Figure 1). Factor β_1 shall be taken as 0.85 for concrete strengths f'_c up to and including 4,000 psi (28 MPa). For strengths above 4,000 psi (28 MPa), β_1 shall be reduced continuously at a rate of 0.05 for each 1,000 psi (7 MPa) of strength in excess of 4,000 psi (28 MPa), but β_1 shall not be taken less than 0.65.

The applicability of the rectangular stress block defined above for the strength computation of high-strength members subject to flexure with or without axial loads has often been questioned. ACI 318-95 does not specify an upper concrete strength limit beyond which the rectangular stress block becomes inapplicable. This is justified for

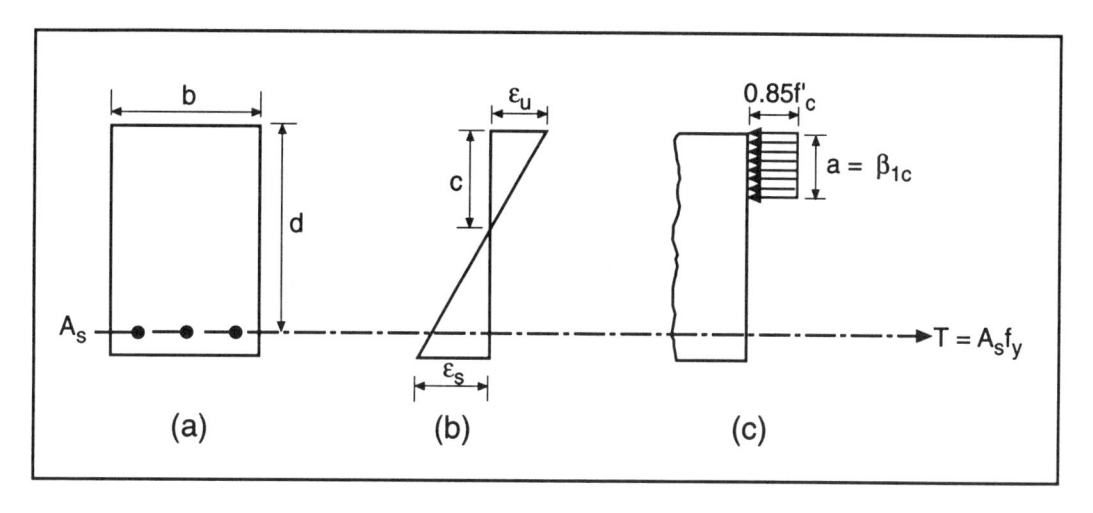

Figure 1. Equivalent Rectangular Stress Block of ACI 318[2]

flexural members not subject to any axial compression. That is because the ACI 318 standard generally requires flexural members to be underreinforced, which translates into shallow neutral axis depths and small concrete compression zones. The behavior of underreinforced concrete flexural members is governed almost entirely by the tension reinforcement. What shape is assumed for the compression concrete stress block becomes largely immaterial.

The ACI rectangular stress block yields acceptable strength prediction in flexure and axial compression as long as $f'_c < 8,000$ psi (55 MPa). This can be seen in Fig. 2 from Ref. 9, which compares 93 tests of eccentrically loaded columns to the strengths predicted by the ACI Code. Because some of the tests are for unreinforced concrete, and others for reinforced, only the strength contributed by concrete has been compared, the steel contribution having been subtracted from both the test and predicted values. In this figure the comparisons are made in terms of ∂ = (Test strength/calculated strength) - 1.0.

For $f'_c > 8,000$ psi (55 MPa), in the absence of substantive enhancement in the current confinement requirements, accurate strength prediction for members subject to combined bending and axial compression would appear to require an adjustment of the ACI rectangular stress block. The importance of confinement is illustrated in Fig. 3 from Ref. 10, which shows the relationship between the parameter $\rho_s f_{yt} / f'_c$ and the ratio of experimentally obtained strength of 111 high-strength concrete columns to that predicted by the ACI Code. ρ_s is the volumetric ratio of transverse reinforcement and f_{yt} is the yield strength of transverse reinforcement. It is clear that columns with a low volumetric ratio of transverse reinforcement may not achieve their strength as predicted by the ACI Code; however, well confined columns can develop strength well in excess of that predicted by the Code. It should be noted that excess strength of columns with relatively high amounts of transverse reinforcement is generally obtained after spalling of cover concrete. This strength enhancement comes as a result of an increase in strength of the confined core concrete.

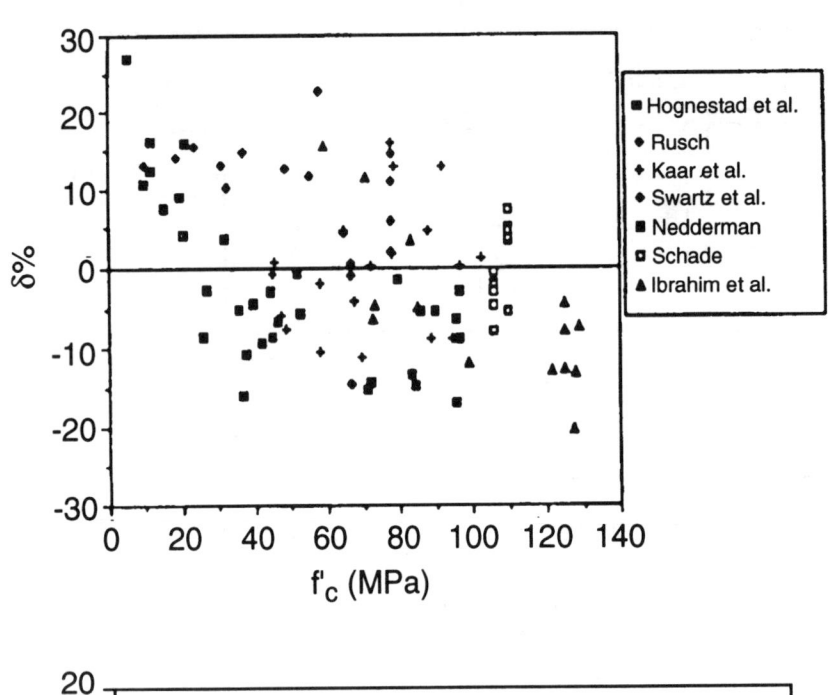

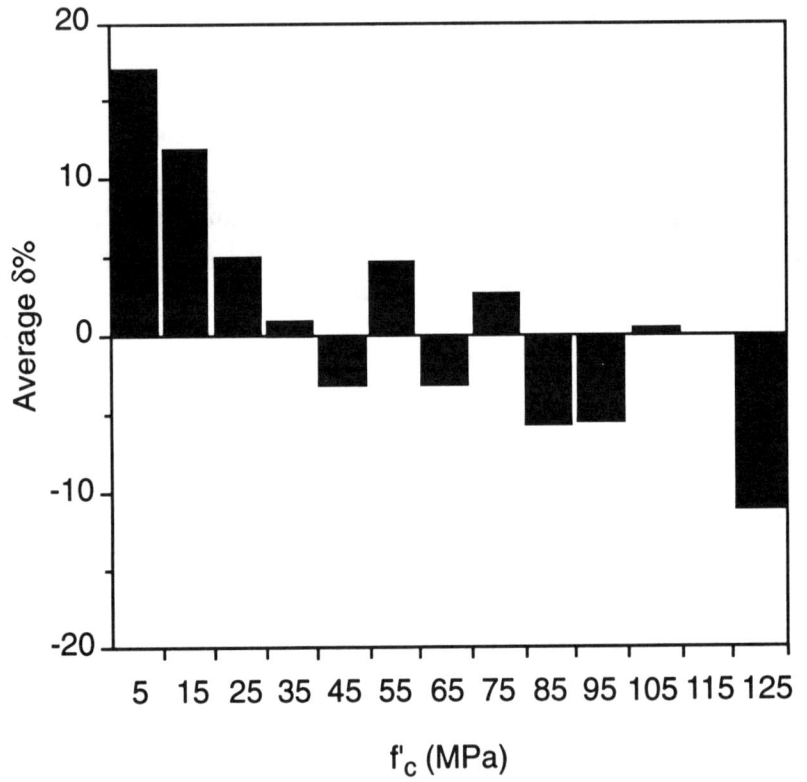

Figure 2. (a) Comparison of Strength from Tests of Eccentrically Loaded Columns with that Given by ACI 318[2], (b) mean values of δ versus Concrete Strength.

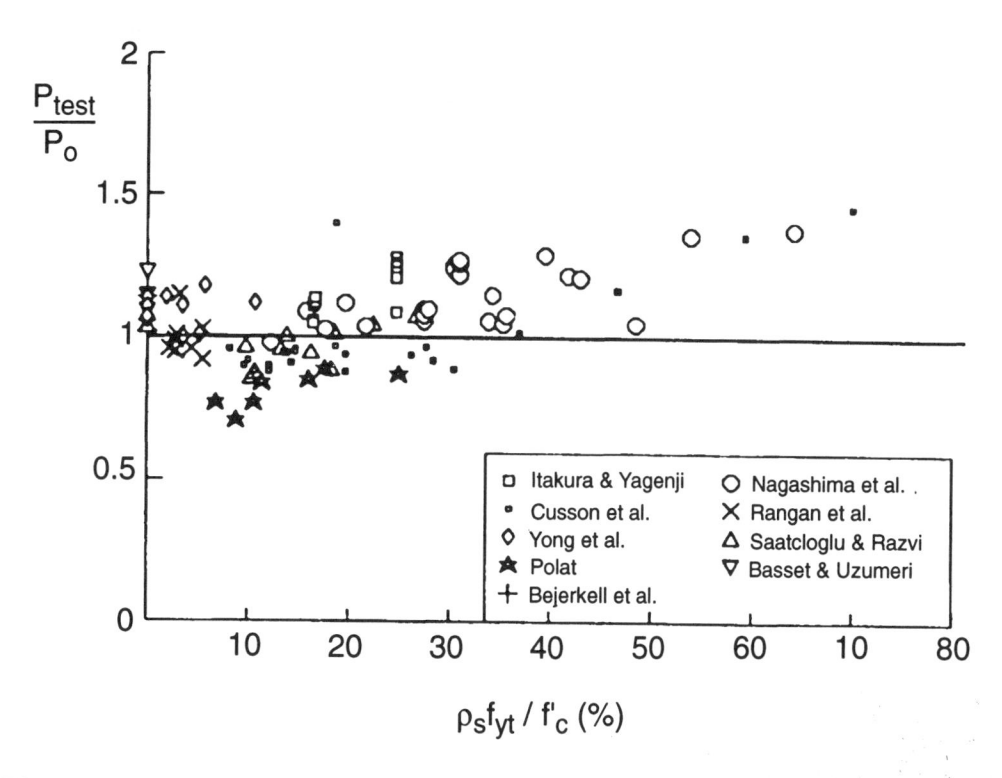

Figure 3. Comparisons of Experimental and Analytical Concentric Tests of Columns

The latest New Zealand Concrete Design Standard[11] has adopted a specific adjustment of the ACI rectangular stress block for high-strength concrete. The latest Canadian Standards Association Standard[12] has adopted a different adjustment. A slight variation of the Canadian adjustment has been proposed for adoption in ACI 318. Other modifications of the ACI rectangular stress block have also been proposed.[13] These different options need to be evaluated, and a proper assessment made. As an alternative to the use of a modified stress block, accurate strength prediction should also be obtainable through sophisticated analysis using the confined core only, ignoring the cover concrete.

MINIMUM REINFORCEMENT FOR FLEXURAL MEMBERS

The provisions for a minimum amount of reinforcement are meant to apply to flexural members that, for architectural or other reasons, are much larger in cross section than required for strength. With a very small amount of tensile reinforcement, the computed moment strength as a reinforced concrete section using cracked section analysis becomes less than that of the corresponding unreinforced concrete section computed from its modulus of rupture. Failure in such a case can be sudden.

To prevent such failure, a minimum amount of tensile flexural reinforcement is required, and should be provided in both positive and negative moment regions.

ACI 318-89 (revised 1992) Section 10.5.1 required that at any section of a flexural member, except as provided in Sections 10.5.2 and 10.5.3, where positive reinforcement is required, the ρ ratio provided must not be less than:

$$\rho_{min} = 200/f_y \tag{4}$$

$$(\rho_{min} = 1.4/f_y)$$

In T-beams and joists where the web is in tension, the ratio ρ was to be computed for this purpose using the width of the web.

The term "positive reinforcement" in the above provision was not clear but probably meant tension reinforcement.

The $200/f_y (1.4/f_y)$ value was originally derived to provide the same 0.5 percent minimum (for mild grade steel) required in older editions of the ACI Code. Indications were that, when concrete strength higher than about 5,000 psi (34 MPa) is used, the $200/f_y (1.4/f_y)$ value may not be sufficient. The minimum amount of flexural reinforcement was therefore changed in ACI 318-95 which requires that at every section of a flexural member where tensile reinforcement is required by analysis, except as provided in 10.5.2, 10.5.3 and 10.5.4, the area A_s provided shall not be less than that given by

$$A_{s,min} = \frac{3\sqrt{f'_c}}{f_y} b_w d \tag{5}$$

$$(A_{s,min} = \frac{\sqrt{f'_c}}{4f_y} b_w d)$$

and not less than $200 \, b_w d/f_y \, (1.4 b_w d/f_y)$.

Section 10.5.2 requires that for a statistically determinate T-section with the flange in tension, the area $A_{s,min}$ shall be equal to or greater than the smaller value given either by

$$A_{s,min} = \frac{6\sqrt{f'_c}}{f_y} b_w d \tag{6}$$

$$(A_{s,min} = \frac{\sqrt{f'_c}}{2f_y} b_w d)$$

or Eq. (5) with b_w set equal to the width of the flange.

Section 10.5.3 (previously 10.5.2) allows that the requirements of 10.5.1 and 10.5.2 need not apply if at every section the area of tensile reinforcement provided is at least one-third greater than that required by analysis.

Section 10.5.4 (previously 10.5.3) provides that for structural slabs and footings of uniform thickness the minimum area of tensile reinforcement in the direction of the span shall be the minimum shrinkage and temperature reinforcement required by 7.12. Maximum spacing of this reinforcement shall not exceed three times the thickness nor 18 in. (457 mm).

SHEAR STRENGTH

ACI 318-95 Chapter 11 on Shear and Torsion restricts the values of $\sqrt{f'_c}$ to no more than 100 psi (25/3 MPa), meaning that the contribution of concrete to the shear or torsional strength of a structural member will not increase any further, once the specified compression strength of concrete goes above 10,000 psi (69 MPa). There is an important exception to this new restriction, however.

Values of $\sqrt{f'_c}$ greater than 100 psi (25/3 MPa) are permitted in computing V_c (nominal shear strength provided by concrete), V_{ci} (nominal shear strength provided by concrete when diagonal cracking results from combined shear and moment), and V_{cw} (nominal shear strength provided by concrete when diagonal cracking results from excessive principal stress in web) for reinforced or prestressed concrete beams and concrete joist construction having minimum web reinforcement equal to $f'_c / 5,000$ ($f'_c / 35$) times, but not more than three times, the amounts required by Section 11.5.5.3, 11.5.5.4 or 11.6.5.2.

Section 11.5.5.3 requires a minimum amount of shear reinforcement in reinforced concrete beams, capable of carrying at least 50 psi (0.35 MPa) of nominal shear stress. Figure 4 illustrates that the amount of shear reinforcement has to be twice as large as soon as the specified compressive strength exceeds 10,000 psi (69 MPa), if the designer is to benefit from the shear strength contributed by the excess compression strength of concrete. The multiplier of two goes up linearly to a value of three, corresponding to a specified compressive strength of 15,000 psi (103 MPa), and remains level for higher strengths.

A limited number of tests of reinforced concrete beams made with high-strength concrete[14, 15] ($f'_c > 8,000$ psi or 55 MPa) suggested that the inclined cracking load increases less rapidly than Eq. 11-3 or 11-6 of ACI 318 would indicate. This was offset by an increased effectiveness of the stirrups compared to the strength predicted by Eqs. 11-17, 11-18 and 11-19. Other tests of high-strength concrete girders with minimum web reinforcement indicated that this amount of web reinforcement may be inadequate to prevent brittle shear failures when inclined cracking occurs.[16]

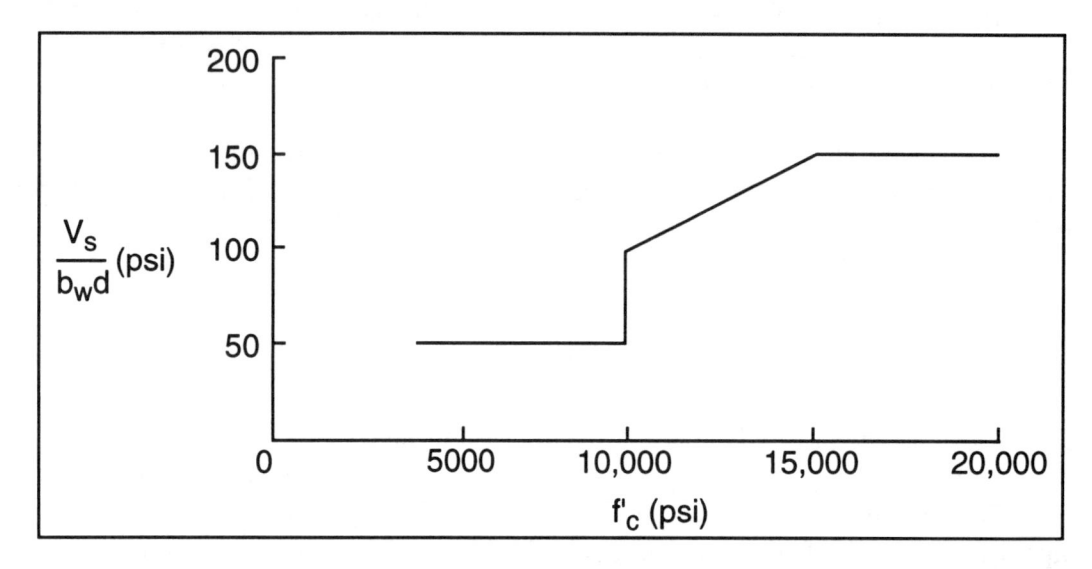

Figure 4. Higher Minimum Shear Reinforcement Requirements for Members with Specified Compressive Strength Exceeding 10,000 psi (69 MPa)[2]

There are no test data on the two-way shear strength of high-strength concrete slabs or torsional strength. Until more practical experience is obtained with beams and slabs built with concretes with strengths greater than 10,000 psi (69 MPa), it is required by code to limit $\sqrt{f'_c}$ to 100 psi (25/3 MPa) in calculations of shear strength and torsional strength.

DEVELOPMENT LENGTH

ACI 318-95 Chapter 12 on Development and Splices of Reinforcement also restricts the value of $\sqrt{f'_c}$ to no more than 100 psi (25/3 MPa), meaning that the required development length of reinforcement embedded in concrete does not decrease any further, once the specified compression strength of the concrete goes above 10,000 psi (69 MPa). This limit was also imposed in view of limited test results on the development of reinforcement embedded in concretes with very high compression strengths. Unlike in Chapter 11, there is no exception to this important new restriction in Chapter 12.

The results of recent research[17,18] have shown that when $\sqrt{f'_c}$ exceeds 100 psi (25/3 MPa), stirrups with a maximum spacing not to exceed a certain value need to be provided over the tension development or lap splice length, as applicable, to assure an adequate level of inelastic deformability before member failure. As a minimum, #3 (10 mm dia.) Grade 60 (414 MPa yield strength) reinforcing bars must be used for these stirrups. Test results indicate that the use of smaller bar sizes may result in fracturing stirrups before achieving an adequate level of inelastic deformability in the member. Research results also show that when $\sqrt{f'_c}$ exceeds 100 psi (25/3 MPa), stirrups with the maximum spacing mentioned above must be provided even when $\sqrt{f'_c}$ used in tension development or lap splice length calculation is limited to 100 psi (0.7 MPa). On the other hand, there is no reason to restrict $\sqrt{f'_c}$ to 100 psi (25/3 MPa) in computing tension splice length or development length when stirrups as required are provided along such length.

HIGH-STRENGTH CONCRETE IN AASHTO SPECIFICATIONS

The only specific reference to high-strength concrete in the AASHTO Standard Specifications[19] is in Section 9.5 which states: "The design of precast prestressed members ordinarily shall be based on f'_c = 5,000 psi (35 MPa). An increase to 6,000 psi (42 MPa) is permissible where, in the Engineer's judgment, it is reasonable to expect that this strength will be obtained consistently. Still higher concrete strengths may be considered on an individual area basis. In such cases, the Engineer shall satisfy himself completely that the controls over materials and fabrication procedures will provide the required strengths."

The 1996 draft AASHTO LRFD Bridge Design Specifications does contain the following restriction: "Concrete strengths above 10.0 ksi shall be used only when physical tests are made to establish the relationship between the concrete strength and other properties."

CONCLUDING REMARKS

The legal codes of virtually all jurisdictions within the United States are based on one of three model codes: the Uniform Building Code (UBC), the BOCA National Building Code (BOCA/NBC), and the Standard Building Code (SBC). The concrete design and construction provisions of all three model codes are based on ACI 318 *Building Code Requirements for Structural Concrete*. There is no upper limit on the strength of concrete, that can be used in structural applications, in ACI 318 or in any of the three model codes. The AASHTO Standard Specifications and LRFD Bridge Design Specifications do contain certain restrictions in that regard. This paper details and discusses the latest ACI 318 provisions that relate specifically to high-strength concrete. It points out two areas in which changes in ACI 318 provisions are needed to ensure continued satisfactory usage of high-strength concrete in non-seismic applications.

REFERENCES

1. American Concrete Institute Committee 363, *State-of-the-Art Report on High-Strength Concrete,* ACI 363R-92, American Concrete Institute, Detroit, MI, September 1992, 55 pp.

2. American Concrete Institute Committee 318, *Building Code Requirements for Reinforced Concrete,* ACI 318-89 (Revised 1992), and *Building Code Requirements for Structural Concrete,* ACI 318-95, American Concrete Institute, Detroit, MI 1992, 1995.

3. International Conference of Building Officials, *Uniform Building Code*, Vol. II, Whittier, CA, 1997.

4. Building Officials and Code Administrators International, *National Building Code,* Country Club Hills, IL, 1996.

5. Southern Building Code Congress International, *Standard Building Code,* Birmingham, AL, 1997.

6. Martinez, S., Nilson, A.H., and Slate, F.O., *Spirally Reinforced High-Strength Concrete Columns,* Research Report No. 82-1, Department of Structural Engineering, Cornell University, Ithaca, New York, August 1982.

7. Cook, J.E., "10,000 psi Concrete," *Concrete International,* Vol. 11, No. 10, October 1989, pp. 67-75.

8. Carrasquillo, L., Nilson, A. H., and Slate, F.O., "Properties of High-Strength Concrete Subjected to Short-Term Loads," *ACI Journal,* Proceedings Vol. 78, No. 3, May-June 1981, pp. 171-178, and Discussion, Proceedings Vol. 79, No. 2, March-April 1982, pp. 162-163.

9. Ibrahim, H., and McGregor, J.G., "Tests of High-Strength Columns under Combined Axial Load and Moment," *ACI Structural Journal,* Vol. 94, No. 1, January/February 1997, pp. 40-48.

10. Razvi, S.R., and Saatcioglu, M., "Strength and Deformability of Confined High-Strength Concrete Columns," ACI Structural Journal, Vol. 91, No. 6, November-December 1994, pp. 678-687.

11. *Concrete Design Standard, NZS 3101: 1995, Part 1* and *Commentary on the Concrete Design Standard, NZS 3101: 1995, Part 2,* Standards Association of New Zealand, Wellington, 1995.

12. *CSA A23.3-94 Design of Concrete Structures,* Canadian Standards Association, Rexdale, Ontario, 1994.

13. Azizinamini, A., Kuska, S., Brungardt, P., and Hatfield, E., "Seismic Behavior of Square High-Strength Concrete Columns," *ACI Structural Journal,* Vol. 91, No. 3, May-June 1994, pp. 336-345.

14. Mphonde, A.G., and Frantz, G.C., "Shear Tests of High- and Low-Strength Concrete Beams without Stirrups," *ACI Journal,* Proceedings Vol. 81, No. 4, July-August 1984, pp. 350-357.

15. Elzanaty, A.H., Nilson, A. H., and Slate, F. O., "Shear Capacity of Reinforced Concrete Beams Using High-Strength Concrete," *ACI Journal,* Proceedings Vol. 83, No. 2, March-April 1986, pp. 290-296.

16. Roller, J.J., and Russell, H.G., "Shear Strength of High-Strength Concrete Beams with Web Reinforcement," *ACI Journal,* Proceedings Vol. 87, No. 2, March-April 1990, pp. 151-198.

17. Azizinamini, A., Pavel, R., Hatfield, E., and Ghosh, S. K., "Behavior of Spliced Reinforcing Bars Embedded in High-Strength Concrete," to be published.

18. Azizinamini, A., Pavel, R., Eligehausen, R., and Ghosh, S. K., "Proposed Modifications to ACI 318 Provisions for Calculating Tension Development and Tension Splice Lengths," to be published.

19. American Association of State Highway and Transportation Officials, *Standard Specifications for Highway Bridges,* Sixteenth Edition, Washington, D.C., 1996.

20. American Association of State Highway and Transportation Officials, *AASHTO LRFD Bridge Design Specifications, U. S. Units,* 1996 Interim Revisions, as approved by the AASHTO Subcommittee on Bridges and Structures, Washington, D.C., 1996.

USE OF HIGH-PERFORMANCE CONCRETE IN A BRIDGE STRUCTURE IN VIRGINIA

Jose Gomez
Research Scientist
Virginia Transportation Research Council
Charlottesville, Virginia, U.S.A.

Tommy Cousins
Associate Professor
Department of Civil Engineering, Virginia Tech
Blacksburg, Virginia, U.S.A.

Celik Ozyildirim
Principal Research Scientist
Virginia Transportation Research Council

ABSTRACT

This paper summarizes current work in high-performance concrete (HPC) in bridge structures by the Virginia Department of Transportation. It focuses on the instrumentation plan to monitor the performance of the HPC structure in Richlands, Virginia. This structure has two 22.6-m spans, designed to be continuous for live load, and has high strength, low permeability concrete in all its components. The prestressed beams were designed with a concrete strength of 69 MPa (10,000 psi) and 15 mm (0.6 in) prestressing strands, at a 51 mm (2 in) spacing. Selected beams will be instrumented to monitor internal concrete temperatures, internal concrete strains, long term deflections, strand transfer length, and strand end slip.

INTRODUCTION

HPC has enhanced specific properties, such as workability, durability, strength, and dimensional stability, resulting in long-lasting, economical structures.[1] Many conventional concrete bridge structures deteriorate rapidly, requiring costly repairs well before their expected service lives are reached. Four major types of environmental distress affect bridge structures and cause early deterioration: corrosion of the reinforcement, alkali-aggregate reactivity, freeze-thaw deterioration, and attack by sulfates.[2] In each case, water or solutions penetrate the concrete and initiate or accelerate the damage. HPCs designed for low permeability resist the infiltration of aggressive liquids, and thus are more durable.

Until 1992, the Virginia Department of Transportation (VDOT) specifications for prestressed concrete required a minimum 28-day design strength of 35 MPa (5,000 psi) and a maximum water-cementitious material ratio (W/CM) of 0.49. The required air content was 4.5±1.5% which was increased by 1.0% when a high-range water-reducing admixture (HRWRA) is added. In 1992 the maximum W/CM was reduced to 0.40, due to concerns with durability. Also, at that time a pozzolan (Class F fly ash, silica fume) or slag was required for alkali-silica resistivity in concrete containing siliceous aggregates and cements with alkali contents exceeding 0.40%. In 1994, VDOT adopted a special provision, for experimentation in some bridges, for low permeability concrete, based on the rapid chloride permeability test, AASHTO T 277 or ASTM C 1202.[3,4] The special provision requires 1,500 coulombs or less for prestressed concrete, 2,500 coulombs or less for the deck, and 3,500 coulombs or less for the substructure. If the test bridges perform satisfactorily, VDOT plans to adopt this specification for use in all bridges.

The initial VDOT HPC program includes seven bridges (Table 1). The first two bridges have been completed. The contracts for the remaining five structures have been awarded and construction is scheduled to begin in the Spring of 1997.

Table 1. Bridges with high-performance concrete.

Location	Length m (ft)	# of Spans	Span Length m (ft)	Beam Type	Beams per Span	Beam Strength Mpa (psi)	Low Perm. Spec.
Rte. 40 over Falling River, Brookneal	97.5 (320)	4	24.4 (80)	IV	5	55 (8000)	Yes
Rte. 629 over Mattoponi River, Walkerton	365.8 (1200)	12	30.5 (100)	IV	5	55 (8000)	No
Telegraph Road over Fairfax Co. Parkway	55.5 (182)	2	91	Steel	11	N/A	Yes
Rte. 10 over Appomattox River, Richmond	654.5 (2147)	22	27.1/29.6 (89/97)	IV	5	55 (8,000)	No
Rte. 250 over Little Tuckahoe Creek, Richmond	16 (53)	1	16 (53)	II	20	48 (7,000)	Yes
Second Street, Wise Co.	27 (89)	1	27 (89)	IV	6	48 (7,000)	Yes
Virginia Avenue over Clinch River, Richlands	45.1 (148)	2	22.6 (74)	III	5	69 (10,000)	Yes

The Richlands bridge is the focus of this paper. It was designed with the low permeability requirements as well as high strength, details of which are shown in Table 2. In addition, the beams were designed with strands 15 mm (0.6 in) in diameter, spaced at 51 mm (2 in), to increase the prestressing force, thereby fully utilizing the high-strength concrete as well as avoiding steel congestion by reducing the number of strands required.

This paper reports on the instrumentation program that has been developed to achieve the following:

1. investigate the accuracy of the prestress loss calculations done for these beams
2. compare the measured transfer lengths of the beams to the recommended equations in the literature

The prestress loss and the subsequent camber and deflection calculations were done according to the AASHTO *Standard Specifications for Highway Bridges*[5] which do not include provisions for the use of HPC. In addition, the relationship between strand end slip and transfer length will be investigated.

The current bridge was built in 1932 (Figure 1) and carries Virginia Avenue over the Clinch River in Richlands, Virginia. The new structure consists of two 22.6 m (74 ft) spans designed to be continuous for live load (Figure 2). The original design for the replacement structure called for seven beams per span, using conventional concrete. The new design, incorporating the higher concrete strength and larger prestressing strands, required 5 beams per span, as shown in Figure 3. Cost savings are expected because of the low permeability requirements that will ensure longevity with minimal maintenance. To support this design, two 9.4 m (31 ft) long AASHTO Type II, cast with 69 MPa (10,000 psi) concrete and 15 mm (0.6 in) prestressing strands spaced at 51 mm (2 in) were cast and tested for transfer and development length of the larger strands. Satisfactory results were obtained and reported previously.[6]

RESEARCH METHODOLOGY

Three beams in the structure will be instrumented and monitored over an extended period. The beams will consist of the center beam, one outer beam, and one beam in between. It is desired that the instrumented beams be on one span. The following will be determined and monitored:

Table 2. Specification details of the Richlands Bridge.

Element	Compressive Strength Mpa (psi)	Chloride Permeability Coulombs
Beams	69 (10,000)	1,500
Deck	41 (6,000)	2,500
Substructure	21 (3,000)	3,500

Figure 1. Existing structure in Richlands, Virginia.

1. internal concrete temperature both short and long term
2. internal concrete strain prior to and after detensioning as well as long term
3. long term deflections
4. transfer length

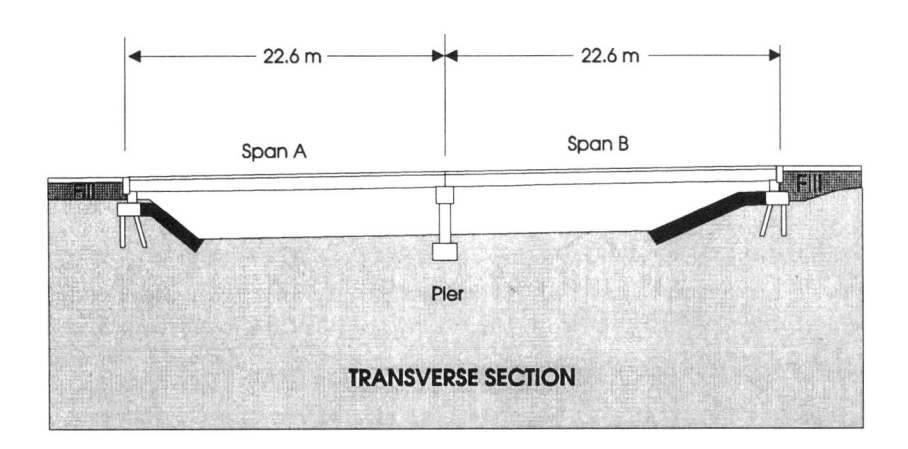

Figure 2. Replacement Structure

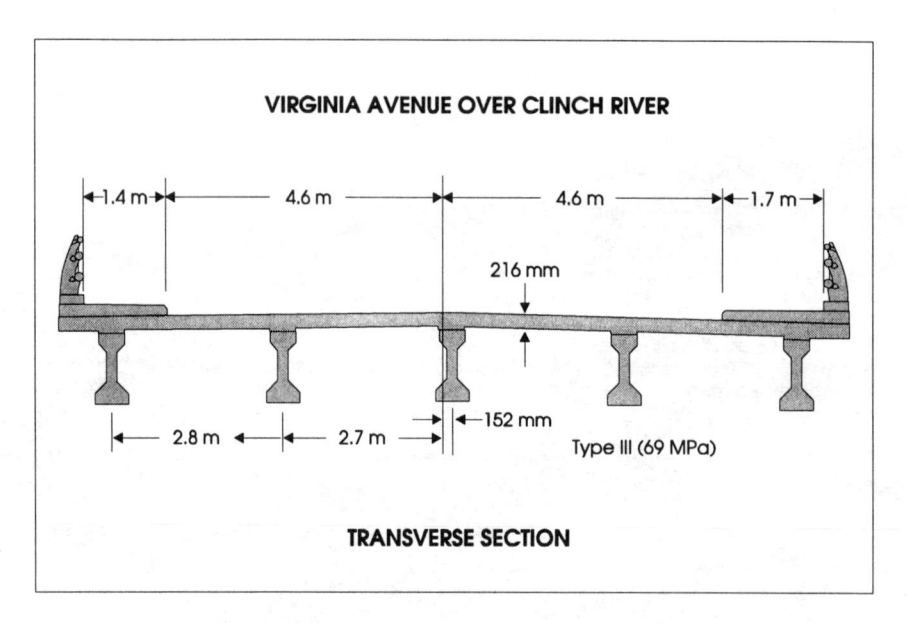

Figure 3. Transverse Section View of the Replacement Structure.

    5.   strand end slip

Additionally, concrete specimens will be prepared from samples of the actual mixes for both the beams and the deck and prepared for testing for creep and shrinkage characteristics.

The beams will be instrumented with thermocouples to monitor concrete temperatures from time of casting through an extended period after placement. Temperatures are measured to determine heat of hydration, temperature for match-cured cylinders, and temperature gradients within the components. Proper temperature management is essential to achieve high early and ultimate strengths and to insure an overall quality product.[6] The thermocouples will be placed at the following locations, as shown in Figure 4:

1. Approximate location of the center of gravity of top flange of beam
2. mid-depth of the beam
3. approximate location of the center of gravity of the bottom flange of beam
4. near the bottom surface of beam

The thermocouples will be placed in each of three beams at the midspan and at one end and in the slab at the midspan of one of the spans, as shown in Figure 4. Data obtained from the thermocouples will be reported elsewhere.

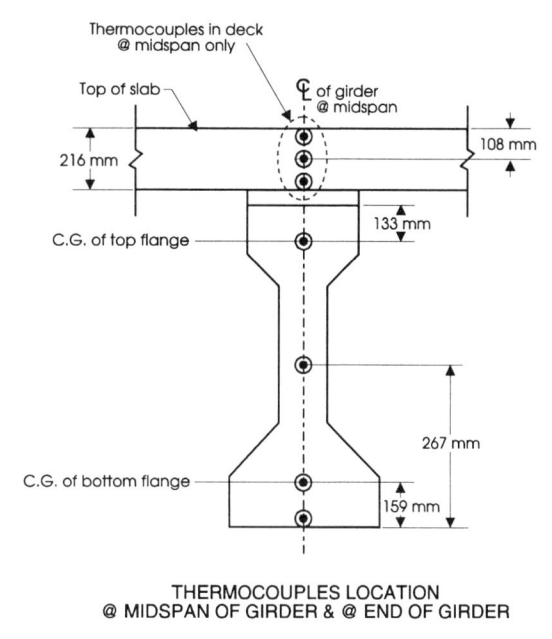

Figure 4. Locations of Thermocouples.

Strain gages will be placed at the midspan of the beams at the center of gravity of the prestressing force to measure the concrete strains at detensioning and continuously thereafter. Long-term strain measurements will be used to determine prestress losses due to elastic shortening, creep, and shrinkage. Three gages, for redundancy of instrumentation, will be spaced longitudinally on a piece of reinforcing steel and placed at the center of gravity of the prestressing force at midspan, as shown in Figures 5.

Long term deflection of the beams will be determined with a precise level and leveling rod. Measurements will be taken at the ends and quarter points of each of the three instrumented beams prior to and just after detensioning. Measurements will be taken after placement at the site and again after placement of the deck. Measurements will be taken periodically thereafter, until the completion of the project.

For end slip determination, 12 strands on both the load and dead ends of each of the three beams will be instrumented with end slip measurement gages as shown in Figure 6. Eight strands across the bottom row, two on the next row up, and two of the draped strands will be measured. Measurements will be taken prior to and after detensioning and again at 1 day.

Transfer length will be determined by measurement of the concrete strains on the surface of the lower flange of the beams. Both ends of the three beams, on both sides (four locations per beam), will be instrumented with brass inserts (Whittemore strain gage points) spaced at 100-mm (4 in) intervals for a total distance of 1800 mm (72 in), as shown in Figure 5. Measurements with a Whittemore mechanical strain gage will be

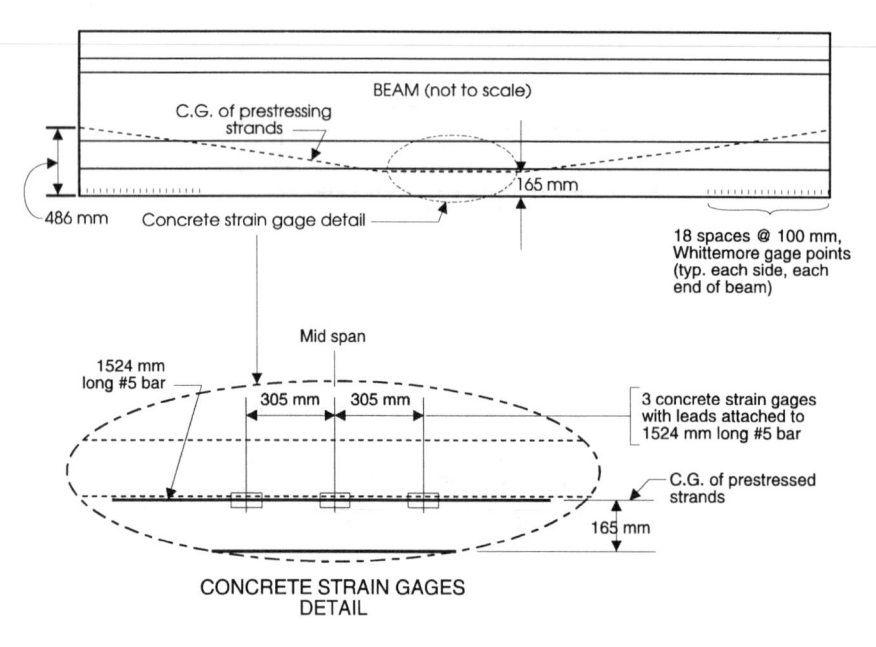

Figure 5. Location of Strain Gages

taken just before and after detensioning, and at one day and 14 days after detensioning. The difference between the initial readings (just prior to detensioning) and the readings taken after detensioning, normalized with respect to the initial readings, yields a strain profile along the face of the beam at the location of the brass studs. Ideally, this strain profile reaches a plateau, indicating that the transfer of prestressing force from the strands to the concrete has occurred.

ANALYSIS OF RESULT

To analyze the accuracy of the prestress loss calculations determined in the design of the beams, data from the measurements of the internal concrete strain as well as vertical girder deflections will be used. The data obtained from these measurements will be tracked for 36 months. The results will be compared to predictions of the AASHTO's *Standard Specifications for Highway Bridges*[5] and other models for prestress loss predictions. Creep and shrinkage properties will also be incorporated in the calculation of losses. Preliminary analysis of the results will be reported.

Transfer length of each of three beams will be determined using concrete surface strain measurements obtained from Whittemore gage points. Transfer length is the embedment length of prestressing strand at the end of a beam necessary to transfer the prestress force to the concrete. By plotting the concrete surface strain measurements as a function of beam length, the transfer length can be determined within a reasonable degree of accuracy.

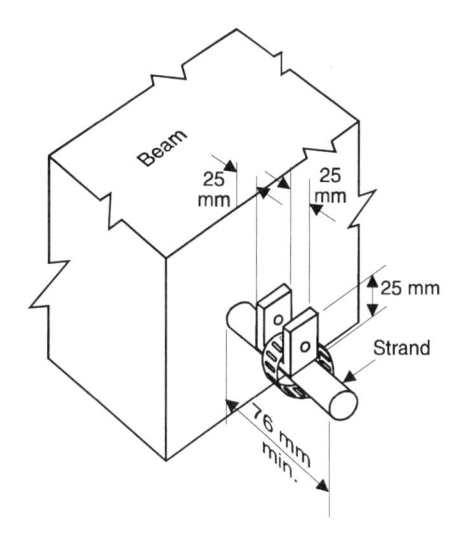

END SLIP INSTRUMENT

Figure 6. Detail of End Slip Gage.

The AASHTO equation for development length of prestressing strand, Equation 9-32, contains a term representing the transfer length.[5] This equation is not a function of concrete strength. The applicability of the equation for use in beams produced from HPC will be investigated by comparing the transfer lengths obtained by direct measurement and via Equation 9-32.

Strand end slip of twelve strands at each end of the three beams will be determined. Strand end slip is the amount of slip, or draw in, that occurs at prestress transfer. Strand end slip and measured transfer lengths will be compared to determine any correlation.

ACKNOWLEDGMENTS

Studies reported here were conducted by the Virginia Transportation Research Council of the Virginia Department of Transportation and sponsored by the Federal Highway Administration. The assistance provided by the FHWA through consultation, review, and use of testing facilities, including the mobile concrete laboratory, is very much appreciated. The authors would also like to recognize Dr. Henry Russell for his guidance and suggestions in the development of the instrumentation plan.

The opinions, findings, and conclusions expressed in this paper are those of the authors and not necessarily those of the sponsoring agency.

REFERENCES

1. Zia, P., Leming, M. L., Ahmad, S. H., Schemmel, J. J., Elliott, R. P., and Naaman, A. E. 1993. *Mechanical behavior of high performance concretes. Volume 1. Summary report.* SHRP-C-361. Strategic Highway Research Program, Washington D. C.

2. Ozyildirim, C. 1993. Durability of Concrete Bridges in Virginia, *ASCE Structures XI Proceedings: Structural Engineering in Natural Hazards Mitigation*, American Society of Civil Engineers, New York, N.Y., pp. 996-1001.

3. *Standard Specifications for Transportation Materials and Methods of Sampling and Testing, Seventeenth Edition*, 1995, American Association of State Highway and Transportation Officials, Inc., Washington, D.C.

4. *1996 Annual Book of ASTM Standards, Volume 4.02, Concrete and Aggregates*, American Society of Testing and Materials, West Conshohocken, PA., 1996.

5. *Standard Specifications for Highway Bridges, Fifteenth Edition*, 1992, American Association of State Highway and Transportation Officials, Inc., Washington, D.C.

6. Ozyildirim, C., Gomez J., and Elnahal, M., 1996. High Performance Concrete Applications In Bridge Structures In Virginia, ASCE Proceedings: Worldwide Advances in Structural Concrete and Masonry, American Society of Civil Engineers, New York, N.Y., pp. 153-163.

LONGITUDINAL SEISMIC RESPONSE OF PRECAST SPLICED GIRDER BRIDGES

Jay Holombo, Graduate Student Researcher
M.J. Nigel Priestley, Professor of Structural Engineering
Frieder Seible, Professor of Structural Engineering
University of California, San Diego
La Jolla, California

ABSTRACT

A research project is underway at the University of California, San Diego (UCSD) in conjunction with the California Department of Transportation (Caltrans) and the Precast/Prestressed Manufacturers Association of California (PCMAC) to cyclically test precast/prestressed high performance concrete (HPC) spliced girder bridges which incorporates Caltrans' rigorous seismic design and detailing requirements. At UCSD, two 40% of full-scale models were tested under fully reversed seismic loading in the longitudinal direction (parallel to the direction of traffic) to verify the adequacy of the newly developed integral capbeam-girder-column details. The first unit incorporated Modified Florida Bulb-Tee girders, while the second unit featured Bathtub girders. Testing of both models was completed in May 1997. Model design, with particular emphasis on seismic design and detailing, model construction and preliminary test results are the subjects of this paper.

INTRODUCTION

In California, bridges with precast elements made up only 3% of the total bridges built in the late 1980's and the early 1990's[1]. An important reason for this statistic is due to seismic concerns. Typical precast bridges constructed in California consist of simply supported girder elements made continuous with a cast-in-place deck, while in-situ construction features full continuity between the column and superstructure. Design engineers have been reluctant to consider the connection between the precast girders and the columns as completely fixed under seismic loading and the resulting substructure of the precast girder bridges have been larger than cast-in-place bridges to compensate for this lack of continuity.

As a consequence, a research project was started by the California Department of Transportation (Caltrans), in conjunction with the Precast Manufacturers of California (PCMAC) to evaluate this problem. Their findings were to replace the existing simply supported configuration with a construction system which incorporates precast high performance concrete (HPC) spliced girders connected with continuous post-tensioning[2]. Splicing the girders allows the system to be continuous for dead load as well as live load, making the system more efficient. Along with the use of precast elements featuring HPC,

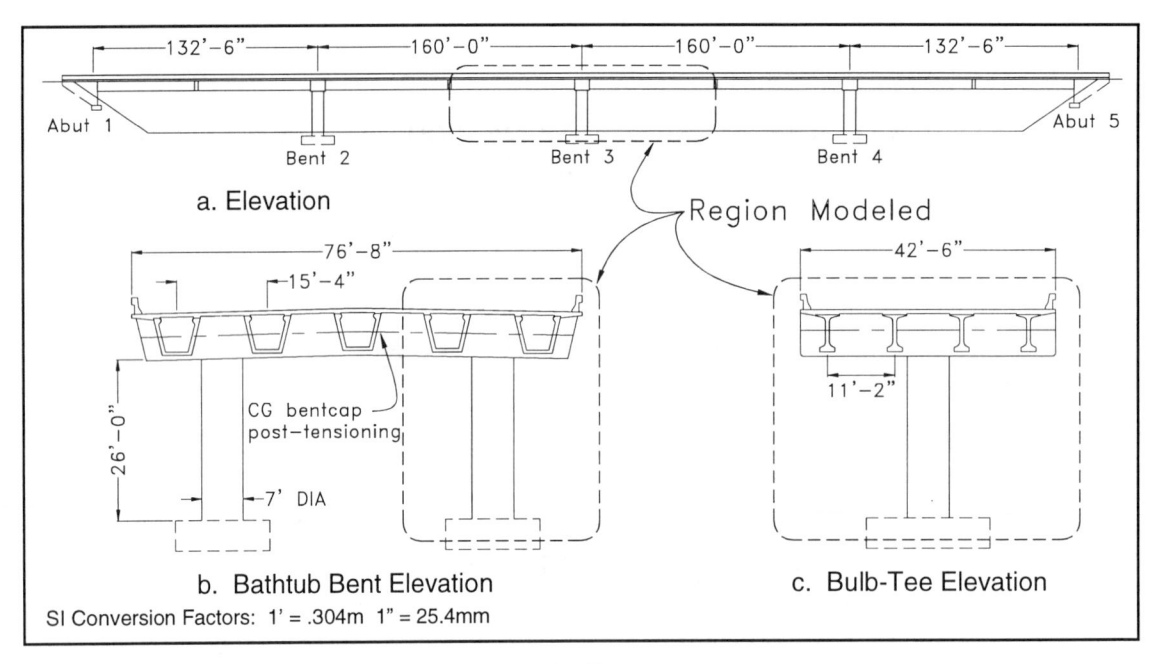

Figure 1. Prototype

this construction scheme allows a substantial reduction of superstructure weight and corresponding seismic mass.

Preliminary design studies showed that designing and detailing an integral superstructure-column connection was possible. However, verification through large scale testing was required[2]. As a result, a research project is currently underway at the University of California San Diego (UCSD), sponsored by Caltrans and PCMAC, to cyclically test precast/prestressed concrete-spliced girder bridges which incorporate Caltrans' rigorous seismic design and detailing requirements. At UCSD, two 40% of full-scale models were tested under fully reversed seismic loading in the longitudinal direction (parallel to the direction of traffic) to verify the adequacy of the newly developed integral capbeam-girder-column details. The first test featured Modified Florida Bulb-Tee girders while the second incorporated Bathtub or "U" shaped girders referred to, in the following text, as the Bulb-Tee Model and the Bathtub Model respectively.

MODEL DESIGN

Prototype

A detailed design study was carried out on the prototype bridges shown in Figure 1. Dimensions and forces for both models were scaled from the prototype. Due to the repetitive nature of the structural configuration, only one bent was needed to capture the seismic behavior. The region studied extends from midspan to midspan, roughly the location of the seismic moment inflection points in the spans adjacent to Bent 3.

Two design studies were carried out on the four span prototype. The Bathtub girder bridge prototype had two columns per bent which allowed flexural pins at the base of the columns. Since the study was concentrating on the longitudinal response, it was assumed that modeling one of the two columns, along with the adjacent girders and contributory bentcap and deck, was sufficient to capture the behavior. The Bulb-Tee girder prototype represents a typical two lane single column bent structure. Hence, full fixity at the bottom of the column was required.

The design 28 day concrete compressive strength for the girders, listed in Table 1, was calculated based on the maximum service compression stress divided by 0.4, as required by Caltrans Bridge Design Specifications[3]. A full live load analysis was performed using HS 20-44 loading for service loads, and a combination of Permit and HS loads was used to determine the ultimate moments and shears. No mild reinforcing was required over the bentcap for the gravity ultimate loads.

Table 1. Prototype Design Concrete Strengths

Item	Design 28 Day Compressive Strength MPa (psi)
Girder Segments	43.8 (6,350) Bulb-Tee 44.8 (6,500) Bathtub
Deck	27.6 (4,000)*
Bentcap	27.6 (4,000)*
Column	22.4 (3,250)*

* Minimum required compressive strength[3].

Model Details

Column - The amount of longitudinal reinforcing in the columns for both models was determined by both gravity and seismic loading. The area of longitudinal steel required was 1% of the gross cross sectional area. However, 2% was used for both model tests to increase the severity of the seismic effects on the superstructure. Both of the model columns were designed to sustain a structural displacement ductility of 4 ($\mu_\Delta = 4$), which is required by Caltrans Design Specifications[3]. Confinement steel in the plastic hinge region was designed based on moment curvature analysis using the confined concrete model developed by Mander et al[4].

The Bathtub Model incorporated a flexural hinge at the base of the column and the corresponding design shear strength was 60% of the Bulb-Tee Model column. The pin was designed using the following Caltrans criteria:

$$V^o \leq 1.4(A_s f_y + P) \tag{1}$$

$$V^o \leq A_{core} \times 0.2 f'_c \tag{2}$$

Where V^o is the column overstrength shear demand, A_s is the pin reinforcement required, A_{core} is the concrete area in the pin and P is the axial load due to dead and seismic loads[3].

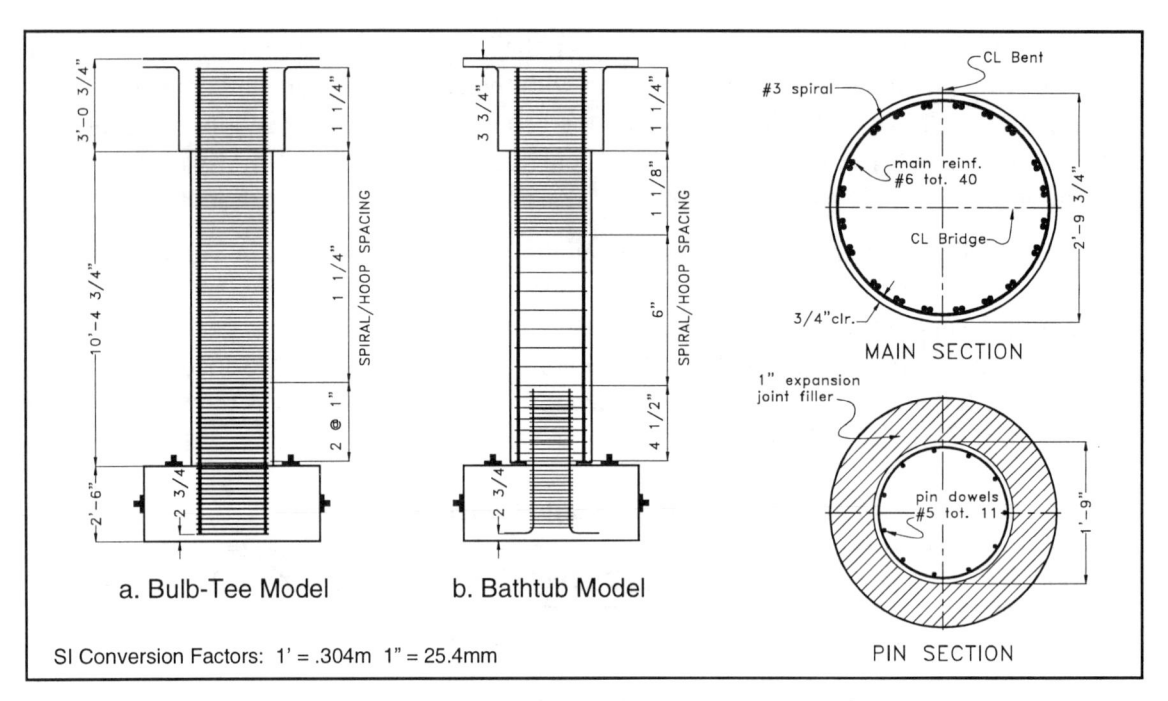

Figure 2. Model Column Details

Bentcap - The bentcaps for both models, shown in Figure 3, were 92mm (3 5/8") deeper than the superstructure to allow the main reinforcement to pass either above and below the precast girders. As a result, only five post-tensioning tendons were required to pass through the precast Bulb-Tee girders.

Bentcap first stage post-tensioning was designed to carry the girders, the fluid deck weight and miscellaneous construction loads with zero tensile stress at the top fiber of the non-composite bent-cap. The bentcap torsional resistance was calculated based on a plastic shear friction model because the distance from the column face to the side of the girders is relatively small (on the order of several inches) and a full spiral crack, assumed to form in most reinforced/prestressed concrete torsion models, can not form in the cap in this small region. The bentcap prestressing provided most of the normal force on the friction plane perpendicular to the bent centerline, between the column and the first adjacent girder[5].

The joint section, which is the same for both models, is shown in Figure 3c. The bentcap was designed wider than the column so the vertical joint shear reinforcement could be placed outside of the column core region in order to reduce congestion. The purpose of the joint stirrups was to help transfer the column tension force up to the top of the joint. The amount of joint shear reinforcement A_{jv} was calculated using the following equation:

$$A_{jv} = 0.125 A_{sc} f_{yc}^{o} / f_{yv} \qquad (3)$$

Where A_{sc} is the area of the column longitudinal steel, f_{yc}^{o} is the ultimate stress of the column reinforcement and f_{yv} is the yield stress of the vertical joint shear reinforcement[5].

The strut and tie mechanism used to develop Equation (3) has an unbalanced horizontal component resisted by the hoops around the column in the joint region. The required volumetric hoop reinforcement ratio can be calculated as:

$$\rho_s = \frac{0.6 A_{sc} f_{yc}^o}{l_a^2 f_{yh}} \qquad (4)$$

Where l_a is the length of the column bar extension in to the joint region, and f_{yh} is the yield stress of the hoops[5]. The provided ratio was $\rho_{s\ provided} = 0.011$ and the required ratio was $\rho_{s\ required} = 0.016$. The difference was made up with the split hairpins shown in Figure 3c. Longitudinal reinforcement was also required to return the horizontal component of the diagonal compression strut to the compression zone of the column. The area of reinforcement required was approximately 1/2 of the vertical area of steel required in Equation (3), and since the joint stirrups continue along the bottom of the joint, the requirement is satisfied.

The Bathtub girders terminate 2 3/8" [60 mm] into the capbeam, as shown in Figure 3d, to allow for easier form-up, and to provide some clamping from the bentcap post-tensioning to improve shear resistance at the bentcap-girder interface. Girder longitudinal soffit reinforcement and the pretensioning strand extended into the cap and were lapped to provide positive moment resistance. Girder post-tensioning ducts were spliced through

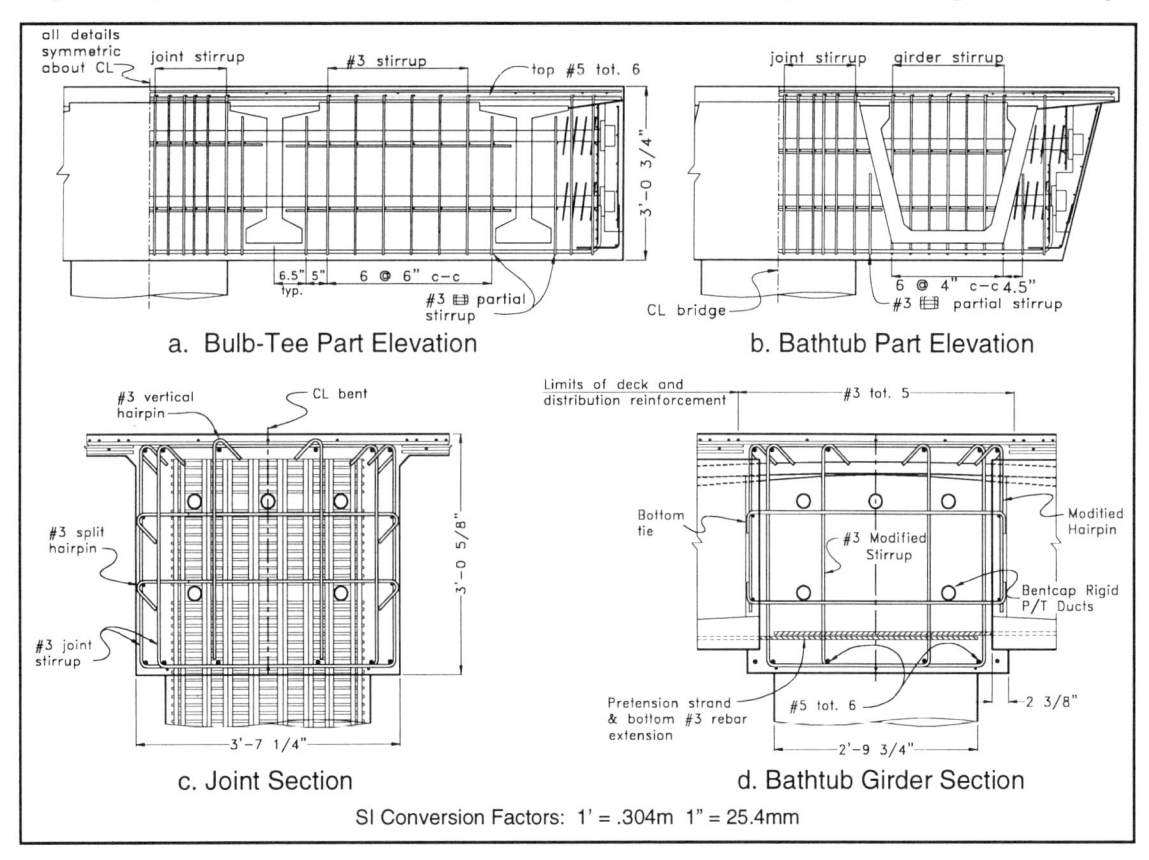

Figure 3. Model Bentcap Details

the capbeam.

Superstructure - The Bulb-Tee Model part-typical section in Figure 4a shows the different girder segments. The pier segments, which pass continuously through the bentcap, had pretensioning in the top flange of the girder to carry the negative moments due to the girder self weight and the reaction from the span segment. Strands placed in the bottom flange for transportation purposes were blanketed along the girder with an exception of the 787mm (2'-7") bonded region at each end, which were used to anchor the strands. Span segments had an identical cross section to the pier segments with the exception of the two fully bonded strands at the bottom of the girder instead of the top. Pretension strands were left extended from the ends of both the pier and span segments and bent into the girder splice.

The Bathtub Model typical section in Figure 4b shows two different sections of the girder segment. The prismatic (midspan) section runs from the face of the end diaphragm to the start of the flare section, which was 1473mm (4'-10") from the end of the girder. The thickness of the web and the soffit vary linearly in the flare section, from zero to 60mm (2 3/8") at the end of the girder. The girder flare not only increases the concrete shear capacity of the superstructure, it also allows the girder post-tensioning ducts to be placed side by side, thus increasing the prestress eccentricity.

Post-tensioning for both models, listed in Table 2, was applied in stages to model the assumed prototype construction sequence. First stage post-tensioning was applied in the lower tendons after the bentcap (and splices in the Bulb-Tee Model) had reached the code required minimum compressive strength of 20.7MPa (3,000psi) and second stage was applied after the deck was cast[3].

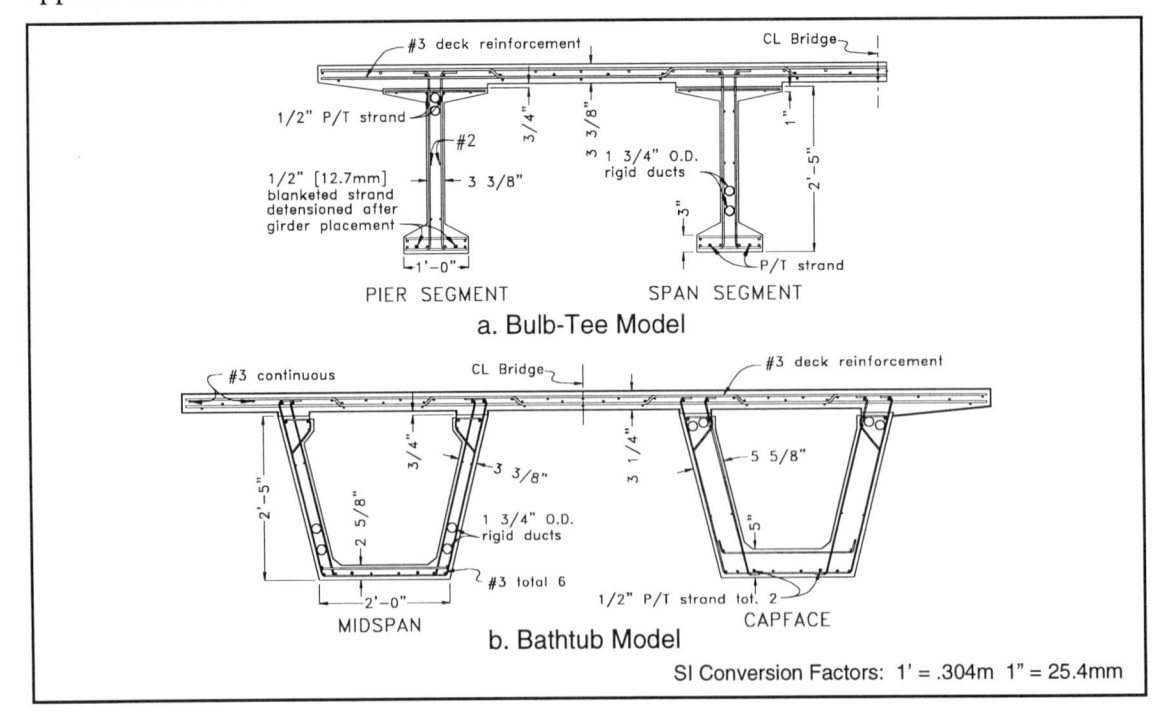

Figure 4. Model Superstructure Details

Table 2. Model Jacking Force (per Tendon)

	Bulb-Tee Model	**Bathtub Model**
1st stage	413 kN (93 kip)	556 kN (125 kip)
2nd stage	551 kN (124 kip)	556 kN (125 kip)

The column plastic moment applied to the superstructure is concentrated at the top of the column, resulting in a non-uniform distribution of moment across the superstructure width. Recognizing this, Caltrans engineers use an "effective width" approach to compute the superstructure capacity to resist plastic hinging. For "T" beam or "I" girder bridge superstructures, an effective width can be calculated as:

$$W_{eff} = D + H_s \qquad (5)$$

Where D is the column diameter and H_s is the height of the superstructure[5]. Equation (5) essentially reduces the contributory superstructure width to the two webs (or girders in the case of the Bulb-Tee Model) adjacent to the column.

TEST SETUP AND LOADING SCHEME

Horizontal Displacement Cycles

The test setup shown in Figure 5a is a 40% scaled representation of the prototype region shown in Figure 1a. Horizontal actuators were placed on both sides of the unit to model the seismic inertia forces acting along the bridge under longitudinal response, while the four vertical actuators at the corners of the unit applied the seismic shear into the superstructure. By assuming that the seismic inflection points remain at the prototype midspans, the vertical actuators were programmed to hold the ends of the test unit to an elevation essentially constant with respect to the bentcap throughout the loading history.

Since the model self weight and required prototype dead load were different multiples of the model scale, and only 1/2 of the span is modeled on either side of the bent, additional forces were needed to correctly model the dead load and prestress secondary moments. As a result, dead load hold-downs were placed 5.49m (18') from the bent centerline. The dead load hold down forces were applied before the deck was poured in order to correctly model strain differential between the deck and the girders, since the prototype girders initially support the deck as a fluid weight.

For both models, initial horizontal force cycles were performed up to the computed first yield of the main steel reinforcement of the column. The experimental stiffness was calculated based on the measured displacement at this force level. The experimental stiffness based on the displacement measured at this force level was used to calculate Δ_y. Three cycles were then carried out at $0.75\Delta_y$, $\mu_\Delta = 1, 1.5, 2, 3, 4, 6$ and 8.

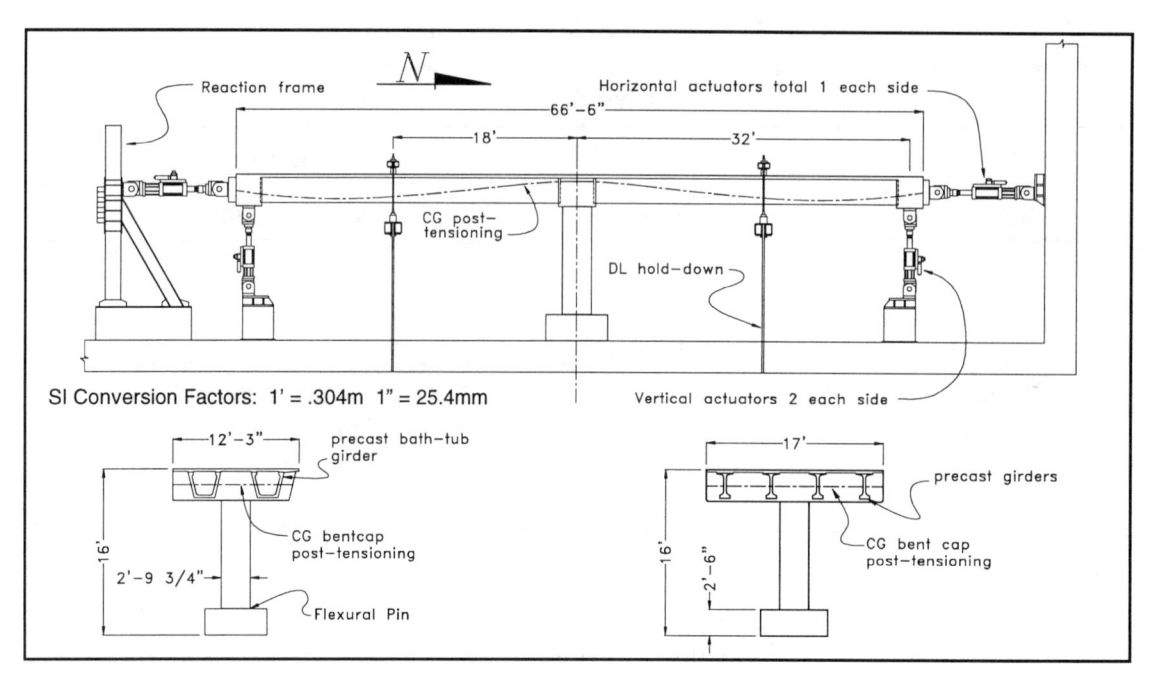

Figure 5. Model Test Setup

Superstructure Capacity Test

Since the structure was designed using capacity design techniques where ductile plastic hinges were to form in the columns, the superstructure/bentcap regions were to remain essentially elastic. Therefore, a second test was performed to evaluate the strength and ductility capacity of the superstructure using the test setup shown in Figure 6a. After the model was returned to approximately zero displacement, the dead load hold-downs and

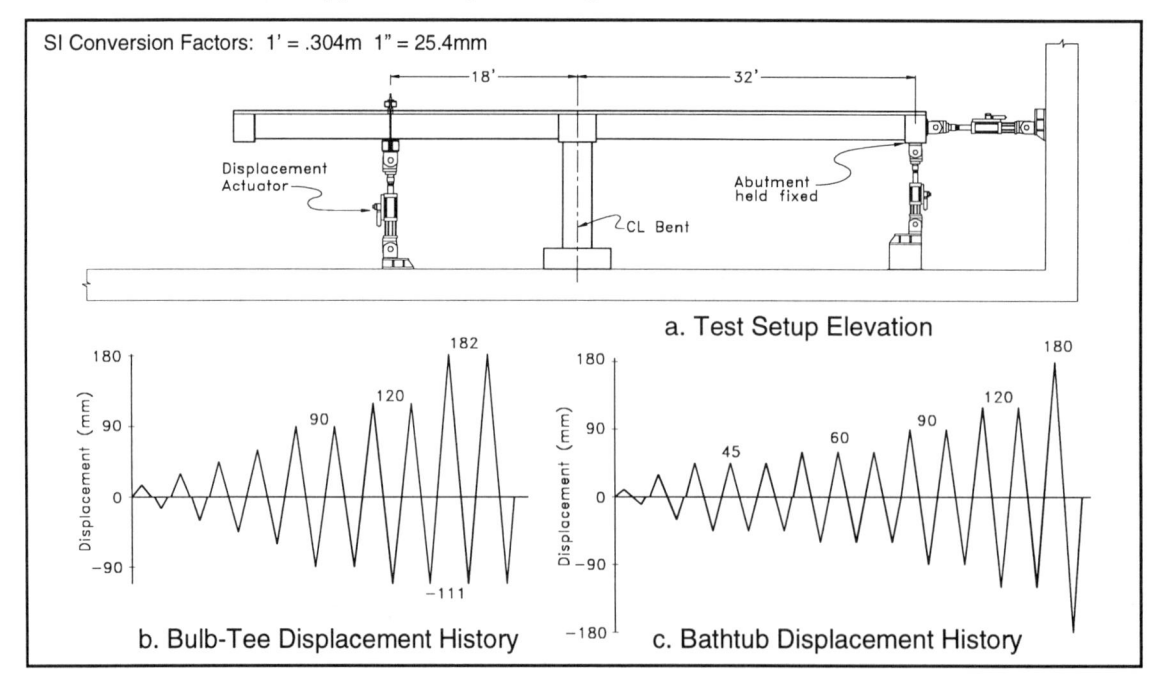

Figure 6. Superstructure Capacity Test Setup

the south actuators were removed. A new actuator(s) (two actuators were used for the Bulb-Tee Model while one was used for the Bathtub Model) was placed 5.47m (18') south from the centerline of bent. The north actuators held the north abutment fixed while the south actuator(s) applied increasing cyclic loads to failure.

Vertical displacements were measured at both the east and west edge of deck at the south actuator (5.49 m (18') south of the bentcap centerline). Displacement cycles, first in the push (up) direction followed by the pull (down) direction were carried out using the loading sequence in Figures 6b and 6c.

PRELIMINARY TEST RESULTS

Horizontal Displacement Cycles

Bulb-Tee Model - At first yield the first deck cracks were observed. The first horizontal joint crack was observed at $\mu_\Delta = 1$ and the first diagonal joint cracking occurred at $\mu_\Delta = 1.5$. Cracking of the girder splice was also observed at this ductility level. General cracking of the column, bentcap and superstructure continued up to $\mu_\Delta = 8$. The superstructure and bentcap cracking shown in Figure 9 closed after removal of the horizontal loading, which was largely due to the prestressing in these regions. During the second cycle at $\mu_\Delta = 8$, popping noises were heard coming from the test unit. It is likely that the noise heard were the 9.5mm (#3) column hoops fracturing. After the third cycle at $\mu_\Delta = 8$, several weld fractures of the 9.5mm (#3) column hoops were observed at the base of the column and the longitudinal bars on the south side of the column had displaced laterally, indicating the onset of buckling and failure of the plastic hinge.

The force displacement response, shown in Figure 7a, reflects increasing lateral strength of the model up to $\mu_\Delta = 8$, which far exceeded the design capacity of $\mu_\Delta = 4$. The measured nominal strength H_i, with correction for P-Δ effects, was exceeded by approximately 27%. H_i was calculated using the following equation:

$$H_i = (M_{i(top)} + M_{i(base)})/L \qquad (6)$$

where M_i is the nominal moment capacity of the column based on a moment curvature analysis at a peak compression strain of 0.004 and L is the clear height of the column. However, the measured peak lateral strength of 1544 kN (344 kip) was 6% less than the peak predicted lateral strength. It appears that the lateral strength was less than predicted due to an apparent sleeving of the column into the joint which increased the effective length of the column.

Bathtub Model - At first yield, the first horizontal joint crack appeared at approximately 1/4 of the joint depth from the bentcap soffit, indicating the activation of the joint shear reinforcement. Diagonal joint shear cracks were first observed at $\mu_\Delta = 1$ on the negative bending side of the bentcap. Cracking intensity and measured lateral force continued to increase up to $\mu_\Delta = 6$. During the second cycle, several banging noises were heard coming from the structure, and the base of the column was first observed sliding along the support. The banging noises were likely due to the pin reinforcement

a. Bulb-Tee Model   b. Bathtub Model

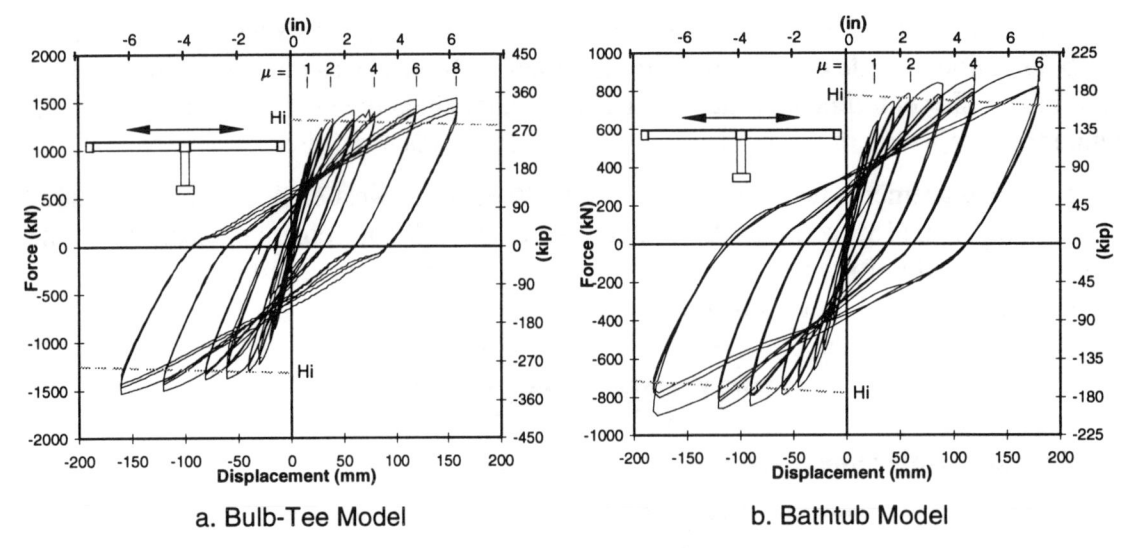

Figure 7. Horizontal Force Displacement Response

fracturing. The sliding displacement at the column base, measured from the peak horizontal push to the peak horizontal pull, was 25 mm (1") during the third cycle. Testing was then stopped at this ductility level due to sliding shear failure of the column base.

The force displacement response shown in Figure 7b reflects increasing lateral strength of the model up to $\mu_\Delta = 6$, which was greater than the design capacity of $\mu_\Delta = 4$. The nominal strength V_i was exceeded by 31% due to the strain hardening of the longitudinal reinforcement and the lightly loaded main column section ($P_{axial} = 0.04 f'_c A_g$, where A_g is the gross concrete area of the column). The peak predicted strength was 4% greater than that measured due to strength degradation of the column pin.

Superstructure Capacity Test

Bulb-Tee Model - At 90 mm of vertical displacement, web shear cracking extended into flexural cracking in the push direction. In the pull direction flexure cracking developed inclined shear extensions. General flexure shear cracking continued until the actuator stroke limits of 182mm (7.17") and 111mm (4.37") had been reached, in push and pull directions respectively. At this level of displacement, no signs of spalling were observed in the girders, which indicated the likely reserve of significant ductility capacity.

The force displacement response shown in Figure 8a shows only minor degradation up to the actuator stroke limits in both directions which corresponds to $\mu_\Delta = 2.7$ and 1.5 in the push and pull directions respectively. The nominal strength was reached in the pull direction and was exceeded in the push direction by approximately 20%. The idealized nominal vertical force was calculated as:

$$V_i = (M_i \pm M_{DL})/L_{cant} \qquad (7)$$

a. Bulb-Tee Model **b. Bathtub Model**

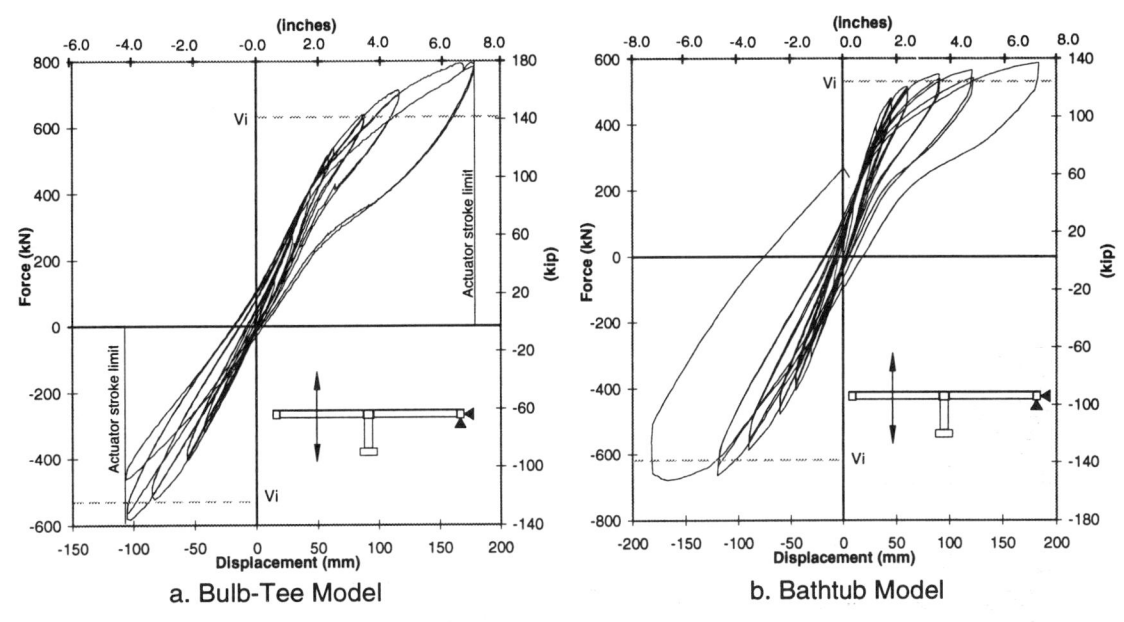

Figure 8. Superstructure Force Displacement Response

Where M_i is the computed nominal moment capacity at the cap face based on a moment curvature analysis and maximum compression strain of 0.004, M_{DL} is the cantilever dead load moment at the capface and L_{cant} is the distance from the displacement actuator to the face of cap.

Bathtub Model - General cracking intensity in the girders and deck increased up to a displacement of 120mm (4.72"). On the push cycle at 120mm (4.72"), the girder soffit pulled out of the bentcap and revealed a wide open crack at the end of the girder. On the reverse cycle, the bottom corners of the girder spalled as the soffit was unable to fit back into the space it previously occupied in the bentcap, as shown in Figure 10. Due to the large inelastic strains in the push direction and to the absence of cover concrete, the 9.5mm (#3) bottom reinforcement in the corners of the girders had buckled during this cycle. At 180mm (7.09") in the push direction, wide open cracking was observed at the girder bentcap interface, accompanied with minor spalling on the top of the deck. Major spalling of the girder soffit occurred in the bottom flange of the girders on the reverse cycle, which extended out approximately 305mm (1'-0") from the face of cap. Spalling along the girder soffit exposed the six 9.5mm (#3) and the two 12.7 mm (1/2") ϕ strands which extended into the bentcap. All of the exposed rebar and strand had buckled during the pull cycle.

The force displacement response shows only minor degradation up to a displacement of 180mm (7.09") in both directions, which corresponds to $\mu_\Delta = 4.4$ and 2.4 in the push and pull directions, respectively, as shown in Figure 8b. The nominal strength was exceeded by approximately 10% in both the push and pull directions.

Figure 9. Bulb-Tee Model at $\mu_\Delta = 6$ (Horizontal Loading)

Figure 10. Bathtub Model (Superstructure Capacity Test)

CONCLUSIONS

Two practical splice-girder designs, which incorporate HPC and an integral superstructure-column connection have been presented. Large scale testing verified the effectiveness of newly developed integral bentcap details under fully reversed simulated seismic loads in the longitudinal direction. Specifically, ductile plastic hinges formed in the columns out to $\mu_\Delta = 8$ and 6 in the Bulb-Tee and Bathtub Models respectively, with only minor strength degradation. Only minor cracking, as shown in Figures 9 and 10, was observed in the bentcap and superstructure, which closed upon removal of the loading. This implies that repair of a prototype superstructure, using the details tested in this project will be essentially cosmetic, after a design level earthquake.

The superstructure capacity tests demonstrated that superstructures which incorporate precast HPC girders can perform in a ductile manner. The Bathtub Model reached $\mu_\Delta = 4.4$ and 2.4 under positive and negative bending respectively, without any special seismic detailing. While failure in the Bulb-Tee Model was not observed, ductile response was recorded up to $\mu_\Delta = 2.7$ and 1.5 under positive and negative bending respectively.

ACKNOWLEDGMENTS

This project was funded by Caltrans with generous donations from PCMAC, Dywidag Systems International, VSL International and the Dayton Superior Corporation.

REFERENCES

1. LoBuono, J., Holombo, J. *Spliced Precast Girders: A New System for High Seismic Areas*. Technical Update, Precast Manufacturers of California, Glendale, California, 1996, 4 pp.

2. Thorkildsen, E., Holombo, J., *Innovative Prestressed Bridges Mark Caltrans Centennial*, PCI Journal, Precast/Prestressed Concrete Institute, Vol. 40, No. 6, November-December, 1995, pp. 34-38.

3. Caltrans, *Bridge Design Specifications Manual*, Department of Transportation, State of California, Sacramento, August, 1986.

4. Mander, J.B., M.J.N. Priestley, R. Park, *Theoretical Stress-Strain Model for Confined Concrete*, Journal of STRUCTURAL ENGINEERING, American Society of Civil Engineering, Vol. 114, No. 8, August, 1988, pp. 1804-1826.

5. Priestley, M.J.N., F. Seible, and M. Calvi, *Seismic Design and Retrofit of Bridges*, John Wiley and Sons, New York, NY 1996, 678 pp.

EVALUATION OF LONG-TERM BEHAVIOR OF HIGH PERFORMANCE PRESTRESSED CONCRETE GIRDERS

John F. Stanton, Professor,
Marc O. Eberhard, Associate Professor,
Paul Barr, Graduate Research Assistant,
Elizabeth A. Fekete, Graduate Research Assistant.

Department of Civil Engineering, University of Washington,
Seattle, Washington, USA.

ABSTRACT

Six high performance concrete girders have been instrumented using vibrating wire strain gages. They have been installed in a new highway bridge, and their long-term behavior will be monitored. Creep and shrinkage of the concrete are also being evaluated in the laboratory using specimens that were made from the same batches of concrete and loaded at the time that the girders were destressed. The objectives are to evaluate the effectiveness of using high performance concrete in bridges and to assess the applicability to high performance concrete of existing methods for predicting prestress losses.

INTRODUCTION

High performance concrete (HPC) is a concrete that provides enhanced performance characteristics for a given application. Additives such as microsilica, fly ash, chemical admixtures, and other materials, individually or in various combinations, are added to conventional concrete to obtain these special characteristics. The Washington State Department of Transportation (WSDOT) is interested in expanding the use of high performance concrete to structural applications. The high strength that can be obtained would be beneficial in precast, prestressed girders to obtain (1) the use of fewer girders per span, (2) longer spans, or (3) girders with reduced height where grade clearance is a problem. Other structural members would be reduced, resulting in less weight. The use of HPC in the deck would enhance the corrosion resistance properties of the concrete to the intrusion of corrosive chemicals.

A new bridge is to be built to carry the eastbound lanes of State Route 18 (SR 18) over SR 516. The WSDOT has designed it assuming the use of HPC in order to test the technology. The bridge will have a roadway width of 38 ft (11.6m), but is to have five lines of girders in place of the usual seven. The three spans will have lengths of 80 ft, 137 ft and 80 ft (24.4, 41.8 and 24.4 m). The girder and deck concretes are to have compressive strengths of 10,000 and 4,000 psi (69 and 28 MPa) respectively.

OBJECTIVES

The Federal Highway Administration (FHWA) has sponsored a number of research programs on HPC around the United States. The SR 18 bridge is also being used as the focus of a study conducted jointly by the WSDOT and the University of Washington. The overall purpose of the program is to obtain information necessary for reliable and more extensive use of high performance concrete in highway structures. The study is divided into two separate phases. The first concentrates on long term behavior and the second, on response to live loads. The specific objectives of the two phases are:

Phase I
1. Instrument the HPC bridge and monitor its long-term behavior.
2. Obtain an extensive array of material test data for the concrete used in the girders.
3. Use the field and laboratory data gathered in (1) and (2) above to compare the long-term behavior of the concrete in the girders with that of the laboratory samples.
4. Estimate the real prestress losses from the field and material data. Compare them with the losses computed using the method contained in the AASHTO LRFD Specifications[1]. Suggest modifications to that procedure if needed.
5. Evaluate the effectiveness of using HPC in prestressed precast concrete bridges.

Phase II
1. Instrument the bridge for (non-destructive) load testing.
2. Load test the bridge to establish the distribution of wheel loads among the five lines of girders. The characteristics of interest are the wide spacing (8'-0") between girders made possible by the use of HPC, the heavy skew (45°) and the partial girder continuity provided by the cast-in-place pier diaphragms.
3. Provide a recommendation to WSDOT for the extent of girder continuity that may be taken into account during design. (At present no continuity is assumed. This assumption is most likely conservative, especially in view of the skew. Savings may be possible if continuity can be shown to exist).

At the time of writing, Phase II has not been started. This paper therefore concentrates on Phase I.

SIGNIFICANCE OF THE WORK

Prestress losses play an important role in the design of prestressed concrete members. In all cases prestress is lost through shortening of the concrete caused by shrinkage and creep, and by relaxation of the prestressing steel. In addition, pretensioned members suffer an instantaneous loss due to elastic shortening of the girder when the strand is released in the prestressing bed. The net camber of, and the stresses in, the girder are the difference between those due to prestressing and those due to gravity loads. For that difference to lie within an acceptable range, the two components must be known accurately. While the gravity loading can usually be determined reasonably well, prestress losses dominate the accuracy with which the effective prestress is known. Inaccurate assessment of the prestress loss can therefore lead to unacceptable stresses, possible cracking, and excessive camber or deflection.

In bridges that are made continuous after the girders are erected, an additional problem arises. The structure is statically indeterminate, so time-dependent effects that occur after the girders are joined will cause a redistribution of moments along the girders. The moment redistribution further exacerbates the difficulties of evaluating the true stresses and deflections.

In high performance concrete several issues combine to increase the difficulties of estimating the prestress losses accurately. First, the losses are relatively large. This is so because the net compressive stress in the bottom flange directly after transfer is high: 4000 psi (28 MPa) in this case. This high stress leads to a large initial elastic shortening loss. Second, most of the methods for estimating prestress losses[2-4] were developed in the 1970s, when concrete technology was radically different from its contemporary state. The losses due to creep and shrinkage may not be well represented by those equations. This is particularly true of high-strength mixes in which the cement paste content may be high.

RESEARCH APPROACH

The approach used here was to place internal instrumentation in the girders and deck slab and to monitor it for three years. Vibrating wire strain (VWS) gages were embedded in the precast girders during production, and more will be installed in the cast-in-place slab when it is cast. Fig. 1 shows a girder, with instrumentation installed, being lifted from the stressing bed.

Figure 1. Instrumented girder being lifted from stressing bed

The VWS gages are being used to monitor long-term strains and were chosen for their long-term reliability and lack of drift. Each gage contains a thermistor, so accurate temperature profiles are also available. Gage locations in the precast girder are shown in Fig. 2. The abbreviations refer to Top Girder (TG), Upper Web (UW), Middle Web (MW), Lower Web (LW), Bottom Left (BL) and Bottom Right (BR). In addition, the camber of the girders is being recorded by several independent means.

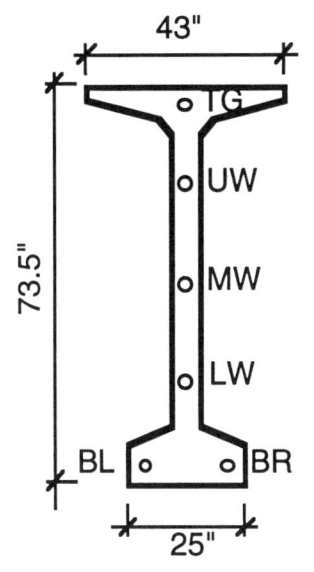

Figure 2. Gage locations

Laboratory samples of the girder concrete were taken. Six creep rigs, each containing either four 6" x 12" (150 x 300 mm) or two 4" x 12" (100 x 300 mm) cylinders, were used to measure creep under a variety of circumstances. Companion drying shrinkage specimens permit the creep and shrinkage deformations to be separated. Compression strength tests, split cylinder tensile tests and elastic modulus tests were conducted at various concrete ages, in order to track their changes with time.

Comparisons are being made between the laboratory and field strain data. The creep rigs are being maintained under constant stress, while the girder concrete is experiencing compressive stresses that inevitably change with time. Furthermore the temperature, humidity and volume/surface ratio for the girders and laboratory samples differ. These differences mean that the two cannot be compared directly. However, most of these characteristics are being varied among the six creep rigs, in order to permit the most meaningful comparisons possible. The purpose is to verify that the properties of the girder concrete can be reliably derived from the cylinder data, so that the prestress losses in future girders can be estimated from material data obtained from laboratory samples of the concrete in question.

The strains and cambers in the girders can be used to calculate the total shortening of the girder at the centroid of the prestressing steel. If the steel were perfectly elastic, the stress in it could be computed directly from this shortening. However, the strand is not perfectly elastic because it is subject to relaxation, and this component of the loss must be estimated from manufacturer's material data in order to establish the change in stress in the strand in the girders. Fortunately, the relaxation of modern "low-relaxation" strand is indeed low, and constitutes only a small fraction of the total prestress loss. Thus moderate relative errors made in estimating it would not cause major errors in the total prestress loss calculation. The measured total shortening, combined with the estimated relaxation loss, can then be used to compute the total stress loss in the strand.

Strand slip-back was also measured during destressing. Until 1996 FHWA had a moratorium on the use of 0.6" diameter strand in federally funded bridges, largely because of the paucity of experimental data to confirm that the existing AASHTO equations for predicting transfer and development lengths apply to it. Although that moratorium has now been lifted, a recent study[5] on 0.5" diameter strand has demonstrated that some scatter exists in bond characteristics. However, if girders are to be made to span further using high performance concrete, larger prestressing forces will be needed and, since the flange dimensions limit the number of strands that can be accommodated, the use of 0.6" (15 mm) diameter strands is essential. Thus the opportunity was taken to measure slip-back and to compute from it the transfer length for the 0.6" (15 mm) strand used here.

FINDINGS

The program of measurement is still at an early stage, since the girders were only erected in May 1997 and, as of July 1997, the deck has not yet been cast. However, several findings have already become apparent.

Concrete Mix Design and Properties.

First, developing a workable mix to provide the required strengths of 7400 psi (51 MPa) at release and 10,000 psi (69 MPa) at 56 days was not a trivial matter. In this case it was undertaken by the precaster. Several different trial designs, using different fly ash and silica fume sources, were necessary to obtain satisfactory results. However when experience had been gained with the girders, the release strength was obtained consistently in 24 hours, using steam curing. The final concrete mix is given in Table 1.

The girder was cured by covering it with an insulating blanket and applying steam heat as necessary to achieve a pre-set time-temperature curve. The controlling thermocouple was located between the gages marked MW and LW in Fig. 2.

Table 1. Concrete Mix Design.

Item	Type	Quantity	Abs. Vol (cu. ft.)	SSD Wt. (lb.)
Coarse aggregate	1/2" to #4, SG = 2.5		11.28	1870
Fine aggregate	blended, SG = 2.3		5.42	890
Cement	Holnam Type III	7.75 sacks	4.09	728
Fly ash		2.3 sacks	1.32	222
Silica Fume	MB Rheomac SF100	0.05 sacks	0.37	50
Water		31.81 gals	4.25	265
Air	entrapped	1%	0.27	0
Superplasticizer	MB Rheobuild 1000			
WR Admixture	MBL 82			
Total			27.00	4025

Note: 1 cu. ft. = 0.0283 m^3, 1 lb. = 0.454 kg, 1 sack = 42.6 kg, 1 gal = 0.00340 m^3

One of the difficulties in developing the mix and the curing program was balancing the requirements for early (release) strength and the 56 day strength of 10,000 psi (69 MPa). The girders are to be made composite with the cast-in-place slab, so the maximum compressive stress after the slab is cast is relatively low, and the 10,000 psi (69 MPa) is not necessary to satisfy compressive stress limits under service conditions. The girders were designed for zero bottom tension under full live load, so the 56 day strength is also not needed to provide tensile stress capacity. Thus the constraints on the mix design could have been eased somewhat by requiring a lower 56 day strength, more closely related to the real needs. Of course, other requirements such as durability would still have to be met regardless of the required concrete compressive strength.

The precaster followed good practice and monitored the concrete temperature using a thermocouple in the girder and a Sure-cure system to ensure that the companion cylinders were cured according to the same temperature history as the girder. The thermistors in the vibrating wire strain gages also recorded temperature, but they showed that a significant thermal gradient existed throughout the girder. A typical plot is shown in Fig. 3. The different curves refer to the gage locations defined in Fig. 2. While the top flange reached a maximum of 140-167°F (60° - 75°C), the bottom flange was typically 45F° (25C°) cooler. This gradient is thought to have occurred because the plant was outdoors and the casting took place in winter, so the ground below the forms acted as a heat sink. The steam heat also rose to the top of the protective blanket. This finding is significant because the bottom flange is the coolest part of the girder, so the concrete there is the least mature and is probably the weakest. Since the Sure-cure thermocouple was usually located just below mid-height of the girder, the true strength in the bottom flange was probably lower than that of the companion cylinders, and may even have been lower than the required compressive strength at release. This could cause problems if the strength shortfall were significant because the highest stress at any time during the life of the girder usually occurs in the bottom flange at release.

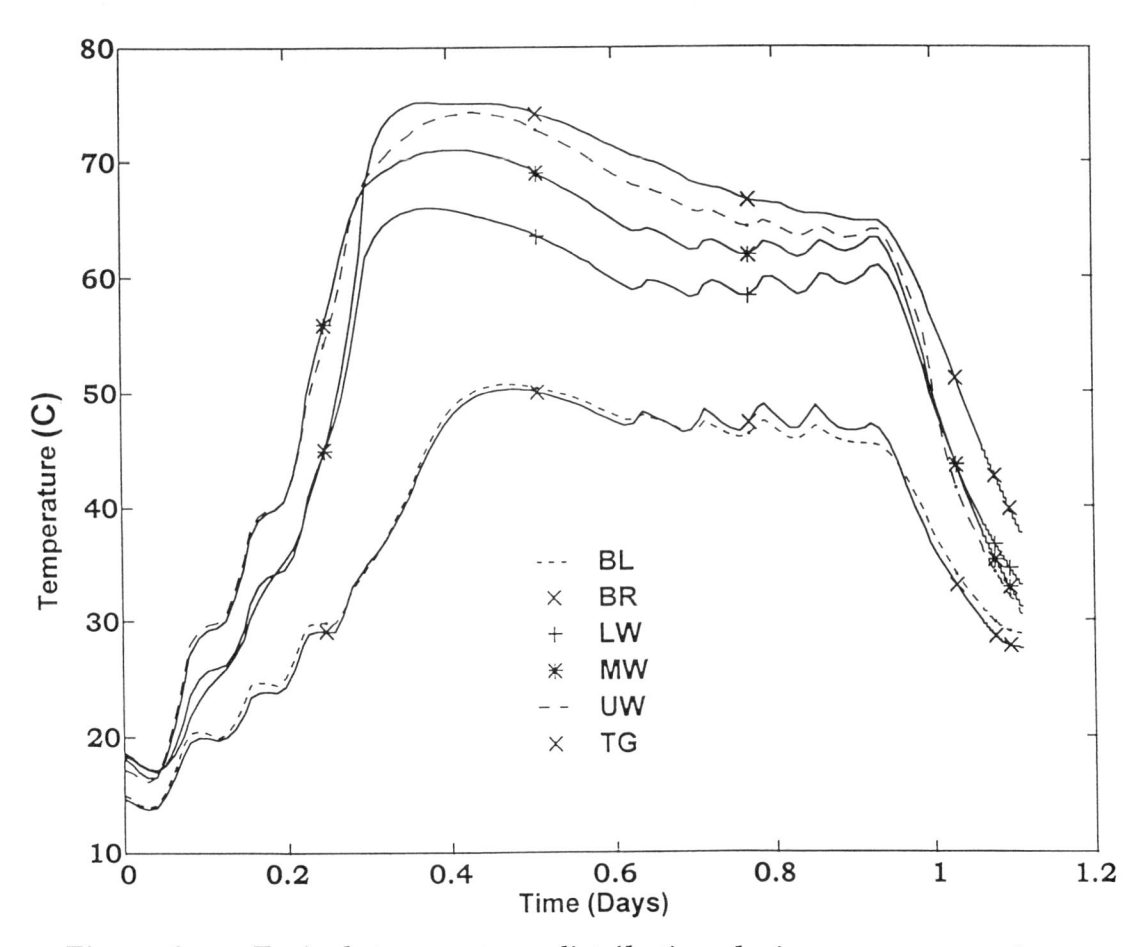

Figure 3.   Typical temperature distribution during concrete curing

Concrete Strains

The vibrating wire strain gages performed very reliably. Apart from two that suffered impact damage shortly after casting, all but four of the original eighty gages are providing consistent, stable readings. They show strains that vary linearly over the cross section, which is the expected pattern. This behavior suggests the absence of drift in the gages, which is obviously essential for long-term measurements.

A typical plot of concrete strain vs. time at mid-span of the 133 ft. long, most highly stressed, girders is shown in Fig. 4. The overall pattern is one of compressive strain that varies over the height of the girder and increases with time. The minor ripples are attributed to daily temperature changes, which are typically smaller at the bottom flange because of shading. The gap in the middle of the data was caused by the need to disconnect the instrumentation during transportation and erection. The close agreement between gages at the same elevation in the girder, combined with the fact that the strains at any time are close to being linearly distributed over the cross-section, constitutes encouraging verification that the readings are reliable.

The strain change at the steel centroid due to instantaneous elastic shortening averaged about 900 micro-strain. This leads to a stress reduction of approximately 25 ksi (172 MPa) in the strand. Three months after casting, this value had approximately doubled due to creep and shrinkage. Continuing creep and shrinkage will cause further prestress loss, but an instantaneous increase in the strand stress of about 9 ksi (60 MPa) is also expected when the slab is cast.

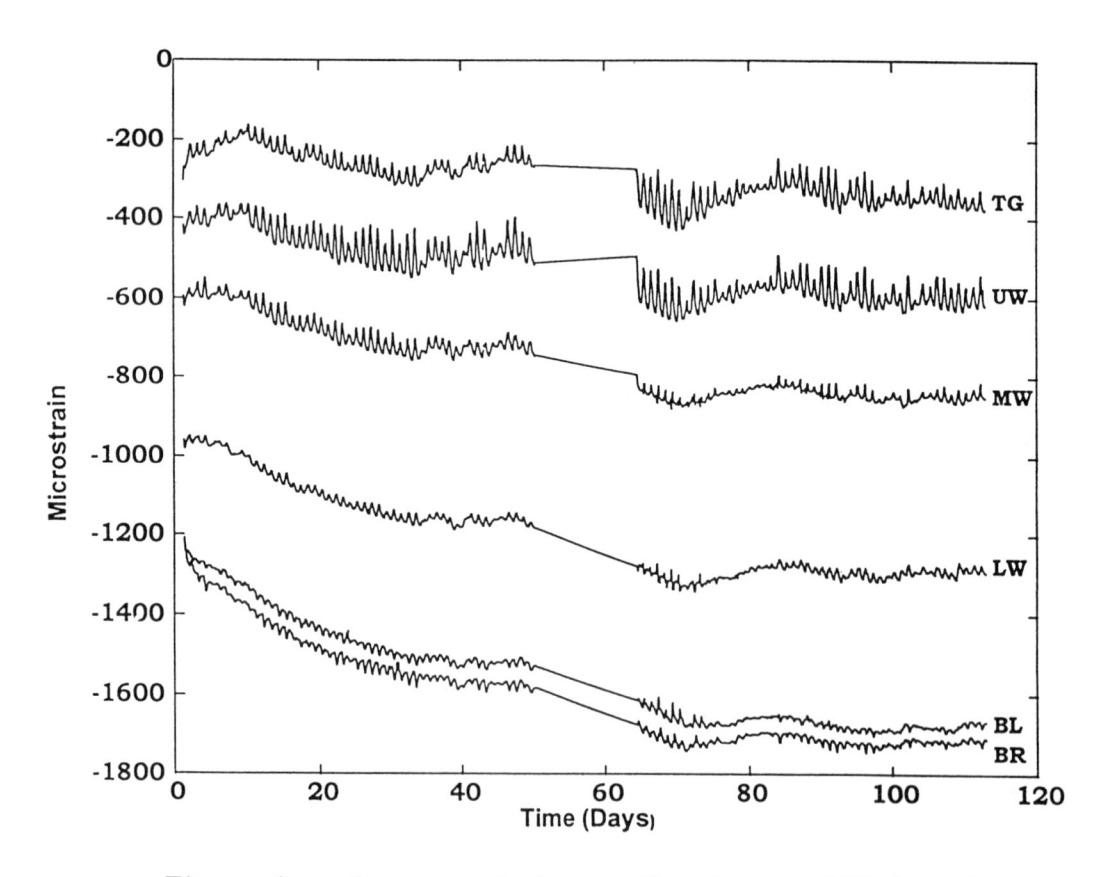

Figure 4.   Concrete strains vs. time in one 133 ft girder

Girder Camber

Camber was also measured, and the values for the long girders are shown in Fig. 5. Camber was recorded in two ways. The first used a stretched wire as a reference elevation and the vertical movement of the mid-span of the girder relative to it was measured with an LVDT. Readings were automated and were taken at the same time intervals as were all the gage readings. This system was installed immediately before the strands were cut. However it suffered damage while the girders were stored in the yard, and cannot be re-installed until the deck is cast for risk of further damage. As a backup, readings were taken using a surveyor's level at various times, and these are the ones shown in Fig. 5. The early readings were influenced by the different support conditions that existed in the stressing bed, in the different storage locations used in the yard and on the bridge piers. Differential heating also played a role. However, since erection, readings have been repeatable to ± 0.040" (1 mm). Those readings have been taken at the same time in the early morning to minimize the influence of differential heating on camber.

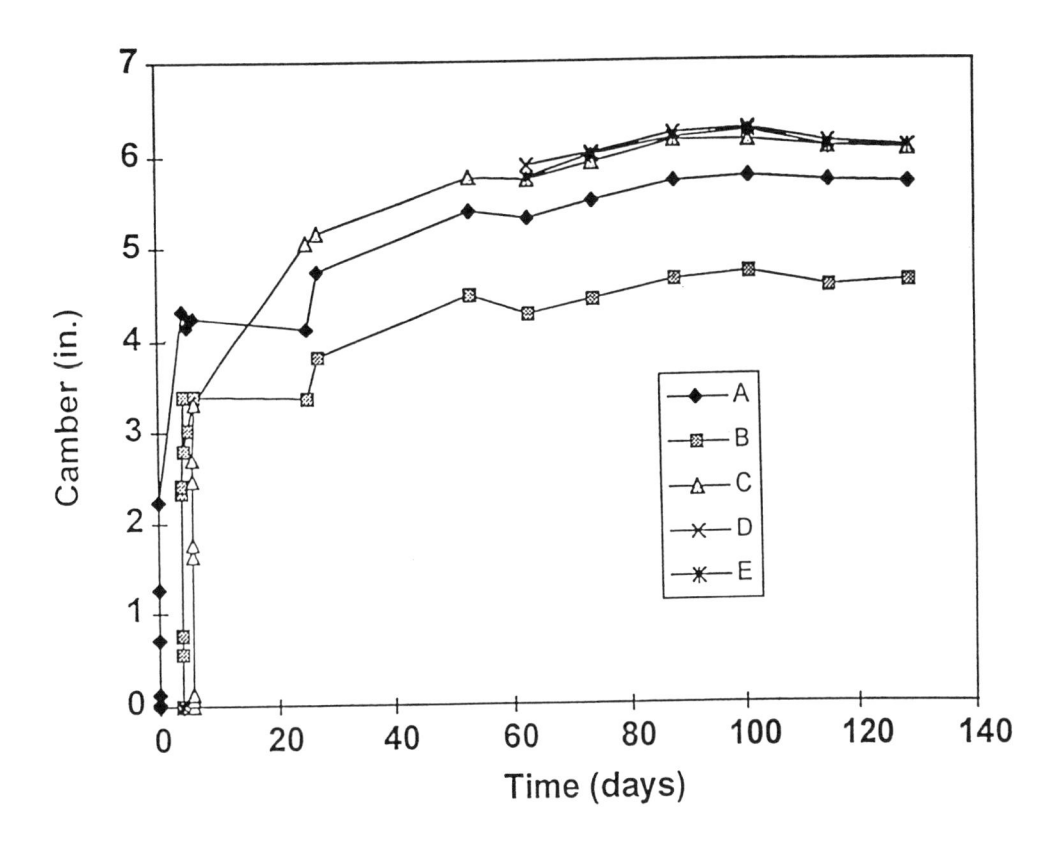

Figure 5.   Camber vs. time in 133 ft girders

Fig. 5 shows that the camber lies between 4.5" and 6.1" (110 and 155 mm). This relatively high value exists because the prestress is high and there have been some delays in casting the deck. Girders 2-C, 2-D and 2-E are behaving almost identically, while girder 2-B has a camber that is approximately 25% smaller than the average of the others. The elastic and creep strains in girder 2-B are also concomitantly smaller. The cause of the difference is not known with certainty, but variations in material properties could have played a role.

Laboratory Creep Data
The results from the creep rigs are preliminary and are illustrated in Fig. 6, which shows total strain vs. time for the rigs containing concrete taken from the same batch as the long center girder, number 2-C. The logistics of loading the creep rigs at the University of Washington in Seattle at the same time as girders were stressed in Spokane were difficult. However the rigs and girders were loaded as closely to simultaneously as possible.

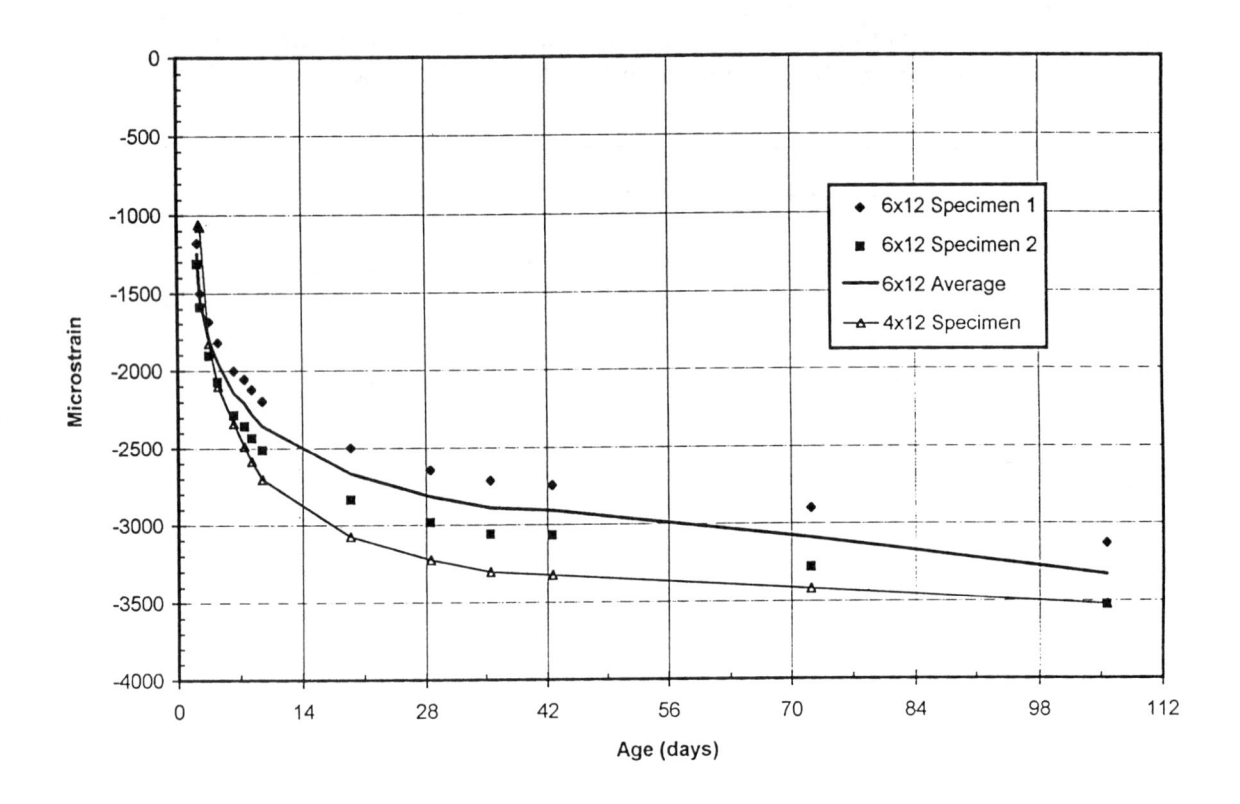

Figure 6. Typical results from creep rigs

Two different cylinder sizes were used (6" x 12" and 4" x 12", or 150 x 300 and 100 x 300 mm), and both sealed and unsealed cylinders were used in each size, in order to obtain data for different volume/surface ratios. The V/S ratio for the girder is approximately 2.9" (74 mm), and for the critical bottom flange it is 3.1" (79 mm), whereas for a 6" x 12" (150 x 300 mm) cylinder with sealed ends it is 1.5" (38 mm). To model the bottom flange of the girder exactly would have required the use of 12" x 24" (300 x 600 mm) cylinders stressed to 4000 psi (28 MPa), for which the necessary creep rigs would have been very large and expensive. Thus, the two cylinder sizes may be used to construct a curve of creep strain vs. V/S ratio, which will be extrapolated to the V/S ratio of the girders. The procedure is not perfect, but it represents the best attempt to model the true behavior within a finite budget.

Fig. 6 shows total strains in the unsealed specimens loaded to 4000 psi (28 MPa), and includes elastic, shrinkage and creep components. The total strain is significantly higher than that in the girders. However, the stress in the bottom flange of the girder is dropping with time, whereas the creep cylinders are being maintained under constant stress. Furthermore, the 4" x 12" (100 x 300 mm) cylinders are creeping more than the 6" x 12" (150 x 300 mm) ones, as would be expected for their different V/S ratios. Insufficient data are yet available to make complete numerical comparisons, but it appears that the factor proposed by PCI[4] for compensating for the V/S ratio may in this case be too small.

It is worth noting that some difficulties were experienced in loading the creep rigs truly concentrically. The rigs were constructed in accordance with ASTM C512, including the use of a spherical head intended to eliminate eccentricity of load. Furthermore, concentric circles had been machined into the plates to aid in centering the specimens. Four sets of gage studs were applied to each cylinder for redundancy. Despite these precautions, the strains obtained by averaging the NS and EW sets of gage studs differed in some rigs, thereby suggesting some eccentricity of load. These issues, combined with the difficulties inherent in reading a demountable mechanical strain gage consistently, suggest that the readings from the laboratory creep specimens may be somewhat less accurate than those from the girders. Discussions with other researchers have revealed similar difficulties in obtaining clean, consistent creep data.

Strand Slip-back and Transfer Length
In the prestressing plant, the strands were stressed by gang-pulling with one jack for the harped web strands and with another for the straight strands in the flange. Destressing was achieved by slowly releasing the hydraulic pressure in the jacks. This procedure led to smooth destressing and to slip-back values that ranged from 0.062" to 0.235" (1.5 to 6.0 mm). These values correspond to transfer lengths of 20" to 74" (500 to 1900 mm) or 33 to 124 d_b. Most values were greater than the 60 d_b predicted by the AASHTO equation for the conditions. Ongoing monitoring of the strand slip, to study the magnitude of the time-dependent slip observed by Logan[5], was not possible for logistical reasons. Almost no flange cracking was observed, in contrast to the relatively severe cracking observed[6] in girders whose strands are destressed by flame-cutting.

SUMMARY

A new bridge is being constructed by the WSDOT, using high performance concrete in the precast, prestressed girders and in the cast-in-place deck. The girders are heavily stressed with 0.6" diameter strands. A research study has been initiated to evaluate the effectiveness of the use of HPC in such bridges. A principal goal is the assessment of the prestress losses, which is one area in which the values are expected to differ from those in similar bridges made from conventional concrete. The evaluation is being conducted by instrumenting the girders and the deck slab, by monitoring those instruments over three years, and by comparing the field data, data from laboratory creep and shrinkage samples of the same concrete, and the results of theoretical predictions of behavior.

CONCLUSIONS

Complete conclusions will not be available until the slab is cast and data collection and analysis are complete. However, the following preliminary conclusions can be drawn:

1. High performance precast prestressed concrete girders are likely to be cured by steam heating. The temperature throughout the girder during curing may differ by as much as 45F° (25C°), depending on conditions. In particular, the bottom flange concrete, in which the critical initial compression stress occurs, may be the coolest. This fact should be taken into account when evaluating the maturity and strength of the concrete in different locations of the girder.

2. The high compressive stresses used in HPC girders are likely to lead to prestress losses that are somewhat larger than those experienced by less highly stressed girders.

3. The instrumentation being used is sufficiently reliable and accurate to permit good computations of the prestress loss as a function of time, which will permit a realistic evaluation of the calculation methods advocated by AASHTO[1] and other agencies.

4. Preliminary comparisons of creep and shrinkage in the field girders and in the laboratory material specimens suggest that the effects of volume/surface ratio may be larger than previously expected.

REFERENCES

1. "Standard Specifications for Highway Bridges", 1st ed. LRFD, American Association of State Highway and Transportation Officials, Washington, DC.

3. Tadros, M.K, Ghali, A. and Dilger, W. (1975). "Time dependent Prestress Loss and Deflection in Prestressed Concrete Members". *PCI Jo.*, 20(3), May-June, 86-98.

4. PCI Committee on Prestress Losses (1975). "Recommendations for Estimating Prestress Losses". *PCI Jo.* 20(4), July-Aug. 43-75.

2. Zia, P., Preston, H.K., Scott, N.L. and Workman, E.B. (1979). "Estimating Prestress Losses". *Concrete International,* 1(6), July, 32-38.

5. Logan, D. (1997) "Acceptance Criteria for Bond Quality of Strand for Pretensioned Prestressed Concrete Applications". *PCI Jo.* 42(4), March-April, 52-90.

6. Kannel, J., French, C.W. and Stolarski, H. (1997). "Release Methodology for Strands to Reduce End Cracking in Pretensioned Concrete Girders". *PCI Jo.* 42(1), Jan.-Feb., 42-55.

DEFORMATION BEHAVIOR OF LONG-SPAN PRESTRESSED HIGH PERFORMANCE CONCRETE BRIDGE GIRDERS

Shawn P. Gross
Graduate Research Assistant / Ph.D. Candidate
Department of Civil Engineering
The University of Texas at Austin
Austin, Texas, U.S.A.

Kenneth A. Byle
Graduate Research Assistant / M.S.E. Candidate
Department of Civil Engineering
The University of Texas at Austin
Austin, Texas, U.S.A.

Ned H. Burns, Ph.D., P.E.
Zarrow Centennial Professor in Engineering
Department of Civil Engineering
The University of Texas at Austin
Austin, Texas, U.S.A.

ABSTRACT

As part of two research projects involving high performance concrete bridges in Texas, camber is being carefully measured on twenty-six long-span prestressed concrete beams. The use of high performance materials, including concrete with compressive strengths in excess of 96.5 MPa (14,000 psi) and 15 mm (0.6 in.) diameter prestressing strand, allow for spans as long as 47.9 m (157 ft.) to be utilized in these structures. As a result, camber and deflection become important issues during the design and construction stages. In this paper, camber measurements on these long-span beams are presented and compared to predicted values.

INTRODUCTION

Camber is an important property of prestressed concrete beams that can greatly influence the construction and service stages of typical highway bridge structures. Inadequate camber prior to casting of the deck slab may make it difficult, if not impossible, for the contractor to satisfy minimum thickness requirements for the deck slab without creating a sagging (downward deflecting) bridge. Camber differences between adjacent girders may cause similar problems during construction. Excessive camber, though less undesirable than inadequate camber, can result in an uneven riding surface for vehicles traveling across the bridge. Considering these and other potential problems, accurate prediction of camber is very important.

FACTORS AFFECTING CAMBER AND DEFLECTION

Parameters Affecting Camber and Deflection

A large number of parameters can affect the camber and deflection behavior of bridge girders. Concrete material properties which influence the deformation behavior include modulus of elasticity, unit weight, and creep. Composite and noncomposite girder section properties, especially the cross-sectional moment of inertia, have a significant impact on camber and deflection. The span length of a girder will affect its camber and deflection behavior dramatically.

Consider the case of a simply-supported highway bridge girder at release of prestress. The camber and deflection theoretically consists of two components. The first is an upward deflection due to the eccentric prestress force (assumed constant along the length). The second is a downward deflection due to the self-weight of the beam.

$$\Delta_p = \frac{PeL^2}{8EI} \tag{1}$$

$$\Delta_w = \frac{5wL^4}{384EI} \tag{2}$$

$$\Delta_{rel} = \Delta_p - \Delta_w \quad \text{(positive indicates upward camber)} \tag{3}$$

The net camber (or deflection) is given by the difference between the two terms (Equation 3) and is plotted in Figure 1 against the span length for typical material and section parameters. Note the sensitivity of the individual components for long spans. Small errors in estimation of material or section parameters, or in prestress force at release, can have a significant impact on the calculated net camber or deflection.

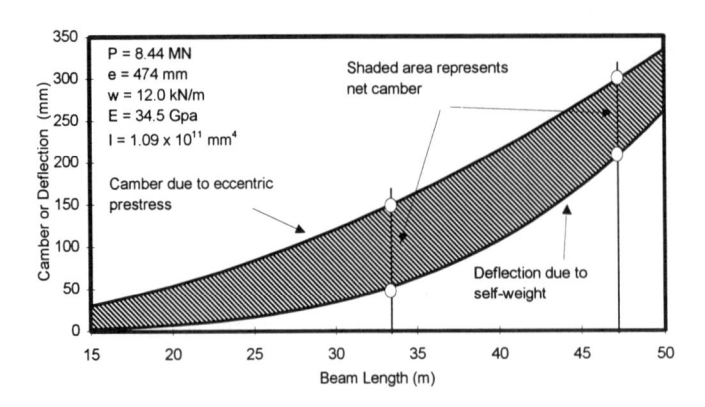

Figure 1. Sensitivity of the components of release camber and deflection to span length

Sources of Camber and Deflection

Camber and deflection in prestressed beams are caused by a wide variety of sources, including:

- pretensioning
- post-tensioning
- self-weight of the member
- weight of the precast deck panels
- weight of the cast-in-place slab
- additional superimposed dead loads
- live loads
- temperature gradients in the member or composite bridge
- early-age thermal effects
- differential shrinkage between the member and the cast-in-place slab

Most of these sources of camber or deflection are related to applied loads or forces. Those sources which are due to sustained loads or forces will be accompanied by a time-dependent camber or deflection due to creep. Some sources are not related to applied loads at all, but are simply phenomena that cause a curvature to develop in the beam. These include temperature gradients and differential shrinkage. Furthermore, some sources are interdependent; camber due to pretensioning or posttensioning is a function of prestress loss at a given time, which depends upon applied loads. Ideally, all of these potential sources should be considered in order to accurately predict the camber or deflection of a beam at a given age.

PROJECT DESCRIPTIONS

Brief project descriptions are presented below. Background information on the two projects is discussed extensively by Ralls[1], and details of the design and fabrication processes are presented by Powell[2] and Patton[3], respectively. Measured material properties obtained as part of extensive quality control programs are given by Carrasquillo and Myers[4].

Louetta Road Overpass (Houston, Texas)

The Louetta Road Overpass on S.H. 249 in Houston consists of two three-span simply-supported highway bridge structures in which all components were constructed using HPC. The Texas U54 beam, an open-top U-shaped cross-section with dimensions shown in Figure 2, was used for the Louetta girders. Beam design lengths, spacing, and required concrete strengths are listed in Table 1. The project was let in February 1994 and was completed in May 1997.

North Concho River/U.S. 87/S.O.R.R. Overpass (San Angelo, Texas)

The San Angelo project, let in June 1995, consists of an 8-span Eastbound HPC bridge adjacent to a 9-span Westbound normal strength concrete bridge. All beams monitored in the two San Angelo bridges were AASHTO Type IV sections, whose dimensions are shown in Figure 2. A two stage process (pretension / post-tension) was used for the fabrication of the Eastbound HPC beams, while a single-stage (pretension) process was used for the Westbound normal strength concrete beams. Beam design lengths, spacing, and required concrete strengths may be found in Table 1. The Westbound normal concrete bridge was completed early in 1997, while the Eastbound HPC bridge is still under construction.

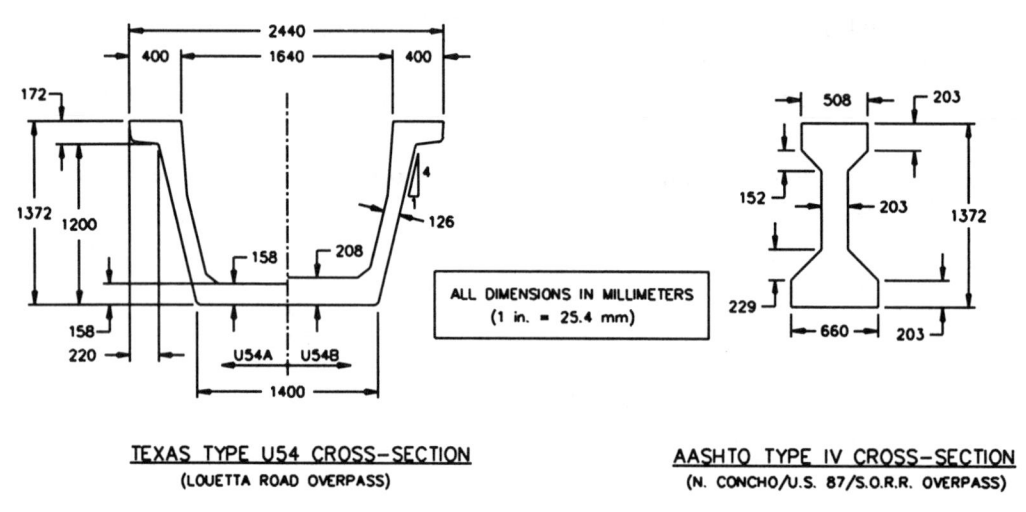

Figure 2. Beam cross-sections for the Louetta Road and N. Concho/U.S. 87/S.O.R.R. Overpasses

Table 1. Summary of the geometric and design material properties of instrumented beams in the Louetta Road and N. Concho/U.S. 87/S.O.R.R. Overpasses

	Num. of Beams Monitored	Cross-section Type	Design Lengths (m)	Spacing (m)	Release Strength (MPa)	Design Strength (MPa)
Louetta HPC Beams (Northbound & Southbound)	12	Texas U54	35.94 - 41.58	3.73 - 4.94	53.1 - 60.7	80.0 - 90.3
San Angelo HPC Beams (Eastbound)	10	AASHTO Type IV	39.30 - 46.74	2.01 - 3.35	55.2 - 60.7	89.6 - 95.1
San Angelo Normal Strength Concrete Beams (Westbound)	4	AASHTO Type IV	39.30	1.73	40.5	55.2
(1 m - 3.281 ft; 1 MPa = 6.895 ksi)							

FIELD MEASUREMENT OF CAMBER AND DEFLECTION

Measurement Systems

Measurement of elastic and time-dependent camber in the field was accomplished by means of two systems: a tensioned wire and ruler system, and a precise surveying system. For the U54 beams in the Louetta Road Overpass and the Type IV beams in the Westbound bridge of the San Angelo project, the tensioned wire system was utilized from fabrication through transportation to the job site. The precise surveying system was then implemented after erection. For the Type IV beams in the Eastbound San Angelo bridge, both systems were used from fabrication through transportation and the precise surveying system will be used for all measurements after erection of the beams.

In the tensioned wire system, bolts were retrofit into each beam at the ends and at midspan. The bolt locations at the ends corresponded to the design bearing locations of the beam. The wire was fixed to a bolt at one end and was free at the other end. A weight was used to tension the wire in the system. At the end where the weight was applied, a grooved roller bearing was attached to the bolt so that the wire could slide freely as it was tensioned. Measurements were taken by first tensioning the wire with the weight, and then reading where the wire crossed on the ruler which was fixed to the retrofit bolts at midspan. A mirror was used to eliminate parallax during readings. A baseline measurement was recorded before release, and all subsequent readings were compared to the baseline reading to determine the camber. The accuracy of the rulers used in this system was 0.3 mm (0.01 in).

In the precise surveying system, beam elevations were recorded at the supports and midspan of each beam using a level and rod. Surveying points were marked on the underside of each beam with paint. This allowed readings to be taken at the same location each time. For readings on the Eastbound San Angelo HPC beams at the precast plant, bolts were retrofit into the bottom flange to be used for measuring relative elevations. To obtain accurate readings, the level was set up as close as possible to the rod and precision rulers like those used in the tensioned wire system were fastened to the rod. Readings under this system were read to 0.5 mm (0.02 in.) accuracy on the rulers.

Both systems worked well and had acceptable repeatability. On the Eastbound San Angelo beams, where both systems were used, there was a 1.5 mm (0.06 in.) average difference between the camber measurements using the two systems. The most significant problem with both systems was difficulty with repeatability in windy conditions.

Measurement Corrections

Analytical corrections were applied to the raw camber measurements to account for thermally induced camber and deflection and the effects of shortened support conditions during storage. Thermally induced camber or deflection is caused by temperature gradients in the beam. Measured temperature gradients from embedded instrumentation[5] were used to calculate an approximate camber or deflection due to the gradient using a procedure outlined by Priestley[6].

The correction for varying support locations adjusted the measured camber to a value that would be read if the span length were the same as the design span length. The theoretical beam weight deflection using the measured support locations was compared to the theoretical beam weight deflection for the design span length, and the difference between the two values was used to adjust the measured camber to the design span length.

Measured Camber Values

Measured camber readings at three stages are reported in Table 2 for all twenty-six beams. Release measurements were recorded immediately after release of prestress, before the beam was moved to storage. Erection measurements were taken shortly after erection of the beam in the bridge and before erection of any panels. Long-term measurements correspond to readings in the completed composite bridge, well after casting of the deck. Where possible, measurements have been corrected for temperature gradients. Support location corrections were not necessary at any of the stages presented here.

Table 2. Measured camber for all instrumented beams

Louetta Road Overpass (Houston)				N. Concho River Overpass (San Angelo)			
Beam	At Release	At Erection	Long-Term[1]	Beam	At Release	At Erection	Long-Term[1]
S14	61.8	98.3	35.6	W14[2]	25.7	39.6	-17.5
S15	62.4	101.9	34.3	W15[2]	21.8	43.4	-16.0
S16	49.1	102.1	51.1	W16[2]	15.5	40.9	-18.8
S24	50.8	84.8	2.3	W17[2]	20.3	37.6	-29.2
S25	44.1	85.9	2.5	E13	36.6	n.a.	n.a.
S26	84.5	131.1	62.7	E14	37.3	n.a.	n.a.
N21	80.5	136.4	77.7	E24	5.1	164.6[1,3]	n.a.
N22	46.4	88.6	16.0	E25	-9.4	154.7[1,3]	n.a.
N23	45.0	83.1	11.4	E26	9.9	152.4[1,3]	n.a.
N31	78.1	132.6	76.7	E33	15.2	163.1[1,3]	n.a.
N32	63.8	107.7	35.1	E34	12.4	153.2[1,3]	n.a.
N33	65.9	99.8	29.0	E35	24.6	178.3[1,3]	n.a.
				E44	13.7	182.9[1,3]	n.a.
				E45	12.4	168.4[1,3]	n.a.

(1 in = 25.4 mm)
1. These measurements were not corrected for temperature movements.
2. Normal strength beams (not HPC)
3. Six pretensioned strands (bonded only at ends of beam) in top flange required for stability have been cut.

COMPARISON OF MEASURED AND PREDICTED CAMBER

Analytical Time-Step Prediction Model

Analytical predictions of camber for all beams were determined using a time-step method. Calculations were performed using a personal computer spreadsheet program developed by the authors. All predictions were performed using known material properties, the actual construction sequence, and transformed section properties. The accuracy of the method was increased by using several small time intervals. Camber was calculated at each time interval by considering the prestress loss, creep parameters, and applied loads at that interval. The effects of differential slab shrinkage were also considered. A plot of the predicted camber response and measured camber response for a typical Louetta HPC beam is presented in Figure 3.

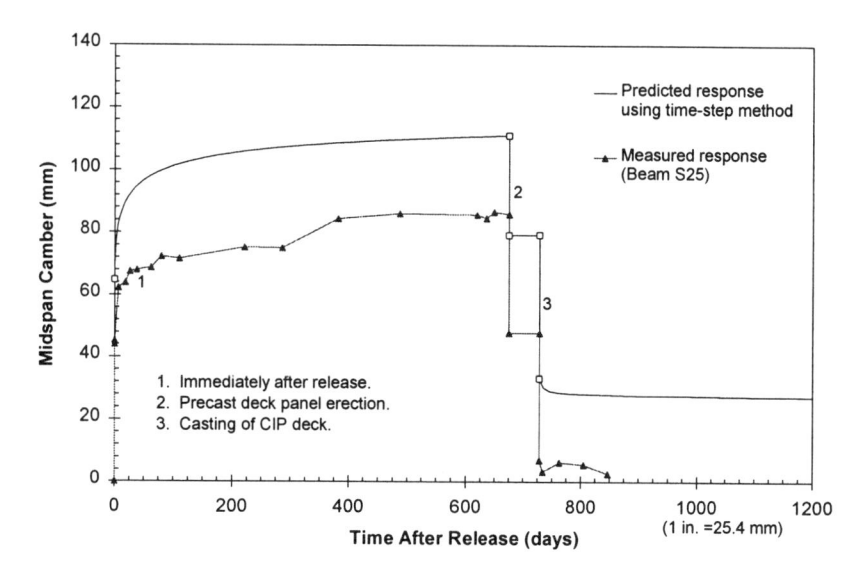

Figure 3. Measured and predicted time-dependent midspan camber for a typical U-beam

Camber at Release

Camber was measured immediately after release of prestress force for all beams. A comparison of the measured and predicted camber responses for all three types of beams is presented in Table 3. Measured release camber was observed to be less than predicted for all 26 beams. The difference between measured and predicted camber measurements ranged from 1 mm (0.04 in.) to 24 mm (0.94 in.), and was rather large on average for all three beam types. Ratios of measured to predicted cambers at release varied tremendously, and in many cases were extremely small. This ratio is clearly a poor indicator of the accuracy of the prediction because the net camber is the algebraic sum of upward and downward components.

Note the wide range of measured cambers observed among the three beam types. The Louetta HPC U54 beams exhibited much greater release cambers than either type of San Angelo beam because the initial prestress force was significantly higher and applied at a very large eccentricity. For either of the San Angelo beam types, the camber due to prestress did not significantly overcome the deflection due to self-weight of the long-span girders. (Note that San Angelo HPC beams were post-tensioned with 20 additional 15 mm (0.6 in.) strands a few weeks after release of pretensioning, resulting in a significant increase in camber.)

Table 3. Summary of Camber at Release

	Measured Release Camber mm	Difference (Measured - Predicted) mm	Ratio (Measured / Predicted)
Louetta HPC U-Beams			
Mean	61.0	-12.2	0.82
Range	44.1 to 84.5	-20.7 to -1.0	0.68 to 0.99
Standard Deviation	14.3	5.8	0.10
San Angelo Normal Type IV Beams			
Mean	20.8	-9.3	0.69
Range	15.5 to 25.7	-14.6 to -4.4	0.51 to 0.85
Standard Deviation	4.2	4.2	0.14
San Angelo HPC Type IV Beams			
Mean	15.8	-16.8	0.39
Range	-9.4 to 37.3	-24.1 to -8.7	-0.64 to 0.77
Standard Deviation	14.1	5.2	0.41
1 mm = 0.039 in.			

Time-Dependent Camber

Time-dependent camber generally followed the form shown in Figure 3. Qualitative comparisons are made here because there is no convenient method of comparing the time-dependent response of all three sets of beams. Measured and predicted camber curves essentially exhibited the same shape between release and erection, with the offset at release remaining relatively constant through the storage period. Where minor variations in the shapes of the two curves did occur, the creep parameters used in the analysis were probably slightly in error. For the San Angelo HPC Type IV beams, the elastic responses due to post-tensioning were somewhat erratic. Predictions of elastic responses to the application of dead loads were reasonably accurate for all beams.

Long-Term Camber

Measured and predicted long-term camber values are compared in Table 4. Long-term measurements represent readings taken a significant time after casting of the composite deck. As with the readings at previous time stages, the measured camber is less than predicted for long-term measurements in the completed composite bridges. The range of long-term camber measurements, reported in Table 2, is 2 to 78 mm (0.1 to 3.1 in.) for the Louetta HPC beams and -29 to -16 mm (-1.1 to -0.6 in.) for the San Angelo normal strength concrete beams. The wide range of values in the Louetta beams is due in large part to the variety of span lengths, spacing, and strand patterns used in the design. Note that all of the monitored San Angelo normal strength concrete beams exhibit a net downward deflection under permanent dead load. For aesthetic reasons, and to assert the confidence of the general public, a net downward deflection under dead load is undesirable. The ratios of measured to predicted camber are again meaningless because the net camber is the algebraic sum of several terms.

Table 4. Comparison of Measured and Predicted Long-Term Camber

	Difference (Measured - Predicted) mm	Ratio (Measured / Predicted)
Louetta HPC U-Beams		
Mean	-19.8	0.56
Range	-29.9 to -5.2	0.07 to 0.92
Standard Deviation	6.7	0.27
San Angelo Normal Type IV Beams		
Mean	-4.7	1.28
Range	-10.3 to -1.4	1.10 to 1.54
Standard Deviation	3.9	0.19
1 mm = 0.039 in.		

DISCUSSION

Analysis of Error in Predicted Camber

In summary, measured camber for all beams was significantly lower than predicted at release and remained less throughout the duration of the measurements, including the erection and long-term stages. The difference was slightly higher or lower at erection and long-term than at release, depending on which set of beams is being discussed. In other words, the shapes of the measured and predicted camber curves were quite similar, with the difference generally appearing as a relatively constant amount over time.

Analysis of potential sources of error for time-dependent stages can be very difficult because so many parameters influence camber at these stages. At release, however, there are no time-dependent effects and a straightforward analysis can theoretically be performed using the parameters found in Equations 1, 2, and 3. Upon examining these equations, potential sources of error include the magnitudes of the self-weight of the member, prestress force, and modulus of elasticity. The span length, eccentricity, and section moment of inertia are assumed to be known accurately. For purposes of discussion, consider that the predicted values each camber or deflection component calculated in Equations 1 and 2 are on the order of 100 to 150 mm (4 to 6 in.). Note that this represents a typical value for most of the beams in this study.

Errors in estimating the self-weight of the member and prestress losses are probably minimal. The concrete unit weight was measured during batching at the precast plant, and prestress losses were monitored on most of the beams in the study[5]. Furthermore, the girder self-weight was increased slightly to account for the extra weight of the reinforcement. As a result, the error in estimation of each can safely be assumed to be on the order of 3 percent or less. In a worst case scenario, each component of camber (or deflection) at release would be in error by 4.5 mm (0.18 in.) and the total error in prediction of camber at release would be 9 mm (0.36 in.).

The effect of an error in estimation of the modulus of elasticity is inversely proportional to the magnitude of the net camber at release, since the modulus is inversely proportional to each component. An underestimation of the modulus by 5 percent would result in an overestimation of the release camber by 5 percent. Modulus of elasticity was also measured on all instrumented beams, and thus despite scatter in the data, it is highly unlikely that the assumed values are in error by more than 5 to 10 percent. In the worst case scenario, predicted camber values would be in error by 10 percent. Furthermore, increases in the predicted modulus of elasticity reduce prestress losses due to elastic shortening, and thus reduce errors in camber estimates due to underestimation of prestress losses.

Again assuming that each camber and deflection component given in Equations 1 and 2 is on the order of 100 to 150 mm (4 to 6 in.), and that the net camber at release is thus on the order 50 mm (2 in.), a reasonable worst-case estimation might be in error by 12 to 15 mm (0.47 to 0.59 in.). This cumulative error, though unlikely to occur, could explain the low observed camber for about half of the beams in the study. However, it cannot explain the deficit of camber at release in beams where the error was as large as 24 mm (0.96 in.).

While some of the error may be attributed to the estimation of these parameters, there is clearly another source or sources of low camber. One possible source may be early-age temperature gradients that develop as the beam is hydrating during the first 24 hours after casting. Temperature measurements during casting show the existence of a significant gradient on the order of 10 °C (18 °F) between the top and bottom of the beam.[5] Since the beam will cool to a uniform temperature over time, the top must shorten more than the bottom. A negative (downward deflection causing) curvature, and resulting deflection are developed. Strain measurements taken during early-ages verify that the top of the beam is shortening more than the bottom.[5]

Since the magnitude of the deflection due to thermal effects would be a function of the length squared, long spans would show a greater lack of camber. The magnitude of this thermal deflection, and the modified equation to determine camber at release would be the following:

$$\Delta_t = \frac{\Phi_t L^2}{8} \qquad (4)$$

$$\Delta_{rel} = \Delta_p - \Delta_w - \Delta_t \qquad (5)$$

Errors in time-dependent camber are largely a function of the error in release camber. Camber growth during storage was predicted with reasonable accuracy, and differences are likely caused by errors in the estimation of creep. Recall that creep is a function of numerous factors, including average relative humidity and volume-to-surface ratio. Creep magnitude is highly variable, and cannot be predicted without some degree of uncertainty.

Impact on Design

As the use of HPC becomes more prevalent in standard highway bridges, longer spans will be accommodated using standard girder cross-sections. As was the case with the San Angelo HPC Type IV beams, some designs will be governed by camber and deflection (serviceability) criteria. Because of the extreme sensitivity inherent in longer spans, predictions will require accurate knowledge of material and section properties. Designers would be wise to allow for a factor of safety against downward deflection under full dead loads, as excessive camber is much less of a problem that insufficient camber. All factors which could contribute to the net camber or deflection of the girder should be considered, including the possibility of an early-age thermal gradient component.

SUMMARY

The following observations have been made in this research program:

1. Prediction of camber and deflection is highly sensitive for long-span beams such as those monitored in this research program.
2. Lower than predicted cambers were observed on all 26 beams measured in this study at release and long-term stages.
3. Errors in estimation of prestress loss, member-self weight, and modulus of elasticity could contribute to low observed camber. It is highly unlikely that they are the only cause of low camber.
4. Thermally induced deflection prior to release may contribute to low camber. This deflection is due to the cooling of the member after hydration, at which a temperature gradient exists, to a uniform temperature state.
5. Errors in prediction of camber growth over time are likely due to differences between the creep model used for prediction and the actual creep history experienced by the member.

ACKNOWLEDGMENTS

The authors wish to thank the joint sponsors of these research projects, The Federal Highway Administration and The Texas Department of Transportation for their support and encouragement. In addition, the authors would like to thank the fabricators and contractors (Williams Brothers Construction Co., Jascon Inc., Texas Concrete Company, and Bexar Concrete Works) involved in the many phases of these projects for their assistance, patience, and interest.

REFERENCES

1. Ralls, M. L., "High Performance Concrete Bridge Construction in Texas," PCI/FHWA International Symposium on High Performance Concrete, Oct. 20-22, 1997.

2. Powell, L. C., "Design and Analysis of Pretensioned/Post-Tensioned Long-Span High Performance Concrete I-Beams," PCI/FHWA International Symposium on High Performance Concrete, Oct. 20-22, 1997.

3. Patton, B., "A Fabricator's Challenges," PCI/FHWA International Symposium on High Performance Concrete, Oct. 20-22, 1997.

4. Carrasquillo, R. L., and Myers, J. J., "Quality Control and Quality Assurance Program for Precast Plant Produced High Performance Concrete U-Beams," PCI/FHWA International Symposium on High Performance Concrete, Oct. 20-22, 1997.

5. Burns, N. H., Gross, S. P., and Byle, K. A., "Instrumentation and Measurements - Behavior of Long-Span Prestressed High Performance Concrete Bridges," PCI/FHWA International Symposium on High Performance Concrete, Oct. 20-22, 1997.

6. Priestley, M. J. N., "Design of Concrete Bridges for Temperature Gradients," ACI Journal, May 1978, pp. 209-217.

RESEARCH AND UTILIZATION OF HIGH PERFORMANCE CONCRETE IN NORTH CAROLINA

Azam Azimi
State Value Management Engineer
North Carolina Department of Transportation
Raleigh, North Carolina, U.S.A.

Paul Zia
Distinguished University Professor, Emeritus
North Carolina State University
Raleigh, North Carolina, U.S.A.

ABSTRACT

High performance concrete (HPC) has been made and used by prestressed concrete producers for many years. The producers use the high early strength of the concrete to maintain the one-day production cycle without, in most cases, taking advantage of the eventual 55 to 69 MPa (8,000 to 10,000 psi) compressive strength in the structural design or the improved durability for a long-term economy.

This paper discusses the structural efficiency and cost effectiveness of using high performance concrete for highway bridge construction including prestressed concrete hollow core slabs as well as girders. The experience in the design and construction of several HPC highway bridges in North Carolina is documented.

INTRODUCTION

Since 1959, the North Carolina Department of Transportation (NCDOT) has provided funding for a cooperative highway research program with the North Carolina State University (NCSU). Through the cooperative program, the faculty and students at NCSU conduct research studies in highway transportation in support of NCDOT in the management, design, construction, and operation of its extensive highway systems. During the past decade, several investigations relative to HPC have been completed at NCSU that provided the essential information for NCDOT to extend its bridge design practices using HPC. This paper summarizes the results of the research studies and describes the experience of NCDOT in the design and construction of several bridges using high performance concrete.

HPC RESEARCH AT NCSU

The NCDOT bridge design standards for prestressed concrete slabs and girders traditionally call for a compressive strength of 34 or 41 MPa (5,000 or 6,000 psi) at 28 days and a corresponding compressive strength of 24 or 28 MPa (3,500 or 4,000 psi) at prestress release. In order to meet the required concrete strength at prestress release and to maintain a one-day production cycle, the prestressed concrete producers usually design their concrete mixtures to produce a compressive strength of well over the required strength at 28 days. Generally this over strength is not utilized in the design of bridge girders.

Recognizing the potential advantages of using high strength concrete in bridge construction, NCDOT initiated a two-year research project in 1985 to study the structural properties of high strength concrete made of the materials which are locally available in North Carolina. The study by Leming et al.[1] demonstrated that compressive strength of concrete in excess of 103 MPa (15,000 psi) at 28 days can be achieved with strict quality control of selection of raw materials and production procedures. With minor changes in standard quality control procedures, it is possible to produce concrete with a compressive strength of over 69 MPa (10,000 psi) on a commercial basis from a "dry batch" concrete plant. Data on the elastic properties, creep, and shrinkage of high strength concrete were obtained with several different materials. In addition, the influence of testing variations such as mold size, specimen preparation and curing regimen was investigated to provide a basis for changes in specifications. Further work of a limited scope was conducted on lightweight concrete and high early strength concrete for use in precast concrete operations.

To determine the structural efficiency and cost effectiveness of using higher strength concrete for highway bridges, a series of parametric studies were initiated in 1986 by Zia et al.[2] Included in the investigation were prestressed concrete hollow core slabs, prestressed concrete AASHTO girders, and reinforced concrete piers. For the hollow core slabs, the concrete compressive strength varied from 34 MPa (5,000 psi) to 97 MPa (14,000 psi). For the girders and piers, concrete strengths varying from 41 MPa (6,000 psi) to 83 MPa (12,000 psi) were considered. It was concluded that the use of high strength concrete in bridges is economical and practical, especially for the purpose of increasing the span range of the current hollow core slab bridges for which a concrete strength of 55 MPa (8,000 psi) would be optimum. For the standard AASHTO girder sections, their maximum spans could be increased by 30 percent if the concrete strength was increased from 41 MPa (6,000 psi) to 83 MPa (12,000 psi). For the box beams, as much as a 20 percent increase in span lengths was obtained. In general, a concrete strength of 69 MPa (10,000 psi) was most desirable for the girder and beam sections. The study of the pier sections indicated clearly the economic benefits of high strength concrete. The cost of carrying a specific level of axial load decreased with increasing concrete strength.

In 1989, a four-year study[3-8] of the mechanical behavior of high performance concretes was initiated at NCSU with the sponsorship of the Strategic Highway Research Program (SHRP) of the National Research Council. This extensive study included a variety of coarse and fine aggregates, cement, chemical and mineral admixtures. Three types of high performance concrete were investigated; namely, concrete of very early strength (VES), high early strength (HES), and very high strength (VHS). A total of 360 trial batches of concrete[4] were produced and evaluated. The tensile and compressive properties, creep, shrinkage, bond, and durability characteristics were determined. The study demonstrated that:

- It was feasible to produce the HPC for highway applications with conventional materials and normal techniques.
- The concrete with 5% entrained air had greatly enhanced frost durability.
- The different types of HPC behaved very much like conventional concrete of comparable strength, except that the HPCs developed their strength characteristics much more rapidly than the conventional concrete.
- The various mechanical properties, such as the modulus of elasticity, the splitting tensile strength, and the flexural tensile strength of the HPCs could all be predicted by the ACI relationships at the appropriate strength levels.
- Shrinkage of the HPCs was considerably less than that of conventional concrete because of the very low W/C ratio used for HPC.
- Creep strains of the different groups of VHS concrete ranged from only 20% to 50% of that of conventional concrete, also because of low W/C ratio and higher compressive strength. Since marine marl is a softer aggregate, the specific creep of the concrete with marine marl was much higher than that with either crushed granite or washed rounded gravel.

Based on the experience of the SHRP research, two additional studies were pursued to develop high early strength concrete with locally available materials in North Carolina and normal production and curing procedures used in precast/prestressed concrete plant. Barcomb[9] produced concrete that would achieve a target compressive strength of 34 MPa (5,000 psi) in 18 hours, and Zia and Hillmann[10] developed three mixture proportions for field production that would achieve a compressive strength of ±48 MPa (7,000 psi) in 18 hours and ± 69 MPa (10,000 psi) in 28 days. Typical results of these two studies are shown in Table 1.

UTILIZATION OF HPC/HSC BY NCDOT

North Carolina, because of its topography and rural atmosphere, has many miles of rural highways. A large number of short- to medium-span timber bridges are part of this rural highway system. In the coming years, due to structural deficiency or functional obsolescence, many of these bridges will require rehabilitation or replacement. A prestressed concrete cored slab bridge could be a cost effective alternative to the current timber design. On the other hand, a long-span prestressed concrete girder bridge could be used economically in many sites throughout the state. Concrete bridges, due to their durability and economy, are widely used in bridge construction in North Carolina. For example, more than 70% of the new bridges designed in North Carolina in 1996 are either prestressed concrete girders or hollow core slabs.

Table 1 – Mixture Proportions of HES Concrete

Material	Barcomb	Zia & Hillmann
Cement kg/m^3 (lbs/cy)	Type I 517 (870)	Type III 496 (836)
Coarse Aggregate, SSD kg/m^3 (lbs/cy)	1046 (1,760)	1023 (1,725)
Fine Aggregate, SSD kg/m^3 (lbs/cy)	618 (1,040)	602 (1,015)
HRWR L/m^3 (oz/cwt)	Daracem 28.25 (84)	Melment 7.4 (23)
AEA L/m^3 (oz/cwt)	MBVR 1.68 (5)	Daravair-R 0.45 (1.4)
Water kg/m^3 (lbs/cy)	119 (200)	173 (292)
W/C	0.34	0.35
Slump mm (in.)	114 (4.5)	140 (5.5)
Air Content %	5.7	4.5
Compressive Strength, MPa (psi) 17 hours 18 hours 28 days	35 (5,020) — 49 (7,040)	— 46 (6,650) 69 (10,050)

Although normal strength concrete is widely used, its low strength-to-weight ratio limits its use in longer span bridges. As mentioned earlier, NCDOT had recognized the potential that HPC could offer, and was accordingly involved in cooperative research on the subject with NCSU.

To incorporate the findings and the recommendations of the research described earlier, NCDOT developed its first HPC specifications of 55 MPa (8,000 psi) in 1989 and reviewed its upcoming bridge projects for a suitable candidate to utilize HPC.

Since then, a total of thirty-six projects have been designed with HPC. These projects are summarized in Table 2. Reduction in size and increase in span-length of the precast prestressed concrete members along with improved durability significantly influenced the choice of HPC for these projects.

Table 2 - NCDOT'S HPC Bridge Projects

Project	Letting Date	County	Concrete Strength @ 28 Days, MPa (psi)	Member Type
1	Jun-91	Iredell	55 (8,000)	21" CS
2	Aug-93	Forsythe	55 (8,000)	45" PCG
3	Apr-94	Catawba	45 (6,500)	54" PCG
4	Jun-94	Wake	55 (8,000)	36" PCG
5	Aug-94	Randolph	45 (6,500)	1'-9" CS
6	Nov-94	Harnett	45 (6,500)	1'-9" CS
7	Feb-95	Edgecombe	49 (7,000)	45" PCG
8	Mar-95	Chatham	45 (6,500)	1'-6", 1'-9" CS
9	Apr-95	Forsythe	55 (8,000)	45", 54" PCG
10	Jun-95	Wake	55(8,000)	54" PCG
11	Jun-95	Ashe	45 (6,500)	1'-9" CS
12	Jul-95	Anson	49 (7,000)	54" PCG
13	Jul-95	Forsythe	55 (8,000)	36" PCG
14	Jul-95	Davie	49 (7,000)	1'-9" CS
15	Aug-95	Warren	45 (6,500)	1'-9" CS
16	Aug-95	Cumberland	49 (7,000)	45" PCG
17	Sep-95	Burke	49 (6,900)	54" PCG
18	Nov-95	Wilson	52 (7,500)	54" PCG
19	Jan-96	Mitchell	49 (7,000)	54" PCG
20	Feb-96	Lee	45 (6,500)	45" PCG
21	Feb-96	Guilford	55 (8,000)	1'-6" CS
22	Feb-96	Alleghany	49 (6,900)	45" PCG
23	Feb-96	Greene	55 (7,800)	54" PCG
24	Jun-96	Gaston	49 (7,000)	45" PCG
25	Jun-96	Martin	45 (6,500)	54" PCG
26	Jul-96	Pitt	55 (8,000)	54" PCG
27	Aug-96	Chatham	55 (8,000)	54" PCG
28	Aug-96	Bertie	42 (6,000)	36" PCG
29	Sep-96	Cumberland	55 (8,000)	54" PCG
30	Oct-96	Yancey	55 (8,000)	1'-9" CS
31	Oct-96	Guilford	45 (6,500)	54" PCG

span bridges carrying US 401 over the Neuse River near Raleigh in Wake County, North Carolina. The longer spans will be 28 m (91.9 ft), and the shorter spans will be 17.5 m (57.4 ft) long. AASHTO Type IV and Type III prestressed concrete girders will be used for these spans respectively. The girders will be designed with HPC of 70 MPa (10,000 psi). The concrete deck will be 42 MPa (6,000 psi). Compared to normal strength concrete, the use of HPC will allow the NCDOT to eliminate one row of girders throughout the length of each bridge. This will also increase the girder spacing to about 3.2 m (10 ft 6 in). The out-to-out width of the north-bound bridge will be 18 m (59 ft) with six girder lines. The out-to-out width of the south-bound bridge will be 14.4 m (47 ft 2 in) with five girder lines. This project is scheduled to be let to contract in July 1998. Currently, NCDOT is also considering the utilization of HPC for two major coastal bridges. These are the Oregon Inlet and Manteo by-pass bridges.

Fig. 1 Completed Hollow Core Slab Bridge with HPC in the Central Span

Fig. 2 Completed Three-span Continuous HPC Prestressed Girder Bridge

CONCLUSION

Since 1990 NCDOT has had good experience in utilizing HPC of up to 55 MPa (8,000 psi) in its bridges. It is anticipated in the very near future to extend the use of higher strength concrete up to 70 MPa (10,000 psi). In addition to providing design solutions and some critical cost savings, it is expected that HPC structures will be maintenance free and the higher impermeability and density of HPC will provide additional resistance to road salts and other environmental pollutants.

ACKNOWLEDGEMENTS

The authors would like to thank NCDOT for its support of HPC research conducted at NCSU. Partial support for the research was also provided by Precast/Prestressed Concrete Institute. The staff of the Bridge Design, Materials, Construction, and Maintenance Units of NCDOT were especially helpful in providing the necessary information for this paper. Cooperation from Gary Concrete Products where the hollow core slab units and the AASHTO girders for the first two bridges were produced are much appreciated. Finally, the authors are indebted to Mrs. Shannon Bishop for her assistance in the preparation of this paper.

REFERENCES

1. Leming, M. L., Tallman, T. E., Altimore, F., Nunez, R., and Salandra, M., "Properties of High Strength Concrete: An Investigation of High Strength Concrete Characteristics Using Materials in North Carolina," Report No. FHWA/NC/88-006, North Carolina Department of Transportation, Raleigh, N.C., July 1988, 202 pp.

2. Zia, P., Schemmel, J. J., and Tallman, T. E., "Structural Applications of High Strength Concrete," Report No. FHWA/NC/89-006, North Carolina Department of Transportation, Raleigh, N.C., June 1989, 330 pp.

3. Zia, P., Leming, M. L., Ahmad, S. H., Schemmel, J. J., Elliot, R. P., and Naaman, A. E., "Mechanical Behavior of High Performance Concretes, Volume 1: Summary Report," SHRP-C-361, Strategic Highway Research Program, National Research Council, Washington, D.C., October 1993, xi, 98 pp.

4. Zia, P., Leming, M. L., Amhad, S. H., Schemmel, J. J., and Elliott, R. P., "Mechanical Behavior of High Performance Concretes, Volume 2: Production of High Performance Concrete," SHRP-C-362, Strategic Highway Research Program, National Research Council, Washington, D.C., November 1993, xi, 92 pp.

5. Zia, P., Leming, M. L., Ahmad, S. H., Schemmel, J. J., and Elliott, R. P., "Mechanical Behavior of High Performance Concretes, Volume 3: Very Early Strength Concrete," SHRP-C-363, Strategic Highway Research Program, National Research Council, Washington, D.C., November 1993, xi, 116 pp.

6. Zia, P., Leming, M. L., Ahmad, S. H., Schemmel, J. J., and Elliott, R. P., "Mechanical Behavior of High Performance Concretes, Volume 4: High Early Strength Concrete," SHRP-C-364, Strategic Highway Research Program, National Research Council, Washington, D.C., December 1993, xi, 179 pp.

7. Zia, P., Ahmad, S. H., Leming, M. L., Schemmel, J. J., and Elliott, R. P., "Mechanical Behavior of High Performance Concretes, Volume 5: Very High Strength Concrete," SHRP-C-365, Strategic Highway Research Program, National Research Council, Washington, D.C., November 1993, xi, 101 pp.

8. Naaman, A. E., Al-khairi, F. M., and Hammound, H., "Mechanical Behavior of High Performance Concretes, Volume 6: High Early Strength Fiber Reinforced Concrete (HESFRC)," SHRP-C-366, Strategic Highway Research Program, National Research Council, Washington, D.C., October 1993, xix, 297 pp.

9. Barcomb, J. J., "High Early Strength Concrete for Precast/Prestressed Concrete Application," MS Thesis, Department of Civil Engineering, North Carolina State University, Raleigh, N.C., 1993, ix, 106 pp.

10. Zia, P., and Hillmann, R. S., "Development of High Early Strength Concrete for Prestressed Concrete Applications," Report No. FHWA/NC/96-002, North Carolina Department of Transportation, Raleigh, N.C., June 1995, xiii, 88 pp.

11. Lane, S.N. and Podolny, W., "The Federal Outlook for High Strength Concrete Bridges," PCI Journal, Vol. 38, No. 3 May/June 1993.

APPLICATION OF HIGH PERFORMANCE CONCRETE IN GILES ROAD BRIDGE, NEBRASKA

XIAOMING HUO
Ph.D. Candidate
University of Nebraska
Omaha, Nebraska, USA

MAHER K. TADROS
Cheryl Prewett Professor
University of Nebraska
Omaha, Nebraska, USA

ABSTRACT

High performance concrete (HPC) with improved durability and strength parameters is used in a three span bridge at 120th and Giles Road in Sarpy County, Nebraska. This bridge is 68.6 m (225 ft) in length and 25.8 m (84.5 ft) in width with 30-degree skew angle. The 1100 mm (43.3 in.) deep girders are designed with a concrete compressive strength of 82.74 MPa (12,000 psi) and the 190 mm (7 1/2 in.) thick deck with a concrete compressive strength of 55.16 MPa (8,000 psi).

INTRODUCTION

High performance concrete (HPC) is gaining popularity for use in building the United States infrastructure. The definition of HPC by Federal Highway Administration (FHWA) specifies eight parameters, four strength parameters and four durability parameters, in determining the grade of HPC. The four strength parameters are compressive strength, shrinkage, creep and modulus of elasticity. The four durability parameters are freeze/thaw durability, scaling resistance, abrasion resistance and chloride permeability. Generally speaking, the features of HPC include low water-cementitious material ratio, high strength, low permeability, and improved durability. With these characteristics, highway bridges built with HPC are expected to have longer service life and less maintenance than those with conventional concrete. HPC also provides greater design flexibility by allowing longer spans and requiring fewer girders with smaller sections. As a result, superstructures and substructures are lighter and more efficient[1,2]. In order to encourage the application of HPC, FHWA initiated an HPC Showcase program under its Strategic Highway Research Program (SHRP). The bridge at 120th Street and Giles Road in Omaha, Nebraska is one of the HPC showcase bridges sponsored by FHWA.

BRIDGE DESCRIPTION

The HPC bridge at 120th Street and Giles Road is a stream overpass with an overall length and width of 68.6 m (225 ft) and 25.8 m (84.5 ft), respectively. The sectional elevation of the bridge is shown in Fig. 1. A bridge in similar geometry, but using conventional concrete had been built in the close vicinity of this HPC bridge on the same road. Having two bridges with almost identical layouts, traffic volume, and environment provides an excellent opportunity to compare the initial and life-cycle economics, as well as, the overall structural and functional performance of HPC in bridges with those of conventional concrete.

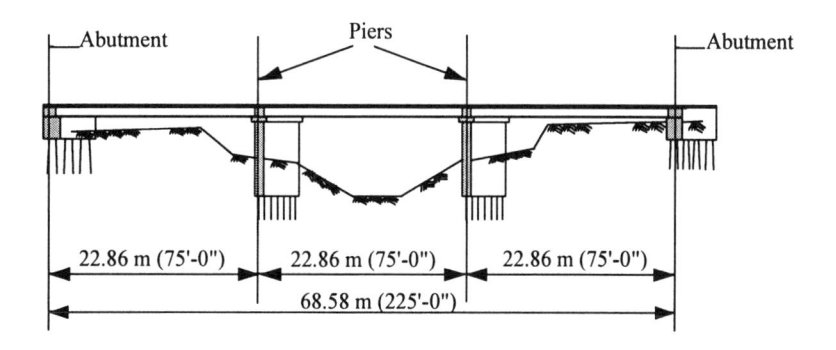

Figure 1. Sectional Elevation of HPC Bridge

Conventional Concrete Bridge Design
The conventional concrete bridge uses 11 Nebraska Type 3A prestressed girder lines with 191 mm (7-1/2 in) thick cast-in-place deck[3]. The depth of a Type 3A girder is 1,143 mm (45 in). Girder continuity for superimposed and live loads is achieved using negative mild reinforcement in the deck. Specified concrete strength for this bridge is 34.47 MPa (5,000 psi) for the girders and 24.13 MPa (3,500 psi) for the deck. The plan view and the typical cross section for this design are shown is Fig. 2 and Fig. 3, respectively.

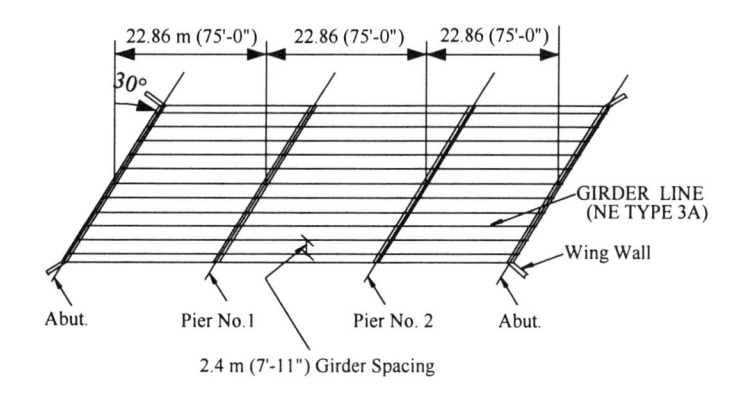

Figure 2. Plan View of Conventional Concrete Bridge

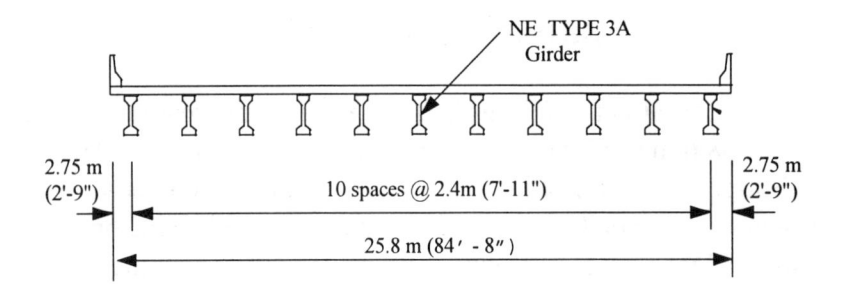

Figure 3. Typical Cross Section of Conventional Concrete Bridge

HPC Bridge

The HPC bridge used a new optimized girder shape NU1100, 1,100 mm (43.3 in) deep, developed by the University of Nebraska. Although these girders are slightly shallower than the Nebraska Type 3A girders used in the conventional concrete bridge, only seven lines of girders were required with the same deck thickness of 190.5 mm (7-1/2 in)[3]. The specified concrete compressive strength is 82.74 MPa (12,000 psi) for the girders and 55.16 MPa (8,000 psi) for the deck. Girder continuity is achieved in a similar way to that of the conventional concrete bridge. The plan view of the HPC bridge is the same as its counterpart with only seven lines of girders instead of eleven. The typical cross section of the HPC bridge is shown in Fig. 4, and the NU1100 girder section is shown in Fig. 5.

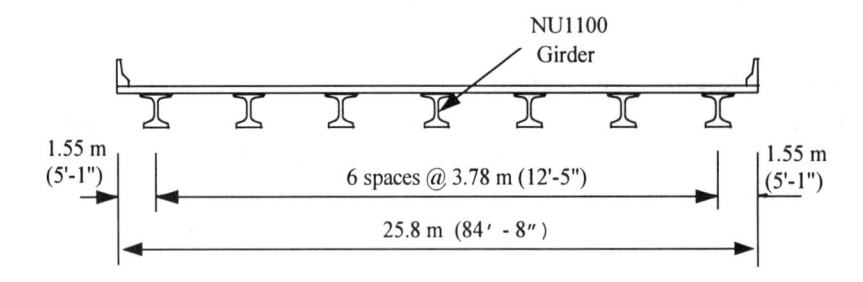

Figure 4. Typical Cross Section of HPC Bridge

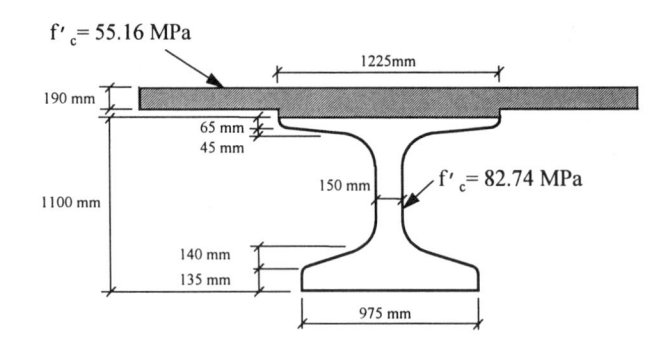

Figure 5. Cross Section of NU1100 Girder Composited with Deck

The NU girder has a very wide and bulky bottom flange which can hold up to 58 strands. For the HPC bridge, a maximum of 30 strands were needed. There were 30 strands at the midspan section and 18 strands at the end sections of the girder. Twelve (12) strands were debonded at the ends. Debonding, rather than draping of the strands, is being promoted nationally as a simpler form of production. Fig. 6 shows the strand details and debond pattern in the sections.

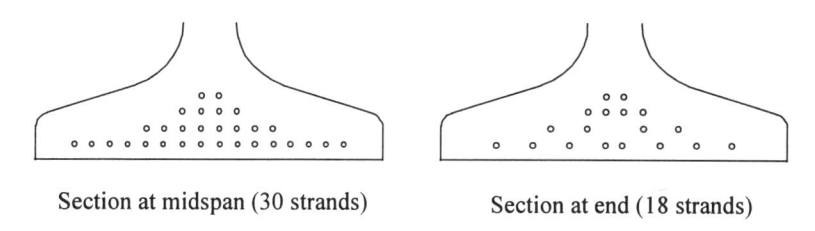

Section at midspan (30 strands)   Section at end (18 strands)

Figure 6. Strand Details in Girder Sections

Influence of High Strength on I-Girder Span Capacity[4]

A comparison between the NU-girder bridge designs with conventional concrete and high strength concrete has been conducted. The bridge studied is a simple span bridge with NU1100 girders. The variables are compressive strength of concrete, type of strands, girder spacing, and number of strands. The analysis results are shown in Fig. 7. Three curves for NU1100 girders are presented in the figure for three cases: conventional concrete with 13 mm (0.5 in.) diameter strands; HPC with 13 mm (0.5 in.) diameter strands; and HPC with 15 mm (0.6 in.) diameter strands. It is shown that using high strength concrete can increase the span capacity tremendously. With 13 mm (0.5 in.) diameter strands, when concrete strength is increased from 41.37 MPa (6,000 psi) to 82.74 MPa (12,000 psi), the girder span capacity can be increased by 20-38% for various girder spacing. Using HPC and 15 mm (0.6 in.) diameter strands, the span capacity is further increased by 7-10%.

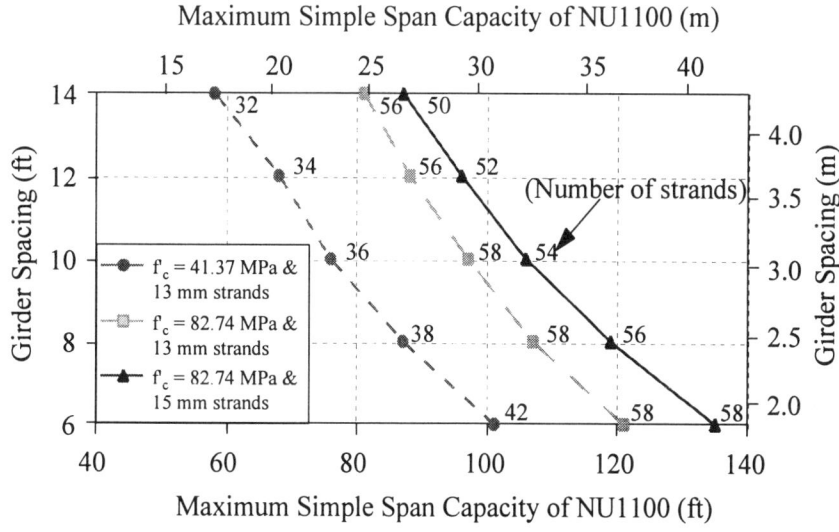

Figure 7. Impact of Using HPC on Simple Span Capacity of NU1100 Girder Bridge

PERFORMANCE CRITERIA OF CONCRETE

As a part of this bridge showcase research project, the research team investigated the viability of obtaining the mixes for the girders and the deck using local aggregates and readily available admixtures. Based on the investigation, and in compliance with the guidelines already established under the SHRP program, the specifications of the mixes were finalized.

Performance Criteria of HPC Girders

For the girder concrete mix, it is specified that:

> Water : Cementitious Ratio .... ≤ 0.28
> Compressive Strength (f'_c) at 56 days   82.74 MPa (12,000 psi)
> Compressive Strength (f'_{ci}) at prestress transfer 37.92 MPa (5,500 psi) [5]
> No air content is specified.

It is well known that compressive strength is inversely related to water/cementitious ratio (w/c), other things being equal. For the currently available Portland cements and additives, the usual mixing and placing methods, and the present curing practices, it has been found that the optimum value of w/c ratio is in the range of 0.22 to 0.38[4,6]. Lower values of w/c ratio are not practical because of the low workability, mixing difficulty, and low quality even with the use of large quantities of superplasticizers.

Performance Criteria of HPC Deck

For the deck concrete mix, it is specified that:

> Compressive strength (f'_c) at 56 days   55.16 MPa (8,000 psi)
> Rapid Chloride Permeability indicator at 56 days   <1800 coulombs
> Air Content $5.0\% \leq x \leq 7.5\%$ [5]

Conventional concrete with relatively high permeability does not provide adequate protection for the reinforcement from deicing salts applied to bridges. Thus, bridges built with conventional concrete require more frequent maintenance and relatively earlier replacement. The major cause of damage is corrosion of reinforcing steel induced by contact with chloride ions from deicing salts or marine environment. The penetration of other ions, especially sulfate, can result in deterioration of the concrete itself. To delay or minimize the deterioration, and thus achieve long-lasting service, concrete with low permeability is needed [6]. Although much effort still needs to be invested, research results thus far have shown that HPC can be used to meet the requirements of low permeability and high durability. The specified Rapid Chloride Permeability (RCP) for the deck mix of this HPC bridge is less than 1800 coulombs at 56 days which is ranked as a low permeability. The actual permeability was measured at 590 coulombs.

MIX DESIGN

Development of concrete mixes for HPC using local material was one of the research objectives. The contractor was provided sample mixes developed by the University of Nebraska, the University of Texas and the University of Minnesota, developed his own mix and provided data on their properties[5]. Tables 1 and 2 show the HPC girder and deck mixes designs that were developed by the contractor. The University researchers and the Nebraska Department of Roads Materials Division engineers then duplicated the contractor's mixes in their laboratories and verified the properties provided by the contractor.

Table 1 - HPC Girder Mix

	Quantity
Cement, Type I (kg/m^3)	445
Fly Ash, Class C (kg/m^3)	119
Silica Fume (kg/m^3)	30
ASTM C33 Sand (kg/m^3)	587
1/2" BRS Limestone (kg/m^3)	1104
Water (kg/m^3)	142
Type-A WR (mL/100 kg)	260
HRWR (mL/100 kg)	1956
W / C	0.24

Note: 1 kg/m^3 = 1.686 lb/yd^3; 1 mL/100 kg = 0.01534 oz/100 lb

Table 2 - HPC Deck Mix

	Quantity
Cement, Type IP (kg/m^3)	445
Fly Ash, Class C (kg/m^3)	44
47-B Sand-Gravel (kg/m^3)	831
1/2" BRS Limestone (kg/m^3)	831
Water (kg/m^3)	151
Air Entrainment (L/m^3)	0.19
Type-A WR (mL/100 kg)	260
HRWR (mL/100kg)	1174
W / C	0.30

Note: 1 kg/m^3 = 1.686 lb/yd^3; 1 mL/100 kg = 0.01534 oz/100 lb

CURING PROCEDURE AND COMPRESSIVE STRENGTH

Curing Procedure

Curing procedures specified for this HPC bridge to ensure quality required that: 1) The curing steam temperatures not exceed 71°C (160°F), and 2) The maximum temperature at the centroid of the bottom flange not exceed 71°C (160°F). Before production of the girders, a trial batch for a 3 m (10 ft) NU1600 girder was cast. The temperature inside of the girder at different points were monitored. Fig. 8 shows the measured temperature during the curing period. The results showed that the temperature at different points of the section were very close to each other and none of them was over 71°C (160°F).

The precast concrete producer monitored the concrete temperatures in the actual production of bridge girders. This experience indicated that lower concrete temperatures resulted in higher ultimate compressive strengths. It was concluded that keeping the concrete temperature in the 27°C - 38°C (80°F - 100°F) range was best for the concrete properties. This was possible to attain as the girders were being produced during a relatively cool period of the year. During hot weather, production may require adding ice to the mixing water. It is interesting to conclude that steam curing of HPC may be an unnecessary expense.

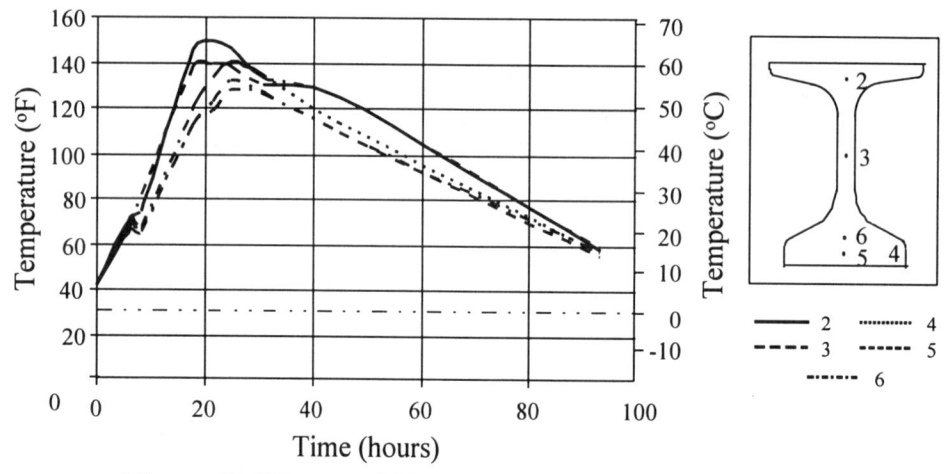

Figure 8. Measured Temperature of the Trial Girder

Compressive Strength

In the Special Provisions of the HPC bridge Specifications, the concrete strength is specified as f'_c = 82.74 MPa (12,000 psi) at 56 days and 37.92 MPa (5,500 psi) at release. The average compressive strength of concrete at release is 58.40 MPa (8,471 psi), which is greater than the required 37.92 MPa (5,500 psi). The acceptance of girders at 56 days is based on the compressive strength of concrete being not less than (82.74) + 1.34s [(12,000) + 1.34s] where s is standard deviation based on 60 cylinders. The contractor was permitted to use 28 day strength at 95% of the 56 day strength as a criterion for acceptance. If the strength requirement was not met, the owner had the option of accepting the girders at a 25% discount or rejecting them. None of the girders produced was discounted or rejected. Test results of field-cured cylinders were required for acceptance. The concrete producer was not satisfied with the requirement of field curing and recommended that standard ASTM curing procedures be used. This issue deserves further investigation[7]. Table 3 summarizes the average cylinder compressive strength of the precast concrete girders at 7, 28 and 56 days for various curing scenarios.

Table 3 - Compressive Strength of HPC Girders

	Field Cured* MPa (psi)	Lab Cured** MPa (psi)	Non-Steam Cured*** MPa (psi)
7 days	72.60 (10,529)	73.35 (10,639)	---
28 days	86.56 (12,554)	93.21 (13,519)	97.61 (14,167)
56 days	96.14 (13,944)	98.42 (14,274)	101.62 (14,739)

*   Cylinders were cured with girders all the time.
**   Cylinders were steam cured with girders and Lab cured with ASTM standards.
*** Cylinders were lab cured following ASTM standards all the time.

Field cured concrete compressive strength at 56 days was 96.14 MPa (13,944 psi). The standard deviation of compressive strength at 56 days was 6.79 MPa (985 psi). The required minimum strength of 82.74 +1.34s = 82.74 + 1.34 (6.79) = 91.84 MPa (12,000 + 1.34s = 12,000 + 1.34 (985) = 13,320 psi), was exceeded since the compressive strength during girder production = 96.14 MPa (13,944 psi).

COST ANALYSIS

The initial cost comparison between the two bridges, similar geometry and function with different concrete and girder section, is shown in Tables 4.

Table 4 Comparison in the Cost between the Conventional and HPC Bridges

Item	Conventional Bridge			HPC Bridge		
	Quantity	Unit price ($)	Total ($)	Quantity	Unit price ($)	Total ($)
Prestressed concrete (m^3)	255	669	170,572	214	971	207,851
Concrete (m^3)	378	320	120,780	395	320	126,420
Epoxy coated steel (kg)	77,733	1.28	99,395	57,696	1.28	73,774
Total			390,747			408,045

Note: 1 m^3 = 1.308 yd^3; 1 kg = 2.205 lb

The cost figures used were taken from the contract bids of each bridge. Only the items which are different in the two bridges are considered. Although the unit cost of HPC material was slightly higher, the overall cost of the superstructure was about the same, not including the labor savings for erecting fewer girder lines. Considering the differences between the various bids at different times, the difference in cost as shown is almost negligible and is expected to be even lower as the contractors get more familiar with HPC. If experience from observation of this bridge indicates longer life cycle and less maintenance, more significant savings would result.

This HPC bridge was completed in July 1996 and opened to traffic in September 1996. It became the first completed HPC bridge in the United States. An aerial view of the bridge is shown in Fig. 9. It demonstrated without a doubt that high performance concrete, which is structurally superior to and more durable than conventional concrete, can be economically used in bridges in the future.

Figure 9. An Aerial View of the HPC Bridge at 120th and Giles Road

LESSONS LEARNED

(1) It is possible to produce precast prestressed concrete girders commercially with concrete strength of 82.74 MPa (12,000 psi), using local materials with or without silica fume but with very larger dozes of superplasticizer.
(2) Concrete strength of 70 MPa (10,000 psi) can be achieved on a daily basis with little added premium.
(3) Due to the high cementitious material content of HPC, the heat of hydration must be controlled. This may create a savings opportunity by eliminating steam curing.

(4) The most important structural property of prestressed concrete girders is the concrete release strength. Higher release strength allows for application of higher prestress and thus translates into higher load resistance. However, this property is the most challenging to achieve.

(5) Strength of CIP deck is not a significant structural requirement. However, it may come as a by-product of low water/cement ratio concrete mixes which is required to achieve low permeability.

(6) Research needs to focus on a CIP deck mix that provides low permeability, low initial shrinkage and low modulus of elasticity, in order to produce the desired high durability.

(7) The deck mix used for the Nebraska bridge was initially more difficult to finish by the contractor as it had no "bleeding" water. The contractor adjusted by applying a vapor retarder.

(8) Curing of the CIP deck concrete was achieved by multiple methods. They included fogging of the freshly placed concrete, spraying of curing compound and finally placement of wet burlap. It is now believed that the two important methods were the curing compound and burlap. Fogging caused delay of construction and was somewhat ineffective due to wind.

(9) Our experience with the HPC precast girders was highly successful. However, our experience with the CIP deck was mixed. Deck performance is a function of many factors, not just material strength or permeability. The structural interaction between the girders and the deck could be one of the factors that affect the performance of deck. Early-age deck shrinkage, which is restrained by the girders, could cause premature transverse deck cracking.

ACKNOWLEDGMENT

The authors acknowledge the financial support of the Federal Highway Administration under its Strategic Highway Research Program (SHRP), the supervision of the Nebraska Department of Roads' Bridge, and Materials and Testing Divisions, and the financial support of the Center for Infrastructure Research at the University of Nebraska-Lincoln. The participation of the following individuals is greatly appreciated: Milo Cress, Federal Highway Administration; Mike Beacham and Claudette Wagner, Nebraska Department of Roads; Karen Bexten, Tadros Associates; Norm Nelson and Ben Recceri, Ready Mixed Concrete Company; and Morris Workman, Wilson Concrete Company.

REFERENCES

1. Aitcin, P.-C. and Neville, A., "High-Performance Concrete demystified", Concrete International, January 1993, pp. 21-26.
2. Ozyildirim, C. "High-Performance Concrete for Transportation Structures", Concrete International, January 1993, pp. 33 - 38.

28-day compressive strength was over 69 MPa (10,000 psi.). The permeability was less than 1,000 coulombs in 56 days. The shrinkage was less than 400 microstrains. The workability of the concrete was good. The successful construction of the floating bridge demonstrates that it is practical and cost effective to use HPC in highway bridges, where high strength, low permeability, low shrinkage and creep, and high abrasion resistance are of benefit. Since then, WSDOT has used HPC for the trunnion and counterweight housings of a bascule bridge, the superstructure and tower of a cable-stayed bridge, and a three-span prestressed concrete I-girder bridge. WSDOT is expecting wider application of HPC in the next few years.

DEVELOPING AN HPC MIX

WSDOT has been using concrete mixes with 28-day compressive strength in the range of 34 MPa (5,000 psi) to 45 MPa (6,500 psi) for precast prestressed, post-tensioning and floating structures for many years. Even though these mixes have not been designated as HPC, they provide high quality concrete for bridge construction.

In the early 1990s, WSDOT had another opportunity to build a new concrete floating bridge across Lake Washington in Seattle. About 38 000 cubic meters (50,000 cubic yards) of concrete were needed to build this bridge. The concrete must be very dense and impermeable, highly abrasion resistant, high strength and relatively crack free to assure the long term performance of this floating structure with a design life of 75 years. WSDOT set out to develop a high performance concrete mix that would meet these requirements. Additionally, the concrete must be readily available locally, have good workability, have low demand on labor skills and not require any special equipment.

Trial concrete mixes were evaluated for achieving an HPC mix that would meet the design and construction requirements. Various parameters were investigated, e.g. effects of different combinations of cementitious materials, water to cementitious materials ratios, chemical admixtures, strength gains, temperature change and so on. A summary of some of the trial mixes is shown in Table 1.

Thermal monitoring of the trial mixes provided the following information:
1. All concrete mixes increased in temperature from 21 degrees C (70 degrees F) to a maximum of 52 degrees C (125 degrees F).
2. Type I cement generated more heat than the Type II cement.
3. With an increasing addition of silica fume, an increase in temperature rise was observed. The maximum temperature was reached in 18 hours for 8% silica fume and 27 hours for 15% silica fume. The rate of temperature decrease was about the same for different silica fume contents.
4. In the mixes with 8% silica fume, it was observed that the Type I and Type II cements reached the same maximum temperatures. However, the rate of temperature increase was lower for Type II cement. The rate of temperature decrease was almost identical.

Table 1 - Trial Mix Proportions by Weight in SI Units
(Weight Per Cubic Meter)

	Type I Cement			Type II Cement						
Ingredients	Mix A	Mix B	Mix C	Mix D	Mix E	Mix F	Mix G	Mix H	Mix I	Mix J
Cement kg	418	418	418	418	418	418	418	418	376	344
Fly Ash kg	--	--	--	--	--	--	42	84	83	57
Silica Fume kg	--	33	63	--	33	63	--	--	42	42
Water kg	147	158	168	147	158	168	161	166	141	138
Fine Aggregate kg	810	780	753	810	780	753	742	687	711	736
Coarse Aggregate kg	1118	1076	1036	1118	1076	1036	1094	1089	1116	736
Water Reducer ml	1636	1636	1636	1636	1636	1636	1636	1636	1636	1636
Superplasticizer ml	2085	2500	3340	2085	2500	3340	--	--	4125	2744
Air Entrainment	None									
Water/Cementitious Mat'l Ratio	0.35	0.35	0.35	0.35	0.35	0.35	0.35	0.33	0.28	0.30
Slump mm	203	171	171	184	146	140	197	171	229	145
Comp. Strength: 7-day MPa	57.6	57.7	57.7	48.7	50.5	46.4	27.9	33.2	51.1	40.1
28-day MPa	71.7	77.2	80.4	66.5	75.3	74.9	45.4	52.4	72.8	69.6
90-day MPa	77.0	83.3	82.8	77.0	84.4	78.4	62.7	71.7	83.2	75.7
Permeability: 28-day Coulomb	No Test	No Test	No Test	4163	778	509	4449	437	332	422

Conversion Factors: 1 kg = 2.20 lb.   100 mm = 3.94 in.
100 ml = 3.38 fl oz.   10 MPa = 1,450 psi

From the trial mixes the contribution of the silica fume to compressive strength and permeability was evident. The optimum benefit was reached at 8% silica fume addition. Generally, the design strength, durability, impermeability and construction requirements could be met using 5-8% silica fume with the normal concrete mixes using Type I or Type II cement and fly ash.

Proper selection of aggregates is important for peak performance. The largest of the coarse aggregates used in the trial mixes was less than 13 mm (1/2 in.) using AASHTO Gradation No. 8.

Trial concrete mixes prepared and tested before or during design are important to make sure that the design requirements are practical and cost effective. The trial mixes also help in preparing the project specifications to guide the contractor in preparing the final mix design.

PROJECT SPECIFICATIONS

Background Information

After evaluating the results of the trial mixes, a demonstration mix design was derived and used to construct full size test sections representative of the reinforcing steel and post-tensioning ducts in the project. The objectives of the test sections were to assess the concrete placement and consolidation techniques, joint preparation and curing procedures to finalize the project specifications. It was a constructability check.

The first HPC specifications were developed for the I-90 Lacey V. Murrow Floating Bridge (LVM). The LVM is 2013 m (6,600 ft.) long, consisting of 20 prestressed concrete pontoons joined to form a continuous structure. A typical pontoon measures 110 m (360 ft.) long, 18.3 m (60 ft.) wide and 5.4 m (17.75 ft.) deep. Figure 1 shows a view of the floating bridge with the City of Seattle in the background. Figure 2 shows a typical cross section of a pontoon. This bridge was completed and opened to traffic in August 1993.

Subsequently, HPC was specified for the trunnion and counterweight housings of the SR 99 First Avenue South Bascule Bridge (FAS) in Seattle. This is a double-leaf bascule bridge crossing the Duwamish River. The span from centerline of trunnion to centerline of Trunnion is 90 m (294 ft.), providing a 44 m (145 ft.) clear navigation channel. It carries 4 lanes of traffic. Figure 3 shows a general view of the bridge and the trunnion and counterweight housings. This bridge was completed and opened to traffic in February 1997.

Figure 1 The Lacey V. Murrow Floating Bridge

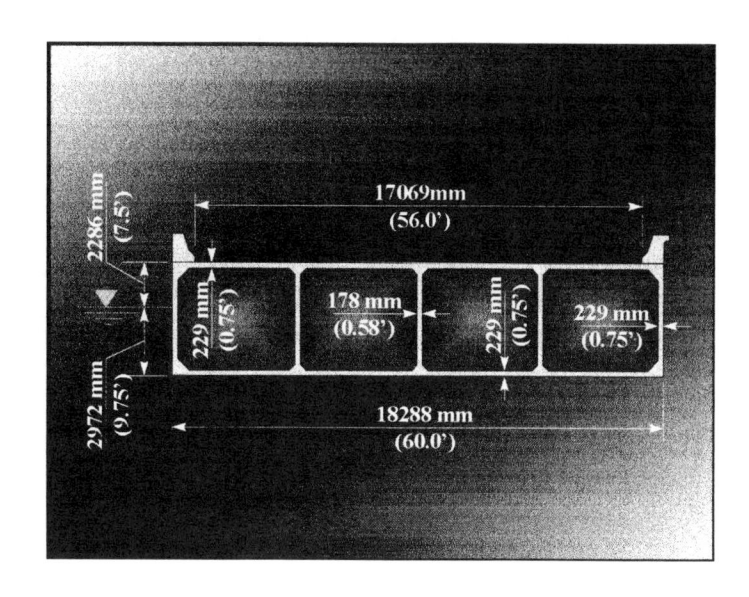

Figure 2 Typical Cross Section of A Pontoon

Figure 3 General View of First Avenue South Bridge

HPC was specified for the superstructure and tower of the SR 509 Theo Foss Waterway Cable-Stayed Bridge (TFW) in Tacoma. It is a two-span cable-stayed structure supported from a single tower. The main span over the waterway is 107 m (350 ft.), while the backspan is 101 m (332 ft.). The roadway width varies from 23 m (74 ft.) to 34 m (112 ft.) Figure 4 gives a general view of the structure. Figure 5 shows a typical cross section of the superstructure. This bridge was completed and opened to traffic in February 1997.

Figure 4 Theo Foss Waterway Cable-Stayed Bridge

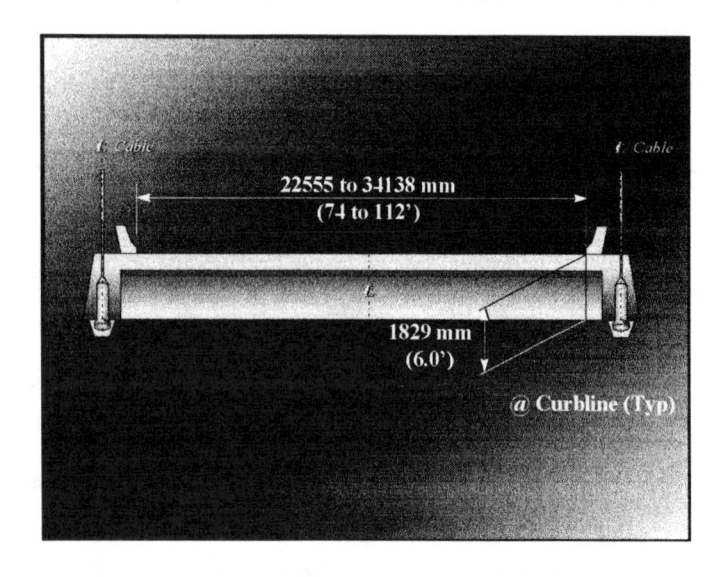

Figure 5 Typical Cross Section of Theo Foss Waterway Bridge

In August 1996, WSDOT signed an agreement with FHWA to conduct a demonstration project using HPC in the precast, prestressed I-girders of the SR 18 Covington Way Bridge (CWB) near Seattle. This is a three-span continuous prestressed girder bridge with a center span of 42 m (137 ft.) and end spans of 24 m (80 ft.). The use of HPC helps to reduce the number of lines of girders from seven to five (See Figure 6). The roadway is 12 m (35 ft.) wide carrying two lanes of traffic. The bridge is now under construction and will be completed in July of 1997.

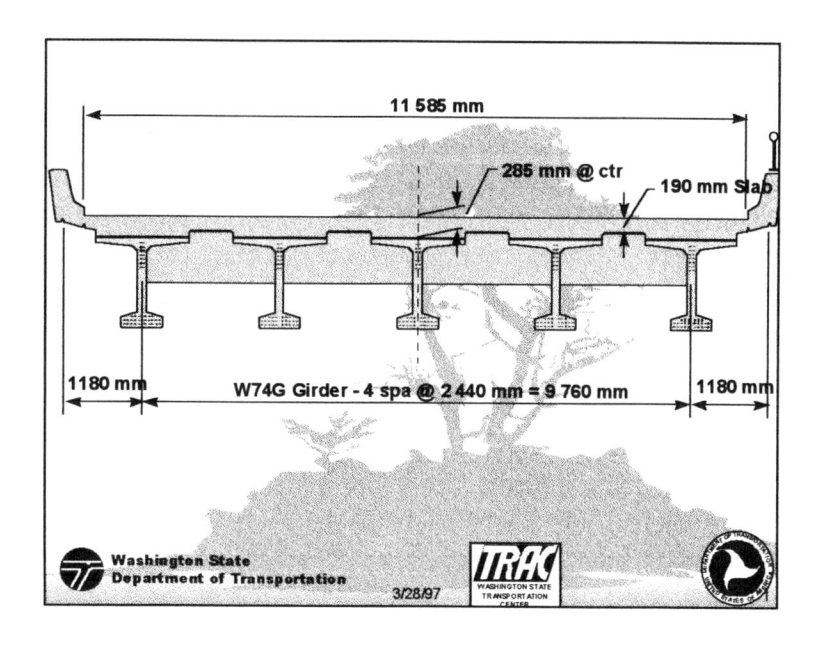

Figure 6 Prestressed Girders of HPC

Special Provisions
The Washington State Standard Specifications for Road, Bridge and Municipal Construction have provisions for contracting agency-provided mix designs for concrete up to 34 MPa (5,000 psi) 28-day strength. For higher strength concrete, the contractors are required to submit, for the state's approval, design mixes that meet the requirements of the special provisions.

Highway structures are generally subjected to environmental and loading conditions which cause the structures to deteriorate. Environmental conditions consist of temperature changes, freeze/thaw cycles, and salt or other aggressive chemicals. Loading conditions consist of traffic, wind, earthquake and other loadings. Eight parameters have been identified to represent HPC long-term performance. These parameters are freeze/thaw durability, scaling resistance, abrasion resistance, chloride permeability, shrinkage, creep, compressive strength, and modulus of elasticity. FHWA has advanced definitions[1] for grades of performance for each of the eight parameters. Each grade represents a measure of resistance to an adverse field condition. Using these grades to represent resistance, a designer can specify a concrete mix design with the necessary performance characteristics to meet the design life of a project.

The HPC performance grade characteristics for the projects are shown in Table 2.

Table 2 Performance Grade Characteristics

Characteristics	LVM	FAS	TFW	CWB
56-Day Compressive Strength MPa (Tested Per AASHTO T 22)	69.0	34.5	48.3	69.0
Chloride Permeability Coulombs (Tested Per AASHTO T 277)	< 1000	< 750	< 750	< 1000
Shrinkage in 56 Days microstrain (Tested Per AASHTO T 157)	< 400	NR	< 400	< 400
Freeze/Thaw Durability (Tested Per AASHTO T 161)	NR	NR	NR	> 80%

LVM = The Lacey V. Murrow Bridge
FAS = The First Avenue South Bascule Bridge
TFW = The Theo Foss Waterway Cable-Stayed Bridge
CWB = The Covington Way Precast Prestressed Girder Bridge.
NR = Not Required

CONTRACTORS' APPROVED CONCRETE MIXES

The approved contractor-provided mix designs were in close conformance with the project specifications. The proportions of the contractors' approved mix design weight per cubic meter are shown in Table 3.

Table 3 Approved Mix Design
(Weight Per Cubic Meter)

Ingredients		LVM	FAS	TFW	CWB
Portland Cement Type/Weight	Type/kg	II/370	II/338	I/371	III/432
Fly Ash	kg	50	71	59	132
Silica Fume	kg	30	36	30	30
Fine Aggregate	kg	770	773	724	528
Coarse Aggregate	kg	1050	1145	997	1109
Water	kg	150	125	153	157
Water Reducer, ASTM C494	ml	965	1764	122	1126
Superplasticizer, ASTM C494	ml	5 065	4851	2419	8316
Air Entrainment		None	None	5%	None
Water/Cementitious Materials Ratio		0.33	0.28	0.33	0.265
Slump	mm	180	178	114	152
56-Day Compressive Strength	MPa	80	96	59	74
Chloride Permeability (56-Day)	Coulombs	790	1250	950	BM
Shrinkage (28-Day)	Microstrain	330	NR	500	BM

NR = Not Required
BM = Being Measured

SOME LESSONS LEARNED

Concrete Mix

Trial mix designs have proved to be very valuable in developing the project specifications and in assuring the proper mix for construction. There are many factors that affect the properties of the concrete mixes. Trial mixes performed by the ready-mix suppliers help to reflect the conditions of the site and to minimize surprises and delays in production. Trial mixes save time and dollars in arriving at an approved design mix that has minimal construction problems. For members with complex geometry and congested components, building test sections prior to the start of construction has proved to be very valuable to the contractors and the owners.

The development or specification of concrete mixes should be part of the project development process. The structural engineers, the materials engineers and the concrete suppliers should communicate and coordinate early on in the structural design process to meet the durability and strength requirements with due considerations for availability and constructability. Fitness for purpose philosophy should be applied to the design and specifications to assure cost effectiveness.

Silica Fume

Silica fume in the range of 5 to 8% of cementitious materials has significant benefit in increasing early compressive strengths and reducing permeability. However, it causes increase in heat of hydration and provides little or no bleed water to the surface, requiring special attention in curing to avoid plastic shrinkage cracking.

Air Entrainment

Use of air entrainment for freeze-thaw resistance of concrete is well known. However, the use of air entrainment in high performance concrete with silica fume is highly controversial. Entrained air reduces the strength of concrete. As much as a 5% decrease in compressive strength may be expected for every percentage point of air entrained. As a result, engineers are trying to limit the use of air entrainment where high strength is desired. One argument is that HPC is of such high density and low permeability that air entrainment is not necessary for freeze-thaw resistance. On the other hand, tests of HPC without air entrainment have shown that silica fume concrete will fail miserably when tested in accordance with ASTM C 666, Freeze-Thaw Durability Tests. Some researchers suggest that the standard ASTM C 666 Freeze-Thaw Test Method is not appropriate for testing HPC with silica fume. To air or not to air should depend on the freeze-thaw rate and number of cycles expected at the site. Perhaps, the testing method should be modified to take into account the climatic conditions to which the HPC structures will be exposed.

Air entrainment was specified for the Theo Foss Waterway Bridge. During the trial mixes, the compressive strength and the permeability of two mix designs, one with air entrainment and one without, were compared. The results are shown in Table 4 below.

Air entrainment reduces the compressive strength by as much as 20%. It has caused some problems during construction when the compressive strengths fall below the specified strengths. Adjustment in entrained air is necessary to bring up the strength level.

Table 4 Mix Designs With and Without Air Entrainment

	Air Entrainment	
	None	5%
56-Day Compressive Strength, MPa	68	59
56-Day Permeability Coulomb	610	950

Permeability

Permeability is a key durability parameter in assuring long term performance of concrete. AASHTO Test Method T277, "Rapid Determination of the Chloride permeability of Concrete," is currently the most widely used method for determining chloride permeability in terms of electric charge in coulombs. It is relatively simple and fast six-hour test method. The results are quite variable, especially when the electric charge drops below 1,000 coulombs. However, this method serves as a quick means to estimate the penetration of chloride into concrete. For all practical purposes, it is quite reasonable.

The correlation between electric charge in coulombs and chloride permeability is given in the following table:

Table 5 Chloride Permeability

Charge Passed, Coulombs	Chloride Permeability
> 4000	High
2000 - 4000	Moderate
1000 - 2000	Low
100 - 1000	Very Low
< 100	Negligible

It may be noted that the tabulated numerical values were developed for concrete mixes without fly ash, slag or silica fume. These values should be used with caution. The applicability of AASHTO T 277 test method to a specific type of concrete can be enhanced by establishing the correlations between this method and long-term chloride ponding tests, such as AASHTO T 259.

Aggregate Gradation

Aggregate gradation is important to optimization of compressive strengths, workability and responsiveness of the mix to vibration. The aggregate gradation proportions are generally left up to the contractor. The contractor is responsible for selecting the proper proportions to meet the contract provisions, the placement conditions and methods of placement. Experience has shown that a maximum aggregate size of less than 13 mm (1/2 inch) will provide a workable HPC mix with high strength.

A plot of Sieve Opening vs Percent Retained graph will give an indication whether the gradation is satisfactory. The percent retained is the percentage of total combined aggregate weight (fine and coarse). The graph should be fairly smooth. Sharp peaks and valleys indicate that the gradation can be adjusted to improve workability. The contractors may also use the Shilstone and Voelker factors to determine workability.

Consolidation of Concrete

Proper consolidation of concrete is key to achieving dense concrete free of surface defects. Internal vibration supplemented with external vibration is important when placing concrete in deep walls with heavy reinforcing steel and post-tensioning ducts. External vibration needs more expensive formwork. But the resulting quality concrete more than offsets the extra cost for better formwork. The contractor does not need to spend labor and time in repairing, reworking and patching concrete.

Very little information is available on the design of forms for external vibration of HPC. The forms must be designed to withstand the lateral fluid pressure of the concrete and the repeated, reversing stresses induced by external vibration. Steel forms are generally preferred, but expensive. Steel forms will be cost effective if they are used several times. Trial placement and consolidation of concrete in test sections provide invaluable information on the design of forms and placement of vibrators.

Curing of Concrete

Silica fume concrete mixes yield very little bleed water to the surface. It is essential to supply the water or moisture to surfaces of flatwork or other unformed surfaces to avoid shrinkage cracking. This is done by fog-spraying immediately after finishing the concrete surfaces. After initial set, the unformed surfaces of the concrete are kept continuously wet with water for not less than fourteen days. This is done by immediately covering the concrete surfaces completely with wet burlap and 2 layers of 6 mil white plastic sheet and keeping the burlap continuously wet during the 14-day curing period. When this is carried out diligently and successfully, a relatively crack-free and trouble-free concrete is achieved. Otherwise numerous cracks show up on the flatwork, which can be problematic. The plastic sheets should be overlapped and held down adequately against wind. The wet burlap should be checked frequently to make sure it is kept wet continuously. As simple as these instructions may seem, they are not always followed diligently in the field.

CONCLUDING REMARKS

High performance concrete is constructable and can be used advantageously in highway bridge construction in cast-in-place and precast applications. It has enhanced durability characteristics and strength parameters not normally attainable by using conventional concrete mixes. The enhanced durability characteristics improve resistance to thermal freeze-thaw cycles and to salts and other chemicals. The enhanced strength parameters include reduced creep and shrinkage, higher modulus of elasticity and higher strength. The bridge engineers have the flexibility to specify the durability and strengths needed to meet the short and long term demands of a project. The bridge engineers can specify high performance concrete to reduce weight and construction cost by using smaller, fewer or longer members; to solve vertical clearance problems by designing shallower members; and to improve durability in marine or other harsh environments.

High performance concrete containing fly ash and silica fume is very cohesive and has good workability when properly proportioned. Enhanced with high range water reducers, the concrete can be mixed with a low water cement ratio and placed with a slump as high as 230 mm (9 inches) with no loss in strength or density. The concrete flows laterally with ease in the forms and can be dropped from a height without segregation.

High performance concrete possesses all the essential elements for structural applications to extend service life and reduce life-cycle cost. However, the successful application of high performance concrete takes team effort from the researchers, owners, structural engineers, inspectors, contractors and suppliers. All participants must be dedicated and committed to quality in the constructed project through cooperation and open communication.

The National Strategic Highway Research Program has identified and developed transportation applications for high performance concrete. The Federal Highway Administration is now funding demonstration projects using high performance concrete around the country. These demonstration projects will gather documented information on constructability, testing, structural performance, and design parameters for wider applications of high performance concrete in the transportation industry.

Washington State along with Missouri, Nebraska, New Hampshire, Texas, and Virginia are participating as "Lead States" in the application of high performance concrete in highway bridge design. Lead state participants are to learn all the dos and don'ts in the demonstration projects and then share their lessons learned with other states to assure successful implementation on the use of high performance concrete. The SR 18 Covington Way Bridge described in this paper is Washington State's demonstration project. The bridge is instrumented and will be monitored for at least three years after completion to evaluate the properties of the concrete and the performance of the bridge.

REFERENCES

1. Goodspeed, C., Vanikar S., Cook, R., "HPC Defined for Highway Structures", Federal Highway Administration, Washington, D.C. January 1995.

APPLICATION OF 100 MPa HIGH-STRENGTH, HIGH-FLUIDITY CONCRETE FOR A PRESTRESSED CONCRETE BRIDGE WITH SPAN-DEPTH RATIO OF 40

Kenro Mitsui and T.Yonezawa
Chief Researchers
Research and Development Institute, Takenaka Corporation
Chiba, JAPAN

M. Tezuka
Chief research Engineer
Technical Research Institute Oriental Construction Co., Ltd.
Tochigi, JAPAN

M. Kinoshita
Chief researcher
Chemical Admixture Division, Takemoto Oil&Fat Co., Ltd.
Aichi, JAPAN

ABSTRACT

High strength and high fluidity concrete with specified design strength of 100 MPa (14527 psi) was applied to a prestressed concrete bridge with span-depth ratio of 40 to ensure the strength and rigidity required for the slender structure. Development of acrylic copolymer based new superplasticizer and undensified silica fume made it possible to utilize cast-in-place high strength concrete with water-cement ratio of 0.20. This paper gives the design and construction of the bridge, special technologies and properties of the high strength concrete including properties of fresh concrete, strength, durability, creep and shrinkage.

INTRODUCTION

Conventional high strength concrete with very low water-cement ratio (W/C) has such high viscosity that in-situ placing becomes impracticable. Also, the fluidity is highly dependent on time, limiting the use of such concrete in the ready-mixed form. Conventional technologies have therefore been incapable of in-situ placing of high strength concrete with W/C of around 0.20. The key technology of utilizing high strength concrete is to give high fluidity as well as to improve strength of concrete.

The authors developed high strength and high fluidity concrete with W/C of 0.20 with compressive strength of 140MPa (20338 psi). The technology of this concrete consisted of two main developments. One was the development of a new super-plasticizer that enables concrete to have extremely high fluidity with limited fluidity loss even with very low W/C. The other was development of a special device to automatically store, convey, batch and charge undensified silica fume, which has stronger effects of increasing fluidity and strength of concrete than densified silica fume.

The CNT Super Bridge, shown in Figure 1, is a simply supported box girder bridge of prestressed concrete with a span of 40.356m (132.41 ft) and a girder depth of 1.02m (3.35 ft). The unique span-girder depth ratio of 40 for a simply supported girder bridge was realized by the use of ultra-high strength concrete with design strength of 100 MPa (14527 psi). The specified strength of 100 MPa for this bridge is the highest for a concrete structure ever cast in Japan. This slenderness of the bridge suggested large vibration under loading even though the bridge was to be made of high strength concrete. A special vibration controller referred to as a tuned bar damper (TBD) was therefore developed to restrict the vibration. Hollow prestressed steel bars were set to prevent cracks at the ends of the girder.

This paper gives the design, construction of the bridge and special technologies, which made it possible to utilize ultra-high strength concrete.

Figure 1. CNT Super Bridge

STRUCTURAL DESIGN OF THE BRIDGE

Outline of the Design
The side elevation and a cross-section of the bridge are shown in Figure 2. The cross-section is a box with a sponson deck slab. Approximately 10m (32.8 ft) from both ends is solid. The bottom part of the sponson deck slab has grooves in the middle, where vibration controllers are mounted. The outline of the bridge is as follows:
Name: CNT Super Bridge, *Location:* Inzai-city, Chiba, Japan
Structure: Post-tensioned prestressed concrete simply supported box girder bridge
Bridge length: 41.216 m (135.23 ft), *Girder length:* 41.126 m (134.93 ft), *Span:* 40.356 m (132.41 ft), *Width of slab:* 1.700 m (5.58 ft), *Girder depth:* 1.020 m (3.35ft),

Design of Girder
Since there is no room for tensioning the prestressing bars at the ends of the girder, the prestressing bars are arranged for alternating one-end tensioning with the other end anchored at the end of the girder. The tensioned ends are anchored on the top edge of the

girder. The arrangement is shown in Figure 3. The bars are bent three-dimensionally in vertical and horizontal directions in the solid portions of the girder.

(a) side elevation (b) cross section

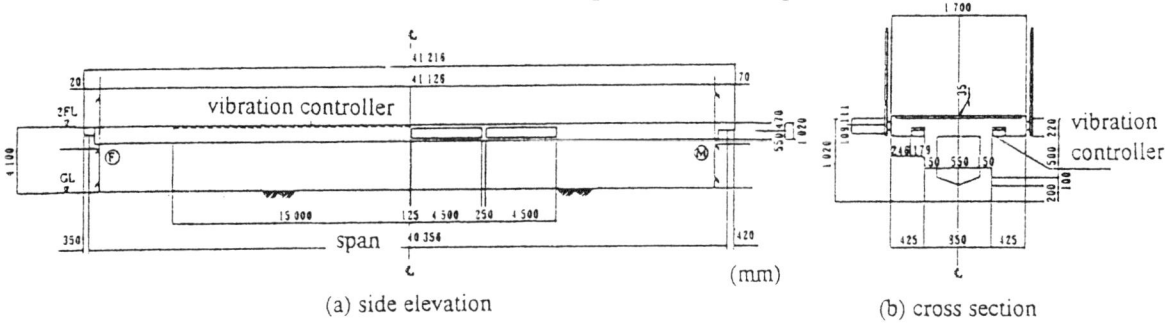

Fig 2: Side elevation and cross section of the bridge

prestressing cable (8-12T 12.7mm)

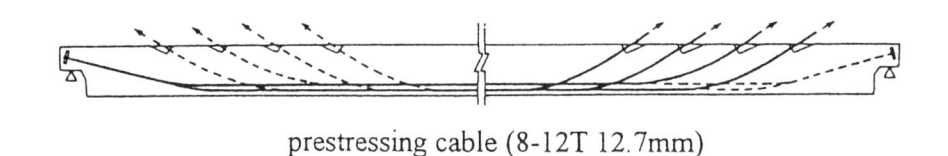

Figure 3: Arrangement of prestressing bars

Loading Conditions and Material Constants

The loading conditions and material constants used are as follows:
Traffic load: 300 kg/m² (555 lb/yd²), *Lateral seismic coefficient:* 0.24.
Specified strength of concrete: 100 MPa (14527 psi)
Strength at the time of tensioning: 65 MPa (9442 psi)
Allowable stress: compressive; 23 MPa (3341 psi), tensile; 2.5 MPa (363 psi)
Modulus of elasticity: 45 GPa (6.54 ksi) (as designed), 35 GPa (5.08 ksi) (at the time of tensioning), Creep *coefficient:* 2.6

Vibration Controller

The vibration controller, TBD, is a form of tuned mass damper of the added-mass type, consisting of flat bars and silicone gel mounted at the bottom of the deck slab. The damper consists of a spring and added mass tuned to the natural frequency of the body structure and damping devices adjusted to optimum values.

The damping mechanism of TBD is shown in Figure 4. The flat bars beneath the girder serve as both the added mass and spring, counteracting the vibration of the girder. Silicone gel inserted between the girder and the flat bar serves as the damping device. The viscosity of this material absorbs the vibration energy and accelerates the damping of the girder vibration. Natural frequency elements of less than 10 Hz are to be restricted with the target frequency being the primary and secondary natural frequencies.

In this TBD system, the flat bars have the primary frequency tuned to the target natural frequency of the girder, as well as a mass of approximately 1/40 of the girder weight. This mass is established with the aim of restricting vibration after the pedestrian load action to around 2 Gal. Two flat bars for the primary natural frequency are mounted on both sides of the girder at midspan. As for the bars for the secondary natural frequency, two each are fixed at two quarters points of the span.

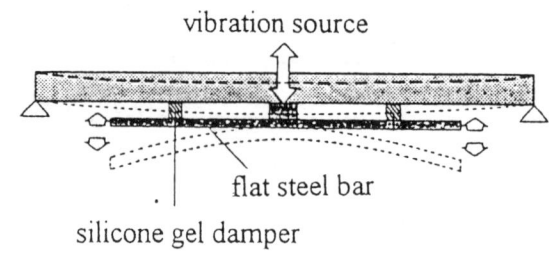

Figure 4: Mechanism of vibration controller

Reinforcement of Girder Ends by Hollow Prestressing Bars

The bridge girder is to rest on the supports, and this can cause cracking of concrete due to the drastic increase of stress at these portions. In addition to sufficient reinforcing bars, hollow prestressing bars were introduced to prevent cracking. The layout of the prestressing bars is shown in Fig. 5. The prestressing system comprises an outer hollow bar and an inner bar for a reaction force as shown in Fig. 6. First, the outer bar is tensioned against the inner bar and fixed with nuts. Then placed in the formwork, and concrete is placed. When the concrete is hardened, the nuts are released to prestress the concrete with the tensile force of the hollow bar, which is transferred to the concrete by the bond between the bar and concrete and the gripping effect of the threaded ends of the bar. The tensioning force on the hollow bars is determined so that the tensile stress near the angle is approximately zero by a finite element analysis of the ends.

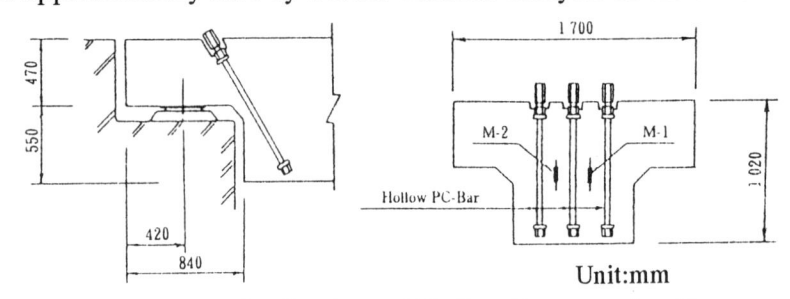

Figure 5. Distribution of Hollow Prestressing Bars

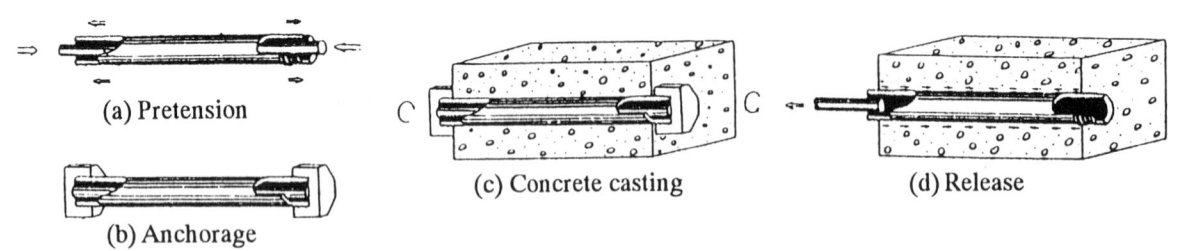

Figure 6. Mechanism of Hollow Prestressing Bars

ULTRA-HIGH STRENGTH CONCRETE

W/C of around 0.20 is required to produce ultra-high strength concrete that meets the design strength of 100 MPa. The ultra-high strength concrete referred to here should have high fluidity to enable very easy placing on site. This concrete was realized by two technologies: 1) an acrylic copolymer based new superplasticizer (SSP) that gives concrete to have much higher fluidity than conventional ones in a very low W/C, and 2) development of a system to use undensified silica fume at a ready-mixed concrete plant.

To evaluate the effect of SSP and undensified silica fume on improving fluidity of high strength concrete, an L-flow test was proposed by the authors[1]. The L-flow meter, shown in Figure 7, is a simple instrument to measure viscosity of high slump concrete. The flow velocity of concrete is calculated from the time that concrete takes to pass through the two sensors. The flow properties of high slump concrete is known be approximated by Bingham liquid, whose constitutive equation is as (1)

$$\tau - \tau_y = \eta \dot{\gamma} \quad (1)$$

where τ: shear stress, τ_y: shear yield stress, η: plastic viscosity, $\dot{\gamma}$: shear strain rate

Concrete with the same slump is expected to have the same shear yield stress. The L-flow velocity is a parameter that represents shear strain rate. Thus it is expected for concrete with the same slump that the inverse of the L-flow velocity is a parameter expressing plastic viscosity of concrete. The larger the L-flow velocity, the smaller the viscosity of concrete.

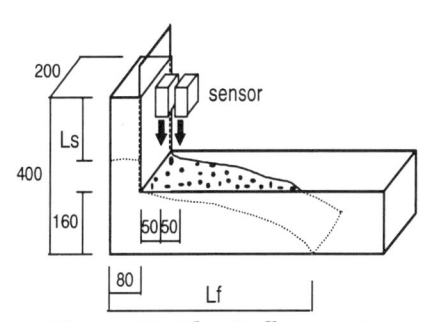

Figure 7: The L-flow meter

Super-Superplasticizer (SSP)

The authors developed a new super superplasticizer (SSP) [3,4], which has an excellent cement dispersing capability in the range of a W/C of lower than 0.25 and a capability to retain high fluidity to allow in-situ placing for a long time. The chemical structure of the SSP is shown in Figure 8. The structure is characterized by comprising an acrylic graft

Figure 8: Chemical structure of SSP

copolymer with polyethylene glycol chains as the graft chains, and containing in the molecule carboxyl groups and sulfonic groups as anion groups of an adequate mole ratio (-COOH/-SO$_3$H=80/20). A number average molecular weight of 3900 by Pullulan conversion was adopted the dispersing capability.

Figures 9 shows the dosage of superplasticizer and L-flow velocity, respectively, when using this superplasticizer (SSP), a conventional naphthalene type superplasticizer (NSF), and a polycarboxylate type superplasticizer (SP). The differences between SSP and conventional superplasticizers are small when the water-binder ratio is 0.30 and SF/(C+SF) is 10%. When the water-binder ratio is 0.20, both the dosage of SSP and the resulting viscosity are lower than the cases of NSF and SP, exhibiting SSP's capabilities suitable for ultra-high strength concrete.

Figure 10 shows loss of flow and L-flow velocity after mixing. The flow of concrete containing SSP remained practically the same for 2 hours after mixing. The L-flow velocity also showed little loss, indicating good fluidity retention effects even with a very low W/C. Concrete containing NSF showed slightly larger loss in flow than in the case of SSP, but the loss of L-flow velocity was much greater than in the case of SSP. It indicates that, in high-strength concrete, the viscosity of concrete would increase over time after mixing while the flow would not show significant loss, and it varies with type of superplasticizer. The slump loss is considered to be caused by the physical agglomeration

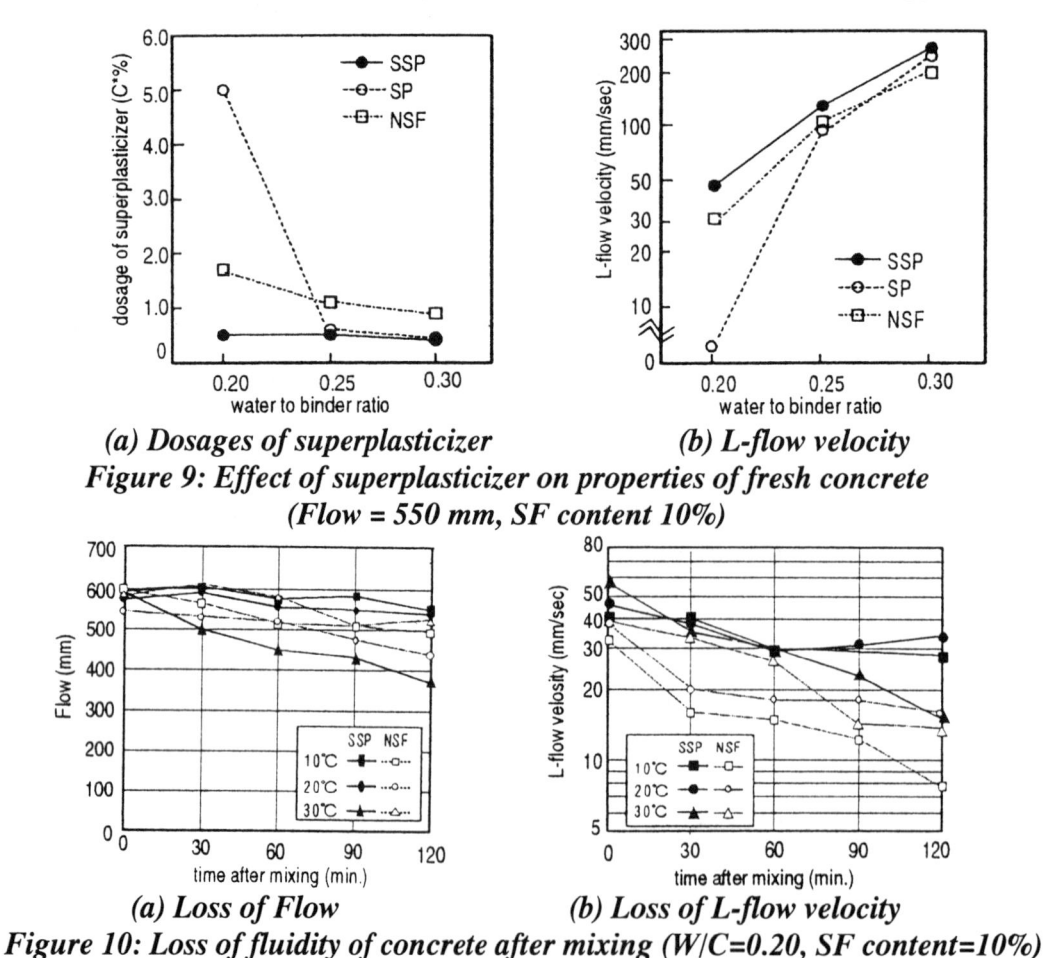

(a) Dosages of superplasticizer *(b) L-flow velocity*
Figure 9: Effect of superplasticizer on properties of fresh concrete (Flow = 550 mm, SF content 10%)

(a) Loss of Flow *(b) Loss of L-flow velocity*
Figure 10: Loss of fluidity of concrete after mixing (W/C=0.20, SF content=10%)

of particles[5]. The steric hindrance of SSP's graft chains is considered to be effective in preventing this physical agglomeration especially with very low W/C.

Utilizing Undensified Silica Fume
Silica fume is generally produced in three forms: undensified, densified, and slurry. Densified and slurry silica fume, which are easy to handle, are generally used in batch plants in Europe and America. However, their effects on increasing the strength and fluidity of high strength concrete are weaker than those of the undensified type.
Figure 11 compares the dosages of superplasticizer required to obtain concrete with slump of 230 mm (9.1 in.), with L-flow velocity by an L-flow meter, using concretes containing silica fume from 14 different factories. Figure 11 reveals that the densified silica fume requires higher dosages of superplasticizer, and causes concrete to have higher viscosity than undensified silica fume.

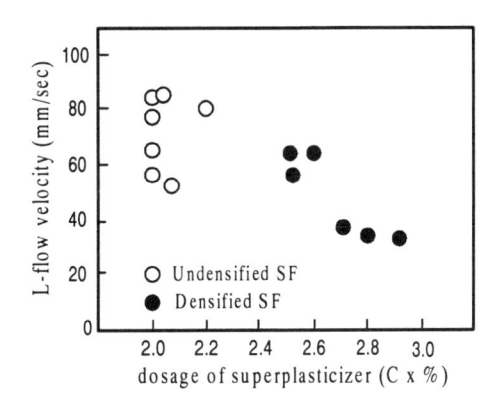

Figure 11: Relationship between dosages Superplasticizer and L-flow velocity of concrete (W/C = 0.28, SF content 10%)

Undensified Silica Fume Supplying Device
Undensified is the most efficient form of silica fume for increasing the strength and fluidity of high strength concrete, but is difficult to handle at batch plants. To overcome these problems, the authors developed a silica fume supplying device that enables undensified silica fume to be stored, conveyed, and weighed at plants. The appearance of the device is shown in Figure 12. The system is controlled automatically from the control room for the plant in connection with the concrete batching plant.

MIXTURE PROPORTION

Materials
The materials used for the bridge are as follows:
Cement: Portland fly ash cement, *Silica fume:* Undensified silica fume (SiO_2 = 93.7%, specific surface area = 15.5 m²/g, specific gravity = 2.39)
Coarse aggregate: crushed quartz schist (specific gravity = 2.63, water absorption = 0.70%, fineness modulus (FM) = 6.63)

Fine aggregate: pit sand (specific gravity = 2.61, water absorption = 1.53%, FM = 2.99)
Superplasticizer: new super-superplasticizer SSP

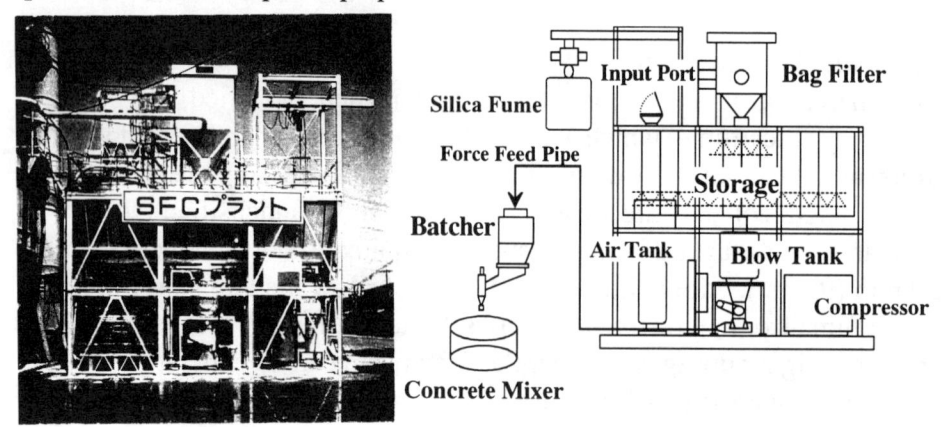

Figure 12: Undensified SF supplying device

Concept of Mixture Proportion

The strength of drilled cores from high strength concrete structures tends to be lower than that of standard cured specimen because of an effect of rapid rising of temperature caused by cement hydration. Consequently, the authors adopted the standpoint that standard-cured specimens do not always represent the strength of concrete in the structure, and decided to proportion the concrete for this bridge by experimentally determining the difference between both in advance.

The concept is expressed by the following equation:

$$f_{cr} \geq f'_{ck} + f_s + K\sigma \qquad (2)$$

where f_{cr} : proportioning strength, f'_{ck} : specified strength, f_s : difference between the strength of concrete in structure and standard-cured specimens at a control age, K: normal deviation, σ : standard deviation of strength

The difference between the strengths of standard-cured specimens and concrete in the structure, f_s, was assumed to be 10 MPa (1453 psi) based on mock-up test and other test results. The standard deviation was assumed to be 5.5 MPa (799 psi) on the basis of the past results regarding ultra-high strength manufactured at the same batcher plant. The normal deviation, K, was set to be 2 by assuming the proportion defective to be 2.3%. As a result of these requirements and trial mixing, f_{cr} of 122 MPa (17723 psi) and a water-binder ratio of 0.20 were specified.

The slump flow was specified to be 600 mm (23.6 in.), and the L-flow velocity, which is an index for viscosity, to be not less than 30 mm/sec (1.2 in/sec). The air content was

Table 1: Mixture Proportion of 100 MPa High-strength Concrete

W/(C+SF)	Unit Weight (kg/m³)					
	Water	Cement	Silica Fume	Fine Agg.	Coarse Agg.	Super-Plasticizer
0.20	135	574	101	748	836	13.5

specified to be 2.0%. The mixture proportion to meet these requirements is given in Table 1.

MOCK-UP TEST

Before the actual application, a construction test was conducted using a full-scale model, to study the quality and placeability of the fresh concrete, as well as the surface texture and strength development of the hardened concrete. The model was made to have identical cross sections and bar arrangement as the bridge, but was 1/10 in length. The filling capability of the concrete was investigated by using transparent forms.

Figure 13 shows the results of compressive strength tests. Both water-cured and field-cured specimens indicated 110 MPa (15980 psi) at 28 day, exceeding the design strength. At 91 days, the water-cured and field-cured specimens had compressive strength of 140 MPa (20338 psi) and 126 MPa (18304 psi), respectively. The drilled cores and the specimens synchronized with the thermal history of the model also reached strength of 115 MPa (16706 psi) at 28 days. Thus it was confirmed that concrete with sufficient strength of the structure was obtainable.

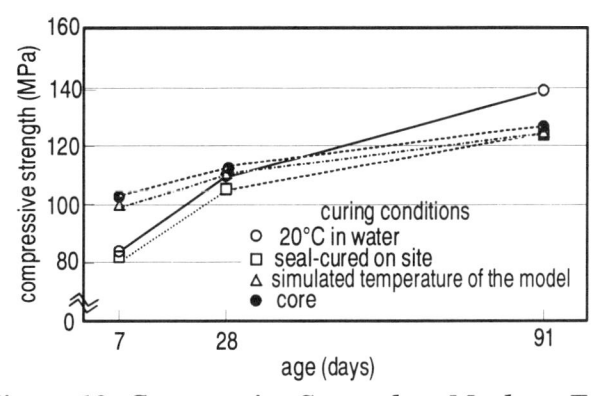

Figure 13: Compressive Strength at Mock-up Test

CONSTRUCTION AND RESULTS

Concrete Placing and Curing

The concrete was deposited with a bucket, and was consolidated with an internal vibrator. Because of no accumulation of bleeding water, the surface of the concrete was kept moist by being sprayed with water from immediately after placing through the final setting, so as to avoid plastic shrinkage cracks due to rapid drying. Wet curing mats were then used up to 14 days. Prestress was applied at 5 days. The strength of standard-cured specimens was 67 MPa (9733 psi) at that time

Properties of Fresh Concrete

Figure 14 shows the results of the tests on the fresh concrete immediately after mixing and upon arrival at the site. Though the transportation from the batcher plant to the site required such a long time as 2 to 2.5 hours, the slump losses were marginal. The loss of slump flow spread met the target value of 600 +/- 50 mm (23.6 +/- 1.97 in.) at the time of placing. The

L-flow velocity, an index for viscosity, showed a reduction of 50 mm/sec (1.97 in) but attained the target value of not less than 30 mm/sec (1.2 in.). These results exhibit that successfully placed under a severe condition of being transported for 2 to 2.5 hours, at which the new superplasticizer's fluidity-retaining effects were fully exerted.

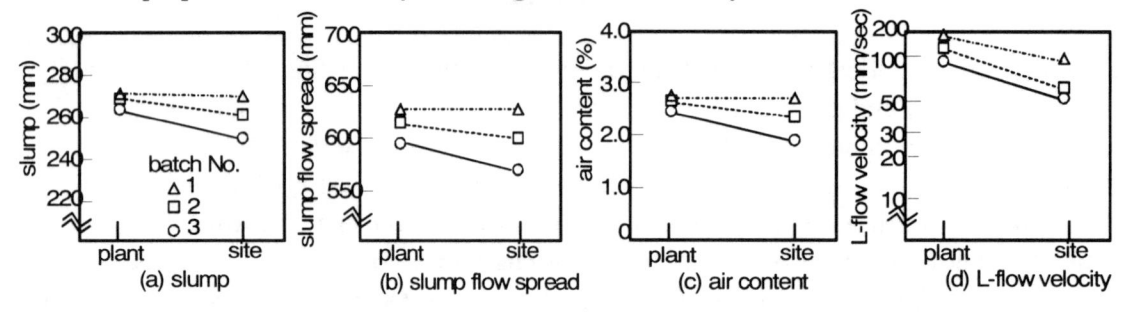

Figure 14: properties of fresh concrete at actual construction

Qualities of Hardened Concrete

Figure 15 shows the results of the compressive strength and modulus of elasticity. Whereas the strength of a standard-cured specimen at 28 days was 127 MPa (18449 psi) and 134 MPa (19466 psi) at 91 days, being 10 MPa (1453 psi) higher than the proportioning strength. The strengths of the adiabatic-cured specimens, to which the same temperature history as that of concrete in the structure was applied, used for the estimation of the strength of concrete in the structure were 112 MPa (16270 psi) at 91 days, sufficiently satisfying the design strength of 100 MPa. The modulus of elasticity was 42 GPa (6.10 ksi) at 91 days.

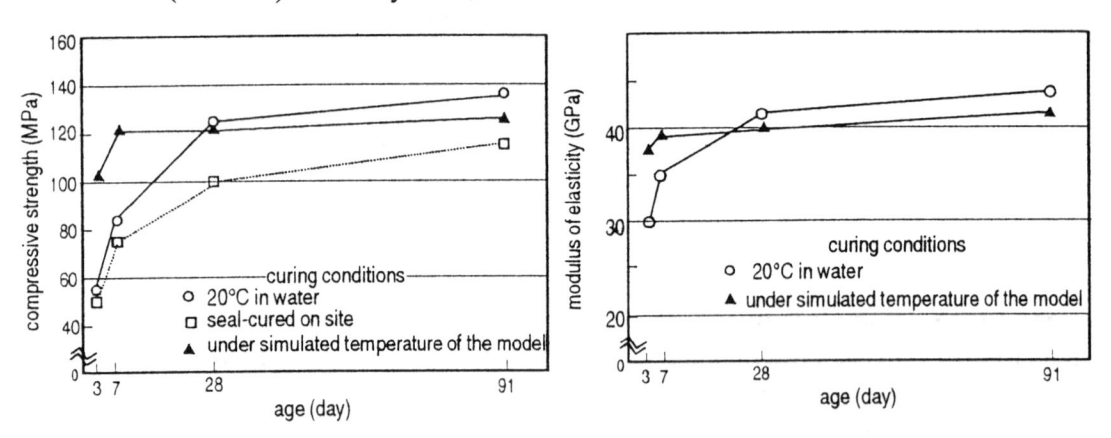

(a) Compressive strength   (b) Modulus of elasticity
Figure 15: Compressive strength and Modulus of elasticity at actual construction

Figure 16 shows the results of freezing and thawing resistance tests in accordance with ASTM C 666A. No reduction in the relative dynamic modulus of elasticity was observed up to 300 cycles even with low air content, 2.0%. This indicates that the ultra-high strength concrete used for the bridge has sufficient freeze-thaw resistance.

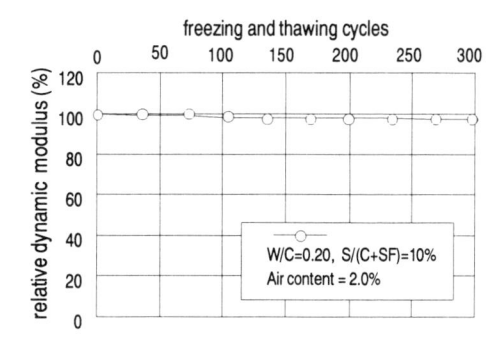

Figure 16: Result of freezing and thawing test

Long Term Deformation Behavior

The long-term behavior of the bridge is monitored to ensure safety and obtain technical data for future reference.

The measuring points for strain in the girder are shown in Figure 17. The specimens for measuring drying shrinkage and creep are prisms 165 x 165 x 500 mm (6.5 x 6.5 x 19.7 in.) in size fabricated simultaneously with the main girder, and placed on the site to equalize the environmental conditions. When the main girder was prestressed, the stress equivalent to the stress produced on the bottom edge of the girder at midspan was acted using prestressing bars on the prismatic specimen to measure the creep. When the tension is reduced by creep and drying shrinkage, the prestressing bars are retensioned to maintain the initial stress. The drying shrinkage and creep strain of the specimen is shown in Fig 18. The drying shrinkage was 220 x 10^{-6} at 380 days. Creep coefficient at 380 days is 1.2 when dividing the creep strain by the elastic strain in Figure 18.

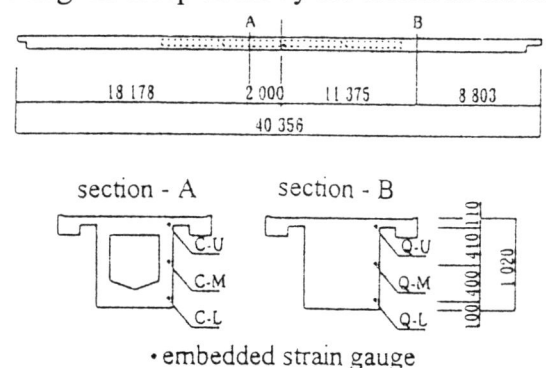

Figure 17: Measuring points of strain in the girder

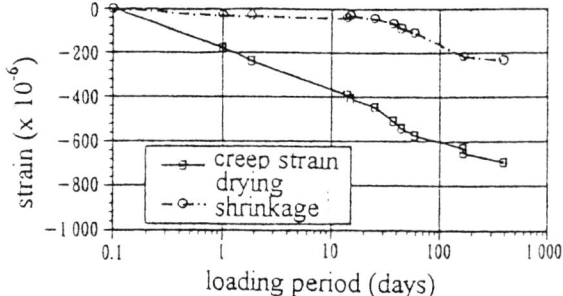

Figure 18: Drying shrinkage and creep strain of the girder

Figure 19 shows the total strain measured in the girder. These strains contain both creep and drying shrinkage strain. Though it is difficult to separate the creep coefficient from these values, the values obtained by dividing total strain by elastic strain are limited to a range of 1 to 2, suggesting fairly small values as long-term deformation. The long-term deformation behavior of this bridge is yet to be concluded. The ongoing measurement will contribute to future investigation.

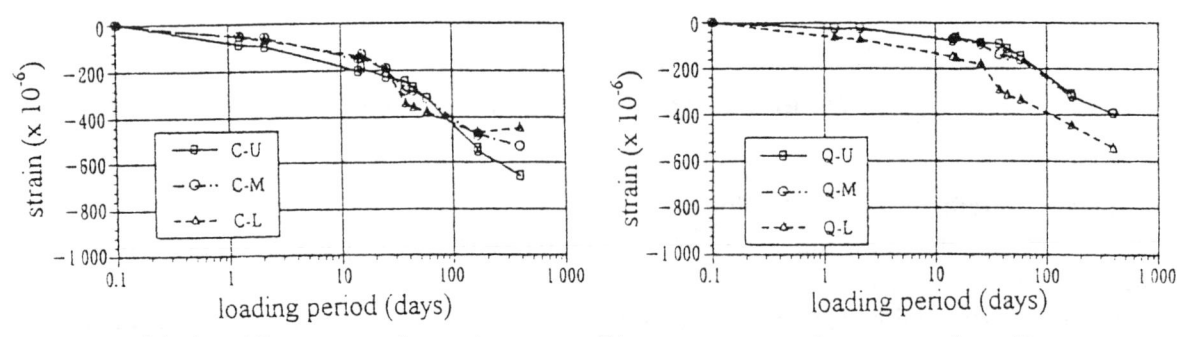

(a) At mid span, section - A   *(b) at a quarter of span, section -B*
Figure 19: Total strain of the girder)

CONCLUSIONS

1) Development of a new super-superplasticizer and the supplying system for undensified silica fume realized cast-in place 100MPa ultra-high strength concrete with high fluidity.
2) The use of ultra high strength concrete, together with vibration controller enabled the very slender concrete bridge with a span-girder depth ratio of 40.

This bridge is not of a large scale, but is a significant structure in that it proposes several new bridge design concepts - the first application of 100 MPa ultra-high strength concrete to a real structure in Japan, and the adoption of vibration controller.

This bridge is considered to contribute to the strengthening of concrete structures, as well as the span extension and lightening of prestressed concrete bridges in Japan.

REFERENCES

1 - Yonezawa, Y., Izumi, I., Mitsui, K. and Okuno, T.; A (1989) Study on the Flowability of High-Strength Concrete using the L-flow test, Proc. of Japan Concrete Institute, Vol.11, 171-176
2 - Sato M., Oura T., Okuno T. and Yonezawa T. (1993) Plant system for silica fume high-strength concrete, Proceedings of the Japan Concrete Institute, Vol.15, No.1, 75-80
3 - Kinoshita, M., Yonezawa, T. and Yuki, Y. (1990) Chemical Structure and Performance of New Type High-Range Water Reducing Agent, Proceedings of Japan Cement Association, No.44, 222-227
4 - Kinoshita M., Yonezawa T. and Mitsui K. (1994), Properties of a new type high range water reducing agent for ultra-high strength concrete, Proc. of Japan Conc. Inst., Vol.16, No.1, 341-346
5 - Hattori, K. (1980). Mechanism of Slump Loss and Its Control, Journal of the Society of Materials Science, Japan, Vol.29, 240-246

VIRGINIA'S BRIDGE STRUCTURES WITH HIGH PERFORMANCE CONCRETE

Celik Ozyildirim and Jose Gomez
Principal Research Scientist and Research Scientist
Virginia Transportation Research Council
Charlottesville, Virginia, U.S.A.

ABSTRACT

The recent high performance concrete bridge structures constructed by the Virginia Department of Transportation (VDOT) are summarized in this paper. Emphasis has been on the development of low-permeability concretes, with high strength as a secondary goal. Laboratory studies and two test programs with AASHTO Type II beams led to field applications.

After the first test program, structures were selected for the use of high-strength and low-permeability concretes in beams with 28-day compressive strengths of 55 MPa. Some of the bridges also included low-permeability concretes throughout the structure. After the second test program, VDOT began a bridge with two new features: concrete beams with 69 MPa strength, and 15-mm diameter prestressing strands spaced at 51 mm.

This paper discusses VDOT's early studies on low-permeability and high- strength concretes, the two test programs, and the seven bridges selected for the 1995-1997 construction seasons. Two of the bridges have been completed, and the others are under construction.

INTRODUCTION

Over the years, hydraulic cement concrete (HCC) has been successfully used in bridge structures. However, many concrete structures have exhibited premature failure requiring costly repairs. Increased traffic and vehicle loads and severe exposures have put more demand on these structures. Limited resources also demand longer service life, prompting the new AASHTO Load and Resistance Factor Design (LRFD) Specification to require a 75-year service life.

When concrete is exposed to the environment, many destructive forces start acting on it and can cause premature failure. The four major types of distress in bridge structures are corrosion of the reinforcement, alkali-silica reactivity, freeze-thaw damage, and sulfate resistance.[1] In each case, water and solutions penetrating into concrete initiate the distress. Corrosion is the most common and costliest distress and has been accelerated by the increase in the use of deicing salts containing chlorides. For longevity, harmful solutions should be kept away from concretes, and this can be achieved by using low-permeability concretes.[2] Such concretes can be categorized as high-performance concrete (HPC), since HPC is expected to provide enhanced workability, durability, strength, and/or dimensional stability.[3] In Virginia, and other parts of the country where cycles of freezing and thawing take place, concretes exposed to the environment must also have the proper air-void system.

To achieve low permeability, pozzolans and slag are used and the water-cementitious material ratio (W/CM) is kept below 0.45.[4] Pozzolanic material and slag generally improve the ultimate strength of conventional portland cement concrete. Sometimes, ultimate strengths higher than those attained by using pozzolans and slag in ordinary mixtures are needed, especially in prestressed beams. In such applications, W/CM is reduced to very low values of about 0.30 or below.[5]

Even though proper selection and proportioning of materials are needed for HPC, it is of utmost importance to pay close attention to proper design and construction practices for success. This paper includes VDOT's early studies on low-permeability and high-strength concretes, the two test programs, and the seven bridges selected for the 1995-1997 construction seasons. Two of the bridges have already been completed, and the others are under construction.

EARLY STUDIES

Permeability

Early studies on low-permeability concretes included preparing concretes with different W/CM and pozzolanic material and subjecting them to two tests.[1,2,4,6] One test is AASHTO T 259 (Resistance of Concrete to Chloride Ion Penetration) in which slabs measuring 300 x 300 mm are ponded with 3% NaCl for 90 days after a 42-day conditioning period, which consists of 14 days of moist curing and then air drying. Powdered samples are obtained from different depths and analyzed for chloride content to indicate the degree of penetration of chlorides into the concrete.

The other test is AASHTO T 277 or ASTM C 1202 (Electrical Indication of Concrete's Ability to Resist Chloride Ion Penetration). In this test cylindrical specimens measuring 100 mm in diameter and 50 mm in thickness are subjected to a potential difference of 60-volt dc for 6 hours. The resulting electrical charge, in coulombs, is related to the resistance of the specimens to chloride ion penetration.[7] Coulombs above 4,000 indicate high chloride permeability; 2,000 to 3,000 moderate; 1,000 to 2,000 low; 100 to 1,000 very low; and less than 100 negligible.

The ponding test is generally accepted as indicating the resistance of concretes to chloride penetration. The rapid permeability test is an indirect test. Quantitative relationships between the coulomb values obtained from the rapid permeability test and the chloride content from the ponding test have been sought. For comparisons, it is important that similar concretes be tested at similar ages. Permeability declines with time, and different concretes have different rates of reduction. Different curing methods or test ages can easily give the appearance of a lack of any relationship.

The permeability of typical concretes used in VDOT's transportation facilities has been determined, using both the rapid permeability and ponding tests. Concretes containing latex or pozzolanic materials (Class F fly ash or silica fume) and slag have varying coulomb values ranging from very low to very high at 28 days but are reduced to a low or very low value with time.[4,8] Portland cement concretes used in bridge decks usually have high values initially, which are reduced to a moderate range with time.[4] The positive influence of latex, pozzolans, and slag has been shown using the coulomb test. Similarly, the resistance of concretes to chloride ion penetration has been shown to increase with the use of latex, pozzolan, or slag.[4,6] Ponding times longer than 90 days were used since 90 days is not sufficient to discern among the different concretes used in bridge structures.

Studies at VDOT have shown that providing a higher curing temperature accelerates the reduction in coulomb values and indicates the long-term coulomb results at 28 days. As a result, in VDOT's proposed low-permeability special provision, specimens are tested at 28 days after moist curing for 1 week at room temperature followed by 3 weeks at 38 C. The special provision requires 1,500 coulombs or less for prestressed concrete, 2,500 coulombs or less for the deck, and 3,500 coulombs or less for the substructure.

Strength
Low-permeability concretes generally contain pozzolans or slag and have a low W/CM.[4,6] These modifications result in high compressive strengths that together with low permeability result in more economical structures, initially by reducing construction costs through increased span lengths and the use of fewer beams and, in the long term, by reducing maintenance costs through increased durability.[9,10]

With the use of HPC mixes in prestressed beams, additional prestressing force can be obtained in the beams. To increase the prestressing force and avoid steel congestion caused by additional strands, it is essential to use larger diameter strands (greater than 13 mm). In 1988, the Federal Highway Administration (FHWA) placed a moratorium on the use of 15-mm strands, which was lifted recently. Additionally, FHWA required a minimum strand spacing of 4 times the nominal strand diameter on prestressing strands for pretensioned applications and a multiplier of 1.6 times AASHTO equation 9-32, the equation used to calculate the development length of fully bonded prestressing strands. Thus, the use of 15-mm strands at 51-mm spacing needed further evaluation.[11]

Test Beams

Before the initiation of the HPC projects, two experimental projects were conducted to support the design of high-strength, low-permeability beams with 15-mm strands. Four prestressed concrete AASHTO Type II beams each containing 10 strands 15-mm in diameter at 51-mm center-to-center spacing, 2 on top and 8 across the bottom, were fabricated at a prestressing plant and tested to failure at the FHWA Structures Laboratory in McLean, Virginia. Two of the beams were prepared with concrete to develop a compressive strength of 69 MPa and the other two to develop a compressive strength of 83 MPa The beams were steam cured to obtain 70% of the compressive strength within 24 hours. They contained slag and had W/CM of about 0.30 or below to achieve high early and ultimate strengths and low permeability. To achieve such low W/CM requires large amounts of cementitious material, proper selection of aggregates, and high dosages of HRWRA. Thorough mixing and good construction practices must be followed during placement, consolidation, and curing. Excessive retardation should be eliminated. To achieve high early strengths, proper temperature management is needed. Beams were steam cured, and a temperature of 70 C was planned within the enclosure. However, the temperature inadvertently approached 85 C, which resulted in temperatures exceeding the boiling point in one of the two beams being monitored. Some of the test specimens stored in the recesses of the forms were damaged due to high heat. The beams were instrumented to measure transfer and development length. Due to high heat, difficulties in uncovering some gage points, and initial operator error, transfer lengths could not be determined. The beams were tested to determine the maximum load-carrying capacity under a concentrated load at midspan. They had a high load-carrying capacity exceeding the theoretical loads, indicative of concrete strengths exceeding 69 MPa.

Due to difficulties in the first test program, and to support the design of a bridge with 15-mm strands at 51-mm spacing, the second test program was initiated. Similar beams were prepared with 69 MPa concrete except that a composite slab was cast when the beams reached 69 MPa. Concretes contained silica fume, and W/CM was 0.28 for the beam and 0.36 for the slab. A concrete block measuring 0.6 x 0.6 x 0.9 m containing 8 untensioned 15-mm strands was prepared with the same concrete used in the beams. The strands were subjected to pull-out testing to evaluate the bond strength. The bond and compressive strengths and transfer lengths were satisfactory; permeability values were very low. At one end of one beam, several strands slipped at loads well below the predicted flexural capacity. This was attributed to poor consolidation at the beam end during casting. One indication of poor consolidation was the exceptionally long transfer length at that end. Thus it was shown that with proper construction practices, 15-mm strands with 51 mm spacing could be achieved. As a result, a bridge was designed to include concrete beams with a compressive strength of 69 MPa and with 15-mm strands.

VDOT HPC BRIDGES

For the 1995-97 construction season, VDOT selected seven HPC bridge structures, as shown in Table 1. The new low permeability specification was planned for five. The first two bridges given in Table 1 are complete. The first bridge had the low-permeability requirement for all the concrete. The beams of the seventh bridge will have a compressive strength of 69 MPa and will contain 15-mm strands.

Table 1. Bridges with High-Performance Concrete

#	Location	L(m)	No. Spans	Span L (m)	Beam Type	Bms/ Span	Bm Str. MPa	Low Perm.
1	Rte. 40	97.5	4	24.4	IV	5	55	Yes
2	Rte. 629	365.8	12	30.5	IV	5	55	No
3	Telegraph Rd.	55.5	2	27.7	Steel			Yes
4	Rte. 10	654.4	22	27/30	IV	5	55	No
5	Rte. 250	16	1	16	II	20	48	Yes
6	Second Str.	27	1	27	IV	6	48	Yes
7	Virginia Ave.	45.1	2	22.6	III	5	69	Yes

First HPC Bridge

The first HPC bridge in Virginia, with both low permeability and high strength in the prestressed beams, was constructed in 1995 on Rte. 40 over the Falling River in Campbell County. The prestressed beams of the first two bridges were fabricated with a release strength of 41 MPa and a 28-day strength of 55 MPa. All of the concrete placed in this structure met the permeability requirement. Mixture proportions and the requirements for air content and strength for the prestressed beams and cast-in-place substructure and deck concretes are given in Table 2. The coarse and fine aggregates were crushed limestone. In the substructure and deck, the coarse aggregate was crushed arch marble, and the fine aggregate natural sand.

A total of 20 AASHTO Type IV beams were placed with a compressive strength of 55 MPa instead of 28 AASHTO Type IV beams with a strength of 41 MPa. VDOT generally uses a minimum 28-day design strength of 35 MPa, and strengths as high as 41 MPa have been specified. The bridge construction costs per square meter of deck surface was $527, less than the 1994 average cost of $624 for 34 bridges in the federal-aid highway system in Virginia.

Table 2. Mixture Proportions and Requirements for the First and Second Bridge

Item	First Bridge Beam	First Bridge Substructure	First Bridge Deck	Second Bridge Beam
Portland Cement (kg/m^3)	446	209	195	303
Silica Fume (kg/m^3)	31	---	---	---
Slag (kg/m^3)	---	139	195	202
Coarse Aggr. (kg/m^3)	994	1056	1052	1157
Coarse Aggr. Size	#67	#57	#57	#68
Fine Aggr. (kg/m^3)	846	772	696	586
W/CM	0.32	0.44	0.40	0.33
Air (%)	5.5±1.5	6±2	6.5±1.5	5.5±1.5
28-d design str. (MPa)	55	21	28	55

Twenty beams were cast at the same plant that cast the test beams, using the same materials. The contractor cured 9 of the beams using steam curing, and the rest were moist cured. Strength and permeability test results for the beams are given in Table 3. In two of the steam-cured batches (B1 and B2) tested by cylinders placed in the form recesses, the average compressive strength was 55.2 MPa within 18 hours of batching. The TMC specimens averaged 57.4 MPa. After steam curing, specimens were kept outdoors. The compressive strength value averaged 67.4 MPa, and the permeability value 272 coulombs at 28 days. At 1 year, the average compressive strength value was the same, and the permeability value was 290 coulombs.

The moist-cured beams were cast on Fridays, and the strands released on Mondays. Two batches (B3 and B4) were tested. The average 3-day compressive strength was 53.9 MPa for moist cured and 61.8 MPa for TMC cylinders. After 3 days of moist curing, specimens were kept outdoors. The compressive strength values of the moist-cured specimens averaged 83.0 MPa, and the permeability values 183 coulombs at 28 days. At 1 year, the average compressive strength was 81.2 MPa, which is slightly less than the 28-day strength attributed to the variability and the curing conditions. The average permeability value at 1 year was 152 coulombs.

Table 3. Properties of Hardened Concrete for Beams of the First Bridge

Prop.	Age	B1	B2	B3	B4
Comp. Str. (MPa)	1d	56.3	54.1		
	1d[a]	58.1	56.7		
	3d			53.9	53.9
	3d[a]			61.2	62.3
	28d	67.9	66.8	83.6	82.5
	56d	68.2	68.0	83.6	84.9
	1 yr	67.8	67.0	81.6	80.9
E (GPa)	28d	41.2	40.3	42.8	42.9
	56d	41.1	43.8	45.9	44.3
Splitting (MPa)	28d	5.2	5.7	6.6	6.3
Flexure (MPa)	28d	6.7	6.0	6.9	6.8
Perm (coulomb)	28d	254	290	178	188
	1yr	280	300	119	184

Note: B1 and B2 were steam cured and then kept outdoors. B3 and B4 were moist cured for 3 days and then kept outdoors, those marked with ([a]) were temperature matched cured specimens (TMC).

The steam-cured specimens had higher early strengths but lower ultimate strengths than the moist-cured specimens, indicating the adverse effects of higher early curing temperatures. The elastic modulus values were in excess of 40 GPa at 28 days, splitting tensile strengths 5.2 MPa, and the flexural strengths 6.0 MPa.

Strength and permeability test results for both the substructure (A31, A32, A33) and deck concretes (A41, A42, A43, A44) are given in Table 4. The values were satisfactory. The coulomb values were in the low and very low range at 28 days when cured the last 3 weeks at 38 C. At 1 year, all values were in the very low range.

Table 4. Properties of Hardened A3 and A4 Concrete for the First Bridge

Property	Age	A31	A32	A33	A41	A42	A43	A44
Comp Str (Mpa)	1d	14.6	13.4	10.2	4.1	2.9		
	3d						11.9	11.4
	7d	28.9	30.5	26.9	40.1	37.5	37.2	33.7
	28d	40.1	42.5	40.0	57.9	55.8	62.4	64.1
	1y	47.4	49.2	46.4	65.6	64.0	73.6	74.5
E (GPa)	28d		33.3	32.6				
Splitting	28d	4.1	4.3	4.0	5.3	4.7	5.2	5.2
Flex	28d	5.8	5.6	5.1	6.0	5.7	7.2	6.9
Perm	28d	1831	1347	1670	1428	1405	1256	1677
Perm[a]	28d	1323	883	1076	696	773	743	898
Perm	1y	815	710	904	705	674	602	782

[a] Last 3 weeks cured at 38 C (100 F).

Second Bridge

In the second bridge, on Rte. 629 over the Mattaponi River, only the beams were specified to be HPC. AASHTO Type IV beams were used with a compressive strength of 55 MPa instead of an equal number of larger AASHTO Type V beams with a strength of 41 MPa. Sixty beams were cast, and all were steam cured. The mixture proportions are given in Table 2. Cementitious material was finely ground Type II cement and slag. The coarse aggregate was crushed granite. The fine aggregate was natural sand. Based on a single cylinder from each of the 60 batches, the average strength at release was 45.1 MPa, with a standard deviation of 4.0 MPa, and 61.4 MPa, with a standard deviation of 3.2 MPa at 28 days. Extra cylinders were available from two batches and were tested for permeability. The values, as an average of two specimens, were 323 and 536 coulombs at 6 weeks. The bridge construction cost per square meter of deck surface was $511.

CONCLUSIONS

Air-entrained HPCs with very low permeability values have been achieved with the use of a pozzolan or slag at W/CM below 0.45. Additionally, workable air-entrained HPCs with high early strength (exceeding 41 MPa at release) and ultimate strength with very low permeability (exceeding 55 MPa at 28 days) have been successfully made on a production basis. Higher early and ultimate strengths (exceeding 55 MPa at release and exceeding 69 MPa at 28 days) were produced using locally available materials for test beams. Proper temperature management is essential to achieve high early and ultimate strengths.

ACKNOWLEDGEMENTS

Studies reported here were conducted by the Virginia Transportation Research Council of the Virginia Department of Transportation, and sponsored by the Federal Highway Administration. The opinions, findings, and conclusions expressed in this paper are those of the authors and not necessarily those of the sponsoring agency.

REFERENCES

1. Ozyildirim, C. 1993. Durability of Concrete Bridges in Virginia. *ASCE Structures XI Proceedings: Structural Engineering in Natural Hazards Mitigation.* American Society of Civil Engineers, New York, pp. 996-1001.

2. Ozyildirim, C. 1993. High-Performance Concrete for Transportation Structures. *Concrete International*, Vol. 15, No. 1, pp. 19-26.

3. Zia, P., Leming, M. L., Ahmad, S. H., Schemmel, J. J., Elliott, R. P., and Naaman, A. E. 1993. *Mechanical Behavior of High Performance Concretes. Volume 1. Summary Report.* SHRP-C-361. Strategic Highway Research Program, Washington D. C.

4. Ozyildirim, C. 1994. Resistance to Penetration of Chlorides into Concrete Containing Latex, Fly Ash, Slag, and Silica Fume. *ACI SP-145, Durability of Concret.*, American Concrete Institute, pp. 503-518.

5. Ozyildirim, C., Gomez J., and Elnahal, M. 1996. High Performance Concrete Applications in Bridge Structures in Virginia. *ASCE Proceedings: Worldwide Advances in Structural Concrete and Masonry.* American Society of Civil Engineers, New York, pp. 153-163.

6. Ozyildirim, C. 1994. Rapid Chloride Permeability Testing of Silica Fume Concrete. *Cement, Concrete, and Aggregate*, ASTM, Vol. 16, No. 1, pp. 53-56

7. Whiting, D. 1981. *Rapid Determination of the Chloride Permeability of Concrete*. FHWA/RD-81/119. Federal Highway Administration, Washington, D. C.

8. Sprinkel, M. M. 1992. Twenty-Year Performance of Latex-Modified Concrete Overlays. *Transportation Research Record 1335*. Transportation Research Board, Washington, D.C., pp. 27-35.

9. Lane, S. N., and Podolny, W. 1993. The Federal Outlook for High Strength Concrete Bridges. *PCI Journal*, Vol. 38, No. 3.

10. Bruce, R. N., Russell, H. G., Roller, J. J., and Martin, B. T. 1994. *Feasibility Evaluation of Utilizing High-Strength Concrete in Design and Construction of Highway Bridge Structures*. LA-FHWA-94-282. Louisiana Transportation Research Center, Baton Rouge.

11. Buckner, C.D. 1994. *An Analysis of Transfer and Development Lengths for Pretensioned Concrete Structures*. FHWA-RD-94-049. Office of Engineering and Highway Operations R & D, Federal Highway Administration, Washington, D.C.

TEXAS HIGH PERFORMANCE CONCRETE BRIDGES IMPLEMENTATION STATUS

Mary Lou Ralls, P.E.
Bridge Construction/Maintenance Engineer
Texas Department of Transportation
Austin, Texas, U.S.A.

Ramon L. Carrasquillo, Ph.D., P.E.
Professor
University of Texas
Austin, Texas, U.S.A.

ABSTRACT

Construction of the first two high performance concrete (HPC) bridges in Texas will be completed in 1997. The Louetta Road Overpass in Houston and the North Concho River, US 87, & South Orient Railroad Overpass in San Angelo, in this paper referred to as the San Angelo HPC (Eastbound) bridge, utilize HPC in the beams, decks, and substructures. This paper describes the bridges and discusses design optimization, concrete mix design and properties, performance-related tests and results to date, and comparison of test results with the HPC definition.

INTRODUCTION

The construction contract for the Louetta Road Overpass, on State Highway 249 in Harris County just outside the northwest city limits of Houston, was awarded to Williams Brothers Construction Company of Houston in February of 1994. In June of 1995, the construction of the San Angelo HPC bridge in Tom Green County began under the direction of Jascon, Inc., of Uvalde and Reece Albert, Inc., of San Angelo. Figures 1 and 2 show plan views of the bridges, Figures 3 and 4 are site photographs (the San Angelo photograph was taken during deck construction), and Table 1 describes member types.

Research projects in cooperation with the Center for Transportation Research at the University of Texas at Austin are part of both projects. The research team is led by Ramon L. Carrasquillo, Ph.D., P.E., and Ned H. Burns, Ph.D., P.E. Funding is through the Federal Highway Administration (FHWA) and the Texas Department of Transportation (TxDOT). The San Angelo research is also funded by the HPC pooled-fund States. Technical assistance has been received from all levels of FHWA, from a National Peer Advisory Group set up for the projects, and from TxDOT's Project Advisors Committees.

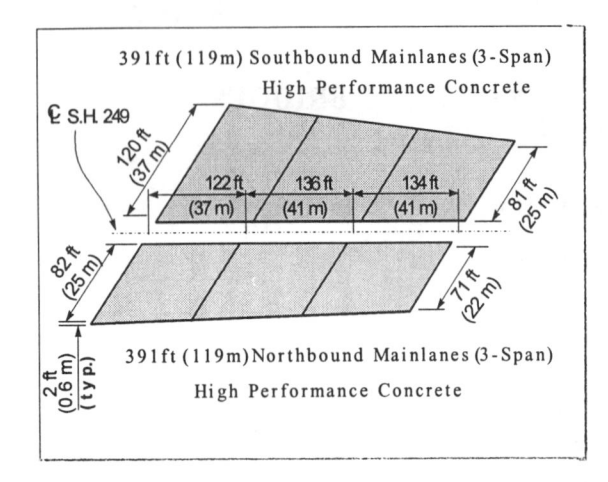

Figure 1. Louetta Plan View

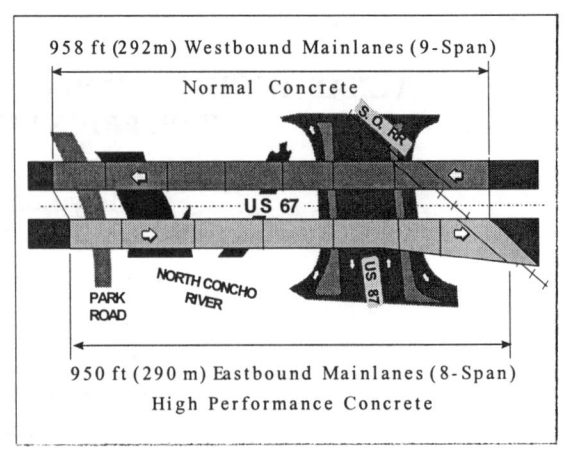

Figure 2. San Angelo Plan View

Figure 3. Louetta Road Overpass

Figure 4. San Angelo HPC Bridge

Table 1 - Bridge Member Types

Member	Louetta	San Angelo
Decks ①	Cast-in-Place Topping & Composite Precast Panels	Cast-in-Place Topping & Composite Precast Panels
Beams ②	54-in. deep Texas U-Beam	54-in. deep AASHTO Ty. IV Beam
Interior Supports ③	Precast Post-Tensioned Concrete Individual Pier Segments	Cast-in-Place Reinforced Concrete Column & Continuous Cap

Note: 1 mm = 0.0394 in.

① Construction or controlled joints in decks at interior supports of continuous units.
② Simple-span beams.
③ Abutments are conventional non-HPC.

STRUCTURAL DESIGN OPTIMIZATION

As part of the research project, the American Association of State Highway and Transportation Officials (AASHTO) bridge specifications[1] were evaluated for use in the design of the HPC bridges. Table 2 gives the primary structural design parameters evaluated. For the high-strength HPC beams, the allowable concrete tensile stresses were increased by 33 percent based on previous research[2], and the concrete modulus of elasticity was held at a constant 41 GPa (6 million psi) to provide adequate stiffness at both release and service condition. AASHTO shear provisions were conservative for the high-strength HPC beams[3].

To utilize concrete strengths in excess of 70 MPa (10,000 psi), it was found that 15-mm (0.6-inch) diameter prestressed strands at 50-mm (2-inch) spacing were needed. However, FHWA had a moratorium on the use of these larger strand in pretensioned applications[4]. Extensive strand transfer and development length tests were conducted on the 15-mm (0.6-inch) diameter prestressed strands at 50-mm (2-inch) spacing[3]. After successful testing, FHWA approved the use of the larger strand on these projects and subsequently lifted the moratorium, except for length multipliers, in 1996.

The decks and interior supports were designed according to AASHTO except that no tension was allowed in the Louetta pier design. Standard HS20 live load with impact was used. Provisions that the researchers are evaluating for HPC include concrete compressive and tensile stresses, concrete modulus of elasticity, and prestress losses.

Table 2 - Structural Design Parameters

Member	Allowable Concrete Compressive Stress	Allowable Concrete Tensile Stress	Concrete Modulus of Elasticity	Shear Steel	Prestress Loss	Strand Transfer & Development Length
Decks: CIP ① Panels	AASHTO AASHTO	AASHTO AASHTO	AASHTO AASHTO	AASHTO AASHTO	N.A. ② AASHTO	N.A. AASHTO
Beams	AASHTO	$10\sqrt{f'_{ci}}$ psi $8\sqrt{f'_c}$ psi	6×10^6 psi	AASHTO	AASHTO	AASHTO ③
Substr.: CIP Precast	AASHTO AASHTO	AASHTO 0 psi	AASHTO AASHTO	AASHTO AASHTO	N.A. AASHTO	N.A. N.A.

① Cast-in-place. Note: 1 Mpa = 145 psi
② Not applicable.
③ Following successful completion of ultimate strength tests[3] and approval by FHWA.

Dimensions of the bridges are given in Table 3, and design concrete compressive strengths are given in Table 4. Additional details may be obtained from Reference 3.

Table 3 - HPC Bridge Dimensions

HPC Dimensions	Louetta Northbound & Southbound	San Angelo Eastbound, Spans 1-5
Span length	121.5 - 135.5 ft	131.0 - 157.0 ft
Beam spacing	11.7 - 15.8 ft	6.6 - 11.0 ft

Deck thickness:
  Cast-in-place
  Precast Panels
  Total deck thickness
	3.75 in.
	3.50 in.
	7.25 in.
		3.50 in.
		4.00 in. [1]
		7.50 in.

[1] Texas standard panel thickness.

Note: 1 mm = 0.0394 in.
1 m = 3.281 ft

Table 4 - Design Compressive Strengths (psi)

Member	Louetta Northbound	Louetta Southbound	San Angelo Eastbound	San Angelo Westbound
CIP decks at 28-day	4000	8000	6000	4000
Panels at Release	6000	6000	4000	4000 [1][4]
Panels at 28 days	8000	8000	6000	5000 [1][4]
Beams at Release	6900-8800	6900-8800	8900-10,800 [2] 4000-6800 [3]	4000-6600 [4]
Beams at 56 days [5]	9800-13,100	9800-13,100	10,900-14,700 [2] 5800-7800 [3]	5000-8900 [4]
Caps at 28 days	No Cap	No Cap	8000	6000 [4]
Columns at 28 days	10,000	10,000	6000	3600 [4]

[1] Texas standard precast panel compressive strengths.
[2] Spans 1-5.
[3] Spans 6-8; underneath geometric constraints controlled design.
[4] Non-HPC.
[5] Beams could be shipped as soon as design strengths were attained.

Note: 1 Mpa = 145 psi

Initial benefits from the use of high-strength HPC beams can be seen in Table 5, which compares Spans 1-5 of the San Angelo Eastbound HPC bridge with normal-concrete beam design. Note that Span 1 of the normal-concrete design requires 7 beams at 1.7 m (5.7 ft) spacing compared to 4 high-strength HPC beams at 3.3 m (11.0 ft) spacing. Also note that an overall reduction of one interior support and 10 beams was obtained through the optimization of the high-strength HPC beam designs in Spans 1-5 of the Eastbound HPC bridge.

Caution is needed, however, when designing long, slender I-shaped beams incorporating high strength HPC. Stability and camber became the controlling parameters in the beam designs for Spans 1-5 of the HPC Eastbound bridge. The result was construction costs of

$510 per square meter ($47 per square ft) of deck area, on the high end of previous bridge construction costs in San Angelo. This compares to an average $258 per square meter ($24 per square ft) of deck area for the Louetta bridges, comparable to other non-HPC U-beam bridges in Houston at that time. The Louetta beams did not have stability and camber concerns.

A more exact design that takes into account all variables is needed when stretching AASHTO Type IV beams past 40 m (130 ft)[5]. Other means of optimization, such as wider beam spacings and more stable and efficient beam shapes, should be considered.

Table 5 - San Angelo HPC Bridge Member Reduction

Span No.	Normal Concrete ①			HPC Eastbound, Spans 1-5				Reduction in No. of Bms.
	Span Length (ft)	No. of Bms.	Bm. Spac. (ft)	Span Length (ft)	No. of Bms.	Bm. Spac. (ft)	Concrete Strength (psi)	
1	131	7	5.7	131	4	11.0	13,600	3
2	114	5	8.5	157	6	6.6	13,500	-1
3	114	5	8.5	150	5	8.3	14,700	0
4	114	5	8.5	149	5	8.3	14,000	0
5	114	5	8.5	- ②	-	-	-	5
6	140	8	5.0	140	5	8.5	10,900	3
Total	121 (avg)	35	7.1 (avg)	145 (avg)	25	8.3 (avg)	13,300 (avg)	10 ③

① Concrete strength < 6500 psi at release, <9000 psi at 28 days. Note: 1 MPa = 145 psi
② Number of spans reduced by 1. 1 m = 3.281 ft
③ Number of beams reduced by 10.

CONCRETE MIX OPTIMIZATION

Rather than including requirements such as modulus of elasticity, flexural strength, and durability for project acceptance, the contract documents specified that the researchers guide the concrete producers in the development of mix designs and quality control/quality assurance procedures to achieve the needed performance characteristics in the field. Standard job control tests were required, and the payment penalty for strengths less than design was waived for the HPC mixes. The producers were encouraged to use local materials and conventional production methods to the maximum extent possible. Additional specimens were cast during construction for the various research tests.

Texas Concrete Company in Victoria, Texas, was the beam fabrication subcontractor for the two HPC bridge projects. Mixture proportions for the HPC beams are shown in Table

6, and fresh concrete properties of the HPC beams are shown in Table 7. The aggregate content for the San Angelo beams was slightly lower and the fine aggregate slightly higher than in the Louetta beams primarily to accommodate the workability of the concrete for casting a different beam shape having a slightly lower slump. Note the low water-cementitious materials ratio, the replacement of approximately a third of the cement with fly ash, and the high maximum slump allowed. Air entrainment is not required in prestressed concrete beams in Texas, and is not needed for durable concrete unless the concrete is exposed to alternate cycles of freezing and thawing under wet conditions (above 90 percent saturation). Steam curing was not necessary to achieve the needed early concrete strengths. During the winter months, very small amounts of steam were used only to prevent the loss of heat from the beams during the early curing until the heat of hydration generated enough temperature that the concrete gained strength.

Table 6 - Mixture Proportions for HPC Beams

Component	Louetta Northbound & Southbound	San Angelo Eastbound, Spans 1-5
Coarse aggregate type	Dolomitic Limestone, 0.5" max, ASTM GR 7 ①	Dolomitic Limestone, 0.5" max, ASTM GR 7 ①
Coarse aggregate quantity	1918 pcy	1863 pcy
Fine aggregate type	Sand FM = 2.60	Sand FM = 2.60
Fine aggregate quantity	1029 pcy	1062 pcy
Water	247 pcy	246 pcy
Cement type	Type III	Type III
Cement quantity	671 pcy	671 pcy
Fly Ash ②	316 pcy	312 pcy
Repl. of Cement, by Wt.	32 %	32 %
Retarder, ASTM Type B	27 oz / cy	28 oz / cy
HRWR, ASTM Type F ③	200 oz / cy	200 oz / cy
Air Entr., ASTM C260	None	None

① Crushed.
② ASTM Class C, with additional requirements per TxDOT's Departmental Mat'ls. Spec. D-9-8900.
③ High-range water-reducer.

Note: 1 MPa = 145 psi
1 mm = 0.0394 in
1 mL/m^3 = 0.0258 oz/cy
1 kg/m^3 = 1.686 pcy

Mixture proportions for the cast-in-place decks are shown in Table 8. Fresh concrete properties of the decks are shown in Table 9. Several deck comparisons are being studied.

Table 7 - Fresh Concrete Properties of HPC Beams

Component	Louetta Northbound & Southbound	San Angelo Eastbound, Spans 1-5
W/(C + FA), by Wt.	0.25	0.25
Total Air	0.9 % ①	0.9 % ①
Slump	8 - 10 in.	6 - 9 in.
Unit Weight	154.0 pcf	*Not Available*

① No air entrainment.

Note: 1 mm = 0.0394 in
1 kg/m^3 = 1.686 pcy

Table 8 - Mixture Proportions for Cast-in-Place Decks

Component	Louetta Northbound	Louetta Southbound	San Angelo Eastbound ①	San Angelo Westbound Spans 1-5	San Angelo Westbound Spans 6-9
Coarse agg.: Type Quantity	limestone, ② 1.5" max 1856 pcy	limestone, ② 1" max 1811 pcy	river gravel, ② 1.25" max 1900 pcy	river gravel, ② 1.25" max 1856 pcy	river gravel, ② 1.25" max 1856 pcy
Fine agg.: Type Quantity	Sand FM=2.54 1243 pcy	Sand FM=2.54 1304 pcy	Sand FM=2.70 1365 pcy	Sand FM=2.70 1240 pcy	Sand FM=2.70 1243 pcy
Water	230 pcy	245 pcy	219 pcy	258 pcy	258 pcy
Cement: Type Quantity	Type I 383 pcy	Type I 473 pcy	Type II 490 pcy	Type II 427 pcy	Type II 611 pcy
Fly Ash ③ Repl. of Cem., by Wt.	148 pcy 28 %	221 pcy 32 %	210 pcy 30 %	184 pcy 30 %	None None
Retarder, ④ ASTM Ty. B	45 oz / cy	22 oz / cy	28 oz / cy	26 oz / cy	26 oz / cy
HRWR, ⑤ ASTM Ty. F	None	122 oz / cy	156 oz / cy	None	None
Air Entr., ASTM C260	2 oz / cy	None	3 oz / cy	3 oz / cy	3 oz / cy

① Under construction in mid-1997.
② Crushed.
③ ASTM Class C, with additional requirements per TxDOT's Departmental Mat'ls. Spec. D-9-8900.
④ The retarder dosage rate varies depending on weather conditions at the time of placement.

Note: 1 MPa = 145 psi
1 mm = 0.0394 in
1 mL/m^3 = 0.0258 oz/cy
1 kg/m^3 = 1.686 pcy

Values given here represent the upper bound amounts used during hot weather conditions.
The retarder is taken out during winter months.
⑤ High-range water-reducer.

Table 9 - Fresh Concrete Properties of Cast-in-Place Decks

Component	Louetta Northbound	Louetta Southbound	San Angelo Eastbound ①	San Angelo Westbound Spans 1-5	San Angelo Westbound Spans 6-9
W/(C+FA), by Wt.	0.43	0.35	0.31	0.42	0.42
Total Air	5.0 %	1.4 % ②	6 %	6 %	6 %
Slump	3 - 4 in.	8 - 9.5 in.	7 - 9 in.	3 - 4 in.	3 - 4 in.
Unit Weight	143.2 pcf	150.2 pcf	149.4 pcf	145.3 pcf	145.6 pcf

① Under construction; values shown are for trial mixes.
② No air entrainment.

Note: 1 mm = 0.0394 in
1 kg/m^3 = 1.686 pcy

The Northbound Louetta cast-in-place HPC deck is 4000 psi concrete (the standard deck design strength in Texas) that is enhanced for durability with 28 percent by weight replacement of the cement with fly ash. It did not require the use of a high-range water-reducer. The Southbound Louetta deck used 8000 psi concrete with 32 percent by weight replacement of the cement with fly ash. Because of its higher specified compressive strength and the low 0.35 water-cementitious materials ratio, a high-range water-reducer was used. This drier mix required much greater attention to fogging and interim and final curing in order to avoid plastic shrinkage cracking. Constructability is a concern when using these low water-cementitious materials ratios in cast-in-place concrete decks.

The San Angelo cast-in-place decks also have comparison studies. The Eastbound HPC deck uses 6000 psi with 30 percent replacement of cement with fly ash, a low 0.31 water-cementitious materials ratio, and 6 percent total air because San Angelo is in a freeze-thaw zone. The Eastbound HPC deck required the use of a high-range water-reducer. The Westbound HPC deck uses 4000 psi with a more typical 0.42 water-cementitious materials ratio and 6 percent total air. In the Westbound bridge, Span 1-5 decks have 30 percent fly ash replacement of cement for enhanced durability, compared to the San Angelo standard of no fly ash in Span 6-9 decks.

All of the HPC components are being optimized according to the requirements for each use. These include optimizing the cementitious paste to reduce early temperature rise, enhance flowability, reduce permeability, and achieve the required strength. The admixtures are being optimized to achieve adequate workability and early strength gain as needed in each specific application. The aggregates are being optimized in terms of maximum size and quantity or content in order to achieve the needed modulus of elasticity, reduce shrinkage and creep, and achieve workability.

PERFORMANCE-RELATED TESTS AND MEASUREMENTS

Nineteen different performance-related tests and measurements are being conducted in the two HPC bridge projects, as shown in Table 10. Some of the field test results to date are shown in Table 11, which gives compressive strength, modulus of elasticity and splitting tensile strength test results. Table 12 gives the permeability test results to date. Analysis of test results continues.

Table 10 - Performance-Related Tests and Measurements

Member	Louetta		San Angelo	
	Northbound	Southbound	Eastbound	Westbound
Cast-in-place decks	1-7, 11-15	1-7, 11-15	1-15	1-15
Precast panels	1-6, 11-14		1-6, 11-14	
Beams	1-6, 11-16		1-7, 11-19	1-6, 11-18 ①
Caps	No Cap		1-3, 6	1-3, 6 ①
Columns	1-3, 6, 11, 13		1-3, 6	1-3, 6 ①

1)	Compressive Strength		① Non-HPC.	
2)	Modulus of Elasticity	11)	Temperatures	
3)	Splitting Tensile	12)	Deflections	
4)	Shrinkage	13)	Concrete Strains	
5)	Creep	14)	Coefficient of Thermal Expansion	
6)	Chloride Ion Penetration	15)	Maturity	
7)	Ponding	16)	Transfer & development length	
8)	Freeze-Thaw	17)	Unstressed strand pull-out	
9)	Abrasion	18)	Strand chemical residue	
10)	Scaling	19)	Strand force prior to release	

Table 11 - Average Field Test Results to Date (psi)

Member	Louetta		San Angelo		
	Northbound	Southbound	Eastbound	Westbound Spans 1-5	Westbound Spans 6-9
CIP decks at 28 days: Compressive Strength Mod. of Elast. Split. Tens. Strength	4890 4460 x 10³ 465	9220 4730 x 10³ 725	①	6370 5510 x 10³ 540	④ 5310 5000 x 10³ 510
Deck panels at 28 days: Compressive Strength Mod. of Elast. Split. Tens. Strength	9680 5900 x 10³ 760		10050 5010 x 10³ 790	④ 8610 4915 x 10³ 690	

		②	③ ④
Beams at Release:			
Time to release	(16-21 hrs)	(19-20 hrs)	(26 hrs)
Compressive Strength	8690	8630	8310
Mod. of Elast.	6000 x 10³	5950 x 10³	5910 x 10³
Beams at 28 days:			
Compressive Strength	13,900	13,450 ⑤	10,200
Mod. of Elast.	6690 x 10³	6550x10³ ⑤	6020 x 10³
Split. Tens. Strength	920	980 ⑤	750
Beams at 56 days:			
Compressive Strength	15,200	14,830	11,130
Mod. of Elast.	7000 x 10³	6580 x 10³	6450 x 10³
Split. Tens. Strength	940	1050	890
Caps at 28 days:			④
Compressive Strength	No Cap	10,410	7100
Mod. of Elast.		6700 x 10³	5370 x 10³
Split. Tens. Strength		845	600
Columns at 28 days:			
Compressive Strength	12,840	8290	5100
Mod. of Elast.	7080 x 10³	5560 x 10³	4740 x 10³
Split. Tens. Strength	930	720	530

① Under construction in mid-1997.
② From 1 beam to date.
③ From 2 beams in Span 1 (coarse agg. = river gravel).
④ Non-HPC.
⑤ At post-tensioning (25 days).

Note: 1 Mpa = 145 psi

Table 12 - Permeability Test Results @ 56 days, Coulombs Passed

Member	Louetta		San Angelo		
	Northbound	Southbound	Eastbound	Westbound Spans 1-5	Westbound Spans 6-9
CIP decks	1730	900	①	2130	3050 ③
Panels	1430		1980	3230 ③	
Beams	560		280②	550 ③	
Caps	No Cap		560	1180③	
Columns	410		580	2260③	

① Under construction in mid-1997.
② Test results from 1 beam to date.
③ Non-HPC.

COMPARISON WITH HPC DEFINITION

Grading System

Efficient use of HPC in transportation structures requires the designer to optimize each aspect of concrete performance including both fresh and hardened concrete. As a result, HPC becomes an engineered component to be used in the design and construction process. In an effort to better understand and describe each aspect of HPC, Goodspeed, et al.,[6] proposed a grading system for cataloging the different properties and characteristics of HPC. Table 13 gives the different grades for the HPC used in the Louetta and San Angelo bridge projects. The design requirement for compressive strength has values in all grades for strength ranging from N.A. to Grade 4 [>97 MPa (>14 ksi)]. Modulus of elasticity values are in Grade 1 [28 - 40 GPa (4 - 6 million psi)] and Grade 2 [40 - 50 GPa (6 - 7.5 million psi)] while the chloride penetration values are in Grade 2 [800 - 2000 coulombs] and Grade 3 [< 800 coulombs].

Table 13 - HPC Grades for Texas HPC Bridge Members

Performance Characteristic	Louetta			San Angelo		
	Decks	Beams	Substr.	Decks	Beams	Substr.
Freeze-thaw durability	*	*	*	②	②	*
Scaling resistance	*	*	*	②	*	*
Abrasion resistance	*	*	*	②	*	*
Chloride penetration	2	3	3	②	3	3
Strength ①	N/A, 2	2-3	3	N/A, 1	N/A,1-4	1-2
Elasticity	1	2	2	②	2	1

① Strength grades are based on design requirements; grades for other performance characteristics are based on test results to date.
② Test either in progress or planned but concrete not cast yet.
* No tests performed.

Although an excellent guideline for describing and comparing the relative performance aspects of HPC, Goodspeed's grading system is not detailed enough for use by the structural design engineer when specifying the required performance criteria of HPC for a job. For specification purposes, engineers need to be more specific when addressing a given concrete property such as strength than just requiring that the strength be within a range as is the case for the grading system. However, Goodspeed's grading system approximates better the needs of the specifying engineer when addressing durability requirements.

Specifications

Engineers could use the listing of concrete properties and performance requirements in Goodspeed's work as a menu of available aspects of concrete that could be optimized in the design. However, engineers need to be provided with guidelines for selecting an acceptable value for any given concrete property that is to be optimized and/or specified. This is the case for such properties as compressive strength, flexural strength, modulus of elasticity, coefficient of thermal expansion, creep, shrinkage and others. An example of guidelines for the selection of the modulus of rupture for HPC is given in Figure 5. As shown, the guideline consists of a band of acceptable values for modulus of rupture for various compressive strengths.

For a specified compressive strength, the lower point on the band corresponds to a modulus that is obtainable without much effort for typical mixes. The upper point on the band is a modulus that can be obtained for that compressive strength, but which requires more engineering of the mix to achieve the higher modulus. Increasing values within the band require increasing engineering of the mix for that compressive strength. Similar graphs can be developed for the other properties/characteristics, thus allowing the designer to optimize the design based on the specific needs.

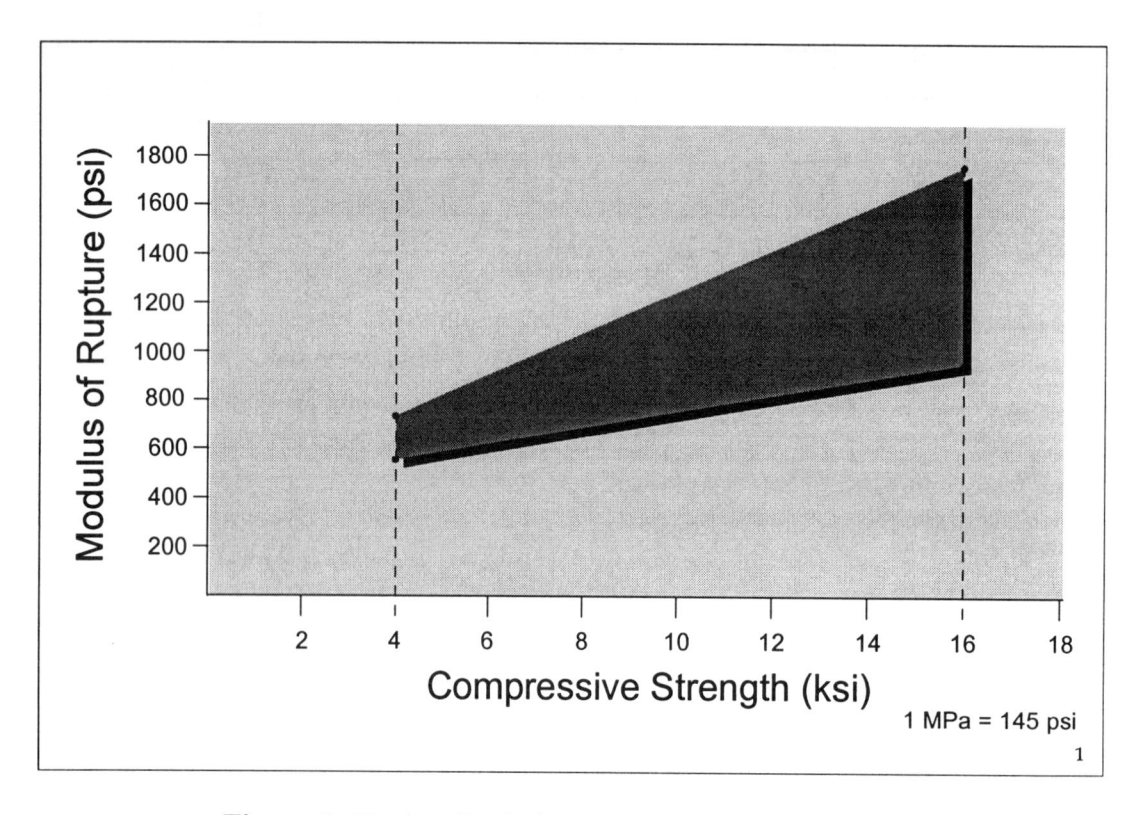

Figure 5. Design Optimization for Modulus of Rupture

Regarding specifying the needed requirements to ensure that HPC has adequate durability and service life, engineers need to be provided with guidelines relating the exposure condition, desired service life and performance testing. Long-term durability of HPC is a function of the concrete's freeze-thaw resistance, sulfate resistance, alkali-aggregate reaction, abrasion resistance, scaling resistance, corrosion protection, and structural details such as clear cover and drainage to avoid ponding. These performance aspects of concrete are addressed by such controlling factors as entrained air, permeability, strength and materials characteristics and selection.

In this approach, the level of required service-life durability performance is specified based on exposure condition in service. A correlation is provided between the exposure condition in service and the measured performance on a specific test, with higher test performance required for more severe exposure conditions. An example correlation between level of performance required for service life and measured test performance is given for freeze-thaw resistance in Table 14.

Table 14 - Example Freeze-Thaw Resistance Performance

Exposure		Test Performance (ASTM C666)
Condition	Degree	
Freeze-thaw > X cycles; wet; seawater; thin cover	Severe	DF > 90 %
Freeze-thaw < X cycles; wet; seawater; thin cover	Moderate	DF > 80 %
Dry	Mild	N.A.

SUMMARY AND CONCLUSIONS

The Louetta and San Angelo bridge research and construction projects are progressing satisfactorily. Project acceptance strength tests are meeting requirements, and data are being collected and analyzed for 19 performance-related tests and measurements to evaluate material and structural behavior of HPC bridges. Design optimization and service life optimization are being evaluated. Significant findings to date include the following.

1) Very low concrete permeabilities are being achieved with an approximate 30 percent replacement of cement with fly ash.
2) For Spans 1-5 of the San Angelo Eastbound bridge, one support and 10 of 35 beams were eliminated by optimizing the design using high-strength HPC beams. Optimizing with wider beam spacings and more stable beam shapes should be considered as an alternative to long slender I-shaped beam spans.
3) Bridge beam release strengths greater than 55 MPa (8000 psi) in 16-21 hrs have been attained with high-strength HPC in the precast fabrication plant.
4) Bridge beam compressive strengths exceeding 90 MPa (13,000 psi) at 28 days and 103 MPa (15,000 psi) at 56 days have been achieved in the precast fabrication plant.
5) Bridge beam modulus of elasticity reached 41 GPa (6 million psi) in 16-21 hrs.
6) Constructability is a concern when using low water-binder ratios in decks.
7) Evaluation of the Texas HPC projects indicates that an alternate approach for the HPC definition would be to specify each performance requirement on the basis of design optimization and service life optimization considering exposure conditions, as discussed herein.

REFERENCES

1. "Interim Specifications - Bridges, 1993," *Standard Specifications for Highway Bridges, Fifteen Edition, 1992*, American Association of State Highway and Transportation Officials, Inc., Washington, D.C.

2. Carrasquillo, R.L., Slate, F.O., and Nilson, A.H., "Properties of High Strength Concrete Subject to Short-Term Loading," *Journal of the American Concrete Institute*, Vol. 78, No. 3, May-June, 1981, pp. 171-179.

3. "SHRP High Performance Concrete Bridge Showcase," *participant notebook*, Federal Highway Administration and Texas Department of Transportation, in cooperation with the University of Texas at Austin, March 25-27, 1996, Houston, Texas.

4. "Prestressing Strand for Pretension Applications -- Development Length Revisited," U.S. Department of Transportation, Federal Highway Administration Memorandum, October 26, 1988.

5. Gross, S.P., Byle, K.A., and Burns, N.H., "Camber of Long-Span High Performance Concrete Girders," unpublished *Technical Memorandum No. 1*, Center for Transportation Research Project 9-589, University of Texas at Austin, October 1996.

6. Goodspeed, C.H., Vanikar, S., and Cook, R.A., "High-Performance Concrete Defined for Highway Structures," *Concrete International*, V. 18, No. 2, February 1996, pp. 62-67.

COLORADO SHOWCASE ON HPC BOX-GIRDER BRIDGE: DEVELOPMENT AND TRANSFER LENGTH TESTS

P.B. Shing, D. Cooke, and D. M. Frangopol
University of Colorado
Boulder, Colorado, U.S.A.

M.A. Leonard, M. L. McMullen, and W. Hutter
Colorado Department of Transportation
Denver, Colorado, U.S.A.

ABSTRACT

A 65.5-m (215-ft.)-long, four-span, bridge in Denver, Colorado, is being replaced by a two-span bridge of precast pretensioned side-by-side box girders. To attain a high span-to-depth ratio for the superstructure, 69-MPa (10,000-psi) high performance concrete is specified for the box girders. The prestressing strand in the girders is 15 mm (0.6 in.) in diameter and spaced at 51 mm (2 in.) on center. As part of this project, three pretensioned box girders were tested to evaluate the transfer and development length. The test results indicate that the AASHTO specifications on transfer and development length are conservative for the box girders.

INTRODUCTION

The Colorado Department of Transportation (CDOT) is carrying out a showcase project sponsored by the Federal Highway Administration (FHWA) on the use of high performance concrete (HPC) in bridges. The project involves the replacement of a 65.5-m (215-ft.)-long, four-span, bridge by a two-span bridge at Interstate 25 over Yale Avenue in Denver, Colorado. One design objective in this project is to minimize the superstructure depths in order to obtain the desired vertical clearance beneath the bridge with a minimum change to existing vertical alignments. Side-by-side 32-m (105-ft.)-long precast box girders with a composite topping slab will be employed to minimize construction costs and optimize erection. To achieve the desired span-to-depth ratio, 69-MPa (10,000 psi) HPC and Grade 270, 15-mm (0.6-in.)-diameter prestressing strands spaced at 51 mm (2 in.) on center are used.

As part of this project, three box girders with 69 MPa (10,000 psi) concrete were tested at the University of Colorado (CU) to determine the transfer and development length of 15-mm (0.6-in.)-diameter prestressing strands at 51-mm (2 in.) spacing. These girders, designed by CDOT, were scaled down versions of the girders to be used in the bridge. The test program and results, including the accompanying material and strand pullout tests, are presented in this paper.

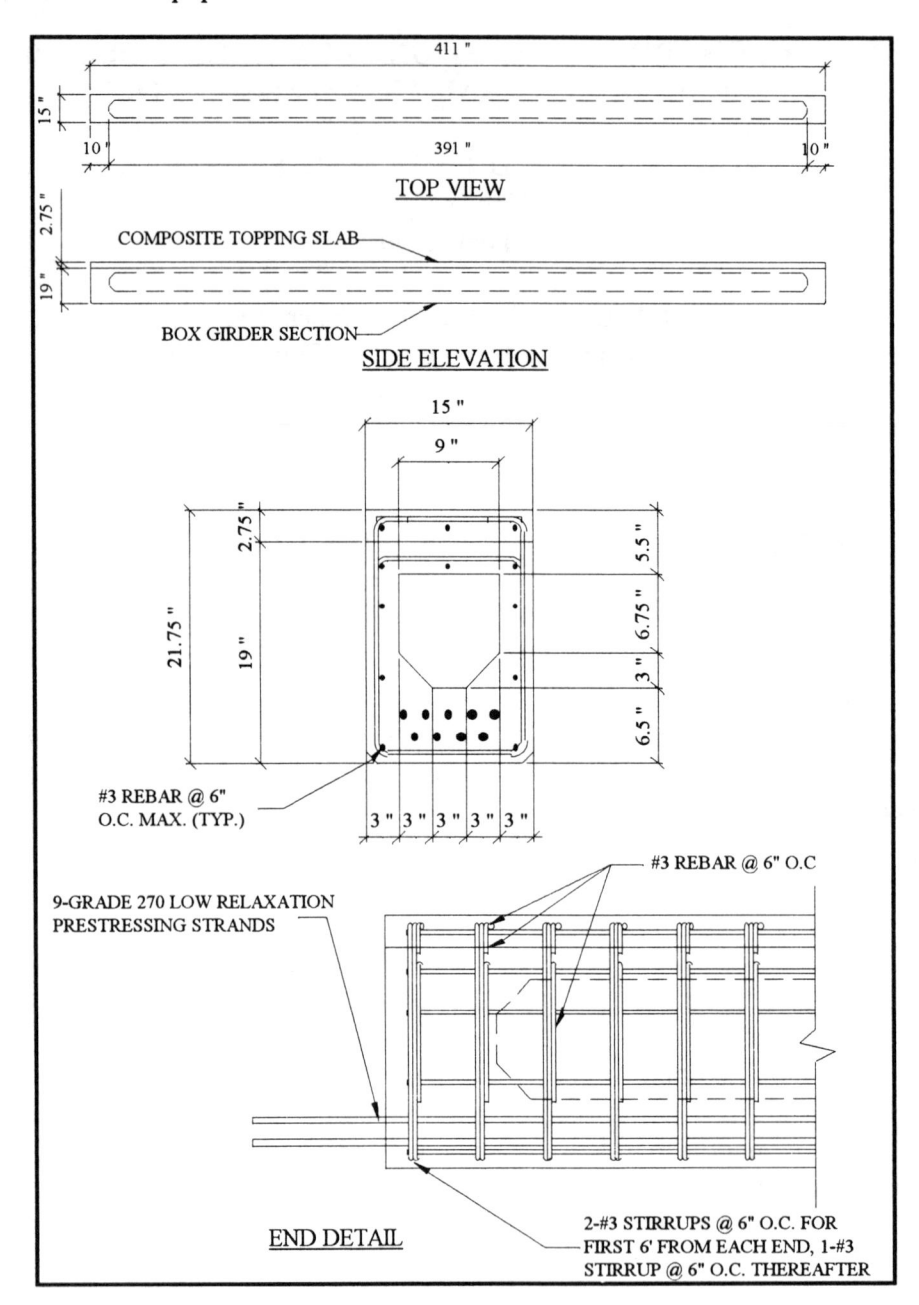

Figure 1 - Test Girder (1 in. = 25.4 mm)

EXPERIMENTAL PROGRAM

Girder Specimens
The girder design is shown in Figure 1. Each girder specimen consisted of a box section with a composite topping slab. The girder section was prestressed with nine Grade 270,

seven-wire, 15-mm (0.6-in.)-diameter strands at 51-mm (2-in.) spacing. The prestress immediately before release was specified to be 1,407 MPa (204 ksi). The girder concrete was specified to have a compressive strength of 45 MPa (6.5 ksi) at prestress release and 69 MPa (10 ksi) at 56 days. The specified concrete compressive strength for the topping slab was 40 MPa (5.8 ksi) at 28 days. Scaling was determined such that the test girders would have the same performance as the actual bridge girders in terms of stresses at the top and bottom of the section, the percentage of the compression region provided by the topping slab, and the strain in the prestressing strands at flexural failure.

The reinforcement ratio of these specimens was 0.0072, which was close to the maximum allowable ratio of 0.0079 stipulated in the AASHTO specifications.[1] This resulted in a design strand strain at failure of approximately 0.011, which reflects the behavior of the actual bridge girders. The moment capacity of the girders calculated with the simplified method in the code,[1] based on specified material properties, is 839 kN-m (619 kip-ft.), and that calculated with the strain compatibility method is 872 kN-m (644 kip-ft.). All three girders were fabricated in the same casting bed. They were steamed cured for approximately eight hours. The maximum temperature measured in the curing process was about 80°C (176 °F). Stress transfer took place two days after casting with flame cut. Each girder was designated numerically 1, 2, or 3, with the ends of the girder labeled east (E) and west (W).

Test Setup and Procedure

Transfer Length Measurement -- Transfer length was determined by measuring strains on girder surface with a Whittemore gage. This gage had a 200 mm (7.87 in.) gage length. Threaded target points were cast into both sides of each girder at the same level as the tendon center of gravity. As shown in Figure 2, these target points were located at 100 mm (3.94 in.) spacing over the first 1600 mm (63 in.) from each end of a girder, and at 200 mm (7.87 in.) spacing in the rest of the girder.

Development Length Tests -- For each test, an estimate of the development length for the member was made. The member was then loaded until the peak load was reached, using a point load at a distance from the end of the member equal to the estimated development length. Based on whether a bond failure or flexural failure occurred, the estimate of development length was revised for the next test. For each test, the specimen was simply supported on specially fabricated supports as shown in Figure 2. Teflon pads were placed between the girder and the support. During each test, load, deflections at midspan and under the load, concrete strain at the tendon center of gravity, and strand slips were measured. Linear voltage differential transducers (LVDT's) were used to measure both strand slips and deflections. Concrete strain at the tendon center of gravity was measured at select load intervals using the mechanical gage.

Other Measurements

Camber -- Bolts were attached to one side of each girder at both ends at the level of the center of gravity of the girder section. A fishing line was stretched tight between these bolts and a reference mark was made on the girder at midspan. The tension in the line was maintained the same in measuring camber. Camber was measured for each girder immediately after prestress release, immediately before and after the topping slab casting, 14, and 28 days after girder casting, and on the day of development length testing.

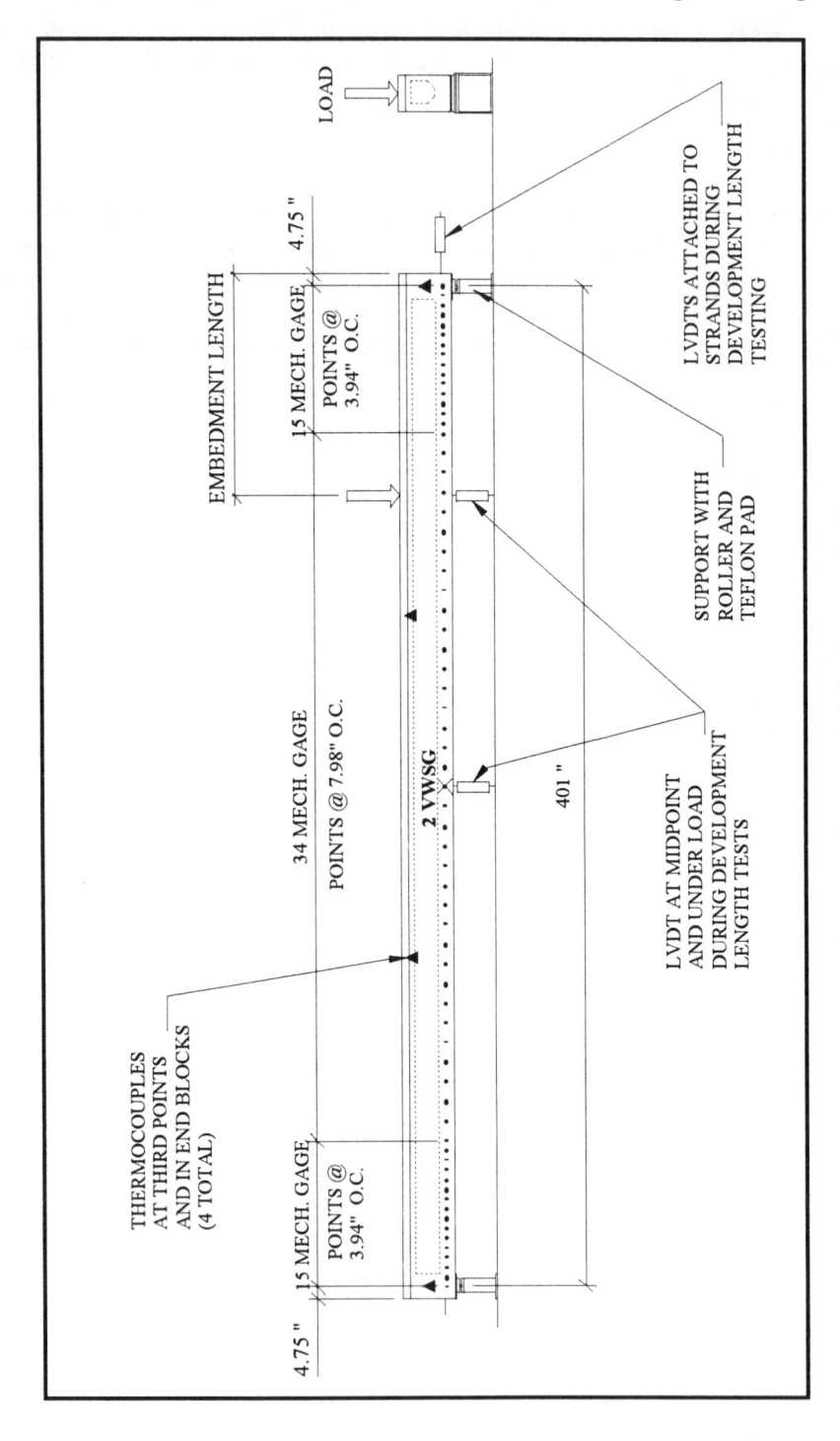

Figure 2 - Test Setup and Instrumentation (1 in. = 25.4 mm)

End Slip at Transfer -- Strand slip measuring devices, developed at FHWA's Turner-Fairbanks Highway Research Center, were attached to the strands near the ends of each girder before prestress release. These devices were essentially a section of aluminum channel which was clamped to the strand at approximately 90 mm (3.5 in.) from the end face of the girder. A Brown & Sharpe Digit-Cal Mark IV caliper was used to measure the distance from one end of the channel to the girder. Measurements were taken immediately before and after prestress release.

Pullout Tests -- Pullout tests were performed. A total of 8 strand samples, approximately 1.83 m (6 ft.) long, were embedded to a depth of 457 mm (18 in.) in a 610 mm (24 in) deep by 914 mm (36 in) long by 610 mm (24 in.) wide concrete block. Two such blocks were cast and allowed to cure for two days before the commencement of the pullout tests. A total of nine strands were tested.

Curing Temperature -- Four thermocouples were placed in each girder, one at each third point of the span in the bottom flange of the girder, and one centered in each end block. These devices were read every 30 minutes for approximately a week after casting.

MATERIAL PROPERTIES

Prestressing Strands

The prestressing strands used in the test girders were manufactured by Insteel Wire Products. All strands were 15-mm (0.6-in.)-diameter, Grade 270, low-relaxation prestressing strand meeting the requirements of ASTM A-416-94. A little rust was observed on the strand surface.

Table 1 - Girder Concrete Mix Design

Material	Quantity kg/m^3 (lb./yd.^{3})
Type III Cement	474 (800)
Water	156 (263)
Coarse Aggregate (3/8" Cooley)	930 (1570)
Fine Aggregate (Sand, Cooley)	782 (1320)
Silica Fume	14 (30)
Water Reducer (Polyheed 997)	2.96 L/m^3 (100 oz./yd.^{3})
Water Reducer (Rheo 1000)	5.91 L/m^3 (120-200 oz./yd.^{3})

Girder Concrete

The concrete mix design for the girder, shown in Table 1, was developed by Rocky Mountain Prestress. Material tests were conducted per ASTM standards wherever applicable on both moist and air cured specimens. The air cured specimens were initially steamed cured together with the girders. Tests to determine compressive strength, modulus of elasticity, modulus of rupture, shrinkage behavior, and split cylinder strength were conducted at the University of Colorado (CU). Additional tests to determine shrinkage and creep behavior were performed by Commercial Testing Laboratories (CTL-Thompson). The air cured specimens yielded higher strengths at an early age due to steam curing and consistently lower strengths after seven days when compared to the moist cured specimens. The compressive strength of concrete is plotted against age in Figure 3.

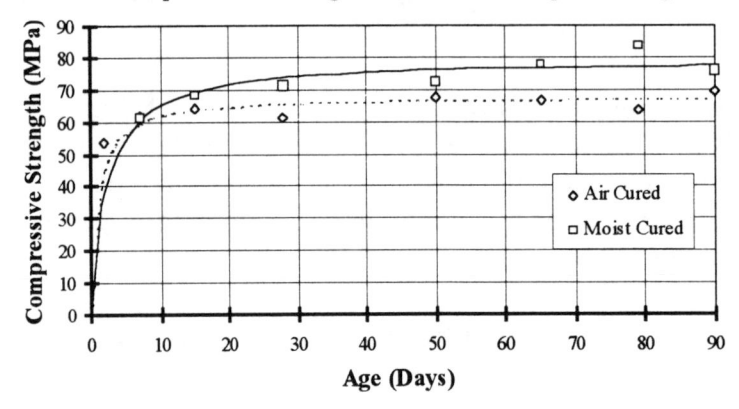

Figure 3 - Girder Concrete Compressive Strength
(1 MPa = 145 psi)

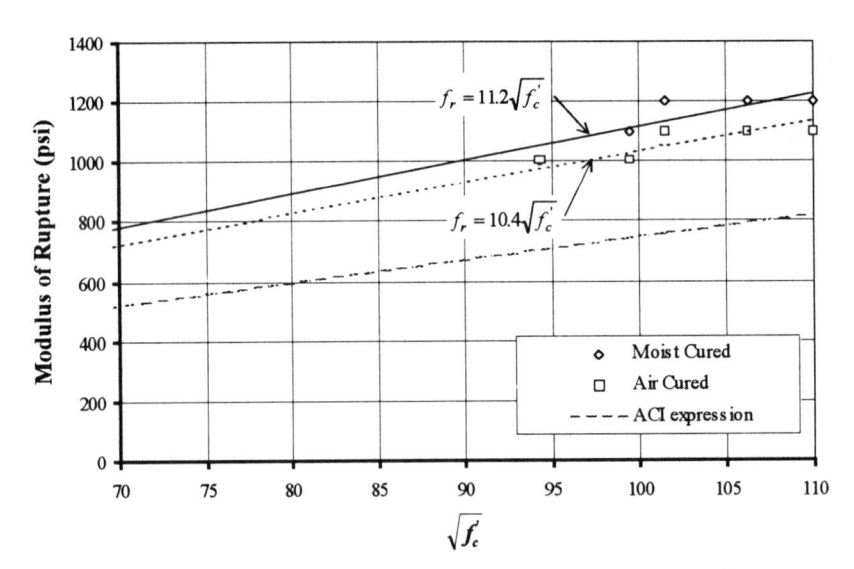

Figure 4 - Modulus of Rupture Data (f'$_c$ in psi)
(1 psi = 6.89 x 10^3 Pa)

The modulus of rupture and modulus of elasticity data are plotted against $\sqrt{f_c'}$ in Figures 4 and 5. Linear fits to these data have been performed and are shown with the equations representing these lines. Also shown are the equations given by ACI 318,[2] which are $f_r = 7.5\sqrt{f_c'}$ and $E_c = 57,000\sqrt{f_c'}$. As can be seen from the graphs, the ACI expression for the modulus of rupture is somewhat conservative, while the expression for the modulus of elasticity yields values that are higher than the test data. In fact, for normal weight concrete with compressive strength between 20.7 MPa (3,000 psi) and 82.7 MPa (12,000 psi), the following equation is recommended for the modulus of elasticity based on research conducted at the Cornell University.[3]

$$E_c = \left(40,000\sqrt{f_c'} + 1,000,000\right)\left(\frac{w_c}{145}\right)^{1.5} \quad (1)$$

in which w_c is the unit weight of the hardened concrete in pcf. It can be seen in Figure 5 that the Cornell expression fits the modulus of elasticity data from the moist cured specimens well.

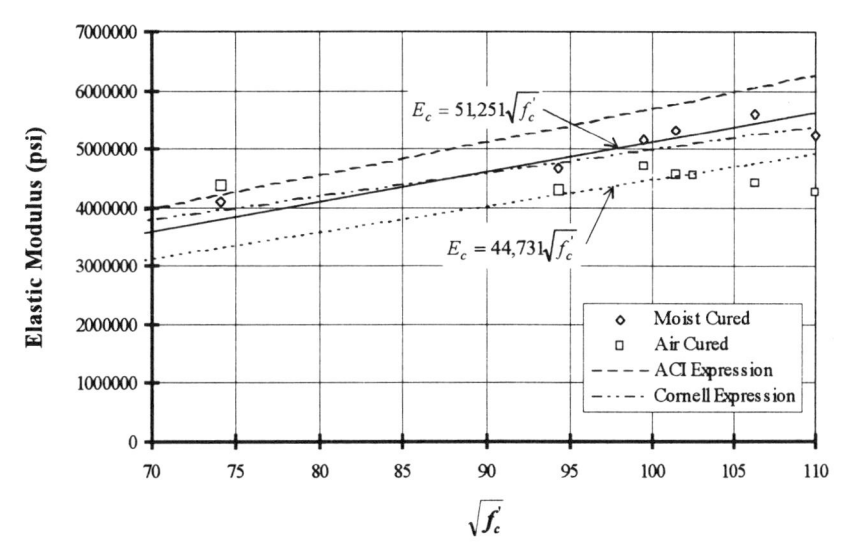

Figure 5 - Modulus of Elasticity Data (f'c in psi)
(1 psi = 6.89 x 10³ Pa)

At CU, the shrinkage of four air cured specimens and two lime-water cured specimens was monitored. The four air cured specimens were initially steam cured together with the girders, cured in lime saturated water for the following two days, and air cured at room temperature thereafter. The two lime-water cured specimens were cured in lime saturated water for the first 28 days and were air cured at room temperature thereafter. The shrinkage data are presented in Figure 6.

Creep tests were performed at CTL-Thompson on 102x203-mm (4x8-in.) cylinders under 19.3 MPa (2,800 psi) compression. Due to measurement problems with the specimens initially cast with the girders, these specimens were cast several weeks after the girders. However, they had the same mix design and curing conditions as the girders. During the creep tests, the specimens were air cured in the laboratory at a temperature of 23°C (73°F) and a humidity of 50%. Loading began two days after casting, corresponding to the day of stress transfer for the girders. The test results are plotted in Figure 7. Creep was also calculated using the following empirical expression.[3]

$$\delta_t = \frac{t^{0.60}}{10+t^{0.60}}\delta_u \qquad (2)$$

in which δ_t is the unit creep strain at time t, t is the time in days after loading, and δ_u is the ultimate creep strain. Based on research at the Cornell University,[3] an ultimate creep strain of 41×10^{-6} per MPa (0.28×10^{-6} per psi) has been suggested for 69 MPa (10 ksi) concrete. The values based on the above formula are plotted in Figure 7 as well. As can be seen from the graph, the creep exhibited by the specimens is much higher than what would be expected with the above formula. Shrinkage measurements were made in the creep tests on specimens cured under the same environment. The shrinkage data are plotted in Figure 6.

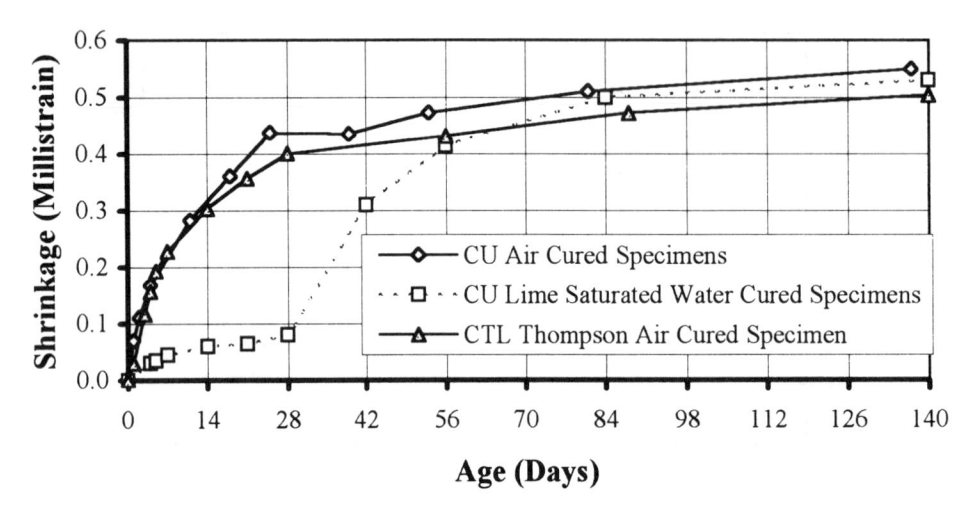

Figure 6 - Average Shrinkage Strain for Girder Concrete

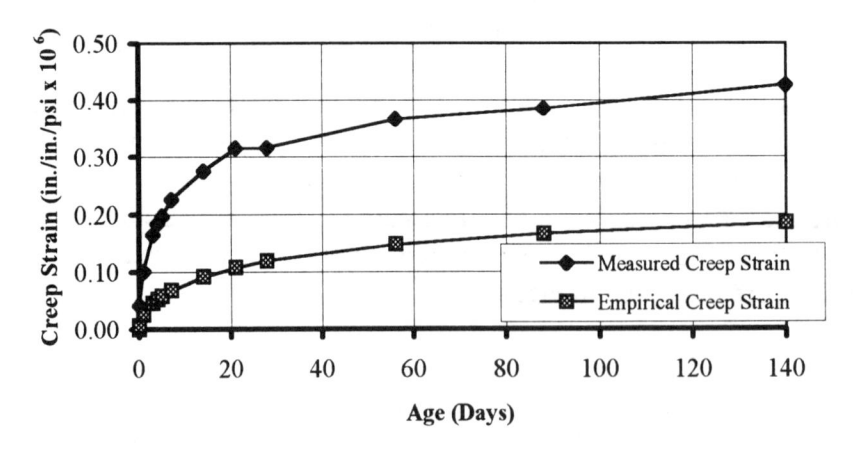

Figure 7 - Unit Creep Strain for Girder Concrete
(1 psi = 6.89 x 10³ Pa)

For the topping slab concrete, air cured cylinders reached a compressive strength of 54 MPa (7.9 ksi) in 72 days, which was considerably higher than the specified strength.

TEST RESULTS

Pullout Test Results

Pullout tests were conducted two days after the casting of the pullout blocks. The concrete compressive strength at this time was 56.5 MPa (8,200 psi). The average load at which first slip occurred was 134 kN (30 kips) and the average maximum pullout load was 215 kN (48 kips). Based on a recent report,[4] the ACI/FHWA transfer and development length requirements are satisfactory for 13-mm (0.5-in)-diameter strand demonstrating a pullout capacity exceeding 160 kN (36 kips). However, no standard has yet been established for 15-mm (0.6-in.)-diameter strand.

Strand Slip at Transfer

Strand slip was measured at the end of each girder right after stress transfer. The elastic shortening which occurred over the strand between the end of the girder and the reference point was calculated and subtracted from the apparent strand slip to give the actual strand slip. The average end slip was 1.49 mm (0.060 in.)

Camber

A time step procedure was used to estimate the camber of the girders, based on the creep and shrinkage properties measured, and the calculated camber is compared to the measured values. Steel relaxation was estimated using an empirical equation for low relaxation strand.[3] The calculated camber at 28 days using these values is 55 mm (2.2 in.), which is significantly higher than the average measured value of 31 mm (1.2 in.). Creep is the prime contributor to increases in camber and it is suspected that the measured creep values might have been too high. Based on the creep estimated with the formula in Eq. (2), the calculated camber is 35 mm (1.4 in.).

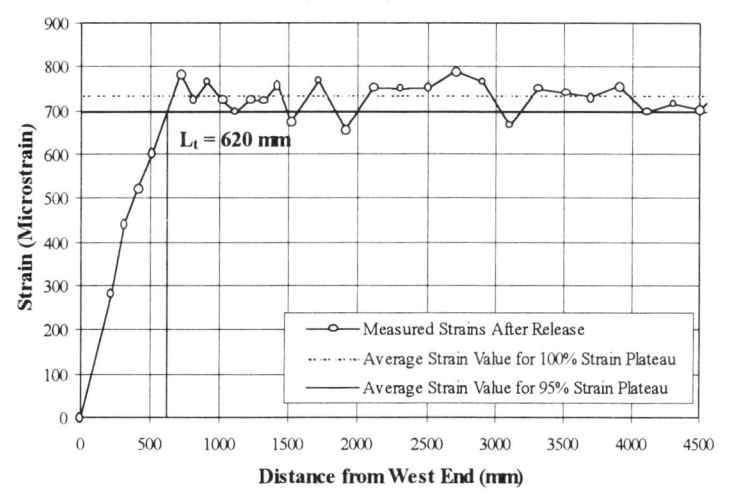

Figure 8 - Transfer Length After Release by 95% Average Maximum Strain Plateau Method (1-W) (1 mm = 0.0394 in.)

Transfer Length

Transfer length was determined using the 95% average maximum strain plateau method.[5] Figure 8 shows the curve used to determine the transfer length at the west end of Girder 1. The average transfer length measured immediately after stress transfer was 593 mm (23.4 in.).

Development Length

Development length testing began 50 days after girder casting. The embedment length at which each test was conducted is shown in Table 2. Only one end was loaded in each test. Both Girders 1 and 2 failed in flexure, while the failure of Girder 3 was caused by bond slips in both tests.

Table 2 - Development Length Tests

Test Designation	Test Date	Embedment Length mm (in.)
1-E	10/2/96	2159 (85)
1-W	10/11/96	2057 (81)
2-W	10/21/96	1918 (76)
2-E	10/28/96	1651 (65)
3-E	11/4/96	1524 (60)
3-W	11/11/96	1497 (59)

Extremely small strand slips were observed in the testing of Girder 2, even though the ultimate load capacities of this girder were governed by the compression failure of the topping slab at both ends and the flexural bond stress was developed in regions away from the transfer zones. The observed slips could be caused by the development of shear cracks in the transfer zone, which led to localized increases in strand stress.

The ultimate load capacities of Girder 3 were governed by strand slips at both ends. Strand slips led to a sudden shear failure at the east end of Girder 3. The approximate strand stress in the east end of Girder 3 calculated from the surface strain of concrete is shown in Figure 9. It can be seen from Figure 9 that the flexural bond stress propagated into the transfer zone. The transition from flexural failure to bond failure occurred between tests 2-E and 3-E. The embedment length for test 2-E was 1651 mm (65 in.), while that for test 3-E was 1524 mm (60 in.). From these observations, it can be concluded that the required development length for these girders should be between 1651 mm (65 in.) and 1524 mm (60 in.). Furthermore, Figure 9 shows that with an embedment length of 1524 mm (60 in.), the flexural bond stress just reached the transfer zone. Hence, it can be concluded that 1524 mm (60 in.) is a reasonable estimate of the required development length.

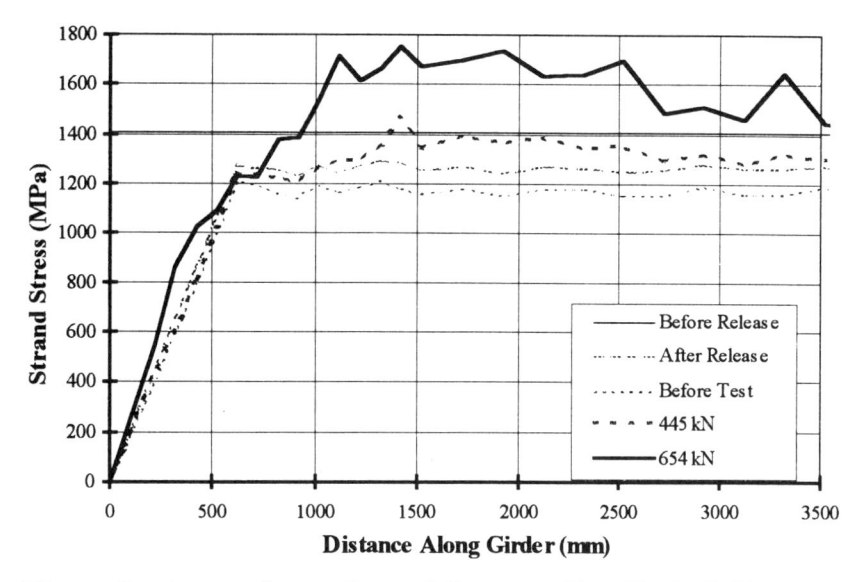

Figure 9 - Approximate Strand Stress at East End of Girder 3
(1 mm = 0.0394 in.; 1 MPa = 145 psi)

Comparison of Experimental Results to Code Specifications

Flexure and shear strengths were calculated for the girders using the AASHTO specifications.[1] The material properties measured at the time of each test were used in these calculations. The calculated strengths are compared to the maximum moment and shear values obtained in the tests in Table 3.

Table 3 - Comparison of Calculated Flexure and Shear Strengths to the Maximum Values obtained in the Tests

Test	M_{test} kN-m (kip-ft.)	$M_{n,AASHTO}$ kN-m (kip-ft.)	$\dfrac{M_{test}}{M_{n,AASHTO}}$	V_{test} kN (kips)	$V_{n,AASHTO}$ kN (kips)
1-E	925 (682)	872 (643)	1.06	432 (97)	628 (141)
1-W	907 (669)	874 (644)	1.04	444 (100)	639 (144)
2-W	967 (714)	874 (645)	1.11	504 (113)	661 (149)
2-E	857 (632)	874 (645)	0.98	521 (117)	703 (158)
3-E	895 (660)	874 (645)	1.02	589 (132)	729 (164)
3-W	848 (626)	874 (645)	0.97	568 (128)	729 (164)

In three of the six tests, the measured maximum moment values are less than the calculated values. In Girder 3, it is the bond failures causing the lower than predicted failure moments. The failure moment for test 2-E is lower than predicted probably due to a premature compression failure of a poorly compacted section of topping slab near the load point. The maximum shear forces measured in all the girder tests are lower than the calculated values. In spite of this, both ends of girder 3 failed, in whole or in part, by shear following strand slips.

The required transfer length and development length for the girders have been calculated in accordance with the AASHTO specifications.[1] The calculated transfer length is 701 mm (27.6 in.), which is about 18% greater than the measured values. The calculated development length is 2332 mm (91.8 in.), which is about 53% longer than the measured.

CONLUSIONS

The average transfer length for the three girders was determined to be 593 mm (23.4 in.). The development length for these girders was determined to be about 1524 mm (60 in.). The AASHTO specifications overestimate the transfer length by 18% and the development length by 53% for these girders. However, these results cannot be generalized to all 15-mm (0.6-in.)-diameter strands as the bond quality may vary depending on the surface condition and manufacturing process of the strand.

ACKNOWLEDGMENTS

This study was sponsored by FHWA. However, opinions expressed in this paper are those of the writers and do not necessarily reflect those of the sponsor or the Colorado Department of Transportation (CDOT). The writers also appreciate the assistance of Nat Jansen of FHWA, David Price of CDOT, and a number of undergraduate students, Sam Scupham, Michael Meiggs, and Ann Grooms, in the experimental work. The constructive comments provided by Gail Kelly of FHWA on this paper are also greatly appreciated.

REFERENCES

1. AASHTO, *Standard Specifications for Highway Bridges,* Fifteen Edition, American Association of State Highway and Transportation Officials, Washington, D.C., 1992.

2. ACI Committee 318, *Building Code Requirements for Reinforced Concrete (ACI 318-89),* American Concrete Institute, Detroit, MI, 1989.

3. Nilson, A.H., *Design of Prestressed Concrete*, Second Edition, John Wiley & Sons, Inc., N.Y., 1987.

4. Logan, D. R., "Acceptance Criteria for Bond Quality of Strand for Pretensioned Prestressed Concrete Applications," *PCI Journal*, V. 42, No. 7, March-April 1997, pp. 52-90.

5. Russell, B.W., and Burns, N.H., "Design Guidelines for Transfer, Development, and Debonding of Large Diameter Seven Wire Strands in Pretensioned Concrete Girders," *Research Report 1210-5F,* Center for Transportation Research, University of Texas, Austin, TX, 1993, 286 pp.

HPC BRIDGE SHOWCASE PROJECT IN ALABAMA

J. Michael Stallings
Gottlieb Associate Professor of Civil Engineering
Auburn University
Auburn University, Alabama, USA

David Pittman
Assistant Professor of Civil Engineering
Auburn University
Auburn University, Alabama, USA

ABSTRACT

A FHWA high performance concrete (HPC) bridge showcase project is just beginning in Alabama. The goal of the project is to encourage the use of HPC in bridges in Alabama and the Southeast by demonstrating the advantages and successful use of HPC in a regional showcase bridge. HPC mixes will be developed to meet a number of durability requirements and provide compressive strengths of 69 MPa (10,000 psi) for prestressed girders and 41 MPa (6,000 psi) for the deck and substructure elements. The long-term performance of the girders and bridge structure will be monitored and short-term live load tests will be performed.

INTRODUCTION

Bridges with precast, prestressed concrete girders and cast-in-place concrete decks dominate new construction in Alabama due to their lower first cost relative to other bridge systems. High performance concrete (HPC) offers a way to reduce the cost while improving the durability of these already popular bridges. But, until designers and bridge contractors become comfortable with this new technology, HPC will not be fully utilized, and confidence can only be provided by successful completion of HPC bridge projects. The project described here will provide a demonstration of HPC technology on a bridge with standard precast, prestressed concrete girders and cast-in-place concrete deck to help build the required confidence. The goal of the project is to encourage further implementation of high performance concrete (HPC) in Alabama and the Southeast by demonstrating the successful use and the advantages of HPC. The project discussed here was just beginning at the time of this writing, so the focus of this paper is an outline of the project scope.

HPC will be used to construct a replacement for the main bridge over Uphapee Creek on Alabama Highway 199. The bridge replacement is necessary due to stream bed scour resulting from sand and gravel mining downstream. HPC will be used in the deck, columns and pier caps, and also in the prestressed girders. Two additional bridges, a relief bridge at Uphapee Creek and a bridge over nearby Bulger Creek, are a part of the same bridge replacement project and will be constructed using Normal Performance Concrete (NPC). The three bridges are near a stone quarry and experience significant heavy truck traffic from sand and gravel trucks as well as log trucks. Short-term and long-term monitoring of the bridges will track their performance and provide information illustrating the benefits of using HPC.

HIGH PERFORMANCE CONCRETE

The terminology "high performance concrete" refers to concretes designed to meet durability requirements and have high compressive strengths. Specifying the compressive strength of a concrete mix along with air entrainment for durability has been common in the past. The proposed project will give the Alabama Department of Transportation (ALDOT) the opportunity to specify durability parameters such as permeability, and resistance to freeze-thaw, scaling and abrasion. This opportunity also presents questions of where to set the durability requirements so that concrete mixes can be designed and placed to meet the requirements without creating significant additional costs. Answering these questions is part of the research effort.

The target properties for the precast, prestressed concrete girders and cast-in-place concrete (includes the bridge deck and substructure) are given in Table 1. The goals for the prestressed girder concrete are a 28-day compressive strength of 69 MPa (10,000 psi), and a compressive strength at release of 55 MPa (8000 psi). The cast-in-place deck, columns and pier cap concrete will have a 56-day compressive strength of 41 MPa (6000 psi). Currently in Alabama, 28-day compressive strengths of 34 MPa (5000 psi) to 45 MPa (6500 psi) are used for prestressed girders and 28 MPa (4000 psi) for decks and substructure elements. The target values for the durability related characteristics of the HPC girder concrete are at the boundary between FHWA Performance Grades 2 and 3 as defined by Goodspeed et al.[1] The cast-in-place concrete meet the definition of Performance Grade 1.

The mix development will begin with a feasibility study at Auburn University using materials readily available in Alabama to determine whether target concrete properties are attainable. Development of the final mix designs will be the bridge contractor's responsibility. Data and experience gained during the feasibility study will be shared with the contractor to assist in the development of the final design. After the Contractor has developed HPC mix designs for the girder and bridge deck concrete (as well as the "normal" concrete (NPC) mix designs to be used as a comparison), the mixtures will be subjected to verification testing to determine values for all the performance characteristics listed in Table 1. The mixture validation process will be conducted on the two HPC mixtures, and a conventional NPC mixture (34 MPa) for comparisons.

Table 1 - Target Performance Characteristics for HPC

Characteristic	Girders	Cast-in-Place
Compressive Strength, at release 28 days 56 days	55 MPa[1] 69 MPa N.A.	N.A.[2] N.A. 41 MPa
Modulus of Elasticity, at release 28 days 56 days	41,000 MPa 48,000 MPa N.A.	N.A. N.A. 28,000 MPa
Freeze-Thaw Durability Factor	> 80%	> 80%
Chloride Ion Permeability	< 800 Coulombs	< 3000 Coulombs
Creep (ASTM C512)	< 45 microstrain/MPa	N.A.
Shrinkage (ASTM C672)	< 400 microstrain	< 800 microstrain
Scaling Resistance Visual Rating	N.A.	< 2
Abrasion Resistance Depth of Wear	N.A.	1 mm at 50 cycles

[1] 6.895 Mpa = 1 ksi.
[2] Does not apply.

After the laboratory verification testing, the Contractor (and Prestress Producer) will construct small test sections (a test girder and a test slab) using the girder and bridge deck HPC mixtures. The purpose of the test sections is to let the Contractor demonstrate his ability to mix, haul, place, consolidate, finish, and cure the HPC mixtures properly.

BRIDGE DESCRIPTION

The HPC bridge over Uphapee Creek will have seven 34.7 m (114 ft) simple spans and a 12.2 m (40 ft) clear roadway width. Five AASHTO BT - 54 girders will be used in each span at a transverse spacing of 2.67 m (8.75 ft). The HPC deck thickness will be 180 mm (7 in.). The spans will be supported on cast-in-place concrete columns and pier caps on either drilled shaft or driven steel pile foundations. The girder designs will take full advantage of the HPC and 15 mm (0.6 in.) diameter prestressing strand. The deck, column and pier cap designs will take full advantage of the improved durability of the HPC, but practical minimum member thickness standards of the ALDOT may limit how well the increased compressive strength can be utilized.

Comparisons of preliminary designs illustrate the benefits of using HPC for this project. Only a few years ago, concrete compressive strengths of 34 MPa (5000 psi) were routinely used by ALDOT in prestressed girder designs. At that strength level, 30.4 m (100 ft) spans each with six BT-54 girders would be required at Uphapee Creek. Recently, concrete strengths used in normal designs have increased from 34 MPa (5000 psi) to 44 MPa (6500 psi). This incremental increase in strength reduces the number of girders required in each of the original 30.4m (100 ft) spans from six to five. Using the 69 MPa (10,000 psi) HPC girders at Uphapee Creek allows an increase in span length from 30.4 m (100 ft) to 34.7 m (114 ft). The increased span lengths reduces the number of spans required from eight to seven, so the cost of one pier is eliminated.

INSTRUMENTATION, MONITORING AND LIVE LOAD TESTING

Both the HPC and NPC bridges will be instrumented and monitored for short-term and long-term performance. This will provide information to future contractors and prestress producers regarding differences in prestress losses, deflections and cambers that affect bridge construction. The monitoring will begin in the casting yard and will continue beyond completion of construction, possibly for several years. Girder strains, deflections and temperature distributions will be monitored.

Data from the instrumented girders will be collected during construction. Measurements of the girder temperatures, stresses and deflections will be made at intervals as required to properly record the load history on the girders for detailed calculations of prestress losses, cambers and deflections. Measurements will be made at critical times such as erection of the instrumented girders, casting of the deck and installation of the barriers.

Live load tests of the completed bridge will be performed before the bridge is opened to traffic. The purpose of the tests will be to measure live load stresses and deflections created by single and side-by-side, static and dynamic vehicle loadings. The live load measurements will provide valuable data regarding the stiffness of the HPC bridge system and the effects of differences in modulus of elasticity between the NPC and HPC concretes.

EVALUATION OF PRESTRESS LOSS AND CAMBER CALCULATIONS

The accuracy of prestress loss and camber calculations is dependent on the accuracy in treatment of factors such as the load history, modulus of elasticity, shrinkage and creep of the concrete. One aspect of the research project is an evaluation of applicability of the methods currently used for these calculations to HPC designs. Computed losses and cambers at various ages through the project will be compared to the measurements made in the bridge monitoring program.

PROJECT SCHEDULE

The bridge construction project is currently scheduled for letting to contract in January, 1998. Estimated times for casting of the prestressed girders and deck are Spring of 1998 and Fall of 1998, respectively. The construction project and HPC technology will be presented in a HPC Showcase in Auburn, Alabama in the Spring of 1999.

SUMMARY STATEMENT

A FHWA HPC bridge showcase project is underway in Alabama. The bridge chosen for the project is a standard precast, prestressed concrete girder bridge with a cast-in-place concrete deck. The goal of the project is to provide an illustration of the advantages of HPC and a demonstration of successful application of HPC. This will provide confidence in the use of HPC in this already popular bridge type and lead to more durable and economical bridges.

ACKNOWLEDGMENTS

The authors gratefully acknowledge the funding and support of this project provided by the Federal Highway Administration, the Alabama Department of Transportation, and the Auburn University Highway Research Center.

REFERENCES

1. Goodspeed, C., Vanikar, S. and Cook, R., "High-Performance Concrete (HPC) Defined for Highway Structures," Federal Highway Administration, Washington, D.C., November 1995, 15 pp.